# UNDERSTANDING PHYSICS

# PART 1



Karen Cummings
*Rensselaer Polytechnic Institute*
*Southern Connecticut State University*

Priscilla W. Laws
*Dickinson College*

Edward F. Redish
*University of Maryland*

Patrick J. Cooney
*Millersville University*

GUEST AUTHOR

Edwin F. Taylor
*Massachusetts Institute of Technology*

ADDITIONAL MEMBERS OF ACTIVITY BASED PHYSICS GROUP

David R. Sokoloff
*University of Oregon*

Ronald K. Thornton
*Tufts University*

*Understanding Physics* is based on *Fundamentals of Physics* by David Halliday, Robert Resnick, and Jearl Walker.

**John Wiley & Sons, Inc.**

This book is dedicated to Arnold Arons, whose pioneering work in physics education and reviews of early chapters have had a profound influence on our work.


| | |
|---|---|
| SENIOR ACQUISITIONS EDITOR | Stuart Johnson |
| SENIOR DEVELOPMENT EDITOR | Ellen Ford |
| MARKETING MANAGER | Bob Smith |
| SENIOR PRODUCTION EDITOR | Elizabeth Swain |
| SENIOR DESIGNER | Kevin Murphy |
| INTERIOR DESIGN | Circa 86, Inc. |
| COVER DESIGN | David Levy |
| COVER PHOTO | © Antonio M. Rosario/The Image Bank/Getty Images |
| ILLUSTRATION EDITOR | Anna Melhorn |
| PHOTO EDITOR | Hilary Newman |

This book was set in 10/12 Times Ten Roman by Progressive and printed and bound by Von Hoffmann Press. The cover was printed by Von Hoffmann Press.

This book is printed on acid free paper. ♾





***Library of Congress Cataloging in Publication Data:***

Understanding physics / Karen Cummings . . . [et al.]; with additional members of the Activity Based Physics Group.
p. cm.
Includes index.
ISBN 0-471-46435-X (pt. 1 : pbk. : acid-free paper)
1. Physics. I. Cummings, Karen. II. Activity Based Physics Group.

QC23.2.U54 2004
530—dc21 2003053481

L.C. Call no. Dewey Classification No. L.C. Card No.
ISBN 0-471-46435-X

Printed in the United States of America

10 9 8 7 6 5 4 3 2 1

# Preface

Welcome to *Understanding Physics*. This book is built on the foundations of the 6th Edition of Halliday, Resnick, and Walker's *Fundamentals of Physics* which we often refer to as HRW 6th. The HRW 6th text and its ancestors, first written by David Halliday and Robert Resnick, have been best-selling introductory physics texts for the past 40 years. It sets the standard against which many other texts are judged. You are probably thinking, "Why mess with success?" Let us try to explain.

## Why a Revised Text?

A physics major recently remarked that after struggling through the first half of his junior level mechanics course, he felt that the course was now going much better. What had changed? Did he have a better background in the material they were covering now? "No," he responded. "I started reading the book before every class. That helps me a lot. I wish I had done it in Physics One and Two." Clearly, this student learned something very important. It is something most physics instructors wish they could teach all of their students as soon as possible. Namely, no matter how smart your students are, no matter how well your introductory courses are designed and taught, your students will master more physics if they learn how to read an "understandable" textbook carefully.

We know from surveys that the vast majority of introductory physics students do not read their textbooks carefully. We think there are two major reasons why: (1) many students complain that physics textbooks are impossible to understand and too abstract, and (2) students are extremely busy juggling their academic work, jobs, personal obligations, social lives and interests. So they develop strategies for passing physics without spending time on careful reading. We address both of these reasons by making our revision to the sixth edition of *Fundamentals of Physics* easier for students to understand and by providing the instructor with more **Reading Exercises** (formerly known as Checkpoints) and additional strategies for encouraging students to read the text carefully. Fortunately, we are attempting to improve a fine textbook whose active author, Jearl Walker, has worked diligently to make each new edition more engaging and understandable.

In the next few sections we provide a summary of how we are building upon HRW 6th and shaping it into this new textbook.

## A Narrative That Supports Student Learning

One of our primary goals is to help students make sense of the physics they are learning. We cannot achieve this goal if students see physics as a set of disconnected mathematical equations that each apply only to a small number of specific situations. We stress conceptual and qualitative understanding and continually make connections between mathematical equations and conceptual ideas. We also try to build on ideas that students can be expected to already understand, based on the resources they bring from everyday experiences.

In *Understanding Physics* we have tried to tell a story that flows from one chapter to the next. Each chapter begins with an introductory section that discusses why new topics introduced in the chapter are important, explains how the chapter builds on previous chapters, and prepares students for those that follow. We place explicit emphasis on basic concepts that recur throughout the book. We use extensive forward and backward referencing to reinforce connections between topics. For example, in the introduction of Chapter 16 on Oscillations we state: "Although your study of simple harmonic motion will enhance your understanding of mechanical systems it is also vital to understanding the topics in electricity and magnetism encountered in Chapters 30-37. Finally, a knowledge of SHM provides a basis for understanding the wave nature of light and how atoms and nuclei absorb and emit energy."

## Emphasis on Observation and Experimentation

Observations and concrete everyday experiences are the starting points for development of mathematical expressions. Experiment-based theory building is a major feature of the book. We build ideas on experience that students either already have or can easily gain through careful observation.

Whenever possible, the physical concepts and theories developed in *Understanding Physics* grow out of simple observations or experimental data that can be obtained in typical introductory physics laboratories. We want our readers to develop the habit of asking themselves: What do our observations, experiences and data imply about the natural laws of physics? How do we know a given statement is true? Why do we believe we have developed correct models for the world?

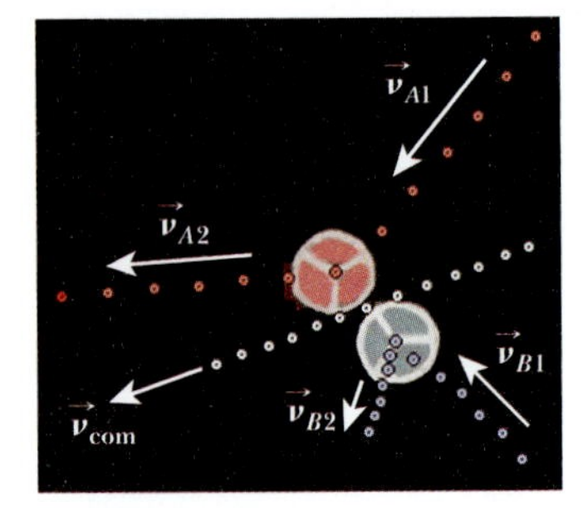


**FIGURE P-1** ■ A video analysis shows that the center of mass of a two-puck system moves at a constant velocity.

Toward this end, the text often starts a chapter by describing everyday observations with which students are familiar. This makes *Understanding Physics* a text that is both relevant to students' everyday lives and draws on existing student knowledge. We try to follow Arnold Arons' principle "idea first, name after." That is, we make every attempt to begin a discussion by using everyday language to describe common experiences. Only then do we introduce formal physics terminology to represent the concepts being discussed. For example, everyday pushes, pulls, and their impact on the motion of an object are discussed before introducing the term "force" or Newton's Second Law. We discuss how a balloon shrivels when placed in a cold environment and how a pail of water cools to room temperature before introducing the ideal gas law or the concept of thermal energy transfer.

The "idea first, name after" philosophy helps build patterns of association between concepts students are trying to learn and knowledge they already have. It also helps students reinterpret their experiences in a way that is consistent with physical laws.

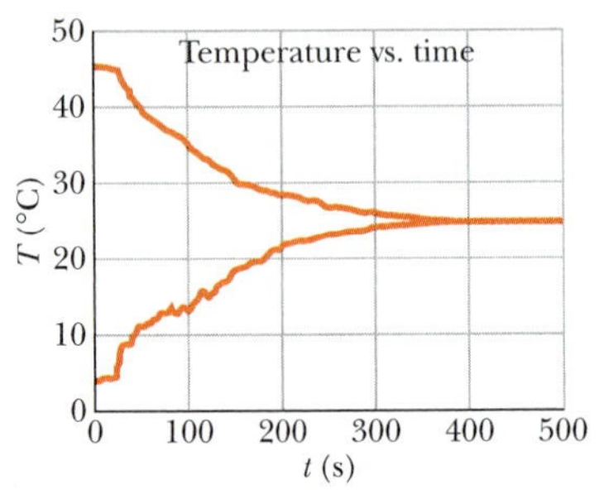


**FIGURE P-2** ■ Electronic temperature sensors reveal that if equal amounts of hot and cold water mix the final temperature is the average of the initial temperatures.

Examples and illustrations in *Understanding Physics* often present data from modern computer-based laboratory tools. These tools include computer-assisted data acquisition systems and digital video analysis software. We introduce students to these tools at the end of Chapter 1. Examples of these techniques are shown in Figs. P-1 and P-2 (on the left) and Fig. P-3 on the next page. Since many instructors use these computer tools in the laboratory or in lecture demonstrations, these tools are part of the introductory physics experience for more and more of our students. The use of real data has a number of advantages. It connects the text to the students' experience in other parts of the course and it connects the text directly to real world experience. Regardless of whether data acquisition and analysis tools are used in the student's own laboratory, our use of realistic rather that idealized data helps students develop an appreciation of the role that data evaluation and analysis plays in supporting theory.

**FIGURE P-3** ■ A video analysis of human motion reveals that in free fall the center of mass of an extended body moves in a parabolic path under the influence of the Earth's gravitational force.

## Using Physics Education Research

In re-writing the text we have taken advantage of two valuable findings of physics education research. One is the identification of concepts that are especially difficult for many students to learn. The other is the identification of active learning strategies to help students develop a more robust understanding of physics.

### Addressing Learning Difficulties

Extensive scholarly research exists on the difficulties students have in learning physics.[1] We have made a concerted effort to address these difficulties. In *Understanding Physics,* issues that are known to confuse students are discussed with care. This is true even for topics like the nature of force and its effect on velocity and velocity changes that may seem trivial to professional physicists. We write about subtle, often counter-intuitive topics with carefully chosen language and examples designed to draw out and remediate common alternative student conceptions. For example, we know that students have trouble understanding passive forces such as normal and friction forces.[2] How can a rigid table exert a force on a book that rests on it? In Section 6-4 we present an idealized model of a solid that is analogous to an inner spring mattress with the repulsion forces between atoms acting as the springs. In addition, we invite our readers to push on a table with a finger and experience the fact that as they push harder on the table the table pushes harder on them in the opposite direction.

**FIGURE P-4** ■ Compressing an innerspring mattress with a force. The mattress exerts an oppositely directed force, with the same magnitude, back on the finger.

### Incorporating Active Learning Opportunities

We designed *Understanding Physics* to be more interactive and to foster thoughtful reading. We have retained a number of the excellent Checkpoint questions found at the end of HRW 6th chapter sections. We now call these questions **Reading Exercises.** We have created many new Reading Exercises that require students to reflect on the material in important chapter sections. For example, just after reading Section 6-2 that introduces the two-dimensional free-body diagram, students encounter Reading Exercise 6-1. This multiple-choice exercise requires students to identify the free-body diagram for a helicopter that experiences three non-collinear forces. The distractors were based on common problems students have with the construction of free-body diagrams. When used in "Just-In-Time Teaching" assignments or for in-class group discussion, this type of reading exercise can help students learn a vital problem solving skill as they read.

---

[1] L. C. McDermott and E. F. Redish, "Resource Letter PER-1: Physics Education Research," *Am. J. Phys.* **67**, 755-767 (1999)

[2] John J. Clement, "Expert novice similarities and instruction using analogies," *Int. J. Sci. Ed. 20*, 1271-1286 (1998)

We also created a set of **Touchstone Examples.** These are carefully chosen sample problems that illustrate key problem solving skills and help students learn how to use physical reasoning and concepts as an essential part of problem solving. We selected some of these touchstone examples from the outstanding collection of sample problems in HRW 6th and we created some new ones. In order to retain the flow of the narrative portions of each chapter, we have reduced the overall number of sample problems to those necessary to exemplify the application of fundamental principles. Also, we chose touchstone examples that require students to combine conceptual reasoning with mathematical problem-solving skills. Few, if any, of our touchstone examples are solvable using simple "plug-and-chug" or algorithmic pattern matching techniques.

**Alternative problems** have been added to the extensive, classroom tested end-of-chapter problem sets selected from HRW 6th. The design of these new problems are based on the authors' knowledge of research on student learning difficulties. Many of these new problems require careful qualitative reasoning. They explicitly connect conceptual understanding to quantitative problem solving. In addition, estimation problems, video analysis problems, and "real life" or "context rich" problems have been included.

The organization and style of *Understanding Physics* has been modified so that it can be easily used with other research-based curricular materials that make up what we call *The Physics Suite*. The *Suite* and its contents are explained at length at the end of this preface.

## Reorganizing for Coherence and Clarity

For the most part we have retained the organization scheme inherited from HRW 6th. Instructors are familiar with the general organization of topics in a typical course sequence in calculus-based introductory physics texts. In fact, ordering of topics and their division into chapters is the same for 27 of the 38 chapters. The order of some topics has been modified to be more pedagogically coherent. Most of the reorganization was done in Chapters 3 through 10 where we adopted a sequence known as *New Mechanics*. In addition, we decided to move HRW 6th Chapter 25 on capacitors so it becomes the last chapter on DC circuits. Capacitors are now introduced in Chapter 28 in *Understanding Physics*.

### The New Mechanics Sequence

HRW 6th and most other introductory textbooks use a familiar sequence in the treatment of classical mechanics. It starts with the development of the kinematic equations to describe constantly accelerated motion. Then two-dimensional vectors and the kinematics of projectile motion are treated. This is followed by the treatment of dynamics in which Newton's Laws are presented and used to help students understand both one- and two-dimensional motions. Finally energy, momentum conservation, and rotational motion are treated.

About 12 years ago when Priscilla Laws, Ron Thornton, and David Sokoloff were collaborating on the development of research-based curricular materials, they became concerned about the difficulties students had working with two-dimensional vectors and understanding projectile motion before studying dynamics.

At the same time Arnold Arons was advocating the introduction of the concept of momentum before energy.[3] Arons argued that (1) the momentum concept is simpler than the energy concept, in both historical and modern contexts and (2) the study

[3] Private Communication between Arnold Arons and Priscilla Laws by means of a document entitled "Preliminary Notes and Suggestions," August 19, 1990; and Arnold Arons, *Development of Concepts of Physics* (Addison-Wesley, Reading MA, 1965)

of momentum conservation entails development of the concept of center-of-mass which is needed for a proper development of energy concepts. Additionally, the impulse-momentum relationship is clearly an alternative statement of Newton's Second Law. Hence, its placement immediately after the coverage of Newton's laws is most natural.

In order to address these concerns about the traditional mechanics sequence, a small group of physics education researchers and curriculum developers convened in 1992 to discuss the introduction of a new order for mechanics.[4] One result of the conference was that Laws, Sokoloff, and Thornton have successfully incorporated a new sequence of topics in the mechanics portions of various curricular materials that are part of the Physics Suite discussed below.[5] These materials include *Workshop Physics*, the *RealTime Physics Laboratory Module in Mechanics*, and the *Interactive Lecture Demonstrations*. This sequence is incorporated in this book and has required a significant reorganization and revisions of HRW 6th Chapters 2 through 10.

The New Mechanics sequence incorporated into Chapters 2 through 10 of understanding physics includes:

- Chapter 2: One-dimensional kinematics using constant horizontal accelerations and vertical free fall as applications.
- Chapter 3: The study of one-dimensional dynamics begins with the application of Newton's laws of motion to systems with one or more forces acting along a single line. Readers consider observations that lead to the postulation of "gravity" as a constant invisible force acting vertically downward.
- Chapter 4: Two-dimensional vectors, vector displacements, unit vectors and the decomposition of vectors into components are treated.
- Chapter 5: The study of kinematics and dynamics is extended to two-dimensional motions with forces along only a single line. Examples include projectile motion and circular motion.
- Chapter 6: The study of kinematics and dynamics is extended to two-dimensional motions with two-dimensional forces.
- Chapters 7 & 8: Topics in these chapters deal with impulse and momentum change, momentum conservation, particle systems, center of mass, and the motion of the center-of-mass of an isolated system.
- Chapters 9 & 10: These chapters introduce kinetic energy, work, potential energy, and energy conservation.

## Just-in-Time Mathematics

In general, we introduce mathematical topics in a "just-in-time" fashion. For example, we treat one-dimensional vector concepts in Chapter 2 along with the development of one-dimensional velocity and acceleration concepts. We hold the introduction of two- and three-dimensional vectors, vector addition and decomposition until Chapter 4, immediately before students are introduced to two-dimensional motion and forces in Chapters 5 and 6. We do not present vector products until they are needed. We wait to introduce the dot product until Chapter 9 when the concept of physical work is presented. Similarly, the cross product is first presented in Chapter 11 in association with the treatment of torque.

[4] The New Mechanics Conference was held August 6-7, 1992 at Tufts University. It was attended by Pat Cooney, Dewey Dykstra, David Hammer, David Hestenes, Priscilla Laws, Suzanne Lea, Lillian McDermott, Robert Morse, Hans Pfister, Edward F. Redish, David Sokoloff, and Ronald Thornton.

[5] Laws, P. W. "A New Order for Mechanics" pp. 125-136, *Proceedings of the Conference on the Introductory Physics Course*, Rensselaer Polytechnic Institute, Troy New York, May 20-23, Jack Wilson, Ed. 1993 (John Wiley & Sons, New York 1997)

### Notation Changes

Mathematical notation is often confusing, and ambiguity in the meaning of a mathematical symbol can prevent a student from understanding an important relationship. It is also difficult to solve problems when the symbols used to represent different quantities are not distinctive. Some key features of the new notation include:

- We adhere to recent notation guidelines set by the U.S. National Institute of Standard and Technology Special Publication 811 (SP 811).
- We try to balance our desire to use familiar notation and our desire to avoid using the same symbol for different variables. For example, $p$ is often used to denote momentum, pressure, and power. We have chosen to use lower case $p$ for momentum and capital $P$ for pressure since both variables appear in the kinetic theory derivation. But we stick with the convention of using capital $P$ for power since it does not commonly appear side by side with pressure in equations.
- We denote vectors with an arrow instead of bolding so handwritten equations can be made to look like the printed equations.
- We label each vector component with a subscript that explicitly relates it to its coordinate axis. This eliminates the common ambiguity about whether a quantity represents a magnitude which is a scalar or a vector component which is not a scalar.
- We often use subscripts to spell out the names of objects that are associated with mathematical variables even though instructors and students will tend to use abbreviations. We also stress the fact that one object is exerting a force on another with an arrow in the subscript. For example, the force exerted by a rope on a block would be denoted as $\vec{F}_{\text{rope}\rightarrow\text{block}}$.

Our notation scheme is summarized in more detail in Appendix A4.

## Encouraging Text Reading

We have described a number of changes that we feel will improve this textbook and its readability. But even the best textbook in the world is of no help to students who do not read it. So it is important that instructors make an effort to encourage busy students to develop effective reading habits. In our view the single most effective way to get students to read this textbook is to assign appropriate reading, reading exercises, and other reading questions after every class. Some effective ways to follow up on reading question assignments include:

1. Employ a method called "Just-In-Time-Teaching" (or JiTT) in which students submit their answers to questions about reading before class using just plain email or one of the many available computer based homework systems (WebAssign or E-Grade for example). You can often read enough answers before class to identify the difficult questions that need more discussion in class;
2. Ask students to bring the assigned questions to class and use the answers as a basis for small group discussions during the class period;
3. Assign multiple choice questions related to each section or chapter that can be graded automatically with a computer-based homework system; and
4. Require students to submit chapter summaries. Because this is a very effective assignment, we intentionally avoided doing chapter summaries for students.

Obviously, all of these approaches are more effective when students are given some credit for doing them. Thus you should arrange to grade all, or a random sample, of the submissions as incentives for students to read the text and think about the answers to Reading Exercises on a regular basis.

## The Physics Suite

In 1997 and 1998, Wiley's physics editor, Stuart Johnson, and an informally constituted group of curriculum developers and educational reformers known as the *Activity Based Physics Group* began discussing the feasibility of integrating a broad array of curricular materials that are physics education research-based. This led to the assembly of an *Activity Based Physics Suite* that includes this textbook. The *Physics Suite* also includes materials that can be combined in different ways to meet the needs of instructors working in vastly different learning environments. The *Interactive Lecture Demonstration Series*[6] is designed primarily for use in lecture sessions. Other *Suite* materials can be used in laboratory settings including the *Workshop Physics Activity Guide,*[7] the *Real Time Physics Laboratory* modules,[8] and *Physics by Inquiry.*[9] Additional elements in the collection are suitable for use in recitation sessions such as the University of Washington *Tutorials in Introductory Physics* (available from Prentice Hall)[10] and a set of *Quantitative Tutorials*[11] developed at the University of Maryland. The *Activity Based Physics Suite* is rounded out with a collection of thinking problems developed at the University of Maryland. In addition to this **Understanding Physics** text, the Physics Suite elements include:

1. **Teaching Physics with the Physics Suite** by Edward F. Redish (University of Maryland). This book is not only the "Instructors Manual" for *Understanding Physics*, but it is also a book for anyone who is interested in learning about recent developments in physics education. It is a handbook with a variety of tools for improving both teaching and learning of physics—from new kinds of homework and exam problems, to surveys for figuring out what has happened in your class, to tools for taking and analyzing data using computers and video. The book comes with a Resource CD containing 14 conceptual and 3 attitude surveys, and more than 250 thinking problems covering all areas of introductory physics, resource materials from commercial vendors on the use of computerized data acquisition and video, and a variety of other useful reference materials. (Instructors can obtain a complimentary copy of the book and Resource CD, from John Wiley & Sons.)
2. **RealTime Physics** by David Sokoloff (University of Oregon), Priscilla Laws (Dickinson College), and Ronald Thornton (Tufts University). *RealTime Physics* is a set of laboratory materials that uses computer-assisted data acquisition to help students build concepts, learn representation translation, and develop an understanding of the empirical base of physics knowledge. There are three modules in the collection: Module 1: Mechanics (12 labs), Module 2: Heat and Thermodynamics (6 labs), and Module 3: Electric Circuits (8 labs). (Available both in print and in electronic form on *The Physics Suite CD.*)

[6]David R. Sokoloff and Ronald K. Thornton, "Using Interactive Lecture Demonstrations to Create an Active Learning Environment." *The Physics Teacher*, **35**, 340-347, September 1997.

[7]Priscilla W. Laws, *Workshop Physics Activity Guide*, Modules 1-4 w/ Appendices (John Wiley & Sons, New York, 1997).

[8]David R. Sokoloff, *RealTime Physics*, Modules 1-2, (John Wiley & Sons, New York, 1999).

[9]Lillian C. McDermott and the Physics Education Group at the University of Washington, *Physics by Inquiry* (John Wiley & Sons, New York, 1996).

[10]Lillian C. McDermott, Peter S. Shaffer, and the Physics Education Group at the University of Washington, *Tutorials in Introductory Physics*, First Edition (Prentice-Hall, Upper Saddle River, NJ, 2002).

[11]Richard N. Steinberg, Michael C. Wittmann, and Edward F. Redish, "Mathematical Tutorials in Introductory Physics," in, *The Changing Role Of Physics Departments In Modern Universities,* Edward F. Redish and John S. Rigden, editors, AIP Conference Proceedings **399**, (AIP, Woodbury NY, 1997), 1075-1092.

3. **Interactive Lecture Demonstrations** by David Sokoloff (University of Oregon) and Ronald Thornton (Tufts University). ILDs are worksheet-based guided demonstrations designed to focus on fundamental principles and address specific naïve conceptions. The demonstrations use computer-assisted data acquisition tools to collect and display high quality data in real time. Each ILD sequence is designed for delivery in a single lecture period. The demonstrations help students build concepts through a series of instructor led steps involving prediction, discussions with peers, viewing the demonstration and reflecting on its outcome. The ILD collection includes sequences in mechanics, thermodynamics, electricity, optics and more. (Available both in print and in electronic form on *The Physics Suite CD.*)
4. **Workshop Physics** by Priscilla Laws (Dickinson College). *Workshop Physics* consists of a four part activity guide designed for use in calculus-based introductory physics courses. Workshop Physics courses are designed to replace traditional lecture and laboratory sessions. Students use computer tools for data acquisition, visualization, analysis and modeling. The tools include computer-assisted data acquisition software and hardware, digital video capture and analysis software, and spreadsheet software for analytic mathematical modeling. Modules include classical mechanics (2 modules), thermodynamics & nuclear physics, and electricity & magnetism. (Available both in print and in electronic form on *The Physics Suite CD.*)
5. **Tutorials in Introductory Physics** by Lillian C. McDermott, Peter S. Shaffer and the Physics Education Group at the University of Washington. These tutorials consist of a set of worksheets designed to supplement instruction by lectures and textbook in standard introductory physics courses. Each tutorial is designed for use in a one-hour class session in a space where students can work in small groups using simple inexpensive apparatus. The emphasis in the tutorials is on helping students deepen their understanding of critical concepts and develop scientific reasoning skills. There are tutorials on mechanics, electricity and magnetism, waves, optics, and other selected topics. (Available in print from Prentice Hall, Upper Saddle River, New Jersey.)
6. **Physics by Inquiry** by Lillian C. McDermott and the Physics Education Group at the University of Washington. This self-contained curriculum consists of a set of laboratory-based modules that emphasize the development of fundamental concepts and scientific reasoning skills. Beginning with their observations, students construct a coherent conceptual framework through guided inquiry. Only simple inexpensive apparatus and supplies are required. Developed primarily for the preparation of precollege teachers, the modules have also proven effective in courses for liberal arts students and for underprepared students. The amount of material is sufficient for two years of academic study. (Available in print.)
7. **The Activity Based Physics Tutorials** by Edward F. Redish and the University of Maryland Physics Education Research Group. These tutorials, like those developed at the University of Washington, consist of a set of worksheets developed to supplement lectures and textbook work in standard introductory physics courses. But these tutorials integrate the computer software and hardware tools used in other Suite elements including computer data acquisition, digital video analysis, simulations, and spreadsheet analysis. Although these tutorials include a range of classical physics topics, they also include additional topics in modern physics. (Available only in electronic form on *The Physics Suite CD.*)
8. **The Understanding Physics Video CD for Students** by Priscilla Laws, et. al.: This CD contains a collection of the video clips that are introduced in *Understanding Physics* narrative and alternative problems. The CD includes a number of QuickTime movie segments of physical phenomena along with the QuickTime player

software. Students can view video clips as they read the text. If they have video analysis software available, they can reproduce data presented in text graphs or complete video analyses based on assignments designed by instructors.

9. **WPTools** by Priscilla Laws and Patrick Cooney: These tools consist of a set of macros that can be loaded with Microsoft Excel software that allow students to graph data transferred from computer data acquisition software and video analysis software more easily. Students can also use the *WPTools* to analyze numerical data and develop analytic mathematical models.
10. **The Physics Suite CD.** This CD contains a variety of the Suite Elements in electronic format (Microsoft Word files). The electronic format allows instructors to modify and reprint materials to better fit into their individual course syllabi. The CD contains much useful material including complete electronic versions of the following:

    *RealTime Physics*

    *Interactive Lecture Demonstrations*

    *Workshop Physics*

    *Activity Based Physics Tutorials*

### A Final Word to the Instructor

Over the past decade we have learned how valuable it is for us as teachers to focus on what most students actually need to do to learn physics, and how valuable it can be for students to work with research-based materials that promote active learning. We hope you and your students find this book and some of the other *Physics Suite* materials helpful in your quest to make physics both more exciting and understandable to your students.

## Supplements for Use with *Understanding Physics*

### Instructor Supplements

1. **Instructor's Solution Manual** prepared by Anand Batra (Howard University). This manual provides worked-out solutions for most of the end-of-chapter problems.
2. **Test Bank** by J. Richard Christman (U. S. Coast Guard Academy). This manual includes more than 2500 multiple-choice questions adapted from HRW 6th. These items are also available in the *Computerized Test Bank* (see below).
3. **Instructor's Resource CD.** This CD contains:
    - The entire *Instructor's Solutions Manual* in both Microsoft Word© (IBM and Macintosh) and PDF files.
    - A *Computerized Test Bank,* for use with both PCs and Macintosh computers with full editing features to help you customize tests.
    - All text illustrations, suitable for classroom projection, printing, and web posting.
4. **Online Homework and Quizzing:** *Understanding Physics* supports WebAssign and eGrade, two programs that give instructors the ability to deliver and grade homework and quizzes over the Internet.

### Student Supplements

1. **Student Study Guide** by J. Richard Christman (U. S. Coast Guard Academy). This student study guide provides chapter overviews, hints for solving selected end-of-chapter problems, and self-quizzes.
2. **Student Solutions Manual** by J. Richard Christman (U. S. Coast Guard Academy). This manual provides students with complete worked-out solutions for approximately 450 of the odd-numbered end-of-chapter problems.

## Acknowledgements

Many individuals helped us create this book. The authors are grateful to the individuals who attended the weekend retreats at Airlie Center in 1997 and 1998 and to our editor, Stuart Johnson and to John Wiley & Sons for sponsoring the sessions. It was in these retreats that the ideas for *Understanding Physics* crystallized. We are grateful to Jearl Walker, David Halliday and Bob Resnick for graciously allowing us to attempt to make their already fine textbook better.

The authors owe special thanks to Sara Settlemyer who served as an informal project manager for the past few years. Her contributions included physics advice (based on her having completed Workshop Physics courses at Dickinson College), her use of Microsoft Word, Adobe Illustrator, Adobe Photoshop and Quark XPress to create the manuscript and visuals for this edition, and skillful attempts to keep our team on task—a job that has been rather like herding cats.

**Karen Cummings:** I would like to say "Thanks!" to: Bill Lanford (for endless advice, use of the kitchen table and convincing me that I really could keep the same address for more than a few years in a row), Ralph Kartel Jr. and Avery Murphy (for giving me an answer when people asked why I was working on a textbook), Susan and Lynda Cummings (for the comfort, love and support that only sisters can provide), Jeff Marx, Tim French and the poker crew (for their friendship and laughter), my colleagues at Southern Connecticut and Rensselaer, especially Leo Schowalter, Jim Napolitano and Jack Wilson (for the positive influence you have had on my professional life) and my students at Southern Connecticut and Rensselaer, Ron Thornton, Priscilla Laws, David Sokoloff, Pat Cooney, Joe Redish, Ken and Pat Heller and Lillian C. McDermott (for helping me learn how to teach).

**Priscilla Laws:** First of all I would like thank my husband and colleague Ken Laws for his quirky physical insights, for the Chapter 11 Kneecap puzzler, for the influence of his physics of dance work on this book, and for waiting for me countless times while I tried to finish "just one more thing" on this book. Thanks to my daughter Virginia Jackson and grandson Adam for all the fun times that keep me sane. My son Kevin Laws deserves special mention for sharing his creativity with us—best exemplified by his murder mystery problem, *A(dam)nable Man,* reprinted here as problem 5-68. I would like to thank Juliet Brosing of Pacific University who adapted many of the Workshop Physics problems developed at Dickinson for incorporation into the alternative problem collection in this book. Finally, I am grateful to my Dickinson College colleagues Robert Boyle, Kerry Browne, David Jackson, and Hans Pfister for advice they have given me on a number of topics.

**Joe Redish:** I would like to thank Ted Jacobsen for discussions of our chapter on relativity and Dan Lathrop for advice on the sources of the Earth's magnetic field, as well as many other of my colleagues at the University of Maryland for discussions on the teaching of introductory physics over many years.

**Pat Cooney:** I especially thank my wife Margaret for her patient support and constant encouragement and I am grateful to my colleagues at Millersville University: John Dooley, Bill Price, Mike Nolan, Joe Grosh, Tariq Gilani, Conrad Miziumski, Zenaida Uy, Ned Dixon, and Shawn Reinfried for many illuminating conversations.

We also appreciate the absolutely essential role many reviewers and classroom testers played. We took our reviewers very seriously. Several reviewers and testers deserve special mention. First and foremost is Arnold Arons who managed to review 29 of the 38 chapters either from the original HRW 6th material or from our early drafts before he passed away in February 2001. Verne Lindberg from the Rochester Institute of Technology deserves special mention for his extensive and very insightful reviews of most of our first 18 chapters. Ed Adelson from Ohio State did a particularly good job reviewing most of our electricity chapters. Classroom tester Maxine Willis from Gettysburg Area High School deserves special recognition for compiling valuable comments that her advanced placement physics students made while class testing Chapters 1-12 of the preliminary version. Many other reviewers and class testers gave us useful comments in selected chapters.

**Class Testers**

Gary Adams
Rensselaer Polytechnic Institute

Marty Baumberger
Chestnut Hill Academy

Gary Bedrosian
Rensselaer Polytechnic Institute

Joseph Bellina,
Saint Mary's College

Juliet W. Brosing
Pacific University

Shao-Hsuan Chiu
Frostburg State

Chad Davies
Gordon College

Hang Deng-Luzader
Frostburg State

John Dooley
Millersville University

Diane Dutkevitch
Yavapai College

Timothy Hayes
Rensselaer Polytechnic Institute

Brant Hinrichs
Drury College

Kurt Hoffman
Whitman College

James Holliday
John Brown University

Michael Huster
Simpson College

Dennis Kuhl
Marietta College

John Lindberg
Seattle Pacific University

Vern Lindberg
Rochester Institute of Technology

Stephen Luzader
Frostburg State

Dawn Meredith
University of New Hampshire

Larry Robinson
Austin College

Michael Roth
University of Northern Iowa

John Schroeder
Rensselaer Polytechnic Institute

Cindy Schwarz
Vassar College

William Smith
Boise State University

Dan Sperber
Rensselaer Polytechnic Institute

Roger Stockbauer
Louisiana State University

Paul Stoler
Rensselaer Polytechnic Institute

Daniel F. Styer
Oberlin College

Rebecca Surman
Union College

Robert Teese
Muskingum College

Maxine Willis
Gettysburg Area High School

Gail Wyant
Cecil Community College

Anne Young
Rochester Institute of Technology

David Ziegler
Sedro-Woolley High School

**Reviewers**

Edward Adelson
Ohio State University

Arnold Arons
University of Washington

Arun Bansil
Northeastern University

Chadan Djalali
University of South Carolina

William Dawicke
Milwaukee School of Engineering

Robert Good
California State University-Hayware

Harold Hart
Western Illinois University

Harold Hastings
Hofstra University

Laurent Hodges
Iowa State University

Robert Hilborn
Amherst College

Theodore Jacobson
University of Maryland

Leonard Kahn
University of Rhode Island

Stephen Kanim
New Mexico State University

Hamed Kastro
Georgetown University

Debora Katz
U. S. Naval Academy

Todd Lief
Cloud Community College

Vern Lindberg
Rochester Institute of Technology

Mike Loverude
California State University-Fullerton

Robert Luke
Boise State University

Robert Marchini
Memphis State University

Tamar More
Portland State University

Gregor Novak
U. S. Air Force Academy

Jacques Richard
Chicago State University

Cindy Schwarz
Vassar College

Roger Sipson
Moorhead State University

George Spagna
Randolf-Macon College

Gay Stewart
University of Arkansas-Fayetteville

Sudha Swaminathan
Boise State University

We would like to thank our proof readers Georgia Mederer and Ernestine Franco, our copyeditor Helen Walden, and our illustrator Julie Horan.

Last but not least we would like to acknowledge the efforts of the Wiley staff; Senior Acquisitions Editor, Stuart Johnson, Ellen Ford (Senior Development Editor), Justin Bow (Program Assistant), Geraldine Osnato (Project Editor), Elizabeth Swain (Senior Production Editor), Hilary Newman (Senior Photo Editor), Anna Melhorn (Illustration Editor), Kevin Murphy (Senior Designer), and Bob Smith (Marketing Manager). Their dedication and attention to endless details was essential to the production of this book.

**Karen Cummings** (Southern Connecticut State University)
**Priscilla W. Laws** (Dickinson College)
**Edward F. Redish** (University of Maryland)
**Patrick J. Cooney** (Millersville University)

with
**David R. Sokoloff** (University of Oregon)
**Ronald K. Thornton** (Tufts University)
**Edwin F. Taylor** (Massachusetts Institute of Technology)

# Contents

# Introduction

The test of all knowledge is experiment. But what is the source of knowledge? Where do the laws that are to be tested come from? . . . Experiment, itself, helps to produce these laws, in the sense that it gives us hints. But also needed is imagination to create from these hints the great generalizations—to guess at the wonderful, simple, but very strange patterns beneath them all, and then to experiment to check again whether we have made the right guess.[1]

[1] R. P. Feynman, *The Feynman Lectures on Physics,* Ch. 1, (Addison-Wesley, Reading, MA, 1964).

## The Nature of Physics and Learning Physics

Welcome to the study of physics. Physics is a process of learning about the physical world by finding ways to make sense of what we observe and measure. As the inspiring teacher Richard Feynman wrote, "Progress in all of the natural sciences depends on this interaction between experiment and theory."[2]

The point here is that to learn physics you must continually compare and contrast your observations to your intuitions and expectations. Sometimes your intuitions will be right, sometimes they'll be partially right, and sometimes they'll be dead wrong. Comparing observations to your intuitions will not only help you learn more physics, it will help you to understand how scientific knowledge is created.

Physics is supposed to help you make sense of the physical world. If a physical phenomenon doesn't make sense at first, keep thinking. Keep analyzing observations and experiments and considering what they mean. Einstein said, "Physics is the refinement of common sense." The key here is on the word "refinement." Physics is more than common sense. It's common sense made consistent by continued reference to both theory and experiment.

In some ways learning physics may seem much simpler than learning biology or chemistry. There are fewer things to consider and the systems we study are simpler. If you write down all the most basic equations you encounter in a physics course there are far fewer to remember than the number of organisms you encounter in a general biology course or the number of reactions you encounter in general chemistry. Also, many physical phenomena seem relatively simple. A system consisting of a ball rolling down an inclined plane or a battery connected to a bulb is a lot simpler than an octopus or the chemical cyclohexane. But many students complain that introductory physics is harder to learn than other sciences. What's going on? One problem is that it is easy to fall into the trap of thinking of physics as a jumble of separate equations to be memorized. *This is not so!* Most equations used in introductory physics courses can be derived from a relatively small number of fundamental relationships.

If you focus your efforts on trying to memorize the properties of hundreds of specific systems you will quickly get overwhelmed. Instead, you should focus on the nature of the scientific process by studying the behavior of a limited number of ideal systems. How can you tell whether a prediction you have made about the behavior of a physical system is correct? How do investigators discover or create "scientific laws?" How can we be sure a law or theory is valid? These questions are critical to solving real-world scientific problems such as how to create a new computer chip, diagnose an illness, or improve the performance of an athlete. Your efforts to *learn fundamental relationships* and to apply them to new scientific problems are the key to understanding physics.

## The Art of Simplifying

In physics, we try to understand the rules that govern the way the natural world behaves. But the natural world is a very complex place. So, we start by considering the simplest system that allows us to observe and explain a type of behavior. For example, when studying motion we start with a small object whose structure and shape we can ignore. We pretend a football is just a tiny blob. We figure out how it moves after being thrown and under the influence of gravity only—pretending that it is in a vacuum and that it never rotates or deforms. These are clearly not good assumptions for a real football! But they provide an excellent starting point for making sense of its basic motion. Over small distances (a few feet), and for reasonably low speeds (below

[2]R. P. Feynman, *The Feynman Lectures on Physics,* Ch. 1, (Addison-Wesley, Reading, MA, 1964).

about 20 miles/hour) the idealized description works very well. As you get up to higher speeds and distances, the effects of the air grow in importance. However, this additional complication is manageable. Once you understand the basic principles of motion, you can add details to your "model" to account for the effects of air and thereby make the situations you understand more realistic and extend the number of cases you can treat.

A typical physicist's initial strategy is to understand simple systems as completely as possible by constructing physical laws that describe them. Once that is accomplished, the next step is to add more and more real-world complexity to the system one step at a time. This is the process investigators use to contribute to the powerful body of knowledge that is physics. This is the process we suggest you also use to construct and extend your knowledge of physics.

**FIGURE I-1** ■ Jason being explicit about all the simplifications he is being asked to make.

## Expect Surprises

You will probably find many surprises in your study of physics, and you don't need to wait until you study relativity or quantum mechanics to do so (though both topics are really interesting and lots of fun). Even the physics phenomena that we present in the early chapters of this book will reveal some facts about our everyday world that many people find surprising. For example, if you take a ball made of lead and a similar ball made of plastic, the lead ball may weigh 20 times as much as the plastic ball. Yet if you stand on a chair that is perched on a sturdy table and drop the two balls at the same time, they fall ten feet to the ground in almost exactly the same time. Why doesn't the lead ball go faster? Or, when an object is immersed in water, it seems to weigh less—and its weight reduction is equal to the weight of the water that it pushed out of the way. What could that water have to do with anything? That water is gone! When you connect two identical bulbs up to a battery, if you connect them in one way they'll both have the same brightness as a single bulb connected to the battery. But, if you connect them in another way, they both get much dimmer. Huh? Why does that happen? This book is full of such surprises.

**FIGURE I-2** ■ Two balls with different masses fall with the same acceleration whenever air drag is negligible.

## Using This Book as a Learning Tool

This textbook is one of many resources that you will need to make use of in order to learn physics. It is very important that you read this textbook on a regular basis and do the *Reading Exercises* at the end of many sections in each chapter. We attempt to present both the experimental results that support theories and some of the reasoning that has gone into the development of theories. However, you will understand the physics only when you make your own observations and are actively engaged in reasoning. So, it is critical that you observe a physical phenomenon directly or ponder the outcome of an experiment that we describe. Then you need to *think* about whether the explanation of the phenomenon we present makes sense. In addition, you must test and refine your understanding of theoretical concepts by applying them to solving problems included at the end of each chapter. Solving problems requires you to use both the physical principles you have learned and the mathematical relationships that describe these principles. Finally, if possible, you will want to test your understanding of physical systems by predicting the outcomes of experiments that you can perform in a basic introductory physics laboratory.

We hope this book will help you enjoy the practice of physics as much as we do.

(Left to right): Priscilla W. Laws, Edward F. Redish, Karen Cummings, and Patrick J. Cooney. Photo by David Hildebrand.

# 1 Measurement

You can watch the Sun set and disappear over a calm ocean, once while lying on the beach, and then once again if you stand up. This is a surprising observation! Furthermore, if you measure the time between the two sunsets, you can approximate the Earth's radius by using an understanding of the shape and motion of the Earth relative to the Sun along with some basic high school mathematics.

**How can such a simple observation be used to measure the Earth?**

*The answer is in this chapter.*

## 1-1 Introduction

Physics is the study of the basic components of the universe and their interactions. The fact that you can use the time difference between sunsets while lying on the beach and then standing to estimate the size of the earth is indeed surprising to most people. It is one example of how the interplay between mathematics, theoretical principles, and observations allow us to develop a deeper understanding of the physical world. In fact, the ongoing quest of physics is to develop a unified set of ideas to explain apparently different phenomena. Scientific theories are only valid if they serve to explain and predict the outcomes of new observations and experiments. Many theories in physics are expressed in mathematical equations, and predictions usually involve quantities that can be measured.

Measurement is the process of associating numbers with physical quantities. In fact, *physical quantities are defined in terms of the procedures used to measure them.* But the numbers that result from measurements are not meaningful unless people who are using and interpreting them know what was measured and what units were used to obtain the numbers. For example, if you were asked to go to a store to buy 27, you would immediately ask 27 of what? If you were told 27 containers of milk, you might ask 27 of what size or unit—pints, quarts, or gallons? Unambiguous communication with others about the results of a scientific measurement requires agreement on (1) the definition of the physical quantity and (2) the basic units used for comparison when the measurements are made.

The focus in this chapter will be on the fundamental physical quantities and measurement processes used to study motion. Later on we introduce additional physical quantities defined for the study of thermal interactions, electricity, magnetism, and light. You will learn about common elements of physical measurements, reasons why precise measurements are highly valued, and the international system of standard basic units that allows scientists all over the world to communicate with each other.

## 1-2 Basic Measurements in the Study of Motion

A long jumper speeds up along a runway, leaps into the air, and then comes to a sudden stop in a sand pit. How can such a motion be described and studied scientifically?

In studying motion, at least three questions come to mind. How far has something moved and in what directions? How long did it take? How much stuff was moved? Let's consider length, time, and mass, the three basic physical quantities used in the study of motion. How are they usually defined? What procedures are used to measure them on an everyday basis?

**Length:** Our "How far?" question involves being able to measure the distance between two points. Suppose you had no measuring instrument. Is there any way you could meaningfully ask and answer the question, "What is the total distance that the jumper ran?" The only approach possible would be to compare this distance to the size of one of your body parts such as your hand or foot. It is not surprising that the hand and the foot have been used throughout history as basic units of measurement. The distance can then be described as a ratio between it and a convenient item chosen to be a length standard.

**Time:** To answer the question, "How long did it take?" you need to be able to measure a time interval. To do this, you define the time between repetitive events as a standard. Historically, repetitive events that have been used as time standards have included the day (the time it takes for the Sun to appear to revolve around the Earth), the year, and the time it takes for a pendulum of a certain length to swing back and

**FIGURE 1-1** ■ A common method of determining mass assumes two objects have the same mass if they balance each other.

forth. A time interval, or time duration, is measured by determining how many years have passed or how many swings of a pendulum have occurred during the interval being measured.

**Mass:** Mass is a measure of "amount of stuff." Throughout recorded history, merchants and scientists have used balances to determine how many units of "standard mass" are needed to balance whatever is being measured. (See Fig. 1-1.) A standard of mass can be a certain object that everyone agrees should be used. Replicas of the standard mass that balance with it can be passed around and used by many people.

The everyday procedures outlined above for measuring length, time, and mass share common elements that characterize all physical quantities.

1. These quantities are defined by the procedures used to measure them.
2. Their measurement always involves the determination of a ratio between a unit, known as a base quantity, and the quantity being measured.
3. Such comparisons can only be made with limited precision.

As you will see, there are often many alternative procedures that can be used to measure the same quantity. Indeed, a major factor in the progress of science and technology has been the discovery of better, more **precise** methods of measurement.

**READING EXERCISE 1-1:** List one common base unit used for time, for length, and for mass not mentioned in the discussion in this section. ■

**READING EXERCISE 1-2:** What is a more precise base unit for length measurement that is reliable over a period of years—a 12-inch ruler or your foot? Explain the reason for your answer. ■

**READING EXERCISE 1-3:** What problems might arise when using the length of the day as a standard unit of time? ■

## 1-3 The Quest for Precision

Using a grocery store spring scale to find an apple's mass is fine for shopping purposes. But a mass can be determined to a far greater precision with a chemical microbalance. At best, the apple's mass can only be determined to the nearest gram, whereas the chemistry lab sample can be determined to the nearest hundred-thousandth of a gram.

Throughout history people have sought to measure physical quantities as precisely as possible, because reducing measurement uncertainties has been of tremendous importance in commerce, navigation, astronomical observation, engineering, and scientific research. For example, in 1707, the British navy lost almost 2000 men when four warships ran aground because navigators were unable to measure longitude with sufficient precision. In 1714, as a result of this mishap and others, the British government offered a prize of £20,000 (current value about $12 million) to anyone who could devise a scheme to measure longitude to within half a degree. John Harrison, a self-educated clockmaker, collected the prize in 1765 after designing a series of elaborate chronometers. His early models were driven by a combination of rust-proof brass and self-lubricating wooden gears that kept time to within 1 second per day.

**HOW CAN TIME MEASUREMENTS BE USED TO DETERMINE LONGITUDE?** Harrison's measurement technique is one of several examples of how a time standard and a knowledge of how fast something is moving are used to measure distance more precisely. In this case, since the Earth turns through 360° on its axis in 24 hours, a precise chronometer can be set so that it reads exactly noon when the Sun is at its highest point in a port with known longitude. Out at sea, the clock time that was set in port will differ from the local solar time by 4 minutes for each degree of longitude difference. Thus, the difference between the observed local noon and the clock reading can then be used to calculate longitude.

Of all the measured quantities, time and other measurements based on time are the most precise. By the end of the 20th century, many of us were wearing inexpensive digital watches driven by the oscillations of quartz crystals. These watches are 1000 times better than John Harrison's chronometer, since they are accurate to within 1 part in $10^8$ or 1 thousandth of a second per day. Atomic clocks, precise to 3 billionths of a second per day, are now being used as time standards in many countries.

**READING EXERCISE 1-4:** A ship embarks from Southampton, England where its clock was set to 12:00:00 at local noon. After 14 days under sail its chronometer reads 12 h 20 min 13 s at the moment the Sun is highest in the sky (local noon). (a) By how many degrees has the ship's longitude changed? (b) Suppose the clock is not precise and has gained 2 minutes out of the 20 160 minutes that have elapsed since it set sail. How far off will the longitude measurement be? (c) The circumference of the Earth is 24 000 nautical miles. Suppose the ship was traveling along the equator. How many miles off course could the ship be if the uncertainty of longitude is 0.5°? ■

## 1-4 The International System of Units

In the past, communication between scientists was complicated by the fact that for every physical quantity there were a multitude of measurement procedures and basic units of comparison. In addition, there are so many physical quantities that it is a problem to organize them. Fortunately, these quantities are not all independent; for example, speed is the ratio of a length to a time. Thus, what we do is pick out—by international agreement—a small number of physical quantities, such as length and time, and assign standards to them alone. We then define all other physical quantities in terms of these *base quantities* and their standards (which we now call *base standards*). Speed, for example, is defined in terms of the base quantities length and time.

**WHY IS IT IMPORTANT TO HAVE A STANDARD SYSTEM OF UNITS THAT IS USED BY ALL SCIENTISTS AND ENGINEERS?** In December 1998, the National Aeronautics and Space Administration launched the *Mars Climate Orbiter* on a scientific mission to collect Martian climate data. Nine months later, on September 23, 1999, the *Orbiter* disappeared while approaching Mars at an unexpectedly low altitude. (See Fig. 1-2). An investigation revealed that the orbital calculations were incorrect due to an error in the transfer of information between the spacecraft's team in Colorado and the mission navigation team in California. One team was using English units such as feet and pounds for a critical calculation, while the other group assumed the result of the calculation was being reported in metric units such as meters and kilograms. This misunderstanding about the units being used cost U.S. taxpayers approximately 125 million dollars.

**FIGURE 1-2** ■ The *Mars Climate Orbiter* failed to go into orbit around Mars and disappeared due to a miscalculation that resulted from confusion about what units were being used.

In 1971, the 14th General Conference on Weights and Measures recognized the need to use standard units for physical quantities. Conference attendees chose seven physical quantities as base quantities and defined a standard unit of measure for each one. Although other sets of physical quantities could be defined, the seven shown in Table 1-1 form the basis of the widely accepted International System of Units. The system is popularly known as the *metric system* or by its abbreviation, SI, which derives from its French name, *Système International.*

All other SI units are known as *derived units* because they can be expressed in terms of the base units. For example, the SI unit for power, called the **watt** (symbol: W), is defined in terms of the base units for mass, length, and time. As you will see in Chapter 9,

$$1 \text{ watt} = 1 \text{ W} = 1 \text{ kg} \cdot \text{m}^2/\text{s}^3. \tag{1-1}$$

The fact that the dozens of units used in different branches of physics can all be derived from a set of seven base units seems incredible and is a profound testimonial to the unity of physics.

To express the very large and very small quantities that we often run into in physics, we use *scientific notation*, which employs powers of 10. In this notation,

$$3\,560\,000\,000 \text{ m} = 3.56 \times 10^9 \text{ m} \tag{1-2}$$

and

$$0.000\,000\,492 \text{ s} = 4.92 \times 10^{-7} \text{ s}. \tag{1-3}$$

Scientific notation on computers sometimes takes on an even briefer look, as in 3.56 E9 and 4.92 E-7, where E stands for "exponent of ten." It is briefer still on some calculators, where E is replaced with an empty space.

**TABLE 1-1**
**The SI Base Units**

| Quantity | Unit Name | Unit Symbol |
|---|---|---|
| Length | meter | m |
| Time | second | s |
| Mass | kilogram | kg |
| Amount of substance | mole | mol |
| Electric current | ampere | A |
| Thermodynamic temperature | kelvin | K |
| Luminous intensity | candela | cd |

**TABLE 1-2**
**Common Prefixes for SI Units**

| Factor | Prefix | Symbol | Factor | Prefix | Symbol |
|---|---|---|---|---|---|
| $10^{12}$ | tera- | T | $10^{-15}$ | femto- | f |
| $10^{9}$ | giga- | G | $10^{-12}$ | pico- | p |
| $10^{6}$ | mega- | M | $10^{-9}$ | nano- | n |
| $10^{3}$ | kilo- | k | $10^{-6}$ | micro- | $\mu$ |
| | | | $10^{-3}$ | milli- | m |
| | | | $10^{-2}$ | centi- | c |
| | | | $10^{-1}$ | deci- | d |

When reporting the results of very large or very small measurements, it is convenient to define prefixes that designate what power of ten a number has. For example, we can use the prefix kilo-, which represents $10^3$, to express $1.0 \times 10^3$ grams as 1.0 kilogram. Some of the most common prefixes used in physics and engineering are listed in Table 1-2. A complete list of SI prefixes is included on the inside front cover. As you can see, each prefix represents a certain power of 10 as a factor. Attaching a prefix to an SI unit has the effect of multiplying it by the associated factor. Thus, we can express a particular electric power as

$$1.27 \times 10^9 \text{ watts} = 1.27 \text{ gigawatts} = 1.27 \text{ GW}, \tag{1-4}$$

or a particular length as

$$2.35 \times 10^{-9} \text{ m} = 2.35 \text{ nanometers} = 2.35 \text{ nm}. \tag{1-5}$$

**FIGURE 1-3** ■ Two different surfaces of a crystal of pure silicon.

Some prefixes, as used in milliliter, centimeter, kilogram, and megabyte, may be familiar to you.

Once we have set up a standard unit—say, for length—we must work out procedures by which any length, be it the distance to a star or the radius of a hydrogen atom, can be expressed in terms of the standard. Rulers, which approximate our length standard, give us one such procedure for measuring length. We can use a ruler to measure another length by counting how many times the standard can be fit, laid end-to-end, to the other length. The count is our assigned length and is given in terms of the standard's unit. However, many of our comparisons must be indirect. You cannot use a ruler, for example, to measure the distance to a star or the radius of an atom. Figure 1-3 shows an image of the surface of a crystal of silicon obtained with a modern scanning probe microscope.

Base standards must be both accessible and invariable. If we define the length standard as the distance between one's nose and the index finger on an outstretched arm, we certainly have an accessible standard—but it will, of course, vary from person to person. The demand for precision in science and engineering pushes us to aim first for invariability. We then exert great effort to make duplicates of the base standards that are accessible to those who need them. In the United States, the National Institute of Standards and Technology (NIST) is responsible for maintaining base standards and researching issues related to measurement.

The topics that we will investigate first, those related to the physics of forces and motion, require that we make measurements of time, length, and mass. Therefore, we begin by discussing the formal SI definitions of these quantities.

## 1-5 The SI Standard of Time

Time has two separate aspects that are important in physics. We may want to note at what moment an event occurred or began, or we may want to know how long the event lasted. These are two very different aspects of the measurement of time. For example, the moment at which your physics teacher walks into the room for class on a given day will be measured differently by different students because their watches will not all be synchronized. However, the measured duration of the class will not be affected by the fact that the watches are not synchronized. Thus, "*When* did it happen?" and "What is its *duration*?" are two different questions.

Any phenomenon that regularly repeats itself is a possible time standard. The Earth's rotation, which determines the length of the day, has been used in this way for centuries. Originally the second was defined as the fraction 1/86 400 of a "mean solar day." Figure 1-4 shows a two-century-old example of a time-keeping instrument used to measure the Earth's rotation in terms of a 20-hour day. A quartz clock, in which a quartz ring is made to vibrate continuously, can be calibrated against Earth's rotation via astronomical observations and used to measure time intervals in the laboratory. However, even this calibration cannot be carried out with the accuracy called for by modern scientific and engineering technology.

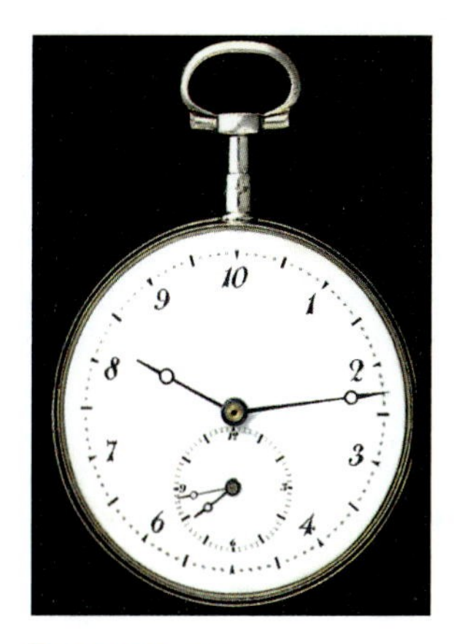

**FIGURE 1-4** ■ When the metric system was proposed in 1792, the hour was redefined to provide a 20-hour day. The idea did not catch on. The maker of this watch wisely provided a small dial that kept both 10-hour and conventional 12-hour time. Do the two dials indicate the same time?

To meet the need for more accuracy in the measurement of time, atomic clocks have been developed that replace the use of Earth's rotation in the definition of our time standard. In 1967, the 13th General Conference on Weights and Measures adopted a standard second based on the radiation absorption characteristics of the cesium-133 atom. Like other atoms, a cesium-133 atom can absorb electromagnetic radiation that has a very precise frequency when the atom makes a transition between two of its well-defined energy states known in technical jargon as "hyperfine levels." The fixed frequency of this external radiation is used to drive a cesium clock. Such a precisely repetitive event is just what is needed for a high-precision timekeeper. Although the technical details of how a cesium clock works is beyond the scope of this text, interested readers can consult the NIST web site at http://www.nist.gov for more information about how the cesium clock is used as a time standard. (See Fig. 1-5.) This new SI standard of time defines the second as follows:

> One second is the duration of 9 192 631 770 periods of the radiation corresponding to the transition between the two hyperfine levels of the ground state of the cesium-133 atom.

An atomic clock at NIST is the standard for Coordinated Universal Time (CUT) in the United States. Its time signals are available from NIST's Web site listed previously. You can also download a Java program from this site that will synchronize your computer's clock to Coordinated Universal Time so you can use your computer as a time standard by which to set other clocks.

Atomic clocks are so consistent that, in principle, two cesium clocks would have to run for 6000 years before their readings would differ by more than 1 second. This amounts to a precision better than 1 part in $10^{11}$. Even such accuracy pales in comparison to that of clocks currently being developed; their precision may be as fine as 1 part in $10^{18}$.

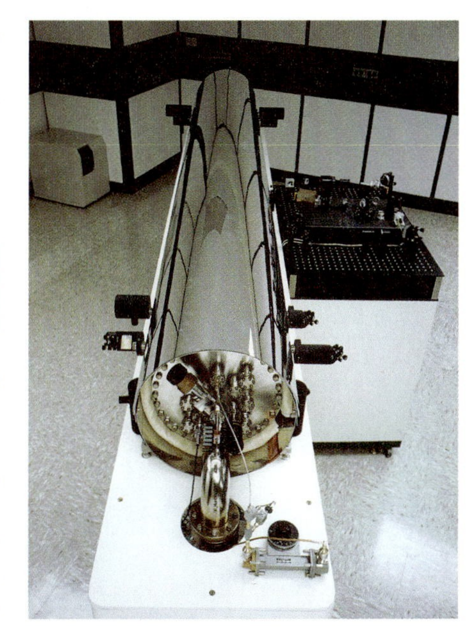

**FIGURE 1-5** ■ The cesium atomic frequency standard at the National Institute of Standards and Technology in Boulder, Colorado. It is the primary standard for the unit of time in the United States. To set your watch by it, call (303) 499-7111, or call (900) 410-8463 or http://tycho.usno.navy.mil/time.html for Naval Observatory time signals.

**READING EXERCISE 1-5:** (a) You and a friend are observing a storm. Each of you has your own watch. Describe under what conditions you will both measure the same time for a flash of lightning. Describe under what conditions you will both measure the same duration of time between the lightning flash and the clap of thunder. (b) Look at Fig. 1-4. Do the 10-hour and 12-hour clocks really show the same time? ■

## TOUCHSTONE EXAMPLE 1-1*: Sunset

Suppose that while lying on a beach watching the Sun set over a calm ocean, you start a stopwatch just as the top of the Sun disappears. You then stand, elevating your eyes by a height $h = 1.70$ m, and stop the watch when the top of the Sun again disappears. If the elapsed time on the watch is $t = 11.1$ s, what is the radius $r$ of Earth?

**SOLUTION** ■ A **Key Idea** here is that just as the Sun disappears, your line of sight to the top of the Sun is tangent to Earth's surface. Two such lines of sight are shown in Fig. 1-6. There your eyes are located at point $A$ while you are lying, and at height $h$ above point $A$ while you are standing. For the latter situation, the line of sight is tangent to Earth's surface at point $B$. Let $d$ represent the distance between point $B$ and the location of your eyes when you are standing, and draw radii $r$ as shown in Fig. 1-6. From the Pythagorean theorem, we then have

$$d^2 + r^2 = (r + h)^2 = r^2 + 2rh + h^2,$$

or

$$d^2 = 2rh + h^2. \quad (1\text{-}6)$$

Because the height $h$ is so much smaller than Earth's radius $r$, the term $h^2$ is negligible compared to the term $2rh$, and we can rewrite Eq. 1-6 as

$$d^2 \approx 2rh. \quad (1\text{-}7)$$

In Fig. 1-6, the angle between the radii to the two tangent points $A$ and $B$ is $\theta$, which is also the angle through which the Sun moves about Earth during the measured time $t = 11.1$ s. During a full day, which is approximately 24 h, the Sun moves through an angle of 360° about Earth. This allows us to write

$$\frac{\theta}{360°} = \frac{t}{24\text{ h}},$$

which, with $t = 11.1$ s, gives us

$$\theta = \frac{(360°)(11.1\text{ s})}{(24\text{ h})(60\text{ min/h})(60\text{ s/min})} = 0.04625°.$$

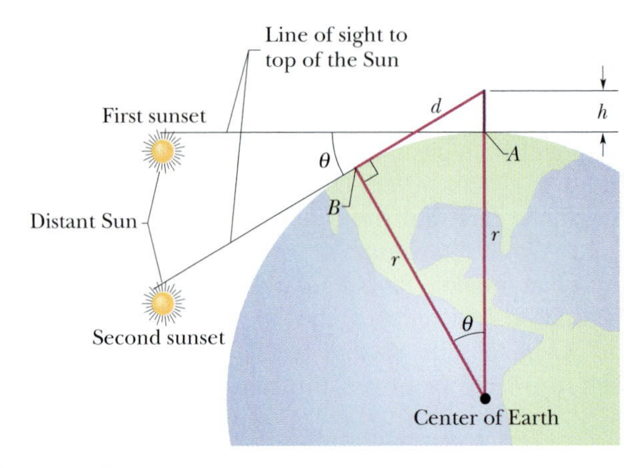


**FIGURE 1-6** ■ Your line of sight to the top of the setting Sun rotates through the angle $\theta$ when you stand up at point $A$, and elevate your eyes by a distance $h$. (Angle $\theta$ and distance $h$ are exaggerated here for clarity.)

Again in Fig. 1-6, we see that $d = r \tan\theta$. Substituting this for $d$ in Eq. 1-7 gives us

$$r^2 \tan^2\theta = 2rh,$$

or

$$r = \frac{2h}{\tan^2\theta}.$$

Substituting $\theta = 0.04625°$ and $h = 1.70$ m, we find

$$r = \frac{(2)(1.70\text{ m})}{\tan^2(0.04625°)} = 5.22 \times 10^6\text{ m}, \quad \text{(Answer)}$$

which is within 20% of the accepted value ($6.37 \times 10^6$ m) for the mean radius of Earth.

*Adapted from "Doubling Your Sunsets, or How Anyone Can Measure the Earth's Size with a Wristwatch and Meter Stick," by Dennis Rawlins, *American Journal of Physics*, Feb. 1979, Vol. 47, pp. 126–128. This technique works best at the equator.

## 1-6 The SI Standards of Length

In 1792, the newly born Republic of France established a new system of weights and measures. Its cornerstone was the meter, defined to be one ten-millionth of the distance from the North Pole to the equator. However, the first prototype of a 1-meter-long rod was short by 0.2 millimeter, because researchers miscalculated the flattening of the Earth due to its rotation. Nonetheless, this shortened length became the standard meter. For practical reasons, the meter came to be defined as the distance between two fine lines engraved near the ends of a special platinum-iridium bar, the **standard meter bar,** which was kept at the International Bureau of Weights and Measures near Paris. Accurate copies of the bar have been sent to standards laboratories throughout the world including NIST.

Eventually, modern science and technology required an even more precise standard. Today, the length standard is based on the speed of light. As you will learn in Chapter 38, one of the landmark discoveries of the 20th century was Einstein's recognition that the speed of light in a vacuum is the same for all observers. Since the speed of light can be measured to very high precision, it was adopted as a defined quantity in 1983. Time measurements with atomic clocks are also very precise, so it made sense to redefine the meter in terms of the time it takes light to travel 1 meter. By defining the speed of light $c$ to be exactly

$$c = 299\,792\,458 \text{ m/s}, \tag{1-8}$$

light would travel 1 meter in a time period equal to 1/299 792 458 of a second. That is, if one takes this speed and multiplies by this time period, then the distance traveled by the light is exactly 1 meter. According to the 17th General Conference on Weights and Measures:

> The meter is the length of the path traveled by light in a vacuum during a time interval of 1/299 792 458 of a second.

This approach of measuring lengths in terms of a speed and time is similar to that taken by John Harrison in the 18th century when he proposed measuring longitude in terms of the angular speed of the Earth's rotation and time.

Defining the standard meter in terms of the time it takes light to travel a meter has not done away with the need for secondary standards like bars of metal with fine lines delineating the beginning and end points of a meter. We currently use the metal bar as a secondary standard against which we can easily compare other objects. Defining the meter in terms of the speed of light simply gives us a more precise way to verify that our secondary standard is correct.

## 1-7 SI Standards of Mass

Currently there are two accepted base units for mass—one suitable for determining large masses and the other for determining masses on an atomic scale.

### The Standard Kilogram

The initial SI standard of mass is a platinum-iridium cylinder (Fig. 1-7) kept at the International Bureau of Weights and Measures near Paris. By international agreement, it is defined as a mass of 1 kilogram. Accurate replicas have been sent to standards laboratories in other countries, and the masses of other bodies can be determined by balancing them against a replica. The United States copy of the standard kilogram is housed in a vault at NIST. It is removed, no more than once a year, for the purpose of checking replicas used elsewhere. Since 1889, the U.S. replica of the standard kilogram has been taken to France twice for comparison with the primary standard.

**FIGURE 1-7** ■ The international 1 kg standard of mass is a cylinder 39 mm in both height and diameter.

### The Atomic Mass Unit

The mass of the known universe is estimated to be $1 \times 10^{53}$ kg. In contrast, the electron, which plays a vital role in chemical bonding, has a mass of $9 \times 10^{-31}$ kg. Obviously, the masses of electrons and atoms can be compared with each other more precisely than they can be compared with the standard kilogram. For this reason, we have a second mass standard. It is the carbon-12 atom, which, by international agreement, has been assigned a mass of 12 **atomic mass units** (u). The relation between the atomic mass unit and the kilogram is

$$1\ \text{u} = 1.660\,538\,73 \times 10^{-27}\ \text{kg}, \tag{1-9}$$

with an uncertainty of $\pm 13$ in the last two decimal places. Scientists can determine the masses of other atoms relative to the mass of carbon-12 with much better precision than they can using a standard kilogram.

We presently lack a reliable way to extend the precision of the atomic mass unit to more common units of mass, such as the kilogram. However, it is not hard to imagine how one might do this. If we had an object made up of carbon-12 atoms and knew the exact number of atoms in the object, than we could build a precise standard kilogram based on the atomic unit. Work on this is currently underway at NIST and other similar institutions.

**READING EXERCISE 1-6:** Describe a procedure for determining the mass of the object that has a mass much less than 1 kilogram. Assume that you have a balance, a replica of a standard kilogram, and a big blob of clay available to you. ■

## 1-8 Measurement Tools for Physics Labs

Institutions like NIST and the International Bureau of Weights and Measures in Paris have many exotic instruments for performing extremely precise measurements. Traditionally, physics students use more common measuring tools in the laboratory, such as meter sticks, vernier calipers (Fig. 1-8), mechanical and electronic balances, digital stopwatches, and multimeters. With careful use, these tools provide adequate precision for studying the time durations and distances investigated in introductory physics laboratories.

**FIGURE 1-8** ■ Vernier calipers are cleverly designed to make length measurements to within 1/10 of a millimeter.

In the past few years, new computer tools have become popular in introductory laboratories and in interactive lecture demonstrations. These tools greatly enhance the speed and precision of measurements while allowing students to make many measurements easily and accurately. These tools include **computer data acquisition systems** (Fig. 1-9) **and video capture and analysis tools.** Data obtained using these new computer tools will be shown throughout this text. These data will be used to provide

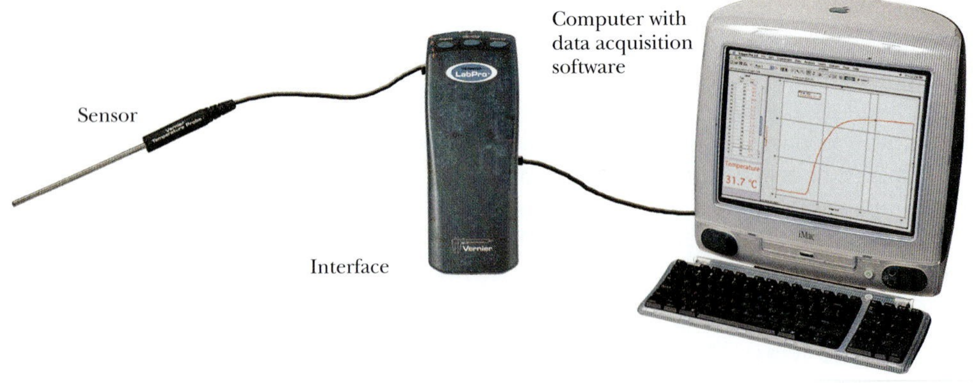


**FIGURE 1-9** ■ The photo shows a computer data acquisition system consisting of a sensor, an interface, a computer, and software for real-time data collection.

experimental evidence to motivate and test various theories presented in this book. You may be replicating some of these experiments in laboratory or lecture sessions.

**FIGURE 1-10** ■ An ultrasonic motion detector.

## Computer Data Acquisition System

When a sensor is attached to a computer through an interface, a very powerful data collection, analysis, and display system is created.* Computers coupled with appropriate software packages are capable of analyzing signals and displaying them on the screen in easily understood formats. Using these capabilities, a graphical representation of data can be displayed in "real time."

A number of different sensors are used in contemporary introductory physics laboratories (Fig. 1-10). These include sensors for the detection of linear and rotational motion (Fig. 1-11), acceleration, force, temperature, pressure, voltage, current, and magnetic field. To determine distances, the most popular motion sensor emits pulses of ultra high frequency sound. Although these ultrasonic pulses are above the range of human hearing, the motion sensor can detect reflections of these pulses after they bounce off objects within the sensor's field of "view."

**FIGURE 1-11** ■ Two electronic interfaces used in popular introductory physics computer data acquisition systems: The LabPro Interface (Vernier Software and Technology) and the Science Workshop 500 Interface (PASCO scientific).

Since the speed of ultrasound in room temperature air is known, the computer motion software can calculate the distance to an object by recording how long the pulse takes to reflect off the object and return to the sensor. This is similar to how a bat "sees," and how some auto-focus cameras determine the distance to an object. This approach to measuring a distance or length is not unlike that used by international standards organizations to define the meter in terms of the speed of light. Since ultrasonic motion detectors can send and receive short pulses up to 50 times a second, the computer software can also make rapid calculations of velocities and accelerations of slowly moving objects "on the fly," and graph them in real time. Sample graphs are shown in Fig. 1-12.

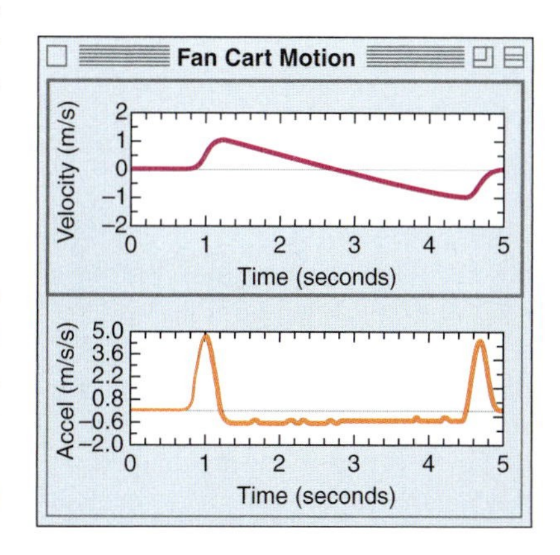


**FIGURE 1-12** ■ Real-time graphs of position, velocity, or acceleration, as a function of time, can be generated by an ultrasonic motion detector.

## Digital Video Capture and Analysis Tools

Software and hardware enable student investigators to digitize images from a video camera, VCR, or videodisc. Once a digital video movie is created, it can be analyzed using video analysis software. Video data are collected by locating items of interest in each frame of a movie as it is displayed on a computer screen. Video analysis is a useful tool for studying one- and two-dimensional motions, electrostatics, and digital simulations of molecular motions. Examples of digital video clips and their analysis will be presented in this text from time to time. (See Figs. 1-13 and 1-14).

---

* These systems go by many names, such as computer-based data collection system, e-measure, CADAA (computer-assisted data acquisition and analysis system), or MBL system (Microcomputer Based Laboratory system).

**FIGURE 1-13** ■ An overlay of five digital video frames showing a ballet dancer moving toward the left while performing a grand jeté.

## 1-9 Changing Units

An American traveling overseas notices a road sign indicating that the distance to the next town is 32 km. She wants to get a feel for how far away the town is, and needs to convert the kilometers to the more familiar units of miles. How would she go about doing that?

We often need to change the units in which a physical quantity is expressed. A good method is called *chain-link conversion*. In this method, we multiply the original measurement by one or more conversion factors. A **conversion factor** is defined as a ratio of units that is equal to 1. For example, because 1 mile and 1.61 kilometers are identical distances, we have

$$\frac{1\ \text{mi}}{1.61\ \text{km}} = 1 \quad \text{and also} \quad \frac{1.61\ \text{km}}{1\ \text{mi}} = 1. \tag{1-10}$$

Thus, the ratios (1 mi)/(1.61 km) and (1.61 km)/(1 mi) can be used as conversion factors. This is *not* the same as writing $1/1.61 = 1$ or $1.61 = 1$; each *number* and its *unit* must be treated together. Because multiplying any quantity by one leaves it unchanged, we can introduce such conversion factors wherever we find them useful. In chain-link conversion, we use the factors to cancel unwanted units. For example, to convert 32 kilometers to miles, we have

$$32\ \text{km} = (32\ \cancel{\text{km}})\left(\frac{1\ \text{mi}}{1.61\ \cancel{\text{km}}}\right) = 20\ \text{mi}. \tag{1-11}$$

Suppose instead that our traveler wanted to know how many feet there are in 32 kilometers. Then two conversion factors would be needed, so that

$$32\ \text{km} = (32\ \cancel{\text{km}})\left(\frac{1\ \cancel{\text{mi}}}{1.61\ \cancel{\text{km}}}\right)\left(\frac{5280\ \text{ft}}{1\ \cancel{\text{mi}}}\right) = 1.05 \times 10^5\ \text{ft}. \tag{1-12}$$

The number of feet is expressed in scientific notation so that the correct number of significant figures can be represented. See the next section and Appendix A for more details on how to represent significant figures properly.

Appendix D and the inside back cover give conversion factors between SI and other systems of units, including many of the non-SI units still used in the United States. However, the conversion factors are written in the style of "1 mi = 1.61 km" rather than the ratios we show here.

It is important to note that the **value of a physical quantity** is actually the product of a number and a unit. Thus, the number associated with a particular physical quantity depends on the unit in which it is expressed. For example, the distance to the trav-

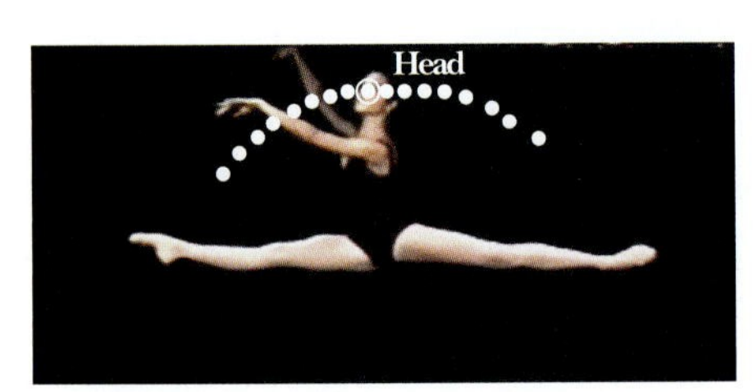


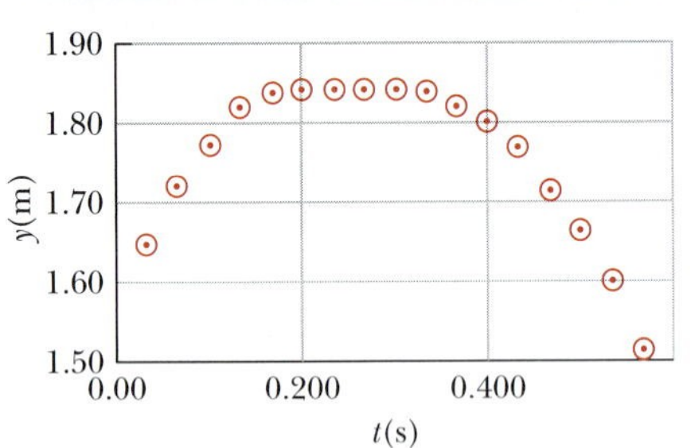


**FIGURE 1-14** ■ A video analysis of the motion of the dancer reveals that while performing the grand jeté depicted in Fig. 1-13, her head is moving in a straight horizontal line between the times 0.180 s and 0.330 s. To observers following the motion of her head, the dancer appears to be floating for this short period of time. How does she accomplish this? In Chapter 8 we describe how video analysis helps us explore this question.

eler's town has a value of 32 km. The numerical component of its value expressed in the unit "kilometers" is 32. However, the value of the distance when expressed in miles is 20 mi, and the numerical component of its value when expressed in miles is 20. Since 20 miles is actually the *same distance* as 32 kilometers, it is meaningful to write 32 km = 20 mi. In this context the equal sign (=) signifies that 32 km is the *same distance* as 20 mi expressed in different units. However, it is totally meaningless to write 32 = 20. Thus it is extremely important to include appropriate units in all calculations.

### TOUCHSTONE EXAMPLE 1-2: Marathon

When Pheidippides ran from Marathon to Athens in 490 B.C.E. to bring word of the Greek victory over the Persians, he probably ran at a speed of about 23 rides per hour (rides/h). The ride is an ancient Greek unit for length, as are the stadium and the plethron: 1 ride was defined to be 4 stadia, 1 stadium was defined to be 6 plethra, and, in terms of a modern unit, 1 plethron is 30.8 m. How fast did Pheidippides run in kilometers per second (km/s)?

**SOLUTION** ■ The **Key Idea** in chain-link conversions is to write the conversion factors as ratios that will eliminate unwanted units. Here we write

$$23 \text{ rides/h} = \left(23 \, \frac{\cancel{\text{rides}}}{\cancel{\text{h}}}\right)\left(\frac{4 \, \cancel{\text{stadia}}}{1 \, \cancel{\text{ride}}}\right)\left(\frac{6 \, \cancel{\text{plethra}}}{1 \, \cancel{\text{stadium}}}\right)\left(\frac{30.8 \, \cancel{\text{m}}}{1 \, \cancel{\text{plethron}}}\right)\left(\frac{1 \text{ km}}{1000 \, \cancel{\text{m}}}\right)\left(\frac{1 \, \cancel{\text{h}}}{3600 \text{ s}}\right)$$

$$= 4.7227 \times 10^{-3} \text{ km/s} \approx 4.7 \times 10^{-3} \text{ km/s}. \quad \text{(Answer)}$$

**READING EXERCISE 1-7:** (a) Explain why it is correct to write 1 min/60 s = 1, but it is not correct to write 1/60 = 1. (b) Use the relevant conversion factors and the method of chain-link conversions to calculate how many seconds there are in a day. ■

## 1-10 Calculations with Uncertain Quantities

### Issue 1: Significant Figures and Decimal Places

In July 1988, in Indianapolis, Indiana, the U.S.'s Florence Griffith Joyner set a world record in the women's 100-meter dash with an official time of 10.49 seconds (Fig. 1-15). The timing in the race is considered good to the nearest 1/100 of a second. Suppose you had been asked to report the time in minutes instead of seconds. If you used a calculator to transform the 10.49 seconds into minutes by multiplying by (1 min)/(60 s), you might report the following by copying all the digits on your display:

$$10.49 \text{ s} = (10.49 \text{ s})\left(\frac{1 \text{ min}}{60 \text{ s}}\right) = 0.174\,833\,333 \text{ min}. \tag{1-13}$$

No matter how precise a measuring instrument is, all measured quantities have uncertainties associated with them. The precision implied by the calculated time in minutes shown above is both meaningless and misleading! We should have rounded the answer to four significant digits, 0.1748 min, so as not to imply that it is more precise than the given data. The given time of 10.49 seconds consists of four digits, called **significant figures.** This tells us we should round the answer to four significant figures. In this text, final results of calculations are often rounded to match the least number of significant figures in the given data. *Significant figures* should not be confused with *decimal places*. Consider the lengths 35.6 mm, 3.56 cm, and 0.0356 m. They all have three significant figures, but they have one, two, and four decimal places, respectively.

**FIGURE 1-15** ■ The late Florence Griffith Joyner set a world's record in the women's 100-meter dash in 1988.

As you work with scientific calculations in data analysis in the laboratory or complete the problems in this text, it is important to pay strict attention to reporting your answer to the same precision as the lowest precision in any of the factors used in your calculation. Information on how to keep track of significant figures and measurement uncertainties in calculations is included in Appendix A, and a table of fundamental constants that have been measured to high precision is in Appendix B.

## Issue 2: Order of Magnitude

In order to make estimations, engineering and science professionals will sometimes round a number to be used in a calculation up or down to the nearest power of ten. This makes the number very easy to use in calculations. The result of this rounding procedure is known as the *order of magnitude* of a number. To determine an order of magnitude, we start by expressing the number of interest in scientific notation. Next, the mantissa is rounded up to 10 or down to 1 depending on which is closest. For example, if $A = 2.3 \times 10^4$, then the order of magnitude of $A$ is $10^4$ (ten to the fourth) since 2.3 is closer to 1 than it is to 10. On the other hand, if $B = 7.8 \times 10^4$, then the order of magnitude of $B$ is $10^5$ (ten to the fifth) since 7.8 is closer to 10 than it is to 1. Order of magnitude estimations are common when detailed or precise data are not required in a calculation or are not known.

**READING EXERCISE 1-8:** Using the method outlined in Appendix A, determine the number of significant figures in each of the following numbers: (a) 27 meters, (b) 27 cows, (c) 0.003 429 87 second, (d) $-1.970\,500 \times 10^{-11}$ coulombs, (e) 5280 ft/mi. (*Note:* By definition there are exactly 5280 feet in a mile.) ■

**READING EXERCISE 1-9:** A popular science book lists the radius of the Earth as 20 900 000 000 ft. (a) How many significant figures does this number have if you apply the method described in Appendix A for determining the number of significant figures? (b) How many significant figures did the author probably intend to report? (c) How could you rewrite this number so that it represents three significant figures? (d) What order of magnitude is the radius of the Earth in feet? ■

### TOUCHSTONE EXAMPLE 1-3: Ball of String

The world's largest ball of string is about 2 m in radius. To the nearest order of magnitude, what is the total length $L$ of the string in the ball?

**SOLUTION** ■ We could, of course, take the ball apart and measure the total length $L$, but that would take great effort and make the ball's builder most unhappy. A **Key Idea** here is that, because we want only the nearest order of magnitude, we can estimate any quantities required in the calculation.

Let us assume the ball is spherical with radius $R = 2$ m. The string in the ball is not closely packed (there are uncountable gaps between nearby sections of string). To allow for these gaps, let us somewhat overestimate the cross-sectional area of the string by assuming the cross section is square, with an edge length $d = 4$ mm. Then, with a cross-sectional area of $d^2$ and a length $L$, the string occupies a total volume of

$$V = (\text{cross-sectional area})(\text{length}) = d^2L.$$

This is approximately equal to the volume of the ball, given by $\frac{4}{3}\pi R^3$, which is about $4R^3$ because $\pi$ is about 3. Thus, we have

$$d^2L = 4R^3,$$

or

$$L = \frac{4R^3}{d^2} = \frac{4(2\text{ m})^3}{(4 \times 10^{-3}\text{ m})^2} \qquad \text{(Answer)}$$

$$= 2 \times 10^6\text{ m} \approx 10^6\text{ m} = 10^3\text{ km}.$$

(Note that you do not need a calculator for such a simplified calculation.) Thus, to the nearest order of magnitude, the ball contains about 1000 km of string!

**READING EXERCISE 1-10:** Suppose you are to calculate the volume of a cube that is $L = 1.4$ cm on a side and you start by calculating the area, $A$, of a face of the cube $A = L^2$ and then calculating $V = AL$. (a) What intermediate value for $A$ should you use in the calculation for $V$? (b) What is the value of the volume to the correct number of significant figures? (c) What value do you get for $V$ if you incorrectly retain only two significant figures after you calculate $A$? ■

**READING EXERCISE 1-11:** Perform the following calculations and express the answers to the correct number of significant figures. (a) Multiply 3.4 by 7.954. (b) Add 99.3 and 98.7. (c) Subtract 98.7 from 99.3. (d) Evaluate the cosine of 3°. (e) If five railroad track segments have an average length of 2.134 meters, what is the total length of these five rails when they lie end to end? ■

**READING EXERCISE 1-12:** Suppose you measure a time to the nearest 1/100 of a second and get a value of 1.78 s. (a) What is the absolute precision of your measurement? (b) What is the relative precision of your measurement? ■

# Problems

## SEC. 1-5 ■ THE SI STANDARD OF TIME

**1. Speed of Light** Express the speed of light, $3.0 \times 10^8$ m/s, in (a) feet per nanosecond and (b) millimeters per picosecond.

**2. Fermi** Physicist Enrico Fermi once pointed out that a standard lecture period (50 min) is close to 1 microcentury. (a) How long is a microcentury in minutes? (b) Using

$$\text{percentage difference} = \left(\frac{\text{actual} - \text{approximation}}{\text{actual}}\right) 100,$$

find the percentage difference from Fermi's approximation.

**3. Five Clocks** Five clocks are being tested in a laboratory. Exactly at noon, as determined by the WWV time signal, on successive days of a week the clocks read as in the following table. Rank the five clocks according to their relative value as good timekeepers, best to worst. Justify your choice.

| Clock | Sun. | Mon. | Tues. | Wed. | Thurs. | Fri. | Sat. |
|---|---|---|---|---|---|---|---|
| A | 12:36:40 | 12:36:56 | 12:37:12 | 12:37:27 | 12:37:44 | 12:37:59 | 12:38:14 |
| B | 11:59:59 | 12:00:02 | 11:59:57 | 12:00:07 | 12:00:02 | 11:59:56 | 12:00:03 |
| C | 15:50:45 | 15:51:43 | 15:52:41 | 15:53:39 | 15:54:37 | 15:55:35 | 15:56:33 |
| D | 12:03:59 | 12:02:52 | 12:01:45 | 12:00:38 | 11:59:31 | 11:58:24 | 11:57:17 |
| E | 12:03:59 | 12:02:49 | 12:01:54 | 12:01:52 | 12:01:32 | 12:01:22 | 12:01:12 |

**4. The Shake** A unit of time sometimes used in microscopic physics is the *shake*. One shake equals $10^{-8}$ s. (a) Are there more shakes in a second than there are seconds in a year? (b) Humans have existed for about $10^6$ years, whereas the universe is about $10^{10}$ years old. If the age of the universe now is taken to be 1 "universe day," for how many "universe seconds" have humans existed?

**5. Astronomical Units** An astronomical unit (AU) is the average distance of Earth from the Sun, approximately $1.50 \times 10^8$ km. The speed of light is about $3.0 \times 10^8$ m/s. Express the speed of light in terms of astronomical units per minute.

**6. Digital Clocks** Three digital clocks $A$, $B$, and $C$ run at different rates and do not have simultaneous readings of zero. Figure 1-16 shows simultaneous readings on pairs of the clocks for four occasions. (At the earliest occasion, for example, $B$ reads 25.0 s and $C$ reads 92.0 s.) If two events are 600 s apart on clock $A$, how far apart are they on (a) clock $B$ and (b) clock $C$? (c) When clock $A$ reads 400 s, what does clock $B$ read? (d) When clock $C$ reads 15.0 s, what does clock $B$ read? (Assume negative readings for prezero times.)

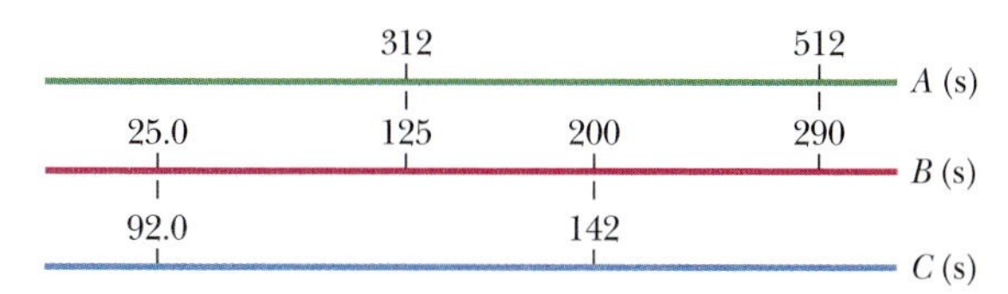


FIGURE 1-16 ■ Problem 6.

**7. Length of Day** Assuming the length of the day uniformly increases by 0.0010 s per century, calculate the cumulative effect on the measure of time over 20 centuries. (Such slowing of Earth's rotation is indicated by observations of the occurrences of solar eclipses during this period.)

**8. Time Zones** Until 1883, every city and town in the United States kept its own local time. Today, travelers reset their watches only when the time change equals 1.0 h. How far, on the average, must you travel in degrees of longitude until your watch must be reset by 1.0 h? (*Hint:* Earth rotates 360° in about 24 h.)

**9. A Fortnight** A fortnight is a charming English measure of time equal to 2.0 weeks (the word is a contraction of "fourteen nights"). That is a nice amount of time in pleasant company but perhaps a painful string of microseconds in unpleasant company. How many microseconds are in a fortnight?

**10. Time Standards** Time standards are now based on atomic clocks. A promising second standard is based on *pulsars*, which are rotating neutron stars (highly compact stars consisting only of neutrons). Some rotate at a rate that is highly stable, sending out a radio beacon that sweeps briefly across Earth once with each rotation, like a lighthouse beacon. Pulsar PSR 1937 + 21 is an example; it rotates once every 1.557 806 448 872 75 ± 3 ms, where the trailing ±3 indicates the uncertainty in the last decimal place (it does *not* mean ±3 ms). (a) How many times does PSR 1937 + 21 rotate in 7.00 days? (b) How much time does the pulsar take to rotate $1.0 \times 10^6$ times, and (c) what is the associated uncertainty?

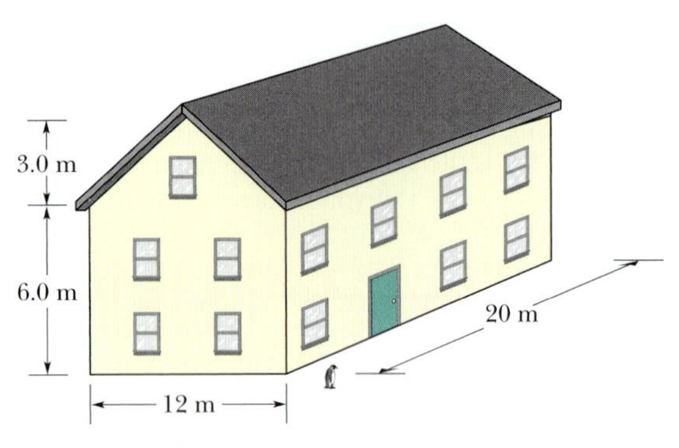


FIGURE 1-18 ■ Problem 18.

## SEC. 1-6 ■ THE SI STANDARDS OF LENGTH

**11. Furlongs** Horses are to race over a certain English meadow for a distance of 4.0 furlongs. What is the race distance in units of (a) rods and (b) chains? (1 furlong = 201.168 m, 1 rod = 5.0292 m, and 1 chain = 20.117 m.)

**12. Types of Barrels** Two types of *barrel* units were in use in the 1920s in the United States. The apple barrel had a legally set volume of 7056 cubic inches; the cranberry barrel, 5826 cubic inches. If a merchant sells 20 cranberry barrels of goods to a customer who thinks he is receiving apple barrels, what is the discrepancy in the shipment volume in liters?

**13. The Earth** Earth is approximately a sphere of radius $6.37 \times 10^6$ m. What are (a) its circumference in kilometers, (b) its surface area in square kilometers, and (c) its volume in cubic kilometers?

**14. Points and Picas** Spacing in this book was generally done in units of points and picas: 12 points = 1 pica, and 6 picas = 1 inch. If a figure was misplaced in the page proofs by 0.80 cm, what was the misplacement in (a) points and (b) picas?

**15. Antarctica** Antarctica is roughly semicircular, with a radius of 2000 km (Fig. 1-17). The average thickness of its ice cover is 3000 m. How many cubic centimeters of ice does Antarctica contain? (Ignore the curvature of Earth.)

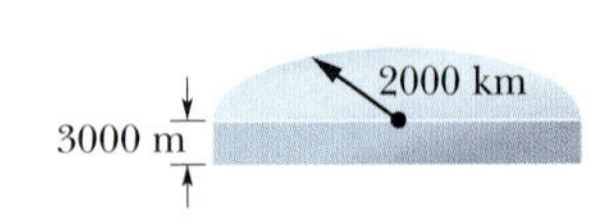


FIGURE 1-17 ■ Problem 15.

**16. Roods and Perches** An old manuscript reveals that a landowner in the time of King Arthur held 3.00 acres of plowed land plus a livestock area of 25.0 perches by 4.00 perches. What was the total area in (a) the old unit of roods and (b) the more modern unit of square meters? Here, 1 acre is an area of 40 perches by 4 perches, 1 rood is 40 perches by 1 perch, and 1 perch is 16.5 ft.

**17. The Acre-Foot** Hydraulic engineers in the United States often use, as a unit of volume of water, the *acre-foot*, defined as the volume of water that will cover 1 acre of land to a depth of 1 ft. A severe thunderstorm dumped 2.0 in. of rain in 30 min on a town of area 26 km$^2$. What volume of water, in acre-feet, fell on the town?

**18. A Doll House** In the United States, a doll house has the scale of 1:12 of a real house (that is, each length of the doll house is $\frac{1}{12}$ that of the real house) and a miniature house (a doll house to fit within a doll house) has the scale of 1:144 of a real house. Suppose a real house (Fig. 1-18) has a front length of 20 m, a depth of 12 m, a height of 6.0 m, and a standard sloped roof (vertical triangular faces on the ends) of height 3.0 m. In cubic meters, what are the volumes of the corresponding (a) doll house and (b) miniature house?

## SEC. 1-7 ■ THE SI STANDARDS OF MASS

**19. Earth's Mass** Earth has a mass of $5.98 \times 10^{24}$ kg. The average mass of the atoms that make up Earth is 40 u. How many atoms are there in Earth?

**20. Gold** Gold, which has a mass of 19.32 g for each cubic centimeter of volume, is the most ductile metal and can be pressed into a thin leaf or drawn out into a long fiber. (a) If 1.000 oz of gold, with a mass of 27.63 g, is pressed into a leaf of 1.000 $\mu$m thickness, what is the area of the leaf? (b) If, instead, the gold is drawn out into a cylindrical fiber of radius 2.500 $\mu$m, what is the length of the fiber?

**21. Mass of Water** (a) Assuming that each cubic centimeter of water has a mass of exactly 1 g, find the mass of one cubic meter of water in kilograms. (b) Suppose that it takes 10.0 h to drain a container of 5700 m$^3$ of water. What is the "mass flow rate," in kilograms per second, of water from the container?

**22. The Thunderstorm** What mass of water fell on the town in Problem 17 during the thunderstorm? One cubic meter of water has a mass of $10^3$ kg.

**23. Iron** Iron has a mass of 7.87 g per cubic centimeter of volume, and the mass of an iron atom is $9.27 \times 10^{-26}$ kg. If the atoms are spherical and tightly packed, (a) what is the volume of an iron atom and (b) what is the distance between the centers of adjacent atoms?

**24. Grains of Sand** Grains of fine California beach sand are approximately spheres with an average radius of 50 $\mu$m and are made of silicon dioxide. A solid cube of silicon dioxide with a volume of 1.00 m$^3$ has a mass of 2600 kg. What mass of sand grains would have a total surface area (the total area of all the individual spheres) equal to the surface area of a cube 1 m on an edge?

## SEC. 1-9 ■ CHANGING UNITS

**25. A Diet** A person on a diet might lose 2.3 kg per week. Express the mass loss rate in milligrams per second, as if the dieter could sense the second-by-second loss.

**26. Cats and Moles** A mole of atoms is $6.02 \times 10^{23}$ atoms. To the nearest order of magnitude, how many moles of atoms are in a large domestic cat? The masses of a hydrogen atom, an oxygen atom, and a carbon atom are 1.0 u, 16 u, and 12 u, respectively. (*Hint:* Cats are sometimes known to kill moles.)

**27. Sugar Cube** A typical sugar cube has an edge length of 1 cm. If you had a cubical box that contained a mole of sugar cubes, what would its edge length be? (One mole = $6.02 \times 10^{23}$ units.)

**28. Micrometer** The micrometer (1 $\mu$m) is often called the *micron*. (a) How many microns make up 1.0 km? (b) What fraction of a centimeter equals 1.0 $\mu$m? (c) How many microns are in 1.0 yd?

**29. Hydrogen** Using conversions and data in the chapter, determine the number of hydrogen atoms required to obtain 1.0 kg of hydrogen. A hydrogen atom has a mass of 1.0 u.

**30. A Gry** A *gry* is an old English measure for length, defined as 1/10 of a line, where *line* is another old English measure for length, defined as 1/12 inch. A common measure for length in the publishing business is a *point*, defined as 1/72 inch. What is an area of 0.50 $gry^2$ in terms of points squared ($points^2$)?

# Additional Problems

**31. Harvard Bridge** Harvard Bridge, which connects MIT with its fraternities across the Charles River, has a length of 364.4 Smoots plus one ear. The unit of one Smoot is based on the length of Oliver Reed Smoot, Jr., class of 1962, who was carried or dragged length by length across the bridge so that other pledge members of the Lambda Chi Alpha fraternity could mark off (with paint) 1-Smoot lengths along the bridge. The marks have been repainted biannually by fraternity pledges since the initial measurement, usually during times of traffic congestion so that the police could not easily interfere. (Presumably, the police were originally upset because a Smoot is not an SI base unit, but these days they seem to have accepted the unit.) Figure 1-19 shows three parallel paths, measured in Smoots (S), Willies (W), and Zeldas (Z). What is the length of 50.0 Smoots in (a) Willies and (b) Zeldas?

**FIGURE 1-19** ■ Problem 31.

**32. Little Miss Muffet** An old English children's rhyme states, "Little Miss Muffet sat on her tuffet, eating her curds and whey, when along came a spider who sat down beside her. . . ." The spider sat down not because of the curds and whey but because Miss Muffet had a stash of 11 tuffets of dried flies. The volume measure of a tuffet is given by 1 tuffet = 2 pecks = 0.50 bushel, where 1 Imperial (British) bushel = 36.3687 liters (L). What was Miss Muffet's stash in (a) pecks, (b) bushels, and (c) liters?

**33. Noctilucent Clouds** During the summers at high latitudes, ghostly, silver-blue clouds occasionally appear after sunset when common clouds are in Earth's shadow and are no longer visible. The ghostly clouds have been called *noctilucent clouds* (NLC), which means "luminous night clouds," but now are often called *mesospheric clouds*, after the *mesosphere*, the name of the atmosphere at the altitude of the clouds.

These clouds were first seen in June 1885, after dust and water from the massive 1883 volcanic explosion of Krakatoa Island (near Java in the Southeast Pacific) reached the high altitudes in the Northern Hemisphere. In the low temperatures of the mesosphere, the water collected and froze on the volcanic dust (and perhaps on comet and meteor dust already present there) to form the particles that made up the first clouds. Since then, mesospheric clouds have generally increased in occurrence and brightness, probably because of the increased production of methane by industries, rice paddies, landfills, and livestock flatulence. The methane works its way into the upper atmosphere, undergoes chemical changes, and results in an increase of water molecules there, and also in bits of ice for the mesospheric clouds.

If mesospheric clouds are spotted 38 min after sunset and then quickly dim, what is their altitude if they are directly over the observer?

**34. Staircase** A standard interior staircase has steps each with a rise (height) of 19 cm and a run (horizontal depth) of 23 cm. Research suggests that the stairs would be safer for descent if the run were, instead, 28 cm. For a particular staircase of total height 4.57 m, how much farther would the staircase extend into the room at the foot of the stairs if this change in run were made?

**35. Large and Small** As a contrast between the old and the modern and between the large and the small, consider the following: In old rural England 1 hide (between 100 and 120 acres) was the area of land needed to sustain one family with a single plough for one year. (An area of 1 acre is equal to 4047 $m^2$.) Also, 1 wapentake was the area of land needed by 100 such families. In quantum physics, the cross-sectional area of a nucleus (defined in terms of the chance of a particle hitting and being absorbed by it) is measured in units of barns, where 1 barn is $1 \times 10^{-28}$ $m^2$. (In nuclear physics jargon, if a nucleus is "large," then shooting a particle at it is like shooting a bullet at a barn door, which can hardly be missed.) What is the ratio of 25 wapentakes to 11 barns?

**36. Cumulus Cloud** A cubic centimeter in a typical cumulus cloud contains 50 to 500 water droplets, which have a typical radius of 10 $\mu$m. (a) How many cubic meters of water are in a cylindrical cumulus cloud of height 3.0 km and radius 1.0 km? (b) How many 1-liter pop bottles would that water fill? (c) Water has a mass per unit volume (or density) of 1000 kg/$m^3$. How much mass does the water in the cloud have?

**37. Oysters** In purchasing food for a political rally, you erroneously order shucked medium-size Pacific oysters (which come 8 to 12 per U.S. pint) instead of shucked medium-size Atlantic oysters (which come 26 to 38 per U.S. pint). The filled oyster container delivered to you has the interior measure of 1.0 m $\times$ 12 cm $\times$ 20 cm, and a U.S. pint is equivalent to 0.4732 liter. By how many oysters is the order short of your anticipated count?

**38. U.K. Gallons** A tourist purchases a car in England and ships it home to the United States. The car sticker advertised that the car's fuel consumption was at the rate of 40 miles per gallon on the open road. The tourist does not realize that the U.K. gallon differs from the U.S. gallon:

$$1 \text{ U.K. gallon} = 4.545\,963\,1 \text{ liters}$$
$$1 \text{ U.S. gallon} = 3.785\,306\,0 \text{ liters.}$$

For a trip of 750 miles (in the United States), how many gallons of fuel does (a) the mistaken tourist believe she needs and (b) the car actually require?

**39. Types of Tons** A ton is a measure of volume frequently used in shipping, but that use requires some care because there are at least three types of tons: A *displacement ton* is equal to 7 barrels bulk, a *freight ton* is equal to 8 barrels bulk, and a *register ton* is equal to 20 barrels bulk. A *barrel bulk* is another measure of volume: 1 barrel bulk = 0.1415 $m^3$. Suppose you spot a shipping order for "73 tons" of M&M candies, and you are certain that the client who sent the order intended "ton" to refer to volume (instead of weight or mass, as discussed in Chapter 6). If the client actually meant displacement tons, how many extra U.S. bushels of the candies will you erroneously ship to the client if you interpret the order as (a) 73 freight tons and (b) 73 register tons? One cubic meter is equivalent to 28.378 U.S bushels.

**40. Wine Bottles** The wine for a large European wedding reception is to be served in a stunning cut-glass receptacle with the interior dimensions of 40 cm × 40 cm × 30 cm (height). The receptacle is to be initially filled to the top. The wine can be purchased in bottles of the sizes given in the following table, where the volumes of the larger bottles are given in terms of the volume of a standard wine bottle. Purchasing a larger bottle instead of multiple smaller bottles decreases the overall cost of the wine. To minimize that overall cost, (a) which bottle sizes should be purchased and how many of each should be purchased, and (b) how much wine is left over once the receptacle is filled?

1 standard
1 magnum = 2 standard
1 jeroboam = 4 standard
1 rehoboam = 6 standard
1 methuselah = 8 standard
1 salmanazar = 12 standard
1 balthazar = 16 standard = 11.356 L
1 nebuchadnezzar = 20 standard

**41. The Corn–Hog Ratio** The *corn-hog ratio* is a financial term commonly used in the pig market and presumably is related to the cost of feeding a pig until it is large enough for market. It is defined as the ratio of the market price of a pig with a mass of 1460 slugs to the market price of a U.S. bushel of corn. The slug is the unit of mass in the English system. (The word "slug" is derived from an old German word that means "to hit"; we have the same meaning for "slug" as a verb in modern English.) A U.S. bushel is equal to 35.238 L. If the corn–hog ratio is listed as 5.7 on the market exchange, what is it in the metric units of

$$\frac{\text{price of 1 kilogram of pig}}{\text{price of 1 liter of corn}}?$$

(*Hint:* See the Mass table in Appendix D.)

**42. Volume Measures in Spain** You can easily convert common units and measures electronically, but you still should be able to use a conversion table, such as those in Appendix D. Table 1-3 is part of a conversion table for a system of volume measures once common in Spain; a volume of 1 fanega is equivalent to 55.501 $dm^3$ (cubic decimeters). (a) Complete the table, using three significant figures. Then express 7.00 almude in terms of (b) medio, (c) cahiz, and (d) cubic centimeters ($cm^3$).

**TABLE 1-3**
**Problem 42**

| | cahiz | fanega | cuartilla | almude | medio |
|---|---|---|---|---|---|
| 1 cahiz = | 1 | 12 | 48 | 144 | 288 |
| 1 fanega = | | 1 | 4 | 12 | 24 |
| 1 cuartilla = | | | 1 | 3 | 6 |
| 1 almude = | | | | 1 | 2 |
| 1 medio = | | | | | 1 |

**43. Pirate Ship** You receive orders to sail due east for 24.5 mi to put your salvage ship directly over a sunken pirate ship. However, when your divers probe the ocean floor at that location and find no evidence of a ship, you radio back to your source of information, only to discover that the sailing distance was supposed to be 24.5 *nautical miles,* not regular miles. Use the Length table in Appendix D to calculate how far horizontally you are from the pirate ship in kilometers.

**44. The French Revolution** For about 10 years after the French revolution, the French government attempted to base measures of time on multiples of ten: One week consisted of 10 days, 1 day consisted of 10 hours, 1 hour consisted of 100 minutes, and 1 minute consisted of 100 seconds. What are the ratios of (a) the French decimal week to the standard week and (b) the French decimal second to the standard second?

**45. Heavy Rain** During heavy rain, a rectangular section of a mountainside measuring 2.5 km wide (horizontally), 0.80 km long (up along the slope), and 2.0 m deep suddenly slips into a valley in a mud slide. Assume that the mud ends up uniformly distributed over a valley section measuring 0.40 km × 0.40 km and that the mass of a cubic meter of mud is 1900 kg. What is the mass of the mud sitting above an area of 4.0 $m^2$ in that section?

**46. Liquid Volume** Prior to adopting metric systems of measurement, the United Kingdom employed some challenging measures of liquid volume. A few are shown in Table 1-4. (a) Complete the table, using three significant figures. (b) The volume of 1 bag is equivalent to a volume of 0.1091 $m^3$. If an old British story has a witch cooking up some vile liquid in a cauldron with a volume of 1.5 chaldrons, what is the volume in terms of cubic meters?

**TABLE 1-4**
**Problem 46**

| | wey | chaldron | bag | pottle | gill |
|---|---|---|---|---|---|
| 1 wey = | 1 | 10/9 | 40/3 | 640 | 120 240 |
| 1 chaldron = | | | | | |
| 1 bag = | | | | | |
| 1 pottle = | | | | | |
| 1 gill = | | | | | |

**47. The Dbug** Traditional units of time have been based on astronomical measurements, such as the length of the day or year. However, one human-based measure of time can be found in Tibet, where the *dbug* is the average time between exhaled breaths. Estimate the number of dbugs in a day.

**48. Tower of Pisa** The following photograph of the Leaning Tower of Pisa was taken from an advertisement found in a 1994 airline magazine. Assume that the photo of the man talking on the telephone to the left has been dubbed in and is not part of the original photograph.

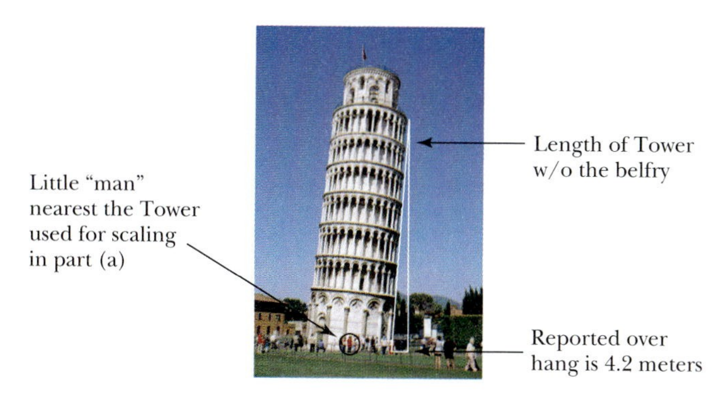


**FIGURE 1-20** ■ Problem 48.

**(a)** Examine the photograph. Take the measurements in centimeters that are needed to find a scale factor that enables you to estimate the length of the tower in meters (i.e., its height if it were standing up straight.) Use only the evidence in the photograph—no other data are allowed. Then estimate the tower length in meters.
**(b)** According to data published in Sir Bannester Fletcher's *A History of Architecture* (U. of London Athlone Press, 1975, p. 470) the diameter of the lower part of the tower is 16.0 m. Using these data, find another scale factor for estimating the length of the tower, and then re-estimate the length of the tower using this new scale factor.
**(c)** Which of the scale factors (a) or (b) do you think will give the best estimate of the length of the tower? Explain the reasons for your answer.
**(d)** Using the scale factor you found in part (b), what is the length of the tower without the belfry or narrow top segment (i.e., just consider the bottom 7 stories)?

**49. Mexican Food** You are to fix dinners for 400 people at a convention of Mexican food fans. Your recipe calls for 2 jalapeño peppers per serving (one serving per person). However, you have only habanero peppers on hand. The spiciness of peppers is measured in terms of the *scoville heat unit* (SHU). On average, one jalapeño pepper has a spiciness of 4000 SHU and one habanero pepper has a spiciness of 300 000 SHU. To salvage the situation, how many (total) habanero peppers should you substitute for the jalapeño peppers in the recipe for the convention?

**50. Big or Small?** Discuss the question: "Is 500 feet big or small?" Before you do so, carry out the following estimates.

**(a)** You are on the top floor of a 500-ft-tall building. A fire breaks out in the building and the elevator stops working. You have to walk down to the ground floor. Estimate how long this would take you. (Your stairwell is on the other side of the building from the fire.)
**(b)** You are hiking the Appalachian Trail on a beautiful fall morning as part of a 10 mi hike with a group of friends. You are walking along a well-tended, level part of the trail. Estimate how long it would take you to walk 500 ft.
**(c)** You are driving on the New Jersey Turnpike at 65 mi/hr. You pass a sign that says "Lane ends 500 feet." How much time do you have in order to change lanes?

**51. Doubling System** Historically the English had a doubling system when measuring volumes; 2 mouthfuls equal 1 jigger, 2 jiggers equal 1 jack (also called a jackpot); 2 jacks equal 1 jill; 2 jills = 1 cup; 2 cups = 1 pint; 2 pints = 1 quart; 2 quarts = 1 pottle; 2 pottles = 1 gallon; 2 gallons = 1 pail. (The nursery rhyme "Jack and Jill" refers to these units and was a protest against King Charles I of England for his taxes on the jacks of liquor sold in the tavern. (See A. Kline, *The World of Measurement,* New York: Simon and Schuster, 1975, pp. 32–39.) American and British cooks today use teaspoons, tablespoons, and cups; 3 teaspoons = 1 tablespoon; 4 tablespoons = 1/4 cup. Assume that you find an old English recipe requiring 3 jiggers of milk. How many cups does this represent? How many tablespoons? You can assume that the cups in the two systems represent the same volume.

**52. Fuel Efficiency** In America, we measure fuel efficiency of our cars by citing the number of miles you can drive on 1 gallon of gas (miles/gallon). In Europe, the same information is given by quoting how many liters of gas it takes to go 100 kilometers (liter/100 kilometers).

**(a)** My current car gets 21 miles/gallon in highway travel. What number (in liter/100 kilometers) should I give to my Swedish friend so that he can compare it to the mileage for his Volvo?
**(b)** The car I drove in England last summer needed 6 liters of gas to go 100 kilometers. How many miles/gallon did it get?
**(c)** If my car has a fuel efficiency, $f$, in miles/gallon, what is its European efficiency, $e$, in liters/100 kilometers? (Write an equation that would permit an easy conversion.)

**53. Pizza Sale** Two terrapins decide to go to Jerry's for a pizza. When they get there they find that Jerry's is having a special:

| | | |
|---|---|---|
| **SPECIAL TODAY:** | one 20″ pizza | $15 |
| **REGULAR PRICE** | one 10″ pizza | $5 |
| | one 20″ pizza | $18 |

Raphael: "Great! Let's get a large one."

Donatello: "Don't be dumb. Let's get three of the small ones for the same price. That'll give us more pizza and be cheaper."

Raphael: "Why would it be a special if it's more than we could get for the regular price? Let's get the large."

Who's right? Which would you buy? What would the difference be if you were buying them at Ledo's (square pizzas)?

**54. Dollar and Penny** A student makes the following argument: "I can prove a dollar equals a penny. Since a dime (10 cents) is one-tenth of a dollar, I can write:

$$10\,¢ = \$0.1.$$

Square both sides of the equation. Since squares of equals are equal,

$$100\,¢ = \$0.1.$$

Since 100 ¢ = \$1 and \$0.01 = 1 ¢, it follows that \$1 = 1 ¢."
What's wrong with the argument?

**55. Scaling Up** Here are two related problems—one precise, one an estimation.

**(a)** A sculptor builds a model for a statue of a terrapin to replace Testudo.* She discovers that to cast her small scale model she needs 2 kg of bronze. When she is done, she finds that she can give it two coats of finishing polyurethane varnish using exactly one small can of varnish. The final statue is supposed to be 5 times as large as the model in each dimension. How much bronze will she need? How much varnish should she buy? (*Hint*: If this seems difficult, you might start by writing a simpler question that is easier to work on before tackling this one.)

FIGURE 1-21 ■ Problem 55.

**(b)** The human brain has 1000 times the surface area of a mouse's brain. The human brain is convoluted, the mouse's is not. How much of this factor is due just to size (the human brain is bigger)? How sensitive is your result to your estimations of the approximate dimensions of a human brain and a mouse brain?

**56. Finding the Right Dose** We know from our dimensional analysis that if an object maintains its shape but changes its size, its area changes as the square of its length and its volume changes as the cube of its length. Suppose you are a parent and your child is sick and has to take some medicine. You have taken this medicine previously and you know its dose for you. You are 5′10″ tall and weigh 180 lb, and your child is 2′11″ tall and weighs 30 lb. Estimate an appropriate dosage for your child's medicine in the following cases. Be sure to discuss your reasoning.

**(a)** The medicine is one that will enter the child's bloodstream and reach every cell in the body. Your dose is 250 mg.
**(b)** The medicine is one that is meant to coat the child's throat. Your dose is 15 ml.

**57. Ping-Pong Ball Packing** Estimate how many Ping-Pong balls it would take to fill your classroom (assuming all the doors and windows are closed).

**58. Feeding the Cougar** When visiting the Como Park Zoo in St. Paul, Minnesota, with my young grandson, we encountered the sign shown at the right on the cage of the mountain lion. The detailed numbers surprised me. The amount of food given to the cat was specified to the tenth of a gram and the average cat's weight was specified to within 10 grams—about 1/3 of an ounce. This seemed to be overly precise. Can you figure out what they were trying to say and what a plausible accuracy might be for those two numbers—the amount of food given and the average cat's weight?

* Testudo is the statue of a terrapin (the university mascot) in front of the main library on the University of Maryland campus.

**COUGAR**
**North America**

| | |
|---|---|
| Natural Diet: | Hoofed animals, small animals |
| Zoo diet: | 1.3608 kg. commercially prepared diet for large cats, six days a week |
| Average Weight: | 90.72 kg. |
| Average Lifespan: | 20 years |

The cougar is also called mountain lion or puma.
It is the only large cat at Como Zoo that purrs.
Cougars are very solitary animals. They are seldom seen by humans.

FIGURE 1-22 ■ Problem 58.

**59. Blowing Off the Units.** Throughout your physics course, your instructor will expect you to be careful with the units in your calculations. Yet, some students tend to neglect them and just trust that they always work out properly. Maybe this real-world example will keep you from such a sloppy habit.

On July 23, 1983, Air Canada Flight 143 was being readied for its long trip from Montreal to Edmonton when the flight crew asked the ground crew to determine how much fuel was already onboard the airplane. The flight crew knew that they needed to begin the trip with 22 300 kg of fuel. They knew that amount in kilograms because Canada had recently switched to the metric system: previously fuel had been measured in pounds. The ground crew could measure the onboard fuel only in liters, which they reported as 7 682 L. Thus, to determine how much fuel was onboard and how much additional fuel must be added, the flight crew asked the ground crew for the conversion factor from liters to kilograms of fuel. The response was 1.77, which the flight crew used (1.77 kg corresponds to 1 L). (a) How many kilograms of fuel did the flight crew think they had? (In this problem, take all the given data as being exact.) (b) How many liters did they ask to be added to the airplane?

Unfortunately, the response from the ground crew was based on pre-metric habits—the number 1.77 was actually the conversion factor from liters to pounds of fuel (1.77 lb corresponds to 1 L). (c) How many kilograms of fuel were actually onboard? (Except for the given 1.77, use four significant figures for other conversion factors.) (d) How many liters of additional fuel were actually needed? (e) When the airplane left Montreal, what percentage of the required fuel did it actually have?

On route to Edmonton, at an altitude of 7.9 km, the airplane ran out of fuel and began to fall. Although the airplane then had no power, the pilot somehow managed to put it into a downward glide. However, the nearest working airport was too far to reach by only gliding, so the pilot somehow angled the glide toward an old nonworking airport.

Unfortunately, the runway at that airport had been converted to a track for race cars, and a steel barrier had been constructed across it. Fortunately, as the airplane hit the runway, the front landing gear collapsed, dropping the nose of the airplane onto the runway. The skidding slowed the airplane so that it stopped just short of the steel barrier, with stunned race drivers and fans looking on. All on board the airplane emerged safely. The point here is this: Take care of the units.

# 2 Motion Along a Straight Line

On September 26, 1993, Dave Munday, a diesel mechanic by trade, went over the Canadian edge of Niagara Falls for the second time, freely falling 48 m to the water (and rocks) below. On this attempt, he rode in a steel chamber with an airhole. Munday, keen on surviving this plunge that had killed other stuntmen, had done considerable research on the physics and engineering aspects of the plunge.

**If he fell straight down, how could he predict the speed at which he would hit the water?**

*The answer is in this chapter.*

## 2-1 Motion

The world, and everything in it, moves. Even a seemingly stationary thing, such as a roadway, moves because the Earth is moving. Not only is the Earth rotating and orbiting the Sun, but the Sun is also moving through space. The motion of objects can take many different forms. For example, a moving object's path might be a straight line, a curve, a circle, or something more complicated. The entity in motion might be something simple, like a ball, or something complex, like a human being or galaxy.

In physics, when we want to understand a phenomenon such as motion, we begin by exploring relatively simple motions. For this reason, in the study of motion we start with **kinematics,** which focuses on describing motion, rather than on **dynamics,** which deals with the causes of motion. Further, we begin our study of kinematics by developing the concepts required to measure motion and mathematical tools needed to describe them in one dimension (or in 1D). Only then do we extend our study to include a consideration of the causes of motion and motions in two and three dimensions. Further simplifications are helpful. Thus, in this chapter, our description of the motion of objects is restricted in two ways:

1. **The motion of the object is along a straight line.** The motion may be purely vertical (that of a falling stone), purely horizontal (that of a car on a level highway), or slanted (that of an airplane rising at an angle from a runway), but it must be a straight line.
2. **The object is effectively a particle** because its size and shape are not important to its motion. By "particle" we mean either: (a) a point-like object with dimensions that are small compared to the distance over which it moves (such as the size of the Earth relative to its orbit around the Sun), (b) an extended object in which all its parts move together (such as a falling basketball that is not spinning), or (c) that we are only interested in the path of a special point associated with the object (such as the belt buckle on a walking person).

We will start by introducing very precise definitions of words commonly used to describe motion like speed, velocity, and acceleration. These definitions may conflict with the way these terms are used in everyday speech. However, by using precise definitions rather than our casual definitions, we will be able to describe and predict the characteristics of common motions in graphical and mathematical terms. These mathematical descriptions of phenomena form the basic vocabulary of physics and engineering.

Although our treatment may seem ridiculously formal, we need to provide a foundation for the analysis of more complex and interesting motions.

**READING EXERCISE 2-1:** Which of the following motions are along a straight line: (a) a string of carts traveling up and down along a roller coaster, (b) a cannonball shot straight up, (c) a car traveling along a straight city street, (d) a ball rolling along a straight ramp tilted at a 45° angle. ■

**READING EXERCISE 2-2:** In reality there are no point particles. Rank the following everyday items from most particle-like to least particle-like: (a) a 2-m-tall long jumper relative to a 25 m distance covered in a jump, (b) a piece of lead shot from a shotgun shell relative to its range of 5 m, (c) the Earth of diameter $13 \times 10^6$ m relative to the approximate diameter of its orbit about the Sun of $3 \times 10^{11}$ m. ■

## 2-2 Position and Displacement Along a Line

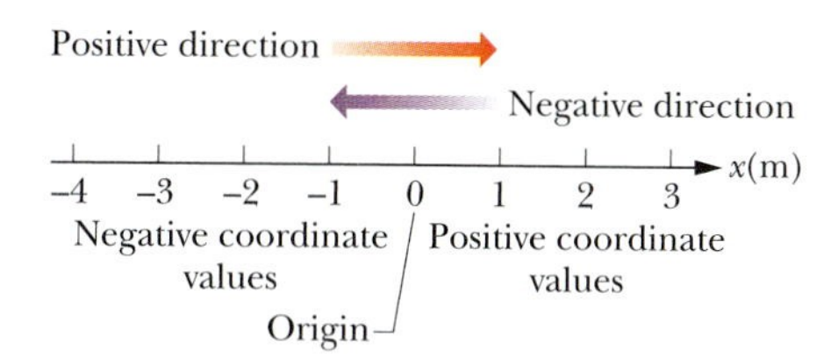


**FIGURE 2-1** ■ Position is determined on an axis that is marked in units of meters and that extends indefinitely in opposite directions.

### Defining a Coordinate System

In order to study motion along a straight line, we must be able to specify the location of an object and how it changes over time. A convenient way to locate a point of interest or an object is to define a coordinate system. Houses in Costa Rican towns are commonly located with addresses such as "200 meters east of the Post Office." In order to locate a house, a distance scale must be agreed upon (meters are used in the example), and a reference point or origin (in this case the Post Office), and a direction (in this case east) must be specified. Thus, in locating an object that can move along a straight line, it is convenient to specify its position by choosing a one-dimensional **coordinate system.** The system consists of a *point of reference known as the origin (or zero point)*, a line that passes through the chosen origin called a *coordinate axis*, one direction along the coordinate axis, chosen as positive and the other direction as negative, and the units we use to measure a quantity. We have labeled the coordinate axis as the $x$ axis, in Fig. 2-1, and placed an origin on it. The direction of increasing numbers (coordinates) is called the **positive direction,** which is toward the right in Fig. 2-1. The opposite direction is the **negative direction.**

Figure 2-1 is drawn in the traditional fashion, with negative coordinates to the left of the origin and positive coordinates to the right. It is also traditional in physics to use meters as the standard scale for distance. However, we have freedom to choose other units and to decide which side of the origin is labeled with negative coordinates and which is labeled with positive coordinates. Furthermore, we can choose to define an $x$ axis that is vertical rather than horizontal, or inclined at some angle. In short, we are free to make choices about how we define our coordinate system.

Good choices make describing a situation much easier. For example, in our consideration of motion along a straight line, we would want to align the axis of our one-dimensional coordinate system along the line of motion. In Chapters 5 and 6, when we consider motions in two dimensions, we will be using more complex coordinate systems with a set of mutually perpendicular coordinate axes. Choosing a coordinate system that is appropriate to the physical situation being described can simplify your mathematical description of the situation. To describe a particle moving in a circle, you would probably choose a two-dimensional coordinate system in the plane of the circle with the origin placed at its center.

### Defining Position as a Vector Quantity

The reason for choosing our standard one-dimensional coordinate axis and orienting it along the direction of motion is to be able to define the position of an object relative to our chosen origin, and then be able to keep track of how its position changes as the object moves. It turns out that the position of an object relative to a coordinate system can be described by a mathematical entity known as a **vector.** This is because, in order to find the position of an object, we must specify both how far and in which direction the object is from the origin of a coordinate system.

> A **VECTOR** is a mathematical entity that has both a magnitude and a direction. Vectors can be added, subtracted, multiplied, and transformed according to well-defined mathematical rules.

There are other physical quantities that also behave like vectors such as velocity, acceleration, force, momentum, and electric and magnetic fields.

However, not all physical quantities that have signs associated with them are vectors. For example, temperatures do not need to be described in terms of a coordinate

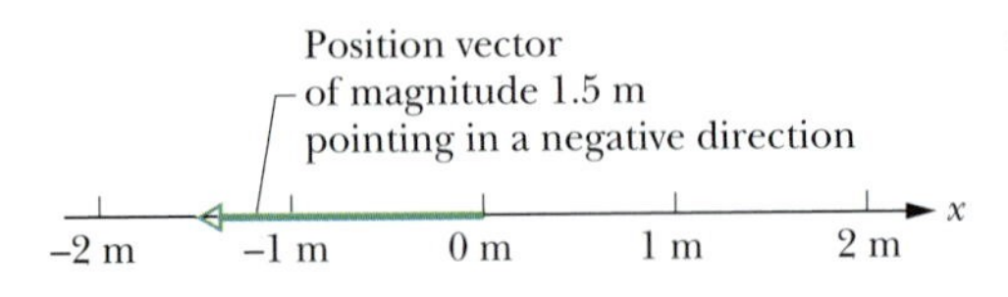


**FIGURE 2-2** ■ A position vector can be represented by an arrow pointing from the origin of a chosen coordinate system to the location of the object.

system, and single numbers, such as $T = -5°C$ or $T = 12°C$, are sufficient to describe them. The minus sign, in this case, does not signify a direction. Mass, distance, length, area, and volume also have no directions associated with them and, although their values depend on the units used to measure them, their values do not depend on the orientation of a coordinate system. Such quantities are called **scalars.**

> A **SCALAR** is defined as a mathematical quantity whose value does not depend on the orientation of a coordinate system and has no direction associated with it.

In general, a one-dimensional vector can be represented by an arrow. The length of the arrow, which is inherently positive, represents the **magnitude** of the vector and the direction in which the arrow points represents the **direction** associated with the vector.

We begin this study of motion by introducing you to the properties of one-dimensional position and displacement vectors and some of the formal methods for representing and manipulating them. These formal methods for working with vectors will prove to be very useful later when working with two- and three-dimensional vectors.

A **one-dimensional position vector** is defined by the location of the origin of a chosen one-dimensional coordinate system and of the object of interest. The **magnitude** of the position vector is a scalar that denotes the distance between the object and the origin. For example, an object that has a position vector of magnitude 5 m could be located at the point +5 m or −5 m from the origin.

On a conventional $x$ axis, the direction of the position vector is positive when the object is located to the right of the origin and negative when the object is located to the left of the origin. For example, in the system shown in Fig. 2-1, if a particle is located at a distance of 3 m to the left of the origin, its position vector has a magnitude of 3 m and a direction that is negative. One of many ways to represent a position vector is to draw an arrow from the origin to the object's location, as shown in Fig. 2-2, for an object that is 1.5 m to the left of the origin. Since the length of a vector arrow represents the magnitude of the vector, its length should be proportional to the distance from the origin to the object of interest. In addition, the direction of the arrow should be from the origin to the object.

Instead of using an arrow, a position vector can be represented mathematically. In order to develop a useful mathematical representation we need to define a **unit vector** associated with our $x$ axis.

> A **UNIT VECTOR FOR A COORDINATE AXIS** is a dimensionless vector that points in the direction along a coordinate axis that is chosen to be positive.

It is customary to represent a unit vector that points along the positive $x$ axis with the symbol $\hat{\imath}$ (i-hat), although some texts use the symbol $\hat{x}$ (x-hat) instead. When considering three-dimensional vectors, the unit vectors pointing along the designated positive $y$ axis and $z$ axis are denoted by $\hat{\jmath}$ and $\hat{k}$, respectively.

These vectors are called "unit vectors" because they have a dimensionless value of one. However, you should not confuse the use of word "unit" with a physical unit. Unit vectors should be shown on coordinate axes as small pointers with no physical units, such as meters, associated with them. This is shown in Fig. 2-3 for the $x$ axis unit vector. Since the scale used in the coordinate system has units, it is essential that the units always be associated with the number describing the location of an object along an axis. Figure 2-3 also shows how the unit vector is used to create a position vector corresponding to an object located at position −1.5 meters on our $x$ axis. To do this we stretch or multiply the unit vector by the magnitude of the position vector, which

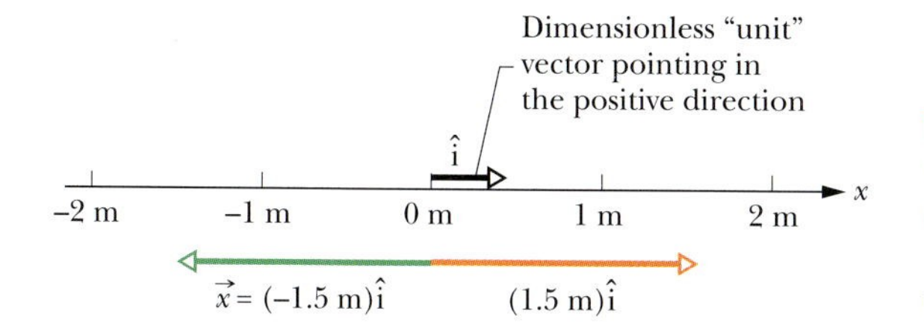


**FIGURE 2-3** ■ Arrows representing: (1) a dimensionless unit vector, $\hat{i}$, pointing in the positive $x$ direction; (2) a vector representing the unit vector multiplied by 1.5 meters; and (3) a vector multiplied by 1.5 meters and inverted by multiplication by $-1$ to create the position vector $\vec{x} = (-1.5\text{ m})\hat{i}$. This position vector has a magnitude of 1.5 meters and points in a negative direction.

is 1.5 m. Note that we are using the coordinate axis to describe a position in meters relative to an origin, so it is essential to include the units with the number. This multiplication of the dimensionless unit vector by 1.5 m creates a 1.5-m-long vector that points in the same direction as the unit vector. It is denoted by $(1.5\text{ m})\hat{i}$. However, the vector we want to create points in the negative direction, so the vector pointing in the positive direction must be inverted using a minus sign. The position vector we have created is denoted as $\vec{x}$. It can be divided into two parts—a vector component and a unit vector,

$$\vec{x} = (-1.5\text{ m})\hat{i}.$$

In this example, the $x$-component of the position vector, denoted as $x$, is $-1.5$ m.

Here the quantity 1.5 m with no minus sign in front of it is known as the magnitude of this position vector. In general, the magnitude is denoted as $|\vec{x}|$. Thus, the one-dimensional position vector for the situation shown in Fig. 2-3 is denoted mathematically using the following symbols:

$$\vec{x} = x\hat{i} = (-1.5\text{ m})\hat{i}.$$

The $x$-component of a position vector, denoted $x$, can be positive or negative depending on which side of the origin the particle is. Thus, in one dimension in terms of absolute values, the vector component $x$ is either $+|x|$ or $-|x|$, depending on the object's location.

In general, a component of a vector along an axis, such as $x$ in this case, is not a scalar since our $x$-component will change sign if we choose to reverse the orientation of our chosen coordinate system. In contrast, *the magnitude of a position vector is always positive, and it only tells us how far away the object is from the origin*, so the magnitude of a vector is always a scalar quantity. The sign of the component ($+$ or $-$) tells us in which direction the vector is pointing. The sign will be negative if the object is to the left of the origin and positive if it is to the right of the origin.

## Defining Displacement as a Vector Quantity

The study of motion is primarily about how an object's location changes over time under the influence of forces. In physics the concept of **change** has an exact mathematical definition.

> **CHANGE** is defined as the difference between the state of a physical system (typically called the final state) and its state at an earlier time (typically called the initial state).

This definition of change is used to define displacement.

> **DISPLACEMENT** is defined as the change of an object's position that occurs during a period of time.

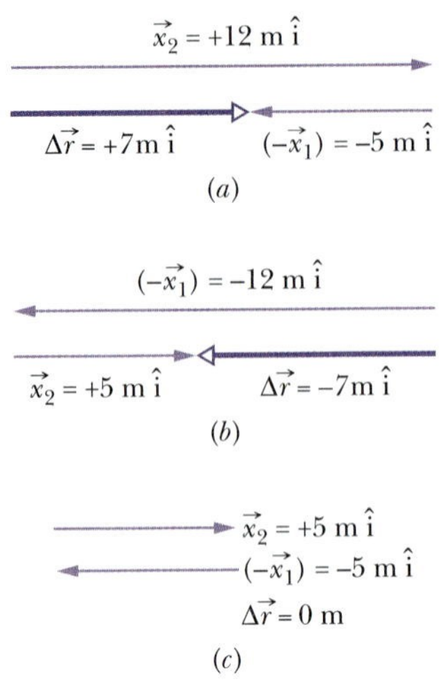

$\vec{x}_2 = +12 \text{ m } \hat{\text{i}}$

$\Delta\vec{r} = +7\text{m } \hat{\text{i}}$ $(-\vec{x}_1) = -5 \text{ m } \hat{\text{i}}$

(*a*)

$(-\vec{x}_1) = -12 \text{ m } \hat{\text{i}}$

$\vec{x}_2 = +5 \text{ m } \hat{\text{i}}$ $\Delta\vec{r} = -7\text{m } \hat{\text{i}}$

(*b*)

$\vec{x}_2 = +5 \text{ m } \hat{\text{i}}$

$(-\vec{x}_1) = -5 \text{ m } \hat{\text{i}}$

$\Delta\vec{r} = 0 \text{ m}$

(*c*)

**FIGURE 2-4** ■ The wide arrow shows the displacement vector $\Delta\vec{r}$ for three situations leading to: (*a*) a positive displacement, (*b*) a negative displacement, and (*c*) zero displacement.

Since position can be represented as a vector quantity, displacement is the difference between two vectors, and thus, is also a vector. So, in the case of motion along a line, an object moving from an "initial" position $\vec{x}_1$ to another "final" position $\vec{x}_2$ at a later time is said to undergo a **displacement** $\Delta\vec{r}$, given by the difference of two position vectors

$$\Delta\vec{r} \equiv \vec{x}_2 - \vec{x}_1 = \Delta x\hat{\text{i}} \quad \text{(displacement vector)}, \tag{2-1}$$

where the symbol $\Delta$ is used to represent a change in a quantity, and the symbol "$\equiv$" signifies that the displacement $\Delta\vec{r}$ is given by $\vec{x}_2 - \vec{x}_1$ because we have *chosen* to define it that way.

As you will see when we begin to work with vectors in two and three dimensions, it is convenient to consider subtraction as the addition of one vector to another that has been inverted by multiplying the vector component by $-1$. We can use this idea of defining subtraction as the addition of an inverted vector to find displacements. Let's consider three situations:

(a) A particle moves along a line from $\vec{x}_1 = (5 \text{ m})\hat{\text{i}}$ to $\vec{x}_2 = (12 \text{ m})\hat{\text{i}}$. Since $\Delta\vec{r} = \vec{x}_2 - \vec{x}_1 = \vec{x}_2 + (-\vec{x}_1)$,

$$\Delta\vec{r} = (12 \text{ m})\hat{\text{i}} - (5 \text{ m})\hat{\text{i}} = (12 \text{ m})\hat{\text{i}} + (-5 \text{ m})\hat{\text{i}} = (7 \text{ m})\hat{\text{i}}.$$

The *positive* result indicates that the motion is in the positive direction (toward the right in Fig. 2-4*a*).

(b) A particle moves from $\vec{x}_1 = (12 \text{ m})\hat{\text{i}}$ to $\vec{x}_2 = (5 \text{ m})\hat{\text{i}}$. Since $\Delta\vec{r} = \vec{x}_2 - \vec{x}_1 = \vec{x}_2 + (-\vec{x}_1)$,

$$\Delta\vec{r} = (5 \text{ m})\hat{\text{i}} - (12 \text{ m})\hat{\text{i}} = (5 \text{ m})\hat{\text{i}} + (-12 \text{ m})\hat{\text{i}} = (-7 \text{ m})\hat{\text{i}}.$$

The negative result indicates that the displacement of the particle is in the negative direction (toward the left in Fig. 2-4*b*).

(c) A particle starts at 5 m, moves to 2 m, and then returns to 5 m. The displacement for the full trip is given by $\Delta\vec{r} = \vec{x}_2 - \vec{x}_1 = \vec{x}_2 + (-\vec{x}_1)$, where $\vec{x}_1 = (5 \text{ m})\hat{\text{i}}$ and $\vec{x}_2 = (5 \text{ m})\hat{\text{i}}$:

$$\Delta\vec{r} = (5 \text{ m})\hat{\text{i}} + (-5 \text{ m})\hat{\text{i}} = (0 \text{ m})\hat{\text{i}}$$

and the particle's position hasn't changed, as in Fig. 2-4*c*. Since displacement involves only the original and final positions, the actual number of meters traced out by the particle while moving back and forth is immaterial.

If we ignore the sign of a particle's displacement (and thus its direction), we are left with the **magnitude** of the displacement. This is the distance between the original and final positions and is always positive. It is important to remember that displacement (or any other vector) has not been completely described until we state its direction.

We use the notation $\Delta\vec{r}$ for displacement because when we have motion in more than one dimension, the notation for the position vector is $\vec{r}$. For a one-dimensional motion along a straight line, we can also represent the displacement as $\Delta\vec{x}$. The magnitude of displacement is represented by surrounding the displacement vector symbol with absolute value signs:

$$\{\text{magnitude of displacement}\} = |\Delta\vec{r}| \quad \text{or} \quad |\Delta\vec{x}|$$

**READING EXERCISE 2-3:** Can a particle that moves from one position with a negative value, to another position with a negative value, undergo a positive displacement? ■

### TOUCHSTONE EXAMPLE 2-1: Displacements

Three pairs of initial and final positions along an $x$ axis represent the location of objects at two successive times: (pair 1) $-3$ m, $+5$ m; (pair 2) $-3$ m, $-7$ m; (pair 3) 7 m, $-3$ m.

(a) Which pairs give a negative displacement?

**SOLUTION** ■ The **Key Idea** here is that the displacement is negative when the final position lies *to the left* of the initial position. As shown in Fig. 2-5, this happens when the final position is *more negative* than the initial position. Looking at pair 1, we see that the final position, $+5$ m, is positive while the initial position, $-3$ m, is negative. This means that the displacement is from left (more negative) to right (more positive) and so the displacement is positive for pair 1. (Answer)

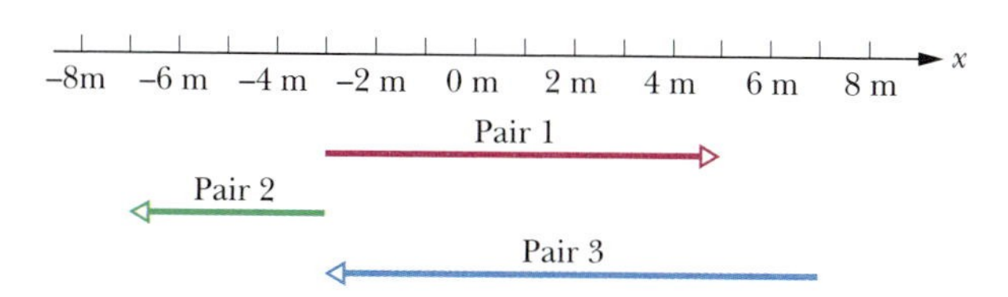


**FIGURE 2-5** ■ Displacement associated with three pairs of initial and final positions along an $x$ axis.

For pair 2 the situation is different. The final position, $-7$ m, lies to the left of the initial position, $-3$ m, so the displacement is negative. (Answer)

For pair 3 the final position, $-3$ m, is to the left of the origin while the initial position, $+7$ m, is to the right of the origin. So the displacement is from the right of the origin to its left, a negative displacement. (Answer)

(b) Calculate the value of the displacement in each case using vector notation.

**SOLUTION** ■ The **Key Idea** here is to use Eq. 2-1 to calculate the displacement for each pair of positions. It tells us the difference between the final position and the initial position, in that order,

$$\Delta\vec{x} = \vec{x}_2 - \vec{x}_1 \quad \text{(displacement).} \tag{2-2}$$

For pair 1 the final position is $\vec{x}_2 = (+5\text{ m})\hat{\mathrm{i}}$ and the initial position is $\vec{x}_1 = (-3\text{ m})\hat{\mathrm{i}}$, so the displacement between these two positions is just

$$\Delta\vec{x} = (+5\text{ m})\hat{\mathrm{i}} - (-3\text{ m})\hat{\mathrm{i}} = (+5\text{ m})\hat{\mathrm{i}} + (3\text{ m})\hat{\mathrm{i}} = (+8\text{ m})\hat{\mathrm{i}}. \quad \text{(Answer)}$$

For pair 2 the same argument yields

$$\Delta\vec{x} = (-7\text{ m})\hat{\mathrm{i}} - (-3\text{ m})\hat{\mathrm{i}} = (-7\text{ m})\hat{\mathrm{i}} + (3\text{ m})\hat{\mathrm{i}} = (-4\text{ m})\hat{\mathrm{i}}. \quad \text{(Answer)}$$

Finally, the displacement for pair 3 is

$$\Delta\vec{x} = (-3\text{ m})\hat{\mathrm{i}} - (+7\text{ m})\hat{\mathrm{i}} = (-3\text{ m})\hat{\mathrm{i}} + (-7\text{ m})\hat{\mathrm{i}} = (-10\text{ m})\hat{\mathrm{i}}. \quad \text{(Answer)}$$

(c) What is the magnitude of each position vector?

**SOLUTION** ■ Of the six position vectors given, one of them—namely $\vec{x}_1 = (-3\text{ m})\hat{\mathrm{i}}$—appears in all three pairs. The remaining three positions are $\vec{x}_2 = (+5\text{ m})\hat{\mathrm{i}}$, $\vec{x}_3 = (-7\text{ m})\hat{\mathrm{i}}$, and $\vec{x}_4 = (+7\text{ m})\hat{\mathrm{i}}$. The **Key Idea** here is that the magnitude of a position vector just tells us *how far* the point lies from the origin without regard to whether it lies to the left or to the right of the origin. Thus the magnitude of our first position vector is 3 m (Answer) since the position specified by $\vec{x} = (-3\text{ m})\hat{\mathrm{i}}$ is 3 m to the left of the origin. It's *not* $-3$ m, because magnitudes only specify distance from the origin, not direction.

For the same reason, the magnitude of the second position vector is just 5 m (Answer) while the magnitude of the third and the fourth are *each* 7 m. (Answer) The fact that the third point lies 7 m to the left of the origin while the fourth lies 7 m to the right doesn't matter here.

(d) What is the value of the $x$-component of each of these position vectors?

**SOLUTION** ■ To answer this question you need to remember what is meant by the component of a vector. The key equation relating a vector in one dimension to its component along its direction is $\vec{x} = x\,\hat{\mathrm{i}}$, where $\vec{x}$ (with the arrow over it) is the vector itself and $x$ (with no arrow over it) is the component of the vector in the direction specified by the unit vector $\hat{\mathrm{i}}$. So the component of $\vec{x} = (-3\text{ m})\hat{\mathrm{i}}$ is $-3$ m, while that of $\vec{x} = (+5\text{ m})\hat{\mathrm{i}}$ is just $+5$ m, and $\vec{x} = (-7\text{ m})\hat{\mathrm{i}}$ has as its component along the $\hat{\mathrm{i}}$ direction ($-7$ m) while for $\vec{x} = (+7\text{ m})\hat{\mathrm{i}}$ it's just ($+7$ m). In other words, the component of a vector in the direction of $\hat{\mathrm{i}}$ is just the signed number (with its units) that multiplies $\hat{\mathrm{i}}$. (Answer)

## 2-3 Velocity and Speed

Suppose a student stands still or speeds up and slows down along a straight line. How can we describe accurately and efficiently where she is and how fast she is moving? We will explore several ways to do this.

## Representing Motion in Diagrams and Graphs

**Motion Diagrams:** Now that you have learned about position and displacement, it is quite easy to describe the motion of an object using pictures or sketches to chart how position changes over time. Such a representation is called a *motion diagram*. For example, Fig. 2-6 shows a student whom we treat as if she were concentrated into a particle located at the back of her belt. She is *standing still* at a position $\vec{x} = (-2.00\text{ m})\hat{\text{i}}$ from a point on a sidewalk that we choose as our origin. Figure 2-7 shows a more complex diagram describing the student in motion. Suppose we see that just as we start timing her progress with a stopwatch (so $t = 0.0$ s), the back of her belt is 2.47 m to the left of our origin. The $x$-component of her position is then $x = -2.47$ m. The student then moves toward the origin, almost reaches the origin at $t = 1.5$ s, and then continues moving to the right so that her $x$-component of position has increasingly positive values. It is important to recognize that just as we chose an origin and direction for our coordinate axis, we also chose an origin in time. If we had chosen to start our timing 12 seconds earlier, then the new motion diagram would show the back of her belt as being at $x = -2.47$ m at $t = 12$ s.

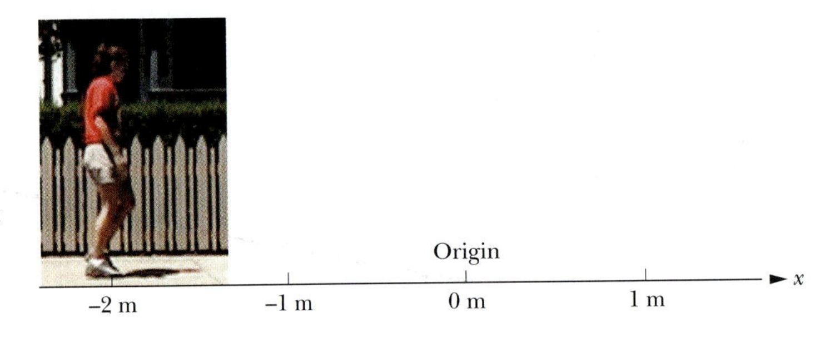


**FIGURE 2-6** ■ A motion diagram of a student standing still with the back of her belt at a horizontal distance of 2.00 m to the left of a spot of the sidewalk designated as the origin.

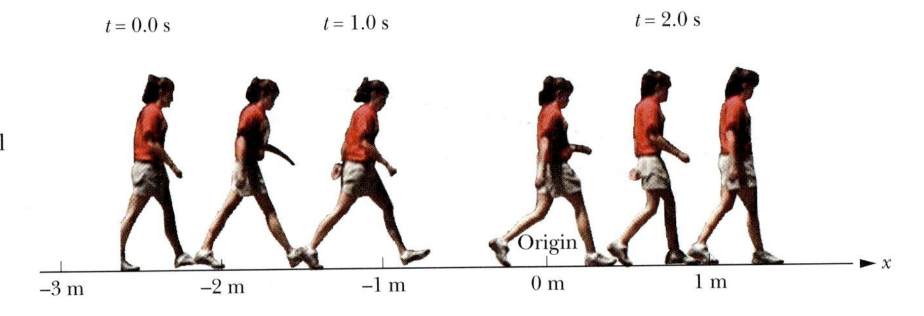


**FIGURE 2-7** ■ A motion diagram of a student starting to walk slowly. The horizontal position of the back of her belt starts at a horizontal distance of 2.47 m to the left of a spot designated as the origin. She is speeding up for a few seconds and then slowing down.

**Graphs:** Another way to describe how the position of an object changes as time passes is with a graph. In such a graph, the $x$-component of the object's position, $x$, can be plotted as a function of time, $t$. This position–time graph has alternate names such as a graph of $x$ as a function of $t$, $x(t)$, or $x$ vs. $t$. For example, Fig. 2-8 shows a graph of the student *standing still* with the back of her belt located at a horizontal position of −2.00 m from a spot on the sidewalk that is chosen as the origin.

The graph of no motion shown in Fig. 2-8 is not more informative than the picture or a comment that the student is standing still for 3 seconds at a certain location. But it's another story when we consider the graph of a motion. Figure 2-9 is a graph of a student's $x$-component of position as a function of time. It represents the same information depicted in the motion diagram in Fig. 2-7. Data on the student's motion are first recorded at $t = 0.0$ s when the $x$-component of her position is $x = -2.47$ m. The student then moves toward $x = 0.00$ m, passes through that point at about $t = 1.5$ s, and then moves on to increasingly larger positive values of $x$ while slowing down.

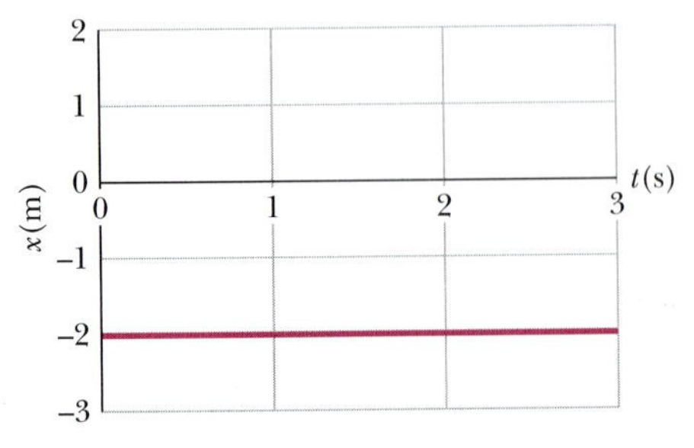


**FIGURE 2-8** ■ The graph of the $x$-component of position for a student who is standing still at $x = -2.0$ m for at least 3 seconds.

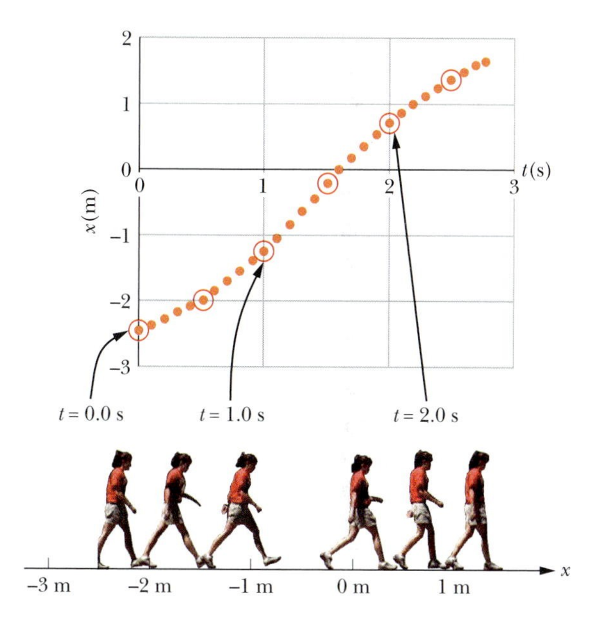


**FIGURE 2-9** ■ A graph that represents how the position component, $x$, of the walking student shown in Fig. 2-7 changes over time. The motion diagram, shown below the graph, is associated with the graph at three points in time as indicated by the arrows.

Although the graph of the student's motion in Fig. 2-9 seems abstract and quite unlike a motion diagram, it is richer in information. For example, the graph allows us to estimate the motion of the student at times between those for which position measurements were made. Equally important, we can use the graph to tell us how fast the student moves at various times, and we deal with this aspect of motion graphs next.

What can motion diagrams and $x$ vs. $t$ graphs tell us about how fast and in what direction something moves along a line? It is clear from an examination of the motion diagram at the bottom of Fig. 2-9 that the student covers the most distance and so appears to be moving most rapidly between the two times $t_1 = 1.0$ s and $t_2 = 1.5$ s. But this time interval is also where the slope (or steepness) of the graph has the greatest magnitude. Recall from mathematics that the average slope of a curve between two points is defined as the ratio of the change in the variable plotted on the vertical axis (in this case the $x$-component of her position) to the change in the variable plotted on the horizontal axis (in this case the time). Hence, on position vs. time graphs (such as those shown in Fig. 2-8 and Fig. 2-9),

$$\text{average slope} \equiv \frac{\Delta x}{\Delta t} = \frac{x_2 - x_1}{t_2 - t_1} \quad \text{(definition of average slope).} \qquad (2\text{-}3)$$

Since time moves forward, $t_2 > t_1$, so $\Delta t$ always has a positive value. Thus, a slope will be positive whenever $x_2 > x_1$, so $\Delta x$ is positive. In this case a straight line connecting the two points on the graph slants upward toward the right when the student is moving along the positive $x$-direction. On the other hand, if the student were to move "backwards" in the direction along the $x$ axis we chose to call negative, then $x_2 < x_1$. In this case, the slope between the two times would be negative and the line connecting the points would slant downward to the right.

## Average Velocity

For motion along a straight line, the steepness of the slope in an $x$ vs. $t$ graph over a time interval from $t_1$ to $t_2$ tells us "how fast" a particle moves. The direction of motion is indicated by the sign of the slope (positive or negative). Thus, this slope or ratio $\Delta x/\Delta t$ is a special quantity that tells us how fast and in what direction something moves. We haven't given the ratio $\Delta x/\Delta t$ a name yet. We do this to emphasize the fact that the ideas associated with figuring out how fast and in what direction something moves are more important than the names we assign to them. However, it is inconvenient not to have a name. The common name for this ratio is **average velocity,** which is defined as the ratio of displacement vector $\Delta\vec{x}$ for the motion of interest to the time interval $\Delta t$ in which it occurs. This vector can be expressed in equation form as

$$\langle\vec{v}\rangle \equiv \frac{\Delta\vec{x}}{\Delta t} = \frac{\Delta x}{\Delta t}\hat{\mathrm{i}} = \frac{x_2 - x_1}{t_2 - t_1}\hat{\mathrm{i}} \quad \text{(definition of 1D average velocity),} \qquad (2\text{-}4)$$

where $x_2$ and $x_1$ are components of the position vectors at the final and initial times. Here we use angle brackets $\langle\ \rangle$ to denote the average of a quantity. Also, we use the special symbol "$\equiv$" for equality to emphasize that the term on the left is equal to the term on the right by definition. The time change is a positive scalar quantity because we never need to specify its direction explicitly. In defining $\langle\vec{v}\rangle$ we are basically multiplying the displacement vector, $\Delta\vec{x}$, by the scalar $(1/\Delta t)$. This action gives us a new vector that points in the same direction as the displacement vector.

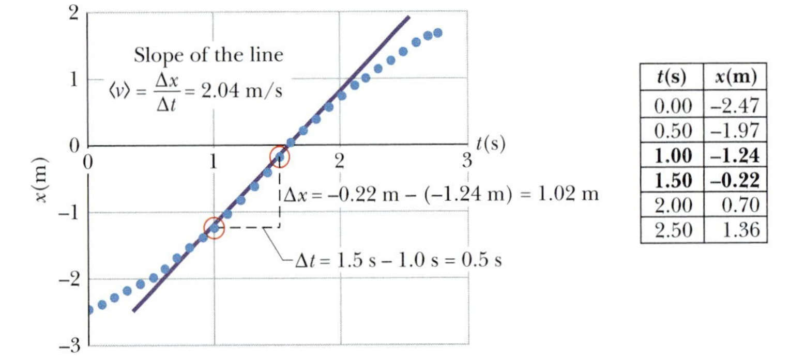


| $t$(s) | $x$(m) |
|---|---|
| 0.00 | −2.47 |
| 0.50 | −1.97 |
| **1.00** | **−1.24** |
| **1.50** | **−0.22** |
| 2.00 | 0.70 |
| 2.50 | 1.36 |

**FIGURE 2-10** ■ Calculation of the slope of the line that connects the points on the curve at $t_1 = 1.0$ s and $t_2 = 1.5$ s. The $x$-component of the average velocity is given by this slope.

Figure 2-10 shows how to find the average velocity for the student motion represented by the graph shown in Fig. 2-9 between the times $t_1 = 1.0$ s and $t_2 = 1.5$ s. The average velocity during that time interval is

$$\langle\vec{v}\rangle \equiv \frac{\Delta x}{\Delta t}\hat{\mathrm{i}} = \frac{x_2 - x_1}{t_2 - t_1}\hat{\mathrm{i}} = \frac{-0.22\text{ m} - (-1.24\text{ m})}{(1.5\text{ s} - 1.0\text{ s})}\hat{\mathrm{i}} = (2.04\text{ m/s})\hat{\mathrm{i}}.$$

The $x$-component of the average velocity along the line of motion, $\langle v_x\rangle = 2.04$ m/s, is simply the slope of the straight line that connects the point on the curve at the beginning of our chosen interval and the point on the curve at the end of the interval. Since our student is speeding up and slowing down, the values of $\langle\vec{v}\rangle$ and $\langle v_x\rangle$ will in general be different when calculated using other time intervals.

## Average Speed

Sometimes we don't care about the direction of an object's motion but simply want to keep track of the distance covered. For instance, we might want to know the total distance a student walks (number of steps times distance covered in each step). Our student could be pacing back and forth wearing out her shoes without having a vector displacement. Similarly, average speed, $\langle s\rangle$, is a different way of describing "how fast" an object moves. Whereas the average velocity involves the particle's displacement $\Delta\vec{x}$, which is a vector quantity, the average speed involves the total distance covered (for example, the product of the length of a step and the number of steps the student took), which is independent of direction. So **average speed** is defined as

$$\langle s \rangle \equiv \frac{\text{total distance}}{\Delta t} \quad \text{(definition of average speed).} \qquad (2\text{-}5)$$

Since neither the total distance traveled nor the time interval over which the travel occurred has an associated direction, average speed does not include direction information. Both the total distance and the time period are always positive, so average speed is always positive too. Thus, an object that moves back and forth along a line can have no vector displacement, so it has zero velocity but a rather high average speed. At other times, while the object is moving in only one direction, the average speed $\langle s \rangle$ is the same as the magnitude of the average velocity $\langle \vec{v} \rangle$. However, as you can demonstrate in Reading Exercise 2-4, when an object doubles back on its path, the average speed is not simply the magnitude of the average velocity $|\langle \vec{v} \rangle|$.

## Instantaneous Velocity and Speed

You have now seen two ways to describe how fast something moves: average velocity and average speed, both of which are measured over a time interval $\Delta t$. Clearly, however, something might speed up and slow down during that time interval. For example, in Fig. 2-9 we see that the student is moving more slowly at $t = 0.0$ s than she is at $t = 1.5$ s, so her velocity seems to be changing during the time interval between 0.0 s and 1.5 s. The average slope of the line seems to be increasing during this time interval. Can we refine our definition of velocity in such a way that we can determine the student's true velocity at any one "instant" in time? We envision something like the almost instantaneous speedometer readings we get as a car speeds up and slows down.

Defining an instant and instantaneous velocity is not a trivial task. As we noted in Chapter 1, the time interval of 1 second is defined by counting oscillations of radiation absorbed by a cesium atom. In general, even our everyday clocks work by counting oscillations in an electronic crystal, pendulum, and so on. We associate "instants in time" with positions on the hands of a clock, and "time intervals" with changes in the position of the hands.

For the purpose of finding a velocity at an instant, we can attempt to make the time interval we use in our calculation so small that it has almost zero duration. Of course the displacement we calculate also becomes very small. So **instantaneous velocity** along a line—like average velocity—is still defined in terms of the ratio of $\Delta\vec{x}/\Delta t$. But we have this ratio passing to a limit where $\Delta t$ gets closer and closer to zero. Using standard calculus notation for this limit gives us the following definition:

$$\vec{v} \equiv \lim_{\Delta t \to 0} \frac{\Delta \vec{x}}{\Delta t} = \frac{d\vec{x}}{dt} \quad \text{(definition of 1D instantaneous velocity).} \qquad (2\text{-}6)$$

> In the language of calculus, the **INSTANTANEOUS VELOCITY** is the rate at which a particle's position vector, $\vec{x}$, is changing with time at a given instant.

In passing to the limit the ratio $\Delta\vec{x}/\Delta t$ is not necessarily small, since both the numerator and denominator are getting small together. The first part of this expression,

$$\vec{v} = v_x \hat{\mathrm{i}} = \lim_{\Delta t \to 0} \frac{\Delta \vec{x}}{\Delta t},$$

tells us that we can find the (instantaneous) velocity of an object by taking the slope of a graph of the position component vs. time at the point associated with that moment in time. If the graph is a curve rather than a straight line, the *slope at a point* is actually the tangent to the line at that point. Alternatively, the second part of the expression, shown in Eq. 2-6,

$$\vec{v} = \frac{d\vec{x}}{dt},$$

indicates that, if we can approximate the relationship between $\vec{x}$ and $t$ as a continuous mathematical function such as $\vec{x} = (3.0 \text{ m/s}^2)t^2$, we can also find the object's instantaneous velocity by taking a derivative with respect to time of the object's position $\vec{x}$. When $\vec{x}$ varies continuously as time marches on, we often denote $\vec{x}$ as a position function $\vec{x}(t)$ to remind us that it varies with time.

**Instantaneous speed,** which is typically called simply **speed,** is just the magnitude of the instantaneous velocity vector, $|\vec{v}|$. Speed is a scalar quantity consisting of the velocity value that has been stripped of any indication of the direction the object is moving, either in words or via an algebraic sign. A velocity of $(+5 \text{ m/s})\hat{\text{i}}$ and one of $(-5 \text{ m/s})\hat{\text{i}}$ both have an associated speed of 5 m/s.

**READING EXERCISE 2-4:** Suppose that you drive 10 mi due east to a store. You suddenly realize that you forgot your money. You turn around and drive the 10 mi due west back to your home and then return to the store. The total trip took 30 min. (a) What is your average velocity for the entire trip? (Set up a coordinate system and express your result in vector notation.) (b) What was your average speed for the entire trip? (c) Discuss why you obtained different values for average velocity and average speed. ■

**READING EXERCISE 2-5:** Suppose that you are driving and look down at your speedometer. What does the speedometer tell you—average speed, instantaneous speed, average velocity, instantaneous velocity—or something else? Explain. ■

**READING EXERCISE 2-6:** The following equations give the position component, $x(t)$, along the $x$ axis of a particle's motion in four situations (in each equation, $x$ is in meters, $t$ is in seconds, and $t > 0$): (1) $x = (3 \text{ m/s})t - (2 \text{ m})$; (2) $x = (-4 \text{ m/s}^2)t^2 - (2 \text{ m})$; (3) $x = (-4 \text{ m/s}^2)t^2$; and (4) $x = -2$ m.
(a) In which situations is the velocity $\vec{v}$ of the particle constant? (b) In which is the vector $\vec{v}$ pointing in the negative $x$ direction? ■

**READING EXERCISE 2-7:** In Touchstone Example 2-2, suppose that right after refueling the truck you drive back to $x_1$ at 35 km/h. What is the average velocity for your entire trip? ■

## TOUCHSTONE EXAMPLE 2-2: Out of Gas

You drive a beat-up pickup truck along a straight road for 8.4 km at 70 km/h, at which point the truck runs out of gasoline and stops. Over the next 30 min, you walk another 2.0 km farther along the road to a gasoline station.

(a) What is your overall displacement from the beginning of your drive to your arrival at the station?

**SOLUTION** ■ Assume, for convenience, that you move in the positive direction along an $x$ axis, from a first position of $x_1 = 0$ to a second position of $x_2$ at the station. That second position must be at $x_2 = 8.4 \text{ km} + 2.0 \text{ km} = 10.4$ km. Then the **Key Idea** here is that your displacement $\Delta x$ along the $x$ axis is the second position minus the first position. From Eq. 2-1, we have

$$\Delta x = x_2 - x_1 = 10.4 \text{ km} - 0 = 10.4 \text{ km} \qquad \text{(Answer)}$$

Thus, your overall displacement is 10.4 km in the positive direction of the $x$ axis.

(b) What is the time interval $\Delta t$ from the beginning of your drive to your arrival at the station?

**SOLUTION** ■ We already know the time interval $\Delta t_{wlk}$ (= 0.50 h) for the walk, but we lack the time interval $\Delta t_{dr}$ for the drive. However, we know that for the drive the displacement $\Delta x_{dr}$ is 8.4 km and the average velocity $\langle v_{dr\,x}\rangle$ is 70 km/h. A **Key Idea** to use here comes from Eq. 2-4: This average velocity is the ratio of the displacement for the drive to the time interval for the drive,

$$\langle v_{dr\,x}\rangle = \frac{\Delta x_{dr}}{\Delta t_{dr}}.$$

Rearranging and substituting data then give us

$$\Delta t_{dr} = \frac{\Delta x_{dr}}{\langle v_{dr\,x}\rangle} = \frac{8.4\text{ km}}{70\text{ km/h}} = 0.12\text{ h}.$$

Therefore,

$$\Delta t = \Delta t_{dr} + \Delta t_{wlk} = 0.12\text{ h} + 0.50\text{ h} = 0.62\text{ h}.$$

(c) What is your average velocity $\langle v_x\rangle$ from the beginning of your drive to your arrival at the station? Find it both numerically and graphically.

**SOLUTION** ■ The **Key Idea** here again comes from Eq. 2-4: $\langle v_x\rangle$ for the entire trip is the ratio of the displacement of 10.4 km for the entire trip to the time interval of 0.62 h for the entire trip. With Eq. 2-4, we find it is

$$\langle v_x\rangle = \frac{\Delta x}{\Delta t} = \frac{10.4\text{ km}}{0.62\text{ h}} = 16.8\text{ km/h} \approx 17\text{ km/h}. \quad \text{(Answer)}$$

To find $\langle v_x\rangle$ graphically, first we graph $x(t)$ as shown in Fig. 2-11, where the beginning and arrival points on the graph are the origin and the point labeled "Station." The **Key Idea** here is that your average velocity in the $x$ direction is the slope of the straight line connecting those points; that is, it is the ratio of the *rise* ($\Delta x$ = 10.4 km) to the *run* ($\Delta t$ = 0.62 h), which gives us $\langle v_x\rangle$ = 16.8 km/h.

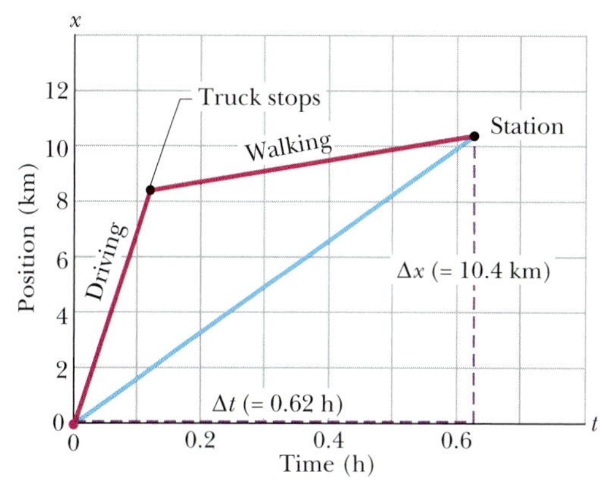


**FIGURE 2-11** ■ The lines marked "Driving" and "Walking" are the position–time plots for the driving and walking stages. (The plot for the walking stage assumes a constant rate of walking.) The slope of the straight line joining the origin and the point labeled "Station" is the average velocity for the trip, from beginning to station.

(d) Suppose that to pump the gasoline, pay for it, and walk back to the truck takes you another 45 min. What is your average speed from the beginning of your drive to your return to the truck with the gasoline?

**SOLUTION** ■ The **Key Idea** here is that your average speed is the ratio of the total distance you move to the total time interval you take to make that move. The total distance is 8.4 km + 2.0 km + 2.0 km = 12.4 km. The total time interval is 0.12 h + 0.50 h + 0.75 h = 1.37 h. Thus, Eq. 2-5 gives us

$$\langle s\rangle = \frac{12.4\text{ km}}{1.37\text{ h}} = 9.1\text{ km/h}. \quad \text{(Answer)}$$

## 2-4 Describing Velocity Change

The student shown in Fig. 2-9 is clearly speeding up and slowing down as she walks. We know that the slope of her position vs. time graph over small time intervals keeps *changing*. Now that we have defined velocity, it is meaningful to develop a mathematical description of how fast velocity changes. We see two approaches to describing velocity change. We could determine velocity change over an interval of displacement magnitude, $|\Delta x|$, and use $\Delta\vec{v}/|\Delta x|$ as our measure. Alternatively, we could use the ratio of velocity change to the interval of time, $\Delta t$, over which the change occurs or ($\Delta\vec{v}/\Delta t$). This is analogous to our definition of velocity.

Both of our proposals are possible ways of describing velocity change—neither is right or wrong. In the fourth century B.C.E., Aristotle believed that the ratio of velocity change to distance change was probably constant for any falling objects. Almost 2000 years later, the Italian scientist Galileo did experiments with ramps to slow down the motion of rolling objects. Instead he found that it was the second ratio, $\Delta\vec{v}/\Delta t$, that was constant.

Our modern definition of acceleration is based on Galileo's idea that $\Delta\vec{v}/\Delta t$ is the most useful concept in the description of velocity changes in falling objects.

Whenever a particle's velocity changes, we define it as having an **acceleration.** The **average acceleration,** $\langle\vec{a}\rangle$, over an interval $\Delta t$ is defined as

$$\langle\vec{a}\rangle = \frac{\vec{v}_2 - \vec{v}_1}{t_2 - t_1} = \frac{\Delta\vec{v}}{\Delta t} \quad \text{(definition of 1D average acceleration).} \tag{2-7}$$

When the particle moves along a line (that is, an $x$ axis in one-dimensional motion),

$$\langle\vec{a}\rangle = \frac{(v_{2x} - v_{1x})}{(t_2 - t_1)}\hat{\text{i}}.$$

It is important to note that an object is accelerated even if all that changes is only the *direction* of its velocity and not its speed. Directional changes are important as well.

## Instantaneous Acceleration

If we want to determine how velocity changes during an instant of time, we need to define **instantaneous acceleration** (or simply **acceleration**) in a way that is similar to the way we defined instantaneous velocity:

$$\vec{a} \equiv \lim_{\Delta t \to 0} \frac{\Delta\vec{v}}{\Delta t} = \frac{d\vec{v}}{dt} \quad \text{(definition of 1D instantaneous acceleration).} \tag{2-8}$$

> In the language of calculus, the **ACCELERATION** of a particle at any instant is the rate at which its velocity is changing at that instant.

Using this definition, we can determine the acceleration by taking a time derivative of the velocity, $\vec{v}$. Furthermore, since velocity of an object moving along a line is the derivative of the position, $\vec{x}$, with respect to time, we can write

$$\vec{a} = \frac{d\vec{v}}{dt} = \frac{d}{dt}\left(\frac{d\vec{x}}{dt}\right) = \frac{d^2\vec{x}}{dt^2} \quad \text{(1D instantaneous acceleration).} \tag{2-9}$$

Equation 2-9 tells us that the instantaneous acceleration of a particle at any instant is equal to the second derivative of its position, $\vec{x}$, with respect to time. Note that if the object is moving along an $x$ axis, then its acceleration can be expressed in terms of the $x$-component of its acceleration and the unit vector $\hat{\text{i}}$ along the $x$ axis as

$$\vec{a} = a_x\hat{\text{i}} = \frac{dv_x}{dt}\hat{\text{i}} \quad \text{so} \quad a_x \equiv \frac{dv_x}{dt}.$$

Figure 2-12*c* shows a plot of the $x$-component of acceleration of an elevator cab. Compare the graph of the $x$-component of acceleration as a function of time ($a_x$ vs. $t$) with the graph of the $x$-component of velocity as a function of time ($v_x$ vs. $t$) in part *b*. Each point on the $a_x$ vs. $t$ graph is the derivative (slope or tangent) of the corresponding point on the $v_x$ vs. $t$ graph. When $v_x$ is constant (at either 0 or 4 m/s), its time derivative is zero and hence so is the acceleration. When the cab first begins to move, the $v_x$ vs. $t$ graph has a positive derivative (the slope is positive), which means that $a_x$ is positive. When the cab slows to a stop, the derivative or slope of the $v_x$ vs. $t$ graph is negative; that is, $a_x$ is negative. Next compare the slopes of the $v_x$ vs. $t$ graphs during the

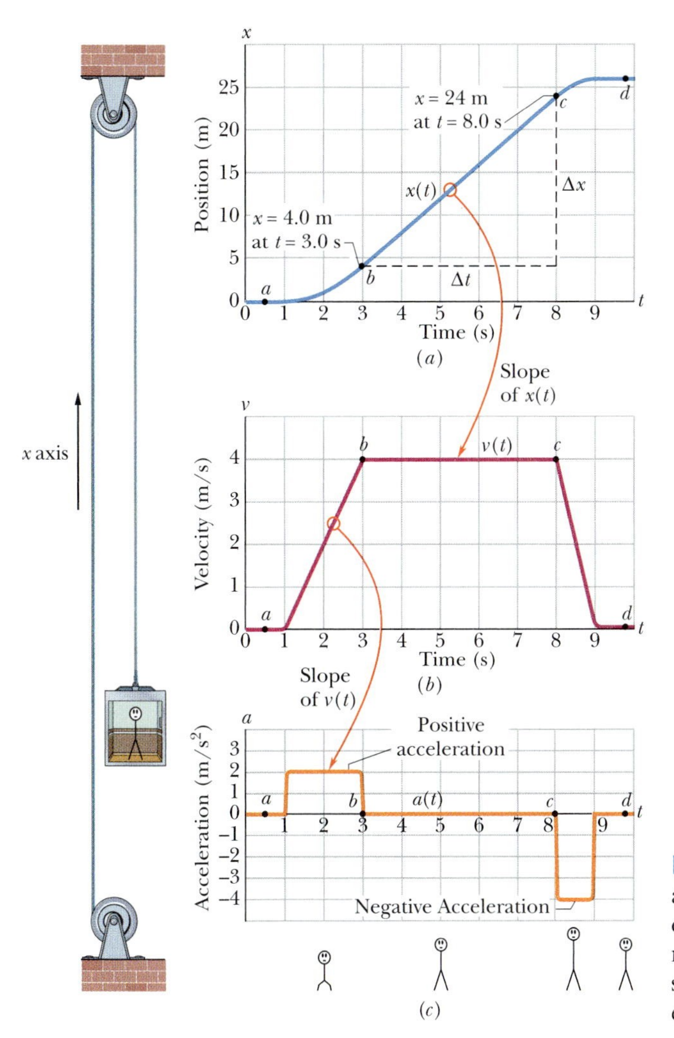


**FIGURE 2-12** ■ (*a*) The $x$ vs. $t$ graph for an elevator cab that moves upward along an $x$ axis. (*b*) The $v_x$ vs. $t$ graph for the cab. Note that it is the derivative of the $x$ vs. $t$ graph ($v_x = dx/dt$). (*c*) The $a_x$ vs. $t$ graph for the cab. It is the derivative of the $v_x$ vs. $t$ graph ($a_x = dv_x/dt$). The stick figures along the bottom suggest times that a passenger might feel light and long as the elevator accelerates downward or heavy and squashed as the elevator accelerates upward.

two acceleration periods. The slope associated with the cab's stopping is steeper, because the cab stops in half the time it took to get up to speed. The steeper slope means that the magnitude of the stopping acceleration is larger than that of the acceleration as the car is speeding up, as indicated in Fig. 2-12*c*.

Acceleration has both a magnitude and a direction and so it is a vector quantity. The algebraic sign of its component $a_x$ represents the direction of velocity change along the chosen $v_x$ axis. When acceleration and velocity are in the same direction (have the same sign) the object will speed up. If acceleration and velocity are in opposite directions (and have opposite signs) the object will slow down.

> It is important to realize that speeding up is not always associated with an acceleration that is positive. Likewise, slowing down is not always associated with an acceleration that is negative. The relative directions of an object's velocity and acceleration determine whether the object will speed up or slow down.

Since acceleration is defined as any change in velocity over time, whenever an object moving in a straight line has an acceleration it is either speeding up, slowing down, or turning around. Beware! In listening to common everyday language, you will probably hear the word acceleration used only to describe speeding up and the word deceleration to mean slowing down. It's best in studying physics to use the more formal definition of acceleration as a vector quantity that describes both the magnitude and direction of *any type of velocity change*. In short, an object is accelerating when it is slowing down as well as when it is speeding up. We suggest avoiding the use of the term deceleration while trying to learn the formal language of physics.

The fundamental unit of acceleration must be a velocity (displacement/time) divided by a time, which turns out to be displacement divided by time squared. Displacement is measured in meters and time in seconds in the SI system described in Chapter 1. Thus, the "official" unit of acceleration is m/s². You may encounter other units. For example, large accelerations are often expressed in terms of "$g$" units where $g$ is directly related to the magnitude of the acceleration of a falling object near the Earth's surface. A $g$ unit is given by

$$1\,g = 9.8\ \text{m/s}^2. \tag{2-10}$$

On a roller coaster, you have brief accelerations up to $3g$, which, in standard SI units, is $(3)(9.8\ \text{m/s}^2)$ or about $29\ \text{m/s}^2$. A more extreme example is shown in the photographs of Fig. 2-13, which were taken while a rocket sled was rapidly accelerated along a track and then rapidly braked to a stop.

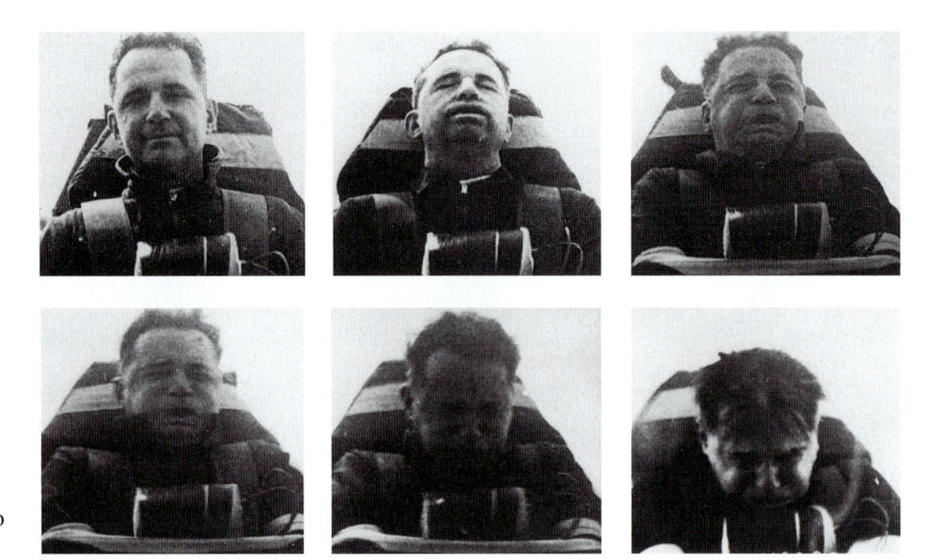

**FIGURE 2-13** ■ Colonel J.P. Stapp in a rocket sled as it is brought up to high speed (acceleration out of the page) and then very rapidly braked (acceleration into the page).

**READING EXERCISE 2-8:** A cat moves along an $x$ axis. What is the sign of its acceleration if it is moving (a) in the positive direction with increasing speed, (b) in the positive direction with decreasing speed, (c) in the negative direction with increasing speed, and (d) in the negative direction with decreasing speed? ■

## TOUCHSTONE EXAMPLE 2-3: Position and Motion

A particle's position on the $x$ axis of Fig. 2-1 is given by

$$x = 4\ \text{m} - (27\ \text{m/s})\,t + (1\ \text{m/s}^3)t^3,$$

with $x$ in meters and $t$ in seconds.

(a) Find the particle's velocity function $v_x(t)$ and acceleration function $a_x(t)$.

**SOLUTION** ■ One **Key Idea** is that to get the velocity function $v_x(t)$, we differentiate the position function $x(t)$ with respect to time. Here we find

$$v_x = -(27\ \text{m/s}) + 3\cdot(1\ \text{m/s}^3)t^2 = -(27\ \text{m/s}) + (3\ \text{m/s}^3)t^2 \quad \text{(Answer)}$$

with $v_x$ in meters per second.

Another **Key Idea** is that to get the acceleration function $a_x(t)$, we differentiate the velocity function $v_x(t)$ with respect to time. This gives us

$$a_x = 2 \cdot 3 \cdot (1 \text{ m/s}^3)t = +(6 \text{ m/s}^3)t, \quad \text{(Answer)}$$

with $a_x$ in meters per second squared.

(b) Is there ever a time when $v_x = 0$?

**SOLUTION** ■ Setting $v_x(t) = 0$ yields

$$0 = -(27 \text{ m/s}) + (3 \text{ m/s}^3)t^2,$$

which has the solution

$$t = \pm 3 \text{ s}. \quad \text{(Answer)}$$

Thus, the velocity is zero both 3 s before and 3 s after the clock reads 0.

(c) Describe the particle's motion for $t \geq 0$.

**SOLUTION** ■ The **Key Idea** is to examine the expressions for $x(t)$, $v_x(t)$, and $a_x(t)$.

At $t = 0$, the particle is at $x(0) = +4$ m and is moving with a velocity of $v_x(0) = -27$ m/s—that is, in the negative direction of the $x$ axis. Its acceleration is $a_x(0) = 0$, because just then the particle's velocity is not changing.

For $0 < t < 3$ s, the particle still has a negative velocity, so it continues to move in the negative direction. However, its acceleration is no longer 0 but is increasing and positive. Because the signs of the velocity and the acceleration are opposite, the particle must be slowing.

Indeed, we already know that it turns around at $t = 3$ s. Just then the particle is as far to the left of the origin in Fig. 2-1 as it will ever get. Substituting $t = 3$ s into the expression for $x(t)$, we find that the particle's position just then is $x = -50$ m. Its acceleration is still positive.

For $t > 3$ s, the particle moves to the right on the axis. Its acceleration remains positive and grows progressively larger in magnitude. The velocity is now positive, and it too grows progressively larger in magnitude.

## 2-5 Constant Acceleration: A Special Case

If you watch a small steel ball bobbing up and down at the end of a spring, you will see the velocity changing continuously. But instead of either increasing or decreasing at a steady rate, we have a very nonuniform pattern of motion. First the ball speeds up and slows down moving in one direction, then it turns around and speeds up and then slows down in the other direction, and so on. This is an example of a nonconstant acceleration that keeps changing in time.

Although there are many examples of nonconstant accelerations, we also observe a surprising number of examples of constant or nearly constant acceleration. As we already discussed, Galileo discovered that if we choose to define acceleration in terms of the ratio $\Delta\vec{v}/\Delta t$, then a falling ball or a ball tossed into the air that slows down, turns around, and speeds up again is always increasing its velocity in a downward direction at the same rate—provided the ball is moving slowly enough that air drag is negligible.

There are many other common motions that involve constant accelerations. Suppose you measure the times and corresponding positions for an object that you suspect has a constant acceleration. If you then calculate the velocities and accelerations of the object and make graphs of them, the graphs will resemble those in Fig. 2-14. Some examples of motions that yield similar graphs to those shown in Fig. 2-14 include: a car that you accelerate as soon as a traffic light turns green; the same car when you apply its brakes steadily to bring it to a smooth stop; an airplane when first taking off or when completing a smooth landing; or a dolphin that speeds up suddenly after being startled.

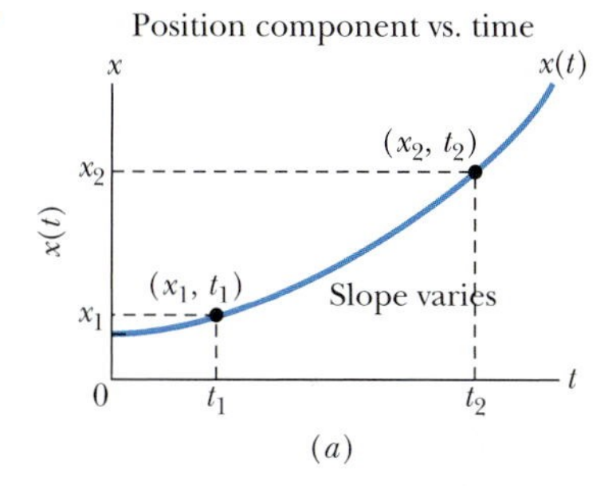


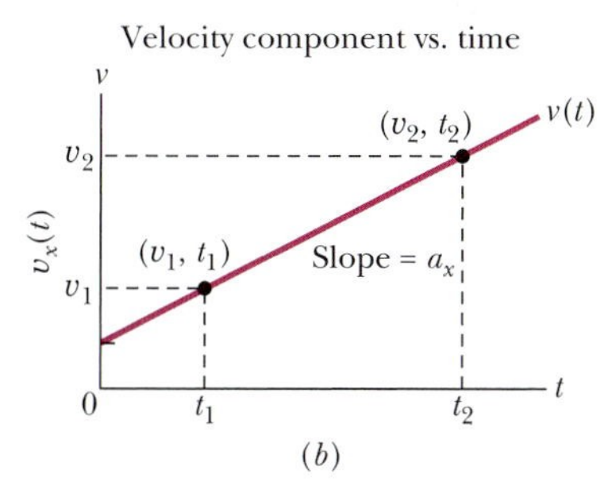


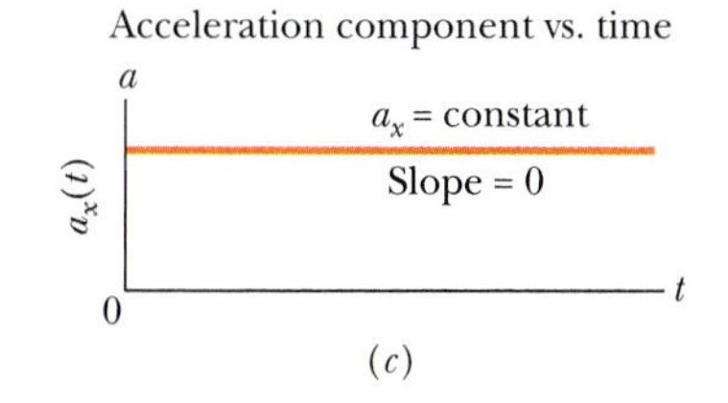


**FIGURE 2-14** ■ (a) The position component $x(t)$ of a particle moving with constant acceleration. (b) Its velocity component $v_x(t)$, given at each point by the slope of the curve in (a). (c) Its (constant) component of acceleration, $a_x$, equal to the (constant) slope of $v_x(t)$.

### Derivation of the Kinematic Equations

Because constant accelerations are common, it is useful to derive a special set of **kinematic equations** to describe the motion of any object that is moving along a line with a constant acceleration. We can use the definitions of acceleration and velocity and an assumption about average velocity to derive the kinematic equations. These equations allow us to use known values of the vector components describing positions, velocities, and accelerations, along with time intervals to predict the motions of constantly accelerated objects.

Let's start the derivation by noting that when the acceleration is constant, the average and instantaneous accelerations are equal. As usual we place our $x$ axis along the line of the motion. We can now use vector notation to write

$$\vec{a} = a_x \hat{\mathrm{i}} = \langle \vec{a} \rangle, \tag{2-11}$$

so that

$$\langle \vec{a} \rangle = \frac{(v_{2x} - v_{1x})}{t_2 - t_1} \hat{\mathrm{i}},$$

where $a_x$ is the component of acceleration along the line of motion of the object. We can use the definition of average acceleration (Eq. 2-7) to express the acceleration component $a_x$ in terms of the object's velocity components along the line of motion, where $v_{2x}$ and $v_{1x}$ are the object's velocity components along the line of motion,

$$a_x = \frac{(v_{2x} - v_{1x})}{t_2 - t_1}. \tag{2-12}$$

This expression allows us to derive the kinematic equations in terms of the vector components needed to construct the actual one-dimensional velocity and acceleration vectors. The subscripts 1 and 2 in most of the equations in this chapter, including Eq. 2-12, refer to initial and final times, positions, and velocities.

If we solve Eq. 2-12 for $v_{2x}$, then the $x$-component of velocity at time $t_2$ is

$$v_{2x} = v_{1x} + a_x(t_2 - t_1) = v_{1x} + a_x \Delta t \quad \text{(primary kinematic } [a_x = \text{constant] equation)},$$

$$\text{or} \quad \Delta v_x = a \Delta t. \tag{2-13}$$

This equation is the first of two primary equations that we will derive for use in analyzing motions involving constant acceleration. Before we move on, we should think carefully about what the expression $t_2 - t_1$ represents in this equation: *It represents the time interval in which we are tracking the motion.*

In a manner similar to what we have done above, we can rewrite Eq. 2-4, the expression for the average velocity along the $x$ axis,

$$\langle \vec{v} \rangle = \langle v_x \rangle \hat{\mathrm{i}} = \frac{\Delta x}{\Delta t} \hat{\mathrm{i}} = \frac{(x_2 - x_1)}{t_2 - t_1} \hat{\mathrm{i}}.$$

Hence, the $x$-component of the average velocity is given by

$$\langle v_x \rangle = \frac{(x_2 - x_1)}{(t_2 - t_1)}.$$

Solving for $x_2$ gives

$$x_2 = x_1 + \langle v_x \rangle (t_2 - t_1). \tag{2-14}$$

In this equation $x_1$ is the $x$-component of the position of the particle at $t = t_1$ and $\langle v_x \rangle$ is the component along the $x$ axis of average velocity between $t = t_1$ and a later time $t = t_2$. Note that unless the velocity is constant, the average velocity component along the $x$ axis, $\langle v_x \rangle$, is not equal to the instantaneous velocity component, $v_x$.

However, we do have a plausible alternative for expressing the average velocity component in the special case when the acceleration is constant. Figure 2-15 depicts the fact that velocity increases in a linear fashion over time for a constant acceleration. It seems reasonable to assume that the component along the $x$ axis of the *average* velocity over any time interval is the average of the components for the in-

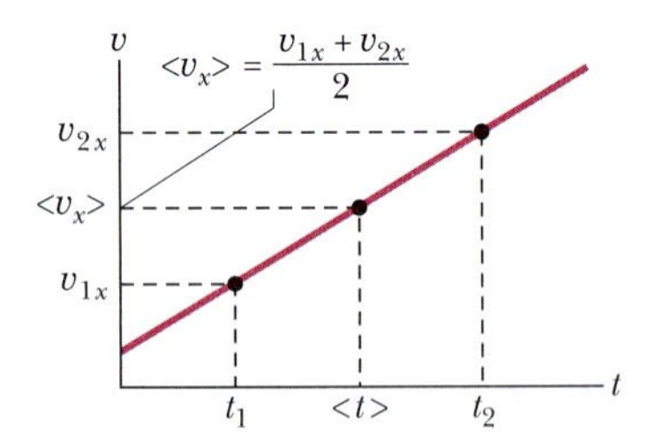


**FIGURE 2-15** When the acceleration is constant, then we assume (without rigorous proof) that the average velocity component in a time interval is the average of the velocity components at the beginning and end of the interval.

stantaneous velocity at the beginning of the interval, $v_{1\,x}$, and the instantaneous velocity component at the end of the interval, $v_{2\,x}$. So we expect that when a velocity increases linearly, the average velocity component over a given time interval will be

$$\langle v_x \rangle = \frac{v_{1\,x} + v_{2\,x}}{2}. \tag{2-15}$$

Using Eq. 2-13, we can substitute $v_{1\,x} + a_x(t_2 - t_1)$ for $v_{2\,x}$ to get

$$\langle v_x \rangle = \frac{1}{2}\left[v_{1\,x} + v_{1\,x} + a_x(t_2 - t_1)\right] = v_{1\,x} + \tfrac{1}{2}a_x(t_2 - t_1). \tag{2-16}$$

Finally, substituting this equation into Eq. 2-14 yields

$$x_2 - x_1 = v_{1\,x}(t_2 - t_1) + \tfrac{1}{2}a_x(t_2 - t_1)^2 \quad \text{(primary kinematic } [a_x = \text{constant}] \text{ equation)}, \tag{2-17}$$

or

$$\Delta x = v_{1\,x}\Delta t + \tfrac{1}{2}a_x \Delta t^2$$

This is our second primary equation describing motion with constant acceleration. Figures 2-14*a* and 2-16 show plots of Eq. 2-17.

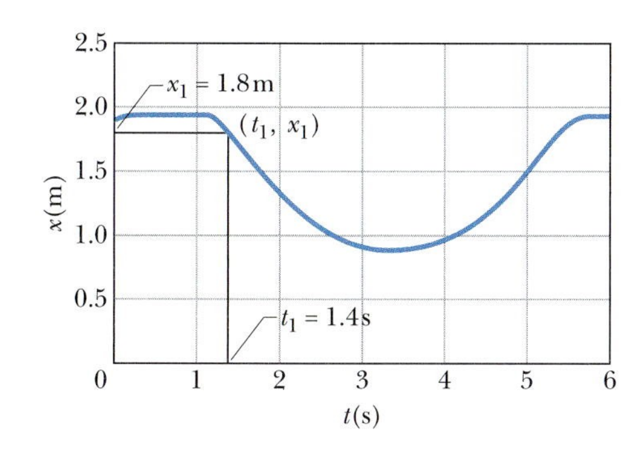


**FIGURE 2-16** A fan on a low-friction cart is turned on at about 1.2 s but isn't thrusting fully until $t_1$ = 1.4 s. Data for the graph were collected with a computer data acquisition system outfitted with an ultrasonic motion detector. Between 1.4 s and about 5.4 s the cart appears to be undergoing a constant acceleration as it slows down, turns around, and speeds up again. Thus, the constant acceleration kinematic equations can be used to describe its motion but only during motion within that time interval. Thus, we can set $t_1$ to 1.4 s and $x_1$ to 1.8 m.

These two equations are very useful in the calculation of unknown quantities that can be used to characterize constantly accelerated motion. There are five or six quantities contained in our primary equations (Eqs. 2-13 and 2-17). The simplest kinematic calculations involve situations in which all but one of the quantities is known in one of the primary equations. In more complex situations, both equations are needed. Typically for a complex situation, we need to calculate more than one unknown. To do this, we find the first unknown using one of the primary equations and use the result in the other equations to find the second unknown. This method is illustrated in the next section and in Touchstone Examples 2-4 and 2-6.

The primary equations above, $v_{2\,x} = v_{1\,x} + a_x(t_2 - t_1) = v_{1\,x} + a_x\Delta t$ (Eq. 2-13), and $x_2 - x_1 = v_{1\,x}(t_2 - t_1) + \frac{1}{2}a_x(t_2 - t_1)^2$ (Eq. 2-17), are derived directly from the definitions of velocity and acceleration, with the condition that the acceleration is constant. These two equations can be combined in three ways to yield three additional equations. For example, solving for $v_{1\,x}$ in $v_{2\,x} = v_{1\,x} + a_x(t_2 - t_1)$ and substituting the result into $x_2 - x_1 = v_{1\,x}(t_2 - t_1) + \frac{1}{2}a_x(t_2 - t_1)^2$ gives us

$$v_{2\,x}^2 = v_{1\,x}^2 + 2a_x(x_2 - x_1).$$

We recommend that you learn the two primary equations and use them to derive other equations as needed. Then you will not need to remember so much. Table 2-1 lists our two primary equations. Note that a really nice alternative to using the two

**TABLE 2-1**
**Equations of Motion with Constant Acceleration**

| Equation Number | Primary Vector Component Equation* |
|---|---|
| 2-13 | $v_{2x} = v_{1x} + a_x(t_2 - t_1)$ |
| 2-17 | $x_2 - x_1 = v_{1x}(t_2 - t_1) + \frac{1}{2}a_x(t_2 - t_1)^2$ |

*A reminder: In cases where the initial time $t_1$ is chosen to be zero it is important to remember that whenever the term $(t_2 - t_1)$ is replaced by just $t$, then $t$ actually represents a *time interval of* $\Delta t = t - 0$ over which the motion of interest takes place.

equations in Table 2-1 is to use the first of the equations (Eq. 2-13) along with the expression for the average velocity component in Eq. 2-15,

$$\langle v_x \rangle = \frac{\Delta x}{\Delta t} = \frac{v_{1x} + v_{2x}}{2} \quad \text{(an alternative "primary" equation),}$$

to derive all the other needed equations. The derivations of the kinematic equations that we present here are not rigorous mathematical proofs but rather what we call plausibility arguments. However, we know from the application of the kinematic equations to constantly accelerated motions that they do adequately describe these motions.

## Analyzing the Niagara Falls Plunge

At the beginning of this chapter we asked questions about the motion of the steel chamber holding Dave Munday as he plunged into the water after falling 48 m from the top of Niagara Falls. How long did the fall take? That is, what is $\Delta t$? How fast was the chamber moving when it hit the water? (What is $\vec{v}$?) As you will learn in Chapter 3, if no significant air drag is present, objects near the surface of the Earth fall at a constant acceleration of magnitude $|a_x| = 9.8\ \text{m/s}^2$. Thus, the kinematic equations can be used to calculate the time of fall and the impact speed.

Let's start by defining our coordinate system. We will take the $x$ axis to be a vertical or up–down axis that is aligned with the downward path of the steel chamber. We place the origin at the bottom of the falls and define up to be positive as shown in Fig. 2-17. (Later when considering motions in two and three dimensions, we will often denote vertical axes as $y$ axes and horizontal axes as $x$ axes, but these changes in symbols will not affect the results of calculations.)

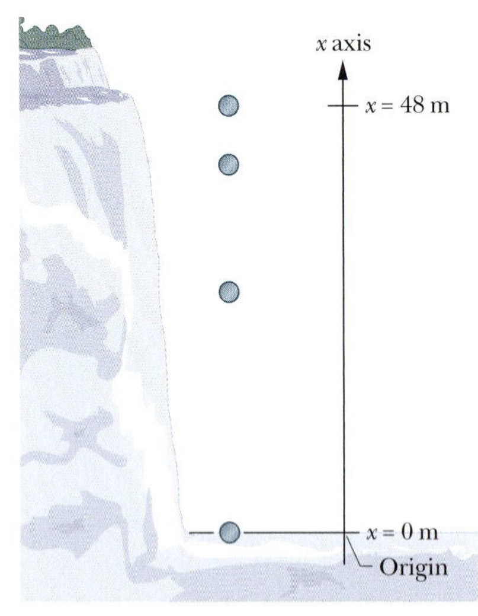


**FIGURE 2-17** ■ A coordinate system chosen to analyze the fall of a steel chamber holding a man who falls 48 m from the top to the bottom of Niagara Falls.

We know that the value of the vertical displacement is given by

$$x_2 - x_1 = (0\ \text{m}) - (+48\ \text{m}) = -48\ \text{m}$$

and that the velocity is getting larger in magnitude in the downward (negative direction). Since the velocity is downward and the object is speeding up, the vertical acceleration is also downward (in the negative direction). Its component along the axis of motion is given by $a_x = -9.8\ \text{m/s}^2$. Finally, we assume that Dave Munday's capsule dropped from rest, so $v_{1x} = 0$ m/s. Thus we can find the time of fall ($\Delta t = t_2 - t_1$) using Eq. 2-17. Solving this equation for the time elapsed during the fall ($t_2 - t_1$) when the initial velocity $v_{1x}$ is zero gives

$$\Delta t = t_2 - t_1 = \sqrt{\frac{2(x_2 - x_1)}{a_x}} = \sqrt{\frac{2(-48\ \text{m})}{-9.8\ \text{m/s}^2}} = 3.13\ \text{s} = 3.1\ \text{s}.$$

This is a fast trip indeed!

Next we can use the time interval of the fall in the other primary kinematic equation, Eq. 2-13, to find the velocity at impact. This gives a component of impact velocity at the end of the fall of

$$v_{2x} = v_{1x} + a_x(t_2 - t_1) = 0 \text{ m/s} + (-9.8 \text{ m/s}^2)(3.13 \text{ s}) = -31 \text{ m/s}.$$

The minus sign indicates that the impact velocity component is negative and is, therefore, in the downward direction. In vector notation, the velocity $\vec{v} = v_x \hat{\text{i}}$ is thus $\vec{v} = (-31 \text{ m/s})\hat{\text{i}}$. Note that this is a speed of about 69 mi/hr. Since the time interval was put into the calculation of velocity of impact as an intermediate value, we retained an extra significant figure to use in the next calculation.

**READING EXERCISE 2-9:** The following equations give the $x$-component of position $x(t)$ of a particle in meters (denoted m) as a function of time in seconds for four situations: (1) $x = (3 \text{ m/s})t - 4 \text{ m}$; (2) $x = (-5 \text{ m/s}^3)t^3 + (4 \text{ m/s})t + 6 \text{ m}$; (3) $x = (2 \text{ m/s}^2)t^2 - (4 \text{ m/s})t$; (4) $x = (5 \text{ m/s}^2)t^2 - 3 \text{ m}$. To which of these situations do the equations of Table 2-1 apply? Explain. ■

## TOUCHSTONE EXAMPLE 2-4: Slowing Down

Spotting a police car, you brake your Porsche from a speed of 100 km/h to a speed of 80.0 km/h during a displacement of 88.0 m, at a constant acceleration.

(a) What is that acceleration?

**SOLUTION** ■ Assume that the motion is along the positive direction of an $x$ axis. For simplicity, let us take the beginning of the braking to be at time $t_1 = 0$, at position $x_1$. The **Key Idea** here is that, with the acceleration constant, we can relate the car's acceleration to its velocity and displacement via the basic constant acceleration equations (Eqs. 2-13 and 2-17). The initial velocity is $v_{1x} = 100 \text{ km/h} = 27.78 \text{ m/s}$, the displacement is $x_2 - x_1 = 88.0 \text{ m}$, and the velocity at the end of that displacement is $v_{2x} = 80.0 \text{ km/h} = 22.22 \text{ m/s}$. However, we do not know the acceleration $a_x$ and time $t_2$, which appear in both basic equations, so we must solve those equations simultaneously.

To eliminate the unknown $t_2$, we use Eq. 2-13 to write

$$t_2 - t_1 = \frac{v_{2x} - v_{1x}}{a_x}, \tag{2-18}$$

and then we substitute this expression into Eq. 2-17 to write

$$x_2 - x_1 = v_{1x}\left(\frac{v_{2x} - v_{1x}}{a_x}\right) + \frac{1}{2}a_x\left(\frac{v_{2x} - v_{1x}}{a_x}\right)^2.$$

Solving for $a_x$ and substituting known data then yields

$$a_x = \frac{v_{2x}^2 - v_{1x}^2}{2(x_2 - x_1)} = \frac{(22.22 \text{ m/s})^2 - (27.78 \text{ m/s})^2}{2(88.0 \text{ m})}$$
$$= -1.58 \text{ m/s}^2. \qquad \text{(Answer)}$$

(b) How much time is required for the given decrease in speed?

**SOLUTION** ■ Now that we know $a_x$, we can use Eq. 2-18 to solve for $t_2$:

$$t_2 - t_1 = \frac{v_{2x} - v_{1x}}{a_x} = \frac{22.22 \text{ m/s} - 27.78 \text{ m/s}^2}{-1.58 \text{ m/s}^2} = 3.52 \text{ s}. \qquad \text{(Answer)}$$

If you are initially speeding and trying to slow to the speed limit, there is plenty of time for the police officer to measure your excess speed.

You can use one of the alternate equations for motion with a constant acceleration, Eq. 2-15, to check this result. The **Key Idea** here is that the distance traveled is just the product of the average velocity and the elapsed time, when the acceleration is constant. The Porsche traveled 88.0 m while it slowed from 100 km/h down to 80 km/h. Thus its average velocity while it covered the 88.0 m was

$$\langle v_x \rangle = \frac{(100 \text{ km/h} + 80 \text{ km/h})}{2}$$
$$= 90 \frac{\text{km}}{\text{h}} \cdot \left(\frac{1000 \text{ m}}{1 \text{ km}}\right) \cdot \left(\frac{1 \text{ h}}{3600 \text{ s}}\right) = 25.0 \text{ m/s},$$

so the time it took to slow down was just

$$t_2 - t_1 = \frac{x_2 - x_1}{\langle v_x \rangle} = \frac{88.0 \text{ m}}{25.0 \text{ m/s}} = 3.52 \text{ s}, \qquad \text{(Answer)}$$

which still isn't enough time to avoid that speeding ticket!

## TOUCHSTONE EXAMPLE 2-5: Motion Data

Suppose that you gave a box sitting on a carpeted floor a push and then recorded its position three times per second as it slid to a stop. The table gives the results of such a measurement. Let's analyze the position vs. time data for the box sliding on the carpet and use curve fitting and calculus to obtain the velocity measurements. We will use Excel spreadsheet software to perform our analysis, but other computer- or calculator-based fitting or modeling software can be used.

**Box Sliding on Carpet**

| t[s] | x[m] |
|---|---|
| 0.000 | 0.537 |
| 0.033 | 0.583 |
| 0.067 | 0.623 |
| 0.100 | 0.659 |
| 0.133 | 0.687 |
| 0.167 | 0.705 |
| 0.200 | 0.719 |
| 0.233 | 0.720 |

(a) Draw a graph of the $x$ vs. $t$ data and discuss whether the relationship appears to be linear or not.

**SOLUTION** ■ The **Key Idea** here is that the relationship between two variables is linear if the graph of the data points lie more or less along a straight line. There are many ways to graph the data for examination: by hand, with a graphing calculator, with a spreadsheet graphing routine, or with other graphing software such as Data Studio (available from PASCO scientific) or Graphical Analysis (available from Vernier Software and Technology). The graph in Fig. 2-18 that shows a curve and so the relationship between position, $x$, and time is not linear.

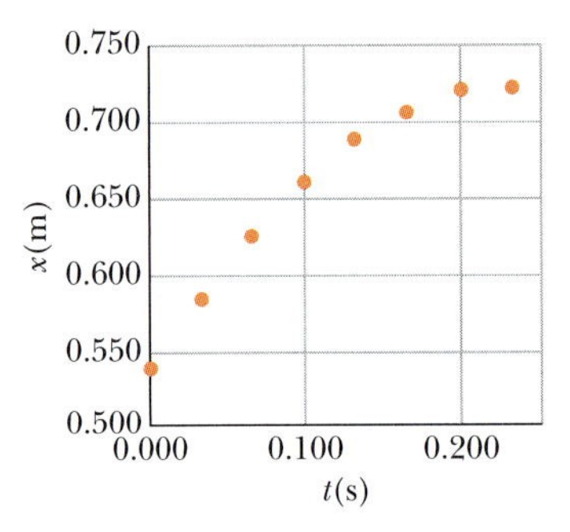


**FIGURE 2-18** ■ Solution to Touchstone Example 2-5(a). A graph of position versus time for a box sliding across a carpet.

(b) Draw a motion diagram of the box as it comes to rest on the carpet.

**SOLUTION** ■ The **Key Idea** here is to use the data to sketch the position along a line at equal time intervals. In Fig. 2-19, the black circles represent the location of the rear of the box at intervals of 1/30 of a second.

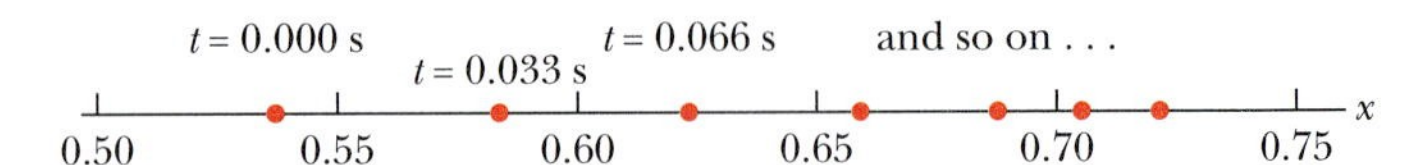


**FIGURE 2-19** ■ Solution to Touchstone Example 2-5(b). A motion diagram for a box sliding across a carpet.

(c) Is the acceleration constant? If so, what is its component along the $x$ axis?

**SOLUTION** ■ The **Key Idea** here is to explore whether or not the relationship between position and time of the box as it slides to a stop can be described with a quadratic (parabolic) function of time as described in Eq. 2-17. This can be done by entering the data that are given into a spreadsheet or graphing calculator and either doing a quadratic model or a fit to the data. The outcome of a quadratic model is shown in Fig. 2-20. The $x$-model column contains the results of calculating $x$ using the equation $x_2 - x_1 = v_{1\,x}(t_2 - t_1) + \frac{1}{2}a_x(t_2 - t_1)^2$ for each of the times in the first column using the initial position, velocity and acceleration data shown in the boxes. The line shows the model data. If the kinematic equation fits the data, then we can conclude that the acceleration component is a constant given by $a_x = -6.6\ \text{m/s}^2$. Thus the acceleration is in the negative $y$ direction.

| $a$ | −6.7 | (m/s²) |
|---|---|---|
| $v_1$ | 1.6 | (m/s) |
| $x_1$ | 0.537 | (m) |

Box Sliding on Carpet

| $t$(s) | $x$-data (m) | $x$-model (m) |
|---|---|---|
| 0.000 | 0.537 | 0.537 |
| 0.033 | 0.583 | 0.587 |
| 0.067 | 0.623 | 0.629 |
| 0.100 | 0.659 | 0.664 |
| 0.133 | 0.687 | 0.691 |
| 0.167 | 0.705 | 0.711 |
| 0.200 | 0.719 | 0.723 |
| 0.233 | 0.720 | 0.728 |

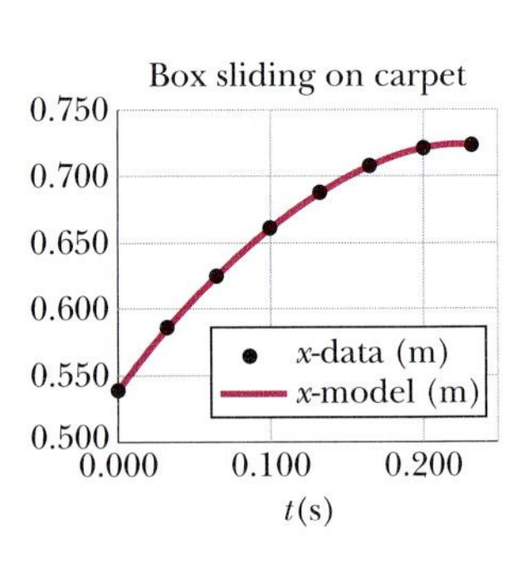


**FIGURE 2-20** ■ Solution to Touchstone Example (c). Data and a graph of position as function of time for a box sliding over carpet. Actual data is compared to a model of what is expected from Eq. 2-17 (assumed constant acceleration). The value of acceleration which produced the best match between the model and actual data is −6.6 m/s².

## TOUCHSTONE EXAMPLE 2-6: Distance Covered

Figure 2-21*b* shows a graph of a person riding on a low-friction cart being pulled along with a bungee cord as shown in Fig. 2-21*a*. Use information from the two graphs and the kinematic equations to determine approximately how far the student moved in the time interval between 1.1 s and 2.0 s.

**SOLUTION** ■ The **Key Idea** is that the initial velocity can be determined from the velocity vs. time graph on the left and the acceleration during the time interval can be determined from the acceleration vs. time graph on the right (or by finding the slope of the velocity vs. time graph on the left during the time interval). Note that the velocity at $t_1 = 1.1$ s is given by $v_{1\,x} \approx 0.4$ m/s. The acceleration during the time interval of interest is given by $a_x \approx 0.4$ m/s$^2$. Since the acceleration is constant over the time interval of interest, we can use the data in Eq. 2-17 to get

$$x_2 - x_1 = v_{1\,x}(t_2 - t_1) + \tfrac{1}{2}a_x(t_2 - t_1)^2$$

$$= (0.4\text{ m/s})(2.0\text{ s} - 1.1\text{ s}) + \tfrac{1}{2}(0.4\text{ m/s}^2)(2.0\text{ s} - 1.1\text{ s})^2$$

$$\approx 0.5\text{ m.} \qquad \text{(Answer)}$$

Half a meter is not very far!

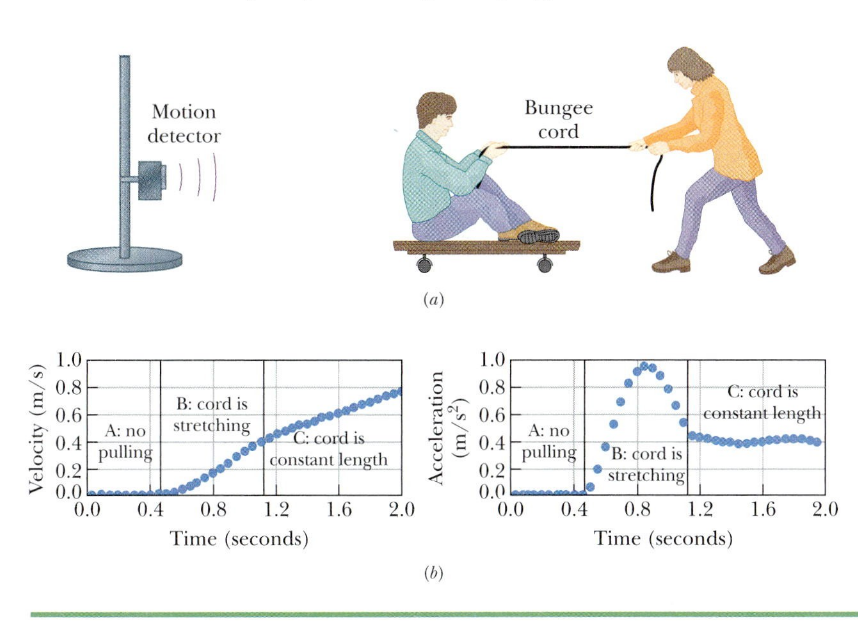


**FIGURE 2-21** ■ (*a*) A person riding on a low-friction cart is pulled by another person who exerts a constant force along a straight line by keeping the length of a bungee cord constant. (*b*) These graphs show velocity and acceleration components vs. time for a rider on a cart. For the first 0.5 s (region A) the cart is at rest. Between 0.5 s and 1.1 s (region B) the cord is beginning to stretch. Between 1.1 s and 2.0 s (region C) a constant force is acting and the acceleration is also constant.

# Problems

*In several of the problems that follow you are asked to graph position, velocity, and acceleration versus time. Usually a sketch will suffice, appropriately labeled and with straight and curved portions apparent. If you have a computer or graphing calculator, you might use it to produce the graph.*

### SEC. 2-3 ■ VELOCITY AND SPEED

**1. Fastball** If a baseball pitcher throws a fastball at a horizontal speed of 160 km/h, how long does the ball take to reach home plate 18.4 m away?

**2. Fastest Bicycle** A world speed record for bicycles was set in 1992 by Chris Huber riding Cheetah, a high-tech bicycle built by three mechanical engineering graduates. The record (average) speed was 110.6 km/h through a measured length of 200.0 m on a desert road. At the end of the run, Huber commented, "Cogito ergo zoom!" (I think, therefore I go fast!) What was Huber's elapsed time through the 200.0 m?

**3. Auto Trip** An automobile travels on a straight road for 40 km at 30 km/h. It then continues in the same direction for another 40 km at 60 km/h. (a) What is the average velocity of the car during this 80 km trip? (Assume that it moves in the positive $x$ direction.) (b) What is the average speed? (c) Graph $x$ vs. $t$ and indicate how the average velocity is found on the graph.

**4. Radar Avoidance** A top-gun pilot, practicing radar avoidance maneuvers, is manually flying horizontally at 1300 km/h, just 35 m above the level ground. Suddenly, the plane encounters terrain that slopes gently upward at 4.3°, an amount difficult to detect visually (Fig. 2-22). How much time does the pilot have to make a correction to avoid flying into the ground?

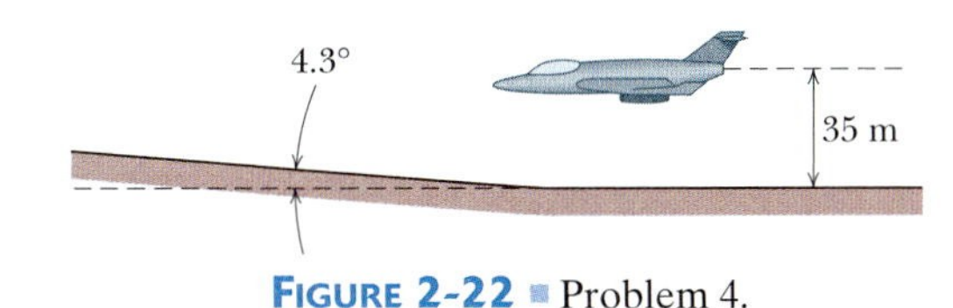


**FIGURE 2-22** ■ Problem 4.

**5. On Interstate 10** You drive on Interstate 10 from San Antonio to Houston, half the *time* at 55 km/h and the other half at 90 km/h. On the way back you travel half the *distance* at 55 km/h and the other half at 90 km/h. What is your average speed (a) from San Antonio to Houston, (b) from Houston back to San Antonio, and (c) for the entire trip? (d) What is your average velocity for the entire trip? (e) Sketch $x$ vs. $t$ for (a), assuming the motion is all in the positive $x$ direction. Indicate how the average velocity can be found on the sketch.

**6. Walk Then Run** Compute your average velocity in the following two cases: (a) You walk 73.2 m at a speed of 1.22 m/s and then run 73.2 m at a speed of 3.05 m/s along a straight track. (b) You walk for 1.00 min at a speed of 1.22 m/s and then run for 1.00 min at 3.05 m/s along a straight track. (c) Graph $x$ vs. $t$ for both cases and indicate how the average velocity is found on the graph.

**7. Position and Time** The position of an object moving along an $x$ axis is given by $x = (3\text{ m/s})t - (4\text{ m/s}^2)t^2 + (1\text{ m/s}^3)t^3$, where $x$ is in meters and $t$ in seconds. (a) What is the position of the object at $t =$ 1, 2, 3, and 4 s? (b) What is the object's displacement between $t_0 = 0$ and $t_4 = 4$ s? (c) What is its average velocity for the time interval from $t_2 = 2$ s to $t_4 = 4$ s? (d) Graph $x$ vs. $t$ for $0 \leq t \leq 4$ s and indicate how the answer for (c) can be found on the graph.

**8. Two Trains and a Bird** Two trains, each having a speed of 30 km/h, are headed at each other on the same straight track. A bird that can fly 60 km/h flies off the front of one train when they are 60 km apart and heads directly for the other train. On reaching the other train it flies directly back to the first train, and so forth. (We have no idea *why* a bird would behave in this way.) What is the total distance the bird travels?

**9. Two Winners** On two *different* tracks, the winners of the 1 kilometer race ran their races in 2 min, 27.95 s and 2 min, 28.15 s. In order to conclude that the runner with the shorter time was indeed faster, how much longer can the other track be in *actual* length?

**10. Scampering Armadillo** The graph in Fig. 2-23 is for an armadillo that scampers left (negative direction of $x$) and right along an $x$ axis. (a) When, if ever, is the animal to the left of the origin on the axis? When, if ever, is its velocity (b) negative, (c) positive, or (d) zero?

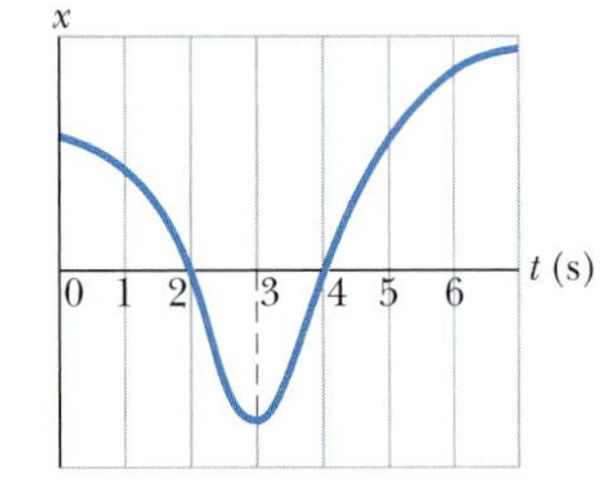


FIGURE 2-23 ■ Problem 10.

**11. Position and Time** (a) If a particle's position is given by $x = 4m - (12\text{ m/s})t + (3\text{ m/s}^2)t^2$ (where $t$ is in seconds and $x$ is in meters), what is its velocity at $t_1 = 1$ s? (b) Is it moving in the positive or negative direction of $x$ just then? (c) What is its speed just then? (d) Is the speed larger or smaller at later times? (Try answering the next two questions without further calculation.) (e) Is there ever an instant when the velocity is zero? (f) Is there a time after $t_3 = 3$ s when the particle is moving in the negative direction of $x$?

**12. Particle Position and Time** The position of a particle moving along the $x$ axis is given in meters by $x = 9.75\text{m} + (1.5\text{ m/s}^3)t^3$ where $t$ is in seconds. Calculate (a) the average velocity during the time interval $t = 2.00$ s to $t = 3.00$ s; (b) the instantaneous velocity at $t = 2.00$ s; (c) the instantaneous velocity at $t = 3.00$ s; (d) the instantaneous velocity at $t = 2.50$ s; and (e) the instantaneous velocity when the particle is midway between its positions at $t = 2.00$ s and $t = 3.00$ s (f) Graph $x$ vs. $t$ and indicate your answers graphically.

**13. Velocity–Time Graph** How far does the runner whose velocity–time graph is shown in Fig. 2-24 travel in 16 s?

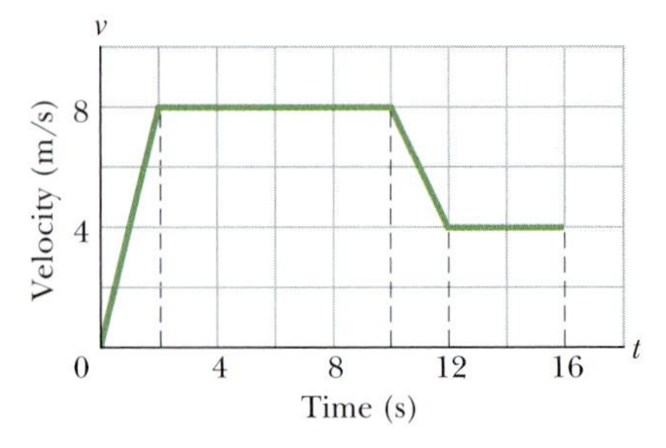

FIGURE 2-24 ■ Problem 13

## SEC. 2-4 ■ DESCRIBING VELOCITY CHANGE

**14. Various Motions** Sketch a graph that is a possible description of position as a function of time for a particle that moves along the $x$ axis and, at $t = 1$ s, has (a) zero velocity and positive acceleration; (b) zero velocity and negative acceleration; (c) negative velocity and positive acceleration; (d) negative velocity and negative acceleration. (e) For which of these situations is the speed of the particle increasing at $t = 1$ s?

**15. Two Similar Expressions** What do the quantities (a) $(dx/dt)^2$ and (b) $d^2x/dt^2$ represent? (c) What are their SI units?

**16. Frightened Ostrich** A frightened ostrich moves in a straight line with velocity described by the velocity–time graph of Fig. 2-25. Sketch acceleration vs. time.

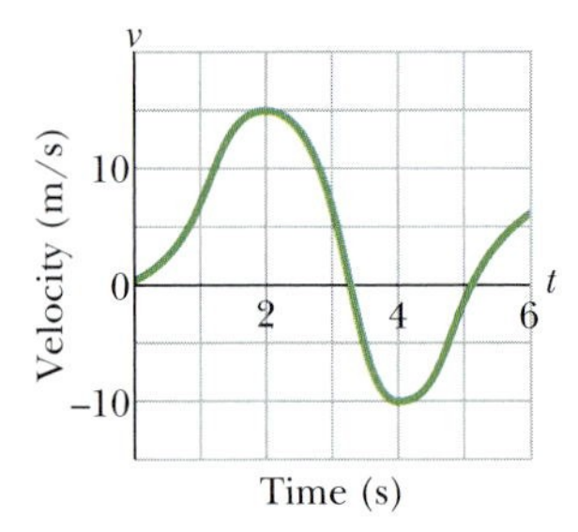


FIGURE 2-25 ■ Problem 16.

**17. Speed Then and Now** A particle had a speed of 18 m/s at a certain time, and 2.4 s later its speed was 30 m/s in the opposite direction. What were the magnitude and direction of the average acceleration of the particle during this 2.4 s interval?

**18. Stand Then Walk** From $t_0 = 0$ to $t_5 = 5.00$ min, a man stands still, and from $t_5 = 5.00$ min to $t_{10} = 10.0$ min, he walks briskly in a straight line at a constant speed of 2.20 m/s. What are (a) his average velocity $\langle\vec{v}\rangle$ and (b) his average acceleration $\langle\vec{a}\rangle$ in the time interval 2.00 min to 8.00 min? What are (c) $\langle\vec{v}\rangle$ and $\langle\vec{a}\rangle$ in the time interval 3.00 min to 9.00 min? (e) Sketch $x$ vs. $t$ and $v$ vs. $t$, and indicate how the answers to (a) through (d) can be obtained from the graphs.

**19. Particle Position and Time** The position of a particle moving along the $x$ axis depends on the time according to the equation $x = ct^2 - bt^3$, where $x$ is in meters and $t$ in seconds. (a) What units must $c$ and $b$ have? Let their numerical values be 3.0 and 2.0. respectively. (b) At what time does the particle reach its maximum positive $x$ position? From $t_0 = 0.0$ s to $t_4 = 4.0$ s, (c) what distance does the particle move and (d) what is its displacement? At $t = 1.0$, 2.0, 3.0, and 4.0 s, what are (e) its velocities and (f) its accelerations?

## SEC. 2-5 ■ CONSTANT ACCELERATION: A SPECIAL CASE

**20. Driver and Rider** An automobile driver on a straight road increases the speed at a constant rate from 25 km/h to 55 km/h in 0.50 min. A bicycle rider on a straight road speeds up at a constant rate from rest to 30 km/h in 0.50 min. Calculate their accelerations.

**21. Stopping a Muon** A muon (an elementary particle) moving in a straight line enters a region with a speed of $5.00 \times 10^6$ m/s and then is slowed at the rate of $1.25 \times 10^{14}$ m/s$^2$. (a) How far does the muon take to stop? (b) Graph $x$ vs. $t$ and $v$ vs. $t$ for the muon.

**22. Rattlesnake Striking** The head of a rattlesnake can accelerate at 50 m/s$^2$ in striking a victim. If a car could do as well, how long would it take to reach a speed of 100 km/h from rest?

**23. Accelerating an Electron** An electron has a constant acceleration of $+3.2$ m/s$^2\hat{\text{i}}$. At a certain instant its velocity is $+9.6$ m/s$\hat{\text{i}}$. What is its velocity (a) 2.5 s earlier and (b) 2.5 s later?

**24. Speeding Bullet** The speed of a bullet is measured to be 640 m/s as the bullet emerges from a barrel of length 1.20 m. Assuming constant acceleration, find the time that the bullet spends in the barrel after it is fired.

**25. Comfortable Acceleration** Suppose a rocket ship in deep space moves with constant acceleration equal to 9.8 m/s$^2$, which gives the illusion of normal gravity during the flight. (a) If it starts from rest, how long will it take to acquire a speed one-tenth that of light, which travels at $3.0 \times 10^8$ m/s? (b) How far will it travel in so doing?

**26. Taking Off** A jumbo jet must reach a speed of 360 km/h on the runway for takeoff. What is the least constant acceleration needed for takeoff from a 1.80 km runway?

**27. Even Faster Electrons** An electron with initial velocity $v_1 = 1.50 \times 10^5$ m/s enters a region 1.0 cm long where it is electrically accelerated (Fig. 2-26). It emerges with velocity $v_2 = 5.70 \times 10^6$ m/s. What is its acceleration, assumed constant? (Such a process occurs in conventional television sets.)

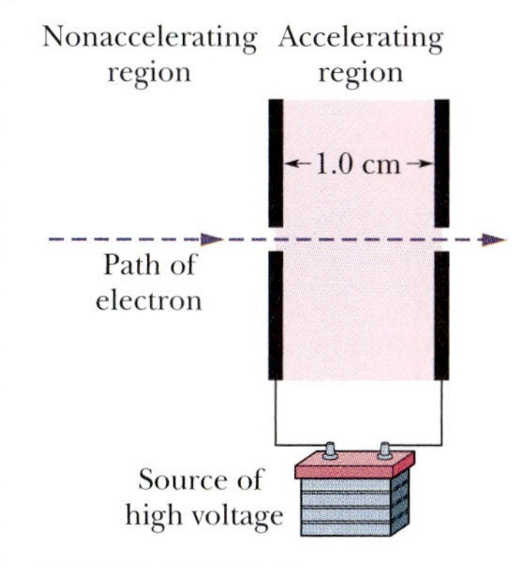


FIGURE 2-26 ■ Problem 27.

**28. Stopping Col. Stapp** A world's land speed record was set by Colonel John P. Stapp when in March 1954 he rode a rocket-propelled sled that moved along a track at 1020 km/h. He and the sled were brought to a stop in 1.4 s. (See Fig. 2-13) In $g$ units, what acceleration did he experience while stopping?

**29. Speed Trap** The brakes on your automobile are capable of slowing down your car at a rate of 5.2 m/s$^2$. (a) If you are going 137 km/h and suddenly see a state trooper, what is the minimum time in which you can get your car under the 90 km/h speed limit? The answer reveals the futility of braking to keep your high speed from being detected with a radar or laser gun.) (b) Graph $x$ vs. $t$ and $v$ vs. $t$ for such a deceleration.

**30. Judging Acceleration** Figure 2-27 depicts the motion of a particle moving along an $x$ axis with a constant acceleration. What are the magnitude and direction of the particle's acceleration?

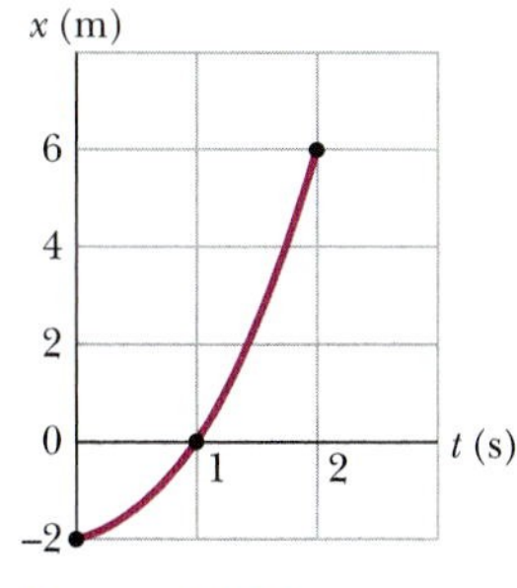


FIGURE 2-27 ■ Problem 30.

**31. Hitting a Wall** A car traveling 56.0 km/h is 24.0 m from a barrier when the driver slams on the brakes. The car hits the barrier 2.00 s later. (a) What is the car's constant acceleration before impact? (b) How fast is the car traveling at impact?

**32. Red and Green Trains** A red train traveling at 72 km/h and a green train traveling at 144 km/h are headed toward one another along a straight, level track. When they are 950 m apart, each engineer sees the other's train and applies the brakes. The brakes slow each train at the rate of 1.0 m/s$^2$. Is there a collision? If so, what is the speed of each train at impact? If not, what is the separation between the trains when they stop?

**33. Between Two Points** A car moving with constant acceleration covered the distance between two points 60.0 m apart in 6.00 s. Its speed as it passes the second point was 15.0 m/s. (a) What was the speed at the first point? (b) What was the acceleration? (c) At what prior distance from the first point was the car at rest? (d) Graph $x$ vs. $t$ and $v$ vs. $t$ for the car from rest ($t_0 = 0$).

**34. Chasing a Truck** At the instant the traffic light turns green, an automobile starts with a constant acceleration $a$ of 2.2 m/s$^2$. At the same instant a truck, traveling with a constant speed of 9.5 m/s, overtakes and passes the automobile. (a) How far beyond the traffic signal will the automobile overtake the truck? (b) How fast will the car be traveling at that instant?

**35. Reaction Time** To stop a car, first you require a certain reaction time to begin braking; then the car slows under the constant braking. Suppose that the total distance moved by your car during these two phases is 56.7 m when its initial speed is 80.5 km/h, and 24.4 in when its initial speed is 48.3 km/h. What are (a) your reaction time and (b) the magnitude of the acceleration?

**36. Avoiding a Collision** When a high-speed passenger train traveling at 161 km/h rounds a bend, the engineer is shocked to see that a locomotive has improperly entered onto the track from a siding and is a distance $D = 676$ m ahead (Fig. 2-28). The locomotive is moving at 29.0 km/h. The engineer of the high-speed train immediately applies the brakes. (a) What must be the magnitude of the resulting constant acceleration if a collision is to be just avoided? (b) Assume that the engineer is at $x = 0$ when, at $t = 0$, he first spots the locomotive. Sketch the $x(t)$ curves representing the locomotive and, high-speed train for the situations in which a collision is just avoided and is not quite avoided.

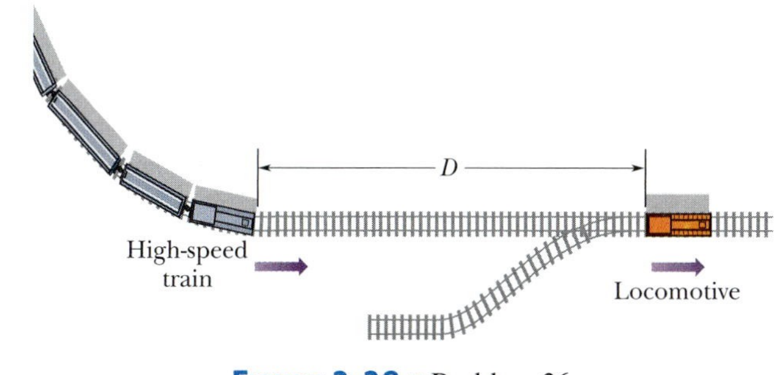


FIGURE 2-28 ■ Problem 36.

**37. Going Up** An elevator cab in the New York Marquis Marriott has a total run of 190 m. Its maximum speed is 305 m/min. Its acceleration (both speeding up and slowing) has a magnitude of 1.22 m/s$^2$. (a) How far does the cab move while accelerating to full speed from rest? (b) How long does it take to make the nonstop 190 m run, starting and ending at rest?

**38. Shuffleboard Disk** A shuffleboard disk is accelerated at a constant rate from rest to a speed of 6.0 m/s over a 1.8 m distance by a player using a cue. At this point the disk loses contact with the cue and slows at a constant rate of 2.5 m/s$^2$ until it stops. (a) How much time elapses from when the disk begins to accelerate until it stops? (b) What total distance does the disk travel?

**39. Electric Vehicle** An electric vehicle starts from rest and accelerates at a rate of 2.0 m/s$^2$ in a straight line until it reaches a speed of 20 m/s. The vehicle then slows at a constant rate of 1.0 m/s$^2$ until it stops. (a) How much time elapses from start to stop? (b) How far does the vehicle travel from start to stop?

**40. Red Car–Green Car** In Fig. 2-29 a red car and a green car, identical except for the color, move toward each other in adjacent lanes and parallel to an $x$ axis. At time $t = 0$, the red car is at $x = 0$ and the green car is at $x = 220$ m. If the red car has a constant velocity of 20 km/h, the cars pass each other at $x = 44.5$ m, and if it has a constant velocity of 40 km/h, they pass each other at $x = 76.6$ m. What are (a) the initial velocity and (b) the acceleration of the green car?

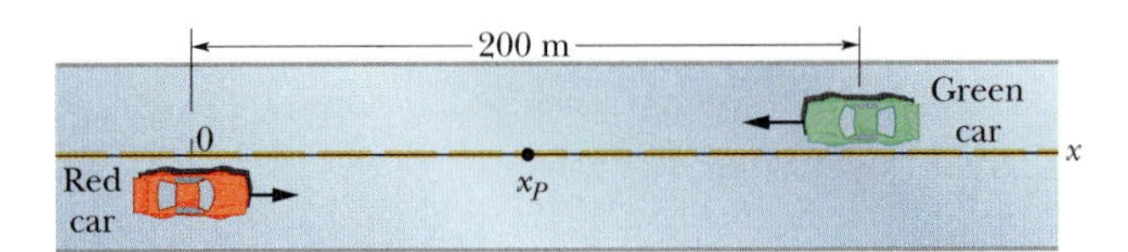


FIGURE 2-29 ■ Problem 40.

**41. Position Function** The position of a particle moving along an $x$ axis is given by $x = (12\ \text{m/s}^2)t^2 - (2\ \text{m/s}^3)t^3$, where $x$ is in meters and $t$ is in seconds. (a) Determine the position, velocity, and acceleration of the particle at $t_3 = 3.0$ s. (b) What is the maximum positive coordinate reached by the particle and at what time is it reached? (c) What is the maximum positive velocity reached by the particle and at what time is it reached? (d) What is the acceleration of the particle at the instant the particle is not moving (other than at $t_0 = 0$)? (e) Determine the average velocity of the particle between $t_0 = 0$ and $t_3 = 3$ s.

# Additional Problems

**42. Kids in the Back!** An unrestrained child is playing on the front seat of a car that is traveling in a residential neighborhood at 35 km/h. (How many mi/h is this? Is this car going too fast?) A small dog runs across the road and the driver applies the brakes, stopping the car quickly and missing the dog. Estimate the speed with which the child strikes the dashboard, presuming that the car stops before the child does so. Compare this speed with that of the world-record 100 m dash, which is run in about 10 s.

**43. The Passat GLX** Test results (*Car & Driver*, February 1993, p. 48) on a Volkswagen Passat GLX show that when the brakes are fully applied it has an average braking acceleration of *magnitude* 8.9 m/s$^2$. If a preoccupied driver who is moving at a speed of 42 mph looks up suddenly and sees a stop light 30 m in front of him, will he have sufficient time to stop? The weight of the Volkswagen is 3 152 lb.

**44. Velocity and Pace** When we drive a car we usually describe our motion in terms of speed or velocity. A speed limit, such as 60 mi/h, is a speed. When runners or joggers describe their motion, they often do so in terms of a *pace*—how long it takes to go a given distance. A 4-min mile (or better, "4 minutes/mile") is an example of a pace.

**(a)** Express the speed 60 mi/h as a pace in min/mi.
**(b)** I walk on my treadmill at a pace of 17 min/mi. What is my speed in mi/h?
**(c)** If I travel at a speed, $v$, given in mi/h, what is my pace, $p$, given in min/mi? (Write an equation that would permit easy conversion.)

**45. Spirit of America** The 9000 lb Spirit of America (designed to be the world's fastest car) accelerated from rest to a final velocity of 756 mph in a time of 45 s. What would the acceleration have been in meters per second? What distance would the driver, Craig Breedlove, have covered?

**46. Driving to New York** You and a friend decide to drive to New York from College Park, Maryland (near Washington, D.C.) on Saturday over the Thanksgiving break to go to a concert with some friends who live there. You figure you have to reach the vicinity of the city at 5 P.M. in order to meet your friends in time for dinner before the concert. It's about 220 mi from the entrance to Route 95 to the vicinity of New York City. You would like to get on the highway about noon and stop for a bite to eat along the way. What does your average velocity have to be? If you keep an approximately constant speed (not a realistic assumption!), what should your speedometer read while you are driving?

**47. NASA Internship** You are working as a student intern for the National Aeronautics and Space Administration (NASA) and your supervisor wants you to perform an indirect calculation of the upward velocity of the space shuttle relative to the Earth's surface just 5.5 s after it is launched when it has an altitude of 100 m. In order to obtain data, one of the engineers has wired a streamlined flare to the side of the shuttle that is gently released by remote control after 5.5 s. If the flare hits the ground 8.5 s after it is released, what is the upward velocity of the flare (and hence of the shuttle) at the time of its release? (Neglect any effects of air resistance on the flare.) *Note*: Although the flare idea is fictional, the data on a typical shuttle altitude and velocity at 5.5 s are straight from NASA!

**48. Cell Phone Fight** You are arguing over a cell phone while trailing an unmarked police car by 25 m; both your car and the police car are traveling at 110 km/h. Your argument diverts your attention from the police car for 2.0 s (long enough for you to look at the phone and yell, "I won't do that!"). At the beginning of that 2.0 s, the police officer begins emergency braking at 5.0 m/s$^2$. (a) What is the separation between the two cars when your attention finally returns? Suppose that you take another 0.40 s to realize your danger and begin braking. (b) If you too brake at 5.0 m/s$^2$, what is your speed when you hit the police car?

**49. Reaction Distance** When a driver brings a car to a stop by braking as hard as possible, the stopping distance can be regarded as the sum of a "reaction distance," which is initial speed multiplied by the driver's reaction time, and a "braking distance," which is the distance traveled during braking. The following table gives typical values. (a) What reaction time is the driver assumed to have? (b) What is the car's stopping distance if the initial speed is 25 m/s?

| Initial Speed (m/s) | Reaction Distance (m) | Braking Distance (m) | Stopping Distance (m) |
|---|---|---|---|
| 10 | 7.5 | 5.0 | 12.5 |
| 20 | 15 | 20 | 35 |
| 30 | 22.5 | 45 | 67.5 |

**50. Tailgating** In this problem we analyze the phenomenon of "tailgating" in a car on a highway at high speeds. This means traveling too close behind the car ahead of you. Tailgating leads to multiple car crashes when one of the cars in a line suddenly slows down. The question we want to answer is: "How close is too close?"

To answer this question, let's suppose you are driving on the highway at a speed of 100 km/h (a bit more than 60 mi/h). The driver ahead of you suddenly puts on his brakes. We need to calculate a number of things: how long it takes you to respond; how far you travel in that time, and how far the other car travels in that time.

**(a)** First let's estimate how long it takes you to respond. Two times are involved: how long it takes from the time you notice something happening till you start to move to the brake, and how long it takes to move your foot to the brake. You will need a ruler to do this. Take the ruler and have a friend hold it from the one end hanging straight down. Place your thumb and forefinger opposite the bottom of the ruler. As your friend releases the ruler suddenly, try to catch it with your thumb and forefinger. Measure how far it falls before you catch it. Do this three times and take the average distance. Assuming the ruler is falling freely without air resistance (not a bad assumption), calculate how much time it takes you to catch it, $t_1$. Now estimate the time, $t_2$, it takes you to move your foot from the gas pedal to the brake pedal. Your reaction time is $t_1 + t_2$.
**(b)** If you brake hard and fast, you can bring a typical car to rest from 100 km/h (about 60 mi/h) in 5 seconds.

1. Calculate your acceleration, $-a_0$, assuming that it is constant.
2. Suppose the driver ahead of you begins to brake with an acceleration $-a_0$. How far will he travel before he comes to a stop? (*Hint*: How much time will it take him to stop? What will be his average velocity over this time interval?)

**(c)** Now we can put these results together into a fairly realistic situation. You are driving on the highway at 100 km/hr and there is a driver in front of you going at the same speed.

1. You see him start to slow immediately (an unreasonable but simplifying assumption). If you are also traveling 100 km/h, how far (in meters) do you travel before you begin to brake? If you can also produce the acceleration $-a_0$ when you brake, what will be the total distance you travel before you come to a stop?
2. If you don't notice the driver ahead of you beginning to brake for 1 s, how much additional distance will you travel?
3. Discuss, on the basis of these calculations, what you think is a safe distance to stay behind a car at 60 mi/h. Express your distance in "car lengths" (about 15 ft). Would you include a safety factor beyond what you have calculated here? How much?

**51. Testing the Motion Detector** A motion detector that may be used in physics laboratories is shown in Fig. 2-30. It measures the distance to the nearest object by using a speaker and a microphone. The speaker clicks 30 times a second. The microphone detects the sound bouncing back from the nearest object in front of it. The computer calculates the time delay between making the sound and receiving the echo. It knows the speed of sound (about 343 m/s at room temperature), and from that it can calculate the distance to the object from the time delay.

**FIGURE 2-30** ■ Problem 51.

**(a)** If the nearest object in front of the detector is too far away, the echo will not get back before a second click is emitted. Once that happens, the computer has no way of knowing that the echo isn't an echo from the second click and that the detector isn't giving correct results any more. How far away does the object have to be before that happens?
**(b)** The speed of sound changes a little bit with temperature. Let's try to get an idea of how important this is. At room temperature (72 °F) the speed of sound is about 343 m/s. At 62 °F it is about 1% smaller. Suppose we are measuring an object that is really 1.5 meters away at 72 °F. What is the time delay $\Delta t$ that the computer detects before the echo returns? Now suppose the temperature is 62 °F. If the computer detects a time delay of $\Delta t$ but (because it doesn't know the temperature) calculates the distance using the speed of sound appropriate for 72 °F, how far away does the computer report the object to be?

**52. Hitting a Bowling Ball** A bowling ball sits on a hard floor at a point that we take to be the origin. The ball is hit some number of times by a hammer. The ball moves along a line back and forth across the floor as a result of the hits. (See Fig. 2-31.) The region to the right of the origin is taken to be positive, but during its motion the ball is at times on both sides of the origin. After the ball has been moving for a while, a motion detector like the one discussed in Problem 51 is started and takes the following graph of the ball's velocity.

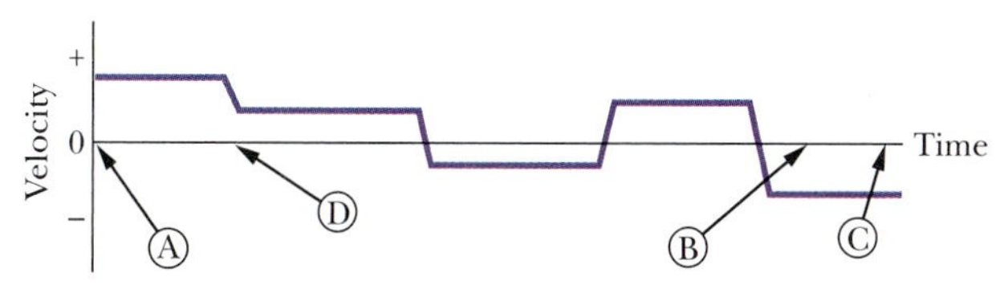


**FIGURE 2-31** ■ Problem 52.

Answer the following questions with the symbols L (left), R (right), N (neither), or C (can't say which). Each question refers only to the time interval displayed by the computer.

**(a)** At which side of the origin is the ball for the time marked A?
**(b)** At the time marked B, in which direction is the ball moving?
**(c)** Between the times A and C, what is the direction of the ball's displacement?
**(d)** The ball receives a hit at the time marked D. In what direction is the ball moving after that hit?

**53. Waking the Balrog** In *The Fellowship of the Ring*, the hobbit Peregrine Took (Pippin for short) drops a rock into a well while the travelers are in the caves of Moria. This wakes a balrog (a bad thing) and causes all kinds of trouble. Pippin hears the rock hit the water 7.5 s after he drops it.

**(a)** Ignoring the time it takes the sound to get back up, how deep is the well?
**(b)** It is quite cool in the caves of Moria, and the speed of sound in air changes with temperature. Take the speed of sound to be 340 m/s (it is pretty cool in that part of Moria). Was it OK to ignore the time it takes sound to get back up? Discuss and support your answer with a calculation.

**54. Two Balls, Passing in the Night*** Figure 2-32 represents the position vs. clock reading of the motion of two balls, *A* and *B*,

---

*From A. Arons, *A Guide to Introductory Physics Teaching* (New York: John Wiley, 1990).

moving on parallel tracks. Carefully sketch the figure on your homework paper and answer the following questions:

**(a)** Along the $t$ axis, mark with the symbol $t_A$ any instant or instants at which one ball is passing the other.
**(b)** Which ball is moving faster at clock reading $t_B$?
**(c)** Mark with the symbol $t_C$ any instant or instants at which the balls have the same velocity.
**(d)** Over the period of time shown in the diagram, which of the following is true of ball $B$? Explain your answer.

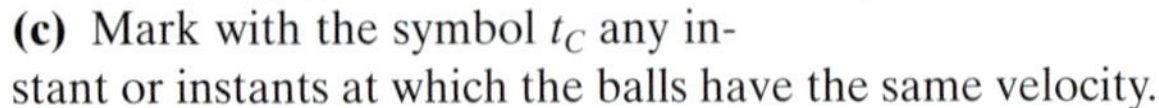

FIGURE 2-32 ■ Problem 54.

1. It is speeding up all the time.
2. It is slowing down all the time.
3. It is speeding up part of the time and slowing down part of the time.

**55. Graph for a Cart on a Tilted Airtrack—with Spring** The graph in Fig. 2-33 below shows the velocity graph of a cart moving on an air track. The track has a spring at one end and has its other end raised. The cart is started sliding up the track by pressing it against the spring and releasing it. The clock is started just as the cart leaves the spring. Take the direction the cart is moving in initially to be the positive $x$ direction and take the bottom of the spring to be the origin.

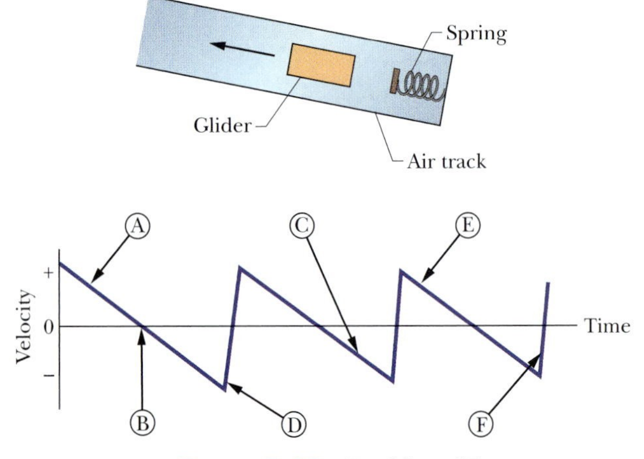


FIGURE 2-33 ■ Problem 55

Letters point to six points on the velocity curve. For the physical situations described below, identify which of the letters corresponds to the situation described. You may use each letter more than once, more than one letter may be used for each answer, or none may be appropriate. If none is appropriate, use the letter N.

**(a)** This point occurs when the cart is at its highest point on the track.
**(b)** At this point, the cart is instantaneously not moving.
**(c)** This is a point when the cart is in contact with the spring.
**(d)** At this point, the cart is moving down the track toward the origin.
**(e)** At this point, the cart has acceleration of zero.

**56. Rolling Up and Down** A ball is launched up a ramp by a spring as shown in Fig. 2-34. At the time when the clock starts, the ball is near the bottom of the ramp and is rolling up the ramp as shown. It goes to the top and then rolls back down. For the graphs shown in Fig. 2-34, the horizontal axis represents the time. The vertical axis is unspecified.

For each of the following quantities, select the letter of the graph that could provide a correct graph of the quantity for the ball in the situation shown (if the vertical axis were assigned the proper units). Use the $x$ and $y$ coordinates shown in the picture. If none of the graphs could work, write N.

**(a)** The $x$-component of the ball's position __________
**(b)** The $y$-component of the ball's velocity __________
**(c)** The $x$-component of the ball's acceleration __________
**(d)** The $y$-component of the normal force the ramp exerts on the ball __________
**(e)** The $x$-component of the ball's velocity __________
**(f)** The $x$-component of the force of gravity acting on the ball __________

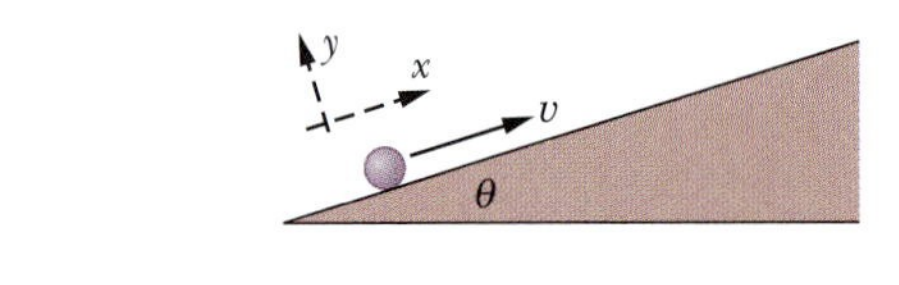


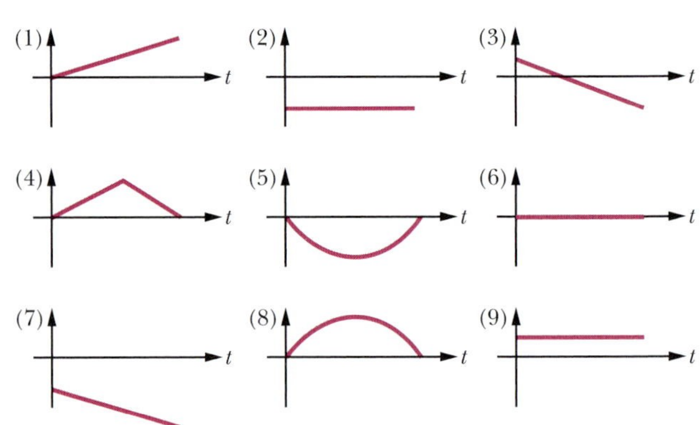


FIGURE 2-34 ■ Problem 56

**57. Model Rocket** A model rocket, propelled by burning fuel, takes off vertically. Plot qualitatively (numbers not required) graphs of $y$, $v$, and $a$ versus $t$ for the rocket's flight. Indicate when the fuel is exhausted, when the rocket reaches maximum height, and when it returns to the ground.

**58. Rock Climber** At time $t = 0$, a rock climber accidentally allows a piton to fall freely from a high point on the rock wall to the valley below him. Then, after a short delay, his climbing partner, who is 10 m higher on the wall, throws a piton downward. The positions $y$ of the pitons versus $t$ during the fall are given in Fig. 2-35. With what speed was the second piton thrown?

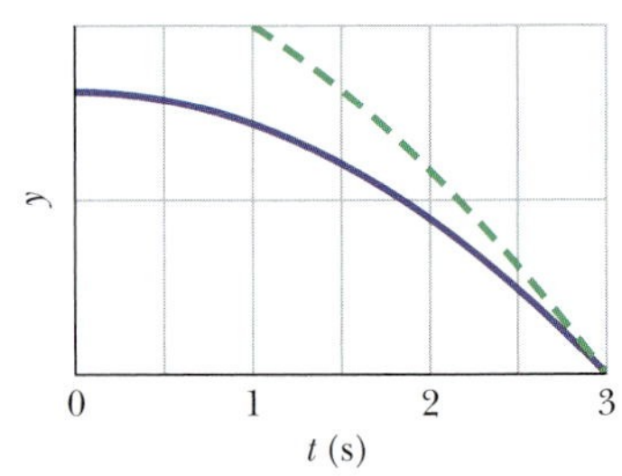


FIGURE 2-35 ■ Problem 58.

**59. Two Trains** As two trains move along a track, their conductors suddenly notice that they are headed toward each other. Figure 2-36 gives their velocities $v$ as functions of time $t$ as the conductors slow the trains.

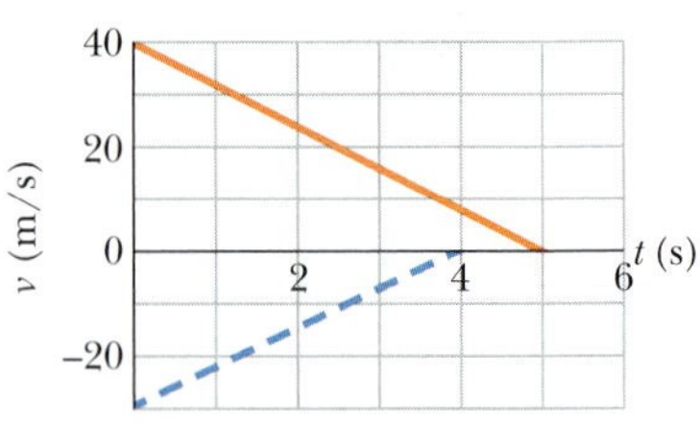


FIGURE 2-36 ■ Problem 59.

The slowing processes begin when the trains are 200 m apart. What is their separation when both trains have stopped?

**60. Runaway Balloon** As a runaway scientific balloon ascends at 19.6 m/s, one of its instrument packages breaks free of a harness and free-falls. Figure 2-37 gives the vertical velocity of the package versus time, from before it breaks free to when it reaches the ground. (a) What maximum height above the break-free point does it rise? (b) How high was the break-free point above the ground?

**FIGURE 2-37** ■ Problem 60.

**61. Position Function Two** A particle moves along the $x$ axis with position function $x(t)$ as shown in Fig. 2-38. Make rough sketches of the particle's velocity versus time and its acceleration versus time for this motion.

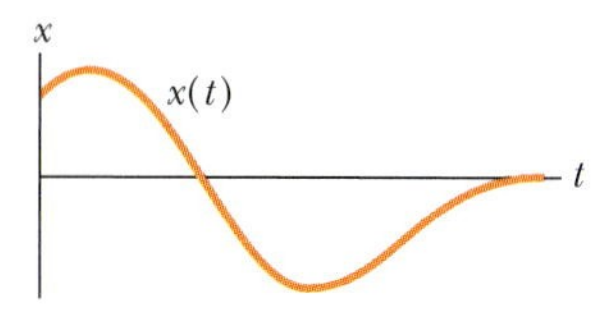


**FIGURE 2-38** ■ Problem 61.

**62. Velocity Curve** Figure 2-39 gives the velocity $v$(m/s) versus time $t$ (s) for a particle moving along an $x$ axis. The area between the time axis and the plotted curve is given for the two portions of the graph. At $t = t_A$ (at one of the crossing points in the plotted figure), the particle's position is $x = 14$ m. What is its position at (a) $t = 0$ and (b) $t = t_B$?

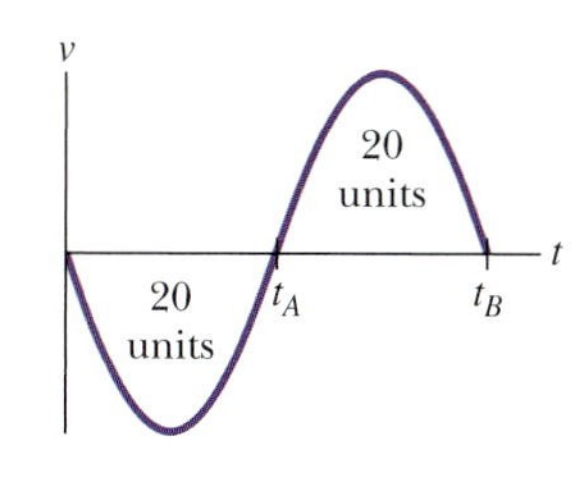


**FIGURE 2-39** ■ Problem 62.

**63. The Motion Detector Rag** This assignment is based on the Physics Pholk Song CD distributed by Pasco scientific. These songs are also available through the Dickinson College Web site at http://physics.dickinson.edu.

**(a)** Refer to the motion described in the first verse of the *Motion Detector Rag*; namely, you are moving for the same amount of time that you are standing. Sketch a position vs. time graph for this motion. Also, describe the shape of the graph in words.

**(b)** Refer to the motion described in the second verse of the *Motion Detector Rag*. In this verse, you are making a "steep downslope," then a "gentle up-slope," and last a flat line. You spend the same amount of time engaged in each of these actions. Sketch a position vs. time graph of this motion. Also, describe what you are doing in words. That is, are you standing still, moving away from the origin (or motion detector), moving toward the origin (or motion detector)? Which motion is the most rapid, and so on?

**(c)** Refer to the motion described in the third verse of the *Motion Detector Rag*. You start from rest and move away from the motion detector at an acceleration of $+1.0$ m/s$^2$ for 5 seconds. Sketch the acceleration vs. time graph to this motion. Sketch the corresponding velocity vs. time graph. Sketch the shape of the corresponding position vs. time graph.

**64. Hockey Puck** At time $t = 0$, a hockey puck is sent sliding over a frozen lake, directly into a strong wind. Figure 2-40 gives the velocity $v$ of the puck vs. time, as the puck moves along a single axis. At $t = 14$ s, what is its position relative to its position at $t = 0$?

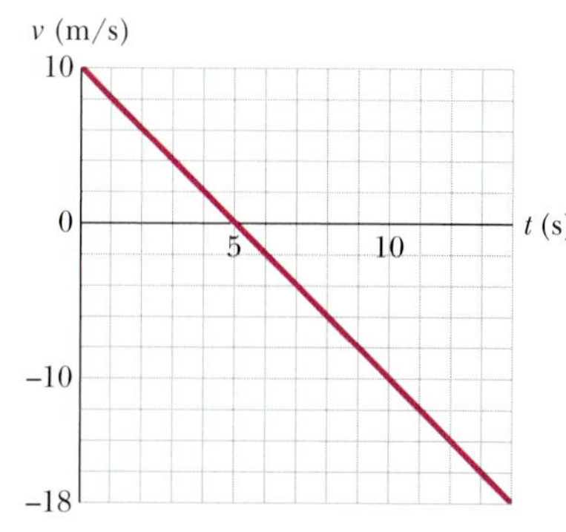


**FIGURE 2-40** ■ Problem 64.

**65. Describing One-Dimensional Velocity Changes** In each of the following situations you will be asked to refer to the mathematical definitions and the concepts associated with the *number line*. Note that being more positive is the same as being less negative, and so on.

**(a)** Suppose an object undergoes a *change* in velocity from $+1$ m/s to $+4$ m/s. Is its velocity becoming *more positive* or *less positive*? What is meant by more positive? Less positive? Is the acceleration positive or negative?

**(b)** Suppose an object undergoes a *change* in velocity from $-4$ m/s to $-1$ m/s. Is its velocity becoming *more positive* or *less positive*? What is meant by more positive? Less positive? Is the acceleration positive or negative?

**(c)** Suppose an object is turning around so that it undergoes a *change* in velocity from $-2$ m/s to $+2$ m/s. Is its velocity becoming *more positive* or *less positive* than it was before? What is meant by more positive? Less positive? Is it undergoing an acceleration while it is turning around? Is the acceleration positive or negative?

**(d)** Another object is turning around so that it undergoes a *change* in velocity from $+1$ m/s to $-1$ m/s. Is its velocity becoming *more positive* or *less positive* than it was before? What is meant by more positive? Less positive? Is it undergoing an acceleration while it is turning around? Is the acceleration positive or negative?

**66. Bowling Ball Graph** A bowling ball was set into motion on a fairly smooth level surface, and data were collected for the total distance covered by the ball at each of four times. These data are shown in the table.

| Average Time (s) | Distance (m) |
|---|---|
| 0.00 | 0.0 |
| 0.92 | 2.0 |
| 1.85 | 4.0 |
| 2.87 | 6.0 |

**(a)** Plot the data points on a graph.

**(b)** Use a ruler to draw a straight line that passes as close as possible to the data points you have graphed.

**(c)** Using methods you were taught in algebra, calculate the value of the slope, $m$, and find the value of the intercept, $b$, of the line you have sketched through the data.

**67. Modeling Bowling Ball Motion** A bowling ball is set into motion on a smooth level surface, and data were collected for the total distance covered by the ball at each of four times. These data are shown in the table in Problem 66. Your job is to learn to use a spreadsheet program — for example, Microsoft Excel—to create a mathematical model of the bowling ball motion data shown. You are to find what you think is the best value for the slope, $m$, and the $y$-intercept, $b$. Practicing with a tutorial worksheet entitled MODTUT.XLS will help you to learn about the process of modeling for a linear relationship. Ask your instructor where to find this tutorial worksheet.

After using the tutorial, you can create a model for the bowling ball data given above. To do this:

**(a)** Open a new worksheet and enter a title for your bowling ball graph.
**(b)** Set the y-label to Distance ($m$) and the x-label to Time ($s$).
**(c)** Refer to the data table above. Enter the measured times for the bowling ball in the Time ($s$) column (formerly x-label).
**(d)** Set the y-exp column to D-data (m) and enter the measured distances for the bowling ball (probably something like 0.00 m, 2.00 m, 4.00 m, and 6.00 m.).
**(e)** Place the symbol $m$ (for slope) in the cell B1. Place the symbol $b$ (for y-intercept) in cell B2.
**(f)** Set the y-theory column to D-model (m) and then put the appropriate equation for a straight line of the form Distance = m*Time + b in cells C7 through C12. Be sure to refer to cells C1 for slope and C2 for y-intercept as absolutes; that is, use $C$1 and $C$2 when referring to them.
**(g)** Use the spreadsheet graphing feature to create a graph of the data in the D-exp and D-theory columns as a function of the data in the Time column.
**(h)** Change the values in cells Cl and C2 until your theoretical line matches as closely as possible your red experimental data points in the graph window.
**(i)** Discuss the meaning of the slope of a graph of distance vs. time. What does it tell you about the motion of the bowling ball?

**68. A Strange Motion** After doing a number of the exercises with carts and fans on ramps, it is easy to draw the conclusion that everything that moves is moving at either a constant velocity or a constant acceleration. Let's examine the horizontal motion of a triangular frame with a pendulum at its center that has been given a push. It undergoes an unusual motion. You should determine whether or not it is moving at either a constant velocity or constant acceleration. (*Note:* You may want to look at the motion of the triangular frame by viewing the digital movie entitled PASCO070. This movie is included on the VideoPoint compact disk. If you are not using VideoPoint, your instructor may make the movie available to you some other way.)

The images in Fig. 2-41 are taken from the 7th, 16th, and 25th frames of that movie.

Data for the position of the center of the horizontal bar of the triangle were taken every tenth of a second during its first second of motion. The origin was placed at the zero centimeter mark of a fixed meter stick. These data are in the table below.
**(a)** Examine the position vs. time graph of the data shown above. Does the triangle appear to have a constant velocity throughout the first second? A constant acceleration? Why or why not?
**(b)** Discuss the nature of the motion based on the shape of the graph. At approximately what time, if any, is the triangle changing direction? At approximately what time does it have the greatest negative velocity? The greatest positive velocity? Explain the reasons for your answers.
**(c)** Use the data table and the definition of average velocity to calculate the average velocity of the triangle at each of the times between 0.100 s and 0.900 s. In this case you should use the position just before the indicated time and the position just after the indicated time in your calculation. For example, to calculate the average velocity at $t_2 = 0.100$ seconds, use $x_3 = 44.5$ cm and $x_1 = 52.1$ cm along with the differences of the times at $t_3$ and $t_1$. *Hint:* Use only times and positions in the gray boxes to get a velocity in a gray box and use only times and positions in the white boxes to get a velocity in a white box.
**(d)** Since people usually refer to velocity as distance divided by time, maybe we can calculate the average velocities as simply $x_1/t_1$, $x_2/t_2$, $x_3/t_3$, and so on. This would be easier. Is this an equivalent method for

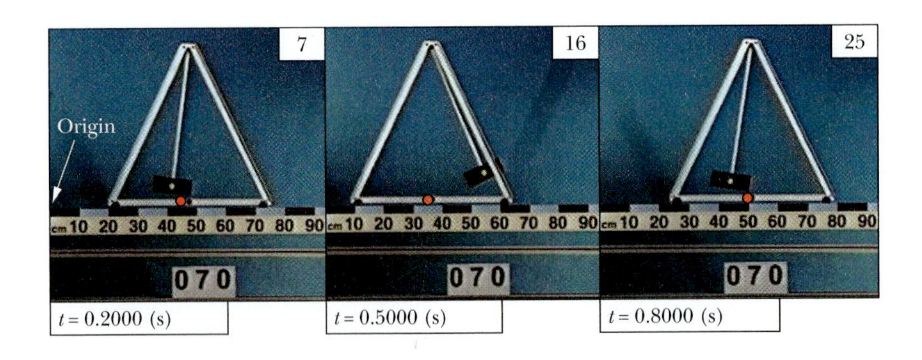


| Fr# | Pr# | $t$(s) | $x$(cm) | <$v$>(cm/s) |
|---|---|---|---|---|
| 1 | 1 | 0.000 | 52.1 | no entry |
| 4 | 2 | 0.100 | 49.9 | −38.0 |
| 7 | 3 | 0.200 | 44.5 | |
| 10 | 4 | 0.300 | 39.1 | |
| 13 | 5 | 0.400 | 35.2 | |
| 16 | 6 | 0.500 | 34.8 | |
| 19 | 7 | 0.600 | 36.9 | |
| 22 | 8 | 0.700 | 43.0 | |
| 25 | 9 | 0.800 | 49.2 | |
| 28 | 10 | 0.900 | 53.6 | |
| 31 | 11 | 1.000 | 54.4 | no entry |

Triangle Motion
$x$(cm): 0, 10, 20, 30, 40, 50, 60
$t$(s): 0.000, 0.500, 1.000

FIGURE 2-41 ■ Problem 68.

finding the velocities at the different times? Try using this method of calculation if you are not sure. Give reasons for your answer.

**(e)** Often, when an oddly shaped but smooth graph is obtained from data it is possible to fit a polynomial to it. For example, a fourth-order polynomial that fits the data is

$$x = \{(-376 \text{ cm/s}^4)t^4 + (719 \text{ cm/s}^3)t^3 - (347 \text{ cm/s}^2)t^2 + (5.63 \text{ cm/s})t + 52.1 \text{ cm}\}$$

Using this polynomial approximation, find the *instantaneous* velocity at $t = 0.700$ s. Comment on how your answer compares to the average velocity you calculated at 0.700 s. Are the two values close? Is that what you expect?

**69. Cedar Point** At the Cedar Point Amusement Park in Ohio, a cage containing people is moving at a high initial velocity as the result of a previous free fall. It changes direction on a curved track and then coasts in a horizontal direction until the brakes are applied. This situation is depicted in a digital movie entitled DSON002. (*Note*: This movie is included on the VideoPoint compact disk. If you are not using VideoPoint, your instructor may make the movie available to you some other way.)

**(a)** Use video analysis software to gather data for the horizontal positions of the tail of the cage in meters as a function of time. Don't forget to use the scale on the title screen of the movie so your results are in meters rather than pixels. Summarize this data in a table or in a printout attached to your homework.

**(b)** Transfer your data to a spreadsheet and do a parabolic model to show that within 5% or better $x = (-7.5 \text{ m/s}^2)t^2 + (22.5 \text{ m/s})t + 2.38$ m. Please attach a printout of this model and graph with your name on it to your submission as "proof of completion."(*Note*: Your judgments about the location of the cage tail may lead to slightly different results.)

**(c)** Use the equation you found along with its interpretation as embodied in the first kinematic equation to determine the horizontal acceleration, $a$, of the cage as it slows down. What is its initial horizontal velocity, $v_1$, at time $t = 0$ s? What is the initial position, $x_1$, of the cage?

**(d)** The movie ends before the cage comes to a complete stop. Use your knowledge of $a$, $v_1$, and $x_1$ along with kinematic equations to determine the horizontal position of the cage when it comes to a *complete* stop so that the final velocity of the cage is given by $v_2 = v = 0.00$ m/s.

**70. Three Digital Movies** Three digital movies depicting the motions of four single objects have been selected for you to examine using a video-analysis program. They are as follows:

PASCO004: A cart moves on an upper track while another moves on a track just below.

PASCO153: A metal ball attached to a string swings gently.

HRSY003: A boat with people moves in a water trough at Hershey Amusement Park.

Please examine the horizontal motion of each object carefully by viewing the digital movies. In other words, just examine the motion in the $x$ direction (and ignore any slight motions in the $y$ direction). You may use VideoPoint, VideoGraph, or World-in-Motion digital analysis software and a spreadsheet to analyze the motion in more detail if needed. Based on what you have learned so far, there is more than one analysis method that can be used to answer the questions that follow. *Note:* Since we are interested only in the nature of these motions (not exact values) you do not need to scale any of the movies. Working in pixel units is fine.

**(a)** Which of these four objects (upper cart, lower cart, metal ball, or boat), if any, move at a constant horizontal velocity? Cite the evidence for your conclusions.

**(b)** Which of these four objects, if any, move at a constant horizontal acceleration? Cite the evidence for your conclusions.

**(c)** Which of these four objects, if any, move at *neither* a constant horizontal velocity nor acceleration? Cite the evidence for your conclusions.

**(d)** The kinematic equations are very useful for describing motions. Which of the four motions, if any, cannot be described using the kinematic equations? Explain the reasons for your answer.

**71. Speeding Up or Slowing Down** Figure 2-42 shows the velocity vs. time graph for an object constrained to move in one dimension. The positive direction is to the right.

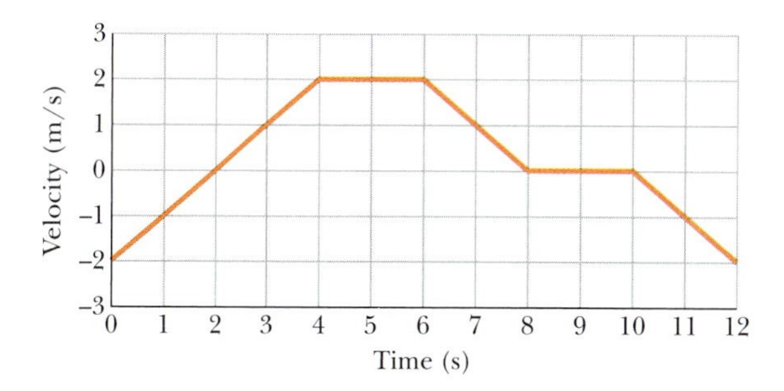


FIGURE 2-42 ■ Problems 71–74.

**(a)** At what times, or during what time periods, is the object speeding up?

**(b)** At what times, or during what time periods, is the object slowing down?

**(c)** At what times, or during what time periods, does the object have a constant velocity?

**(d)** At what times, or during what time periods, is the object at rest?

If there is no time or time period for which a given condition exists, state that explicitly.

**72. Right or Left** Figure 2-42 shows the velocity vs. time graph for an object constrained to move along a line. The positive direction is to the right.

**(a)** At what times, or during what time periods, is the object speeding up and moving to the right?

**(b)** At what times, or during what time periods, is the object slowing down and moving to the right?

**(c)** At what times, or during what time periods, does the object have a constant velocity to the right?

**(d)** At what times, or during what time periods, is the object speeding up and moving to the left?

**(e)** At what times, or during what time periods, is the object slowing down and moving to the left?

**(f)** At what times, or during what time periods, does the object have a constant velocity to the left?

If there is no time or time period for which a given condition exists, state that explicitly.

**73. Constant Acceleration** Figure 2-42 shows the velocity vs. time graph for an object constrained to move along a line. The positive direction is to the right.

**(a)** At what times, or during what time periods, is the object's acceleration zero?

**(b)** At what times, or during what time periods, is the object's acceleration constant?
**(c)** At what times, or during what time periods, is the object's acceleration changing?

If there is no time or time period for which a given condition exists, state that explicitly.

**74. Acceleration to the Right or Left** Figure 2-42 shows the velocity vs. time graph for an object constrained to move along a line. The positive direction is to the right.

**(a)** At what times, or during what time periods, is the object's acceleration increasing and directed to the right?
**(b)** At what times, or during what time periods, is the object's acceleration decreasing and directed to the right?
**(c)** At what times, or during what time periods, does the object have a constant acceleration to the right?
**(d)** At what times, or during what time periods, is the object's acceleration increasing and directed to the left?
**(e)** At what times, or during what time periods, is the object's acceleration decreasing and directed to the left?
**(d)** At what times, or during what time periods, does the object have a constant acceleration to the left?

If there is no time or time period for which a given condition exists, state that explicitly.

# 3 Forces and Motion Along a Line

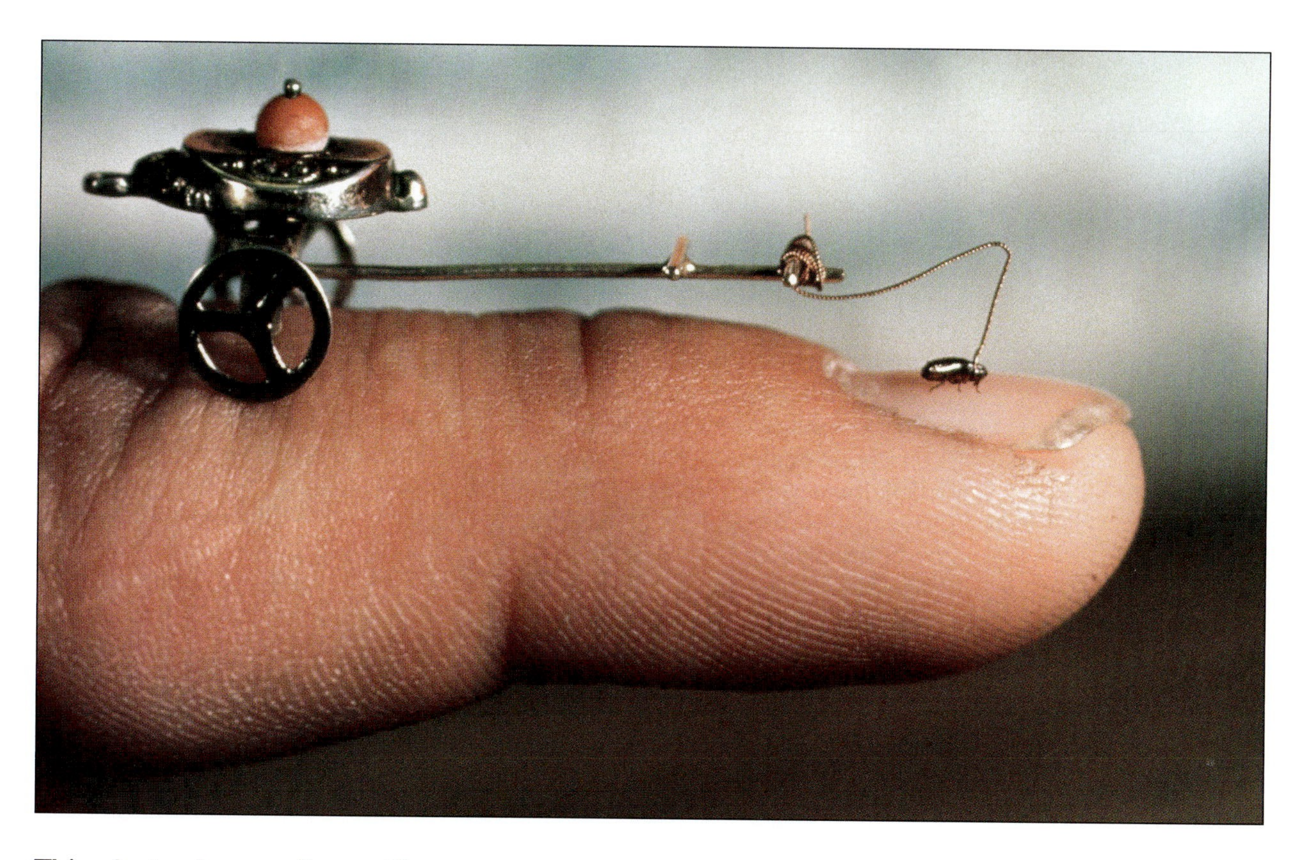

This photo shows a flea pulling a toy cart. In 1996 and 1997 Maria Fernanda Cardoso, a contemporary Colombian artist, created a circus of trained fleas and toured with them. Cardoso used a thin wire to attach Brutus, "the strongest flea on Earth," to a toy train car. She then used sound and carbon dioxide to induce Brutus to hop. Videos show that when Brutus hops, the train car jerks through a distance of about one centimeter. This is an amazing feat because the mass of the toy train car is 160,000 times greater than that of a flea.

**How is it possible for a flea to pull 160 000 times its mass?**

*The answer is in this chapter.*

**FIGURE 3-1** ▪ Isaac Newton (1642–1727) was the primary developer of the laws of classical mechanics.

## 3-1 What Causes Acceleration?

As part of our study of the kinematics of one-dimensional motion, we have introduced definitions of position, velocity, and acceleration. We have used these definitions to describe motion scientifically with graphs and equations. We now turn our attention to **dynamics**—the study of causes of motion. The central question in dynamics is: What causes a body to change its velocity or accelerate as it moves?

Everyday experience tells us that under certain circumstances an object can change its velocity when you interact with it with a push or pull of some sort. We call such a push or pull a **force.** For example, the velocity of a pitched baseball can suddenly change direction when a batter hits it, and a train can slow down when the engineer applies the brakes. However, at times an obvious interaction with an object does not cause a velocity change. Hitting or pushing on a massive object such as a brick wall does not cause it to move. To make matters more complex, many objects seem to undergo velocity changes even when no obvious interaction is present—a car rolls to a stop when you take your foot off the accelerator, and a falling object speeds up.

The laws of motion that relate external interactions between objects to their accelerations were first developed by Isaac Newton, pictured in Fig. 3-1. These laws lie at the heart of our modern interpretation of classical mechanics. Newton's laws are not absolute truths to be found in nature. Instead, they are part of a logically consistent conceptual framework that has emerged from the historical development of concepts, definitions, and measurement procedures.

Newton's laws have attained universal acceptance because they agree with countless observations made by scientists during the past 300 years. They have enabled us to learn about the fundamental nature of gravitational, electrical, and magnetic interactions. Engineers use the laws of motion and a knowledge of forces to predict precisely what motions will occur in the design of industrial-age devices such as engines, bridges, roadways, airplanes, and power plants.

In this chapter we begin our study of the causes of motion along a straight line. In chapters that follow we will extend this study to motions in two and eventually three dimensions.

## 3-2 Newton's First Law

In order to start thinking about what causes changes in an object's velocity, let's set up a thought experiment in which a small object sitting on a level surface is given a swift kick. How would you describe its motion in everyday language? Perhaps you might say something like, "The object speeds up quickly during the kick, but afterward, it begins to slow down as it slides or rolls along the surface, and eventually it comes to a stop." What caused the object to speed up (to change velocity) in the first place? The force of your kick did that. But after the kick is over, what caused the object to slow down? Before Newton's *Principia* was published in 1689, most scientists believed that the natural state of motion is rest and that a sliding object slows down and stops because there is no force to keep it moving.

Let's try to figure out whether this belief that a force is needed to keep an object in motion makes sense by looking at the outcome of an experiment. In the experiment, an object is given a kick and then its velocity is measured as a function of time as it slows down. In particular, the velocities of a plastic box and a small cart are measured as the object moves on different level surfaces—a rough carpet and a smooth track. In each case, the velocity of the slowing object is recorded by a motion detector attached to a microcomputer-based laboratory system. Figure 3-2 shows the experimental setup for two situations of interest—a cart rolling on a track and a plastic box sliding to a stop along a carpet.

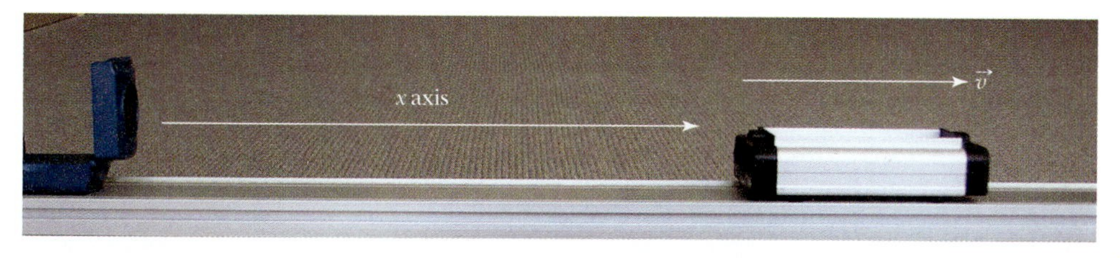

**FIGURE 3-2** ■ Two objects are moving away from a motion detector. The cart on a level track is slowing down very little (top panel), and a plastic box sliding on a carpet is slowing to a stop much sooner than the cart on the track (bottom panel).

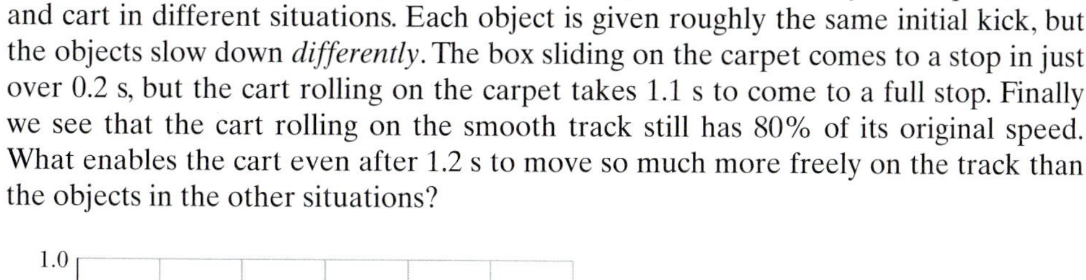

Figure 3-3 shows what happens to the $x$-components of velocity of the plastic box and cart in different situations. Each object is given roughly the same initial kick, but the objects slow down *differently*. The box sliding on the carpet comes to a stop in just over 0.2 s, but the cart rolling on the carpet takes 1.1 s to come to a full stop. Finally we see that the cart rolling on the smooth track still has 80% of its original speed. What enables the cart even after 1.2 s to move so much more freely on the track than the objects in the other situations?

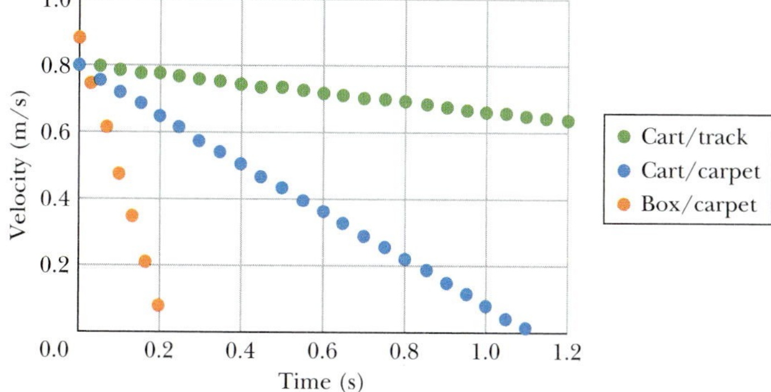


**FIGURE 3-3** ■ An overlay graph of the $x$-component of velocity vs. time for objects slowing to a stop in three different situations. Although the rate of velocity decrease is linear in each case, the slowing rate is distinctly different for each object/surface combination. *Note*: Data for position vs. time were obtained using an ultrasonic motion detector. In each case, the position vs. time data were fit very accurately with a quadratic function and the first time derivative of each $x$ vs. $t$ fit equation was used to determine instantaneous velocity vs. time equations. Each of these $v$ vs. $t$ equations was plotted at the times that position data were recorded.

Let's return to the question that motivated the experiment: Is a force required to keep an object moving at a constant velocity? At first glance, the answer is yes, since the object of interest slows down after the kick in each case. But wait a minute! After the kick, the *rate of slowing* is different in each case. This suggests that the slowing is caused by different forces between the object and the surface over which it moves. We associate the longer slowing time with a smaller frictional force exerted on the object by the surface. A reasonable inference is that it doesn't require a force to keep an object moving at a constant velocity. Rather, forces are present that are causing it to slow down. So what is the natural state of motion in the absence of forces?

Imagine what would happen if we could make the surface that the cart and plastic box move on smoother and smoother or minimize the horizontal friction forces on an object by using an air track, hovercraft, or moving it in outer space. The object would move farther and farther. What if we could observe an object in motion that has no interactions with its surroundings and hence no forces on it? Our experiment suggests that it could move forever at a constant velocity. This was Newton's answer to this question and is embodied in his First Law of Motion, expressed here in contemporary English rather than 17th-century Latin:

**NEWTON'S FIRST LAW:** Consider a body on which no force acts. If the body is at rest, it will remain at rest. If the body is moving, it will continue to move with a constant velocity.

What is force? Clearly Newton is defining force here to be an agent acting on a body that changes its velocity. In the absence of force, a body's velocity will not change. We can state this definition of force more formally.

> **FORCE** is that which causes the velocity of an object to change.

Newton's First Law and his definition of force seem sensible when applied to an object at rest or moving at a constant velocity in a typical physics laboratory. However, in order to measure the velocity of an object, we must choose a coordinate system or reference frame to measure the positions as a function of time. As you saw in Chapter 2, these measurements are needed to calculate velocities and accelerations.

Can we expect Newton's First Law to hold in any reference frame? It turns out that Newton's First Law doesn't hold in all frames of reference. For example, consider what happens to an object in a frame of reference that is accelerating. It is common to see pencils and other small objects that were at rest in a car's frame of reference spontaneously begin to roll around on a dashboard when a car suddenly speeds up or slows down. In this case, Newton's First Law doesn't appear to hold. For this reason, Newton's First Law is often called the law of inertia. Reference frames in which it holds are called inertial frames. Thus, any accelerating frame of reference, in which resting objects appear to start moving spontaneously such as those in a vehicle that is speeding up, slowing down, or turning, is a noninertial frame. Newton's First Law only holds in inertial reference frames. As we develop Newton's other laws of motion, we will restrict ourselves to working in inertial reference frames in which the first law is valid.

**READING EXERCISE 3-1:** Consider the graph shown in Fig. 3-3. (a) Roughly how many seconds does it take the cart rolling on the rough carpet to come to a complete stop? (b) Assuming the cart traveling on the smooth track has a speed of 0.8 m/s at $t = 0.0$ s, what percent of its initial speed does the cart rolling on the track still have just as the cart on the carpet has come to rest? ■

**READING EXERCISE 3-2:** (a) Describe a noninertial reference frame that you have been "at rest" in. (b) What observations did you make in that frame to lead you to conclude that it was noninertial? ■

## 3-3 A Single Force and Acceleration Along a Line

We will simplify our investigation of force and the changes in velocity that it produces by first considering situations in which a single force acts on an object in an inertial reference frame. After we study how a single force affects the motion of an object, we will investigate what happens to an object's motion when two or more forces are acting on an object along its line of motion.

Consider the motion of a person riding on a low-friction cart that can roll easily under the influence of a force. A steady pulling force is applied to the cart and rider. The force acts along the line of the cart's motion. The person who is pulling maintains a steady force on the cart and rider by keeping a short piece of bungee cord stretched to a constant length as shown in Fig. 3-4. By directing a motion detector toward the back of the cart rider, we can track the motion with a computer data acquisition system. If the pulling force is the only significant force on the rider in the direction of his motion, then the results displayed in Fig. 3-5 lead us to make the following observation.

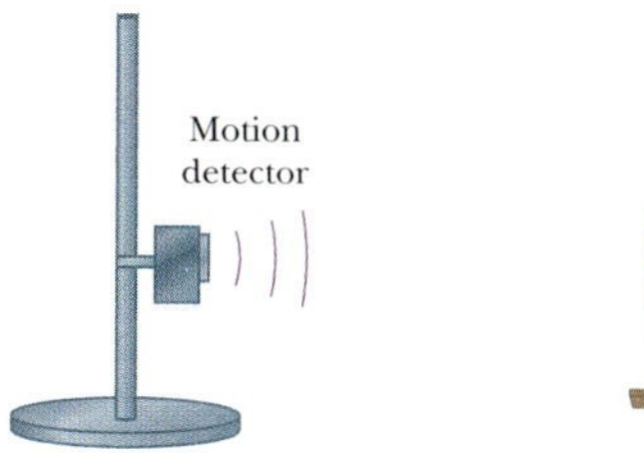


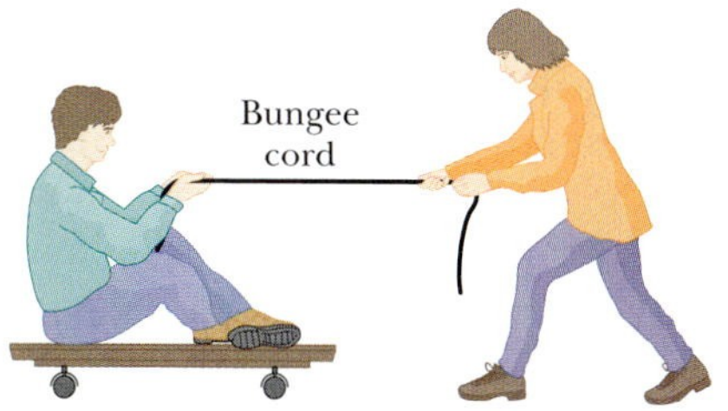


**FIGURE 3-4** ■ A person riding on a low-friction cart is pulled by another person who exerts a constant force along a straight line by keeping the length of a bungee cord constant.

**OBSERVATION:** A constant force acting on an object causes it to move along a straight line with a constant acceleration that is in the same direction as the force.

This observation has been verified many times for different objects moving under the influence of a constant push or pull when friction forces are small.

Many people believe that a constant force will cause a body to move at a constant velocity. This common belief stems from everyday experiences such as driving a car along a highway or sliding a heavy box along a floor. It takes a steady flow of gasoline to move the car at a constant velocity. Thus, the experimental result that a constant force causes a constant acceleration, as shown in Fig. 3-5, is surprising. Remember that we have designed our experiment to apply a single force to a low-friction cart so that there are no significant friction forces acting. Later in this chapter, we will discuss how contact forces, involving a direct push or pull or friction between surfaces like those experienced by a sliding box or a car moving along a highway, can cancel each other to yield zero *net force* on an object. This can then lead us to situations in which pushing or pulling forces, when counteracted by friction forces, do indeed cause bodies to move at a constant velocity.

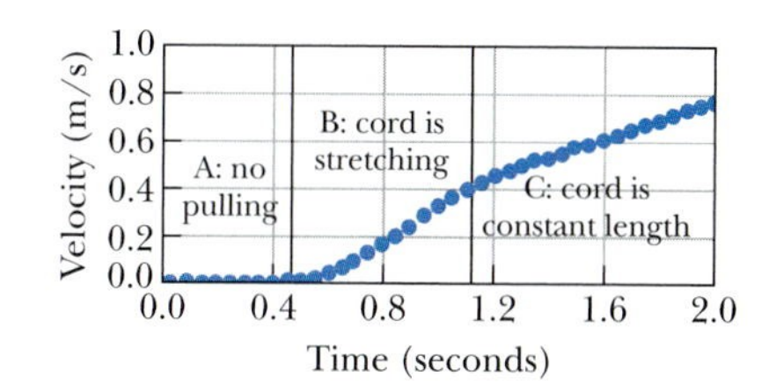


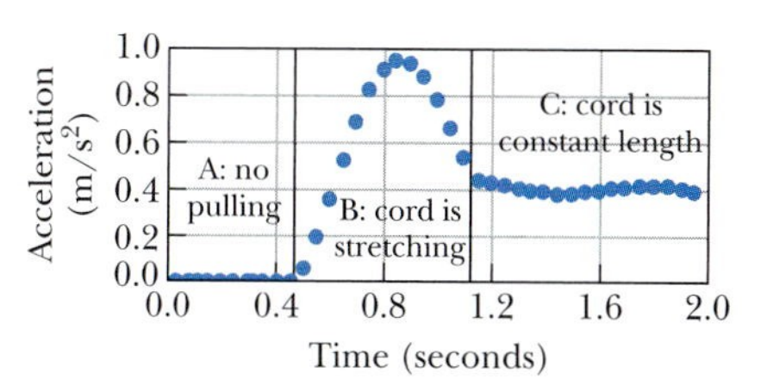


**FIGURE 3-5** ■ These graphs show velocity and acceleration components vs. time for a rider on a cart. For the first 0.5 s (region A) the cart is at rest. Between 0.5 s and 1.1 s (region B) the cord is beginning to stretch. Between 1.1 s and 2.0 s (region C) a constant force is acting and the acceleration is observed to be constant as well.

**READING EXERCISE 3-3:** (a) Describe an experience you have had in which applying what seems like a steady force to an object did *not* cause it to accelerate. (b) Describe a situation in which an object accelerated when you applied what seemed to be a steady force to it. *Note:* You can experiment with applying a steady force to some objects readily available to you. ■

## 3-4 Measuring Forces

As we discussed in Chapter 1, in order to allow us to communicate with others precisely and unambiguously, we need to define a standard unit and a scale for force just as we did for distance, mass, and time. Since all physical quantities are defined by the procedures developed for measuring them, we must start by defining a procedure for measuring our standard unit of force. Our qualitative definition of force is that it is an interaction that causes acceleration, so our standard method for measuring force involves measuring how much acceleration a given force imparts to a standard object. We need to decide, as an international community of scientist and engineers, what the standard object we accelerate will be. It turns out that what we have chosen to use is the international standard kilogram discussed in Chapter 1. The SI unit of force is the newton.

**DEFINITION OF THE STANDARD FORCE UNIT:** One newton of force is defined to be the force necessary to impart an acceleration of 1 $m/s^2$ to the international standard kilogram.

This definition of the newton assumes, of course, that all other forces experienced by the standard mass are small enough to be neglected. To measure any other force in newtons, we simply need to measure the acceleration of our standard object in a low-friction setting and compare its acceleration to 1 m/s$^2$.

Other units of force that are still used in the United States are summarized in Appendix D. These include the dyne, the pound, and the ton.

In order to measure a force in standard units, we must allow it to accelerate a 1 kg object that is free to move without experiencing significant friction forces. For practical reasons we have chosen to measure force by accelerating a low-friction cart on a smooth, level track instead of the actual standard kilogram. We start by adjusting the cart's mass so that it balances with a facsimile of the international standard kilogram.

Next we set up an ultrasonic motion detector with a computer data acquisition system to measure the change in position of the cart as a function of time as it accelerates. The computer data acquisition software can then be used to calculate velocity and acceleration values as a function of time from the position data.

Suppose that someone pulls our low friction cart along a track by means of a spring attached to one end of the cart. Assuming the spring is not yet stretched so far as to be permanently deformed, then the farther it is stretched the greater the size or magnitude of the pull force. The different strengths of pull impart different accelerations to our cart. For a certain strength of pull we find that we can impart an acceleration of 1 m/s$^2$ to the cart—measured by the computer data acquisition system. Of course, this length of the spring is by definition acting on the standard cart with a force of 1 N.

How could we exert a force on the cart of 2 N, 3 N, and so on? We can pull harder on the spring so it stretches enough to cause the cart to accelerate at 2 m/s$^2$. The process can be repeated to yield an acceleration of 3 m/s$^2$, and so on as illustrated in Fig. 3-6.

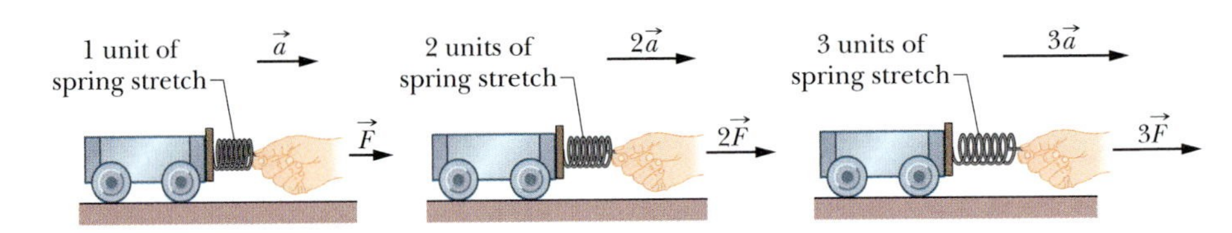


**FIGURE 3-6** ■ An experiment in which a spring is used to apply steady forces to a 1 kg cart. First the spring is stretched enough to yield an acceleration of 1 m/s$^2$, so by definition the force applied to the 1 kg cart is 1 N. As the spring is stretched more and more, the forces on the cart become larger and accelerations of 2 m/s$^2$ and then 3 m/s$^2$ can be created.

Thus, a force can be measured by the acceleration it produces on a standard 1 kg object.

Acceleration is a vector quantity that has both a magnitude and direction. Is force also a vector quantity? Does it have a direction as well as a size associated with it? In order to answer the question of whether force is a vector quantity, consider the following question: Is a force of 1 N directed to the right different from a force of 1 N directed to the left? If so, how? The answer is "yes," these forces are different. A force directed to the right will cause an object to accelerate to the right, and a force directed to the left will cause an object to accelerate to the left. Thus, a force has both a magnitude and a direction associated with it. As we discussed in Section 2-2, to qualify as a vector, a force must also have certain other properties that we have not yet specified. However, it is reasonable for now to assume that force behaves like a vector.

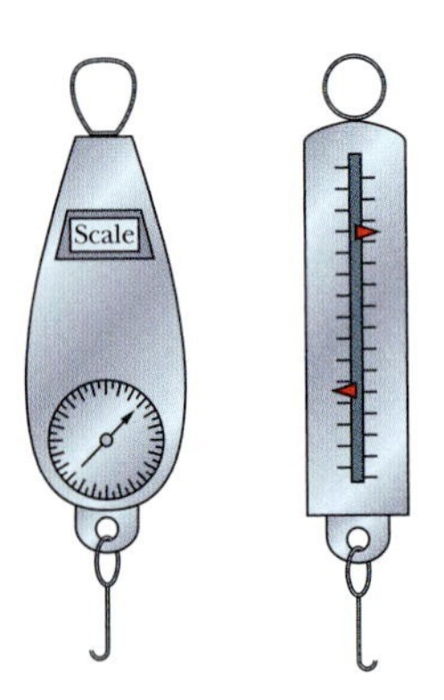


**FIGURE 3-7** ■ Two types of spring scales that can be calibrated to measure forces in newtons by relating the gravitational force exerted by the Earth on a 1 kg weight to the amount of spring stretch.

Measuring force by setting up a system for measuring the acceleration of a standard object is very impractical. Most investigators take advantage of the fact that elastic devices such as springs, rubber bands, and electronic strain gauges (used in the electronic force sensor in Fig. 3-8) stretch more and more as greater forces are exerted on them. These devices can be calibrated "properly" by using the "official" method for measuring force. We can designate a 1 newton force as that which causes our standard mass to accelerate at 1 m/s$^2$ and record the amount of stretch or the electronic reading for the new device. Then we can designate a 2 newton force as that which leads to an acceleration of 2 m/s$^2$ and record the response of the new device and so on for other forces. More often, a secondary calibration can be performed by comparing the readings of a given force-measuring device to that of another force-measuring device that has already been properly calibrated. Spring scales, like those shown in Fig. 3-7, are very popular devices for measuring force. This popularity stems from the fortunate fact that the amount by which a spring stretches is directly proportional to the magnitude of the force acting on the spring—provided the spring is not overstretched. This proportionality was discovered in the 17th century by Robert Hooke, and will be discussed more formally in Chapter 9. The proportionality between spring stretch and force is a convenient property, but not necessary. We could just as well use a nonlinear device such as a piece of bungee cord.

**FIGURE 3-8** ■ An electronic force sensor that can be used with a computer data acquisition system. When the hook at the bottom is pushed or pulled, a metal element is compressed or flexed. This is detected by an electronic strain gauge, which puts out a voltage proportional to force.

**READING EXERCISE 3-4:** A typical rubber band does not obey Hooke's law. However, it can be used as a force scale if not stretched to its limit. Describe how you might use a properly calibrated spring scale, like one of those shown in Fig. 3-7, to create a device that uses the elasticity of a rubber band to measure force. ■

## 3-5 Defining and Measuring Mass

We know from experience that if we push steadily on a wheelbarrow it is much harder to get it moving when it's full than when it's empty. We also know that it is much harder to lift a wheelbarrow when it's full. We can summarize these observations with the statement that a large amount of stuff is harder to move than a small amount of stuff. But how do we measure how much larger "an amount of stuff" on a loaded wheelbarrow is than on an unloaded one? Suppose we pile our wheelbarrow with a huge mound of hay and try to lift it or pull on it. What happens if we replace the hay with a relatively small lead brick? How much hay is the same amount of stuff as a small lead brick? How do we know?

In Section 1-2 we introduced the term mass as a measure of "amount of stuff" and stated that quantities are defined by the procedures used to measure them. In the last section we defined force in terms of basic procedures for measuring it. In this section we will do the same for mass. We introduce two quite different procedures for measuring mass based on two questions: How hard is it to lift a certain pile of stuff? And how hard is it to accelerate the pile of stuff with a standard force?

### Measuring Gravitational Mass

As we mentioned in Chapter 1, the most common historical procedure for measuring mass is to compare the effect of the gravitational forces on two objects using a balance. As early as 5000 B.C.E., ancient Egyptians used the equal arm balance for comparing masses to a standard mass (Fig. 3-9).

**FIGURE 3-9** ■ An old fashioned balance is used to measure gravitational mass using 1 and then 2 replicas of a standard 1 kg mass. So the sphere has a gravitational mass of 1 kg. The cart loaded with extra mass has a gravitational mass of 2 kg.

We assume that two objects have the same mass if they balance with each other. If two masses balance, they are experiencing the same gravitational force. The mass of replicas of the standard 1 kg mass are adjusted using a balance. We can create a mass scale by assuming that masses add so that two replicas of the standard 1 kg mass have a combined mass of 2 kg, and so on. We can also create masses that are fractions of a

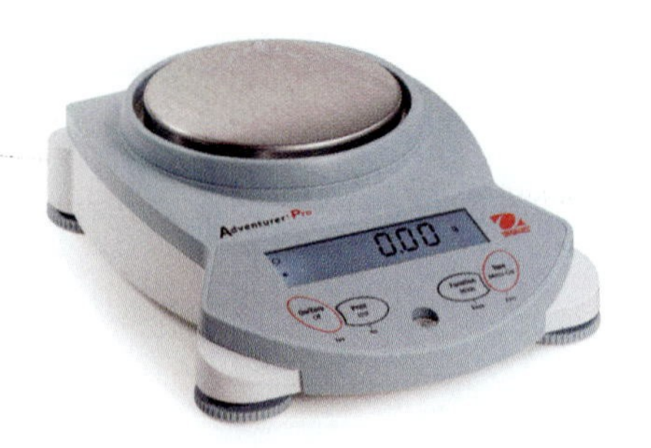


**FIGURE 3-10** ■ A modern electronic balance uses an internal electronic strain gauge to measure gravitational mass. Although the principle on which it works is not obvious, it gives the same result as spring scales do.

kilogram. For example, we can create 1/2 kg masses by creating two less massive objects that balance with each other, but combine to balance with a standard 1 kg mass. Because this procedure for determining a mass involves balancing gravitational forces, we call this type of mass *gravitational mass*.

In modern laboratories, triple beam balances, spring scales, and electronic scales (Fig. 3-10) are used instead of the old-fashioned balance for measuring gravitational mass. As the Earth attracts a mass hanging from a spring, the spring will stretch. A mass on an electronic scale causes an electrical strain gauge to compress.

## Measuring Inertial Mass

As we mentioned, another "measure" of how much stuff we have is to observe how hard it is to get an object moving, or accelerate it, with a known force. We know that by definition a 1 N force will cause a standard 1 kg mass to accelerate at 1 m/s$^2$. In general, when $m = 1$ kg, the magnitude of the acceleration is the same as that of the force. What happens to the relationship between a single force and acceleration when the mass is different from the standard mass?

If we set up a system to measure acceleration and force, such as the computer interface system shown in Fig. 3-11 with an accelerometer and an electronic force sensor attached firmly together, we can study how mass affects the relationship between force and acceleration. We do this by pushing and pulling in a horizontal direction on the force sensor–accelerometer system. We can then tape some additional mass on the system and repeat this procedure.

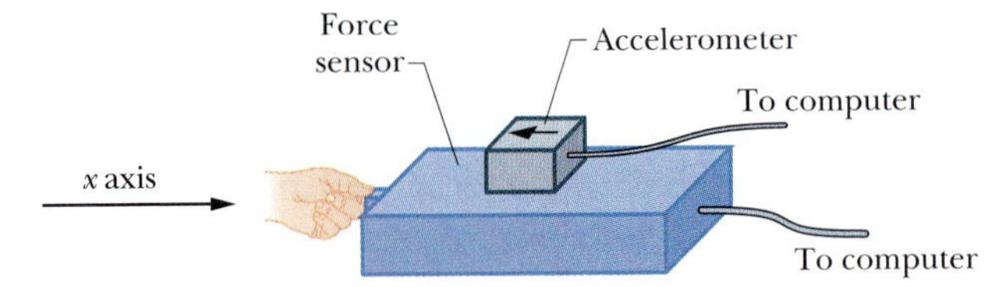


**FIGURE 3-11** ■ Setup showing an electronic accelerometer tracking the acceleration as a function of the forces of a push or pull on a system consisting of itself, a force sensor, and additional mass. The system is held firmly by the hook that is attached to the sensitive area of the force sensor. It is then pushed and pulled horizontally in mid-air with gentle but rather erratic motions.

Figure 3-12 shows graphs of both the $x$-component of force vs. time and the $x$-component of acceleration vs. time for a system that has a *gravitational mass* of 150 g. We find that the force and acceleration components are directly proportional to each other on a moment-by-moment basis. The evidence for this is the fact that the graphs of force vs. time and acceleration vs. time have the same basic shape and are zero at the same times. By the same "graph shape" we mean that if the force is twice as large at one time than another, then so is the acceleration.

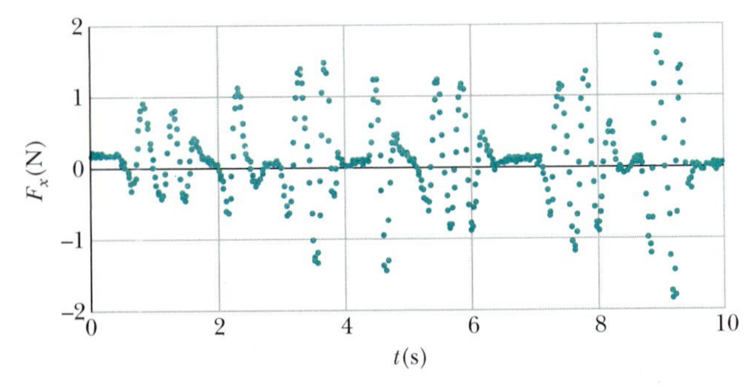


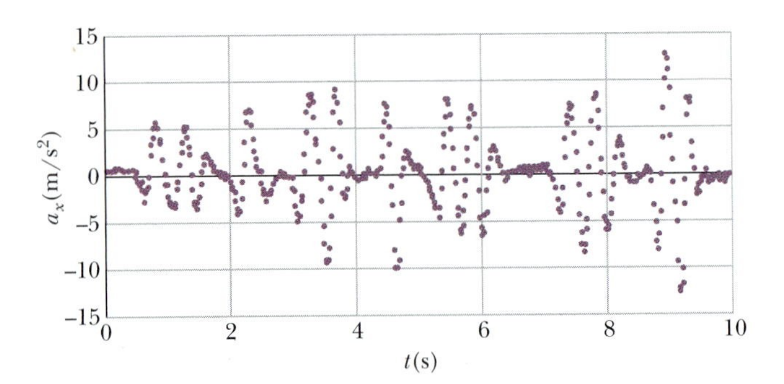


**FIGURE 3-12** ■ Graphs of measured force and acceleration as a function of time. An accelerometer is attached to a force sensor as shown in Fig. 3-11. The combination is being pushed and pulled. Signals from these sensors were sent to a computer via a computer data acquisition interface. The similarity in the shapes of the real-time computer graphs reveal a moment-by-moment proportionality between force and acceleration.

In Figure 3-13 we use the same data displayed in Fig. 3-12 for the 150 g *gravitational mass* to graph the $x$-component of force as a function of the $x$-component of acceleration. The fact that this new graph is a straight line that passes through the origin is additional evidence that there is a direct proportionality between force and acceleration. The constant of proportionality is given by the slope of the graph.

> We define the **INERTIAL MASS** of a system as the constant of proportionality between acceleration and the force that causes it.

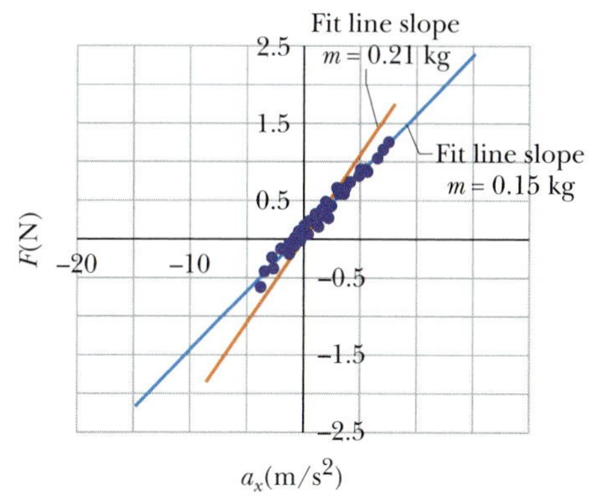


**FIGURE 3-13** ■ Graphs of the horizontal force vs. acceleration components for the accelerometer–force sensor system (shown in Fig. 3-11) as it is being pushed and pulled in a horizontal direction. Two system masses were used, 0.150 kg and 0.200 kg. The resulting slopes for the $F_x$ vs. $a_x$ graphs show a proportionality between force and acceleration with the constant of proportionality being equal, within the limits of experimental uncertainty, to the gravitational mass of the system in each case.

Indeed, we see that if we now do the experiment shown in Fig. 3-13 with a 200 g gravitational mass we get a larger slope. This indicates that when there is more mass it takes more force to get the same acceleration. Perhaps the most interesting feature of Fig. 3-13 is that the *inertial masses* measured as the slopes of the $F_x$ vs. $a_x$ graphs are the *same* as the values of the *gravitational masses* measured with a balance—at least within the limits of experimental uncertainty.

The inertial mass of an object tells us how much it resists acceleration, whereas the gravitational mass is a measure of how hard the Earth pulls on an object. Sophisticated experiments involving precise measurements of the gravitational forces between two objects in a laboratory using a device known as a Cavendish balance have shown that there is no difference between the two types of mass to within less than one part in $10^{12}$. Since the two types of mass seem to have the same values, we will drop the distinction between them and just refer to mass.

## 3-6 Newton's Second Law for a Single Force

The general relationship between force, mass, and acceleration discussed in Section 3-5 is known as Newton's Second Law. By pulling together conclusions we have reached so far, we will state this law for the case of a single force that acts alone along a line. We will then proceed, in this chapter and those that follow, to show that this law is also valid when more than one force acts and when forces act in two dimensions.

How is the acceleration of a body related to its mass and the force acting on it? The experimental evidence presented in Figs. 3-12 and 3-13 shows that the acceleration of a body is *directly proportional* to the force acting on it. The experimental results in Fig. 3-13 show acceleration to be *inversely proportional* to the mass of the body. That is, looking at the graph, we can see that for a given force acting on a body, the acceleration imparted to it is less when the body's mass is large than when it is small. Combining these two relations with our definition of the unit force as producing a unit acceleration of a unit mass, we can summarize what we now know in the single equation

$$\vec{a} = \frac{\vec{F}}{m}. \tag{3-1}$$

The arrows shown in Eq. 3-1 serve as a reminder that we believe that both force and acceleration are vector quantities that have magnitude and direction. The force on a body and acceleration caused by it are in the same direction. Mass is a scalar quantity that does not have a direction associated with it. Newton's Second Law can also be put in words:

> **NEWTON'S SECOND LAW FOR A SINGLE FORCE:** When a single force acts on an object, it will cause the object to accelerate in the direction of the force. The amount of acceleration is given by the acting force divided by the object's mass.

Because it is easier to write, the most common way to refer to Newton's Second Law is in the form

$$\vec{F} = m\vec{a}. \tag{3-2}$$

If the force lies along the $x$ axis, then $\vec{F} = F_x\,\hat{\text{i}}$ and $\vec{a} = a_x\,\hat{\text{i}}$. So we can also express the Second Law in terms of the force and acceleration components as $F_x = ma_x$.

Equations 3-1 and 3-2 represent an interesting combination of definitions and a law of nature. In both Eqs. 3-1 and 3-2, the equality sign does not mean that the two sides of an equation are the same physical quantities or that force is defined as the product of mass and acceleration. Rather, Eq. 3-2 provides a method for *predicting* the acceleration of an object when its mass and the force acting on it are known. Alternatively, Eq. 3-2 tells us that a measurement of acceleration and mass can be used to *determine* the force on a body that is causing it to accelerate.

For standard SI units, $\vec{F} = m\vec{a}$. tells us that

$$1\text{ N} = (1\text{ kg})(1\text{ m/s}^2) = 1\text{ kg}\cdot\text{m/s}^2. \tag{3-3}$$

Force units common in other systems of units are given in Appendix D.

So far we have been studying the relationship between motion and force under very limited circumstances. We have restricted our study to forces acting along a line in an inertial reference frame. We have also restricted ourselves to observations in which we think that the applied force acting on an object, such as a low-friction cart, is the only significant interaction the object is experiencing. By applying this rather unrealistic set of restrictions, we were able to formulate initial definitions of force and mass. We then combined these definitions with observations to develop two of Newton's three laws of motion.

As we already suggested, Newton's first two laws are not simply valid by definition in the way that $\vec{v} \equiv d\vec{x}/dt$ or $\vec{a} \equiv d\vec{v}/dt$ are. Rather, they represent a combination of definitions and natural laws. Can we refine these laws so they are valid in more complicated situations that describe forces and motion along a line? In particular, what happens when more than one force is acting at the same time? How do forces combine? What other forces besides the forces we apply can act on a body? What evidence is there that these forces are real? When forces are obviously due to interactions between two or more bodies, does a body acted upon also exert forces on the body acting on it? How are these related? The rest of this chapter will be devoted to dealing with these questions. Chapters 5 and 6 will deal with how to use Newton's laws to predict motions that result from forces that act in two and three dimensions.

## The Flea Pulling a Train

Let us return to the question we asked at the beginning of the chapter. How can a jumping flea with a tiny mass pull an object that is 160,000 times more massive? When it comes to jumping, insects have a big advantage over larger animals. The strength of their legs increases as the square of the diameter of their legs while the mass that they push off with goes as the cube of their body dimensions. Thus the ratio of the mass they lift with their legs to their cross-sectional area is much smaller than it is for a large animal. While a world-class high jumper can barely jump his or her own height, a flea can jump up to 150 times its own height. So a 2-mm-tall flea can jump to a height of about 30 cm.

The flea's secret is that he can launch himself at a high speed. Suppose Brutus, whose mass is only about $2 \times 10^{-3}$ g, starts a 30-cm high hop that takes him up and forward at the same time. Our flea will be moving at a pretty high horizontal speed. Using kinematic equations, we can estimate its initial hopping speed to be over 2 m/s.

Before the flea completes his hop, it will be rudely interrupted as he comes to the end of the wire. The wire begins to stretch and the wire then pulls Brutus to a sudden stop. But the force that Brutus exerts on his end of the wire while he is being stopped will be transmitted along the wire to the train. This causes the train, which has a mass of 32 g, to jerk forward. While Brutus is falling down, the friction in its wheels causes it to roll to a stop also.

Because Brutus is not pulling with a steady force, it is difficult to make detailed calculations of the motion of the train he is pulling on. Instead, in Touchstone Example 3-1 we calculate what happens to a man who pulls steadily on a pair of real passenger cars.

**READING EXERCISE 3-5:** A student sitting on a skateboard is pulled with a horizontal force to the left of magnitude 26 N and accelerates at 0.42 m/s$^2$. (a) Write the expressions for force and acceleration in vector notation using the unit vector. (b) What is the combined mass of the student and skateboard? (c) The mass of a student and her skateboard is measured using a European bathroom scale calibrated to read in kilograms. What is the scale reading? ■

**READING EXERCISE 3-6:** Consider your answers to Reading Exercise 3-5. (a) Which mass measurement is a determination of inertial mass, the one made in part (b) or part (c)? Explain. (b) What assumption did you make in determining your answer to part (c)? ■

## TOUCHSTONE EXAMPLE 3-1: Pulling a Train

John Massis is shown in the photo pulling two passenger cars by applying a steady force to them at an angle of about 30° with respect to the horizontal. Assume instead that Massis had pulled the two cars of mass $8.0 \times 10^4$ kg with a horizontal force of $2.0 \times 10^3$ N. If there was no friction in the rails, what speed would the cars have after Massis moves them a distance of 1.0 m from their resting location?

In 1974, John Massis of Belgium managed to move two passenger cars belonging to New York's Long Island Railroad. He did so by clamping his teeth down on a bit that was attached to the cars with a rope and then leaning backward while pressing his feet against the railway ties. The cars together weighed about 80 tons, which is almost 1000 times more than the man's mass.

**SOLUTION** ■ A **Key Idea** here is that, from Newton's Second Law, the constant horizontal pulling force on the cars that Massis exerts causes a constant horizontal acceleration of the cars. Because the force is constant, and the motion is assumed to be one-dimensional, we can use the kinematic equations to find the horizontal velocity component $v_{2x}$ at location $x_2$ (where $x_2 - x_1 = +1.0$ m).

Place an $x$ axis along the direction of motion, as shown in Fig. 3-14. We know that the initial velocity component along the horizontal axis $v_{1x}$ is 0, and that the displacement $x_2 - x_1$ is $+1.0$ m. However, we need to find the $x$-component of acceleration, $a_x$.

We can relate the $x$-component of the acceleration of the cars, $a_x$, to the pulling force on the cars from the rope by using Newton's Second Law. If we assume there are no friction forces, we can note that a single pulling force acting along the *horizontal axis* in Fig. 3-14 is

$$F_x^{\text{pull}} = Ma_x \tag{3-2}$$

where $M$ is the mass of the cars and $F_x^{\text{pull}}$ and $a_x$ are the $x$-components of the force and acceleration vectors.

In Fig. 3-14, we see that Massis is pulling in the $x$-direction, so $F_x^{\text{pull}} = 2.0 \times 10^3$ N. Since the mass of the railroad cars, $M$, is $8.0 \times 10^4$ kg we can find $a_x$ by rearranging Eq. 3-2 and substituting for $F_x^{\text{pull}}$ and $M$. The acceleration component becomes

$$a_x = \frac{F_x^{\text{pull}}}{M} = \frac{2.0 \times 10^3\ \text{N}}{8.0 \times 10^4\ \text{kg}} = 0.025\ \text{m/s}^2.$$

Next we use Eq. 2-13

$$v_{2x} = v_{1x} + a_x(t_2 - t_1) = v_{1x} + a_x\Delta t \tag{3-4}$$

to find the velocity of the train after it has moved 1.0 m. Since $v_{1x} = 0$, we find that $v_{2x} = a_x\Delta t$ in this case.

To find $\Delta t$, we can use the fact that the train's average velocity is

$$\langle v_x \rangle = \frac{\Delta x}{\Delta t} = \frac{v_{1x} + v_{2x}}{2},$$

then solve this equation for $\Delta t$. Again using $v_{1x} = 0$, this yields $\Delta t = \Delta x/(v_{2x}/2)$. By substituting $\Delta t$ back into Eq. 3-4, we find that $v_{2x} = a_x \Delta x/(v_{2x}/2)$ or, more simply,

$$v_{2x} = \sqrt{2a_x \Delta x} = \sqrt{(2)(0.025\text{m/s}^2)(1.0\text{ m})} = 0.22\text{m/s}. \quad \text{(Answer)}$$

We assumed in this calculation that the force Massis exerted on the railroad cars was horizontal. Actually his pull was not quite horizontal. This made his job harder. Can you see why?

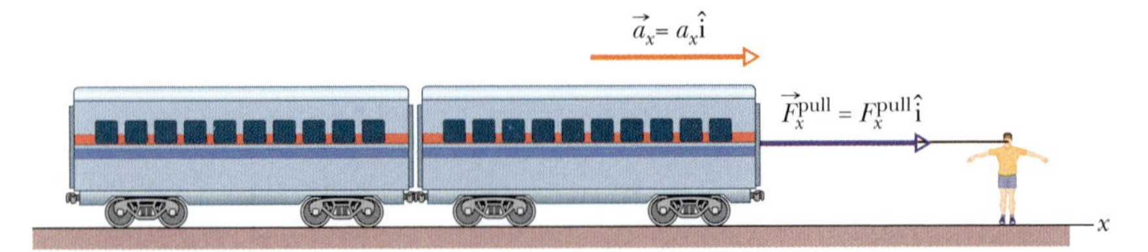


**FIGURE 3-14** ■ Force diagram for the passenger cars attached to a rope. The rope is pulled by Massis with his teeth. We assume that Massis was pulling horizontally in a positive $x$-direction.

## 3-7 Combining Forces Along a Line

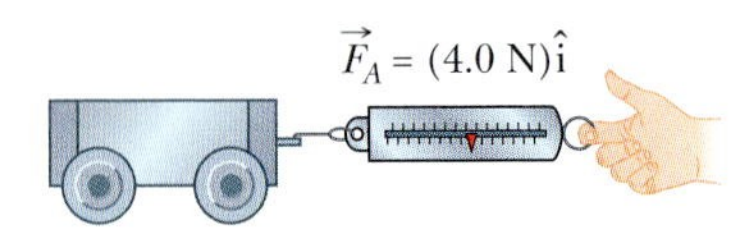


**FIGURE 3-15** ■ Pulling to the right on a low-friction cart with a force of +4.0 N. If the cart has a mass of 0.50 kg, then it will accelerate to the right at +8.0 m/s$^2$.

We have discussed how a single applied force, such as a push or pull, affects the motion of an object. Now let's go one step further and think about what happens if a second applied force also acts on a body.

Suppose that you have a spring attached to a low friction cart like that shown in Fig. 3-15. You pull on the cart using the spring, keeping the spring constantly stretched to produce a constant force of magnitude 4.0 N to the right. Since you are applying a constant force to the cart, it will speed up with a constant acceleration. That is, the cart's velocity will increase at a constant rate.

What do you think would happen if a friend simultaneously pulled on the cart in the same manner, with the same magnitude of force, but in the opposite direction as shown in Fig. 3-16? Would the cart still accelerate? Clearly the answer is "no." How is the motion of the cart affected if you and your friend each apply a 2.0 N force to the cart in the *same direction*? Measurement reveals two things. First, the acceleration produced by a single 4.0 N force is twice that produced by a single 2.0 N force. Second, a single 4.0 N force produces the same acceleration as two 2.0 N forces applied in the same direction, as shown in Fig. 3-17.

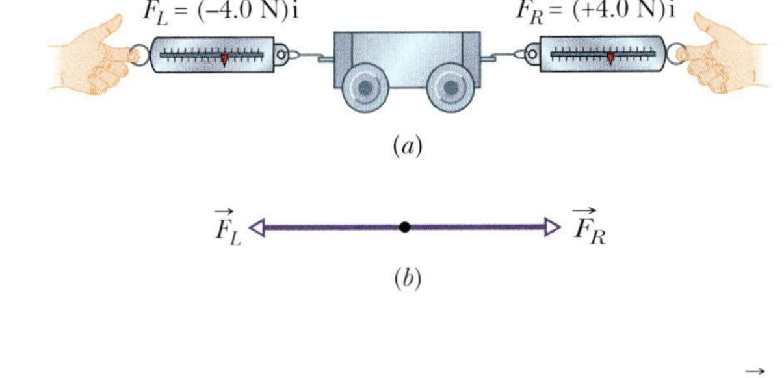


**FIGURE 3-16** ■ (*a*) Pulling to the right on a low-friction cart with a force as someone pulls to the left with the same magnitude of force. Thus, $\vec{F}_R + \vec{F}_L = 0$, so the forces cancel and the cart doesn't move. (*b*) A simple diagram representing the forces acting on the cart. Such a diagram is called a free-body diagram.

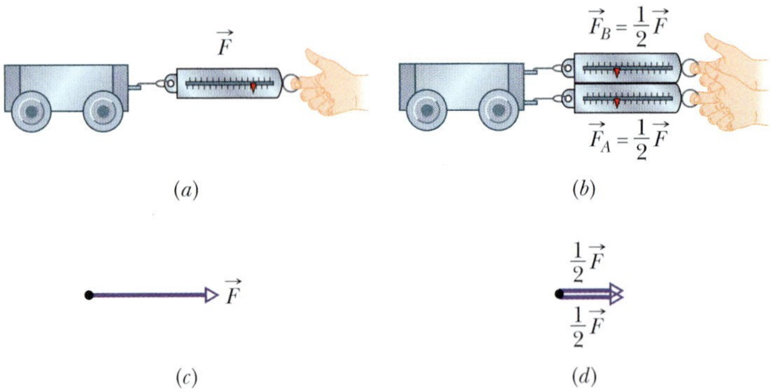

**FIGURE 3-17** ■ (*a*) and (*b*): Pulling to the right on a low-friction cart with one force yields the same acceleration on it as two forces pulling to the right do when each has half the magnitude of the single force. (*c*) and (*d*): Free-body diagrams of situations (*a*) and (*b*).

These observations indicate that keeping track of the magnitudes and directions of all the forces acting on an object is very important if we want to be able to make predictions about the object's subsequent motion. A special type of diagram, called a **free-body diagram,** is an especially useful technique for doing this. Figures 3-16*b* and 3-17*c* and *d* show free-body diagrams that represent various situations.

We construct a free-body diagram by representing each object we are investigating as a point. For example, in Fig. 3-17, we are interested in the motion of the cart (not the hand) and so we represent the cart as a point. We then draw a force vector (as an arrow) for each force acting on the object. We place the tail of each force vector (arrow) on the point and draw the vector in the direction of the force. The relative magnitude of the forces is represented by the relative lengths of the arrows. Hence, the two equal force vectors in Fig. 3-17 are shown to have the same length. Finally, we label the force vectors so that we know which force each arrow represents.

Free-body diagrams help us to translate pictures or statements of a situation into mathematical expressions. That is, they help us to generate mathematical expressions in which we treat a force as a vector quantity with both a magnitude and a direction. As we discussed in Section 3-4, by normal convention, a horizontal force directed to the right has a positive $x$-component and one directed to the left has a negative $x$-component. Thus, each of the one-dimensional vectors we discussed can be represented as the combination of its magnitude and direction as follows:

Two 4.0 N forces acting in opposite directions:

$$\vec{F}_A = F_{Ax}\hat{\mathrm{i}} = (+4.0\ \mathrm{N})\hat{\mathrm{i}} \qquad \vec{F}_B = F_{Bx}\hat{\mathrm{i}} = (-4.0\ \mathrm{N})\hat{\mathrm{i}}$$

Two 2.0 N forces acting in the same direction:

$$\vec{F}_A = F_{Ax}\hat{\mathrm{i}} = (+2.0\ \mathrm{N})\hat{\mathrm{i}} \qquad \vec{F}_B = F_{Bx}\hat{\mathrm{i}} = (+2.0\ \mathrm{N})\hat{\mathrm{i}}$$

The plus or minus sign carried with the vector components to denote the direction of vectors makes it easy for us to remember in what direction the forces and acceleration point along a chosen $x$ axis. The signs make it possible to combine forces mathematically using the rules of vector mathematics. As long as we denote direction with signs as we did above, we can determine the combined effect of multiple forces acting on an object simply by adding up the force components acting along a single line. For example, in the case of our two forces that are applied in opposite directions, we can determine the combined force, usually called the vector sum of the forces or **net force,** by calculating the vector sum, so that

$$\vec{F}^{\,\mathrm{net}} = \vec{F}_A + \vec{F}_B = F_{Ax}\,\hat{\mathrm{i}} + F_{Bx}\,\hat{\mathrm{i}} = (F_{Ax} + F_{Bx})\,\hat{\mathrm{i}} = (+4.0 - 4.0\ N)\,\hat{\mathrm{i}} = 0.$$

The net force or vector sum of the forces for the situation depicted in Fig. 3-17*b* can be calculated as

$$\vec{F}^{\,\mathrm{net}} = (F_{Ax} + F_{Bx})\hat{\mathrm{i}} = [+2.0\ \mathrm{N} + 2.0\ \mathrm{N}]\hat{\mathrm{i}} = (+4.0\ N)\hat{\mathrm{i}}.$$

When the forces do not have the same magnitude or direction, we can still use vector sums. For instance, consider a 3 newton force to the right, denoted by its $x$-component of +3 N, and a 2 newton force to the left, denoted by its $x$-component of –2 N. These two forces combine to give a net force of

$$\vec{F}^{\,\mathrm{net}} = \vec{F}_A + \vec{F}_B = (+3\ \mathrm{N} - 2\ \mathrm{N})\hat{\mathrm{i}} = (+1\ \mathrm{N})\hat{\mathrm{i}}.$$

This means that part of the influence of the 3 N force to the right is counteracted by the application of a 2 N force to the left. In the end, an object with these two forces acting on it behaves as if only a 1 N force, directed to the right, is present. So,

> When two or more forces act on a body, we can find their net force or resultant force by adding the individual forces as vectors taking direction into account.

A single force with the magnitude and direction of the net force has the same effect on the body as all the individual forces together. This fact is called the **principle of superposition for forces.** The world would be quite strange if, for example, you and a friend were to pull on the cart in the same direction, each with a force of 5 N, and yet somehow the net pull was 20 N.

In this book a net force is represented with the vector symbol $\vec{F}^{\text{net}}$. Instead of what was previously given, the proper statement of Newton's First and Second Laws should now be rephrased in terms of net forces.

> **NEWTON'S FIRST LAW:** Consider a body on which no net force acts so that $\vec{F}^{\text{net}} = 0$. If the body is at rest, it will remain at rest. If the body is moving, it will continue to move with a constant velocity.

This statement means that there may be multiple forces acting on a body, but if the net force (the vector sum of the forces) is zero, then the body will not accelerate. Remember, this doesn't mean that the object is stationary. It simply means that the object will not speed up or slow down.

We can also rewrite Newton's Second Law in terms of net force.

> **NEWTON'S SECOND LAW FOR MULTIPLE FORCES:** The acceleration of a body is the *net* force acting on the body divided by the body's mass.

This statement can be expressed mathematically by replacing the force in Eq. 3-1 with net force, so that

$$\vec{a} = \frac{\vec{F}^{\text{net}}}{m} \quad \text{(Newton's Second Law).} \tag{3-5}$$

Once again, because it is easier to write down, a common way to write Newton's Second Law for multiple forces in vector form is

$$\vec{F}^{\text{net}} = m\vec{a} \quad \text{or} \quad F_x^{\text{net}} = ma_x,$$

where

$$\vec{F}^{\text{net}} = \vec{F}_A + \vec{F}_B + \vec{F}_C \cdots \vec{F}_N.$$

Hence, if we want to know the acceleration of an object on which more than one force acts, we can find it using the following procedure:

1. Draw a free-body diagram for the object of interest.
2. Determine the net force acting on the object.
3. Take the ratio of the net force to the mass of the object.

This procedure is used in the examples that follow. You will find it useful in completing many of the end-of-chapter problems as well.

**READING EXERCISE 3-7:** The figure shows two horizontal forces moving a cart along a frictionless track. Suppose a third horizontal force $\vec{F}_C$ could act on the cart. What are the magnitude and direction of $\vec{F}_C$ when the cart is (a) not moving and (b) moving to the left with a constant speed of 5 m/s?

$\vec{F}_A = (-1\text{N})\hat{\text{i}}$ $\quad$ $\vec{F}_B = (+3\text{ N})\hat{\text{i}}$

**READING EXERCISE 3-8:** The figures that follow show overhead views of four situations in which two forces accelerate the same cart along a frictionless track. Rank the situations according to the magnitudes of (a) the net force on the cart and (b) the acceleration of the cart, greatest first.

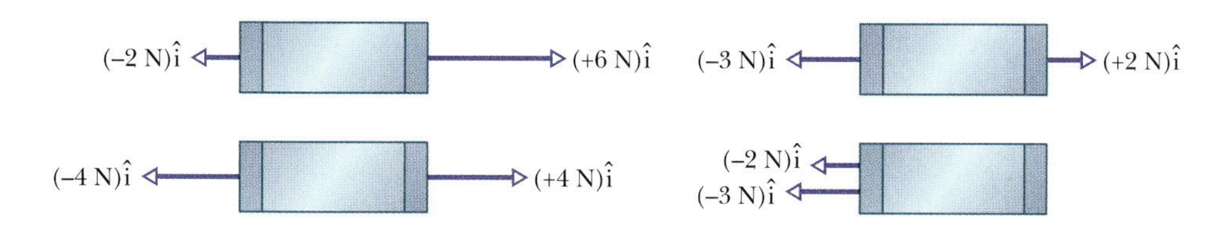


## TOUCHSTONE EXAMPLE 3-2: Three Forces

In the overhead view of Fig. 3-18, a 2.0 kg cookie tin is accelerated at 3.0 m/s² in the direction shown by $\vec{a}$, over a frictionless horizontal surface. The acceleration is caused by three horizontal forces, only two of which are shown: $\vec{F}_A$ with a magnitude of 10 N and $\vec{F}_B$ with a magnitude of 20 N. Choose a coordinate system and then use it to express the third force $\vec{F}_C$ in unit-vector notation.

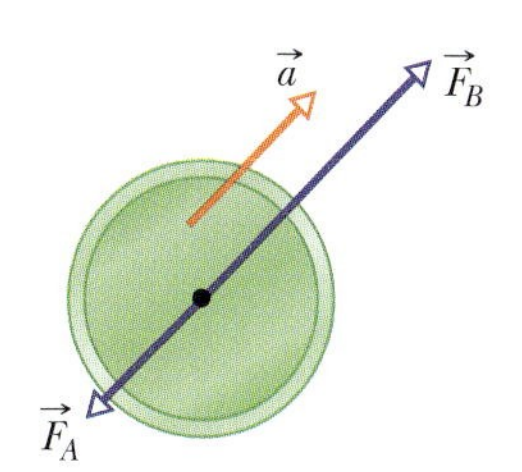


**FIGURE 3-18** ■ Three forces act to produce an acceleration in the direction shown. Only two of the three forces causing this acceleration are included in this picture.

**SOLUTION** ■ The **Key Idea** here is that the net force $\vec{F}^{\,\text{net}}$ on the tin is the sum of the three forces and is related to the acceleration $\vec{a}$ of the tin via Newton's Second Law ($\vec{F}^{\,\text{net}} = m\vec{a}$). Thus,

$$\vec{F}_A + \vec{F}_B + \vec{F}_C = m\vec{a},$$

which gives us

$$\vec{F}_C = m\vec{a} - \vec{F}_A - \vec{F}_B. \tag{3-6}$$

A second **Key Idea** is that this is a one-dimensional problem for which two of the forces and the acceleration are all along the same line. This means that the third force must also lie along the line of the acceleration. Thus we are able to choose a coordinate system in which the three forces lie along a single axis. If we choose our $x$ axis to align with these forces, we have

$$\vec{F}_C = m\vec{a} - \vec{F}_A - \vec{F}_B = (ma_x)\hat{\text{i}} - F_{A\,x}\hat{\text{i}} - F_{B\,x}\hat{\text{i}}$$
$$= (ma_x - F_{A\,x} - F_{B\,x})\,\hat{\text{i}}.$$

Choosing the positive direction to be in the direction of the acceleration, components $a_x$ and $F_{B\,x}$ are positive and the component $F_{A\,x}$ is negative. Thus $\vec{a} = (+3.0\text{ m/s}^2)\hat{\text{i}}$, $\vec{F}_B = (+20\text{ N})\hat{\text{i}}$, and $\vec{F}_A = (-10\text{ N})\hat{\text{i}}$.

Then, substituting known data and factoring $\hat{\text{i}}$ out of the equation, we find

$$\vec{F}_C = [(2.0\text{ kg})(3.0\text{ m/s}^2) - (-10\text{ N}) - (+20\text{ N})]\hat{\text{i}} = (-4\text{ N})\hat{\text{i}}. \quad \text{(Answer)}$$

## 3-8 All Forces Result from Interaction

Careful observation of everyday motions should convince you that objects do not spontaneously speed up, slow down, or change direction. Clearly, pushes, pulls, bumps, winds, interactions with a surface during sliding motion, and so on, will influence an object's velocity by changing the object's speed, direction, or both. According to Newton's Second Law, changes in velocity (accelerations) occur only when the object experiences forces. Forces are always due to the presence of one or more other objects.

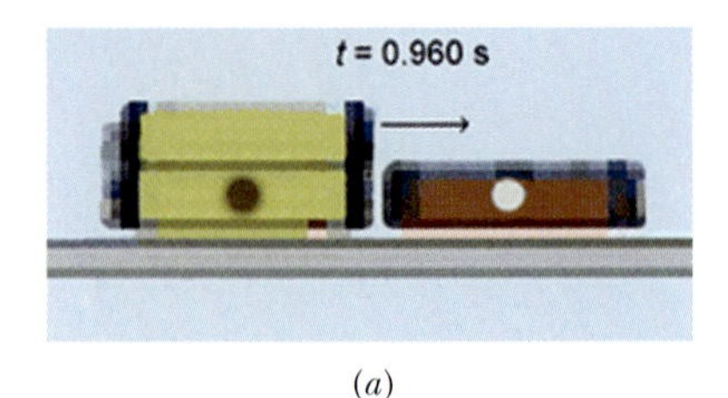


(a)

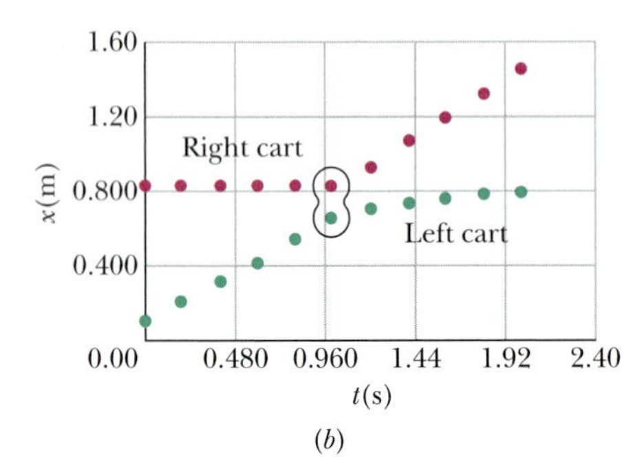


(b)

**FIGURE 3-19** ■ Two low-friction carts are outfitted with neodymium magnets that repel each other. Initially the cart on the left bears down on the stationary cart on the right. (*a*) A video frame shows the carts interacting briefly at about 0.960 s but never touching. (*b*) Graphs of position vs. time for the two carts were obtained using video analysis. An examination of the changes in the slopes for each cart representing their velocity components enables us to deduce that the carts undergo velocity changes due to forces acting in opposite directions. A force to the left on the large cart slows it down. A force to the right on the small stationary cart starts it moving to the right.

In the course of your study of physics, you will be reading and hearing about dozens of forces. Adjectives such as net, combined, total, friction, contact, collision, normal, tension, spring, gravitational, electrostatic, magnetic, atomic, molecular, and so on, will be bandied about. It turns out that currently there are only four fundamental forces that are known: gravitational, electromagnetic, weak nuclear, and strong nuclear. However, essentially all of the types of forces introduced in this book (including "friction forces," "contact forces," and "collision forces") are actually fundamental forces (either electromagnetic or gravitational.) These other descriptive adjectives are used only to help us understand the physical situation in which various forces occur.

For example, in the first experiment we presented in this chapter, we tracked the motion of objects that roll or slide to a stop on different horizontal surfaces after a kick. Figure 3-3 showed that the rate of decrease of velocity for each of these objects was a constant. In other words, each object experienced a constant acceleration until it came to rest (or, in the case of the cart on the track, until data were no longer collected). What causes the objects to slow down? Consider Newton's Second Law and our definition of force (as an agent that causes an acceleration). We must conclude that each of the objects experienced a force. The only obvious interactions are interactions with the surface along which it was sliding or rolling. We call this type of contact interaction a friction force or, informally, friction. We found that in these cases the direction of the friction force on an object is opposite to the direction of the object's motion. We know this because the object slows down.

Forces like pushes, pulls, those experienced in a collision, and the friction force on a sliding or rolling object that moves over a surface are called **contact forces** because the objects involved appear to touch. The interactions that cause contact forces are ultimately due to a superposition of many small electromagnetic forces between the electrons and protons that the materials "in contact" are made of. Thus, contact forces are ultimately electromagnetic forces!

Another important contact force is a pull force exerted through a string, rope, cable, or rod attached to an object. This type of pulling force has a special name. It is called **tension.** Tension is always a pull force. Hence, the direction of a tension force is always the direction in which one would *pull* the object with a string or rope. The fundamental nature and origin of tension forces and frictional forces are discussed in more detail in Chapter 6.

Many other forces seem much less obvious than contact forces because they act at a distance. Electromagnetic and gravitational forces are capable of acting over large distances. But as you will learn in this chapter and later in this book, the source of these invisible or noncontact forces are not totally mysterious. If an everyday object experiences an "invisible force," we are always able to find it interacting with other objects that have some combination of electrical charges, magnets, electrical currents, or masses. An example of this is shown in Fig. 3-19, where two carts interact "at-a-distance" by means of magnetic forces that act in opposite directions.

**READING EXERCISE 3-9:** Consider Fig. 3-3 depicting the results of measurements on the motion of objects just after they have been given swift kicks along a positive $x$ axis. Assume that the cart and the box both have the same mass of 0.5 kg. (a) What is the acceleration of the box on the carpet? Is it positive or negative? (b) What is the acceleration of the cart on the track? Is it positive or negative? ■

**READING EXERCISE 3-10:** Consider your answers to Reading Exercise 3-9. Assume that the cart and the box both have the same mass of 0.5 kg. (a) In each case is the friction force on the object constant or changing as the box or cart slows down? Cite evidence for your answer. (b) What is the magnitude of the friction force on the box due to its interaction with the carpet? Does it point to the right or the left? (c) What is the magnitude of the friction force on the cart due to the combined interaction of the cart wheels with the track and the cart axle? Does it point to the right or the left? ■

## 3-9 Gravitational Forces and Free Fall Motion

We now consider the forces the Earth exerts on objects near its surface. These forces are called "gravitational forces." Since we don't want to complicate our exploration with air resistance, we will limit ourselves to considering the motions of bodies that are relatively dense, small, and smooth like balls and coins. Also, assume that these objects are not moving at high speeds—say, in excess of about 5 m/s. In Chapter 6 we discuss situations where air resistance is a significant factor, and in Chapter 14, we explore the question of how masses such as galaxies and planets exert gravitational forces on each other in more general circumstances.

**FIGURE 3-20** ■ This strobe photo shows a feather and an apple, undergoing free fall in a vacuum. The time interval between each exposure and the next is constant. The feather and the apple appear to be speeding up at the same rate, as evidenced by the increase in distance between the successive images.

### The Gravitational Acceleration Constant

We know that any object dropped near the Earth's surface falls, but the fall is so rapid that we can't easily describe it. Does the object suddenly speed up to a natural velocity and then fall at that rate or does it keep speeding up? A casual observation tells the story. Imagine lying on the floor while someone drops an apple on your forehead from different heights. The impact of the apple will feel harder when the apple is dropped from a greater height, so the apple must keep speeding up. The strobe photo in Fig. 3-20 confirms that an apple and a feather falling in a vacuum keep speeding up.

At the end of Section 2-5 we asserted without any evidence that if there are no other forces on an object, it would move *downward* with a magnitude of acceleration $a = 9.8 \text{ m/s}^2$. But how do we know this? Back in the early 17th century Galileo rolled small balls of different masses down a ramp to slow their falling rates. He found that the velocities of all the balls increased at the same rate.

Today we can use modern technology such as ultrasonic motion detectors, video analysis, and strobe photos to make high-speed measurements of the position and time of an object falling straight down like the tossed ball in Fig. 3-21. For example, Figure 3-22 shows an analysis of a video clip of a small plastic ball shot vertically into the air with a spring-loaded launcher. The graph of the $y$-component of velocity vs. time was produced assuming that the $y$ axis is pointing up. The graph shows that the ball is changing its velocity at the same constant rate when the ball is moving upward, turning around and moving downward. The measured acceleration component is the slope of the $V_y$ vs. $t$ graph, and a linear fit to the graph yields a vertical gravitational acceleration component of $-9.8 \text{ m/s}^2$. Within the limits of experimental uncertainty, we obtain the same result if we use a lead ball instead of a plastic one or for that matter any other object that doesn't experience much air resistance.

The magnitude of the acceleration we measured, denoted as $|\vec{a}|$ or a, is known as the **gravitational acceleration constant** given by $a = 9.8 \text{ m/s}^2$. If air resistance is significant, as is the case for a feather or sheet of paper falling through air, we will not obtain the gravitational acceleration constant from an experiment like the one we just described. However, the fact that the gravitational acceleration is independent of an object's mass, density, or shape can be verified by removing air in the vicinity of a falling object. In Fig. 3-20, a feather and an apple are shown accelerating downward at the same rate in a vacuum in spite of the fact that they have very different masses, shapes, and sizes.

**FIGURE 3-21** ■ A ball of arbitrary mass is tossed in the air near the surface of the Earth. What is its acceleration?

### Gravitational Force and Mass Revisited

Since acceleration requires a net force, and the Earth doesn't need to touch an object to make it accelerate, we conclude that "gravity" is a *noncontact* force. Another piece of evidence that this force of attraction exists is that if you hang an object vertically from the spring force scale we developed in Section 3-4, the spring will be stretched. The stretch of the spring implies that there is something pulling down on the object and that the spring stretches just enough to pull the object up with the same magni-

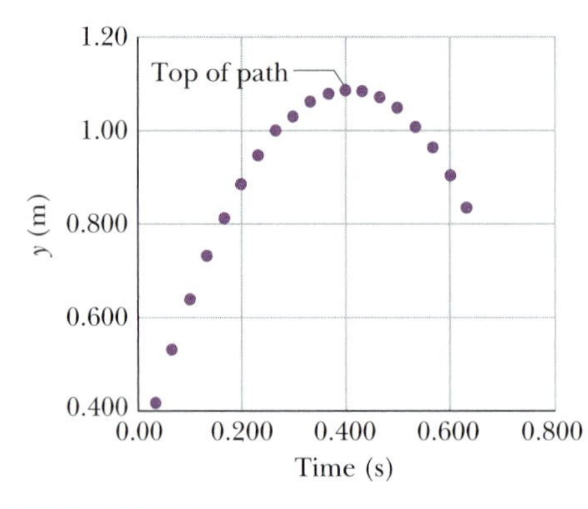


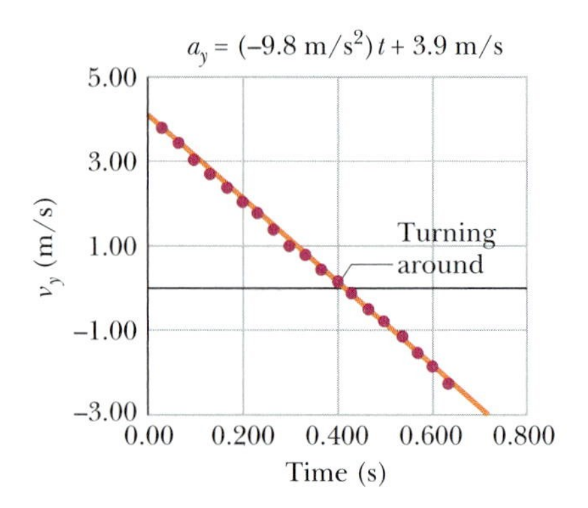


**FIGURE 3-22** ■ Video analysis software is used to perform a frame-by-frame analysis of a digital movie depicting the motion of a small tossed ball. Graphs of position vs. time and the calculated average velocity vs. time are shown. A fit of the velocity vs. time graph reveals that the ball undergoes a constant acceleration in the downward direction of magnitude 9.8 m/s$^2$.

tude of force. If we use a spring scale like that shown in Fig. 3-23 to measure the gravitational force on an object, we find that it is directly proportional to the mass. This is not surprising since we know by experience that a bigger mass is harder to lift. We can express the proportionality between mass and gravitational force $\vec{F}^{\text{grav}}$ in terms of the force magnitude as

$$F^{\text{grav}} = mg, \qquad \text{so that } g = \frac{F^{\text{grav}}}{m} \tag{3-7}$$

where this constant of proportionality $g$ is defined as the **local gravitational strength.** The magnitude of the gravitational force $F^{\text{grav}}$ is commonly referred to as **weight.** Up to an altitude of 16 km or so, $g$ can be expressed to two significant figures as

$$g = 9.8 \text{ N/kg} \qquad \text{(the Earth's local gravitational strength).}$$

Newton's Second Law predicts that, for an object of mass $m$ that has no other forces on it except the gravitational force, the object will fall with an acceleration of magnitude

$$a = F^{\text{grav}}/m = 9.8 \text{ m/s}^2 \qquad \text{(gravitational acceleration constant).} \tag{3-8}$$

Thus we see that $a$ and $g$ have the same value and *different* but dimensionally equivalent units. We use m/s$^2$ when describing the gravitational acceleration $a$. We use the units N/kg when describing the local gravitational strength, $g$.

Equation 3-7 tells us that the Earth pulls harder on a larger mass, whereas Eq. 3-8 tells us the larger mass is harder to accelerate. These two mass-dependent effects cancel each other! Thus, near the Earth's surface,

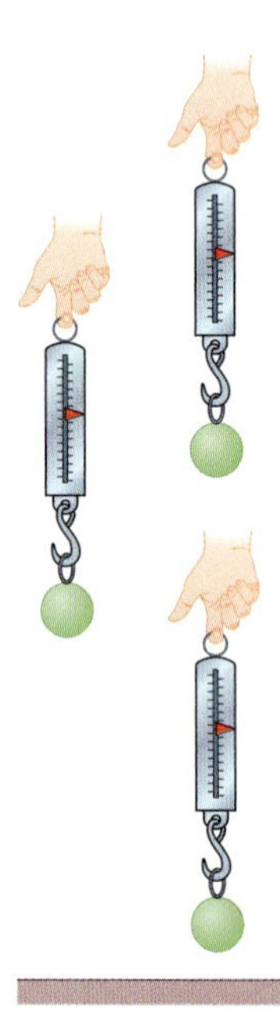

**FIGURE 3-23** ■ Depiction of a spring scale used to determine the gravitational force on an object near the surface of the Earth. The scale reading is essentially the same at all heights reasonably near the Earth's surface (including those found in high flying passenger jets).

> The *magnitude* of the acceleration of any falling object is that of the **GRAVITATIONAL ACCELERATION CONSTANT** 9.8 m/s$^2$, independent of the mass of the falling object.

## Other Properties of the Local Gravitational Force

So what are the characteristics of the gravitational force of attraction exerted by the Earth on objects near its surface? Does this force change as time passes or if the position of the object changes? The answers to these questions become clear if we consider an object hanging vertically from our spring force scale at different times and places.

*Time Dependence*: What we see when performing force measurements with a spring scale is that, for a given object, the amount that the spring stretches changes very little as time passes. Hence, we conclude that, at least over the span of a human life, *the force of gravitational attraction does not appear to be changing over time.*

*Height Dependence*: How does the gravitational force on a given object change with its height above the Earth's surface or its location? The stretch of a spring scale is approximately the same if we are standing at sea level, on top of a table, on top of a tall building, on top of a mountain, or inside of a high-flying passenger jet. This idea is pictured in Fig. 3-23. The gravitational force actually decreases with distance from the surface of the Earth, but the percentage change over the range of elevations that we have described is not measurable to the two significant figures that we have been using to describe $g$. There turns out to be a slight dependence on location and height. But for all heights and locations where people normally travel, the magnitude of the gravitational force of attraction the Earth exerts on another object is the same to two significant figures.

*Direction*: The direction of the gravitational force is apparently down. Since the Earth is approximately spherical, if we look at the Earth's gravitational force from the perspective of outer space, its direction changes from place to place. It will be different in Australia than in the United States.

## Using the Kinematic Equations

Because the constant force of gravity near the surface of the Earth imparts a constant acceleration to objects on which it acts, the kinematic equations of motion derived in Chapter 2 (Table 2-1) can be used to describe free fall near the Earth's surface, but only as long as there are no other nonconstant forces present. The kinematic equations that you worked with in the last chapter describe motion along a line with constant acceleration.

Though the value of $g$ does vary slightly with latitude and elevation, you may safely use a value of 9.8 m/s$^2$ (or 32 ft/s$^2$) in free fall calculations near the Earth's surface as long as air resistance is considered negligible. For many calculations, 10 m/s$^2$ is a convenient approximation, since it varies by only 2% from the more precise value.

When we introduced one-dimensional motion in Chapter 2, we noted that when thinking about the motion of objects, we have freedom to choose our coordinate system. However, to make communicating about these ideas easier, for now we will continue to use a *vertical y* axis that points up as shown in Fig. 3-24. In this coordinate system

$$\vec{F}^{\,\text{grav}} = -mg\,\hat{\text{j}} = ma_y\,\hat{\text{j}}, \tag{3-9}$$

where $a_y$ is the $y$-component of the falling object's acceleration and $\hat{\text{j}}$ the dimensionless unit vector associated with the $y$ axis.

We can easily construct kinematic equations to describe the relationships between vertical vector components for a freely falling object close to the Earth's surface. We simplify writing the equations in Table 2-1 in Chapter 2 by: (1) replacing position component $x$ with the symbol $y$; (2) adding the subscript $y$ to the velocity component to remind us that it is the component of velocity along the $y$ axis; (3) replacing the component of acceleration along the vertical axis that was denoted $a_x$ with $a_y = -g$. *Note:* We have chosen upward to be positive.

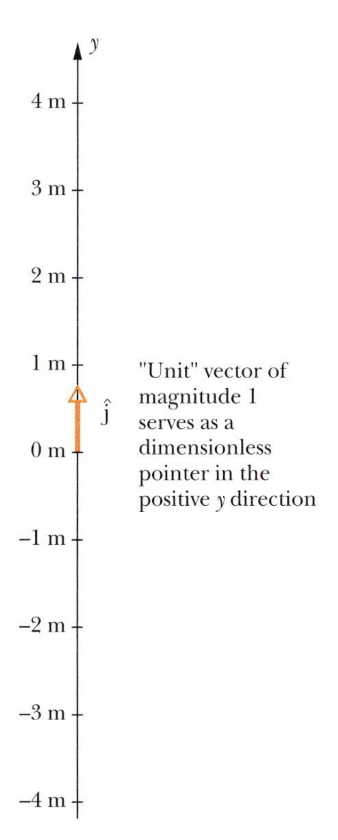


**FIGURE 3-24** ■ It is customary to designate a vertical axis as the $y$ axis, reserving the term $x$ axis for the horizontal direction. The upward direction is typically given as positive. The unit vector is labeled $\hat{\text{j}}$ rather than $\hat{\text{i}}$, and it points upward in the positive $y$ direction.

**READING EXERCISE 3-11:** Suppose that you throw an object upward and can ignore air resistance to the motion of the object. At the highest point in this motion, the object's velocity is instantaneously zero as it reverses direction. Does this mean that the object's acceleration is zero at that point? Explain how your answer is consistent with: (a) the definitions of instantaneous velocity and acceleration that you have learned; (b) the graphs in Fig. 3-22. ■

**READING EXERCISE 3-12:** Rewrite the equations in Table 2-1 so they describe the motion of an object in vertical free fall. Use a conventional coordinate system with the $y$ axis pointing up. ■

## TOUCHSTONE EXAMPLE 3-3: Model Rocket

A model rocket with a mass of 0.50 kg is fired vertically from the ground. Assume that it is streamlined enough that air resistance can be ignored. Suppose it ascends under the influence of a constant net force of 2.0 N acting in a vertical direction and travels for 6.0 s before its fuel is exhausted. Then it keeps moving as a particle-like object in free fall as it continues upward, turns around, and falls back down.

(a) How high is the rocket when it runs out of fuel? What is its velocity at that time?

**SOLUTION** ■ The net force on the rocket is a combination of the upward thrust of the rocket engine and the downward pull of the Earth. The **Key Idea** here is that we can use Newton's Second Law and our knowledge of the constant net force to find the rocket's constant acceleration and then use a kinematic equation to find out how high it will go in 6.0 seconds with that constant acceleration.

Using Eq. 3-5 for Newton's Second Law we get

$$\vec{a} = \frac{\vec{F}^{\,\text{net}}}{m} = \frac{(2.0\text{ N})\,\hat{\text{j}}}{0.50\text{ kg}} = (+4.0\text{ m/s}^2)\,\hat{\text{j}} = a_y\,\hat{\text{j}}.$$

Thus the vertical component of acceleration is $a_y = +4.0\text{ m/s}^2$. The elapsed time since take-off is given by $t_2 - t_1 = 6.0$ s. Since the rocket is fired at the ground level, $y_1 = 0.0$ m and $v_{1\,y} = 0.0$ m/s. Thus we can put numbers in the primary kinematic equation (Eq. 2-17) to get the height of the rocket at the time the fuel has run out,

$$(y_2 - y_1) = v_{1y}(t_2 - t_1) + \tfrac{1}{2}a_y(t_2 - t_1)^2$$
$$= 0.0\text{ m} + \tfrac{1}{2}(4.0\text{ m/s}^2)(6.0\text{ s})^2 = 72\text{ m}. \quad \text{(Answer)}$$

We need to find the $y$-component of velocity just as the rocket's fuel runs out. This is given by the other primary kinematic equation (Eq. 2-13) with respect to time to get

$$v_{2\,y} = v_{1\,y} + a_y\Delta t = 0.0\text{ m/s} + (4.0\text{ m/s}^2)(6.0\text{ s}) = 24\text{ m/s}. \quad \text{(Answer)}$$

(b) What is the total height that the rocket rises?

**SOLUTION** ■ The rocket is now at 72 m above the ground, moving upward with a velocity component of 24 m/s. We need to know how much higher it will go when the only significant force acting on it is the gravitational pull of the Earth. Let's do this by using only the primary kinematic equations for free fall with $a_y = -g$ so that

$$v_{2\,y} = v_{1\,y} - g\Delta t \quad \text{(Eq. 2-13)}$$

$$\text{and} \quad (y_2 - y_1) = v_{1\,y}(t_2 - t_1) - \tfrac{1}{2}g(t_2 - t_1)^2. \quad \text{(Eq. 2-17)}$$

The **Key Idea** here is to use Eq. 2-13 to find the time it takes the rocket to go from its new initial velocity of 24 m/s to its "final" velocity of 0 m/s and then use Eq. 2-17 to find the additional distance moved in the upward direction. Solving $v_{2\,y} = v_{1\,y} - g\Delta t$ for the elapsed time $\Delta t$ gives

$$\Delta t = \frac{v_{2\,y} - v_{1\,y}}{-g} = \frac{(0 - 24)\text{ m/s}}{-9.8\text{ m/s}^2} = 2.45\text{ s}.$$

Solving Eq. 2-17 for the additional rise of the rocket using $\Delta t = 2.45$ s gives

$$(y_2 - y_1) = v_{1\,y}\Delta t - \tfrac{1}{2}g(\Delta t)^2$$
$$= (24\text{ m/s})(2.45\text{ s}) - \tfrac{1}{2}9.8(2.45\text{ s})^2 = 29.4\text{ m}.$$

When added to the previous rise of the rocket under thrust we get

$$\text{maximum height} = 72\text{ m} + 29\text{ m} = 101\text{ m}. \quad \text{(Answer)}$$

(c) What is the net force on the rocket when it continues upward as a free fall particle? As it turns around? When it is traveling back toward the ground?

**SOLUTION** ■ A **Key Idea** here is that the only force on the rocket in free fall is the gravitational force. A second **Key Idea** is that this force is the same whether the rocket is moving up, turning around, or falling down. Its magnitude is given by the mass of the rocket times the gravitational constant $g$. In vector notation, the force is

$$\vec{F} = m\vec{a} = m(-g)\,\hat{\text{j}} = (0.50\text{ kg})(-9.8\text{ N/kg})\,\hat{\text{j}} = (-4.9\text{ N})\,\hat{\text{j}}.$$

## 3-10 Newton's Third Law

Newton's first two laws of motion describe what happens to a *single* object that has forces acting on it. We made the claim in Section 3-8 that for every object that experiences a force there is another object causing that force. Further, we claimed that interactions between two objects always seem to go two ways. We begin this section with a discussion of observations Newton made of the two-way interaction between hanging magnets. We can then state Newton's Third Law, which deals with the relationship between the forces objects exert on each other. We end the section by presenting experimental evidence for the validity of Newton's Third Law using measurements of contact forces.

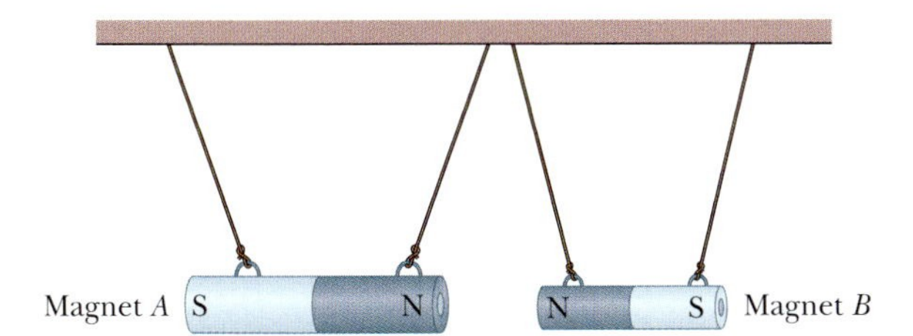


## Qualitative Considerations

Suppose that you hang two strong magnets side by side from long strings with their north poles facing each other as shown in Fig. 3-25. Many of us have observed that the north poles of magnets repel one another. If you were to hold the two north poles very close to each other and let go of the magnets, they would start to accelerate away from each other. The fact that *both* magnets are repelled and begin to accelerate implies that *each* magnet has a force acting on it. If you were to do this with magnets of the same mass, you would observe that the magnitudes of the two accelerations are identical. Observations of the accelerations of the magnets suggest that they are experiencing magnetic forces that have the same magnitude but are oppositely directed. (Actually, to get good measurements we should either mount our magnets on low-friction carts or hang them from long strings so the strings don't exert net horizontal forces on the magnets.)

This notion of equal and opposite forces is familiar to us in the case of contact forces. If you push on a wall it pushes back. This doesn't hurt if you push gently, but if you punch a wall hard it hurts very much. Newton hypothesized that any time two objects interact in such a way that a force is exerted on one of them, there is *always* a force that is equal in magnitude exerted in the opposite direction on the other object. This hypothesis is called Newton's Third Law, and we can state it simply in modern language.

> **NEWTON'S THIRD LAW:** If one object is exerting a force on a second object, then the second object is also exerting a force back on the first object. The two forces have exactly the same magnitude but act in opposite directions.

The most significant idea contained in Newton's Third Law is that *forces always exist in pairs*. It is very important that we realize we are talking about two *different forces* acting on two *different objects*.

In trying to visualize the application of this concept in the situation involving the magnets, it is helpful to draw a force vector at the center of each magnet showing the horizontal force it is experiencing from the other magnet. (Drawing the net force vector at the center of the object on which it acts is another of our many idealizations. The rod-shaped magnets are not really point particles, and each part of one magnet may be exerting forces on each part of the other and vice versa. However, in this situation, it turns out that assuming the rods are particle-like leads us to the same conclusions that treating them like rods would.)

Figure 3-26 shows the force diagrams for the two magnets discussed above, assuming that there are no other forces acting on them. The force exerted on object $A$ by object $B$ is denoted $\vec{F}_{B\to A}$ and the force exerted on object $B$ by object $A$ is denoted $\vec{F}_{A\to B}$. This notation allows us to write an equation that summarizes Newton's Third Law as follows:

$$\vec{F}_{B\to A} = -\vec{F}_{A\to B} \quad \text{(Newton's Third Law in equation form).} \qquad (3\text{-}10)$$

The order of the letters in the subscripts on the force is very important because they tell us which object the force is acting on and the origin of the force. The first letter

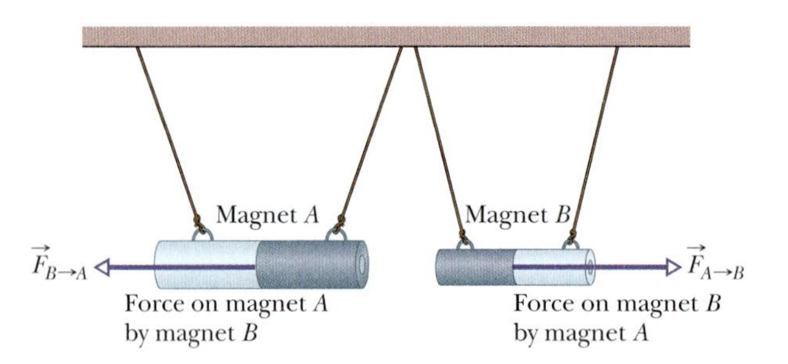


**FIGURE 3-26** ■ We can draw an interaction force vector at the center of each magnet (assuming that Newton's Third Law describes the interactions between two magnets and that the magnets are particle-like in their mutual interaction).

denotes the object that exerts the force and the second letter denotes the object that feels the force. We call the forces shown in Eq. 3-10 between the two interacting magnets a **third-law force pair.** In situations where Newton's laws apply, we believe that if any two bodies are interacting, a third-law force pair is always present.

## Experimental Verification for Contact Forces

We have developed Newton's Third Law in a qualitative fashion by doing a thought experiment. No measurements were taken to verify the law quantitatively. We have asserted that it holds whenever two bodies interact with each other. Now, let's consider whether the Third Law applies to objects that interact via contact (touching) forces. This time we will make measurements to verify the Third Law in a quantitative fashion.

Suppose two people hook the ends of two force sensors together as shown in Fig. 3-27 and have a back-and-forth tug of war. What happens?

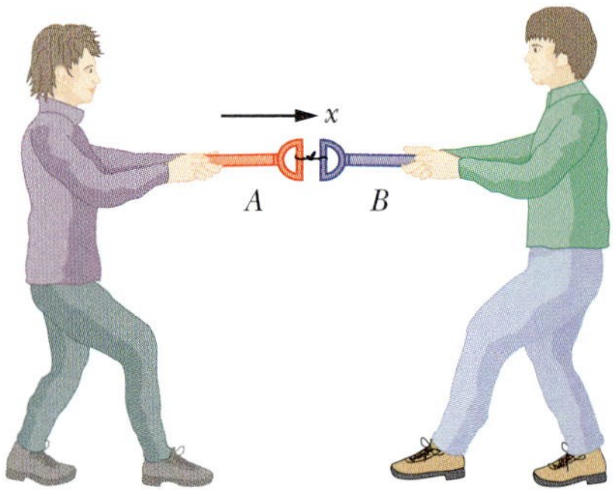

If we interface these force sensors to a computer for data collection, the result would look something like what is shown in Fig. 3-28. This graph verifies that on a moment-by-moment basis the force ($\vec{F}_{B\to A}$) exerted on the person on the left by the person on the right is equal in magnitude but opposite in direction to the force ($\vec{F}_{A\to B}$) exerted on the person on the right by the person on the left.

**FIGURE 3-27** ■ Two people are playing tug-of-war with electronic force sensors hooked together.

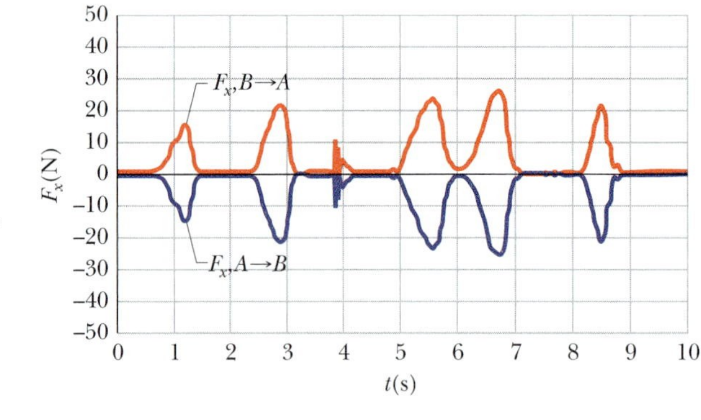


**FIGURE 3-28** ■ A graph of the measured force vs. time for two people playing tug-of-war for 10 seconds. A computer data acquisition system was used to collect and display the data at a rate of 100 readings per second. $\vec{F}_{B\to A}$ is exerted on the person on the left and $\vec{F}_{A\to B}$ on the person on the right.

We have considered magnetic forces (one form of electromagnetic force) and contact forces (another form of electromagnetic force). Does Newton's Third Law also apply to cases where masses are very different and when gravitational forces are present? Is it true for high-speed collisions? Is it true when one object is stationary and the other is not moving at first? For example, is it true for a very heavy truck traveling at high speed that collides head-on with a small car that is at rest? Is it true for a baseball in free fall interacting with the Earth? The answer to all these questions is "yes." We will return to them in Chapter 7, where we discuss the experimental evidence that Newton's Third Law can help us predict the outcomes of collisions between objects.

**READING EXERCISE 3-13:** Suppose that the magnet on the left in Fig. 3-25 is replaced by a steel paper clip that is not magnetized. (a) What can you say about the force that the initially unmagnetized paper clip exerts on the magnet on the right compared to the force the magnet on the right exerts on the paper clip? (b) Do you think Newton's Third Law holds? Explain. ■

## TOUCHSTONE EXAMPLE 3-4: Pushing Two Blocks

In Fig. 3-29*a*, a constant horizontal force $\vec{F}^{\text{app}}$ of magnitude 20 N is applied to block *A* of mass $m_A = 4.0$ kg, which pushes against block *B* of mass $m_B = 6.0$ kg. The blocks slide over a frictionless surface, along an *x* axis.

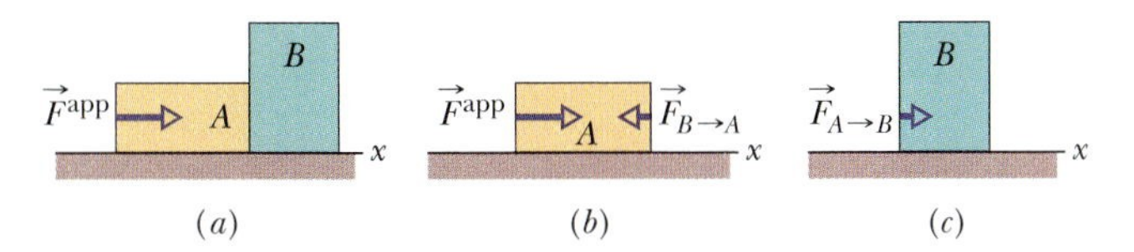


**FIGURE 3-29** ■ (*a*) A constant horizontal force $\vec{F}^{\text{app}}$ is applied to block *A*, which pushes against block *B*. (*b*) Two horizontal forces act on block *A*: applied force $\vec{F}^{\text{app}}$ and force $\vec{F}_{B\to A}$ from block *B*. (*c*) Only one horizontal force acts on block *B*: force $\vec{F}_{A\to B}$ from block *A*.

(a) What is the acceleration of the blocks?

**SOLUTION** ■ We shall first examine a solution with a serious error, then a dead-end solution, and then a successful solution.

*Serious Error*: Because force $\vec{F}^{\text{app}}$ is applied directly to block *A*, we use Newton's Second Law to relate that force to the acceleration $\vec{a}$ of block *A*. Because the motion is along the *x* axis, we use that law for *x*-components ($F_x^{\text{net}} = ma_x$), writing it as

$$F_x^{\text{net}} = m_A a_x.$$

However, this is seriously wrong because $\vec{F}^{\text{app}}$ is not the only horizontal force acting on block *A*. There is also the force $\vec{F}_{B\to A}$ from block *B* (as shown in Fig. 3-29*b*).

*Dead-End Solution*: Let us now include force $\vec{F}_{B\to A}$ by writing, again for the *x* axis,

$$F_x^{\text{app}} + F_{B\to A\,x} = m_A a_x$$

where $F_x^{\text{app}}$ is positive, but $F_{B\to Ax}$ is negative. However, $F_{B\to Ax}$ is a second unknown, so we cannot solve this equation for the desired acceleration $a_x$.

*Successful Solution*: The **Key Idea** here is that, because of the direction in which force $\vec{F}^{\text{app}}$ is applied, the two blocks form a rigidly connected system. We can relate the net force *on the system* to the acceleration *of the system* with Newton's Second Law. Here, once again for the *x* axis, we can write that law as

$$F_x^{\text{app}} = (m_A + m_B)a_x,$$

where now we properly apply $\vec{F}^{\text{app}}$ to the system with total mass $m_A + m_B$. Solving for $a_x$ and substituting known values, we find

$$a_x = \frac{F_x^{\text{app}}}{m_A + m_B} = \frac{20 \text{ N}}{4.0 \text{ kg} + 6.0 \text{ kg}} = 2.0 \text{ m/s}^2.$$

Thus, the acceleration of the system and of each block is in the positive direction of the *x* axis and has the magnitude 2.0 m/s$^2$.

(b) What is the force $\vec{F}_{A\to B}$ on block *B* from block *A* (Fig. 3-29*c*)?

**SOLUTION** ■ The **Key Idea** here is that we can relate the net force on block *B* to the block's acceleration with Newton's Second Law. Here we can write that law, still for components along the *x* axis, as

$$F_{A\to B\,x} = m_B a_x,$$

which, with known values, gives

$$F_{A\to B\,x} = (6.0 \text{ kg})(2.0 \text{ m/s}^2) = 12 \text{ N}.$$

Thus, force $\vec{F}_{A\to B}$ is in the positive direction of the *x* axis and has a magnitude of 12 N.

## TOUCHSTONE EXAMPLE 3-5: Pulling Two Blocks

Two blocks connected by a string are being pulled to the right across a horizontal frictional surface by another string, as shown in Fig. 3-30. The strings are horizontal and their masses are negligible compared to those of the blocks. The tension in the rightmost string is a constant 35 N. Find the tension in the other string if the mass of the left block is four times that of the right block.

**SOLUTION** ■ Note that we do not need to consider the tension in the rope between the objects when we treat them as a system, so as in the previous touchstone example, we choose to apply Newton's Second Law to the two-block system to find its acceleration. Although we don't know the masses of the individual blocks, we do know that $(m_{\text{left}}/m_{\text{right}}) = 4$. If we set $m_{\text{right}} = m$, then Newton's Second Law tells us that

$$F_x^{\text{app}} = (m + 4m)a_x.$$

Solving for $a_x$, we learn that

$$a_x = \frac{F_x^{\text{app}}}{5m}.$$

To find the tension, $T$, in the string between the two blocks, the **Key Idea** is to shift our attention from the two-block system to just the left-hand block. Its acceleration is the same as that of the two-block system, since they are joined by a string of constant length.

The second **Key Idea** is that the magnitude of the net force acting on the left block is equal to the tension, $T_A$, in the string joining the two blocks and that this force is directed to the right; that is, $\vec{F}_A = T_A\,\hat{\text{i}}$. Applying Newton's Second Law to the left block yields

$$F_{Ax} = T_A = 4ma_x.$$

But we've already seen that $a_x = F_x^{\text{app}}/5m$. Combining these two results tells us that

$$T_A = 4ma_x = \frac{4mF_x^{\text{app}}}{5m} = \tfrac{4}{5}F_x^{\text{app}}$$

$$= \tfrac{4}{5}(35\text{ N}) \qquad \text{(Answer)}$$

$$= 28\text{ N}.$$

We now see that $T_A$ depends only on $|\vec{F}^{\text{app}}|$ and on the *ratio* of the two masses; we did not need to know the individual masses to solve the problem.

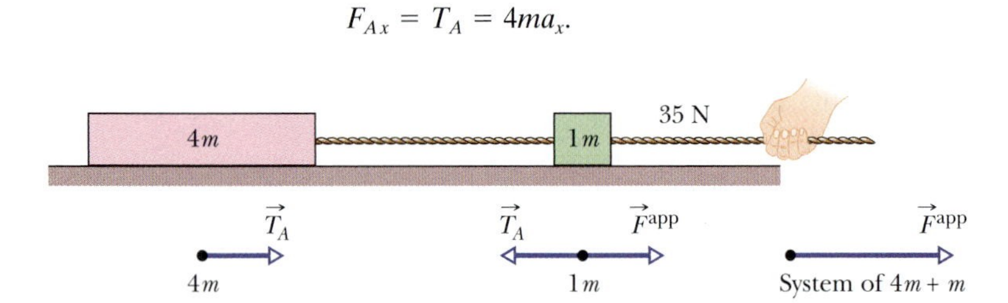


FIGURE 3-30

## TOUCHSTONE EXAMPLE 3-6: Raising Bricks

Figure 3-31 shows a man raising a load of bricks from the ground to the first floor of a building using a rope hung over a pulley. Suppose the load of bricks weighs 900 N and the man weighs 1200 N. What is the maximum upward acceleration that the man can give to the load of bricks by pulling downward on his side of the rope?

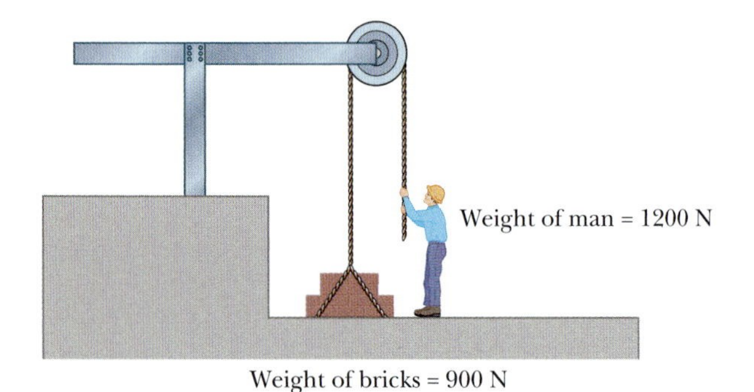


FIGURE 3-31

**SOLUTION** ■ One **Key Idea** here is that whatever force the man exerts downward on the rope, the rope in turn exerts upward on the man. Thus, if the man is not to accelerate upward, the maximum force he can exert downward on the rope is 1200 N. If he exceeds this, then he will experience a *net* upward force and accelerate upward. Thus the tension in the rope cannot exceed 1200 N.

Another **Key Idea** is that the tension in the rope is the same on each side of the pulley; ideal pulleys, such as this one, change the *direction* of the forces that ropes exert on objects but do not affect the *magnitude* of those forces. Thus the maximum *upward* force that the rope can exert on the load of bricks is 1200 N. Since gravity exerts a constant 900 N *downward* on the bricks, this limits the maximum vertical force on the bricks to

$$F_y^{\text{net max}} = F_{\text{rope}\to\text{bricks}\,y}^{\text{max}} + F_{\text{bricks}\,y}^{\text{grav}}$$

$$= 1200\text{ N} - 900\text{ N}$$

$$= 300\text{ N}.$$

Newton's Second Law then tells us that the maximum upward acceleration of the bricks is

$$a_y^{\text{max}} = \frac{F_y^{\text{net max}}}{m_{\text{bricks}}} = \frac{300\text{ N}}{(900\text{ N}/g)}$$

$$= \tfrac{1}{3}g$$

$$= 3.26\text{ m/s}^2. \qquad \text{(Answer)}$$

To find the mass of the bricks here, we have used the fact that the weight of the brick is equal to their mass times the local value of $g = 9.80\,\text{N/kg} = 9.80\,\text{m/s}^2$.

## 3-11 Comments on Classical Mechanics

A word of caution—classical mechanics does not apply to all situations. For instance, physicists know that if the speeds of the interacting bodies are very large—an appreciable fraction of the speed of light—we must replace Newtonian mechanics with *Einstein's special theory of relativity*. In addition, if the interacting bodies are molecules, atoms, or electrons within atoms, there are situations in which we must replace classical mechanics with *quantum mechanics*. Physicists now view Newtonian mechanics as a special case of these two more comprehensive theories. Still, classical mechanics is a very important special case of these other theories because it applies to the motion of objects ranging in size from that of large molecules to that of astronomical objects such as galaxies and galactic clusters. The domain of Newton's laws encompasses our everyday world including the translational, rotational and vibrational motions of cars, ships, airplanes, elevators, steam engines, our bodies, fluids, glaciers, the atmosphere, and oceans.

In the next chapter, we will introduce elements of vector mathematics that allow us to extend and apply Newtonian mechanics to more realistic situations involving motions in two dimensions. In spite of the limitations of Newton's laws, you will see throughout our study of basic classical physics that these laws are extraordinarily powerful in helping us describe, understand, and predict events involving motion in our everyday world and beyond.

# Problems

### SEC. 3-6 ■ NEWTON'S SECOND LAW FOR A SINGLE FORCE

**1. Stopping a Neutron** When a nucleus captures a stray neutron, it must bring the neutron to a stop within the diameter of the nucleus by means of the *strong force*. That force, which "glues" the nucleus together, is approximately zero outside the nucleus. Suppose that a stray neutron with an initial speed of $1.4 \times 10^7$ m/s is just barely captured by a nucleus with diameter $d = 1.0 \times 10^{-14}$ m. Assuming that the strong force on the neutron is constant, find the magnitude of that force. The neutron's mass is $1.67 \times 10^{-27}$ kg.

**2. Riding the Elevator** A 50 kg passenger rides in an elevator that starts from rest on the ground floor of a building at $t = 0$ and rises to the top floor during a 10 s interval. The acceleration of the elevator as a function of the time is shown in Fig. 3-32, where positive values of the acceleration mean that it is directed upward. Give the magnitude and direction of the following forces: (a) the maximum force on the passenger from the floor, (b) the minimum force on the passenger from the floor, and (c) the maximum force on the floor from the passenger.

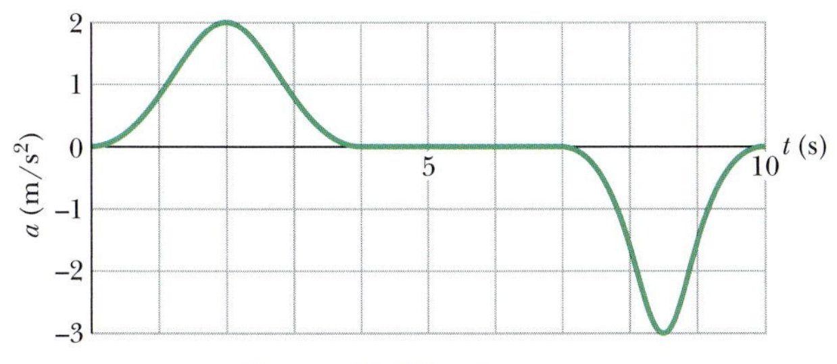


FIGURE 3-32 ■ Problem 2.

**3. Sunjamming** A "sun yacht" is a spacecraft with a large sail that is pushed by sunlight. Although such a push is tiny in everyday circumstances, it can be large enough to send the spacecraft outward from the Sun on a cost-free but slow trip. Suppose that the spacecraft has a mass of 900 kg and receives a push of 20 N. (a) What is the magnitude of the resulting acceleration? If the craft starts from rest, (b) how far will it travel in 1 day and (c) how fast will it then be moving?

**4. Stopping a Salmon** The tension at which a fishing line snaps is commonly called the line's "strength." What minimum strength is needed for a line that is to stop a salmon of weight 85 N in 11 cm if the fish is initially drifting at 2.8 m/s? Assume a constant acceleration.

**5. Rocket Sled** An experimental rocket sled can be accelerated at a constant rate from rest to 1600 km/h in 1.8 s. What is the magnitude of the required net force if the sled has a mass of 500 kg?

**6. Stopping a Car** A car with a mass of 1300 kg is initially moving at a speed of 40 km/h when the brakes are applied and the car is brought to a stop in 15 m. Assuming that the force that stops the car is constant, find (a) the magnitude of that force and (b) the time required for the change in speed. If the initial speed is doubled and the car experiences the same force during the braking, by what factors are (c) the stopping distance and (d) the stopping time multiplied? (There could be a lesson here about the danger of driving at high speeds.)

**7. Rocket and Payload** A rocket and its payload have a total mass of $5.0 \times 10^4$ kg. How large is the force produced by the engine (the thrust) when (a) the rocket is "hovering" over the launchpad just after ignition, and (b) the rocket is accelerating upward at 20 m/s$^2$?

**8. Car Wreck** A car traveling at 53 km/h hits a bridge abutment. A passenger in the car moves forward a distance of 65 cm (with respect to the road) while being brought to rest by an inflated air bag. What magnitude of force (assumed constant) acts on the passenger's upper torso, which has a mass of 41 kg?

**9. The Fall** An 80 kg man drops to a concrete patio from a window only 0.50 m above the patio. He neglects to bend his knees on landing, taking 2.0 cm to stop. (a) What is his average acceleration from when his feet first touch the patio to when he stops? (b) What is the magnitude of the average stopping force?

**10. Starship** An interstellar ship has a mass of $1.20 \times 10^6$ kg and is initially at rest relative to a star system. (a) What constant acceleration is needed to bring the ship up to a speed of $0.10c$ (where $c$ is the speed of light, $3.0 \times 10^8$ m/s) relative to the star system in 3.0 days? (b) What is that acceleration in $g$ units? (c) What force is required for the acceleration? (d) If the engines are shut down when $0.10c$ is reached (the speed then remains constant), how long does the ship take (start to finish) to journey 5.0 light-months, the distance that light travels in 5.0 months?

**11. Force vs. Time** Figure 3-33 gives, as a function of time $t$, the force component $F_x$ that acts on a 3.00 kg ice block, which can move only along the $x$ axis. At $t = 0$, the block is moving in the positive direction of the axis, with a speed of 3.0 m/s. What are its (a) speed and (b) direction of travel at $t = 11$ s?

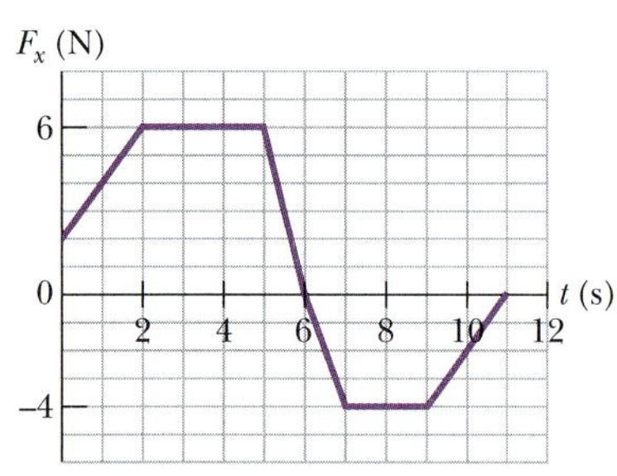


FIGURE 3-33 ■ Problem 11.

**12. Variable Force** A 2.0 kg particle moves along an $x$ axis, being propelled by a variable force directed along that axis. Its position is given by

$$x = 3.0\text{ m} + (4.0\text{ m/s})t + ct^2 - (2.0\text{ m/s}^3)t^3,$$

with $x$ in meters and $t$ in seconds. The factor $c$ is a constant. At $t = 3.0$ s, the force on the particle has a magnitude of 36 N and is in the negative direction of the axis. What is $c$? (Include units.)

## SEC. 3-8 ■ ALL FORCES RESULT FROM INTERACTION

**13. Two People Pull** Two people pull with 90 N and 92 N in opposite directions on a 25 kg sled on frictionless ice. What is the sled's acceleration magnitude?

**14. Take Off** A Navy jet (Fig. 3-34) with a mass of $2.3 \times 10^4$ kg requires an airspeed of 85 m/s for liftoff. The engine develops a maximum force of $1.07 \times 10^5$ N, but that is insufficient for reaching takeoff speed in the 90 m runway available on an aircraft carrier. What minimum force (assumed constant) is needed from the catapult that is used to help launch the jet? Assume that the catapult and the jet's engine each exert a constant force over the 90 m distance used for takeoff.

FIGURE 3-34 ■ Problem 14.

**15. Loaded Elevator** An elevator and its load have a combined mass of 1600 kg. Find the tension in the supporting cable when the elevator, originally moving downward at 12 m/s, is brought to rest with constant acceleration in a distance of 42 m.

**16. Four Penguins** Figure 3-35 shows four penguins that are being playfully pulled along very slippery (frictionless) ice by a curator. The masses of three penguins and the tension in two of the cords are given. Find the penguin mass that is not given.

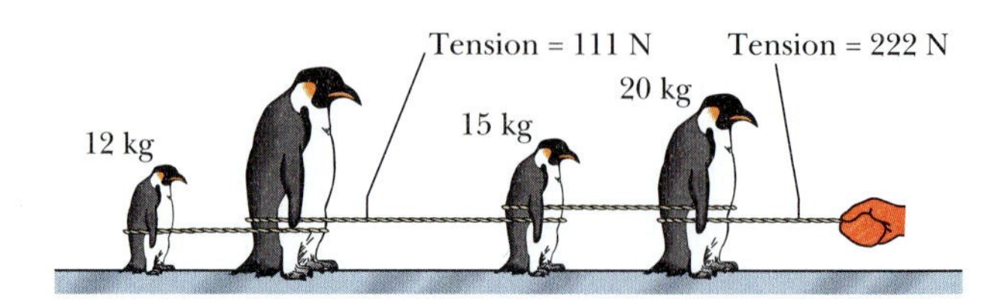


FIGURE 3-35 ■ Problem 16.

**17. Elevator** An elevator with a mass of 2840 kg is given an upward acceleration of 1.22 m/s² by a cable. (a) Calculate the tension in the cable. (b) What is the tension when the elevator is slowing at the rate of 1.22 m/s² but is still moving upward?

**18. Three Blocks** In Fig. 3-36 three blocks are connected and pulled to the right on a horizontal frictionless table by a force with a magnitude of $T_3 = 65.0$ N. If $m_A = 12.0$ kg, $m_B = 24.0$ kg, and $m_C = 31.0$ kg, calculate (a) the acceleration of the system and the magnitudes of the tensions (b) $T_1$ and (c) $T_2$ in the interconnecting cords.

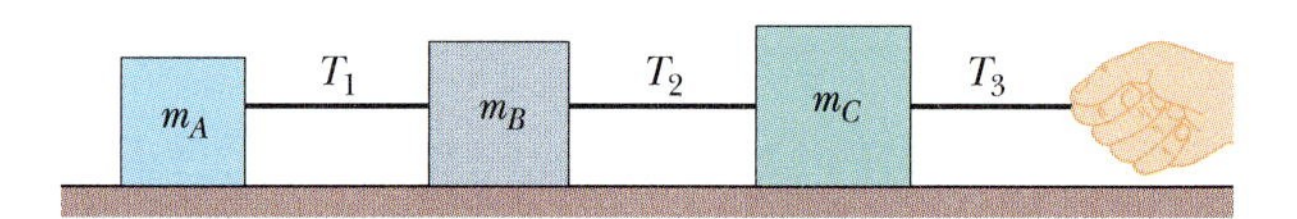


FIGURE 3-36 ■ Problem 18.

**19. Hot-Air Balloon** A hot-air balloon of mass $M$ is descending vertically with downward acceleration of magnitude $a$. How much mass (ballast) must be thrown out to give the balloon an upward acceleration of magnitude $a$ (same magnitude but opposite direction)? Assume that the upward force from the air (the lift) does not change because of the decrease in mass.

**20. Lamp in Elevator** A lamp hangs vertically from a cord in a descending elevator that slows down at 2.4 m/s². (a) If the tension in the cord is 89 N, what is the lamp's mass? (b) What is the cord's tension when the elevator ascends with an upward acceleration of 2.4 m/s²?

**21. Two Forces, Two Blocks** In Fig. 3-37 forces act on blocks $A$ and $B$, which are connected by string. Force $\vec{F}_A = (12\text{ N})\hat{i}$ acts on block $A$, with mass 4.0 kg. Force $\vec{F}_B = (24\text{ N})\hat{i}$ acts on block $B$, with mass 6.0 kg. What is the tension in the string?

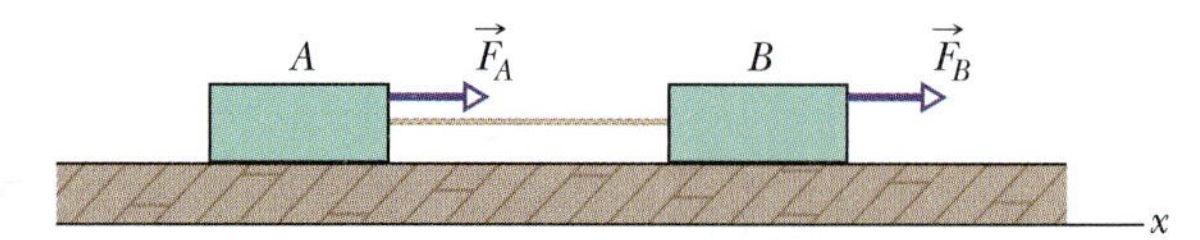


FIGURE 3-37 ■ Problem 21.

**22. Coin Drop** An elevator cab is pulled directly upward by a single cable. The elevator cab and its single occupant have a mass of 2000 kg. When that occupant drops a coin, its acceleration relative to the cab is 8.00 m/s² downward. What is the tension in the cable?

**23. Links** In Fig. 3-38, a chain consisting of five links, each of mass 0.100 kg, is lifted vertically with a constant acceleration of 2.50 m/s².

Find the magnitudes of (a) the force on link 1 from link 2, (b) the force on link 2 from link 3, (c) the force on link 3 from link 4, and (d) the force on link 4 from link 5. Then find the magnitudes of (e) the force $\vec{F}$ on the top link from the person lifting the chain and (f) the *net* force accelerating each link.

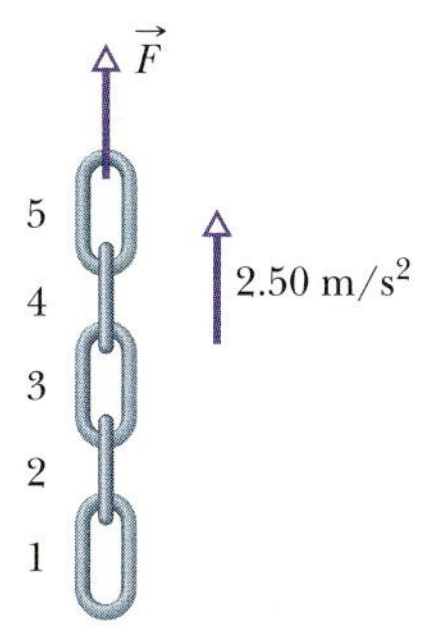


FIGURE 3-38 ▪ Problem 23.

## SEC. 3-9 ▪ GRAVITATIONAL FORCES AND FREEFALL MOTION

**24. Raindrops** Raindrops fall 1700 m from a cloud to the ground. (a) If they were not slowed by air resistance, how fast would the drops be moving when they struck the ground? (b) Would it be safe to walk outside during a rainstorm?

**25. Falling Rock** A rock is dropped from a 100-m-high cliff. How long does it take to fall (a) the first 50 m and (b) the second 50 m?

**26. Long Drop** The Zero Gravity Research Facility at the NASA Lewis Research Center includes a 145 m drop tower. This is an evacuated vertical tower through which, among other possibilities, a 1 m diameter sphere containing an experimental package can be dropped. (a) How long is the sphere in free fall? (b) What is its speed just as it reaches a catching device at the bottom of the tower? (c) When caught, the sphere experiences an average acceleration of $25g$ as its speed is reduced to zero. Through what distance does it travel while stopping?

**27. Leaping Armadillo** A startled armadillo leaps upward, rising 0.544 m in the first 0.200 s. (a) What is its initial speed as it leaves the ground? (b) What is its speed at the height of 0.544 m? (c) How much higher does it go?

**28. Ball Thrown Downward** A ball is thrown *down* vertically with an initial *speed* of $v_1$ from a height of $h$. (a) What is its speed just before it strikes the ground? (b) How long does the ball take to reach the ground? What would be the answers to (c) part a and (d) part b if the ball were thrown *upward* from the same height and with the same initial speed? Before solving any equations, decide whether the answers to (c) and (d) should be greater than, less than, or the same as in (a) and (b).

**29. Boat and Key** A key falls from a bridge that is 45 m above the water. It falls directly into a model boat, moving with constant velocity, that is 12 m from the point of impact when the key is released. What is the speed of the boat?

**30. Downward-Speeding Ball** A ball is thrown vertically downward from the top of a 36.6-m-tall building. The ball passes the top of a window that is 12.2 m above the ground 2.00 s after being thrown. What is the speed of the ball as it passes the top of the window?

**31. Drips** Water drips from the nozzle of a shower onto the floor 200 cm below. The drops fall at regular (equal) intervals of time, the first drop striking the floor at the instant the fourth drop begins to fall. Find the locations of the second and third drops when the first strikes the floor.

**32. Hang Time** A basketball player, standing near the basket to grab a rebound, jumps 76.0 cm vertically. How much (total) time does the player spend (a) in the top 15.0 cm of this jump and (b) in the bottom 15.0 cm? Does this help explain why such players seem to hang in the air at the tops of their jumps?

**33. Air Express** A hot-air balloon is ascending at the rate of 12 m/s and is 80 m above the ground when a package is dropped over the side.

**(a)** How long does the package take to reach the ground?
**(b)** With what speed does it hit the ground?

**34. Other-Worldly Pitch** A ball is shot vertically upward from the surface of a planet in a distant solar system. A plot of $y$ versus $t$ for the ball is shown in Fig. 3-39, where $y$ is the height of the ball above its starting point and $t = 0$ at the instant the ball is shot. What are the magnitudes of (a) the free-fall acceleration on the planet and (b) the initial velocity of the ball?

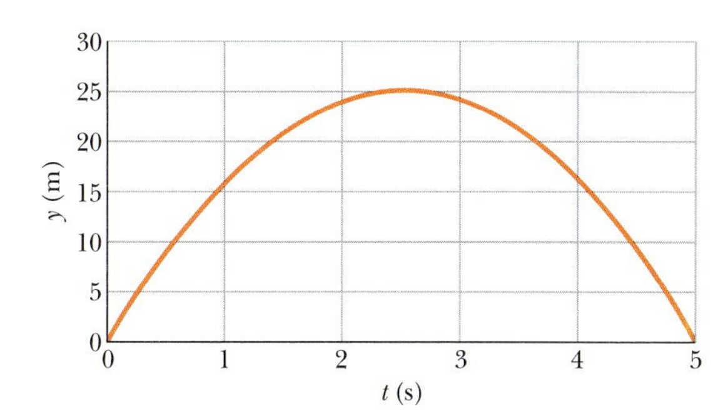


FIGURE 3-39 ▪ Problem 34.

**35. Reaction Time** Figure 3-40 shows a simple device for measuring your reaction time. It consists of a cardboard strip marked with a scale and two large dots. A friend holds the strip *vertically,* with thumb and forefinger at the dot on the right in Fig. 3-40. You then position your thumb and forefinger at the other dot (on the left in Fig. 3-40), being careful not to touch the strip. Your friend releases the strip, and you try to pinch it as soon as possible after you see it begin to fall. The mark at the place where you pinch the strip gives your reaction time. (a) How far from the lower dot should you place the 50.0 ms mark? (b) How much higher should the marks for 100, 150, 200, and 250 ms be? (For example, should the 100 ms marker be two times as far from the dot as the 50 ms marker? Can you find any pattern in the answers?)

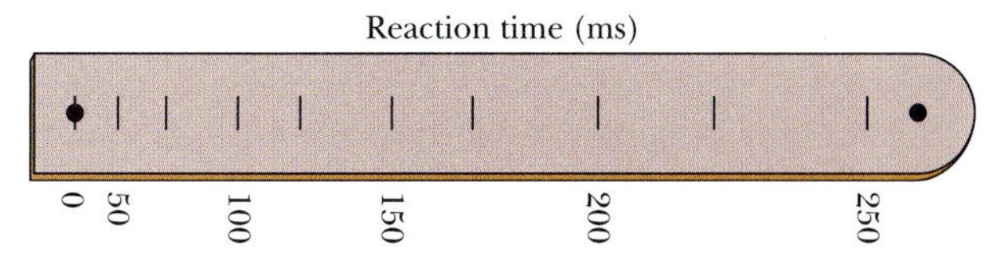


FIGURE 3-40 ▪ Problem 35.

**36. Juggling** A certain juggler usually tosses balls vertically to a height $H$. To what height must they be tossed if they are to spend twice as much time in the air?

**37. Dropping a Wrench** At a construction site a pipe wrench struck the ground with a speed of 24 m/s. (a) From what height was it inadvertently dropped? (b) How long was it falling? (c) Sketch graphs of $y$, $v_y$, and $a_y$ vs. $t$ for the wrench.

**38. Two Stones** A stone is dropped into a river from a bridge 43.9 m above the water. Another stone is thrown vertically down 1.00 s after the first is dropped. Both stones strike the water at the same time. (a) What is the initial speed of the second stone?

(b) Plot velocity vs. time on a graph for each stone, taking zero time as the instant the first stone is released.

**39. Callisto** Imagine a landing craft approaching the surface of Callisto, one of Jupiter's moons. If the engine provides an upward force (thrust) of 3260 N, the craft descends at constant speed; if the engine provides only 2200 N, the craft accelerates downward at 0.39 m/s$^2$. (a) What is the weight of the landing craft in the vicinity of Callisto's surface? (b) What is the mass of the craft? (c) What is the magnitude of the free-fall acceleration near the surface of Callisto?

**40. Rising Stone** A stone is thrown vertically upward. On its way up it passes point $A$ with speed $v$, and point $B$, 3.00 m higher than $A$, with speed $\frac{1}{2}v$. Calculate (a) the speed $v$ and (b) the maximum height reached by the stone above point $B$.

**41. Parachuting** A parachutist bails out and freely falls 50 m. Then the parachute opens, and thereafter she slows at 2.0 m/s$^2$. She reaches the ground with a speed of 3.0 m/s. (a) How long is the parachutist in the air? (b) At what height does the fall begin?

**42. Space Ranger's Weight** Compute the weight of a 75 kg space ranger (a) on Earth, (b) on Mars, where $g = 3.8$ m/s$^2$, and (c) in interplanetary space, where $g = 0$. (d) What is the ranger's mass at each of these locations?

**43. Different *g*'s** A certain particle has a weight of 22 N at a point where $g = 9.8$ m/s$^2$. What are its (a) weight and (b) mass at a point where $g = 4.9$ m/s$^2$? What are its (c) weight and (d) mass if it is moved to a point in space where $g = 0$?

## SEC. 3-10 ■ NEWTON'S THIRD LAW

**44. A Child Stands Then Jumps** A 29.0 kg child, with a 4.50 kg backpack on his back, first stands on a sidewalk and then jumps up into the air. Find the magnitude and direction of the force on the sidewalk from the child when the child is (a) standing still and (b) in the air. Now find the magnitude and direction of the *net* force on Earth due to the child when the child is (c) standing still and (d) in the air.

**45. Sliding Down a Pole** A firefighter with a weight of 712 N slides down a vertical pole with an acceleration of 3.00 m/s$^2$, directed downward. What are the magnitudes and directions of the vertical forces (a) on the firefighter from the pole and (b) on the pole from the firefighter?

**46. Block A, Block B** In Fig. 3-41*a*, a constant horizontal force $\vec{F}_a$ is applied to block $A$, which pushes against block $B$ with a 20.0 N force horizontally to the right. In Fig. 3-41*b*, the same force $\vec{F}_a$ is applied to block $B$; now block $A$ pushes on block $B$ with a 10.0 N force horizontally to the left. The blocks have a total mass of 12.0 kg. What are the magnitudes of (a) their acceleration in Fig. 3-41*a* and (b) force $\vec{F}_a$?

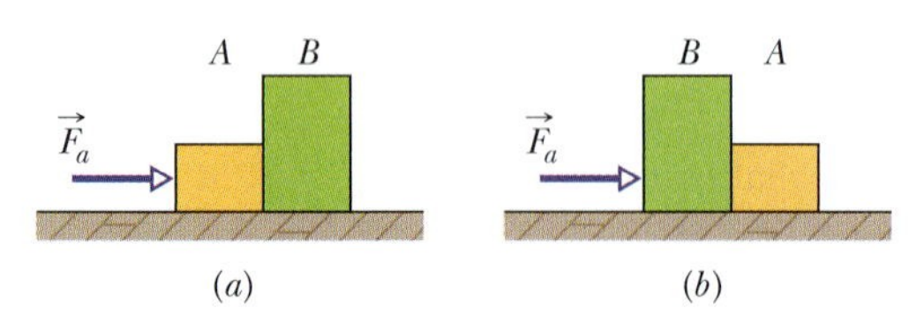


FIGURE 3-41 ■ Problem 46.

**47. Two Blocks** Two blocks are in contact on a frictionless table. A horizontal force is applied to the larger block, as shown in Fig. 3-42. (a) If $m_A = 2.3$ kg, $m_B = 1.2$ kg, and $F = 3.2$ N, find the magnitude of the force between the two blocks. (b) Show that if a force of the same magnitude $F$ is applied to the smaller block but in the opposite direction, the magnitude of the force between the blocks is 2.1 N, which is not the same value calculated in (a). (c) Explain the difference.

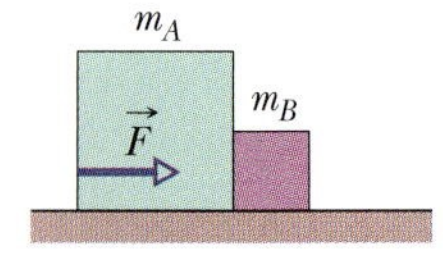


FIGURE 3-42 ■ Problem 47.

**48. Parachuting Two** An 80 kg person is parachuting and experiencing a downward acceleration of 2.5 m/s$^2$. The mass of the parachute is 5.0 kg. (a) What is the upward force on the open parachute from the air? (b) What is the downward force on the parachute from the person?

**49. Getting Down** An 85 kg man lowers himself to the ground from a height of 10.0 m by holding onto a rope that runs over a frictionless pulley to a 65 kg sandbag. With what speed does the man hit the ground if he started from rest?

**50. Climbing a Rope** A 10 kg monkey climbs up a massless rope that runs over a frictionless tree limb and back down to a 15 kg package on the ground (Fig. 3-43). (a) What is the magnitude of the least acceleration the monkey must have if it is to lift the package off the ground? If, after the package has been lifted, the monkey stops its climb and holds onto the rope, what are (b) the magnitude and (c) the direction of the monkey's acceleration, and (d) what is the tension in the rope?

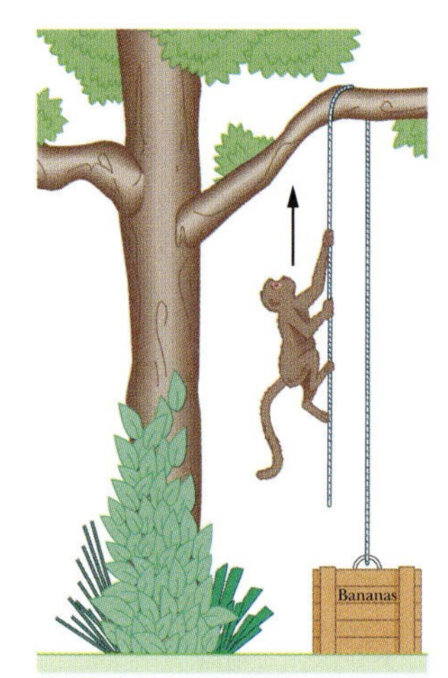


FIGURE 3-43 ■ Problem 50.

**51. Bosun's Chair** Figure 3-44 shows a man sitting in a bosun's chair that dangles from a massless rope, which runs over a massless, frictionless pulley and back down to the man's hand. The combined mass of man and chair is 95.0 kg. With what force magnitude must the man pull on the rope if he is to rise (a) with a constant velocity and (b) with an upward acceleration of 1.30 m/s$^2$? (*Hint:* A free-body diagram can really help.)

FIGURE 3-44 ■ Problem 51.

**52. Girl and Sled** A 40 kg girl and an 8.4 kg sled are on the frictionless ice of a frozen lake, 15 m apart but connected by a rope of negligible mass. The girl exerts a horizontal 5.2 N force on the rope. (a) What is the acceleration of the sled? (b) What is the acceleration of the girl? (c) How far from the girl's initial position do they meet?

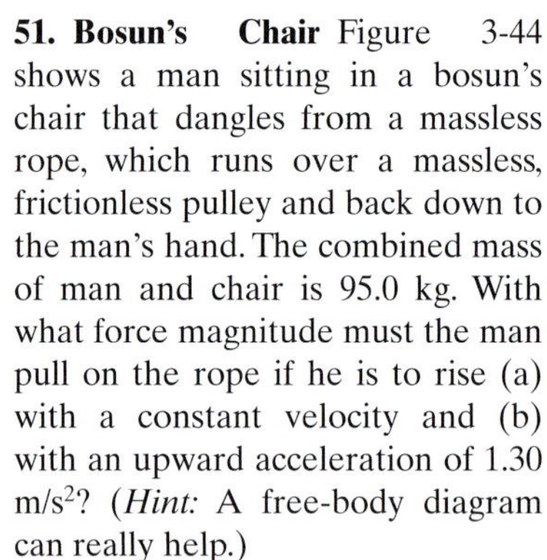

# Additional Problems

**53. Why Bother with N1?** Newton's First Law states that an object will move with a constant velocity if nothing acts on it. This seems to contradict our everyday experience that a moving object comes to a rest unless something acts on it to keep it going. Does this everyday experience contradict Newton's First Law? If it does not, explain how this experience is consistent with Newton's First Law. If it does, explain why we bother to teach Newton's First Law anyway.

**54. When Does N3 Hold?** Newton's Third Law says that objects that touch each other exert forces on each other. These forces satisfy the rule:

*If object A exerts a force on object B, then object B exerts a force back on object A and the two forces are equal in magnitude but opposite in direction.*

Consider the following three situations concerning two identical cars and a much heavier truck.

**(a)** One car is parked and the other car crashes into it.
**(b)** One car is parked and the truck crashes into it.
**(c)** The truck is pushing the car, because the car's engine cannot start. The two are touching and the truck is speeding up.

For each situation, do you think Newton's Third Law holds or does not hold? Explain your reasons for saying so.

**55. Why Bother with N2?** Newton's Second Law written in equation form states

$$\vec{a} = \frac{\vec{F}^{\text{ net}}}{m}.$$

Your roommate says "That's silly. Everyone knows it takes a force to keep something moving at a constant velocity, even when there's no acceleration." Do you agree with your roommate? If so, explain why physics classes bother to teach the law. If you disagree, how would you try to convince your roommate of the error of his/her ways?

**56. Weight vs. Force** A Frenchman, filling out a form, writes "78 kg" in the space marked poids (weight). However weight is a force and kg is a mass unit. What do the French (among others) have in mind when they use mass to report their weight? Why don't they report their weight in newtons? How many newtons does this Frenchman weigh? How many pounds?

**57. Amy Is Pulled** A student named Amy is being pulled across a smooth floor with a big rubber band that is stretched to a constant length. In one case she is riding on a low-friction cart and in the other case she is sliding along the floor. A motion detector is set up to track her motion in each case. The position–time graphs of her motion are shown in Fig. 3-45.

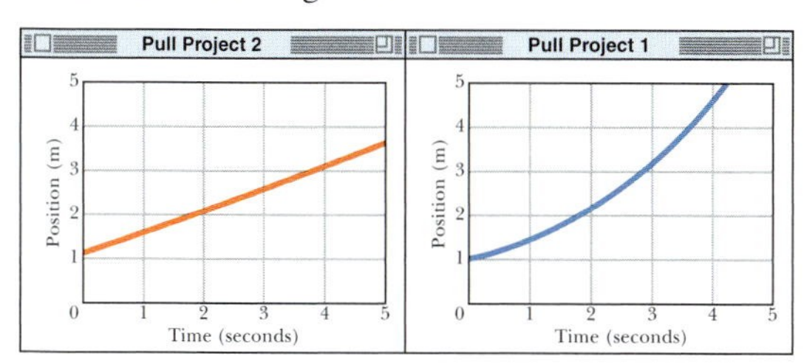


**FIGURE 3-45** ■ Problem 57.

**(a)** Which graph depicts motion at a constant velocity? Pull Project 1 (on the right) or Pull Project 2 (on the left)? Explain.
**(b)** Which graph depicts motion at a roughly constant acceleration? Explain.
**(c)** Which graph demonstrates that something pulled with a constant force moves with a constant velocity? Explain.
**(d)** Which graph demonstrates that something pulled with a constant force moves with a constant acceleration? Explain.
**(e)** Which graph is most likely to show Amy's motion when she is rolling on the cart? Please justify your answer.
**(f)** Explain why it is possible to get two different types of motion even though Amy is being pulled with a constant force in both cases.

**58. Inertial vs. Gravitational Mass** Suppose you have the following equipment available: an electronic balance, a motion detector and an electronic force sensor attached to a computer-based laboratory system. You would like to determine the mass of a block of ice that can slide smoothly along a very level table top without noticeable friction.

**(a)** Describe how you would use some of the equipment to find the *gravitational* mass of the ice.
**(b)** Describe how you would use some of the equipment to find the *inertial* mass of the ice.
**(c)** Which of the two types of masses can be measured in outer space where gravitational forces are very small?

**59. Free Fall Acceleration** Your roommate peeks over your shoulder while you are reading a physics text and notices the following sentence: "In free fall the acceleration is always $g$ and always straight downward regardless of the motion." Your roommate finds this peculiar and raises three objections:

**(a)** If I drop a balloon or a feather, it doesn't fall nearly as fast as a brick.
**(b)** Not everything falls straight down; if I throw a ball it can go sideways.
**(c)** If I hold a wooden ball in one hand and a steel ball in the other, I can tell that the steel ball is being pulled down much more strongly than the wooden one. It will probably fall faster.

How would you respond to these statements? Discuss the extent to which they invalidate the quoted statement. If they don't invalidate the statement, explain why.

**60. Velocity and Force Graphs** In the following situations friction is small and can be ignored. Consider whether the net or combined force on a small cart needs to be positive, negative, or zero to create the following motions. Sketch graphs that show the shapes of the velocity and force functions in each case. Use the format shown in Fig. 3-46. By convention, an object moving away from the origin has a positive velocity. (Draw a separate set of velocity vs. time and force vs. time graph for each of part (a) through (d).)

**(a)** The cart is moving away from the origin at a constant velocity.
**(b)** The cart moves toward the origin, speeding up at a steady rate until it reaches a constant velocity after 3 s.
**(c)** The cart moves toward the origin, slowing down at a steady rate, turns around after 2 s, and then moves away from the origin, speeding up at the *same* steady rate.

**(d)** The cart moves away from the origin, slows down for 3 s, and then speeds up for 3 seconds.

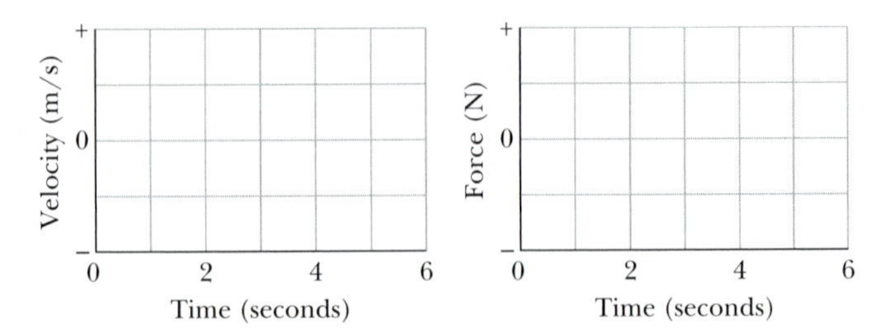


FIGURE 3-46 ■ Problem 60.

**61. Toy Cars (a)** Suppose a toy car moves along a horizontal line without friction and a constant force is applied to the car toward the left.

FIGURE 3-47 ■ Problem 61.

Sketch a set of axes like those shown in Fig. 3-48, and sketch the shape of the acceleration–time graph of the car using a solid line

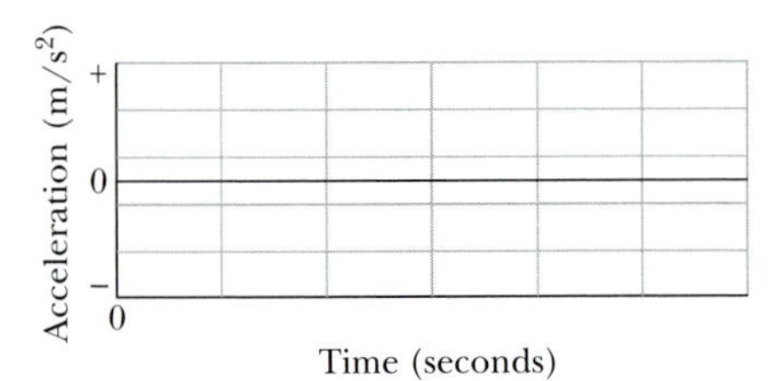


FIGURE 3-48 ■ Problem 61.

**(b)** What if two more identical cars are piled/glued on top of the first car and the same constant force is applied to the three cars? Use a dashed line to sketch the acceleration–time graph of the "triple-car." Explain any differences between this graph and the acceleration–time graph of the single car.

**62. Rocket Thrust and Acceleration** A wise being has placed a standard physics coordinate system in outer space far away from any massive bodies. A specially designed space cylinder that experiences no gravitational or frictional forces is moving along the $x$ axis of this coordinate system. It has two identical rocket engines on each end. These engines can apply thrust forces that act in opposite directions but have equal magnitudes as shown in Fig. 3-49. Diagram A has engines on both ends on, diagram B has all engines off, diagram C has only the left engines on, and diagram D has only the right engines on.

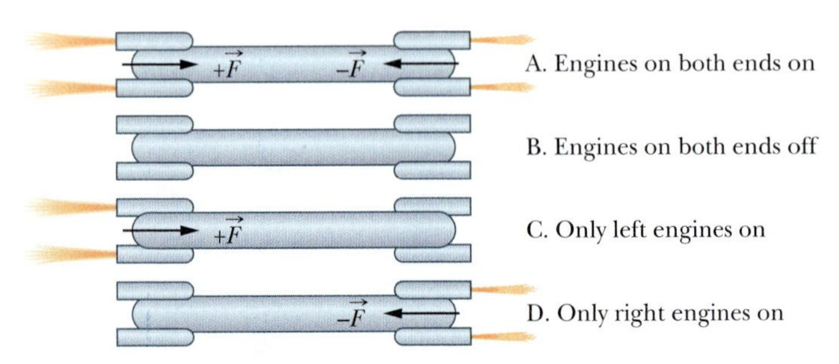


FIGURE 3-49 ■ Problem 62.

Choose all the force combinations (**A** through **D**) which could *keep the rocket moving* as described in each statement below. You may use a choice more than once or not at all. If you think that none is correct, answer choice **E.**

**(a)** Which force combinations could keep the rocket moving toward the right and speeding up at a steady rate (constant acceleration)?
**(b)** Which force combinations could keep the rocket moving toward the right at a steady (constant) velocity?
**(c)** The rocket is moving toward the right. Which force combinations could slow it down at a steady rate (constant acceleration)?
**(d)** Which force combinations could keep the rocket moving toward the left and speeding up at a steady rate (constant acceleration)?
**(e)** The rocket was started from rest and pushed until it reached a steady (constant) velocity toward the right. Which force combinations could keep the rocket moving at this velocity?
**(f)** The rocket is slowing down at a steady rate and has an acceleration to the right. Which force combinations could account for this motion?
**(g)** The rocket is moving toward the left. Which force combinations could slow it down at a steady rate (constant acceleration)?

**63. Two Carts** Two low-friction carts $A$ and $B$ have masses of 2.5 kg and 5.0 kg, respectively. Initially a student is pushing them with an applied force of $\vec{F}_B = -20.0$ N, which is exerted on cart $B$ as shown in Fig. 3-50*a*.

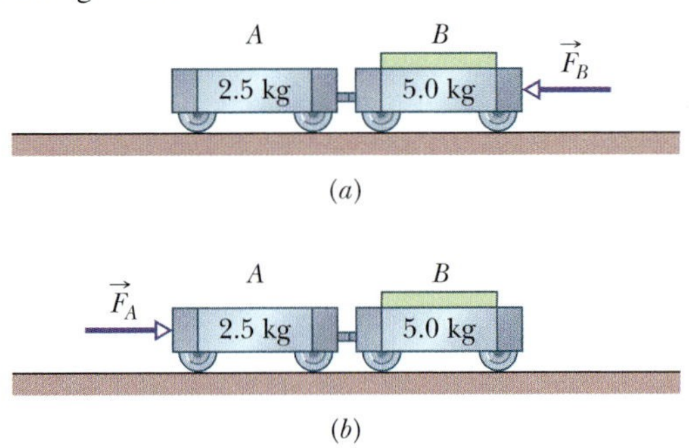


FIGURE 3-50 ■ Problem 63.

**(a)** Find the magnitude and direction of the interaction forces between the two carts $\vec{F}_{B\to A}$ and $\vec{F}_{A\to B}$ where $\vec{F}_{B\to A}$ represents the force on cart $A$ due to cart $B$ and $\vec{F}_{A\to B}$ represents the force on cart $B$ due to cart $A$.
**(b)** If the student pushes on cart $A$ with an applied force of $\vec{F}_A = +20.0$ N instead, as shown in part (b) of Fig. 3-50, determine the magnitude and direction of the interaction forces between the two carts $\vec{F}_{B\to A}$ and $\vec{F}_{A\to B}$ for this situation.
**(c)** Explain why the interaction forces are different in the two cases. *Hint:* If you consider the two carts together as a system with mass 7.5 kg, what is the acceleration of each of carts $A$ and $B$? What does the *net* force on cart $A$ have to be to result in this acceleration?

**64. Spring Scale One** The spring scale in Fig. 3-51 reads 10.5 N. The cart moves toward the right with an acceleration of 3.5 m/s$^2$.

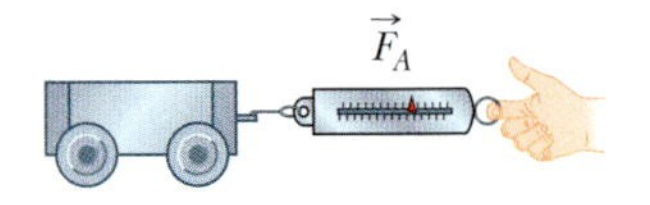

FIGURE 3-51 ■ Problem 64.

**(a)** Suppose a second spring scale is combined with the first and *acts in the same direction* as shown in Fig. 3-52. The spring scale $\vec{F}_A$ still reads 10.5 N. The cart now moves toward the right with an acceleration of 4.50 m/s$^2$. What is the *net* force on the cart? What does spring scale $\vec{F}_B$ read? Show your calculations and explain.

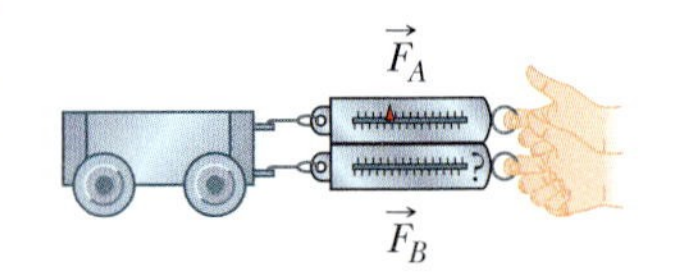

FIGURE 3-52 ■ Problem 64.

**(b)** Suppose a second spring scale is combined with the first and acts in the opposite direction as shown in Fig. 3-53. The spring scale $\vec{F}_A$ still reads 10.5 N.

FIGURE 3-53 ■ Problem 64.

The cart now moves toward the right with an acceleration of 2.50 m/s$^2$. What is the *net* force on the cart? What does spring scale $\vec{F}_B$ read? Show your calculations and explain.
**(c)** Which of Newton's first two laws apply to the situations in this problem?

**65. Spring Scale Two** Two forces are applied to a cart with two different spring scales as shown in Fig. 3-54. The spring scale $\vec{F}_A$ reads 15 N.

FIGURE 3-54 ■ Problem 65.

**(a)** The cart had an initial velocity of 0.00 m/s when the two forces were applied. It remains at rest after the combined forces are applied. What is the *net* force on the cart? What does spring scale $\vec{F}_B$ read? Show your calculations and explain.
**(b)** The cart had an initial velocity of +0.75 m/s and so it was moving to the right when the two forces were applied. It continues moving to the right at that same velocity after the combined forces are applied. What is the *net* force on the cart? What does spring scale $\vec{F}_B$ read? Show your calculations and explain.
**(c)** The cart had an initial velocity of −0.39 m/s and so it was moving to the left when the two forces were applied. It continues moving to the left at that same velocity after the combined forces are applied. What is the *net* force on the cart? What does spring scale $\vec{F}_B$ read? Show your calculations and explain.

**66. Fire Ladder** A physics student is standing on one of the steps of the fire ladder behind a building on campus doing a physics experiment. From there she drops a stone (without giving it any initial velocity) and notes that it takes approximately 2.45 s to hit the ground. The second time she throws the stone vertically upward and notes that it takes approximately 5.16 s for it to hit the ground.

**(a)** Calculate the height above the parking lot from which she releases the first stone.
**(b)** Calculate the initial velocity with which she has thrown the second stone upward.
**(c)** How high above the parking lot did the second stone rise before it started falling again?

**Star Trek Problem Problems 67 and 68 both involve the following:** "You are at the helm of the starship *Defiant* (*NCC-1764*), currently in orbit around the planet Iconia, near the Neutral Zone. Your mission: to rendezvous with a supply vessel at the other end of this solar system . . . You direct the impulse drive to be set at full power for leisurely half-light-speed travel . . . which should bring you to your destination in a few hours."* Assume that the diameter of the Iconian solar system is 100 Astronomical Units (an AU is the mean radius of the Earth's orbit about the Sun: 1AU = $1.49 \times 10^{11}$ m).

*Krauss, Lawrence, *The Physics of Star Trek* (New York: Harper Perennial, 1996), p. 3.

**67. Can You Stand the G-Forces?** In order to minimize the *g*-forces on you, suppose you decide to accelerate with a constant acceleration such that you reach half the speed of light ($c/2 = 1.5 \times 10^8$ m/s) at the midpoint of your trip and then start slowing down so you are at rest just in time to dock with the supply vessel at the other end of this solar system.

**(a)** Draw a single motion diagram showing the speeding-up and slowing-down processes.
**(b)** In a coordinate system in which you move along the positive $x$ axis, what is the direction and magnitude of your initial acceleration? In other words, is your acceleration positive or negative?
**(c)** In a coordinate system in which you move along the positive $x$ axis, what is the direction and magnitude of your acceleration while you are slowing down for your rendezvous with the supply vessel? In other words, is your acceleration positive or negative? (*Hint*: The answer to part (b) and a symmetry argument can save you some effort.)
**(d)** How long will your overall trip take?
**(e)** If the *Defiant* has a mass of $M = 2.850 \times 10^8$ kg, what is the thrust force (in newtons) needed to accelerate your starship?
**(f)** The amount of force you feel being impressed on you by the back of your seat as the starship picks up speed is proportional to your acceleration. A common way to measure typical forces you might feel is to calculate *g*-forces. This is done by comparing the acceleration you experience to the acceleration you would experience while falling freely close to the surface of the Earth. Thus, you can find *g*-forces by dividing your acceleration by 9.8 m/s$^2$. What *g*-forces would you experience while accelerating in the *Defiant*?
**(g)** The maximum sustained *g*-force that a human can stand is about 3 g. What would happen to you during your leisurely acceleration to half the speed of light?

**68. How Long Would a Trip Take If the Forces Were Bearable?** Let's take the trip at a more reasonable acceleration of 3 g.

**(a)** What would your acceleration be in m/s$^2$?
**(b)** How long would it take you, starting from rest, to get halfway (i.e., $d$ = 50 AU) across the Iconian solar system at this 3 g acceleration?
**(c)** What would your maximum speed be (i.e., the speed when you pass the $d$ = 50 AU mark)?
**(d)** How long would it take you to slow down at a 3 g acceleration for docking with the supply vessel? What is the total trip time? Is this feasible?

**69. The Demon Drop** The Demon Drop is a popular ride at the Cedar Point Amusement Park in Ohio. It allows four people to get into a little cage and fall freely for a while. Physics professor Bob Speers of Firelands College in Huron, Ohio, took a video tape of the drop. It is called DSON001. Use VideoPoint and Excel to analyze and develop a mathematical model that describes the fall.

**(a)** Include a printout of your spreadsheet model along with the answers to questions (b) through (e).
**(b)** According to your model, what is the equation you think describes the vertical position of the bottom of the cage as a function of time?
**(c)** According to your model, what is the acceleration of the cage?
**(d)** Can you find values of initial position and velocity that allow you to obtain a good agreement between your model graph and the graph of the data using the accepted value of the free fall acceleration close to the surface of the Earth of $\vec{a} = -g = -9.8$ m/s$^2$?

**(e)** Suppose a group of four people with an average mass of 65 kg each are put in the Demon Drop cage of mass $2.0 \times 10^3$ lb. What is the force on the whole falling system consisting of the cage and the people? Be sure to indicate the direction of the force.

**70. Force, Acceleration, and Velocity Graphs (a)** A force is applied to an object that experiences very little friction. This force causes the object to move resulting in the acceleration vs. time graph shown in Fig. 3-55. Draw a set of graph axes with the same number of time units as that shown in the acceleration graph and carefully sketch the *shape* of a possible graph of the force vs. time for the object.

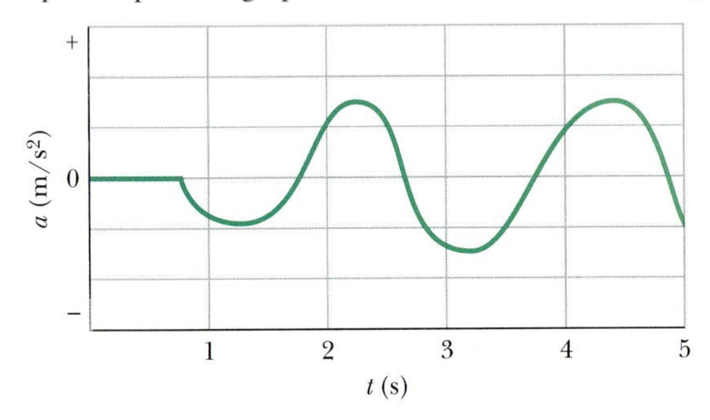


**FIGURE 3-55** ■ Problem 70.

**(b)** A force is applied to an object that experiences very little friction. This force causes the object to move resulting in the velocity vs. time graph shown in Fig. 3-56. Draw a set of axes with the same number of time units as that shown in the velocity graph and carefully sketch the shape of a possible graph of acceleration vs. time for the object.

**(c)** Refer to the velocity vs. time graph shown in part (b) and the acceleration vs. time graph you sketched. Draw a set of graph axes with the same number of time units as that shown in the velocity graph and carefully sketch the shape of a possible graph of force vs. time for the object.

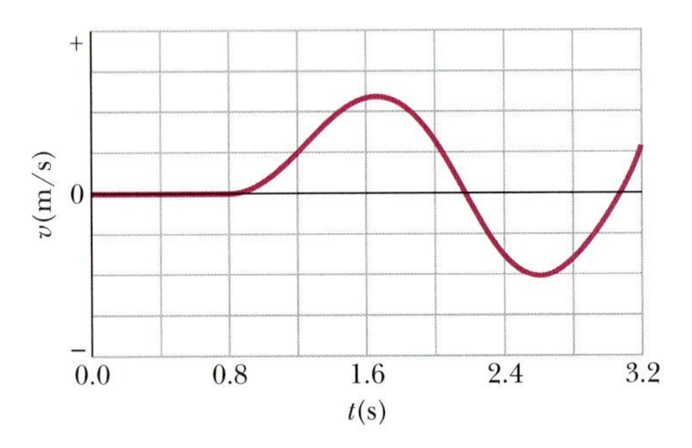


**FIGURE 3-56** ■ Problem 70.

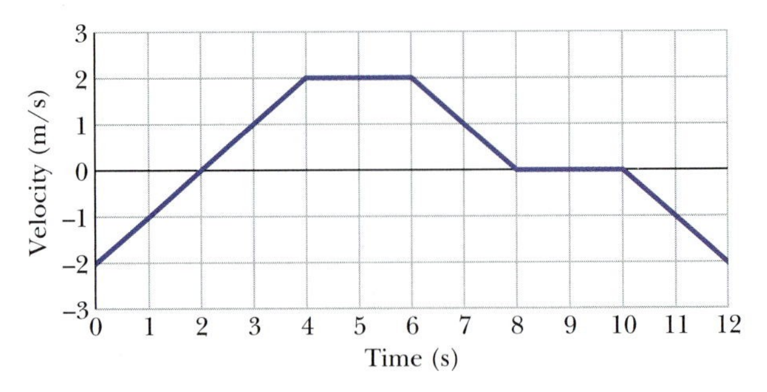


**FIGURE 3-57** ■ Problems 71 and 72.

**71. Force from Velocity One** Figure 3-57 shows the velocity vs. time graph for an object constrained to move along a line. The positive direction is to the right.

**(a)** At what times, or during what time periods, is the net force acting on the object zero?

**(b)** At what times, or during what time periods, is the net force acting on the object constant and nonzero.

**(c)** At what times, or during what time periods, is the net force acting on the object changing?

In each case, explain your reasoning. Describe how your reasoning is consistent or inconsistent with Newton's Laws of Motion. If there is no time or time period for which a given condition exists, state that explicitly.

**72. Force from Velocity Two** Figure 3-57 shows the velocity vs. time graph for an object constrained to move along a line. The positive direction is to the right.

**(a)** At what times, or during what time periods, is the net force on the object increasing and directed to the right?

**(b)** At what times, or during what time periods, is the net force on the object decreasing and directed to the right?

**(c)** At what times, or during what time periods, is the net force on the object constant and directed to the right?

**(d)** At what times, or during what time periods, is the net force on the object increasing and directed to the left?

**(e)** At what times, or during what time periods, is the net force on the object decreasing and directed to the left?

**(f)** At what times, or during what time periods, is the net force on the object constant and directed to the left?

In each case, explain your reasoning. Describe how your reasoning is consistent or inconsistent with Newton's Laws of Motion. If there is no time or time period for which a given condition exists, state that explicitly.

# 4 Vectors

For two decades, spelunking teams crawled, climbed, and squirmed through 200 km of Mammoth Cave and the Flint Ridge cave system, seeking a connection. The photograph shows Richard Zopf pushing his pack through the Tight Tube, far inside the Flint Ridge system. After 12 hours of "caving" along a labyrinthine route, Zopf and six others waded through a stretch of chilling water and found themselves in Mammoth Cave. Their breakthrough established the Mammoth-Flint cave system as the longest cave in the world.

**How can their final point be related to their initial point other than in terms of the actual route they covered?**

---

*The answer is found in this chapter.*

## 4-1 Introduction

As you already learned in Chapters 2 and 3, it is useful to use vectors to represent several of the physical quantities that were used in our study of one-dimensional motion. These quantities include position, displacement, velocity, acceleration, and force. In Chapters 5 and 6, vector mathematics will be used in conjunction with Newton's Laws to study two-dimensional motions such as that of objects that move horizontally while falling (projectile motion), circular motion, motion when friction forces are present, and motions on inclined surfaces.

In order to study motion in two dimensions, you must learn to represent and add two-dimensional vectors both graphically and mathematically. This is not as simple as it is in one dimension. For example, you learned in Chapter 2 that when an $x$ axis (or $y$ axis) is assigned to describe a particle-like object moving along a line, the sign (+ or −) of the $x$-component of its velocity vector indicates the direction of motion. However, if the particle is not moving along a straight line, then keeping track of the changes in the direction of its velocity is not just a matter of using a single plus or minus sign.

We will start our general consideration of vectors and vector operations by extending the definitions of vectors developed in Chapter 2 to two dimensions. We only discuss three-dimensional vectors very briefly in this chapter. However, we will return to them later in the book.

## 4-2 Vector Displacements

In order to define velocity and acceleration in more than one dimension, we need to start with the general definition of displacement. As is true in one dimension, it is useful to represent displacement vectors in two or three dimensions by arrows. For example, if a particle changes its position by moving from point $A$ point $C$, it turns out that its displacement, $\Delta\vec{r}$, can be represented by an arrow that points directly from $A$ to $C$, as shown in Fig. 4-1$a$. Remember, a displacement vector tells us nothing about the actual path that the particle takes. Thus both the curved path and the two straight paths from $A$ to $B$ and $B$ to $C$, shown in Fig. 4-1$a$, can lead a particle from point $A$ to point $C$, so the displacement vector between $A$ and $C$ is the same in both cases.

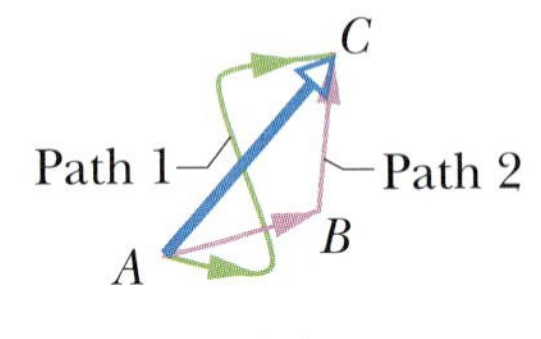


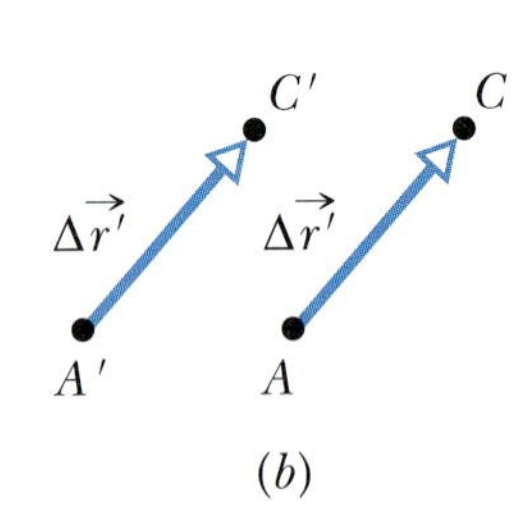


**FIGURE 4-1** ■ ($a$) A displacement vector for a particle's motion between points $A$ and $C$ can be represented by an arrow pointing from $A$ to $C$. Since displacement depends only on the relative locations of $A$ and $C$, both paths shown result in the same displacement. ($b$) The vectors pointing from $A$ to $C$ and from $A'$ to $C'$ also represent the same displacement, $\Delta\vec{r}$, since displacement represents a change in position rather than the positions themselves, so $\Delta\vec{r} = \Delta\vec{r}'$.

In Figure 4-1$b$, the arrows pointing from $A$ to $C$ and from $A'$ to $C'$ have the same magnitude and direction. Thus, they represent identical displacement vectors because they signify the same *change of position* for the particle. Thus a displacement vector that is shifted in space without changing its magnitude (length) and direction is the same vector.

The fact that a displacement vector represents only the overall effect of a motion, and not its detailed path, can lead to miscommunication. The Foxtrot cartoon, Fig. 4-2, shows what can happen when one person assumes that getting from point $A$ to point $B$ is what counts, while another cares how one gets there. Figure 4-3 is a vector diagram of the path Jason takes vs. the path Peter wants him to take.

**READING EXERCISE 4-1:** A soccer field has goals on its north and south end. Consider the following displacement of a soccer ball.

*The ball is initially sitting in the center of the field. It is kicked toward the west. After traveling 3 m, it is kicked toward the north. It travels 6 m before a player stops it.*

Which of the following displacements, if any, are identical to the displacement described above?

(1) The ball is initially sitting directly in front of the south goal. It is kicked toward the east. After traveling 9 m it is kicked toward the north. It travels 6 m before a player kicks it due west. After it travels for 12 m, another player stops it.
(2) The ball is initially sitting in the center of the field. It is kicked toward the east. After traveling 3 m, it is kicked toward the south. It travels 6 m before a player stops it. ■

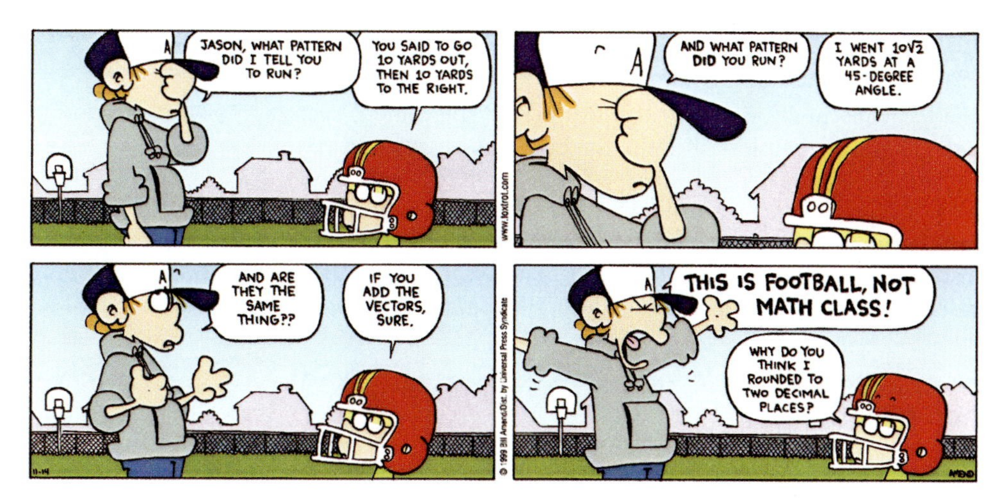


**FIGURE 4-2** ■ Foxtrot cartoon in which Jason is using physics to focus on the path independence of displacement, while his brother Peter is interested in both Jason's displacement and his actual path. 

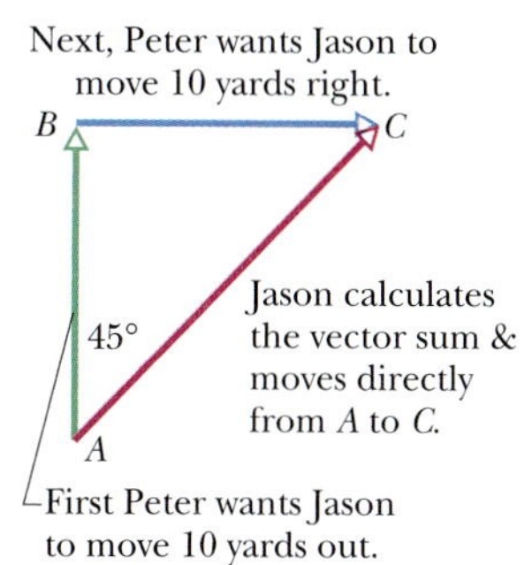


**FIGURE 4-3** ■ Depiction of a vector sum of two successive displacements.

## TOUCHSTONE EXAMPLE 4-1: Jogging in a Circle

Sara is running laps on a circular track. Each full lap is 400 m. If she starts at the northernmost point on the track, initially going east, find her displacement (magnitude and direction) after she has run (a) 100 m, (b) 200 m, (c) 300 m, and (d) 400 m.

**SOLUTION** ■ The **Key Idea** here is that Sara's displacement is a vector whose *magnitude* is the *straight-line* distance from her starting position to her current position. The *direction* of her displacement vector *points* straight from her starting point to her current location.

(a) When Sara has gone 100 m, she has gone one-quarter of the way around the track, as shown in Fig. 4-4*a*. As you can see there, her current position is the same as it would have been if she had gone a distance of one radius of the track due south and then the same distance due east. Since the angle between lines $AP$ and $PB$ in Fig. 4-4*a* is 90°, we can use the Pythagorean theorem to find the straight-line distance from $A$ to $B$. This distance is just

$$d_{AB} = \sqrt{R^2 + R^2} = \sqrt{2R^2} = \sqrt{2}R.$$

Since the track has a circumference of 400 m, its radius is $R = 400\text{ m}/(2\pi) = 63.6$ m. So the *magnitude* of Sara's displacement after she has run the 100 m from $A$ to $B$ is

$$d_{AB} = (\sqrt{2})(400\text{ m})/(2\pi) = 90.0\text{ m}. \qquad \text{(Answer)}$$

The *direction* of her displacement is seen from Fig. 4-4*a* to be due southeast. (Answer)

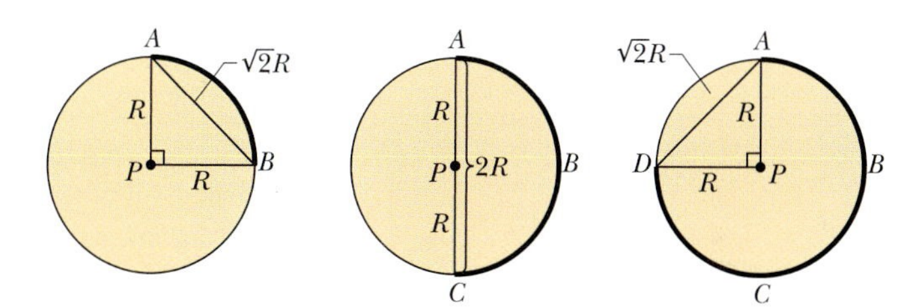


**FIGURE 4-4** ■ (*a*) After Sara has run 100 m from $A$ to $B$. (*b*) After she has run 200 m from $A$ to $C$. (*c*) After she has run 300 m from $A$ to $D$.

(b) When Sara has run 200 m from her starting point, we can see in Fig. 4-4*b* that she has covered exactly half the circumference of the track. This places her a distance $2R = 127$ m away from her starting point and due south of it. (Answer)

(c) Now Sara has covered three-quarters of the track's circumference, as shown in Fig. 4-4*c*. Her distance from her starting point is *the same* as it was when she had run only one-quarter of the way around the track, so now the *magnitude* of her displacement is once again $\sqrt{2}R = 90.0$ m. But the *direction* of her displacement from her starting point is due south*west*.

(d) Now that Sara has run 400 m, she has "come full circle" and returned to her starting point. Since her current position is the same as her starting position, her displacement from her starting position is now *zero*. (Answer)

## 4-3 Adding Vectors Graphically

The basic method for graphical addition of displacement vectors involves considering a single vector that describes the final outcome of two displacements. For example, in Fig. 4-2 big brother, coach Peter, wanted Jason to get from point $A$ to point $C$ by undergoing first one displacement by moving outward (forward along the field) for 10 yards from point $A$ to $B$, and then moving to the right for 10 yards from point $B$ to $C$. Instead Jason used the rules of vector addition to go directly from $A$ to $C$ by traveling a distance of $10\sqrt{2}$ yards at an angle of 45 degrees with respect to the outward direction as shown in Fig. 4-3.

### Addition

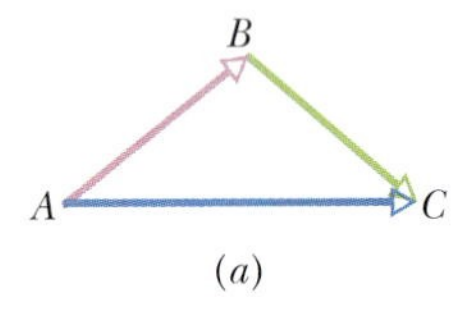


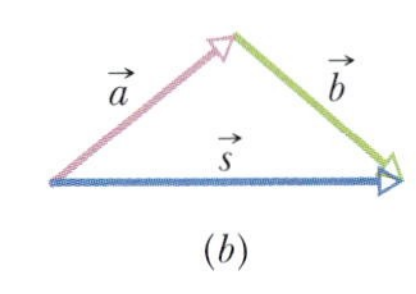


**FIGURE 4-5** ■ (*a*) $AC$ is the sum of the vectors $AB$ and $BC$. (*b*) The same vectors with alternate labels $\vec{a}$, $\vec{b}$ and $\vec{s}$.

Suppose that, as in the vector diagram of Fig. 4-5*a*, a particle moves from $A$ to $B$ and then later from $B$ to $C$. We can represent its overall displacement (no matter what its actual path) with two successive displacement vectors, $AB$ and $BC$. The *net* displacement of these two displacements is a single displacement from $A$ to $C$. We call $AC$ the **vector sum** (or **resultant**) of the vectors $AB$ and $BC$. This sum is not the usual algebraic sum.

In Figure 4-5*b*, we redraw the vectors of Figure 4-5*a* and relabel them in the way that we shall use from now on—namely, with an arrow over a symbol, as in $\vec{a}$. In adding two or more vectors, it's OK to move them to make the addition simpler, as long as the length of each vector and its orientation don't change. Recall from Chapter 2 that if we want to indicate only the magnitude or size of the vector (a quantity that lacks a sign or direction), we shall use the absolute value symbol, as in $|\vec{a}|$, or drop the arrow, as in $a$.

We can represent the relationship among the three vectors in Figure 4-5*b* with the vector equation

$$\vec{s} = \vec{a} + \vec{b}, \tag{4-1}$$

which says that the vector $\vec{s}$ is the vector sum or resultant of vectors $\vec{a}$ and $\vec{b}$. The symbol + in $\vec{s} = \vec{a} + \vec{b}$ and the words "sum" and "add" have different meanings for vectors than they do in algebra because they involve both magnitude and direction.

Figure 4-5 suggests a general procedure for adding two vectors $\vec{a}$ and $\vec{b}$ graphically: (1) On paper, sketch vector $\vec{a}$ to some convenient scale and at the proper angle. (2) Sketch vector $\vec{b}$ to the same scale, with its tail at the head of vector $\vec{a}$, again at the proper angle. (3) The vector sum $\vec{s}$ is the vector that extends from the tail of $\vec{a}$ to the head of $\vec{b}$.

Vector addition, defined in this way, has two important properties. First, the order of addition does not matter. That is,

$$\vec{a} + \vec{b} = \vec{b} + \vec{a} \quad \text{(commutative law).} \tag{4-2}$$

Second, when there are more than two vectors, we can group them in any order as we add them, so

$$\left(\vec{a} + \vec{b}\right) + \vec{c} = \vec{a} + \left(\vec{b} + \vec{c}\right) \quad \text{(associative law).} \tag{4-3}$$

A vector is not simply any entity that has both magnitude and direction. In fact, the rules for vector addition and the associative and commutative properties of vector addition are defining characteristics of vectors.

### Subtraction

Subtracting one vector from another can be considered as the addition of one vector to the **additive inverse** of the other. The additive inverse (sometimes known as the "negative of a vector") is simply the vector we must add to the original vector to get zero. If we want to define the additive inverse of a vector $\vec{b}$, denoted as $-\vec{b}$, we can start with the understanding that $\vec{b} + (-\vec{b})$ should equal zero. Using the graphical method of adding vectors that we discussed above, this demands that the vector $-\vec{b}$ has the same magnitude as $\vec{b}$, but points in the opposite direction so that the two vectors cancel, as shown in Fig. 4-6. Thus, adding vector $\vec{b}$ to its additive inverse gives

$$\vec{b} + (-\vec{b}) = 0.$$

Finding an additive inverse is commonly referred to as **inverting** a vector. Adding $-\vec{b}$ has the effect of subtracting $\vec{b}$. We use this property to define the difference between any two vectors such as $\vec{d} = \vec{a} - \vec{b}$ as

$$\vec{d} = \vec{a} - \vec{b} = \vec{a} + (-\vec{b}) \quad \text{(vector subtraction as a form of addition).} \tag{4-4}$$

That is, we find the difference vector $\vec{d}$ by adding the vector $-\vec{b}$ to the vector $\vec{a}$. Figure 4-7 shows how this is done geometrically.

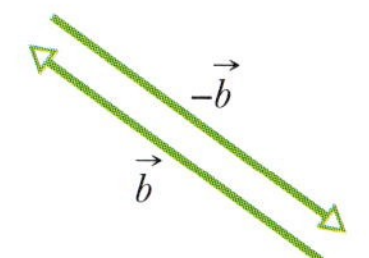


**FIGURE 4-6** ■ The vectors $\vec{b}$ and $-\vec{b}$ have the same magnitude and opposite directions.

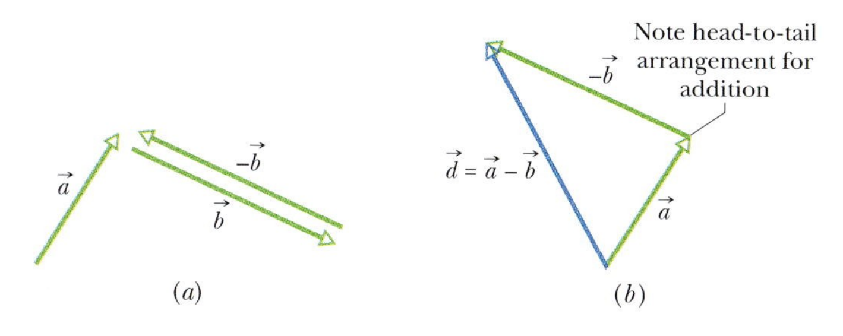


**FIGURE 4-7** ■ Consider the vectors $\vec{a}$ and $\vec{b}$. To subtract vector $\vec{b}$ from vector $\vec{a}$, create vector $-\vec{b}$ from vector $\vec{b}$ as shown in Fig. 4-6. Then add vector $-\vec{b}$ to vector $\vec{a}$.

As another example, consider a car that speeds up (accelerates) along a straight road. The change in the car's velocity is given by $\Delta\vec{v} = \vec{v}_2 - \vec{v}_1$ and the car's average acceleration by $\langle\vec{a}\rangle = \Delta\vec{v}/\Delta t$. The use of vectors to depict how $\Delta\vec{v}$ can be found by vector subtraction is illustrated in Fig. 4-8.

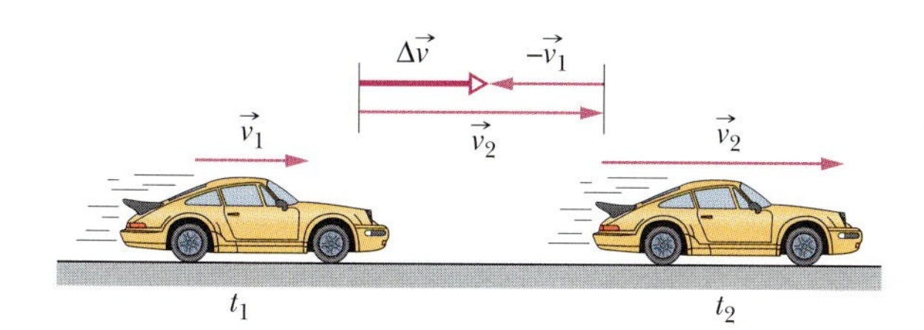


**FIGURE 4-8** ■ Diagram showing how to find the change in velocity of a race car by taking a one-dimensional vector difference in which $\vec{v}_2$ is added to $-\vec{v}_1$ to get $\Delta\vec{v}$.

As in scalar algebra, we can move a term that includes a vector symbol from one side of a vector equation to the other, but we must change its sign. For example, if we are given $\vec{d} = \vec{a} - \vec{b}$ and need to solve for $\vec{a}$, we can rearrange the equation as $\vec{a} = \vec{d} + \vec{b}$. Remember, although we have used displacement vectors here, the rules for addition and subtraction hold for vectors of all kinds, whether they represent velocities, accelerations, forces, or any other vector quantity. However, we can add only vectors of the same kind. For example, we can add two displacements, or two velocities, but adding a displacement and a velocity makes no sense. In the arithmetic of scalars, that would be like trying to add 21 s and 12 m.

**READING EXERCISE 4-2:** The magnitudes of displacements $\vec{a}$ and $\vec{b}$ are 3 m and 4 m, respectively, and $\vec{c} = \vec{a} + \vec{b}$. Considering various orientations of $\vec{a}$ and $\vec{b}$, what are (a) the maximum possible magnitude for $\vec{c}$ and (b) the minimum possible magnitude? ■

## 4-4 Rectangular Vector Components

You have learned a method to find the vector sum or resultant of two vectors that do not point along the same line. Many times it is useful to do the opposite and **decompose** or **resolve** a vector into two or more vectors, which can be added to create the original vector. For example, consider what happens to the motion of a particle-like object when two forces that are not acting in the same direction are applied to it. It turns out that the object will accelerate as if a single force that is the vector sum of the forces is acting on it. In cases where a force vector can be resolved into two or more vectors, it is possible to break down even complex situations into simpler one-dimensional ones, so the skill of resolving vectors is very powerful. We will begin by considering how to describe a two-dimensional vector in a rectangular coordinate system as the sum of two one-dimensional vectors.

### Resolving a Vector

It is typical to describe two dimensional vectors in a coordinate system in which the $x$ and $y$ axes are drawn in the plane of the page. We choose axes that are parallel to the edges of the paper as shown in Fig. 4-9$a$.

We already know how to represent, add, and subtract vectors that are parallel to an $x$ axis or $y$ axis. For this reason, it is convenient to decompose our vector into two component vectors—one parallel to the $x$ axis and the other parallel to the $y$ axis. In this case, the vector $\vec{a}$ is the sum of two component vectors $\vec{a}_x$ and $\vec{a}_y$ as shown in Fig. 4-9$a$. Therefore, $\vec{a} = \vec{a}_x + \vec{a}_y$.

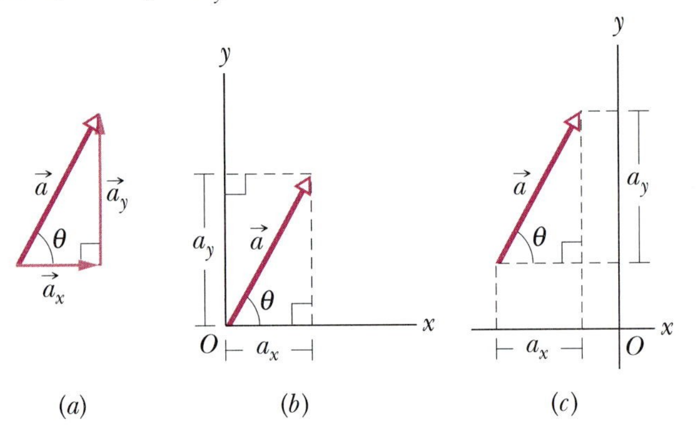


**FIGURE 4-9** ■ ($a$) The component vectors form the legs of a right triangle whose vector sum is the original vector. ($b$) The components $a_x$ and $a_y$ of vector $\vec{a}$ are determined by projections of the tail and tip of the vector on each axis. ($c$) The values of the components are unchanged if the vector is shifted, as long as its magnitude and orientation are the same.

As you can see in Fig. 4-9$a$, we have chosen the length of the component vectors $\vec{a}_x$ and $\vec{a}_y$ so that they conveniently add up to form vector $\vec{a}$. Since $\vec{a}_x = a_x\hat{\text{i}}$ and $\vec{a}_y = a_y\hat{\text{j}}$ (Section 3-2), we can say that we have expressed the component vectors in terms of their components $a_x$ and $a_y$. Note that these components have *no arrows*. We define the **rectangular component** of a vector to be the projection of the vector on an axis. In Fig. 4-9$a$, for example, $a_x$ is the component of vector $\vec{a}$ on (or along) the $x$ axis and $a_y$ is the component along the $y$ axis. The practical way to get the projection or component of a vector along an axis is to draw lines from the two ends of the vector perpendicular to that axis, as shown in Fig. 4-9$b$ or $c$.

In order to understand the idea of a vector component as the projection of the vector onto an axis, think about taking a distant spotlight and shining it onto the vector. The component (or projection) is the length of the shadow that is cast by the vector on one of the axes. For example, take a light and shine it straight down on the vector shown in Fig. 4-9$b$ from the top of the page. The shadow of the vector will fall along the $x$ axis. The length of the shadow is the $x$-component of this vector as shown in Fig. 4-9$b$. We would make the $y$-component of the vector visible as the length of a shadow by shining a light on the vector from the right side of the page. Then the shadow of the vector falls along the $y$ axis and its length is the vector's $y$-component.

Figure 4-9 illustrates that the components (projections) of the vector do not change if we simply move the vector around within our coordinate system. In other words, when you shift a vector without changing its direction, its components, which are lengths, do not change. The length and direction of the projection on an axis tells what the vector component is. The projection of a vector on an $x$ axis is called its **$x$-component** and is denoted as $a_x$. The projection on the $y$ axis is called its **$y$-component** and is denoted $a_y$. We call the process of finding the components of a vector in a chosen coordinate system **resolving the vector.**

## Positive and Negative Components

The components of a vector can be positive or negative depending on the overall orientation of the vector we are resolving relative to the coordinate system we have chosen. In a standard coordinate system, we indicate this by designating components that point up or to the right as positive; then those that point down or to the left are negative. Graphically, small arrowheads on each component can represent its direction. For example, in Fig. 4-9, $a_x$ and $a_y$ are both positive because $\vec{a}$ extends in the positive direction of both axes. (Note the small arrowheads on the components, to indicate their direction.) If we were to reverse vector $\vec{a}$, then both components would be negative and their arrowheads would point toward negative $x$ and $y$. Resolving a different vector $\vec{b}$ shown in Fig. 4-10 yields a positive component $b_x$ and a negative component $b_y$ if we stick with the standard coordinate system.

## Using Sines and Cosines to Find Components

In general, a two-dimensional vector has two components. As Figs. 4-9$b$ and $c$ imply, we can find the value of the components of $\vec{a}$ in Fig. 4-9$b$ using the sine and cosine relations. Since

$$\cos\theta = \frac{\text{adjacent side}}{\text{hypotenuse}} \quad \text{and} \quad \sin\theta = \frac{\text{opposite side}}{\text{hypotenuse}},$$

for the right triangle in Fig. 4-9$a$, the magnitude of the vector $\vec{a}$ is the hypotenuse, and

$$\cos\theta = \frac{a_x}{a} \quad \text{and} \quad \sin\theta = \frac{a_y}{a},$$

where $\theta$ is the angle that the vector $\vec{a}$ makes with the positive direction of the $x$ axis. Remember, the symbols $a$ and $|\vec{a}|$ provide alternate notations for the magnitude of $\vec{a}$, and $a_x$ and $a_y$ are the $x$- and $y$-components of $\vec{a}$, respectively. Rearranging these relationships, we find

$$a_x = a\cos\theta \quad \text{and} \quad a_y = a\sin\theta. \tag{4-5}$$

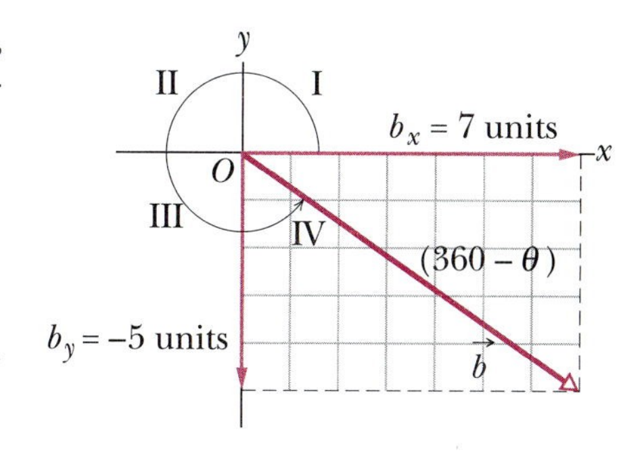


**FIGURE 4-10** ■ The component of $\vec{b}$ on the $x$ axis is positive, and the component on the $y$ axis is negative.

## Reconstructing a Vector from Components

Look at Fig. 4-9$a$ again. It shows that $\vec{a}$ and its $x$- and $y$-components form a right triangle. That means that we can reconstruct a vector $\vec{a}$ from its components. Graphically, we can arrange the components head to tail and then find $\vec{a}$ by completing a right triangle with the vector forming the hypotenuse, from the tail of one component to the head of the other component. We can also get the magnitude of $\vec{a}$ algebraically by using the Pythagorean theorem. That is,

$$a = |\vec{a}| = \sqrt{a_x^2 + a_y^2}.$$

Once a vector has been resolved into its components along a set of axes, the components themselves can be used in place of the vector. For example, $\vec{a}$ in Fig. 4-9*b* is represented (completely determined) by $|\vec{a}|$ and $\theta$. It can also be completely determined by its components $a_x$ and $a_y$. Both pairs of values contain the same information. If we know a vector in *component notation* ($a_x$ and $a_y$) and want it in *magnitude-angle notation* ($a$ and $\theta$), we can use the equations

$$a = |\vec{a}| = \sqrt{a_x^2 + a_y^2} \quad \text{and} \quad \theta = \tan^{-1} = \left(\frac{a_y}{a_x}\right) \tag{4-6}$$

to transform the components into a magnitude and direction. Thus, it is common to represent a two-dimensional vector by ordered rectangular components such as $a_x$ and $a_y$. However, in studying circular motion, it is more convenient to use polar coordinates $(a, \theta)$ to describe a vector. Finding $\theta$ using the inverse tangent must be done with care since the calculated value of $\theta$ must be replaced with $\theta + \pi$ if the $\vec{a}$ vector is in the second (II) or third (III) quadrants as shown in Fig. 4-10.

In the more general three-dimensional case, when using rectangular coordinates, we need to consider another axis, called the $z$ axis, that is mutually perpendicular to the other two axes. In three dimensions the components $a_x$, $a_y$, and $a_z$ can be used to represent a vector in a rectangular coordinate system. If a spherical coordinate system is used instead, then a magnitude and two angles (say, $|\vec{a}|$, $\theta$, and $\phi$) can be used to represent a vector. Three-dimensional vectors are used in the study of rotational motion. However, you will not be using three-dimensional vectors in the next few chapters.

**READING EXERCISE 4-3:** In the figures that follow, which of the indicated methods for combining the $x$- and $y$-components of vector $\vec{a}$ are correct?

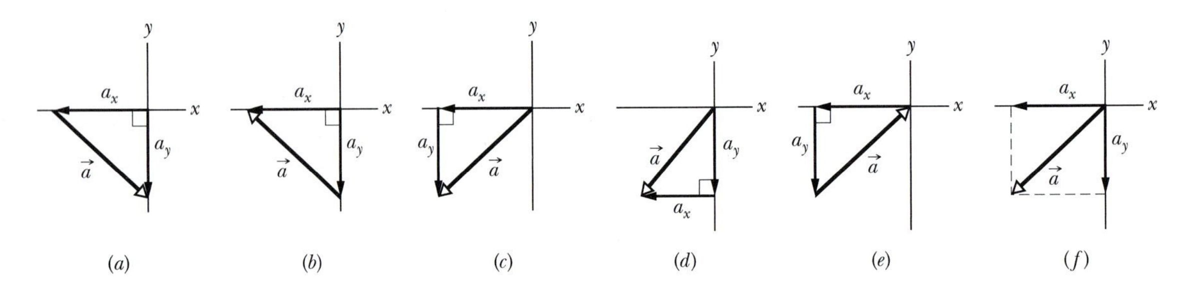


■

**READING EXERCISE 4-4:** Consider the following standard vector. Which vectors have been correctly repositioned so their components are the same as those of the original vector?

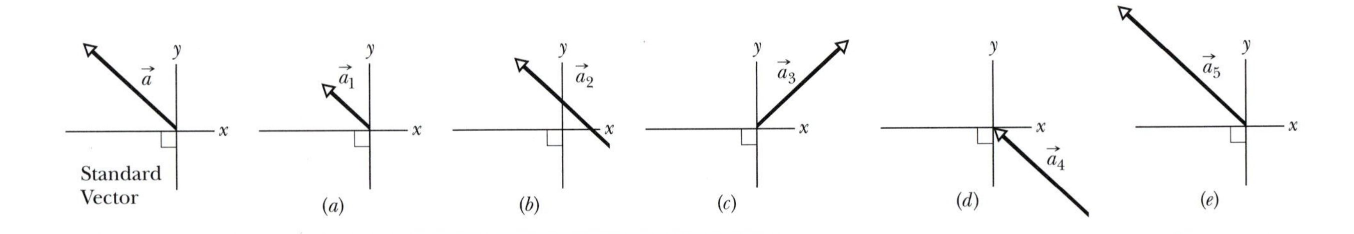


■

## TOUCHSTONE EXAMPLE 4-2: Spelunking

The 1972 team that connected the Mammoth-Flint cave system went from Austin Entrance in the Flint Ridge system to Echo River in Mammoth Cave (Fig. 4-11*a*). Their horizontal travel (parallel to the Earth's surface) was a net 1.0 km westward and 4.2 km southward. What was their horizontal displacement vector from start to finish?

**SOLUTION** ■ The **Key Idea** here is that we have the components of a two-dimensional vector, and we need to find each vector's magnitude and direction to specify the displacement vector. We first choose a two-dimensional coordinate axis and then draw the $x$- and $y$-components of displacement as in Fig. 4-11*b*. The components ($d_x$ = 1.0 km west and $d_y$ = 4.2 km south) form the legs of a horizontal right triangle. The team's horizontal displacement forms the hypotenuse of the triangle, and its magnitude $d$ is given by the Pythagorean theorem:

$$|\vec{d}| = \sqrt{(1.0\text{ km})^2 + (4.2\text{ km})^2} = 4.3\text{ km}.$$

Also from Fig. 4-11*b*, we see that this horizontal displacement is directed south of due west by an angle $\theta$ given by

$$\tan\theta = \frac{dy}{dx} = \frac{4.2\text{ km}}{1.0\text{ km}},$$

so

$$\theta = \tan^{-1}\frac{4.2\text{ km}}{1.0\text{ km}} = 77°. \qquad \text{(Answer)}$$

In summary, the team's horizontal displacement vector had a magnitude of 4.3 km and was at an angle of 77° south of west. The team also traveled a net distance of 25 m upward. The net vertical motion was insignificant compared to the horizontal motion, so we ignored it. However, the relatively small net vertical displacement was of no comfort to the team. They had to climb up and down countless times to get through the cave. The route they actually covered was quite different from the horizontal displacement vector, which merely points in a straight line from start to finish.

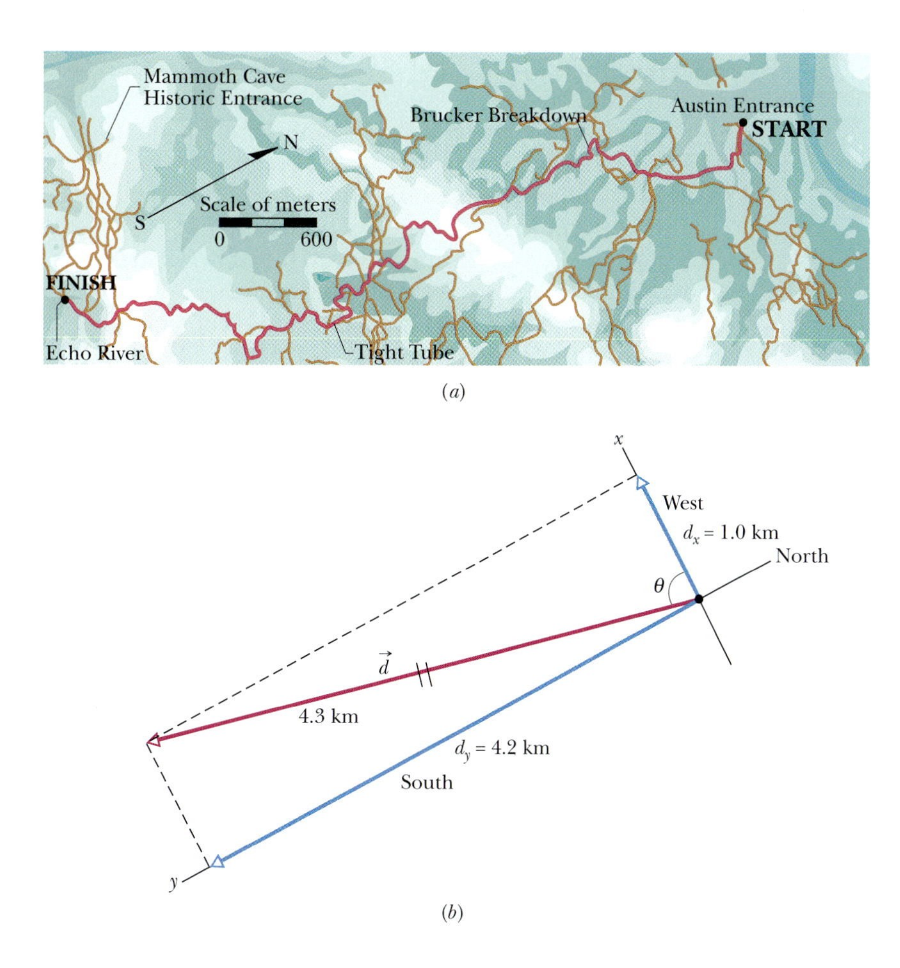


**FIGURE 4-11** ■ (*a*) Part of the Mammoth-Flint cave system, with the spelunking team's route from Austin Entrance to Echo River indicated in red. (*b*) The components of the team's horizontal displacement $\vec{d}$. They are to scale, but at a different scale than in part (*a*). (Adapted from a map by the Cave Research Foundation.)

## 4-5 Unit Vectors

In Section 2-2, we defined a unit vector as a dimensionless vector that points in the direction along a coordinate axis that is chosen to be positive. Its sole purpose is to point—that is, to specify a direction. The unit vectors that point in the positive directions of the $x$, $y$, and $z$ axes are labeled $\hat{i}$, $\hat{j}$, and $\hat{k}$ (Fig. 4-12), where the hat ^ (or caret) is used to note that these vectors are special.

The arrangement of axes in Fig. 4-12 is called a **right-handed coordinate system** because it can be constructed using the thumb and fingers of the right hand. There are several legitimate ways to construct a right-handed coordinate system using the right hand. One method is depicted in Fig. 4-12. The system remains right-handed if it is rotated rigidly to any new orientation. If we used the left hand to construct a coordinate system, $\hat{i}$ would point in the *opposite* direction than it does in Fig. 4-12 while the relative orientations of $\hat{j}$ and $\hat{k}$ would remain unchanged as in Fig. 4-13. Since the use of a right-handed system is standard in the scientific community, we use it exclusively in this book.

Unit vectors are very useful for expressing three-dimensional vectors; for example, we can express any vector in terms of the coordinate system in Fig. 4-12 as

$$\vec{a} = a_x\hat{i} + a_y\hat{j} + a_z\hat{k}. \tag{4-7}$$

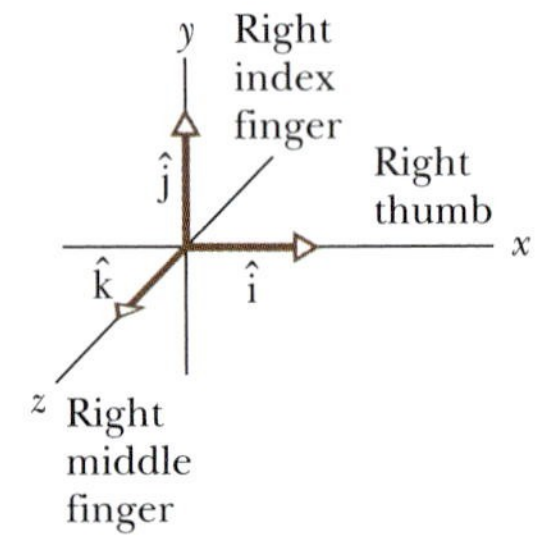

**FIGURE 4-12** ■ Unit vectors $\hat{i}$, $\hat{j}$, and $\hat{k}$ define the directions of a standard right-handed coordinate system. This system gets its name from the fact that the positive directions of the $x$, $y$, and $z$ axes can be determined by the directions of fingers on a right hand. For example when the thumb, index finger, and middle finger of a right hand are arranged so they are at right angles to each other, they determine the directions of the positive $x$, $y$, and $z$ axes, respectively.

The quantities $a_x\hat{i}$, $a_y\hat{j}$, and $a_z\hat{k}$ are vectors called the **component vectors** of $\vec{a}$. The quantities $a_x$, $a_y$, and $a_z$ are called, respectively, the ***x*-component, *y*-component,** and ***z*-component** of $\vec{a}$ (or, as before, simply its components along the axes).

*Note*: The components $a_x$, $a_y$, and $a_z$ are sometimes referred to in other books and articles by different names. They have been called "vector components" since the subscripts $x$, $y$, and $z$ reveal what unit vectors can be used to construct the vectors that lie along each axis. Also they have incorrectly been called "scalar components." *However, they are not scalars.* Real scalars do not change when the coordinate axes are rotated, and the $x$-component, $y$-component, and $z$-component of a vector can change whenever a coordinate axis is rotated.

**READING EXERCISE 4-5:** (a) Using the procedure outlined in Fig. 4-12 and your left hand, sketch a left-handed coordinate system that depicts the positive $x$, $y$, and $z$ axes and put the unit vectors, $\hat{i}$, $\hat{j}$, $\hat{k}$, in place. (b) Describe how the left-handed system differs from the right-handed one. ■

## 4-6 Adding Vectors Using Components

There are many different physical situations in which you will need to be able to add vectors in order to understand what is going on. For example, in Chapter 3 we found that we needed to add together force vectors to understand the motion of an object on which more than one force acts. Although we can add vectors geometrically using a sketch, or, if we have a vector-capable calculator, we can add them directly on the screen, perhaps the most practical method for finding the sum, $\vec{s}$, of two vectors is to combine their components, axis by axis.

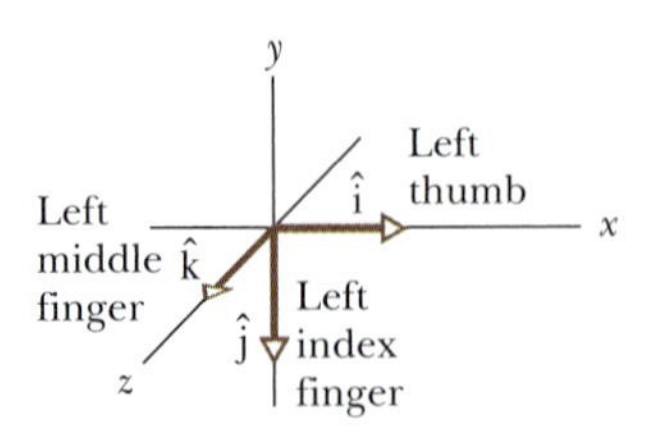

**FIGURE 4-13** ■ A nonstandard left-handed coordinate system that is constructed using the same procedures shown in Fig. 4-12, using the left hand.

To start, consider the mathematical statement for a vector sum in two dimensions

$$\vec{s} = \vec{a} + \vec{b}. \tag{4-8}$$

This statement implies that the vector $\vec{s}$ is the same as the vector $(\vec{a} + \vec{b})$. If this is so, then we can derive the relationships between the components of $\vec{s}$ and those of $\vec{a}$ and $\vec{b}$ mathematically as follows:

$$\vec{s} = (s_x\hat{\mathrm{i}} + s_y\hat{\mathrm{j}}) = \vec{a} + \vec{b} = (a_x\hat{\mathrm{i}} + a_y\hat{\mathrm{j}}) + (b_x\hat{\mathrm{i}} + b_y\hat{\mathrm{j}}) = (a_x + b_x)\hat{\mathrm{i}} + (a_y + b_y)\hat{\mathrm{j}}.$$

Thus, each component of $\vec{s}$ must be the same as the corresponding component of $(\vec{a} + \vec{b})$:

$$s_x = a_x + b_x, \tag{4-9}$$

$$s_y = a_y + b_y. \tag{4-10}$$

In other words, to add vectors $\vec{a}$ and $\vec{b}$, we must first resolve the vectors into their components. Next we must combine these components—taking direction (and thus sign) into account—axis by axis. This gives us the components of the sum vector $\vec{s}$. This is shown in Fig. 4-14. Once we get to this point, we have to make a choice about how to express the result. We can either:

**(a)** express $\vec{s}$ in unit-vector notation as $\vec{s} = s_x\hat{\mathrm{i}} + s_y\hat{\mathrm{j}}$, or

**(b)** combine the components of $\vec{s}$ to get $\vec{s}$ itself and express the vector in magnitude-angle notation, where $|\vec{s}| = \sqrt{s_x^2 + s_y^2}$ and $\tan\theta = s_y/s_x$.

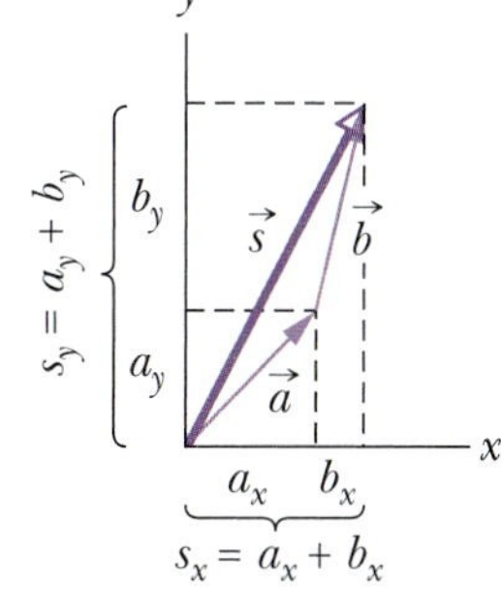


**FIGURE 4-14** ■ Diagram showing how a vector sum can be constructed by adding component vectors.

This procedure for adding vectors by components also applies to vector subtractions. Recall that a subtraction such as $\vec{d} = \vec{a} - \vec{b}$ can be rewritten as an addition $\vec{d} = \vec{a} + (-\vec{b})$. To subtract, we simply add $\vec{a}$ and $-\vec{b}$ by components to get

$$d_x = a_x - b_x, \tag{4-11}$$

and

$$d_y = a_y - b_y, \tag{4-12}$$

where

$$\vec{d} = d_x\hat{\mathrm{i}} + d_y\hat{\mathrm{j}}.$$

**READING EXERCISE 4-6:** (a) In the figure, what are the signs of the $x$-components of $\vec{d}_1$ and $\vec{d}_2$? (b) What are the signs of the $y$-components of $\vec{d}_1$ and $\vec{d}_2$? (c) What are the signs of the $x$- and $y$-components of $\vec{d}_1 + \vec{d}_2$?

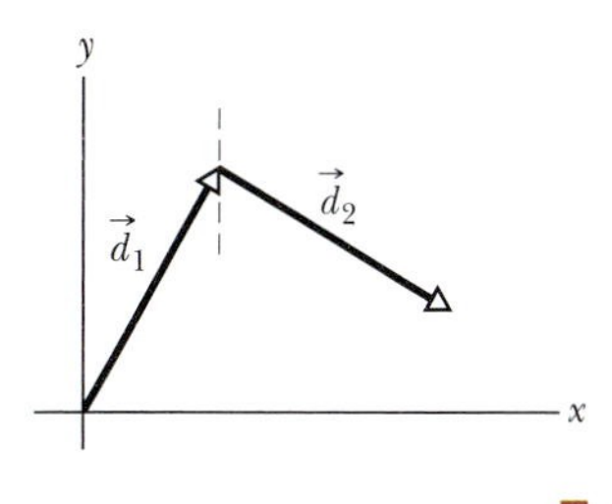


## TOUCHSTONE EXAMPLE 4-3: Three Vectors

Figure 4-15$a$ shows the following three vectors:

$$\vec{a} = (4.2\text{ m})\hat{\mathrm{i}} - (1.5\text{ m})\hat{\mathrm{j}},$$

$$\vec{b} = (-1.6\text{ m})\hat{\mathrm{i}} + (2.9\text{ m})\hat{\mathrm{j}},$$

and

$$\vec{c} = (-3.7\text{ m})\hat{\mathrm{j}}.$$

What is their vector sum $\vec{r}$, which is also shown?

**SOLUTION** ■ The **Key Idea** here is that we can add the three vectors by components, axis by axis. For the $x$ axis, we add the $x$-components of $\vec{a}$, $\vec{b}$, and $\vec{c}$ to get the $x$-component of $\vec{r}$:

$$r_x = a_x + b_x + c_x$$
$$= 4.2\text{ m} - 1.6\text{ m} + 0 = 2.6\text{ m}.$$

Similarly, for the $y$ axis,

$$r_y = a_y + b_y + c_y$$
$$= -1.5\text{ m} + 2.9\text{ m} - 3.7\text{ m} = -2.3\text{ m}.$$

Another **Key Idea** is that we can combine these components of $\vec{r}$ to write the vector in unit-vector notation:

$$\vec{r} = (2.6\text{ m})\hat{\mathrm{i}} - (2.3\text{ m})\hat{\mathrm{j}},$$

where $(2.6\text{ m})\hat{\text{i}}$ is the vector component of $\vec{r}$ along the $x$ axis and $-(2.3\text{ m})\hat{\text{j}}$ is that along the $y$ axis. Figure 4-15$b$ shows one way to arrange these vector components to form $\vec{r}$. (Can you sketch the other way?)

A third **Key Idea** is that we can also answer the question by giving the magnitude and an angle for $\vec{r}$. From Eq. 4-6, the magnitude is

$$r = \sqrt{(2.6\text{ m})^2 + (-2.3\text{ m})^2} \approx 3.5\text{ m}, \qquad \text{(Answer)}$$

and the angle (measured from the positive direction of $x$) is

$$\theta = \tan^{-1}\left(\frac{-2.3\text{ m}}{2.6\text{ m}}\right) = -41°, \qquad \text{(Answer)}$$

where the minus sign means that the angle is in the fourth quadrant.

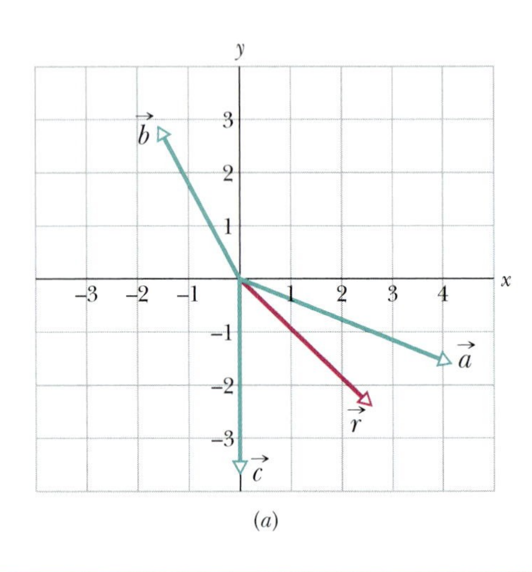


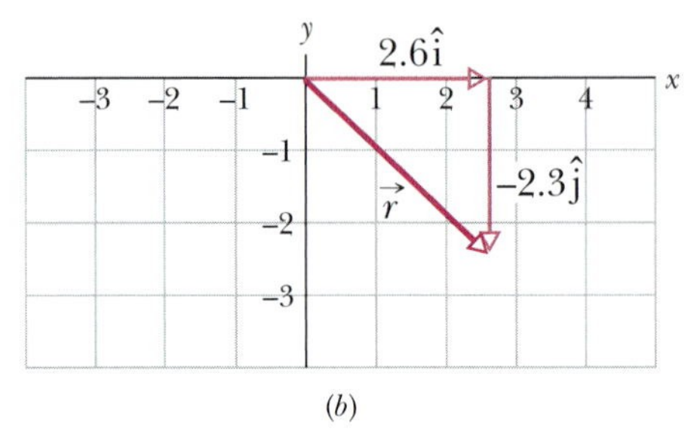


**FIGURE 4-15** ■ Vector $\vec{r}$ is the vector sum of the other three vectors.

## 4-7 Multiplying and Dividing a Vector by a Scalar

We have encountered situations in which we need to multiply or divide a vector by a scalar. For example, we must divide a one-dimensional force vector by a scalar mass to predict the acceleration of an object. Conversely, if we measure an object's acceleration vector in one dimension, we need to multiply the vector by its mass to determine the net force acting on the object. The use of a time interval, $\Delta t$, to create a velocity vector from a displacement vector is another example of a scalar being divided into a vector. According to the rules of mathematics, if we multiply or divide a vector by a scalar we should get a new vector. Dividing a vector by a scalar $s$, can always be transformed into a multiplication. This is because dividing by $s$ is the same as multiplying by $1/s$.

As shown in Fig. 4-16, multiplication of a vector by a scalar simply changes the magnitude of a vector without changing the "line" it lies along.

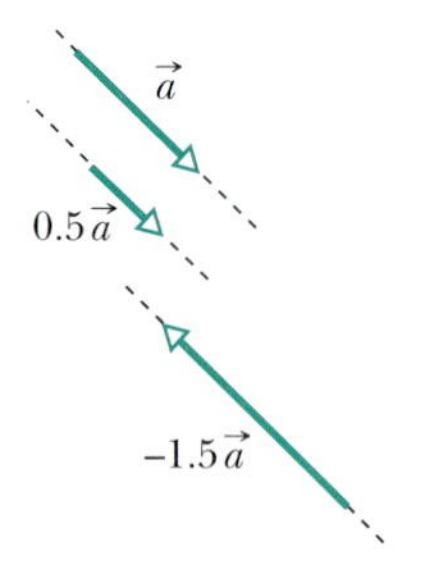


**FIGURE 4-16** ■ The product of a scalar and a vector results in a new vector that is still pointing along the same line but has a new length.

Although we have used familiar one-dimensional examples, it turns out that these rules for the multiplication of a vector by a scalar also work in two and three dimensions when using either graphical or component representations of vectors. The multiplication or division is distributive over addition so that the product of a scalar $e$ and a vector $\vec{V}$, expressed in terms of its rectangular coordinates, can be expressed as

$$e\vec{V} = e(\vec{V}_x + \vec{V}_y) = e(V_x\hat{\text{i}} + V_y\hat{\text{j}}) = eV_x\hat{\text{i}} + eV_y\hat{\text{j}}.$$

There are two other types of vector multiplication that are commonly used in physics, but these both involve the product of two vectors. We call one the dot product, introduced in Section 9-8, and we call the other the cross product, introduced in Section

12-4.* These vector-vector products will be explained in later chapters when they will be needed for the study of work, energy, rotational, and magnetic phenomena.

**READING EXERCISE 4-7:** Use the rules governing multiplication and division of a vector by a scalar to sketch the indicated product vectors with proper magnitude and directions. Note that the average velocity $\langle \vec{v} \rangle$ and the force $\vec{F}$ are given by

$$\langle \vec{v} \rangle = \frac{\Delta \vec{r}}{\Delta t} \quad \text{and} \quad \vec{F} = m\vec{a}.$$

Using the vectors in the diagram: (a) Multiply the acceleration vector $\vec{a}$ by $m = 3$ kg and sketch the vector representing the force acting on the particle. (b) Divide the displacement vector $\Delta\vec{r}$ by the scalar time interval $\Delta t = 0.5$ s to sketch a vector describing a particle-like object's average velocity vector in cm/s.

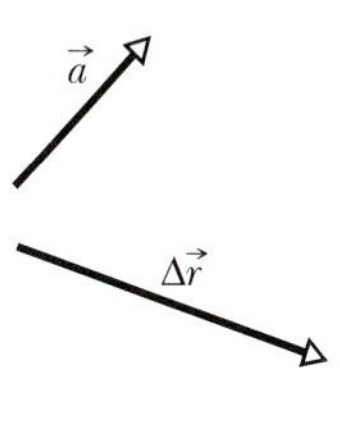


■

**READING EXERCISE 4-8:** Use the rules governing multiplication and division of a vector by a scalar to calculate the indicated product vectors in terms of its rectangular components and unit vectors. Don't forget to include units! Note that the average velocity $\langle \vec{v} \rangle$ and the force $\vec{F}$ are given by

$$\langle \vec{v} \rangle = \frac{\Delta \vec{r}}{\Delta t} \quad \text{and} \quad \vec{F} = m\vec{a}.$$

Multiply the acceleration vector $\vec{a} = a_x\hat{\text{i}} + a_y\hat{\text{j}} = (1.8\ \text{m/s}^2)\hat{\text{i}} + (1.0\ \text{m/s}^2)\hat{\text{j}}$ by $m = 3$ kg to calculate the vector representing the force acting on the particle. (b) Divide the displacement vector $\Delta\vec{r} = \Delta r_x\hat{\text{i}} + \Delta r_y\hat{\text{j}} = (3.2\ \text{m})\hat{\text{i}} + (-0.8\ \text{m})\hat{\text{j}}$ by the time interval $\Delta t = 0.5$ s to calculate the vector describing a particle-like object's average velocity.

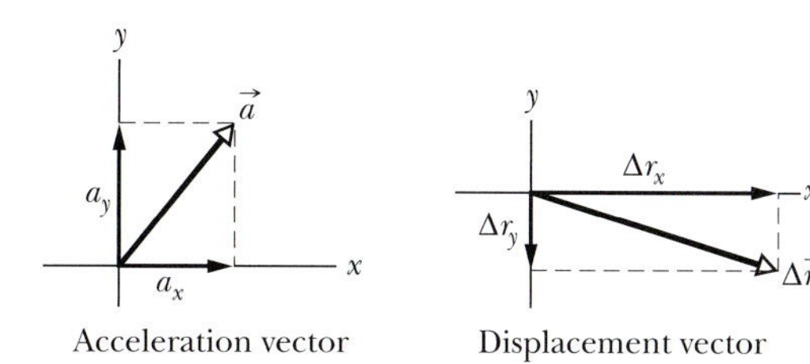


■

## 4-8 Vectors and the Laws of Physics

So far, in every figure that includes a coordinate system, the $x$ and $y$ axes are parallel to the edges of the book page. Thus, when a vector $\vec{a}$ is included, its components $a_x$ and $a_y$ are also parallel to the edges as in Fig. 4-17*a*. However, there are times when it is more convenient to choose a tilted coordinate system. For example, in studying the motion of a cart rolling down an inclined plane, it is easier to rotate the coordinate system so that one of the axes is aligned with the motion. If we choose to rotate the axes (but not the vector $\vec{a}$) through an angle $\phi$ in the $x$-$y$ plane as in Fig. 4-17*b*, the components will have new values—call them $a'_x$ and $a'_y$. Since there are an infinite number of choices of $\phi$, there are an infinite number of different pairs of components for $\vec{a}$.

*The dot product is sometimes called the scalar product because even though it involves a multiplication of two vectors, this product is a scalar. This name can cause confusion since it does not represent the multiplication of a vector by a scalar, which we just discussed. See Section 9-8 for details.

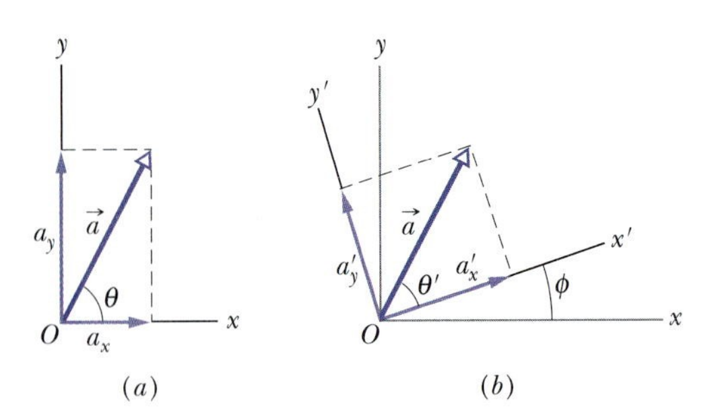


FIGURE 4-17 ■ (*a*) The vector $\vec{a}$ and its components. (*b*) The same vector, with the axes of the coordinate system rotated through an angle $\phi$.

Which then is the "right" pair of components? The answer is that they are all equally valid mathematically. Each pair (with its axes) just gives us a different way of describing the same vector $\vec{a}$. All produce the same magnitude. In Fig. 4-17 we have

$$|\vec{a}| = \sqrt{a_x^2 + a_y^2} = \sqrt{a_x'^2 + a_y'^2}. \tag{4-13}$$

Although the direction that the vector points in space does not change with coordinate rotation, the angle used to relate it to a new coordinate system is changed to

$$\theta' = \theta - \phi. \tag{4-14}$$

In Section 2-2, we defined a scalar as a mathematical quantity whose value does not depend on the orientation of a coordinate system. The magnitude of a vector is a true scalar since it does not change when the coordinate axis is rotated. However, the angles $\theta$ and $\theta'$, as well as the rectangular components $(a_x, a_y)$ and $(a'_x, a'_y)$, *are not scalars*.

The point is that we really do have great freedom in choosing a coordinate system, because the mathematical relations among vectors (including, for example, vector addition) do not depend on the location of the origin or the orientation of the axes.

In Chapter 5, you will use the definitions of position, displacement, velocity, acceleration, and force vectors developed in Chapters 2 and 3 along with what we have learned in this chapter to study motion in two dimensions.

# Problems

## SEC. 4-3 ■ ADDING VECTORS GRAPHICALLY

**1. Two Displacements** Consider two displacements, one of magnitude 3 m and another of magnitude 4 m. Show how the displacement vectors may be combined to get a resultant displacement of magnitude (a) 7 m, (b) 1 m, and (c) 5 m.

**2. Bank Robbery** A bank in downtown Boston is robbed (see the map in Fig. 4-18. To elude police, the robbers escape by helicopter, making three successive flights described by the following displacements: 32 km, 45° south of east; 53 km, 26° north of west; 26 km, 18° east of south. At the end of the third flight they are captured. In what town are they apprehended? (Use the geometrical method to add these displacements on the map.)

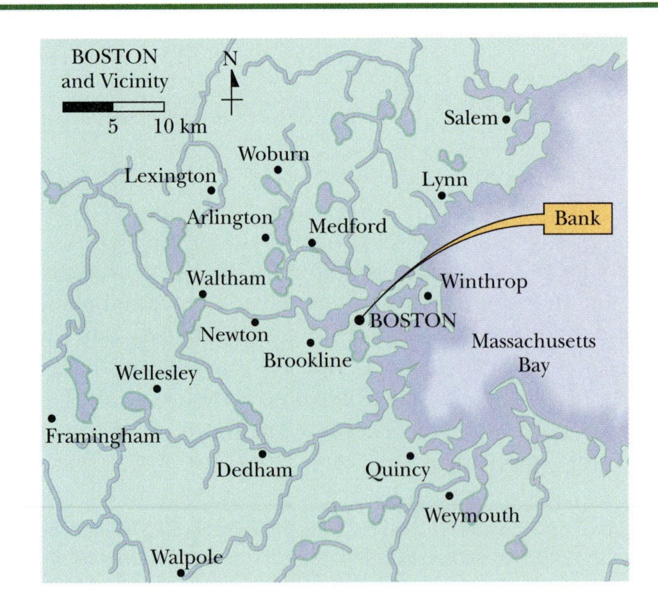


FIGURE 4-18 ■ Problem 2.

**3. Velocity Vector Changes** The motion of three objects is shown in the motion diagrams (*a*), (*b*), and (*c*) of Fig. 4-19. In each case the object is shown at three equally spaced times. A circle with no arrow indicates a velocity of zero magnitude, such as the final velocity in diagram (*a*). Indicate for each part which number is next to the arrow on the right side of the diagram that best shows the *direction* of the change in velocity. (*Hint*: Use the techniques developed in Section 4-3 to draw vectors representing the difference in velocity in each case.) *Note*: This exercise is adapted from a conceptual exercise developed by Dennis Albers of Columbia College.

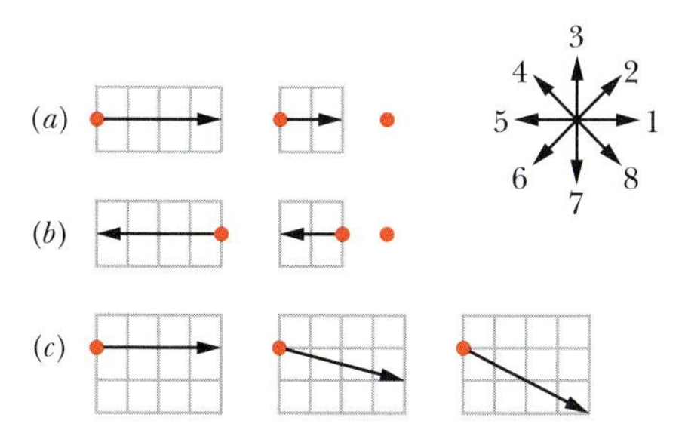


FIGURE 4-19 ■ Problem 3.

**4. The Pea Shooter** A pea leaves a pea shooter at a speed of 5.4 m/s. It makes an angle of +30° with respect to the horizontal.

**(a)** Calculate the $x$-component of the pea's initial velocity.
**(b)** Calculate the $y$-component of the pea's initial velocity.
**(c)** Write an expression for the pea's velocity, $\vec{v}$, using unit vectors for the $x$ direction and the $y$ direction.

## SEC. 4-4 ■ RECTANGULAR VECTOR COMPONENTS

**5. Components** What are (a) the $x$-component and (b) the $y$-component of a vector $\vec{a}$ in the $xy$ plane if its direction is 250° counterclockwise from the positive direction of the $x$ axis and its magnitude is 7.3 m?

**6. Radians and Degrees** Express the following angles in radians: (a) 20.0°, (b) 50.0°, (c) 100°. Convert the following angles to degrees: (d) 0.330 rad, (e) 2.10 rad, (f) 7.70 rad.

**7. Magnitude and Angle** The $x$-component of vector $A$ is $-25.0$ m and the $y$-component is $+40.0$ m. (a) What is the magnitude of $\vec{A}$? (b) What is the angle between the direction of $\vec{A}$ and the positive direction of $x$?

**8. Displacement Vector** A displacement vector $\vec{r}$ in the $xy$ plane is 15 m long and directed as shown in Fig. 4-20. Determine (a) the $x$-component and (b) the $y$-component of the vector.

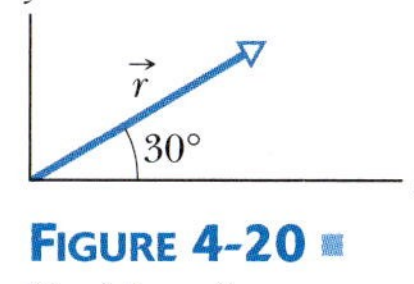


FIGURE 4-20 ■ Problem 8.

**9. Rolling Wheel** A wheel with a radius of 45.0 cm rolls without slipping along a horizontal floor (Fig. 4-21). At time $t_1$, the dot $P$ painted on the rim of the wheel is at the point of contact between the wheel and the floor. At a later time $t_2$, the wheel has rolled through one-half of a revolution. What are (a) the magnitude and (b) the angle (relative to the floor) of the displacement of $P$ during this interval?

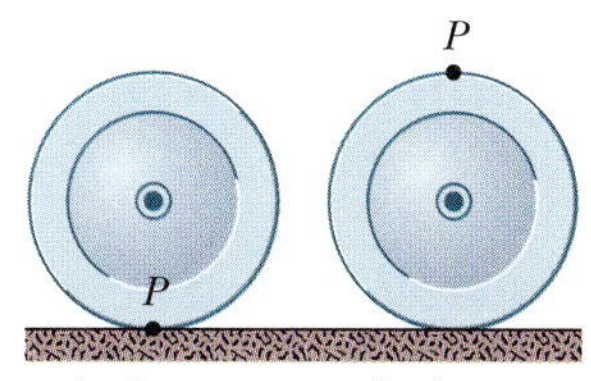


FIGURE 4-21 ■ Problem 9.

**10. Rock Faults** Rock *faults* are ruptures along which opposite faces of rock have slid past each other. In Fig. 4-22 points $A$ and $B$ coincided before the rock in the foreground slid down to the right. The net displacement $\overrightarrow{AB}$ is along the plane of the fault. The horizontal component of $\overrightarrow{AB}$ is the *strike-slip AC*. The component of $\overrightarrow{AB}$ that is directly down the plane of the fault is the *dip-slip AD*. (a) What is the magnitude of the net displacement $\overrightarrow{AB}$ if the strike-slip is 22.0 m and the dip-slip is 17.0 m? (b) If the plane of the fault is inclined 52.0° to the horizontal, what is the vertical component of $\overrightarrow{AB}$?

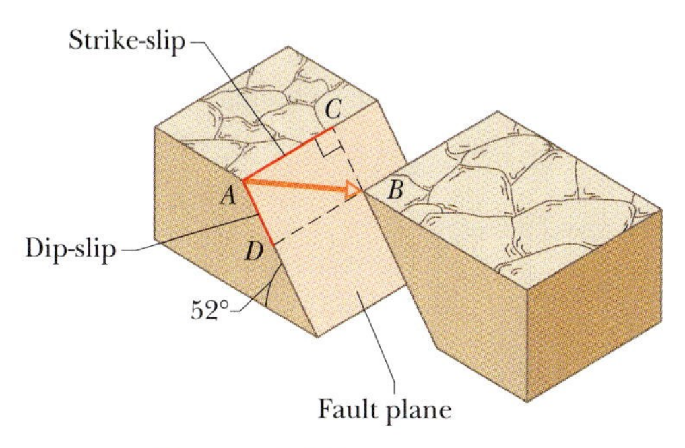


FIGURE 4-22 ■ Problem 10.

**11. A Room** A room has dimensions 3.00 m (height) × 3.70 m × 4.30 m. A fly starting at one corner flies around, ending up at the diagonally opposite corner. (a) What is the magnitude of its displacement? (b) Could the length of its path be less than this magnitude? (c) Greater than this magnitude? (d) Equal to this magnitude? (e) Choose a suitable coordinate system and find the components of the displacement vector in that system. (f) If the fly walks rather than flies, what is the length of the shortest path it can take? (*Hint*: This can be answered without calculus. The room is like a box. Unfold its walls to flatten them into a plane.)

## SEC. 4-6 ■ ADDING VECTORS USING COMPONENTS

**12. The Drive** A car is driven east for a distance of 50 km, then north for 30 km, and then in a direction 30° east of north for 25 km. Sketch the vector diagram and determine (a) the magnitude and (b) the angle of the car's total displacement from its starting point.

**13. A Walk** A woman walks 250 m in the direction 30° east of north, then 175 m directly east. Find (a) the magnitude and (b) the angle of her final displacement from the starting point. (c) Find the distance she walks. (d) Which is greater, that distance or the magnitude of her displacement?

**14. Another Walk** A person walks in the following pattern: 3.1 km north, then 2.4 km west, and finally 5.2 km south. (a) Sketch the vector diagram that represents this motion. (b) How far and (c) in what direction would a bird fly in a straight line from the same starting point to the same final point?

**15. Unit-vector** (a) In unit-vector notation, what is the sum of

$$\vec{a} = (4.0\text{ m})\hat{\text{i}} + (3.0\text{ m})\hat{\text{j}} \quad \text{and} \quad \vec{b} = (-13.0\text{ m})\hat{\text{i}} + (7.0\text{ m})\hat{\text{j}}?$$

What are (b) the magnitude and (c) the direction of $\vec{a} + \vec{b}$ (relative to $\hat{\text{i}}$)?

**16. Find the Components** Find the (a) $x$- (b) $y$- and (c) $z$-components of the sum $\vec{r}$ of the displacements $\vec{c}$ and $\vec{d}$ whose components in meters along the three axes are $c_x = 7.4$, $c_y = -3.8$, $c_z = -6.1$; $d_x = 4.4$, $d_y = -2.0$, $d_z = 3.3$.

**17. Two Vectors** Vector $\vec{a}$ has a magnitude of 5.0 m and is directed east. Vector $\vec{b}$ has a magnitude of 4.0 m and is directed 35° west of north. What are (a) the magnitude and (b) the direction of $\vec{a} + \vec{b}$? What are (c) the magnitude and (d) the direction of $\vec{b} - \vec{a}$? (e) Draw a vector diagram for each combination.

**18. For the Vectors** For the vectors

$$\vec{a} = (3.0 \text{ m})\hat{i} + (4.0 \text{ m})\hat{j} \quad \text{and} \quad \vec{b} = (5.0 \text{ m})\hat{i} + (-2.0 \text{ m})\hat{j},$$

give $\vec{a} + \vec{b}$ in (a) unit-vector notation, and as (b) a magnitude and (c) an angle (relative to $\hat{i}$). Now give $\vec{b} + \vec{a}$ in (d) unit-vector notation, and as (e) a magnitude and (f) an angle.

**19. Two Vectors Two** Two vectors are given by

$$\vec{a} = (4.0 \text{ m})\hat{i} - (3.0 \text{ m})\hat{j} + (1.0 \text{ m})\hat{k}$$

and

$$\vec{b} = (-1.0 \text{ m})\hat{i} + (1.0 \text{ m})\hat{j} + (4.0 \text{ m})\hat{k}.$$

In unit-vector notation, find (a) $\vec{a} + \vec{b}$, (b) $\vec{a} - \vec{b}$, and (c) a third vector $\vec{c}$ such that $\vec{a} - \vec{b} + \vec{c} = 0$.

**20. Two Vectors Three** Here are two vectors:

$$\vec{a} = (4.0 \text{ m})\hat{i} - (3.0 \text{ m})\hat{j} \quad \text{and} \quad \vec{b} = (6.0 \text{ m})\hat{i} + (8.0 \text{ m})\hat{j}.$$

What are (a) the magnitude and (b) the angle (relative to $\hat{i}$) of $\vec{a}$? What are (c) the magnitude and (d) the angle of $\vec{b}$? What are (e) the magnitude and (f) the angle of $\vec{a} + \vec{b}$; (g) the magnitude and (h) the angle of $\vec{b} - \vec{a}$; and (i) the magnitude and (j) the angle of $\vec{a} - \vec{b}$? (k) What is the angle between the directions of $\vec{b} - \vec{a}$ and $\vec{a} - \vec{b}$?

**21. Three Vectors** Three vectors $\vec{a}$, and $\vec{b}$, and $\vec{c}$ each have a magnitude of 50 m and lie in an $xy$ plane. Their directions relative to the positive direction of the $x$ axis are 30°, 195°, and 315°, respectively. What are (a) the magnitude and (b) the angle of the vector $\vec{a} + \vec{b} + \vec{c}$, and (c) the magnitude and (d) the angle of $\vec{a} - \vec{b} + \vec{c}$? What are (e) the magnitude and (f) the angle of a fourth vector $\vec{d}$ such that $(\vec{a} + \vec{b}) - (\vec{c} + \vec{d}) = 0$.

**22. Four Vectors** What is the sum of the following four vectors in (a) unit-vector notation and (b) magnitude-angle notation? For the latter, give the angle in both degrees and radians. Positive angles are counterclockwise from the positive direction of the $x$ axis; negative angles are clockwise.

$\vec{E}$: 6.00 m at +0.900 rad $\qquad \vec{F}$: 5.00 m at −75.0°

$\vec{G}$: 4.00 m at +1.20 rad $\qquad \vec{H}$: 6.00 m at −210°

**23. Two Vectors Four** The two vectors $\vec{a}$ and $\vec{b}$ in Fig. 4-23 have equal magnitudes of 10.0 m. Find (a) the $x$-component and (b) the $y$-component of their vector sum $\vec{r}$, (c) the magnitude of $\vec{r}$, and (d) the angle $\vec{r}$ makes with the positive direction of the $x$ axis.

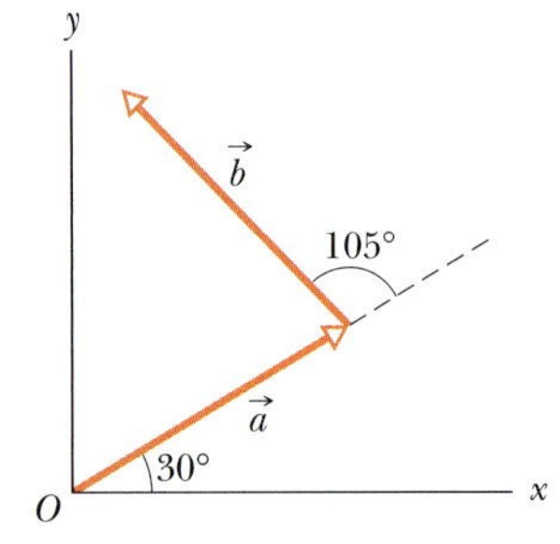


FIGURE 4-23 ■ Problem 23.

**24. The Sum** In the sum $\vec{A} + \vec{B} = \vec{C}$, vector $\vec{A}$ has a magnitude of 12.0 m and is angled 40.0° counterclockwise from the $+x$ direction, and vector $\vec{C}$ has a magnitude of 15.0 m and is angled 20.0° counterclockwise from the $-x$ direction. What are (a) the magnitude and (b) the angle (relative to $+x$) of $\vec{B}$?

**25. Prove** Prove that two vectors must have equal magnitudes if their sum is perpendicular to their difference.

**26. The Sum of Four** Find the sum of the following four vectors in (a) unit-vector notation, and as (b) a magnitude and (c) an angle relative to $+x$.

$\vec{P}$: 10.0 m, at 25.0° counterclockwise from $+x$

$\vec{Q}$: 12.0 m, at 10.0° counterclockwise from $+y$

$\vec{R}$: 8.00 m, at 20.0° clockwise from $-y$

$\vec{S}$: 9.00 m, at 40.0° counterclockwise from $-y$

**27. Prove by Components** Two vectors of magnitudes $a$ and $b$ make an angle $\theta$ with each other when placed tail to tail. Prove, by taking components along two perpendicular axes, that

$$r = \sqrt{a^2 + b^2 + 2ab \cos \theta}$$

gives the magnitude of the sum $\vec{r}$ of the two vectors.

**28. The Sum of Four Again** What is the sum of the following four vectors in (a) unit-vector notation, and as (b) a magnitude and (c) an angle? Positive angles are counterclockwise from the positive direction of the $x$ axis; negative angles are clockwise.

$\vec{A} = (2.00 \text{ mi})\hat{i} + (3.00 \text{ mi})\hat{j} \qquad \vec{B}$: 4.00 m, at +65.0°

$\vec{C} = (-4.00 \text{ m})\hat{i} - (6.00 \text{ m})\hat{j} \qquad \vec{D}$: 5.00 m, at −235°

**29. A Cube** (a) Using unit vectors, write expressions for the four body diagonals (the straight lines from one corner to another through the center) of a cube in terms of its edges, which have length $a$. (b) Determine the angles that the body diagonals make with the adjacent edges. (c) Determine the length of the body diagonals in terms of $a$.

**30. Oasis** Oasis $B$ is 25 km due east of oasis $A$. Starting from oasis $A$, a camel walks 24 km in a direction 15° south of east and then walks 8.0 km due north. How far is the camel then from oasis $B$?

**31. A Plus B** If $\vec{B}$ is added to $\vec{A}$, the result is $6.0\hat{i} + 1.0\hat{j}$. If $\vec{B}$ is subtracted from $\vec{A}$, the result is $-4.0\hat{i} + 7.0\hat{j}$. What is the magnitude of $\vec{A}$?

**32. If-Then** If $\vec{d}_1 + \vec{d}_2 = 5\vec{d}_3$, $\vec{d}_1 - \vec{d}_2 = 3\vec{d}_3$, and $\vec{d}_3 = 2\hat{i} + 4\hat{j}$, then what are (a) $\vec{d}_1$ and (b) $\vec{d}_2$?

**33. Sailing** A sailboat sets out from the U.S. side of Lake Erie for a point on the Canadian side, 90.0 km due north. The sailor, however, ends up 50.0 km due east of the starting point. (a) How far and (b) in what direction must the sailor now sail to reach the original destination?

## SEC. 4-7 ■ MULTIPLYING AND DIVIDING A VECTOR BY A SCALAR

**34. Vector by Scalar** The three vectors in Fig. 4-24 have magnitudes $a$ = 3.00 m, $b$ = 4.00 m, and $c$ = 10.0 m. What are (a) the

$x$-component and (b) the $y$-component of $\vec{a}$; (c) the $x$-component and (d) the $y$-component of $\vec{b}$; and (e) the $x$-component and (f) the $y$-component of $\vec{c}$? If $\vec{c} = p\vec{a} + q\vec{b}$, what are the values of (g) $p$ and (h) $q$?

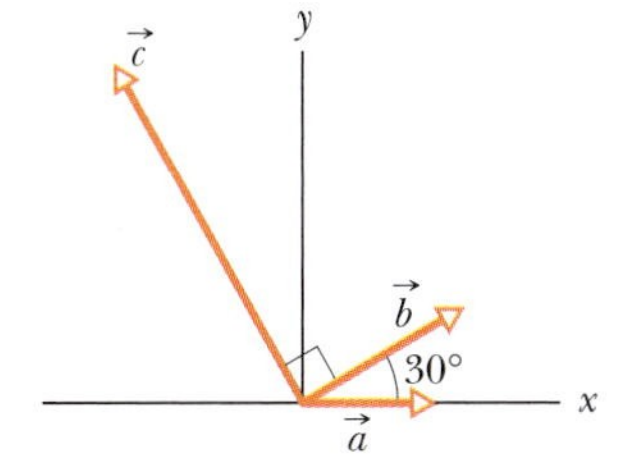


FIGURE 4-24 ■ Problem 34.

**35. Five Times** A vector $\vec{d}$ has a magnitude 3.0 m and is directed south. What are (a) the magnitude and (b) the direction of the vector $5.0\vec{d}$? What are (c) the magnitude and (d) the direction of the vector $-2.0\vec{d}$?

**36. The Sum Is a Third** Vector $\vec{A}$, which is directed along an $x$ axis, is to be added to vector $\vec{B}$, which has a magnitude of 7.0 m. The sum is a third vector that is directed along the $y$ axis, with a magnitude that is 3.0 times that of $\vec{A}$. What is that magnitude of $\vec{A}$?

# Additional Problems

**37. Explorer** An explorer is caught in a whiteout (in which the snowfall is so thick that the ground cannot be distinguished from the sky) while returning to base camp. He was supposed to travel due north for 5.6 km, but when the snow clears, he discovers that he actually traveled 7.8 km at 50° north of due east. (a) How far and (b) in what direction must he now travel to reach base camp?

**38. Bowling Balls** In each case below, sketch the velocity vector. Find the magnitude and direction of motion with respect to the $x$ axis of the coordinate system:

**(a)** $\vec{v} = (2.45 \text{ m/s})\hat{i} + (3.67 \text{ m/s})\hat{j}$
**(b)** $\vec{v} = (-2.45 \text{ m/s})\hat{i} + (5.20 \text{ m/s})\hat{j}$

**39. Lawn Chess** In a game of lawn chess, where pieces are moved between the centers of squares that are each 1.00 m on edge, a knight is moved in the following way: (1) two squares forward, one square rightward; (2) two squares leftward, one square forward; (3) two squares forward, one square leftward. What are (a) the magnitude and (b) the angle (relative to "forward") of the knight's overall displacement for the series of three moves?

**40. Fire Ant** A fire ant, searching for hot sauce in a picnic area, goes through three displacements along level ground: $\vec{d}_1$ for 0.40 m southwest (that is, at 45° from directly south and from directly west), $\vec{d}_2$ for 0.50 m due east (that is, directly east), $\vec{d}_3$ for 0.60 m at 60° north of east (that is 60.0° toward the north from due east). Let the positive $x$ direction be east and the positive $y$ direction be north. What are (a) the $x$-component and (b) the $y$-component of $\vec{d}_1$? What are (c) the $x$-component and (d) the $y$-component of $\vec{d}_2$? What are (e) the $x$-component and (f) the $y$-component of $\vec{d}_3$?

What are (g) the $x$-component, (h) the $y$-component, (i) the magnitude, and (j) the direction of the ant's net displacement? If the ant is to return directly to the starting point, (k) how far and (l) in what direction should it move?

**41. A Heavy Object** A heavy piece of machinery is raised by sliding it 12.5 m along a plank oriented at 20.0° to the horizontal, as shown in Fig. 4-25. (a) How high above its original position is it raised? (b) How far is it moved horizontally?

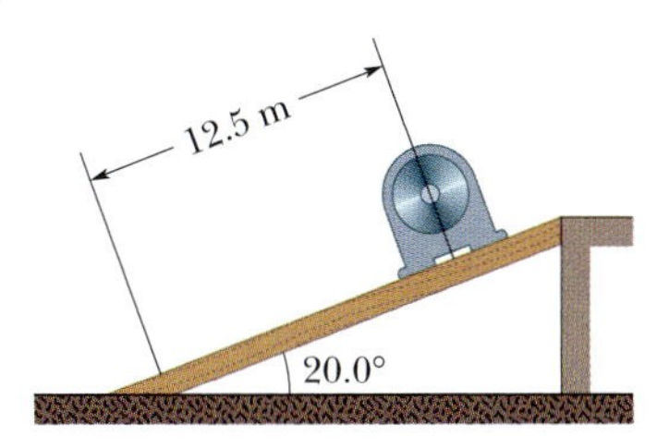


FIGURE 4-25 ■ Problem 41.

**42. Two Beetles** Two beetles run across flat sand, starting at the same point. Beetle 1 runs 0.50 m due east, then 0.80 m at 30° north of due east. Beetle 2 also makes two runs; the first is 1.6 m at 40° east of due north. What must be (a) the magnitude and (b) the direction of its second run if it is to end up at the new location of beetle 1?

**43. Four Moves** You are to make four straight-line moves over a flat desert floor, starting at the origin of an $xy$ coordinate system and ending at the $xy$ coordinates (−140 m, 30 m). The $x$-component and $y$-component of your moves are the following, respectively, in meters: (20 and 60), then ($b_x$ and −70), then (−20 and $c_y$), then (−60 and −70). What are (a) component $b_x$ and (b) component $c_y$? What are (c) the magnitude and (d) the angle (relative to the positive direction of the $x$ axis) of the overall displacement?

**44. Vector C** The magnitude and angle of $\vec{A}$, which lies in an $xy$ plane, are 4.00 and 130°, respectively. What are the components (a) $A_x$ and (b) $A_y$? Vector $\vec{B}$ also lies in the $xy$ plane, and it has components $B_x = -3.86$ and $B_y = -4.60$. What is $\vec{A} + \vec{B}$ in (c) magnitude-angle notation and (d) unit-vector notation? In (e) unit-vector notation and (f) magnitude-angle notation, find $\vec{C}$ such that $\vec{A} - \vec{C} = \vec{B}$. (g) Which of the vector diagrams in Fig. 4-26 correctly show the relationship between those three vectors?

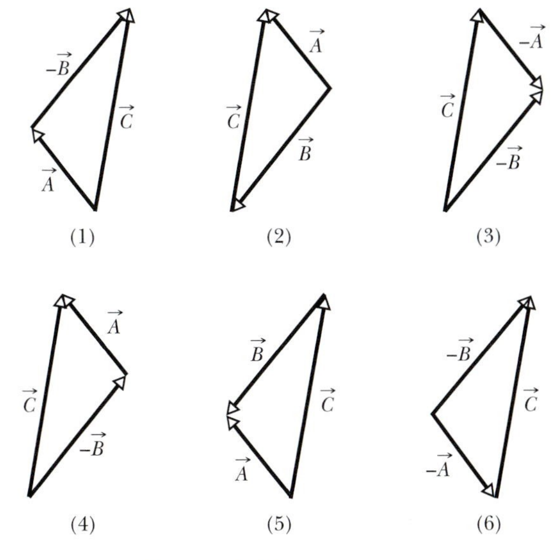


FIGURE 4-26 ■ Problem 44.

**45. Twice the Magnitude** A vector $\vec{B}$, with a magnitude of 8.0 m, is added to a vector $\vec{A}$, which lies along an $x$ axis. The sum of these two vectors is a third vector that lies along the $y$ axis and has a magnitude that is twice the magnitude of $\vec{A}$. What is the magnitude of $\vec{A}$?

**46. To Reach a Point** A person desires to reach a point that is 3.40 km from her present location and in a direction that is 35.0° north of east. However, she must travel along streets that are oriented either north–south or east–west. What is the minimum distance she could travel to reach her destination?

**47. A Golfer** A golfer takes three putts to get the ball into the hole. The first putt displaces the ball 3.66 m north, the second 1.83 m southeast, and the third 0.91 m southwest. What are (a) the magnitude and (b) the direction of the displacement needed to get the ball into the hole on the first putt?

**48. Protestor's Sign** A protester carries his sign of protest 40 m along a straight path, then 20 m along a perpendicular path to his left, and then 25 m up a water tower. (a) Choose and describe a coordinate system for this motion. In terms of that system and in unit-vector notation, what is the displacement of the sign from start to end? (b) The sign then falls to the foot of the tower. What is the magnitude of the displacement of the sign from start to this new end?

**49. Rotated Coordinate System** In Fig. 4.27, a vector $\vec{a}$ with a magnitude of 17.0 m is directed 56.0° counterclockwise from the $+x$ axis, as shown. What are the components (a) $a_x$ and (b) $a_y$ of the vector? A second coordinate system is inclined by 18.0° with respect to the first. What are the components (c) $a'_x$ and (d) $a'_y$ in this primed coordinate system?

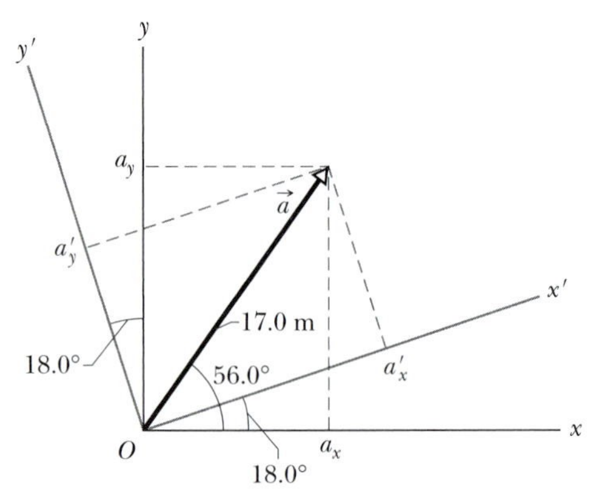


FIGURE 4-27 ■ Problem 49.

**50. Shifted Coordinate System** Consider how the components of a vector in the plane change if I change the reference point. Suppose I start with a coordinate system with an origin at $O$. An arbitrary vector $\vec{r} = x\hat{\mathrm{i}} + y\hat{\mathrm{j}}$ with coordinates $(x, y)$ specifies a point in this system. Suppose also that I have another point $O'$ specified in this coordinate system by a vector $\vec{A} = A_x\hat{\mathrm{i}} + A_y\hat{\mathrm{j}}$. If I change my origin to $O'$ (without rotating the axes), what would the coordinates be for the point specified by $\vec{r}$?

**51. A New System** $\vec{A}$ has the magnitude 12.0 m and is angled 60.0° counterclockwise from the positive direction of the $x$ axis of an $xy$ coordinate system. Also, $\vec{B} = (12.0\text{ m})\hat{\mathrm{i}} + (8.00\text{ m})\hat{\mathrm{j}}$ on that same coordinate system. We now rotate the system, counterclockwise about the origin by 20.0°, to form an $x'y'$ system. On this new system, what are (a) $\vec{A}$ and (b) $\vec{B}$, both in unit-vector notation?

# 5 Net Force and Two-Dimensional Motion

In 1922, one of the Zacchinis, a famous family of circus performers, was the first human cannon ball to be shot across an arena into a net. To increase the excitement, the family gradually increased the height and distance of the flight until, in 1939 or 1940, Emanuel Zacchini soared over three Ferris wheels and through a horizontal distance of 69 m.

**How could he know where to place the net, and how could he be certain he would clear the Ferris wheels?**

*The answer is in this chapter.*

## 5-1 Introduction

In this chapter, we will apply Newton's Second Law of motion developed in Chapter 3 to the analysis of familiar two-dimensional motions. We start with an exploration of projectile motion. When a particle-like object is launched close to the surface of the Earth with a horizontal component of velocity and allowed to fall freely, we say it undergoes projectile motion. The second two-dimensional motion we consider is uniform circular motion in which a particle-like object moves in a circle at a constant speed. This motion can be produced by twirling a ball attached to a string in a circle or by watching a point on the edge of a spinning wheel.

You will use the vector algebra introduced in Chapter 4 and extend the concepts introduced in Chapters 2 and 3 to two dimensions. As you work with this chapter, you will want to review relevant sections in these chapters.

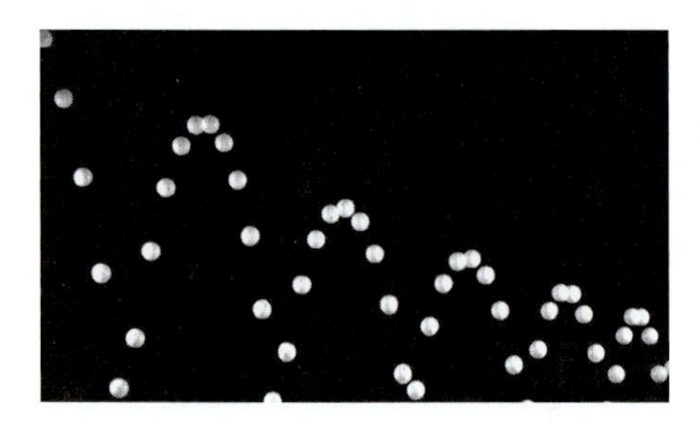

**FIGURE 5-1** ■ A stroboscopic photograph of a golf ball bouncing off a hard surface. Between impacts, the ball undergoes projectile motion that is characterized by curved paths.

## 5-2 Projectile Motion

Projectile motion occurs near the Earth's surface whenever a ball rolls off a table, a basketball arcs toward a basket, a hailstone rolls off a steep roof, or a ball bounces (Fig. 5-1). All of these motions have curved paths. But why are the paths curved, and what sort of curves are they? In Section 3-2 we presented data on objects moving horizontally that provided evidence that in the absence of forces, moving objects tend to continue moving at a constant velocity (Newton's First Law). In Section 3-9 we presented data indicating that near the Earth's surface, objects fall freely in a vertical direction with an acceleration of magnitude 9.8 $m/s^2$. Projectile motions from basketballs to hailstones all involve a combination of horizontal and vertical motions. Galileo was the first to discover how to treat two-dimensional projectile motion as a combination of horizontal and vertical motions.

**FIGURE 5-2** ■ Galileo (1564–1642) was the first scientist to deduce that projectile motion could be analyzed as a combination of two independent linear motions.

### Galileo's Hypothesis

In his *Dialog Concerning Two New Sciences* published in 1632, Galileo (Fig. 5-2) observes:

> "... we have discussed the properties of uniform motion and of motion naturally accelerated along planes of all inclinations. I now propose to set forth those properties which belong to a body whose motion is compounded of two other motions, namely, one uniform and one naturally accelerated. . . . This is the kind of motion seen in a moving projectile; its origin I conceive to be as follows:
>
> Imagine any particle projected along a horizontal plane without friction; then we know . . . that this particle will move along this same plane with a motion which is uniform and perpetual. . . . But if the plane is limited and elevated, then the moving particle . . . will on passing over the edge of the plane acquire, in addition to its previous uniform and perpetual motion, a downward propensity due to its own weight; so that the resulting motion which I call projection is compounded of one which is uniform and horizontal and of another which is vertical and naturally accelerated."

Galileo went on to predict that the curve that describes projectile motion is the parabola. He deduces this using a construction (Fig. 5-3) which shows how uniform horizontal motion and uniformly accelerated vertical motion with displacements increasing in proportion to the square of time can be combined or superimposed to form a parabola. The superposition is not unlike that introduced in Chapter 3 with regard to the combination of forces. However, in this case the quantities must be at right angles to each other.

## Experimental Evidence for Galileo's Hypothesis

Galileo's hypothesis was based on both observations and reasoning. By observing balls accelerating slowly on inclines and balls in free fall, he knew that their positions increased as the square of time. He determined the constant velocity of a ball rolling along a level ramp. Knowing the height of the ramp, he could predict how long the vertical falling motion should take and hence how far in a horizontal direction the ball should travel before it hit the floor, assuming the horizontal velocity was undisturbed by the introduction of the vertical falling motion. Since Galileo did not have contemporary technology such as strobe photography or video analysis at his disposal, his approach to understanding projectile motion was extraordinary.

It is instructive to confirm Galileo's hypothesis regarding a body moving off the edge of a ramp by using a digital video camera to record this motion, as in Fig. 5-4. What would Galileo have observed? If he had drawn the horizontal position component of the ball along the line of its original motion as shown in Fig. 5-4, he would see that the horizontal distance the ball traveled each time period remains constant. That is, the horizontal velocity is constant both before and after the ball reaches the edge of the table. Thus, the horizontal motion must be independent of the falling motion.

Since our image shows that the ball has fallen a distance of approximately $(y_2 - y_1) \approx -0.85$ m in a time interval of $(t_2 - t_1) = 0.40$ s, we see that these quantities are consistent with the kinematic equation 2-17 given by $(y_2 - y_1) = v_{1y}(t_2 - t_1) + {}^1\!/_2\, a_y(t_2 - t_1)^2$. For this situation, the $y$-component of velocity equals zero at time $t_1$ (so that $v_{1y} = 0.0$ m/s) with the vertical acceleration component given by $a_y = -9.8$ m/s. This calculation suggests that the vertical motion is independent of the horizontal motion.

Perhaps the most compelling evidence for the independence of the vertical motion is a stroboscopic photograph of a ball that is released electronically just at the moment that a projectile is shot horizontally, as shown in Fig. 5-5. The vertical position components of these two balls are identical.

In summary, using new technology we can easily conclude, as Galileo did about 400 years ago, that

> The *horizontal and vertical motions of a projectile* (at right angles to each other) *are independent*, and the path of such a projectile can be found by combining its horizontal and vertical position components.

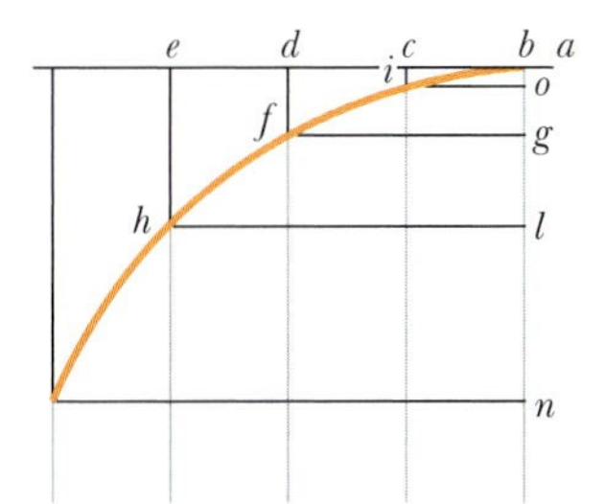


**FIGURE 5-3** ■ Galileo's diagram showing how a parabola can be formed by a set of linearly increasing horizontal coordinates ($b \rightarrow e$) and a set of vertical coordinates that increase as the square of time ($o \rightarrow n$).

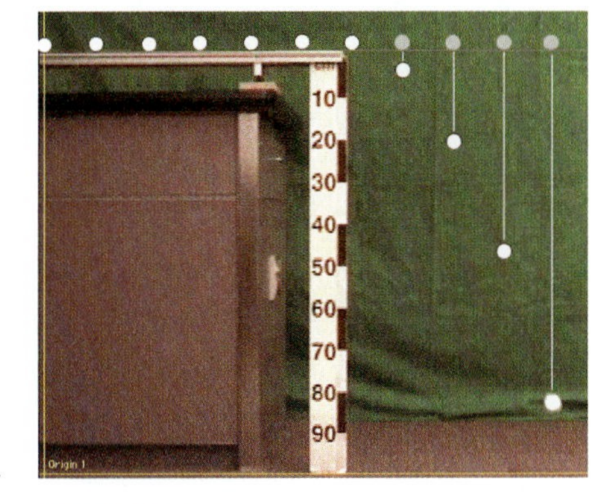


**FIGURE 5-4** ■ Performing Galileo's thought experiment with modern equipment. This digital video image shows the location of a golf ball in the last frame, along with white markers left by the software showing the ball's location every tenth of a second as it rolls off the edge of a table and falls toward the floor.

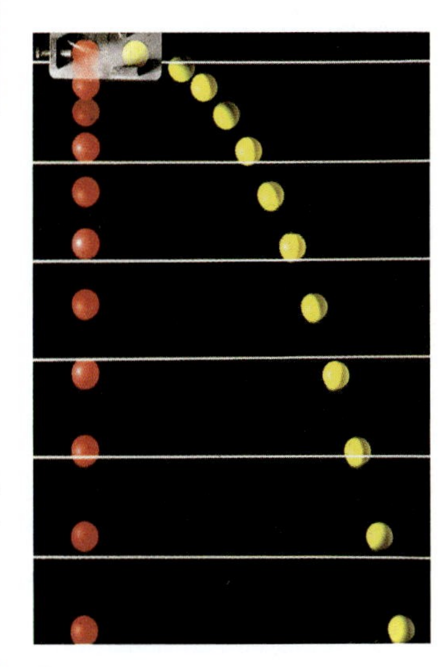

**FIGURE 5-5** ■ One ball is released from rest at the same instant that another ball is shot horizontally to the right.

## Ideal Projectile Motion

When Galileo wrote about projectiles in the *Two Sciences*, he made it quite clear that the moving object should be heavy and that friction should be avoided. In the next section, as we use vector mathematics to consider projectile motion, we will limit ourselves to "ideal" situations. We also assume the only significant force on an object is a constant gravitational force acting vertically downward. For example, the bouncing golf ball is undergoing ideal projectile motion between bounces because it is moving slowly enough that air resistance forces are negligible. On the other hand, flying airplanes and ducks are not ideal projectiles because their sustained flight depends on getting lift forces from air. Examples of ideal and nonideal projectile motion are shown in Fig. 5-6. Ideal projectile motion can be defined as follows:

> A particle-like object undergoes **IDEAL PROJECTILE MOTION** if the only significant force that acts on it is a constant gravitational force.

FIGURE 5-7 ■ The vertical component of this skateboarder's velocity is changing. However, during the entire time he is in the air, the skateboard stays underneath him, allowing him to land on it.

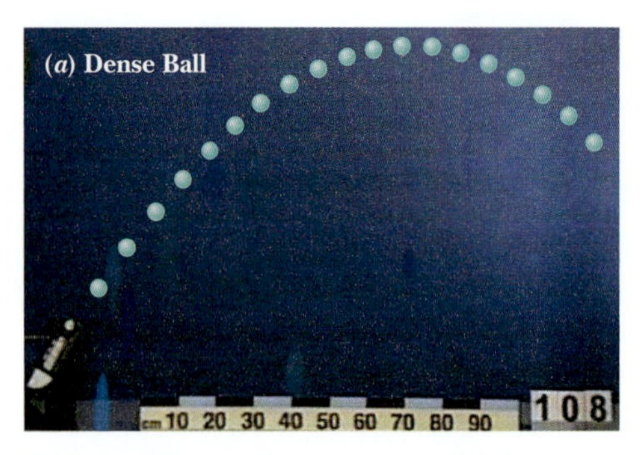


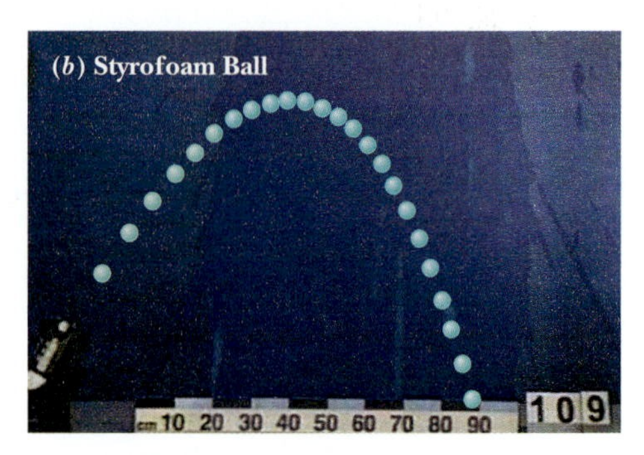


FIGURE 5-6 ■ Video analysis software is used to trace the paths of two small balls of the same size shot from a projectile launcher. The dots mark the location of the ball every 1/30 of a second. (*a*) The left video frame shows the path of a dense plastic ball. (*b*) The right video frame shows a Styrofoam ball path that is not as long or symmetric because it is influenced by air drag forces.

**READING EXERCISE 5-1:** (*a*) Consider the light-colored falling golf ball on the right in Fig. 5-5. Does its horizontal velocity change its vertical acceleration and the vertical velocities it normally acquires in free fall along a straight vertical line? Explain. (*b*) Does the fact that the light-colored golf ball on the right is falling have any effect on the rate that it is moving in the *x* direction? Explain. ■

**READING EXERCISE 5-2:** How does the fact that the skateboarder in Fig. 5-7 has a vertical acceleration and, therefore, vertical velocity affect his horizontal velocity while he is "flying" above his skateboard? ■

## TOUCHSTONE EXAMPLE 5-1: Golf Ball

In Figure 5-4, a golf ball is rolling off a tabletop which is 1.0 meter above the floor. The golf ball has an initial horizontal velocity component of $v_{1x} = 1.3$ m/s. Its location has been marked at 0.100 s intervals.

(a) How long should the ball take to fall on the floor from the time it leaves the edge of the table?

**SOLUTION** ■ The **Key Idea** here is that the golf ball undergoes ideal projectile motion. Therefore, its horizontal and vertical motions are independent and can be considered separately (we need not consider the actual curved path of the ball). We choose a coordinate system in which the origin is at floor level and upward is the positive direction. Then, the first vertical position component of interest in our fall is $y_1 = 1.0$ m and our last vertical position of interest is $y_2 = 0.0$ m. We can use kinematic equation 2-17 to describe the descent along the $y$ axis, so

$$(y_2 - y_1) = v_{1y}(t_2 - t_1) + \tfrac{1}{2}a_y(t_2 - t_1)^2.$$

Noting that at the table's edge the vertical velocity is zero, so $v_{1y} = 0.00$ m/s, we can solve our equation for $(t_2 - t_1)$ to get

$$t_2 - t_1 = \sqrt{\frac{2(y_2 - y_1)}{a_y}},$$

where in free fall the only force is the gravitational force, and so

$$a_y = -g = -9.8 \text{ m/s}^2$$

and

$$(y_2 - y_1) = (0.0 \text{ m} - 1.0 \text{ m}) = -1.0 \text{ m}.$$

This gives

$$(t_2 - t_1) = \sqrt{\frac{2(y_2 - y_1)}{-g}} = \sqrt{\frac{2(-1.0 \text{ m})}{-9.8 \text{ m/s}^2}} = 0.45 \text{ s}.$$

(b) How far will the ball travel in the horizontal direction from the edge of the table before it hits the floor?

**SOLUTION** ■ The **Key Idea** is that the ball will hit the floor after it has fallen for $(t_2 - t_1) = 0.45$ s. Since its horizontal velocity component doesn't change, its average velocity is the same as its instantaneous velocity. We can use the definition of average velocity to calculate the horizontal distance it travels, since

$$v_{1x} = \langle v_x \rangle = \frac{(x_2 - x_1)}{t_2 - t_1};$$

we can solve this equation for $x_2 - x_1$ to get our distance:

$$(x_2 - x_1) = v_{1x}(t_2 - t_1) = (1.3 \text{ m/s})(0.45 \text{ s}) = 0.59 \text{ m}.$$

An examination of Fig. 5-4 shows that this distance is reasonable.

## 5-3 Analyzing Ideal Projectile Motion

Now that we have established the independence of the horizontal and vertical components of the motion of an ideal projectile, we can analyze these motions mathematically. To do this we use the vector algebra we have developed to express the effect of the gravitational force on velocity components along horizontal and vertical axes.

### Velocity Components and Launch Angle

A ball rolling off a level ramp has a launch angle of zero degrees with respect to the horizontal. In general, projectiles are launched at some angle $\theta$ with respect to the horizontal as shown in Fig 5-6($a$). Suppose a projectile is launched at an angle $\theta_1$ relative to the horizontal direction with an initial magnitude of velocity $|\vec{v}_1| = v_1$. If we use a standard rectangular coordinate system, then we can use the definitions of sine and cosine to find the initial $x$- and $y$-components of velocity at time $t_1$ (Fig. 5-8). These components are

$$v_{1x} = v_1 \cos \theta_1 \quad \text{and} \quad v_{1y} = v_1 \sin \theta_1. \tag{5-1}$$

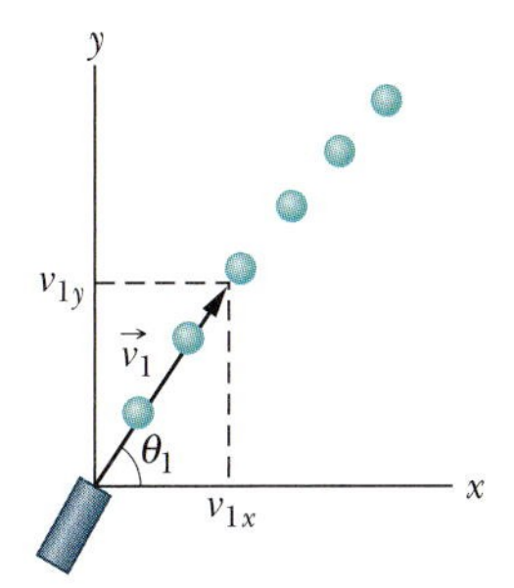


**FIGURE 5-8** ■ If a projectile is launched at an initial angle $\theta_1$ with respect to the horizontal with a magnitude $v_1$, its components can be calculated by using the trigonometric functions.

Alternatively, if the components of the initial velocity are known, we can rearrange Eq. 5-1 to determine the initial velocity vector, $\vec{v}_1$, and the angle of launch, $\theta_1$,

$$\vec{v}_1 = v_{1x}\hat{\mathrm{i}} + v_{1y}\hat{\mathrm{j}} \quad \text{and} \quad \tan\theta_1 = \frac{v_1 \sin \theta_1}{v_1 \cos \theta_1} = \frac{v_{1y}}{v_{1x}}. \tag{5-2}$$

Solving for the launch angle at time $t_1$ gives

$$\theta_1 = \tan^{-1}\left(\frac{v_1 \sin \theta_1}{v_1 \cos \theta_1}\right) = \tan^{-1}\left(\frac{v_{1y}}{v_{1x}}\right). \tag{5-3}$$

During two-dimensional motion, an ideal projectile's position vector $\vec{r}$ and velocity vector $\vec{v}$ change continuously, but its acceleration vector $\vec{a}$ is constant—its value doesn't change and it is always directed vertically downward.

### Position and Velocity Versus Time

In our standard coordinate system an ideal projectile experiences no net force in the $x$ direction ($\vec{F}_x = (0)\hat{\mathrm{i}}$), so its horizontal acceleration stays constant at zero during its flight. The projectile experiences a constant gravitational force $\vec{F}_y = -mg\hat{\mathrm{j}}$ in the $y$ direction, so the $y$-component of its acceleration is a constant with a value of $-g$. Since zero acceleration is a form of constant acceleration, the acceleration is constant in *both* directions. The object's projectile motion acceleration is given by

$$\vec{a}_x = 0\hat{\mathrm{i}} \quad \text{and} \quad \vec{a}_y = -g\hat{\mathrm{j}} \quad \text{(projectile motion accelerations).}$$

From Chapter 2 (Eq. 2-17) we know that (with $a_x = 0$) the changes in the object's position components are given by

$$x_2 - x_1 = v_{1x}(t_2 - t_1) \quad \text{and} \quad y_2 - y_1 = v_{1y}(t_2 - t_1) + \tfrac{1}{2}a_y(t_2 - t_1)^2 \quad \text{(position change),} \tag{2-17}$$

while the object's velocity component changes are given by Eq. 2-13 as

$$v_{2x} - v_{1x} = a_x(t_2 - t_1) \quad \text{and} \quad v_{2y} - v_{1y} = a_y(t_2 - t_1) \quad \text{(velocity change).} \tag{2-13}$$

If we call the time that we start measuring $t_1 = 0$, then the equations are somewhat simplified. However, independent of what time we start measuring, these two equations describe the motion of the object during a specified *interval* of time $t_2 - t_1$. They are our primary equations of motion and are valid for every situation involving constant acceleration, including cases where the acceleration is zero.

## The Horizontal Motion

Suppose that at an initial time $t = t_1$ the projectile has a position component along the $x$ axis of $x_1$ and a velocity component of $v_{1x}$. We must now use the notation $\vec{v}_1, v_{1x}, v_{1y}$ so we can distinguish the initial velocity vector $\vec{v}_1$ from its $x$- and $y$-components. We must also specify which component of acceleration, $a_x$ or $a_y$, we are using in an equation. Using our new notation, we can find the $x$-component of the projectile's horizontal displacement $x_2$ at any later time $t_2$ using $x_2 - x_1 = v_{1x}(t_2 - t_1) + \frac{1}{2}a_x(t_2 - t_1)^2$. However, since there are *no forces* and hence *no acceleration* in the horizontal direction, the $x$-component of acceleration is zero. Noting that the only part of the initial velocity that affects the horizontal motion is its horizontal component $v_{1x}$, we can write this as:

$$x_2 - x_1 = v_{1x}(t_2 - t_1) \quad \text{(horizontal displacement).} \tag{5-4}$$

Because the horizontal component of the object's initial velocity is given by $v_{1x} = v_1 \cos\theta_1$, this equation for the displacement in the $x$ direction can also be written as

$$x_2 - x_1 = (v_1 \cos\theta_1)(t_2 - t_1). \tag{5-5}$$

But the ratio of the displacement to the time interval over which it occurs, $(x_2 - x_1)/(t_2 - t_1)$, is just the $x$-component of average velocity $\langle v_x \rangle$. If the average velocity in the $x$ direction is constant, this means that the instantaneous and average velocities in the $x$ direction are the same. Thus Eq. 5-5 becomes

$$\langle v_x \rangle = \frac{x_2 - x_1}{t_2 - t_1} = v_{1x} = v_1 \cos\theta_1 = \text{a constant.} \tag{5-6}$$

Experimental verification of the constancy of the horizontal velocity component for ideal projectile motion is present in Figs. 5-4 and 5-5. As additional verification we can draw a graph of the $x$-component of position as a function of time for the projectile path depicted in Fig. 5-6$a$. This is shown in Fig. 5-9.

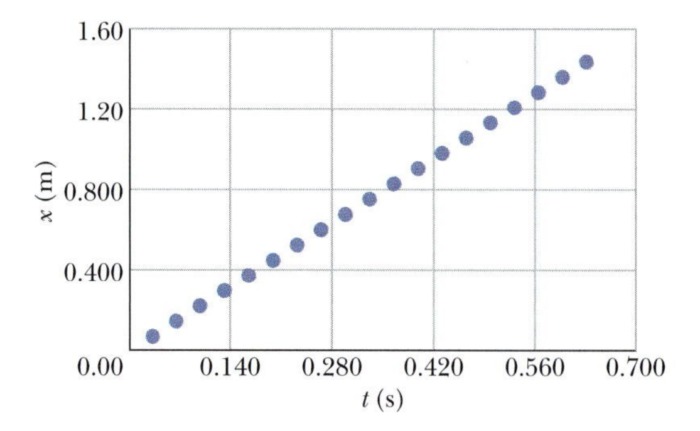


**FIGURE 5-9** ■ A graph constructed from a video analysis of the motion of the projectile depicted in Fig. 5-6$a$. The $x$-component of position is plotted as a function of time. The coordinate system is chosen so that the initial value $x_1$ is zero at the launcher muzzle. The linearity of the graph confirms that the projectile's $x$-component of velocity is a constant given by the slope of the line so that $\langle v_x \rangle = v_{1x} = 2.3$ m/s.

## The Vertical Motion

The vertical motion of an ideal projectile was discussed in Section 3-9 for a particle close to the surface of the Earth. Most important is that the acceleration resulting from the attractive gravitational force that the Earth exerts on the object is constant and directed downward. We denote its magnitude as $g$ and recall that it has a value of 9.8 m/s$^2$. If, as usual, we take upward to be the positive $y$ direction, then since the gravitational force points downward we can replace the $y$-component of acceleration, $a_y$, for position change along the $y$ axis with $-g$. This allows us to rewrite our primary kinematic equation shown in Eq. 2-17 $[(x_2 - x_1) = v_{1\,x}(t_2 - t_1) + \frac{1}{2}a_x(t_2 - t_1)^2]$ for motion along a $y$ axis as

$$y_2 - y_1 = v_{1\,y}(t_2 - t_1) + \tfrac{1}{2}(a_y)(t_2 - t_1)^2 \qquad \text{where } a_y = -g.$$

Note that we used only the $y$-component of the initial velocity in this equation. This is because the vertical acceleration affects only the vertical velocity and position components. Making the substitution $v_{1\,y} = v_1 \sin\theta_1$ from Eq. 5-1, we get

$$y_2 - y_1 = (v_1 \sin\theta_1)(t_2 - t_1) + \tfrac{1}{2}a_y(t_2 - t_1)^2 \qquad \text{where } a_y = -g. \tag{5-7}$$

Similarly, $v_2 - v_1 = a_x(t_2 - t_1)$ (primary equation 2-13) can be rewritten as

$$v_{2\,y} - v_{1\,y} = a_y(t_2 - t_1),$$

or

$$v_{2\,y} - (v_1 \sin\theta_1) = a_y(t_2 - t_1) \qquad \text{where } a_y = -g. \tag{5-8}$$

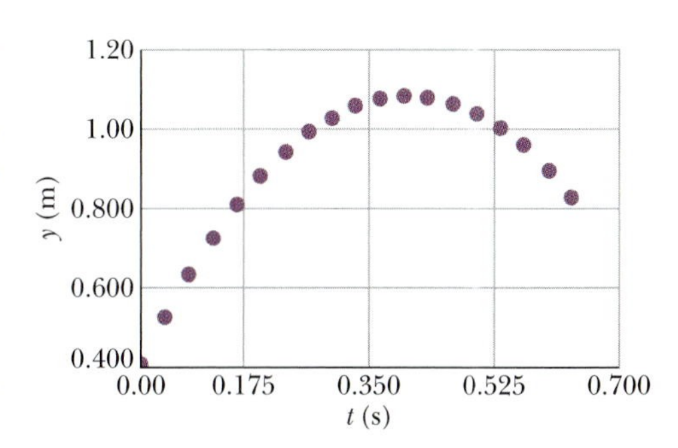


**FIGURE 5-10** ■ A graph constructed from a video analysis of the motion of the projectile depicted in Fig. 5-6*a*. The $y$-component of position is plotted as a function of time. The coordinate system is chosen so that the initial value of the vertical position $y_1$ is zero at the launcher muzzle. A fit to the curve is parabolic and gives a $y$-component of acceleration of $-9.7$ m/s$^2$ and an initial vertical velocity component of $+3.5$ m/s.

As is illustrated in Fig. 5-10, the vertical velocity component behaves just like that for a ball thrown vertically upward. At the instant the velocity is zero, the object must be at the highest point on its path, since the object then starts moving down. The magnitude of the velocity becomes larger with time as the projectile speeds up as it moves back down.

The velocity components for a projectile shot from a small vertical cannon mounted on a cart are shown in Fig. 5-11. Note that the horizontal velocity component does not change. The vertical component decreases in magnitude and becomes zero at the top of the path. This magnitude starts increasing again as the projectile descends and is finally "recaptured" by the cannon.

In the experimental results presented in this section, the effect of the air on the motion was negligible. Thus, we ignored air resistance and performed a mathematical analysis for ideal projectiles. Ignoring the effects of air works well for a compact, dense object such as a marble or a bowling ball, provided it is not launched at very high speeds. However, air resistance cannot be ignored for a less dense object like a crumpled piece of paper thrown rapidly or the styrofoam ball shown in Fig. 5-6*b*. Remember that the ideal projectile motion equations 5-4 through 5-8 have been derived assuming that resistance is negligible. These ideal projectile equations are summarized in Table 5-1. We shall discuss details of the effect of the air on motion in Chapter 6.

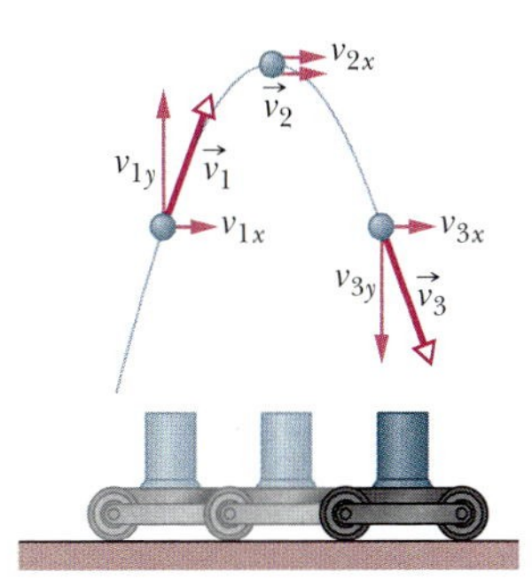


**FIGURE 5-11** ■ Diagram showing sketches of three video frames. The actual movie recorded the motion of a cannon that shoots a projectile while moving. The projectile path and velocity vector components were calculated by modeling data acquired using digital video analysis software.

**TABLE 5-1**
**Kinematic Equations for Ideal Projectile Motion**

| Quantity | Horizontal | Vertical |
|---|---|---|
| Forces | $\vec{F}_x = (0)\hat{i}$ | $\vec{F}_y = (-mg)\hat{j}$ (Eq. 3-7) |
| Acceleration Components | $a_x = 0$ | $a_y = -g = -9.8$ m/s$^2$ |
| Velocity components at $t_1$ | $v_{1x} = v_1 \cos\theta_1$ (Eq. 5-1) | $v_{1y} = v_1 \sin\theta_1$ (Eq. 5-1) |
| Position component Change between $t_1$ and $t_2$ | $x_2 - x_1 = v_{1x}(t_2 - t_1)$ (Eq. 5-4) | $y_2 - y_1 = v_{1y}(t_2 - t_1) + \frac{1}{2}a_y(t_2 - t_1)^2$ (Eq. 5-7) |
| Velocity component Change between $t_1$ and $t_2$ | $v_{2x} - v_{1x} = 0$ | $v_{2y} - v_{1y} = a_y(t_2 - t_1)$ (Eq. 5-8) |

*Note*: Since $g$ (the local gravitational field strength) is always positive, it is not a vector component. There fore, if we follow the standard practice of defining the negative $y$ axis as down, we must put an explicit minus sign in front of it.

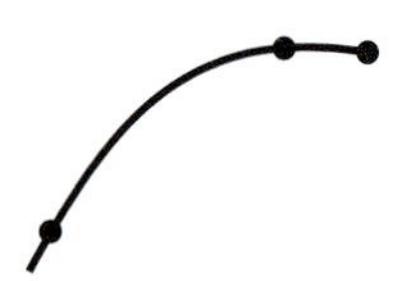

**READING EXERCISE 5-3:** Consider three points along an ideal projectile's path in space as shown in the figure and make three separate sketches of: (a) the force vectors showing the net force on the projectile at each point, (b) the acceleration vectors showing the acceleration of the projectile at each point, (c) the approximate horizontal and vertical components of velocity at each of the three points along with the velocity vector that is determined by these components. ■

**READING EXERCISE 5-4:** Consider the projectile launch shown in Fig. 5-6*a*. (a) According to the data in Figs. 5-9 and 5-10 the initial $x$- and $y$-components of the velocity are $v_{1x} = 2.3$ m/s and $v_{1y} = +3.5$ m/s . Use these values to find the launch angle of the projectile. (b) Use a protractor to measure the launch angle as indicated by the angle of the launcher shown in Fig. 5-6*a*. How do your calculated and measured launch angles compare? They should be approximately the same. ■

**READING EXERCISE 5-5:** A fly ball is hit to the outfield. During its flight (ignore the effects of the air), what happens to its (a) horizontal and (b) vertical components of velocity? What are the (c) horizontal and (d) vertical components of its acceleration during its ascent and its descent, and at the topmost point of its flight? ■

## TOUCHSTONE EXAMPLE 5-2: Rescue Plane

In Fig. 5-12, a rescue plane flies at 198 km/h (= 55.0 m/s) and a constant elevation of 500 m toward a point directly over a boating accident victim struggling in the water. The pilot wants to release a rescue capsule so that it hits the water very close to the victim.

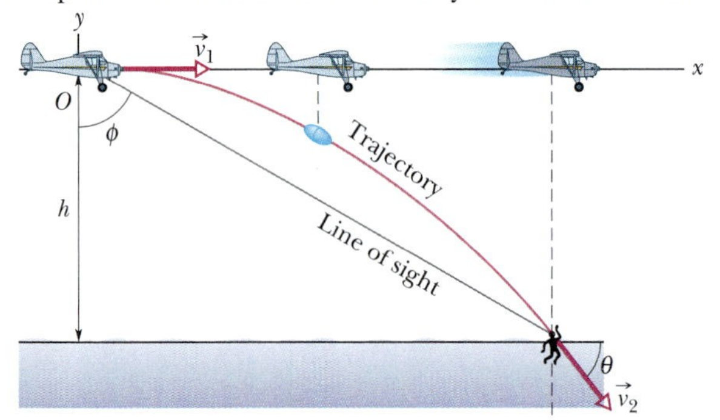


**FIGURE 5-12** ■ A plane drops a rescue capsule while moving at constant velocity in level flight. While the capsule is falling, its horizontal velocity component remains equal to the velocity of the plane.

(a) What should be the angle $\phi$ of the pilot's line of sight to the victim when the release is made?

**SOLUTION** ■ The **Key Idea** here is that, once released, the capsule is a projectile, so its horizontal and vertical motions are independent and can be considered separately (we need not consider the actual curved path of the capsule). Figure 5-12 includes a coordinate system with its origin at the point of release, and we see there that $\phi$ is given by

$$\phi = \tan^{-1}\frac{(x_2 - x_1)}{h}, \tag{5-9}$$

where $x_2$ is the horizontal coordinate of the victim at release (and of the capsule when it hits the water), and $h$ is the elevation of the plane. That elevation is 500 m, so we need only $x_2$ in order to find $\phi$. We should be able to find $x_2$ with Eq. 5-4: $x_2 - x_1 = v_{1x}(t_2 - t_1)$. This can be written as

$$x_2 - x_1 = (v_1 \cos\theta_1)(t_2 - t_1), \tag{5-10}$$

where $\theta_1$ is the angle between the initial velocity $\vec{v}_1$ and the positive $x$ axis. For this problem, $\theta_1 = 0°$.

We know $x_1 = 0$ because the origin is placed at the point of release. Because the capsule is *released* and not shot from the plane, its initial velocity $\vec{v}_1$ is equal to the plane's velocity. Thus, we know also that the initial velocity has magnitude $v_1 = 55.0$ m/s and angle $\theta_1 = 0°$ (measured relative to the positive direction of the $x$ axis). However, we do not know the elapsed time $t_2 - t_1$ the capsule takes to move from the plane to the victim.

To find $t_2 - t_1$, we next consider the vertical motion and specifically Eq. 5-7:

$$y_2 - y_1 = (v_1 \sin\theta_1)(t_2 - t_1) - \tfrac{1}{2}g(t_2 - t_1)^2. \tag{5-11}$$

Here the vertical displacement $y_2 - y_1$ of the capsule is $-500$ m (the negative value indicates that the capsule moves *downward*). Putting this and other known values into Eq. 5-7 gives us

$$-500\text{ m} = (55.0\text{ m/s})(\sin 0°)(t_2 - t_1) - \tfrac{1}{2}(9.8\text{ m/s}^2)(t_2 - t_1)^2.$$

Solving for $t_2 - t_1$, we find $t_2 - t_1 = 10.1$ s. Using that value in Eq. 5-10 yields

$$x_2 - 0\text{ m} = (55.0\text{ m/s})(\cos 0°)(10.1\text{ s}),$$

or

$$x_2 = 555.5\text{ m}.$$

Then Eq. 5-9 gives us

$$\phi = \tan^{-1}\frac{555.5\text{ m}}{500\text{ m}} = 48°. \qquad \text{(Answer)}$$

(b) As the capsule reaches the water, what is its velocity $\vec{v}_2$ in unit-vector notation and as a magnitude and an angle?

**SOLUTION** ■ Again, we need the **Key Idea** that during the capsule's flight, the horizontal and vertical components of the capsule's velocity are independent of each other.

A second **Key Idea** is that the horizontal component of velocity $v_x$ does not change from its initial value $v_{1x} = v_1 \cos\theta_1$ because there is no horizontal acceleration. Thus, when the capsule reaches the water,

$$v_{2x} = v_{1x} = v_1 \cos\theta_1 = (55.0\text{ m/s})(\cos 0°) = 55.0\text{ m/s}.$$

A third **Key Idea** is that the vertical component of velocity $v_y$ changes from its initial value $v_{1y} = v_1 \sin\theta_1$ because there is a vertical acceleration. Using Eq. 5-8 and the capsule's time of fall $t_2 - t_1 = 10.1$ s, we find that when the capsule reaches the water,

$$\begin{aligned} v_{2y} &= v_1 \sin\theta_1 - g(t_2 - t_1) \\ &= (55.0\text{ m/s})(\sin 0°) - (9.8\text{ m/s}^2)(10.1\text{ s}) \\ &= -99.0\text{ m/s}. \end{aligned}$$

Thus, when the capsule reaches the water, it has the velocity

$$\vec{v}_2 = (55.0\text{ m/s})\hat{i} - (99.0\text{ m/s})\hat{j}. \qquad \text{(Answer)}$$

Using either the techniques developed in Section 4-4, or a vector-capable calculator, we find that the magnitude of the final velocity $v_2$ and the angle $\theta_2$ are

$$v_2 = 113\text{ m/s} \quad \text{and} \quad \theta_2 = -61°. \qquad \text{(Answer)}$$

## TOUCHSTONE EXAMPLE 5-3: Ballistic Zacchini

Figure 5-13 illustrates the flight of Emanuel Zacchini over three Ferris wheels, located as shown, and each 18 m high. Zacchini is launched with speed $v_1 = 26.5$ m/s, at an angle $\theta_1 = 53°$ up from the horizontal and with an initial height of 3.0 m above the ground. The net in which he is to land is at the same height.

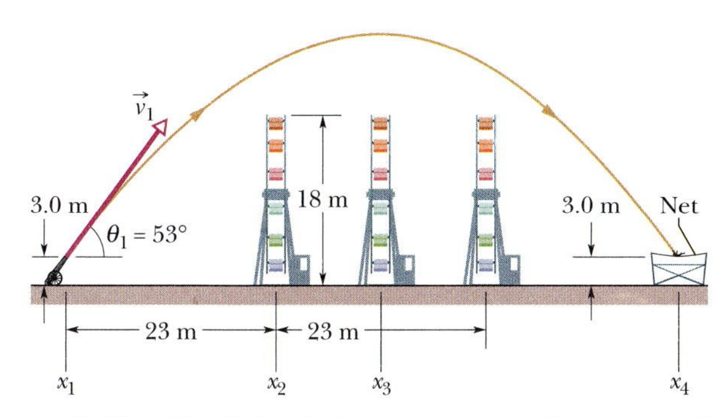


**FIGURE 5-13** ■ The flight of a human cannonball over three Ferris wheels and into a net.

(a) Does he clear the first Ferris wheel?

**SOLUTION** ■ A **Key Idea** here is that Zacchini is a human projectile, so we can use the projectile equations. To do so, we place the origin of an $xy$ coordinate system at the cannon muzzle. Then $x_1 = 0$ and $y_1 = 0$ and we want his height $y_2$ when $x_2 = 23$ m. However, we do not know the elapsed time $t_2 - t_1$ when he reaches that height. To relate $y_2$ to $x_2$ without $t_2 - t_1$, we can solve Eq. 5-5 for $t_2 - t_1$, which gives us

$$(t_2 - t_1) = \frac{x_2 - x_1}{(v_1 \cos\theta_1)}.$$

Then we can replace $t_2 - t_1$ everywhere that it appears in Eq. 5-7,

$$y_2 - y_1 = (v_1 \sin\theta_1)(t_2 - t_1) - \tfrac{1}{2}g(t_2 - t_1)^2.$$

We note that $(v_1 \sin\theta_1)/(v_1 \cos\theta_1) = \tan\theta_1$ to obtain:

$$\begin{aligned} y_2 - y_1 &= (\tan\theta_1)(x_2 - x_1) - \frac{g(x_2 - x_1)^2}{2(v_1 \cos\theta_1)^2} \\ &= (\tan 53°)(23\text{ m}) - \frac{(9.8\text{ m/s}^2)(23\text{ m})^2}{2(26.5\text{ m/s})^2(\cos 53°)^2} \\ &= 20.3\text{ m}. \end{aligned}$$

Since he begins 3.0 m off the ground, he clears the first Ferris wheel by about 5.3 m.

(b) If he reaches his maximum height when he is over the middle Ferris wheel, what is his clearance above it?

**SOLUTION** ■ A **Key Idea** here is that the vertical component $v_y$ of his velocity is zero when he reaches his maximum height. We can combine Eqs. 5-7 and 5-8 to relate $v_y$ and his height $y_3 - y_1$ to obtain:

$$v_{3y}^2 = v_{1y}^2 - 2g(y_3 - y_1) = (v_1 \sin\theta_1)^2 - 2g(y_3 - y_1) = 0.$$

Solving for $y_3 - y_1$ gives us

$$y_3 - y_1 = \frac{(v_1 \sin\theta_1)^2}{2g} = \frac{(26.5\text{ m/s})^2(\sin 53°)^2}{(2)(9.8\text{ m/s}^2)} = 22.9\text{ m},$$

which means that he clears the middle Ferris wheel by 7.9 m.

(c) How far from the cannon should the center of the net be positioned?

**SOLUTION** ■ The additional **Key Idea** here is that, because Zacchini's initial and landing heights are the same, the horizontal distance from cannon muzzle to net is the value of $x_4 - x_1$ where time $t = t_4$ is when $y_4$ is once again zero. Then, since $y_1 = 0$ Eq. 5-7 becomes

$$0 = y_4 - y_1 = (v_1 \sin\theta_1)(t_4 - t_1) - \tfrac{1}{2}g(t_4 - t_1)^2.$$

Since $t_4 - t_1 \neq 0$ we can divide both sides of this equation by $t_4 - t_1$ and solve for the time he is airborne:

$$t_4 - t_1 = \frac{2v_1 \sin\theta_1}{g}.$$

Substituting this time interval in $x_4 - x_1 = (\vec{v}_1 \cos\theta_1)(t_4 - t_1)$ (Eq. 5-5) gives us the total horizontal distance he traveled:

$$x_4 - x_1 = \frac{2v_1^2}{g}\sin\theta_1\cos\theta_1 = \frac{2(26.5\text{ m/s})^2}{9.8\text{ m/s}^2}\sin(53°)\cos(53°)$$

$$= 69\text{ m}.$$

We can now answer the questions that opened this chapter: How could Zacchini know where to place the net, and how could he be certain he would clear the Ferris wheels? He (or someone) did the calculations as we have here. Although he could not take into account the complicated effects of the air on his flight, Zacchini knew that the air would slow him, and thus decrease his range from the calculated value. So, he used a wide net and biased it toward the cannon. He was then relatively safe whether the effects of the air in a particular flight happened to slow him considerably or very little. Still, the variability of this factor of air effects must have played on his imagination before each flight.

Zacchini still faced a subtle danger. Even for shorter flights, his propulsion through the cannon was so severe that he underwent a momentary blackout. If he landed during the blackout, he could break his neck. To avoid this, he had trained himself to awake quickly. Indeed, not waking up in time presents the only real danger to a human cannonball in the short flights today.

## 5-4 Displacement in Two Dimensions

We conclude this chapter with an exploration of objects that move in a circle at a constant speed. Before we begin this exploration, we need to learn more about finding displacement vectors in two dimensions—a task we began in Sections 2-2, 4-2, and 4-3.

How can we track the motion of a particle-like object that moves in a two-dimensional plane instead of being constrained to move along a line? As was the case for tracking motion along a line (treated in Chapter 2), it is useful to define a **position vector,** $\vec{r}$, that extends from the origin of a chosen coordinate system to the object. But this time, we have to choose a two-dimensional coordinate system. However, we can use the vector algebra introduced in the last chapter to resolve (decompose) the position vector into component vectors. This allows us to treat the two-dimensional motions using the techniques developed to describe one-dimensional motions.

### Using Rectangular Coordinates

If we decide to use rectangular coordinates, then we denote the rectangular components of $\vec{r}$ as $x$ and $y$. We can use the unit-vector notation of Section 4-5 to resolve the position vector into

$$\vec{r} = x\hat{\text{i}} + y\hat{\text{j}}, \tag{5-12}$$

where $x\hat{\text{i}}$ and $y\hat{\text{j}}$ are the rectangular component vectors. Note that for a rectangular coordinate system, the components $x$ and $y$ are the same as the coordinates of the object's location $(x, y)$.

The coordinates $x$ and $y$ specify the particle's location along the coordinate axes relative to the origin. For instance, Fig. 5-14 shows a particle with rectangular coordinates $(-3\text{ m}, 2\text{ m})$ that has a position vector given by

$$\vec{r} = (-3\text{ m})\hat{\text{i}} + (2\text{ m})\hat{\text{j}}.$$

The particle is located 3 meters from the $y$ axis in the $-\hat{\text{i}}$ direction and 2 meters from the $x$ axis in the $+\hat{\text{j}}$ direction.

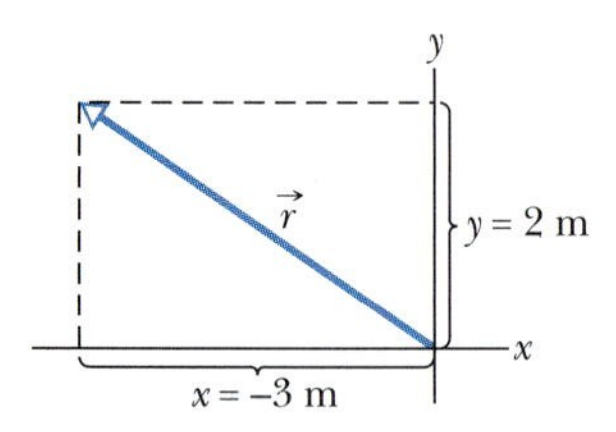


**FIGURE 5-14** ■ The position vector $\vec{r}$ for a particle that has coordinates $(-3\text{ m}, 2\text{ m})$.

As a particle moves, its position vector changes in such a way that the vector always extends to the particle from the reference point (the origin). If the position vector changes—say, from $\vec{r}_1$ to $\vec{r}_2$ during a certain time interval—then the particle's **displacement** $\vec{r}$ during that time interval is

$$\Delta\vec{r} = \vec{r}_2 - \vec{r}_1. \tag{5-13}$$

Using unit-vector notation for $\vec{r}_2$ and $\vec{r}_1$, we can rewrite this displacement as

$$\Delta\vec{r} = (x_2\hat{\mathrm{i}} + y_2\hat{\mathrm{j}}) - (x_1\hat{\mathrm{i}} + y_1\hat{\mathrm{j}}).$$

By grouping terms having the same unit vector, we get

$$\Delta\vec{r} = (x_2 - x_1)\,\hat{\mathrm{i}} + (y_2 - y_1)\,\hat{\mathrm{j}}, \tag{5-14}$$

where the components $(x_1, y_1)$ correspond to position vector $\vec{r}_1$ and components $(x_2, y_2)$ correspond to position vector $\vec{r}_2$. We can also rewrite the displacement by substituting $\Delta x$ for $(x_2 - x_1)$, and $\Delta y$ for $(y_2 - y_1)$, so that

$$\Delta\vec{r} = \Delta x\,\hat{\mathrm{i}} + \Delta y\,\hat{\mathrm{j}}. \tag{5-15}$$

This expression is another example of a very important aspect of motion in more than one dimension. Notice that the coordinates of the displacement in each direction $(\Delta x, \Delta y)$ depend only on the change in the object's position in that one direction, and are independent of changes in position in the other directions. In other words, we don't have to simultaneously consider the change in the object's position in every direction. We can break the motion into two parts (motion in the $x$ direction and motion in the $y$ direction) and consider each direction separately. For motion in three dimensions we could add a $z$ direction and do equivalent calculations.

## Using Polar Coordinates

In dealing with circular motions or rotations it is often useful to describe positions and displacements in polar coordinates. In this case we locate a particle using $r$ or $|\vec{r}|$ which represents the magnitude of its position vector $\vec{r}$ and its angle $\phi_1$ which is measured in a counterclockwise direction from a chosen axis. The relationship between two-dimensional rectangular coordinates and polar coordinates is shown in Fig. 5-15. The transformation between these coordinate systems is based on the definitions of sine and cosine, so that

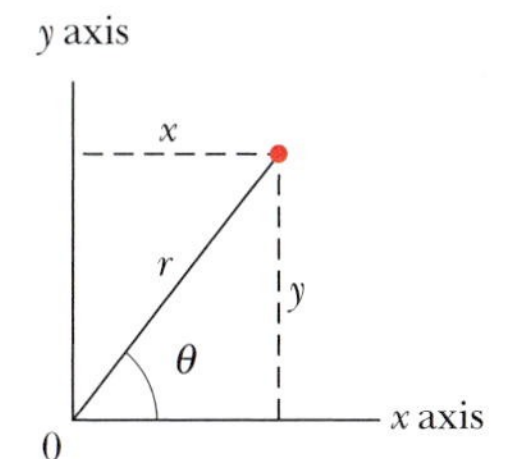


**FIGURE 5-15** ■ Polar coordinates provide an alternative way to locate a particle that is confined to move in two dimensions. These coordinates are especially useful in the description of circular motion where the distance of a particle from the origin does not change.

$$x = r\cos(\theta) \quad \text{and} \quad y = r\sin(\theta). \tag{5-16}$$

Conversely,

$$r = \sqrt{x^2 + y^2} \quad \text{and} \quad \theta = \tan^{-1}\left(\frac{y}{x}\right). \tag{5-17}$$

In circular motion the distance of a particle from the center of the circle defining its motion does not change. So the magnitude of displacement $\Delta r$ or $|\Delta\vec{r}|$ depends only on the change in angle and is given by

$$\Delta r = r(\theta_2 - \theta_1) \quad \text{(small angle circular displacement magnitude).} \tag{5-18}$$

for small angular differences. This expression should look familiar from mathematics classes. There you learned that the relationship between arc length, $s$, the radius, $r$, and angular displacement, $\Delta\theta$ is $s = r\Delta\theta$.

**READING EXERCISE 5-6:** (a) If a wily bat flies from $x, y$ coordinates $(-2\text{ m}, 4\text{ m})$ to coordinates $(6\text{ m}, -2\text{ m})$, what is its displacement $\Delta\vec{r}$ in rectangular unit-vector notation? (b) Is $\Delta\vec{r}$ parallel to one of the two coordinate axes? If so, which axis?

## TOUCHSTONE EXAMPLE 5-4: Displacement

In Fig. 5-16, the position vector for a particle is initially

$$\vec{r}_1 = (-3.0\text{ m})\hat{i} + (4.0\text{ m})\hat{j}$$

and then later is

$$\vec{r}_2 = (9.0\text{ m})\hat{i} + (-3.5\text{ m})\hat{j}.$$

What is the particle's displacement $\Delta\vec{r}$ from $\vec{r}_1$ to $\vec{r}_2$?

**SOLUTION** The **Key Idea** is that the displacement $\Delta\vec{r}$ is obtained by subtracting the initial position vector $\vec{r}_1$ from the later position vector $\vec{r}_2$. That is most easily done by components:

$$\begin{aligned}\Delta\vec{r} &= \vec{r}_2 - \vec{r}_1 \\ &= [9.0 - (-3.0)](\text{m})\hat{i} + [-3.5 - 4.0](\text{m})\hat{j} \\ &= (12\text{ m})\hat{i} + (-7.5\text{ m})\hat{j}.\end{aligned} \quad \text{(Answer)}$$

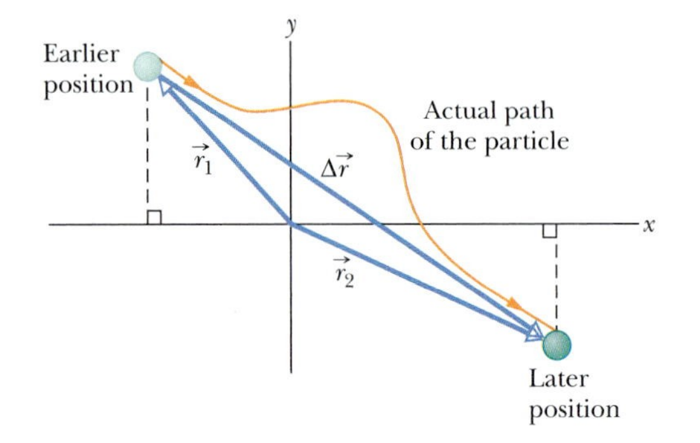


FIGURE 5-16 The displacement $\Delta\vec{r} = \vec{r}_2 - \vec{r}_1$ extends from the head of the initial position vector $\vec{r}_1$ to the head of a later position vector $\vec{r}_2$ regardless of what path is actually taken.

## TOUCHSTONE EXAMPLE 5-5: Rabbit's Trajectory

A rabbit runs across a parking lot on which a set of coordinate axes has, strangely enough, been drawn. The coordinates of the rabbit's position as functions of time $t$ are given by

$$x = (-0.31\text{ m/s}^2)t^2 + (7.2\text{ m/s})t + 28\text{ m} \quad (5\text{-}19)$$

and

$$y = (0.22\text{ m/s}^2)t^2 + (-9.1\text{ m/s})t + 30\text{ m}. \quad (5\text{-}20)$$

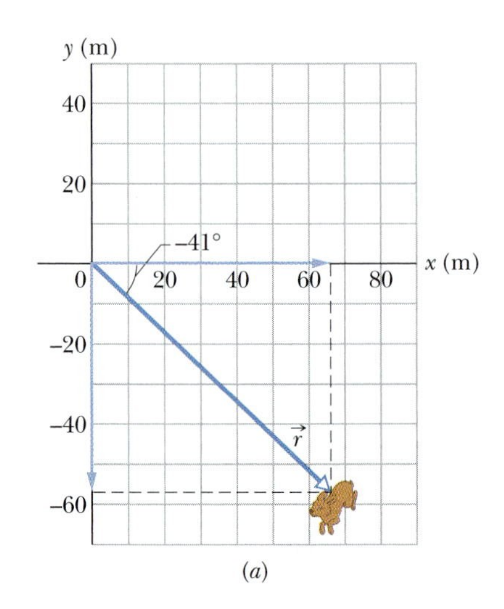


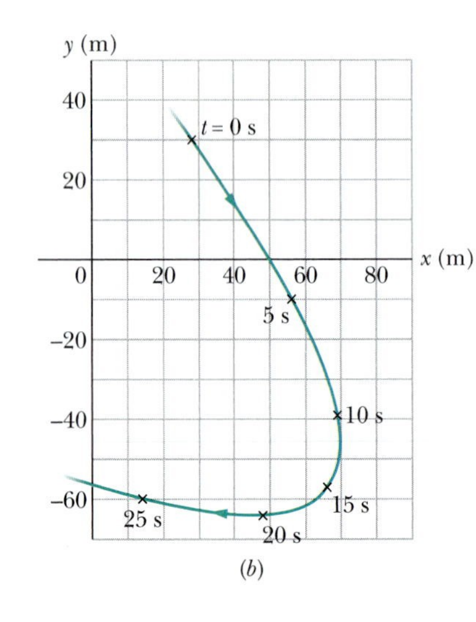


FIGURE 5-17 (a) A rabbit's position vector $\vec{r}$ at time $t = 15$ s. The components of $\vec{r}$ are shown along the axes. (b) The rabbit's path and its position at five values of $t$.

(a) At $t = 15$ s, what is the rabbit's position vector $\vec{r}$ in unit-vector notation and as a magnitude and an angle?

**SOLUTION** The **Key Idea** here is that the $x$ and $y$ coordinates of the rabbit's position, as given by Eqs. 5-19 and 5-20, are the components of the rabbit's position vector $\vec{r}$. Thus, we can write

$$\vec{r}(t) = x(t)\hat{i} + y(t)\hat{j}. \quad (5\text{-}21)$$

(We write $\vec{r}(t)$ rather than $\vec{r}$ because the components are functions of $t$, and thus $\vec{r}$ is also.)

At $t = 15$ s, the components of the position vector are

$$x = (-0.31 \text{ m/s}^2)(15 \text{ s})^2 + (7.2 \text{ m/s})(15 \text{ s}) + 28 \text{ m} = 66 \text{ m},$$

and $$y = (0.22 \text{ m/s}^2)(15 \text{ s})^2 + (-9.1 \text{ m/s})(15 \text{ s}) + 30 \text{ m} = -57 \text{ m}.$$

Thus, at $t = 15$ s,

$$\vec{r} = (66 \text{ m})\hat{\text{i}} - (57 \text{ m})\hat{\text{j}},$$

which is drawn in Fig. 5-17$a$.

To get the magnitude and angle of $\vec{r}$, we can use a vector-capable calculator, or we can be guided by the Pythagorean theorem to write

$$r = \sqrt{x^2 + y^2} = \sqrt{(66 \text{ m})^2 + (-57 \text{ m})^2} = 87 \text{ m},$$

and from trigonometric definition,

$$\theta = \tan^{-1}\frac{y}{x} = \tan^{-1}\left(\frac{-57 \text{ m}}{66 \text{ m}}\right) = -41°$$

(Although $\theta = 139°$ has the same tangent as $-41°$, study of the signs of the components of $\vec{r}$ rules out 139°.)

(b) Graph the rabbit's path for $t = 0$ to $t = 25$ s.

**SOLUTION** ■ We can repeat part (a) for several values of $t$ and then plot the results. Figure 5-17$b$ shows the plots for five values of $t$ and the path connecting them. We can also use a graphing calculator to make a *parametric graph*; that is, we would have the calculator plot $y$ versus $x$, where these coordinates are given by Eqs. 5-19 and 5-20 as functions of time $t$.

## 5-5 Average and Instantaneous Velocity

We have just shown that when tracking motions occurring in more than one dimension, position and displacement vectors can be resolved into rectangular component vectors. Can this also be done with velocity vectors? As is the case for motion in one dimension, if a particle moves through a displacement $\Delta\vec{r}$ in a time interval $\Delta t$ (as shown in Fig. 5-18), then its **average velocity** $\langle\vec{v}\rangle$ is defined as

$$\text{average velocity} \equiv \frac{\text{displacement}}{\text{time interval}},$$

or using familiar symbols

$$\langle\vec{v}\rangle \equiv \frac{\Delta\vec{r}}{\Delta t}. \tag{5-22}$$

This tells us the direction of $\langle\vec{v}\rangle$ must be the same as the displacement $\Delta\vec{r}$. Using our new definition of displacement in two dimensions (Eq. 5-15), we can rewrite this as

$$\langle\vec{v}\rangle = \frac{\Delta x\,\hat{\text{i}} + \Delta y\,\hat{\text{j}}}{\Delta t} = \frac{\Delta x}{\Delta t}\hat{\text{i}} + \frac{\Delta y}{\Delta t}\hat{\text{j}}. \tag{5-23}$$

This equation can be simplified by noting that $\Delta x/\Delta t$ is defined in Chapter 2 as the component of average velocity in the $x$ direction. If we use appropriate definitions for average velocity components in the $y$ direction, then

$$\langle\vec{v}\rangle = \frac{\Delta x}{\Delta t}\hat{\text{i}} + \frac{\Delta y}{\Delta t}\hat{\text{j}} = \langle v_x\rangle\hat{\text{i}} + \langle v_y\rangle\hat{\text{j}}. \tag{5-24}$$

Here we are using the same notation introduced in Chapter 2 to denote the average velocity components. We use subscripts $x$ and $y$ to distinguish each of the average velocity components. Also the angle brackets $\langle\ \rangle$ are used to distinguish average velocity from instantaneous velocity.

As we mentioned in Chapter 2, when we speak of the **velocity** of a particle, we usually mean the particle's **instantaneous velocity** $\vec{v}$. $\vec{v}$ represents the limit the aver-

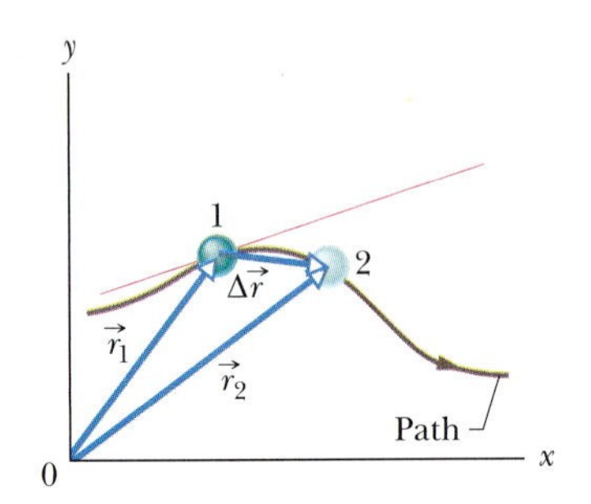


**FIGURE 5-18** ■ The displacement $\Delta\vec{r}$ of a particle during a time interval $\Delta t$ from position 1 with position vector $\vec{r}_1$ at time $t_1$ to position 2 with position vector $\vec{r}_2$ at time $t_2$.

age velocity $\langle\vec{v}\rangle$ approaches as we shrink the time interval $\Delta t$ to zero. Using the language of calculus, we can also write $\vec{v}$ as the derivative

$$\vec{v} = \lim_{\Delta t \to 0} \frac{\Delta\vec{r}}{\Delta t} = \frac{d\vec{r}}{dt}. \tag{5-25}$$

If we substitute $\vec{r} = x\,\hat{\text{i}} + y\,\hat{\text{j}}$, then in unit-vector notation:

$$\vec{v} = \frac{d}{dt}(x\hat{\text{i}} + y\hat{\text{j}}) = \frac{dx}{dt}\hat{\text{i}} + \frac{dy}{dt}\hat{\text{j}}.$$

This equation can be simplified by recognizing that $dx/dt$ is $v_x$ and so on. Thus,

$$\vec{v} = v_x\hat{\text{i}} + v_y\hat{\text{j}}, \tag{5-26}$$

where $v_x$ is the component of $\vec{v}$ along the $x$ axis and $v_y$ is the component of $\vec{v}$ along the $y$ axis. The direction of $\vec{v}$ is tangent to the particle's path at the instant in question.

Figure 5-19 shows a velocity vector $\vec{v}$ of a moving particle and its $x$- and $y$-components. *Caution*: When a position vector is drawn as in Fig. 5-14, it is represented by an arrow that extends from one point (a "here") to another point (a "there"). However, when a velocity vector is drawn as in Fig. 5-19, it does *not* extend from one point to another. Rather, it shows the direction of travel of a particle at that instant, and the length of the arrow is proportional to the velocity magnitude. Since the unit for a velocity is a distance per unit time and is *not* a length, you are free to define a scale to use in depicting the relative magnitudes of a set of velocity vectors. For instance, each 2 cm of length on a velocity vector on a diagram could represent a velocity magnitude of 1 m/s.

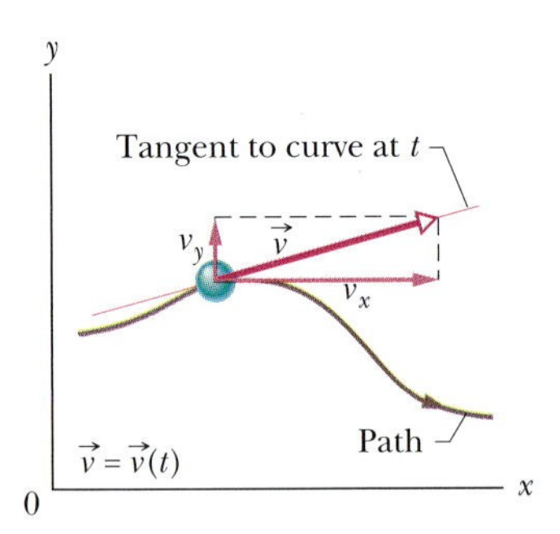


**FIGURE 5-19** ■ A particle moves in a curved path. At a time $t$, the velocity $\vec{v}$ of a particle is shown along with its components $v_x$ and $v_y$. The velocity vector points in the same direction as the tangent to the curve that traces the particle's motion.

Equations 5-24 and 5-26, developed in this section, show that the component of velocity of the object in one direction, such as the horizontal or $x$ direction, can be considered completely separately from the component of velocity of the object in another direction, such as the vertical or $y$ direction.

We assume that a curve that traces out a particle's motion is continuous. Mathematically, the tangent to the curve, its slope, and the instantaneous velocity are different names for the same quantity. Thus, we see that the velocity vector in Fig. 5-19 points along the tangent line that describes the slope of the graph at that point.

**READING EXERCISE 5-7:** The figure below shows a circular path taken by a particle about an origin. If the instantaneous velocity of the particle is $\vec{v} = (2\text{m/s})\hat{\text{i}} - (2\text{m/s})\hat{\text{j}}$, through which quadrant is the particle moving when it is traveling (a) clockwise and (b) counterclockwise around the circle? For both cases, draw $\vec{v}$ on the figure.

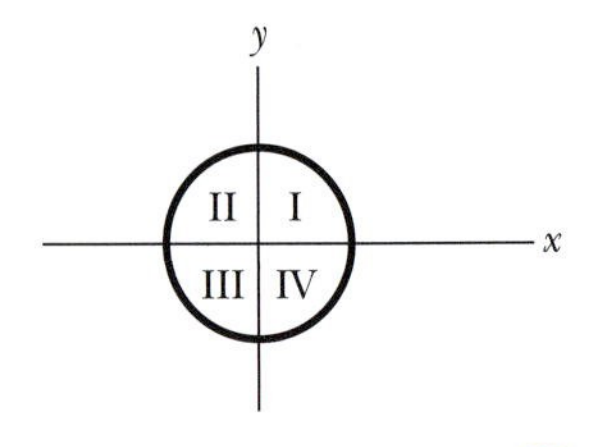


## TOUCHSTONE EXAMPLE 5-6: A Rabbit's Velocity

For the rabbit in Touchstone Example 5-5, find the velocity $\vec{v}$ at time $t = 15$ s, in unit-vector notation and as a magnitude and an angle.

**SOLUTION** ■ There are two **Key Ideas** here: (1) We can find the rabbit's velocity $\vec{v}$ by first finding the velocity components. (2) We can find those components by taking derivatives of the components of the rabbit's position vector. Applying $\vec{v} = v_x\hat{\text{i}} + v_y\hat{\text{j}}$ (Eq. 5-26) with $v_x = dx/dt$ to the expression for the rabbit's $x$ position from Touchstone Example 5-5 (Eq. 5-19), we find the $x$-component of $\vec{v}$ to be

$$v_x = \frac{dx}{dt} = \frac{d}{dt}[(-0.31 \text{ m/s}^2)t^2 + (7.2 \text{ m/s})t + 28 \text{ m})]$$
$$= (-0.62 \text{ m/s}^2)t + 7.2 \text{ m/s}. \quad (5\text{-}27)$$

At $t = 15$ s, this gives $v_x = -2.1$ m/s. Similarly, since $v_y = dy/dt$, using the expression for the rabbit's $y$ position from Touchstone Example 5-5 (Eq. 5-20), we find that the $y$-component is

$$v_y = \frac{dy}{dt} = \frac{d}{dt}[(0.22 \text{ m/s}^2)t^2 + (-9.1 \text{ m/s})t + 30 \text{ m}]$$
$$= (0.44 \text{ m/s}^2)t - 9.1 \text{ m/s}. \quad (5\text{-}28)$$

At $t = 15$ s, this gives $v_y = -2.5$ m/s. Thus, by Equation 5-26,

$$\vec{v} = -(2.1 \text{ m/s})\hat{\text{i}} - (2.5 \text{ m/s})\hat{\text{j}}, \quad \text{(Answer)}$$

which is shown in Fig. 5-20, tangent to the rabbit's path and in the direction the rabbit is running at $t = 15$ s.

To get the magnitude and angle of $\vec{v}$, either we use a vector-capable calculator or we use the Pythagorean theorem and trigonometry to write

$$v = \sqrt{v_x^2 + v_y^2} = \sqrt{(-2.1 \text{ m/s})^2 + (-2.5 \text{ m/s})^2}$$
$$= 3.3 \text{ m/s}, \quad \text{(Answer)}$$

and

$$\theta = \tan^{-1}\frac{v_y}{v_x} = \tan^{-1}\left(\frac{-2.5 \text{ m/s}}{-2.1 \text{ m/s}}\right)$$
$$= \tan^{-1} 1.19 = -130°. \quad \text{(Answer)}$$

(Although 50° has the same tangent as −130°, inspection of the signs of the velocity components indicates that the desired angle is in the third quadrant, given by 50° − 180° = −130°.)

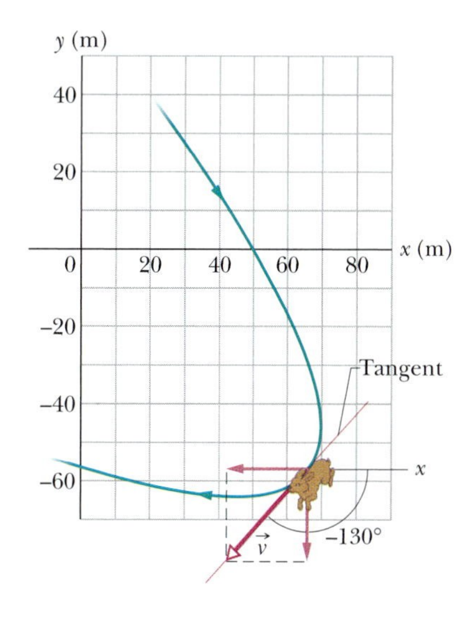


**FIGURE 5-20** ■ The rabbit's velocity $\vec{v}$ at $t = 15$ s. The velocity vector is tangent to the path at the rabbit's position at that instant. The components of $\vec{v}$ are shown.

## 5-6 Average and Instantaneous Acceleration

We have just shown that velocity vectors can be resolved into rectangular component vectors. Can this also be done with acceleration vectors? As is the case for motion in one dimension, if a particle-like object undergoes a velocity change from $\vec{v}_1$ to $\vec{v}_2$ in a time interval $\Delta t$, its **average acceleration** $\langle\vec{a}\rangle$ during $\Delta t$ is

$$\langle\vec{a}\rangle \equiv \frac{\text{change in velocity}}{\text{time interval}},$$

or

$$\langle\vec{a}\rangle = \frac{\vec{v}_2 - \vec{v}_1}{t_2 - t_1} = \frac{\Delta\vec{v}}{\Delta t}. \quad (5\text{-}29)$$

If we shrink $\Delta t$ to zero about some instant, then in the limit the average acceleration $\langle\vec{a}\rangle$ approaches the **instantaneous acceleration** (or just **acceleration**) $\vec{a}$ at that instant. That is,

$$\vec{a} = \lim_{\Delta t \to 0} \frac{\Delta\vec{v}}{\Delta t} = \frac{d\vec{v}}{dt}. \quad (5\text{-}30)$$

If the velocity changes in *either* magnitude *or* direction (or both), the particle is accelerating. For example, a particle that moves in a circle at a constant speed (velocity magnitude) is always changing direction and hence accelerating. We can write this equation in unit-vector form by substituting for $\vec{v} = v_x\hat{\text{i}} + v_y\hat{\text{j}}$ to obtain

$$\vec{a} = \frac{d}{dt}(v_x\hat{\text{i}} + v_y\hat{\text{j}}) = \frac{dv_x}{dt}\hat{\text{i}} + \frac{dv_y}{dt}\hat{\text{j}}.$$

We can rewrite this as

$$\vec{a} = a_x\hat{\text{i}} + a_y\hat{\text{j}}, \tag{5-31}$$

where the components of $\vec{a}$ in two dimensions are given by

$$a_x = \frac{dv_x}{dt} \quad \text{and} \quad a_y = \frac{dv_y}{dt}. \tag{5-32}$$

Thus, we can find the components of $\vec{a}$ by differentiating the components of $\vec{v}$. As is the case for multidimensional position, displacement, and velocity vectors, an acceleration vector can be resolved mathematically into component vectors in a rectangular coordinate system.

As we saw in Chapter 2, the algebraic sign of an acceleration component (plus or minus) represents the direction of velocity change. Speeding up is *not* always associated with a positive acceleration component. Just as we discussed for one-dimensional motion, if the velocity and acceleration components along a given axis have the *same sign* then they are in the same direction. In this case, the object will *speed up*. If the acceleration and velocity components have *opposite signs*, then they are in opposite directions. Under these conditions, the object will *slow down*. So slowing down is not always associated with an acceleration that is negative. It is the relative directions of an object's velocity and acceleration that determine whether the object will speed up or slow down.

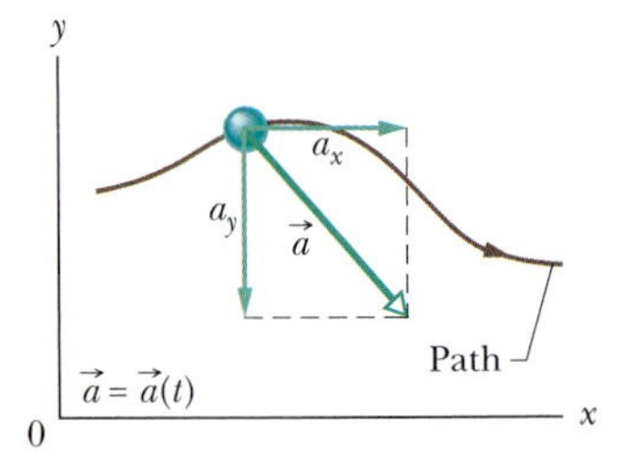


**FIGURE 5-21** ■ A two-dimensional acceleration $\vec{a}$ of a particle at a time $t$ is shown along with its $x$- and $y$-components.

Figure 5-21 shows an acceleration vector $\vec{a}$ and its components for a particle moving in two dimensions. Again, when an acceleration vector is drawn as in Fig. 5-21, although its tail is located at the particle, the vector arrow does *not* extend from one position to another. Rather, it shows the direction of acceleration for the particle, and its length represents the acceleration magnitude. The length can be drawn to any convenient scale.

**READING EXERCISE 5-8:** In Fig. 5-21 the particle is moving along a curved trajectory and its acceleration $\vec{a}$ is *not* tangent to the curve of the particle's trajectory. Under what circumstances, if any, would $\vec{a}$ be tangent to the trajectory? Under what circumstances if any, could $\vec{a}$ be perpendicular to a tangent to the trajectory? In order to answer these questions, you might want to examine the direction of the components of $\vec{a}$ and use what you learned in Chapter 2 about the relationship between velocity and acceleration. ■

## TOUCHSTONE EXAMPLE 5-7: Rabbit's Acceleration

For the rabbit in Touchstone Examples 5-5 and 5-6, find the acceleration $\vec{a}$ at time $t = 15$ s, in unit-vector notation and as a magnitude and an angle.

**SOLUTION** ■ There are two **Key Ideas** here: (1) We can find the rabbit's acceleration $\vec{a}$ by first finding the acceleration components. (2) We can find those components by taking derivatives of the rabbit's velocity components. Applying $a_x = dv_x/dt$ (Eq. 5-32) to $v_x = (-0.62\text{ m/s}^2)t + 7.2\text{ m/s}$ (Eq. 5-27 giving the rabbit's $x$ velocity in Touchstone Example 5-6), we find the $x$-component of $\vec{a}$ to be

$$a_x = \frac{dv_x}{dt} = \frac{d}{dt}[(-0.62\text{ m/s}^2)t + 7.2\text{ m/s})] = -0.62\text{ m/s}^2.$$

Similarly, applying $a_y = dv_y/dt$ (Eq. 5-32) to the rabbit's $y$ velocity from Touchstone Example 5-6 (Eq. 5-28) yields the $y$-component as

$$a_y = \frac{dv_y}{dt} = \frac{d}{dt}[(0.44\text{ m/s}^2)t - 9.1\text{ m/s}] = 0.44\text{ m/s}^2.$$

We see that the acceleration does not vary with time (it is a constant) because the time variable $t$ does not appear in the expression for either acceleration component. Therefore, by Eq. 5-31,

$$\vec{a} = (-0.62\text{ m/s}^2)\hat{\text{i}} + (0.44\text{ m/s}^2)\hat{\text{j}}, \qquad \text{(Answer)}$$

which is shown superimposed on the rabbit's path in Fig. 5-22.

To get the magnitude and angle of $\vec{a}$, either we use a vector-capable calculator or we use the Pythagorean theorem and trigonometry. For the magnitude we have

$$a = \sqrt{a_x^2 + a_y^2} = \sqrt{(-0.62\ \text{m/s}^2)^2 + (0.44\ \text{m/s}^2)^2}$$
$$= 0.76\ \text{m/s}^2. \quad \text{(Answer)}$$

For the angle we have

$$\theta = \tan^{-1}\frac{a_y}{a_x} = \tan^{-1}\left(\frac{0.44\ \text{m/s}^2}{-0.62\ \text{m/s}^2}\right) = -35°.$$

However, this last result, which is what would be displayed on your calculator if you did the calculation, indicates that $\vec{a}$ is directed to the right and downward in Fig. 5-22. Yet, we know from the components above that $\vec{a}$ must be directed to the left and upward. To find the other angle that has the same tangent as $-35°$, but is not displayed on a calculator, we add 180°:

$$-35° + 180° = 145°. \quad \text{(Answer)}$$

This *is* consistent with the components of $\vec{a}$. Note that $\vec{a}$ has the same magnitude and direction throughout the rabbit's run because, as we noted previously, the acceleration is constant.

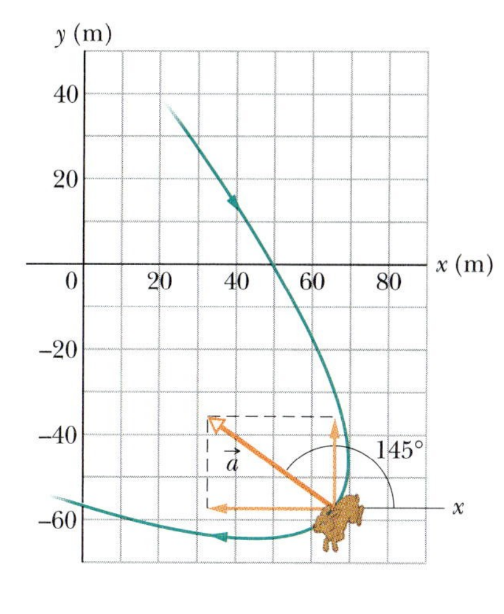


**FIGURE 5-22** ■ The acceleration $\vec{a}$ of the rabbit at $t = 15$ s. The rabbit happens to have this same acceleration at all points along its path.

## TOUCHSTONE EXAMPLE 5-8: Changing Velocity

A particle with velocity $\vec{v}_1 = (-2.0\ \text{m/s})\hat{\text{i}} + (4.0\ \text{m/s})\hat{\text{j}}$ at $t = 0$ undergoes a constant acceleration $\vec{a}$ of magnitude $a = 3.0\ \text{m/s}^2$ at an angle $\theta = 130°$ from the positive direction of the $x$ axis. What is the particle's velocity $\vec{v}_2$ at $t = 5.0$ s, in unit-vector notation and as a magnitude and an angle?

**SOLUTION** ■ We first note that this is two-dimensional motion, in the $xy$ plane. Then there are two **Key Ideas** here. One is that, because the acceleration is constant, Eq. 2-13 ($v_{2x} = v_{1x} + a_x\Delta t$) applies. The second is that, because Eq. 2-13 applies only to straight-line motion, we must apply it separately for motion parallel to the $x$ axis and motion parallel to the $y$ axis. That is, we must find the velocity components $v_x$ and $v_y$ from the equations

$$v_{2x} = v_{1x} + a_x\,\Delta t \quad \text{and} \quad v_{2y} = v_{1y} + a_y\,\Delta t.$$

In these equations, $v_{1x}$ $(= -2.0\ \text{m/s})$ and $v_{1y}$ $(= 4.0\ \text{m/s})$ are the $x$- and $y$-components of $\vec{v}_1$, and $a_x$ and $a_y$ are the $x$- and $y$-components of $\vec{a}$. To find $a_x$ and $a_y$, we resolve $\vec{a}$ either with a vector-capable calculator or with trigonometry:

$$a_x = a\cos\theta = (3.0\ \text{m/s}^2)(\cos 130°) = -1.93\ \text{m/s}^2,$$
$$a_y = a\sin\theta = (3.0\ \text{m/s}^2)(\sin 130°) = +2.30\ \text{m/s}^2.$$

When these values are inserted into the equations for $v_x$ and $v_y$, we find that, at time $t = 5.0$ s,

$$v_{2x} = -2.0\ \text{m/s} + (-1.93\ \text{m/s}^2)(5.0\ \text{s}) = -11.65\ \text{m/s},$$
$$v_{2y} = 4.0\ \text{m/s} + (2.30\ \text{m/s}^2)(5.0\ \text{s}) = 15.50\ \text{m/s}.$$

Thus, at $t = 5.0$ s, we have, after rounding,

$$\vec{v}_2 = (-12\ \text{m/s})\hat{\text{i}} + (16\ \text{m/s})\hat{\text{j}}. \quad \text{(Answer)}$$

We find that the magnitude and angle of $\vec{v}_2$ are

$$v_2 = \sqrt{v_{2x}^2 + v_{2y}^2} = 19.4\ \text{m/s} \approx 19\ \text{m/s}, \quad \text{(Answer)}$$

and

$$\theta_2 = \tan^{-1}\frac{v_{2y}}{v_{2x}} = 127° \approx 130°. \quad \text{(Answer)}$$

Check the last line with your calculator. Does 127° appear on the display, or does $-53°$ appear? Now sketch the vector $v$ with its components to see which angle is reasonable.

## 5-7 Uniform Circular Motion

If a single Olympic event best captures the motions described in this chapter, it's the hammer throw. In this event an athlete spins a heavy steel ball attached to a wire rope with a handle in a circle. An athlete gains maximum distance by swinging a 16 lb ham-

**FIGURE 5-23** ■ A Scotsman twirls a massive steel ball (called a "hammer") in a circle at a highland games competition. When he releases the hammer, it will travel in a direction that is tangent to its original circular path and undergo projectile motion.

mer repeatedly around his head while standing still to build up speed (Fig. 5-23). Finally, the athlete rotates quickly with the hammer before releasing it at the front of a throwing circle. Once the athlete releases the hammer, it begins to travel along a line tangent to the circle in which it was spinning and undergoes projectile motion.

Circular motion, like that of the spinning hammer, is another motion in two dimensions that can be analyzed using Newton's Second Law. Examples of motion that are approximately circular include the revolution of the Earth around the Sun, a race car zooming around a circular track, an electron moving near the center of a large electromagnet, and a stone tied to the end of a string that is twirled in a circle above one's head. In all of these cases if the object's speed is constant, we define its motion as **uniform circular motion.**

> A particle that travels around a circle or a circular arc at constant (*uniform*) speed is said to be undergoing **UNIFORM CIRCULAR MOTION**.

Not all circular motion is uniform. For example, a Hot Wheels® car or roller coaster cart doing a loop-the-loop slows down near the top of the loop and speeds up near the bottom. Analyzing loop-the-loop motions is more complex than analyzing uniform circular motion. For this reason, we start with the ideal case of uniform circular motion.

## Centripetal Force

Consider an object, such as an ice hockey puck, that glides along a frictionless surface. Newton's First Law tells us that you cannot change either its direction or its speed without exerting a force on it. Giving the puck a kick along its line of motion will change its speed but not its direction. How can you have the opposite effect? How can you change the puck's direction without changing its speed? To do this you have to kick perpendicular to its direction of motion. This is an important statement regarding the accelerations that result when we apply a force.

> If a nonzero net force acts on an object, at any instant it can be decomposed into a component along the line of motion and a component perpendicular to the motion. The component of the net force that is *in line* with the object's motion produces only changes in the *magnitude* of the object's velocity (its speed). The component of the net force that is *perpendicular* to the line of motion produces only changes in the *direction* of the object's velocity.

What happens if you give the puck a series of short kicks but adjust their directions constantly so the kicks are always perpendicular to the current direction of motion? Does this lead to circular motion? We can use Newton's Second Law to answer this question. To help visualize the net force needed to maintain uniform circular motion, we consider a similar situation to that of the puck. Imagine twirling a ball at the end of a string in a perfectly horizontal circle of radius $r$. In order to keep the ball moving in a circle, you are constantly changing the direction of the force that the string exerts on the spinning ball. That is, there must constantly be a component of the net force that is perpendicular to the motion. The situation is complicated by the fact that there are actually two forces on the ball as shown in Fig. 5-24—the string force and the gravitational force.

To apply Newton's Second Law to the analysis of this motion, we must find the net force on the ball by taking the vector sum of the two forces acting on it. We start by resolving the string force into horizontal and vertical components, as shown in Fig. 5-24. Since the ball is rotating horizontally and so does not move up or down in a vertical direction, the net vertical force on it must be zero. Since the vertical force

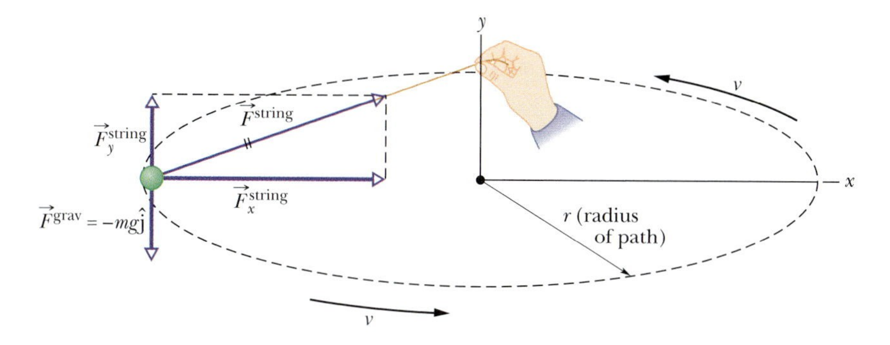


**FIGURE 5-24** ■ If you spin a ball in a circle, the net force on the ball consists of a central or centripetal force $\vec{F}_x^{\text{string}}$ that lies in a horizontal plane. It points toward the center of the circular path.

components cancel, the net force is just the horizontal component of the string force vector, $\vec{F}_x^{\text{string}}$. If you carefully consider the situation depicted in Fig. 5-24, you should be convinced of two important points. First, the direction of the net force on the ball ($\vec{F}_x^{\text{string}}$) is always perpendicular to the line of motion of the ball. This means that the force results only in changes in the direction of the object's velocity. It does not cause changes in the object's speed. Second, the direction of the net force, $\vec{F}_x^{\text{string}}$, is constantly changing so that it always points toward the center of the circle in which the ball moves. We use the adjective **centripetal** to describe any force with this characteristic. The word *centripetal* comes from Latin and means "center-seeking."

> Centripetal is an adjective that *describes* any force or superposition of forces that is directed toward the center of curvature of the path of motion.

It is important to note that the horizontal component of the string force $\vec{F}_x^{\text{string}}$ *is* the centripetal force involved in the ball's motion. There is *not* another force, "the centripetal force," that must be added to the free-body diagram in Fig. 5-24.

If you suddenly let go of the string, the ball will fly off along a straight path that is tangent to the circle at the moment of release. This "linear flying off" phenomenon provides evidence that you *cannot maintain circular motion without a centripetal or center-seeking force.*

If we use a polar coordinate system with its origin at the center of the circular path, we can consider the centripetal force to be a kind of anti-radial force that points *inward* rather than outward in the direction of the circle's radius vector $\vec{r}$.

## Centripetal Acceleration

A very simple example of uniform circular motion is shown in Fig. 5-25. There, an air hockey puck moves around in a circle at constant speed $v$ while tied to a string looped around a central peg. Is this accelerated motion? We can predict that it is for two reasons:

- First, there is a net force on the puck due to the force exerted by the string. There is no vertical acceleration, so the upward force of the air jets and the downward gravitational force must cancel each other. According to Newton's Second Law, if there is a net force on an object there must be an acceleration.
- Second, although the puck moves with constant speed, the *direction* of the puck velocity is continuously changing. Recall that acceleration is related to the change in *velocity* (not speed), so we conclude that this motion is indeed accelerated motion.

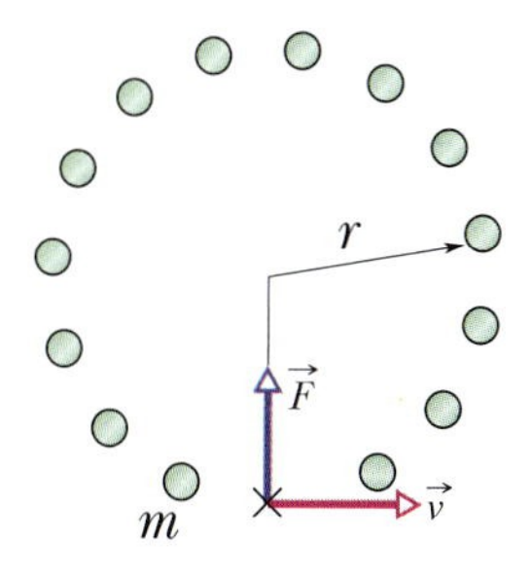


**FIGURE 5-25** ■ A sketch of the locations of an air hockey puck of mass $m$ moving with constant speed $v$ in a circular path of radius $r$ on a horizontal frictionless air table. The centripetal force on the puck, $\vec{F}^{\text{cent}}$, is the pull from the string directed inward toward the center of the circle traced out by the path of the puck.

What direction is this acceleration? Newton's Second Law tells us that the acceleration of an object of mass $m$ is in the *same direction* as the force causing it and is

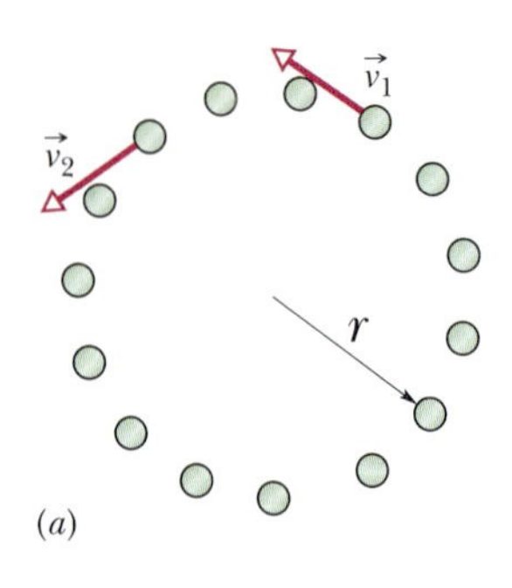


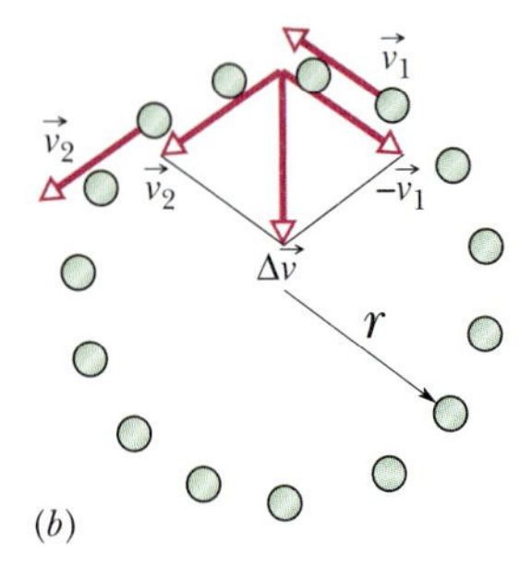


**FIGURE 5-26** ■ (*a*) Sketch based on a video of an object moving counterclockwise on an airtable with constant speed in a circular path with the velocity vectors corresponding to its location at times $t_1$ and $t_2$. (*b*) Shows vectors $\vec{v}_2$ and $-\vec{v}_1$ and their sum $\Delta\vec{v}$ at the location of the object at an average time of $(t_1 + t_2)/2$.

given by Eq. 3-1, which is $\vec{a} = \vec{F}/m$. This suggests that, in uniform circular motion, the acceleration should also be directed radially inward. An acceleration that is directed radially inward is called a **centripetal acceleration.**

Next we will use the general definition of acceleration, a knowledge of geometry, and Newton's Second Law to show that uniform circular motion requires a centripetal acceleration of constant magnitude that depends on the radius of the path of a rotating object as well as its speed.

*Proof that the acceleration is centripetal and has a constant magnitude*: Let's start by considering the definition of average acceleration in Eq. 5-29,

$$\langle\vec{a}\rangle \equiv \frac{\vec{v}_2 - \vec{v}_1}{t_2 - t_1} = \frac{\Delta\vec{v}}{\Delta t} \quad \text{(average acceleration).}$$

Since $\Delta t$ is a scalar, the acceleration must have the same direction as the difference between the two velocity vectors, $\vec{v}_2 - \vec{v}_1$. As usual the difference between the velocity vectors is actually the vector sum of $\vec{v}_2$ and the additive inverse of $\vec{v}_1$. Since the speed is constant (as indicated by the equally-spaced, "frame-by-frame" position markers in Fig. 5-26), the length of the velocity vectors and their additive inverses are the same at times $t_1$ and $t_2$. Another consequence of the speed being constant is that halfway between times $t_1$ and $t_2$, the puck is also midway between the two positions. If we place the tails of $\vec{v}_2$ and $-\vec{v}_1$ arrows at the midpoint between the two locations, we find that the vector sum points toward the center of the circular path taken by the object. This is shown in Fig. 5-26. So the acceleration is indeed centripetal (center-seeking). Furthermore, we could have created the same construction using any two points corresponding to other times that have the same *difference* $\Delta t$. It is obvious that the direction of the velocity change would be different, but it would still point toward the center. Furthermore, the magnitude of the $\Delta v$ vector, and hence the acceleration magnitude, would be constant.

How does the centripetal acceleration depend on speed and path radius? We will prove that the magnitude of the acceleration of an object in uniform circular motion is given by

$$a = \frac{v^2}{R} \quad \text{(magnitude of centripetal acceleration),} \tag{5-33}$$

where $R$ is the radius of the circular path of the object and $|\vec{v}|$ or $v$ represents its speed. Here we start our proof by considering an object in Fig. 5-27, which happens to be moving in a circle in a counterclockwise direction at a constant speed. We choose to describe its motion in polar coordinates. We define $\theta_1$ as zero at time $t_1$ and denote its location as $\vec{r}_1$. The object then moves at constant speed $v$ through an angle $\theta_2 = \theta$ to a new location $\vec{r}_2$ at time $t_2$. Note that, in circular motion, velocity vectors are always perpendicular to their position vectors. This means that the angle between position vectors $\vec{r}_1$ and $\vec{r}_2$ is the same as the angle between velocity vectors $\vec{v}_1$ and $\vec{v}_2$. Furthermore, we note that if $\Delta\vec{v} = \vec{v}_2 - \vec{v}_1$ then $\vec{v}_2 = \Delta\vec{v} + \vec{v}_1$. Also, $\vec{r}_2 = \Delta\vec{r} + \vec{r}_1$. Thus the triangles shown in Fig. 5-27 are similar. According to Eq. 5-15 for small angles, $\Delta r = r(\theta_2 - \theta_1) = r\theta$. So we can write the ratio of magnitudes as

$$\frac{\Delta v}{v} = \frac{\Delta r}{R}.$$

We can solve the similar triangle ratios for the change in velocity and substitute into the expression that defines the magnitude of acceleration in terms of the magnitude of velocity change over a change in time to get

$$a = \frac{\Delta v}{\Delta t} = \frac{(\Delta r)(v)}{(\Delta t)(R)}.$$

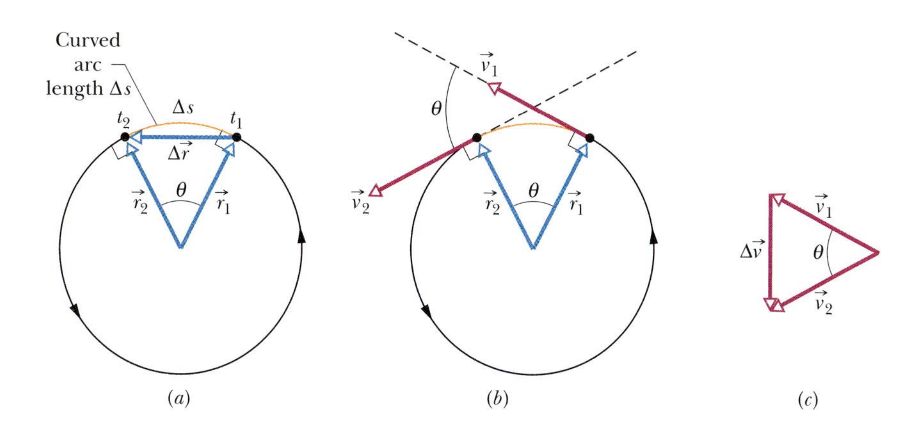


**FIGURE 5-27** ■ Between times $t_1$ and $t_2$, a particle: (*a*) moves from location $\vec{r}_1$ to $\vec{r}_2$. The enclosed arc length $\Delta s$ is curved and slightly longer than the magnitude of the vector displacement $|\Delta\vec{r}|$. (*b*) The velocity vector, which is always perpendicular to the position vector, changes direction but not magnitude. (*c*) Because the velocity vector is always perpendicular to the position vectors, the angle $\theta$ between them is the same as the angle between the position vectors.

However, the speed $v$ is given by the arc length $\Delta s$ along the circular path divided by the time interval, so that $v \approx \Delta s/\Delta t$. In the limit where the change in time is very small, the change in arc length $\Delta s$ and the magnitude of the displacement $\Delta r$ are essentially the same, so

$$v \approx \frac{\Delta s}{\Delta t} \approx \frac{\Delta r}{\Delta t} \quad \text{(when the time interval becomes small).}$$

So, we can replace $\Delta r/\Delta t$ in the expression for acceleration with the particle speed $v$ to get

$$a = \frac{v^2}{R} \quad \text{(centripetal acceleration).} \tag{5-34}$$

## Determining Average Speed, Period, and Frequency of Rotation

Often, we want to know how long it will take an object undergoing uniform circular motion to complete an entire revolution. For example, we might want to know how long it takes a race car on a circular track to complete one lap. This calculation is simplified because objects in uniform circular motion are moving at constant speeds. This means that the magnitude of velocity is constant, and therefore average and instantaneous values are the same. We can then use the relation for the speed of an object,

$$v = \langle v \rangle = \frac{\text{distance traveled}}{\text{time for the travel}}.$$

The distance traveled in one revolution is just the circumference of the circle ($2\pi r$). The time for a particle to go around a closed path exactly once has a special name. It is called the *period of revolution*, or simply the **period** of the motion. The period is represented with the symbol $T$, so,

$$v = \langle v \rangle = \frac{\text{distance traveled}}{\text{time for the travel}} = \frac{2\pi r}{T} \quad \text{(average speed).}$$

Recalling that the magnitudes of the average and instantaneous velocities are the same since the object is moving at constant speed, we can solve this expression for the period:

$$T = \frac{2\pi r}{v} \quad \text{(period of revolution).} \tag{5-35}$$

Another way to describe the motion of a particle moving in a circle is to cite the number of revolutions that the particle makes in a specific amount of time. This number of revolutions in a given time is known as the **frequency,** $f$, of revolution. From the definitions we have given for period and frequency, they are related by the expression

$$f = \frac{1}{T} \quad \text{(frequency)}.$$

## Examples of Centripetal Forces

What are the forces involved in uniform circular motion? Suppose you were to undergo two different types of uniform circular motion—traveling in a tight circle while driving a car and orbiting the Earth in the space shuttle. You are experiencing centripetal forces that cause you to undergo a centripetal acceleration. In one case you feel that you are being rammed against the car door. In the other case you feel "weightless." Let us examine the forces involved in these two examples of uniform circular motion more closely.

*Rounding a curve in a car*: You are sitting in the center of the rear seat of a car moving at a constant high speed along a flat road. When the driver suddenly turns left, rounding a corner in a circular arc, you slide across the seat toward the right and then jam against the car door for the rest of the turn. What is going on?

While the car moves in the circular arc, it is in uniform circular motion. That is, it has an acceleration that is directed toward the center of the circle. By Newton's Second Law, $\vec{F} = m\vec{a}$, a force must cause this acceleration. Moreover, the force must also be directed toward the center of the circle. Thus, it is a centripetal force, where the adjective (centripetal) indicates the direction. In this example, the centripetal force is a frictional force on the tires from the road. Without it, the turn would not be possible. For example, imagine what would happen if you hit a patch of low-friction ice while trying to make such a turn.

If you are to move in uniform circular motion along with the car, there must also be a centripetal force on you as well. In our case, apparently the frictional force on you from the seat was not great enough to make you go in a circle with the car. Thus, the seat slid beneath you, until the right door of the car jammed into you. Then its push on you provided the needed centripetal force on you, and you joined the car's uniform circular motion. For you, the centripetal force is the push from the car door.

*Orbiting the Earth*: This time you are a passenger in the space shuttle *Atlantis*. As it (and you) orbit Earth, you float through your cabin. What is going on?

Both you and the shuttle are in uniform circular motion and have accelerations directed toward the center of the orbital circle. Again by Newton's Second Law, centripetal forces must cause these accelerations. This time the centripetal forces are gravitational pulls (the pull on you and the pull on the shuttle) by Earth, radially inward, toward the center of the Earth. You feel weightless, even though the Earth is pulling on you, because both you and the space shuttle are accelerating at the same rate. This is like feeling lighter when you descend in the elevator discussed in Section 2-4.

*Differences between centripetal forces*: In both car and shuttle, you are in uniform circular motion, acted on by a centripetal force—yet your sensations in the two situations are quite different. In the car, jammed up against the door, you are aware of being compressed by the door. In the orbiting shuttle, however, you are floating around with no sensation of any force acting on you. Why this difference? The difference is due to the nature of the two centripetal forces. In the car, the centripetal force is due to the push on the part of your body touching the car door. You can sense the compression on that part of your body. In the shuttle, the centripetal force is due to Earth's gravitational pull on every atom of your body. Thus, there is no compression (or pull) on any one part of your body and no sensation of force acting on you. (The sensation is said to be one of "weightlessness," but that description is tricky. The Earth's pull on you has certainly not disappeared and, in fact, is only a little less than it would be when you are on the ground.)

Recall also the example of a centripetal force shown in Fig. 5-25. There a hockey puck moves around a circle at constant speed $v$ while tied to a string looped around a central peg. In this case the centripetal force is the radially inward pull on the puck from the string. Without that force, the puck would go off in a straight line instead of moving in a circle.

In the examples discussed above, the source of the centripetal force was different in each situation. The frictional force, the push of the right door of the car, the gravitational attraction of Earth, and the pull of a string, were all centripetal forces that we considered. This is an important point that was made earlier but is worth repeating. A centripetal force *is not* a new kind of force. It is not an additional force. The name merely indicates the direction in which the force acts. Under the right circumstances, a frictional force, a gravitational force, the force from a car door or a string, or any other kind of force can be centripetal. However, for any situation:

> A centripetal force accelerates a body by changing the direction of the body's velocity without changing the body's speed.

From Newton's Second Law with the centripetal acceleration given by $a = v^2/r$, we can determine what magnitude of the centripetal force, $\vec{F}^{\text{cent}}$, is needed to keep an object moving in a circle at a constant speed $v$. This is given by

$$F^{\text{cent}} = ma = \frac{mv^2}{r} \quad \text{(magnitude of centripetal force).} \tag{5-36}$$

Because the speed $v$ and radius $r$ are constant, so are the magnitudes of the acceleration and the force. However, the directions of the centripetal acceleration and force change continuously so as to always point toward the center of a circle. Therefore, unlike the situation for ideal projectile motion, the motions in a chosen $x$ and $y$ direction cannot be treated as independent of each other.

## Centripetal Versus Linear Forces and Accelerations

We find it interesting to contrast the forces involved in projectile motion with those of uniform circular motion. In both situations, a particle experiences a net force and accelerates. In projectile motion, the net force is linear with only a vertical component in rectangular coordinates. The acceleration results from a change in the magnitude of vertical velocity vectors and no change in direction.

In uniform circular motion, we can describe the forces in polar coordinates. In this case, we have a constant force pointing inward antiparallel to the $r$ axis and no force in the direction of increasing $\theta$. But since the $r$ axis is changing direction at a constant rate, the magnitude of the velocity is the same and its acceleration is due to direction changes.

> In summary, linear accelerations are due purely to changes in the magnitude of the velocity, whereas uniform circular accelerations are due purely to *changes in the direction* of the velocity.

**READING EXERCISE 5-9:** Some people say that a centripetal force throws objects outward. For instance, in the example of the car rounding a turn in the discussion above, some might say that the passenger is thrown right when the car turns left. Is it true that centripetal forces throw objects outward? If so, explain how. If not, explain what is really going on. ■

## TOUCHSTONE EXAMPLE 5-9: Little Casey Jones

Little Casey Jones is getting an electric train set for Christmas. In the box, he finds 8 pieces of track—4 pieces of straight track and 4 quarter-circle tracks. The straight tracks are 50 cm long, and the quarter-circle tracks will form a circle of radius 50 cm if they are all put together. Casey assembles them into the figure shown at the right in Fig. 5-28. Casey sets the engine on the track and brings it up to a constant speed. The engine has a mass $M$, and once it is up to speed, it takes 2.0 seconds to make one circuit of the track (at the speed Casey likes to use).

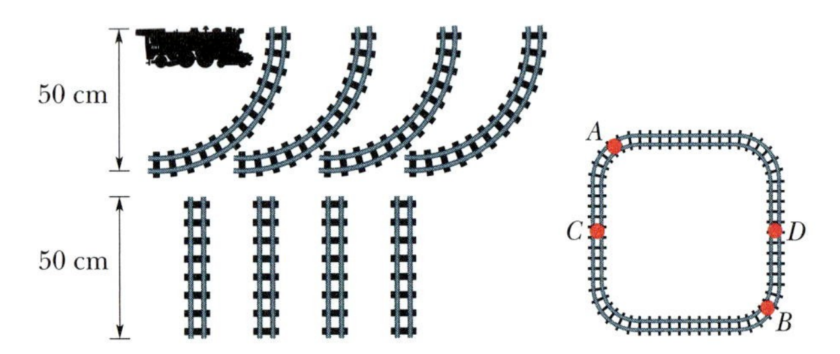


**FIGURE 5-28** ■ Little Casey Jones' new train set.

(a) Without actually calculating values, compare the instantaneous velocities of the engine $\vec{v}_A$, $\vec{v}_B$, $\vec{v}_C$, and $\vec{v}_D$ when it is at points $A$, $B$, $C$, and $D$.

**SOLUTION** ■ Since the engine is traveling at a constant speed, this tells us that the magnitudes of these four velocity vectors are all the same; $|\vec{v}_A| = |\vec{v}_B| = |\vec{v}_C| = |\vec{v}_D| = v$, the speed at which the engine is moving.

However, each of these four velocity vectors has a different *direction* from each of the other three. A **Key Idea** here is that at each instant in time, the engine's velocity vector points in the direction that the engine is moving. We are not told whether the engine is going around the track in a clockwise or a counterclockwise sense. Let's say it is going around in a clockwise sense, as shown in Fig. 5-29.

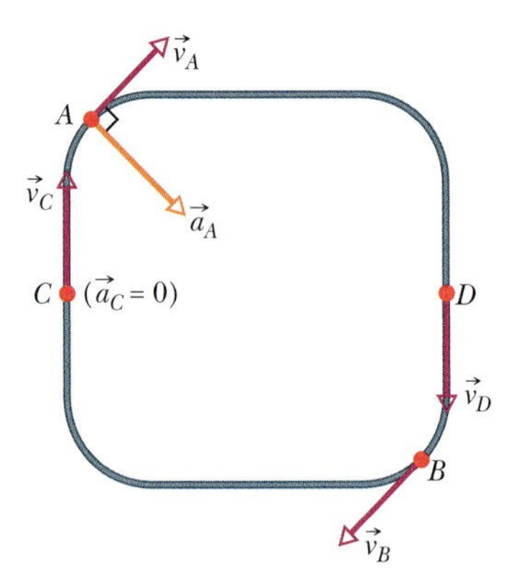


**FIGURE 5-29** ■ The directions of the engine's velocity and acceleration at a number of points around the track.

A second **Key Idea** is that the velocity vector is always tangent to the path that the object is following. Thus, $\vec{v}_C$ would point upward along the track; $\vec{v}_A$ would point upward and to the right, tangent to the curved track; $\vec{v}_D$ would point downward, antiparallel to $\vec{v}_C$; and $\vec{v}_B$ would point downward and to the left, antiparallel to $\vec{v}_A$. The directions of these four vectors are shown in Fig. 5-29. (Answer)

(b) Calculate the average velocity of the engine, $\langle\vec{v}\rangle_{CD}$, for the time interval it takes to go from point $C$ to point $D$.

**SOLUTION** ■ The **Key Idea** here is that the average velocity is given by Eq. 5-22:

$$\langle\vec{v}\rangle_{CD} = \frac{\Delta\vec{r}_{CD}}{\Delta t_{CD}} = \frac{\vec{r}_D - \vec{r}_C}{t_D - t_C}.$$

From the dimensions of the track given in Fig. 5-28, we see that point $D$ is 50 cm (from the straight segment) + 100 cm (from the two curved segments added together) = 150 cm to the right of point $C$. We also see that the distance *along the track* from $C$ to $D$ is one-half of the length of one full circuit of the track. Since the engine is traveling at a constant speed, this means that it must take one-half of the time for one full circuit to go from $C$ to $D$. Thus $\Delta t_{CD} = \frac{1}{2}(2.0\text{ s}) = 1.0\text{ s}$. So the magnitude of the engine's average velocity here is

$$|\langle\vec{v}\rangle_{CD}| = \frac{|\Delta\vec{r}_{CD}|}{\Delta t_{CD}} = \frac{150\text{ cm}}{1.0\text{ s}} = 150\text{ cm/s},$$

directed *to the right* in Fig. 5-29. (Answer)

(c) Calculate the instantaneous acceleration $\vec{a}_A$ and $\vec{a}_C$ of the engine when it is at points $A$ and $C$.

**SOLUTION** ■ The **Key Idea** here is that the instantaneous acceleration is $\vec{a} = d\vec{v}/dt$. The acceleration will be zero only when both the magnitude and the direction of the velocity remain constant. If *either* the magnitude *or* the direction of the velocity are changing, then the acceleration will not be zero.

As the engine goes around the track, we are told that the *magnitude* of its velocity remains constant. On a straight track segment, such as at point $C$, the direction of travel of the engine is also constant, so $\vec{a}_C = 0$. (Answer)

On a curve, such as point $A$, the engine's direction of travel is changing, since it is turning right. In this case since the train is momentarily moving in a circle, we can use the expression for centripetal acceleration (Eq. 5-34) to determine that $|\vec{a}_A| = v_A^2/r_A$ and that $\vec{a}_A$ is directed down and to the right, as shown in Fig. 5-29.

To calculate the magnitude of $\vec{a}_C$, we note that the total length of track is $2\pi r + 4L = (2\pi \times 50\text{ cm}) + (4 \times 50\text{ cm}) = 514\text{ cm}$, so that

$$v_A = \frac{514\text{ cm}}{2\text{ s}} = 257\text{ cm/s},$$

and since the radius of the circle the train is momentarily moving in is 50 cm,

$$a_A = \frac{v_A^2}{r_A} = \frac{(257\text{ cm/s})^2}{50\text{ cm}} = 1322\text{ cm/s}^2,$$

directed inward toward the center of the curve. (Answer)

# Problems

## SEC. 5-3 ■ ANALYZING IDEAL PROJECTILE MOTION

*In some of these problems, exclusion of the effects of the air is unwarranted but helps simplify the calculations.*

**1. Rifle and Bullet** A rifle is aimed horizontally at a target 30 m away. The bullet hits the target 1.9 cm below the aiming point. What are (a) the bullet's time of flight and (b) its speed as it emerges from the rifle?

**2. A Small Ball** A small ball rolls horizontally off the edge of a tabletop that is 1.20 m high. It strikes the floor at a point 1.52 m horizontally away from the edge of the table. (a) How long is the ball in the air? (b) What is its speed at the instant it leaves the table?

**3. Baseball** A baseball leaves a pitcher's hand horizontally at a speed of 161 km/h. The distance to the batter is 18.3 m. (Ignore the effect of air resistance.) (a) How long does the ball take to travel the first half of that distance? (b) The second half? (c) How far does the ball fall freely during the first half? (d) During the second half? (e) Why aren't the quantities in (c) and (d) equal?

**4. Dart** A dart is thrown horizontally with an initial speed of 10 m/s toward point $P$, the bull's-eye on a dart board. It hits at point $Q$ on the rim, vertically below $P$ 0.19 s later. (a) What is the distance $PQ$? (b) How far away from the dart board is the dart released?

**5. An Electron** An electron, with an initial horizontal velocity of magnitude $1.00 \times 10^9$ cm/s, travels into the region between two horizontal metal plates that are electrically charged. In that region, it travels a horizontal distance of 2.00 cm and has a constant downward acceleration of magnitude $1.00 \times 10^{17}$ cm/s$^2$ due to the charged plates. Find (a) the time required by the electron to travel the 2.00 cm and (b) the vertical distance it travels during that time. Also find the magnitudes of the (c) horizontal and (d) vertical velocity components of the electron as it emerges.

**6. Mike Powell** In the 1991 World Track and Field Championships in Tokyo, Mike Powell (Fig. 5-30) jumped 8.95 m, breaking the 23-year long-jump record set by Bob Beamon by a full 5 cm. Assume that Powell's speed on takeoff was 9.5 m/s (about equal to that of a sprinter) and that $g = 9.80$ m/s$^2$ in Tokyo. How much less was Powell's horizontal range than the maximum possible horizontal range (neglecting the effects of air) for a particle launched at the same speed of 9.5 m/s?

**FIGURE 5-30** ■ Problem 6.

**7. Catapulted** A stone is catapulted at time $t_1 = 0$, with an initial velocity of magnitude 20.0 m/s and at an angle of 40.0° above the horizontal. What are the magnitudes of the (a) horizontal and (b) vertical components of its displacement from the catapult site at $t_2 = 1.10$ s? Repeat for the (c) horizontal and (d) vertical components at $t_3 = 1.80$ s, and for the (e) horizontal and (f) vertical components at $t_4 = 5.00$ s.

**8. Golf Ball** A golf ball is struck at ground level. The speed of the golf ball as a function of the time is shown in Fig. 5-31, where $t = 0$ at the instant the ball is struck. (a) How far does the golf ball travel horizontally before returning to ground level? (b) What is the maximum height above ground level attained by the ball?

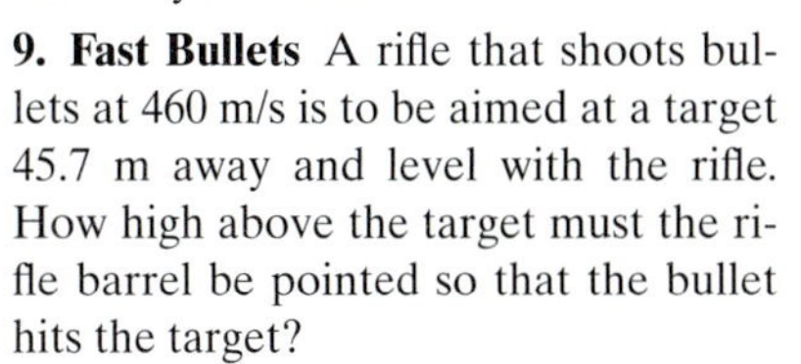

**9. Fast Bullets** A rifle that shoots bullets at 460 m/s is to be aimed at a target 45.7 m away and level with the rifle. How high above the target must the rifle barrel be pointed so that the bullet hits the target?

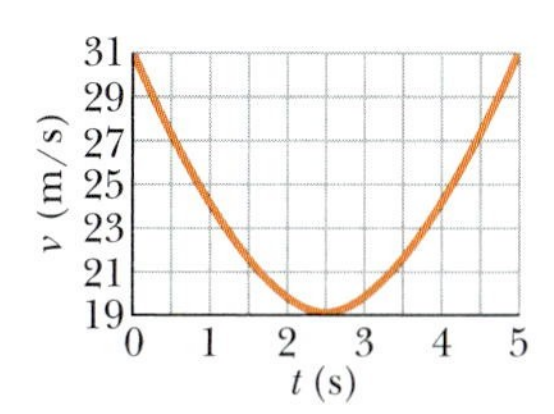


**FIGURE 5-31** ■ Problem 8.

**10. Slow-Pitch** The pitcher in a slow-pitch softball game releases the ball at a point 3.0 ft above ground level. A stroboscopic plot of the position of the ball is shown in Fig. 5-32, where the readings are 0.25 s apart and the ball is released at $t = 0$. (a) What is the initial speed of the ball? (b) What is the speed of the ball at the instant it reaches its maximum height above ground level? (c) What is that maximum height?

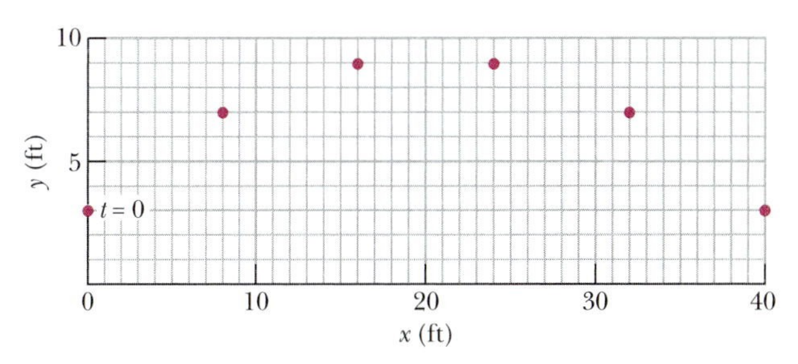


**FIGURE 5-32** ■ Problem 10.

**11. Maximum Height** Show that the maximum height reached by a projectile is $y^{max} = (v_1 \sin \theta_1)^2/2g$.

**12. You Throw a Ball** You throw a ball toward a wall with a speed of 25.0 m/s and at an angle of 40.0° above the horizontal (Fig. 5-33). The wall is 22.0 m from the release point of the ball. (a) How far above the release point does the ball hit the wall? (b) What are the horizontal and vertical components of its velocity as it hits the wall? (c) When it hits, has it passed the highest point on its trajectory?

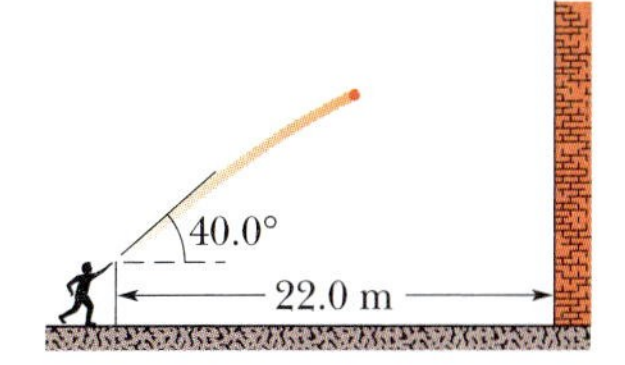


**FIGURE 5-33** ■ Problem 12.

**13. Shot into the Air** A ball is shot from the ground into the air. At a height of 9.1 m. Its velocity is observed to be $\vec{v} = (7.6 \text{ m/s})\hat{\text{i}} + (6.1 \text{ m/s})\hat{\text{j}}$ ($\hat{\text{i}}$ horizontal, $\hat{\text{j}}$ upward). (a) To what maximum height does the ball rise? (b) What total horizontal distance does the ball travel? What are (c) the magnitude and (d) the direction of the ball's velocity just before it hits the ground?

**14. Two Seconds Later** Two seconds after being projected from ground level, a projectile is displaced 40 m horizontally and 53 m vertically above its point of projection. What are the (a) horizontal and (b) vertical components of the initial velocity of the projectile? (c) At the instant the projectile achieves its maximum height above ground level, how far is it displaced horizontally from its point of projection?

**15. Football Player** A football player punts the football so that it will have a "hang time" (time of flight) of 4.5 s and land 46 m away. If the ball leaves the player's foot 150 cm above the ground, what must be (a) the magnitude and (b) the direction of the ball's initial velocity?

**16. Launching Speed** The launching speed of a certain projectile is five times the speed it has at its maximum height. Calculate the elevation angle $\theta_1$ at launching.

**17. Airplane and Decoy** A certain airplane has a speed of 290.0 km/h and is diving at an angle of 30.0° below the horizontal when the pilot releases a radar decoy (Fig. 5-34). The horizontal distance between the release point and the point where the decoy strikes the ground is 700 m. (a) How long is the decoy in the air? (b) How high was the released point?

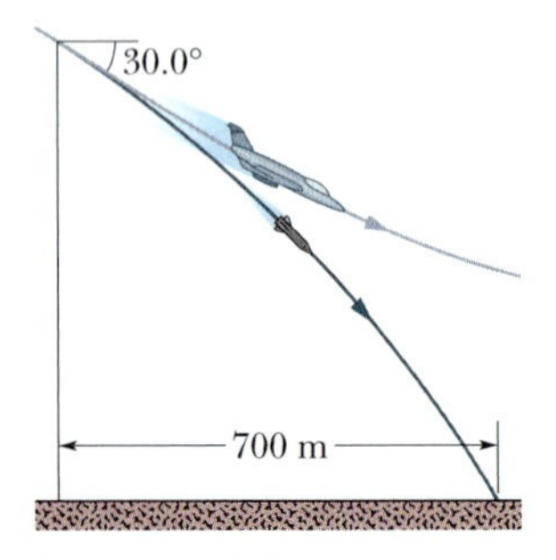


FIGURE 5-34 ■ Problem 17.

**18. Soccer Ball** A soccer ball is kicked from the ground with an initial speed of 19.5 m/s at an upward angle of 45°. A player 55 m away in the direction of the kick starts running to meet the ball at that instant. What must be his average speed if he is to meet the ball just before it hits the ground? Neglect air resistance.

**19. Stairway** A ball rolls horizontally off the top of a stairway with a speed of 1.52 m/s. The steps are 20.3 cm high and 20.3 cm wide. Which step does the ball hit first?

**20. Volleyball** For women's volleyball the top of the net is 2.24 m above the floor and the court measures 9.0 m by 9.0 m on each side of the net. Using a jump serve, a player strikes the ball at a point that is 3.0 m above the floor and a horizontal distance of 8.0 m from the net. If the initial velocity of the ball is horizontal, (a) what minimum magnitude must it have if the ball is to clear the net and (b) what maximum magnitude can it have if the ball is to strike the floor inside the back line on the other side of the net?

**21. Airplane** An airplane, diving at an angle of 53.0° with the vertical, releases a projectile at an altitude of 730 m. The projectile hits the ground 5.00 s after being released. (a) What is the speed of the aircraft? (b) How far did the projectile travel horizontally during its flight? What were the (c) horizontal and (d) vertical components of its velocity just before striking the ground?

**22. Tennis Match** During a tennis match, a player serves the ball at 23.6 m/s, with the center of the ball leaving the racquet horizontally 2.37 m above the court surface. The net is 12 m away and 0.90 m high. When the ball reaches the net, (a) does the ball clear it and (b) what is the distance between the center of the ball and the top of the net? Suppose that, instead, the ball is served as before but now it leaves the racquet at 5.00° below the horizontal. When the ball reaches the net, (c) does the ball clear it and (d) what now is the distance between the center of the ball and the top of the net?

**23. The Batter** A batter hits a pitched ball when the center of the ball is 1.22 m above the ground. The ball leaves the bat at an angle of 45° with the ground. With that launch, the ball should have a horizontal range (returning to the *launch* level) of 107 m. (a) Does the ball clear a 7.32-m-high fence that is 97.5 m horizontally from the launch point? (b) Either way, find the distance between the top of the fence and the center of the ball when the ball reaches the fence.

**24. Detective Story** In a detective story, a body is found 4.6 m from the base of a building and 24 m below an open window. (a) Assuming the victim left that window horizontally, what was the victim's speed just then? (b) Would you guess the death to be accidental? Explain your answer.

**25. Football Kicker** A football kicker can give the ball an initial speed of 25 m/s. Within what two elevation angles must he kick the ball to score a field goal from a point 50 m in front of goalposts whose horizontal bar is 3.44 m above the ground? (If you want to work this out algebraically, use $\sin^2\theta + \cos^2\theta = 1$ to get a relation between $\tan^2\theta$ and $1/\cos^2\theta$, substitute, and then solve the resulting quadratic equation.)

## SEC. 5-4 ■ DISPLACEMENT IN TWO DIMENSIONS

**26. Position Vector for an Electron** The position vector for an electron is $\vec{r} = (5.0\text{ m})\hat{i} - (3.0\text{ m})\hat{j}$. (a) Find the magnitude of $\vec{r}$. (b) Sketch the vector on a coordinate system.

**27. Watermelon Seed** A watermelon seed has the following coordinates: $x = -5.0$ m and $y = 8.0$ m. Find its position vector (a) in unit-vector notation and as (b) a magnitude and (c) an angle relative to the positive direction of the $x$ axis. (d) Sketch the vector on a coordinate system. If the seed is moved to the coordinates (3.00 m, 0 m), what is its displacement (e) in unit-vector notation and as (f) a magnitude and (g) an angle relative to the positive direction of the $x$ axis?

**28. Radar Station** A radar station detects an airplane approaching directly from the east. At first observation, the range to the plane is 360 m at 40° above the horizon. The airplane is tracked for another 123° in the vertical east–west plane, the range at final contact being 790 m. See Fig. 5-35. Find the displacement of the airplane during the period of observation.

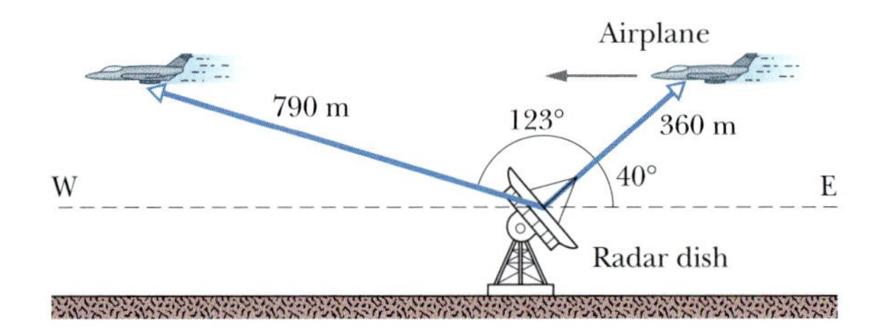


FIGURE 5-35 ■ Problem 28.

**29. Position Vector for a Proton** The position vector for a proton is initially $\vec{r}_1 = (5.0\text{ m})\hat{i} + (-6.0\text{ m})\hat{j}$ and then later is $\vec{r}_2 = (-2.0\text{ m})\hat{i} + (6.0\text{ m})\hat{j}$. (a) What is the proton's displacement vector, and (b) to what axis (if any) is that vector parallel?

**30. Kidnapped** You are kidnapped by armed political-science majors (who are upset because you told them that political science is not a real science). Although blindfolded, you can tell the speed of their car (by the whine of the engine), the time of travel (by mentally counting off seconds), and the direction of travel (by turns along the rectangular street system). From these clues, you know that you are taken along the following course: 50 km/h for 2.0 min, turn 90° to the right, 20 km/h for 4.0 min, turn 90° to the right, 20 km/h for 60 s, turn 90° to the left, 50 km/h for 60 s, turn 90° to the right, 20 km/h for 2.0 min, turn 90° to the left, 50 km/h for 30 s. At that point, (a) how far are you from your starting point and (b) in what direction relative to your initial direction of travel are you?

**31. Drunk Skunk** Figure 5-36 shows the path taken by my drunk skunk over level ground, from initial point $i$ to final point $f$. The angles are $\theta_1 = 30.0°$, $\theta_2 = 50.0°$, and $\theta_3 = 80.0°$, and the distances are $d_1 = 5.00$ m, $d_2 = 8.00$ m, and $d_3 = 12.0$ m. In magnitude-angle notation, what is the skunk's displacement from $i$ to $f$?

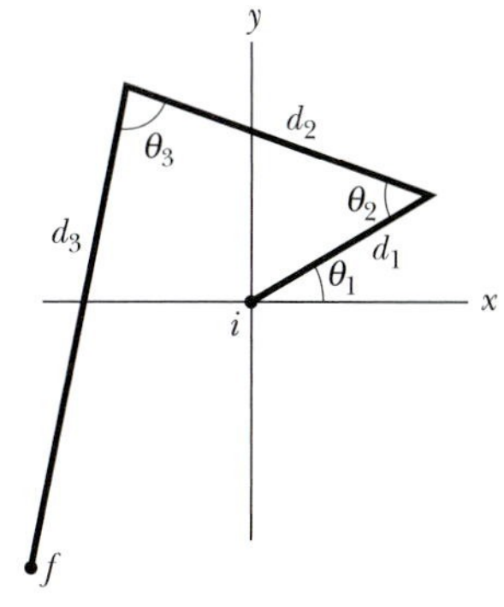


**FIGURE 5-36** ■ Problem 31.

## SEC. 5-5 ■ AVERAGE AND INSTANTANEOUS VELOCITY

**32. Squirrel Path** Figure 5-37 gives the path of a squirrel moving about on level ground, from point $A$ (at time $t_1 = 0$), to points $B$ (at $t_2 = 5.00$ min), $C$ (at $t_3 = 10.0$ min), and finally $D$ (at $t_4 = 15.0$ min). Consider the average velocities of the squirrel from point $A$ to each of the other three points. (a) Of those three average velocities, which has the least magnitude, and what is the average velocity in magnitude-angle notation? (b) Which has the greatest magnitude, and what is the average velocity in magnitude-angle notation?

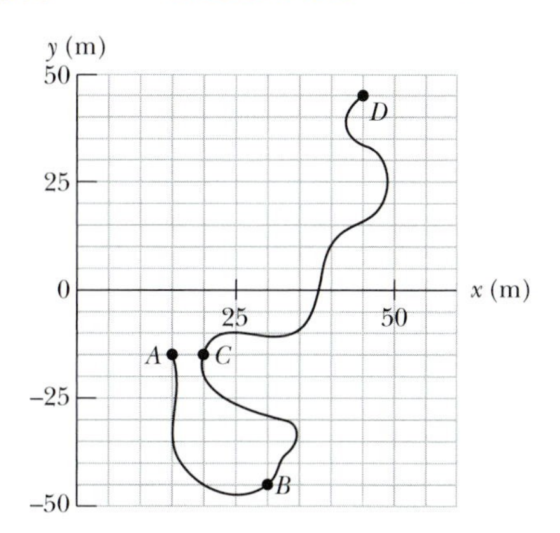


**FIGURE 5-37** ■ Problem 32.

**33. Train** A train moving at a constant speed of 60.0 km/h moves east for 40.0 min. then in a direction 50.0° east of north for 20.0 min, and finally west for 50.0 min. What is the average velocity of the train during this trip?

**34. Ion's Position** An ion's position vector is initially $\vec{r}_1 = (5.0 \text{ m})\hat{\text{i}} + (-6.0 \text{ m})\hat{\text{j}}$, and 10 s later it is $\vec{r}_2 = (-2.0 \text{ m})\hat{\text{i}} + (8.0 \text{ m})\hat{\text{j}}$. What is its average velocity during the 10 s?

**35. Electron's Position** The position of an electron is given by $\vec{r}(t) = [(3.00 \text{ m/s})\, t]\,\hat{\text{i}} + [(-4.00 \text{ m/s}^2)\, t^2]\hat{\text{j}}$. (a) What is the electron's velocity $\vec{v}(t)$? At $t = 2.00$ s, what is $\vec{v}$ (b) in unit-vector notation and as (c) a magnitude and (d) an angle relative to the positive direction of the $x$ axis?

**36. Oasis** Oasis $A$ is 90 km west of oasis $B$. A camel leaves oasis $A$ and during a 50 h period walks 75 km in a direction 37° north of east. The camel then walks toward the south a distance of 65 km in a 35 h period after which it rests for 5.0 h. (a) What is the camel's displacement with respect to oasis $A$ after resting? (b) What is the camel's average velocity from the time it leaves oasis $A$ until it finishes resting? (c) What is the camel's average speed from the time it leaves oasis $A$ until it finishes resting? (d) If the camel is able to go without water for five days (120 h), what must its average velocity be after resting if it is to reach oasis $B$ just in time?

**37. Jet Ski** You are to ride a jet-cycle over a lake, starting from rest at point 1: First, moving at 30° north of due east:

**1.** Increase your speed at 0.400 m/s² for 6.00 s.

**2.** With whatever speed you then have, move for 8.00 s.

**3.** Then slow at 0.400 m/s² for 6.00 s.

Immediately next, moving due west:

**4.** Increase your speed at 0.400 m/s² for 5.00 s.

**5.** With whatever speed you then have, move for 10.0 s.

**6.** Then slow at 0.400 m/s² until you stop.

In magnitude-angle notation, what then is your average velocity for the trip from point 1?

## SEC. 5-6 ■ AVERAGE AND INSTANTANEOUS ACCELERATION

**38. A Proton** A proton initially has $\vec{v}_1 = (4.0 \text{ m/s})\hat{\text{i}} + (-2.0 \text{ m/s})\hat{\text{j}}$ and then 4.0 s later has $\vec{v}_2 = (-2.0 \text{ m/s})\hat{\text{i}} + (-2.0 \text{ m/s})\hat{\text{j}}$. For that 4.0 s. what is the proton's average acceleration $\langle\vec{a}\rangle$ (a) in unit-vector notation and (b) as a magnitude and a direction?

**39. Particle in *xy* Plane** The position $\vec{r}$ of a particle moving in an $xy$ plane is given by $\vec{r}(t) = [(2.00 \text{ m/s}^3)t^3 - (5.00 \text{ m/s})\, t]\hat{\text{i}} + [(6.00 \text{ m}) - (7.00 \text{ m/s}^4)\, t^4]\hat{\text{j}}$. Calculate (a) $\vec{r}$, (b) $\vec{v}$, and (c) $\vec{a}$ for $t = 2.00$ s.

**40. Iceboat** An iceboat sails across the surface of a frozen lake with constant acceleration produced by the wind. At a certain instant the boat's velocity is $\vec{v}_1 = (6.30 \text{ m/s})\hat{\text{i}} + (-8.42 \text{ m/s})\hat{\text{j}}$. Three seconds later, because of a wind shift, the boat is instantaneously at rest. What is its average acceleration for this 3 s interval?

**41. Particle Leaves Origin** A particle leaves the origin with an initial velocity $\vec{v}_1 = (3.00 \text{ m/s})\hat{\text{i}}$ and a constant acceleration $\vec{a} = (-1.00 \text{ m/s}^2)\hat{\text{i}} + (-0.500 \text{ m/s}^2)\hat{\text{j}}$. When the particle reaches its maximum $x$ coordinate, what are (a) its velocity and (b) its position vector?

**42. Particle *A* Particle *B*** Particle $A$ moves along the line $y = 30$ m with a constant velocity $\vec{v}$ of magnitude 3.0 m/s and directed parallel to the positive $x$ axis (Fig. 5-38). Particle $B$ starts at the origin with zero speed and constant acceleration $\vec{a}$ (of magnitude 0.40 m/s²) at the same instant that particle $A$ passes the $y$ axis. What angle $\theta$ between $\vec{a}$ and the positive $y$ axis would result in a collision between these two particles? (If your computation involves an equation with a term such as $t^4$, substitute $u = t^2$ and then consider solving the resulting quadratic equation to get $u$.)

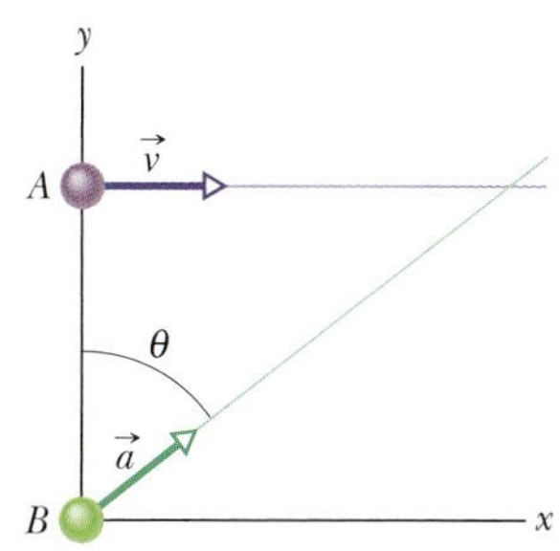


**FIGURE 5-38** ■ Problem 42.

**43. Particle Starts from Origin** A particle starts from the origin at $t = 0$ with a velocity of $\vec{v}_1 = (8.0 \text{ m/s})\hat{\text{j}}$ and moves in the $xy$ plane with a constant acceleration of $\vec{a} = (4.0 \text{ m/s}^2)\hat{\text{i}} + (2.0 \text{ m/s}^2)\hat{\text{j}}$. At the instant the particle's $x$ coordinate is 29 m, what are (a) its $y$ coordinate and (b) its speed?

**44. The Wind and a Pebble** A moderate wind accelerates a smooth pebble over a horizontal $xy$ plane with a constant acceleration

$$\vec{a} = (5.00 \text{ m/s}^2)\hat{\text{i}} + (7.00 \text{ m/s}^2)\hat{\text{j}}.$$

At time $t = 0$, its velocity is $(4.00 \text{ m/s})\hat{\text{i}}$. In magnitude-angle notation, what is its velocity when it has been displaced by 12.0 m parallel to the $x$ axis?

**45. Particle Acceleration** A particle moves so that its position as a function of time is $\vec{r}(t) = (1\text{ m})\hat{\text{i}} + [(4\text{ m/s}^2)\,t^2]\,\hat{\text{j}}$. Write expressions for (a) its velocity and (b) its acceleration as functions of time.

## SEC. 5-7 ■ UNIFORM CIRCULAR MOTION

**46. Sprinter** What is the magnitude of the acceleration of a sprinter running at 10 m/s when rounding a turn with a radius of 25 m?

**47. Sprinter on Circular Path** A sprinter runs at 9.2 m/s around a circular track with a centripetal acceleration of magnitude 3.8 m/s$^2$. (a) What is the track radius? (b) What is the period of the motion?

**48. Rotating Fan** A rotating fan completes 1200 revolutions every minute. Consider the tip of a blade, at a radius of 0.15 m. (a) Through what distance does the tip move in one revolution? What are (b) the tip's speed and (c) the magnitude of its acceleration? (d) What is the period of the motion?

**49. An Earth Satellite** An Earth satellite moves in a circular orbit 640 km above Earth's surface with a period of 98.0 min. What are (a) the speed and (b) the magnitude of the centripetal acceleration of the satellite?

**50. Merry-Go-Round** A carnival merry-go-round rotates about a vertical axis at a constant rate. A passenger standing on the edge of the merry-go-round has a constant speed of 3.66 m/s. For each of the following instantaneous situations, state how far the passenger is from the center of the merry-go-round, and in which direction. (a) The passenger has an acceleration of 1.83 m/s$^2$, east. (b) The passenger has an acceleration of 1.83 m/s$^2$, south.

**51. Astronaut** An astronaut is rotated in a horizontal centrifuge at a radius of 5.0 m. (a) What is the astronaut's speed if the centripetal acceleration has a magnitude of 7.0g? (b) How many revolutions per minute are required to produce this acceleration? (c) What is the period of the motion?

**52. TGV** The fast French train known as the TGV (Train à Grande Vitesse) has a scheduled average speed of 216 km/h. (a) If the train goes around a curve at that speed and the magnitude of the acceleration experienced by the passengers is to be limited to 0.050$g$, what is the smallest radius of curvature for the track that can be tolerated? (b) At what speed must the train go around a curve with a 1.00 km radius to be at the acceleration limit?

**53. Object on the Equator** (a) What is the magnitude of the centripetal acceleration of an object on Earth's equator due to the rotation of Earth? (b) What would the period of rotation of Earth have to be for objects on the equator to have a centripetal acceleration with a magnitude of 9.8 m/s$^2$?

**54. Supernova** When a large star becomes a *supernova*, its core may be compressed so tightly that it becomes a *neutron star*, with a radius of about 20 km (about the size of the San Francisco area). If a neutron star rotates once every second, (a) what is the speed of a particle on the star's equator and (b) what is the magnitude of the particle's centripetal acceleration? (c) If the neutron star rotates faster, do the answers to (a) and (b) increase, decrease, or remain the same?

**55. Ferris Wheel** A carnival Ferris wheel has a 15 m radius and completes five turns about its horizontal axis every minute. (a) What is the period of the motion? What is the centripetal acceleration of a passenger at (b) the highest point and (c) the lowest point, assuming the passenger is at a 15 m radius?

**56. A Particle at Constant Speed** A particle $P$ travels with constant speed on a circle of radius $r = 3.00$ m (Fig. 5-39) and completes one revolution in 20.0 s. The particle passes through $O$ at time $t = 0$. State the following vectors in magnitude-angle notation (angle relative to the positive direction of $x$). With respect to $O$, find the particle's position vector at the times $t$ of (a) 5.00 s, (b) 7.50 s, and (c) 10.0 s. (d) For the 5.00 s interval from the end of the fifth second to the end of the tenth second, find the particle's displacement. (e) For the same interval, find its average velocity. Find its velocity at (f) the beginning and (g) the end of that 5.00 s interval. Next, find the acceleration at (h) the beginning and (i) the end of that interval.

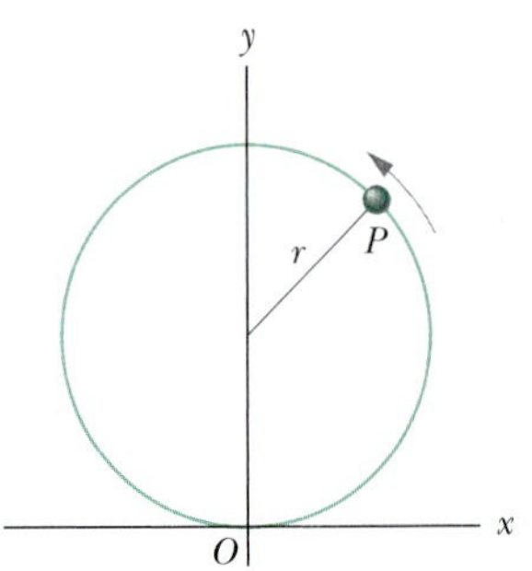


FIGURE 5-39 ■ Problem 56.

**57. Stone on a String** A boy whirls a stone in a horizontal circle of radius 1.5 m and at height 2.0 m above level ground. The string breaks, and the stone flies off horizontally and strikes the ground after traveling a horizontal distance of 10 m. What is the magnitude of the centripetal acceleration of the stone while in circular motion?

**58. Cat on a Merry-Go-Round** A cat rides a merry-go-round while turning with uniform circular motion. At time $t_1 = 2.00$ s, the cat's velocity is

$$\vec{v}_1 = (3.00\text{ m/s})\hat{\text{i}} + (4.00\text{ m/s})\hat{\text{j}},$$

measured on a horizontal $xy$ coordinate system. At time $t_2 = 5.00$ s, its velocity is

$$\vec{v}_2 = (-3.00\text{ m/s})\hat{\text{i}} + (-4.00\text{ m/s})\hat{\text{j}}.$$

What are (a) the magnitude of the cat's centripetal acceleration and (b) the cat's average acceleration during the time interval $t_2 - t_1$?

**59. Center of Circular Path** A particle moves horizontally in uniform circular motion, over a horizontal $xy$ plane. At one instant, it moves through the point at coordinates (4.00 m, 4.00 m) with a velocity of $(-5.00\text{ m/s})\hat{\text{i}}$ and an acceleration of $(12.5\text{ m/s}^2)\hat{\text{j}}$. What are the coordinates of the center of the circular path?

# Additional Problems

**60. Keeping Mars in Orbit** Although the planet Mars orbits the Sun in a Kepler ellipse with an eccentricity of 0.09, we can approximate its orbit by a circle. If you have faith in Newton's laws then you must conclude that there is an invisible centripetal force holding Mars in orbit. The data on the orbit of Mars around the sun are shown in the Fig. 5-40.

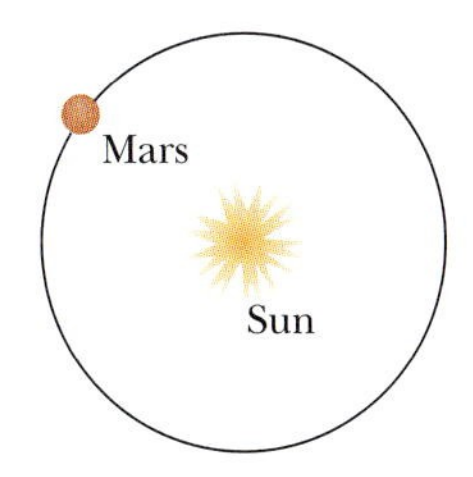


$m_{\text{Sun}} = 2.00 \times 10^{30}$ kg

$m_{\text{Mars}} = 0.107$ Earth masses or
$= 6.42 \times 10^{23}$ kg

$d_{\text{Mars}} = 1.523$ AU $= 2.28 \times 10^{11}$ m =
distance from the sun (= radius of circular orbit)

$\langle v \rangle = 24.13$ km/s (mean orbital speed)

**FIGURE 5-40** ■ Problem 60.

(a) Calculate the magnitude of the centripetal force needed to hold Mars in its circular orbit. Please use the proper number of significant figures. (b) What is the direction of the force as Mars orbits around the Sun? (c) What is the most likely source of this force? (d) Could this force have anything in common with the force that attracts objects to the Earth?

**61. Playing Catch** A boy and a girl are tossing an apple back and forth between them. Figure 5-41 shows the path the apple followed when watched by an observer looking on from the side. The apple is moving from left to right. Five points are marked on the path. Ignore air resistance. (a) Make a copy of this figure. At each of the marked points, draw an arrow that indicates the magnitude and direction of the force on the apple when it passes through that point. (b) Make a second copy of the figure. This time, at each marked point, place an arrow indicating the magnitude and direction of the apple's velocity at the instant it passes that point. (c) Did you change your answer to the first question after solving the second? If so, explain what you were thinking at first and why you changed it.

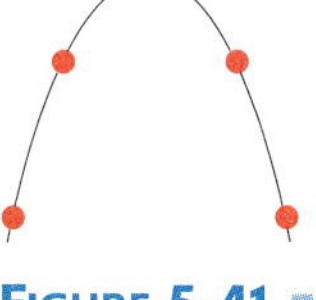

**FIGURE 5-41** ■ Problem 61.

**62. The Cut Pendulum** A pendulum (i.e., a string with a ball at the end) is set swinging by holding it at the point marked A in Fig. 5-42*a* and releasing it. The $x$ and $y$ coordinates are shown with the origin at the crossing point of the axes and the positive directions indicated by the arrowheads. (a) During one swing, the string breaks exactly at the bottom-most point of the swing (the point labeled $B$ in the figure) as the ball is moving from $A$ to $B$ toward $C$. Make a copy of this figure. Using solid lines, sketch on the figure the path of the ball after the string has broken. Sketch qualitatively the $x$ and $y$ coordinates of the ball and the $x$- and $y$-components of its velocity on graphs like those shown in Fig. 5-42*b*. Take $t = 0$ to be the instant the string breaks. (b) During a second trial, the string breaks again, but this time at the top-most point of the swing (the point labeled $C$ in the figure). Using dashed lines, sketch on the figure the path of the ball after the string has broken. Sketch qualitatively the $x$ and $y$ coordinates of the ball and the $x$- and $y$-components of its velocity on graphs like those shown in Fig. 5-42*b*. Take $t = 0$ to be the instant the string breaks.

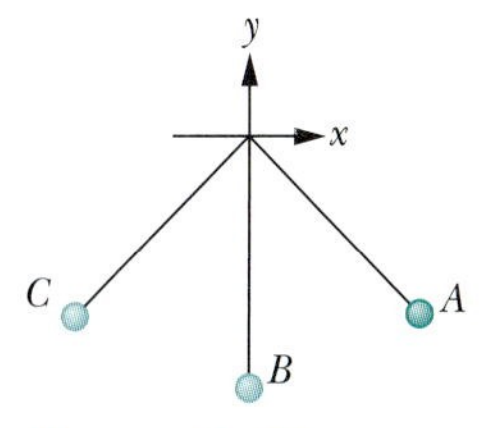


**FIGURE 5-42*a*** ■ Problem 62.

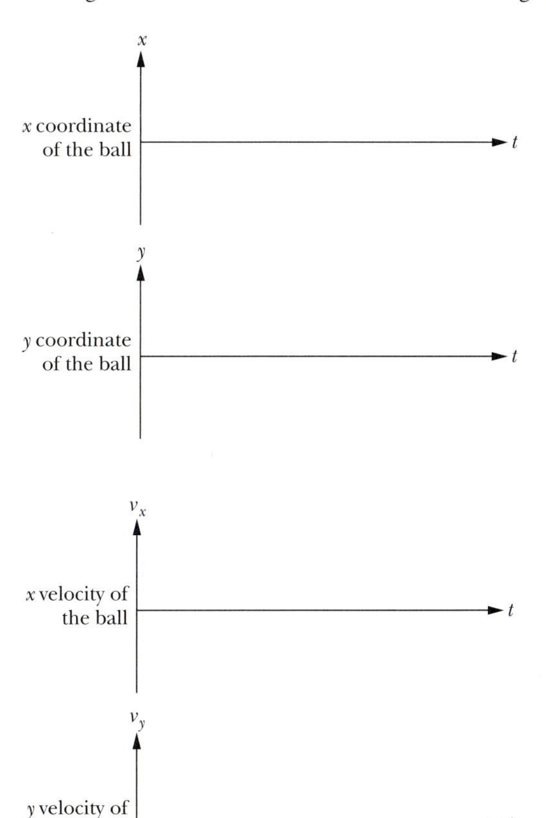


**FIGURE 5-42*b*** ■ Problem 62.

**63. Projectile Graphs** A pop-gun is angled so that it shoots a small dense ball through the air as shown in Fig. 5-43$a$.

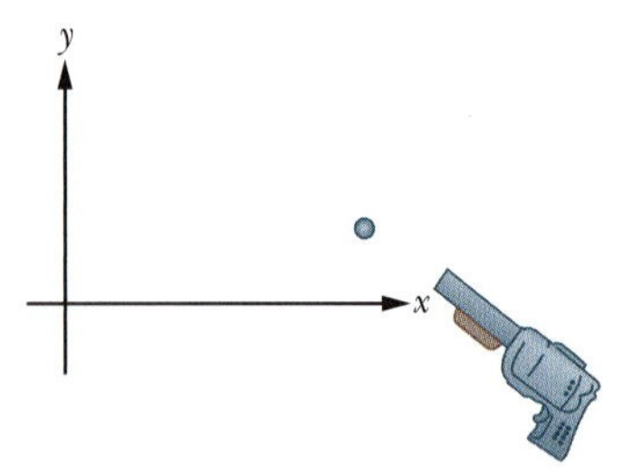


FIGURE 5-43$a$ ■ Problem 63.

**(a)** Sketch the path that the ball will follow on the figure. For the graphs shown in Fig. 5-43$b$, the horizontal axis represents the time. The vertical axis is unspecified. For each of the following quantities, select the letter of the graph that could provide a correct graph of the quantity for the ball in the situation shown (if the vertical axis were assigned the proper units). Use the $x$ and $y$ coordinates shown in the picture. The arrow heads point in the positive direction. If none of the graphs could work, write N. The time graphs begin just after the ball leaves the gun.

**(b)** $y$ coordinate
**(c)** $x$-component of the velocity
**(d)** $y$-component of the net force
**(e)** $y$-component of the velocity
**(f)** $x$ coordinate
**(g)** $y$-component of the acceleration
**(h)** $x$-component of the net force

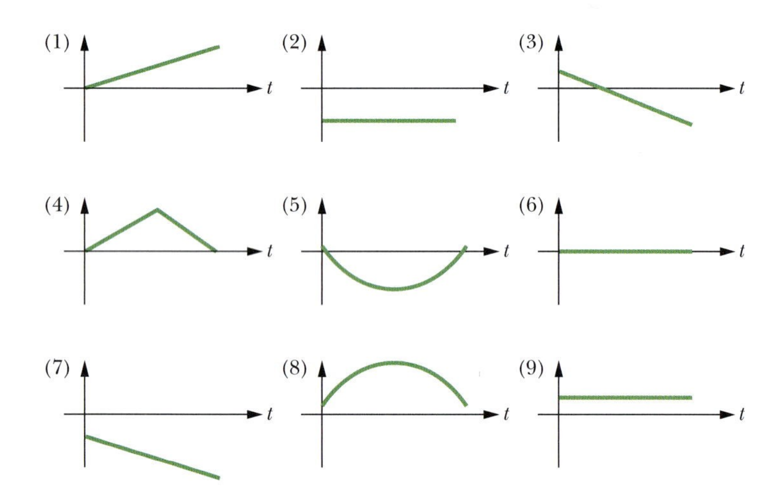


FIGURE 5-43$b$ ■ Problem 63.

**64. Shoot and Drop** In the demonstration discussed in Section 5-2, two identical objects were dropped, one straight down and the other shot off to the side by a spring. Both objects seemed to hit the ground at about the same time. Explain why this happens in terms of the physics we have learned. Does it matter how fast we shoot the one launched sideways? How would the outcome of this experiment change if the objects had different masses? (*Hint*: See Fig. 5-5.)

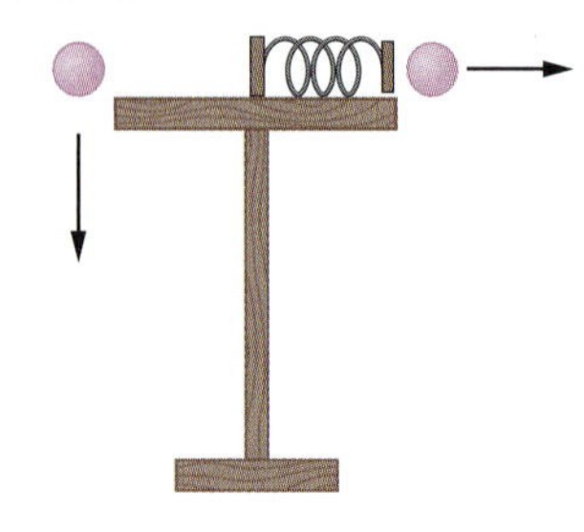

FIGURE 5-44 ■ Problem 64.

**65. Billiards over the Edge** Two identical billiard balls are labeled $A$ and $B$. Maryland Fats places ball $A$ at the very edge of the table. He places ball $B$ at the other side. He strikes ball $B$ with his cue so that it flies across the table and off the edge. As it passes $A$, it just touches ball $A$ lightly, knocking it off. Figure 5-45$a$ shows the balls just at the instant they have left the table. Ball $B$ is moving with a speed $v_1$, and ball $A$ is essentially at rest.

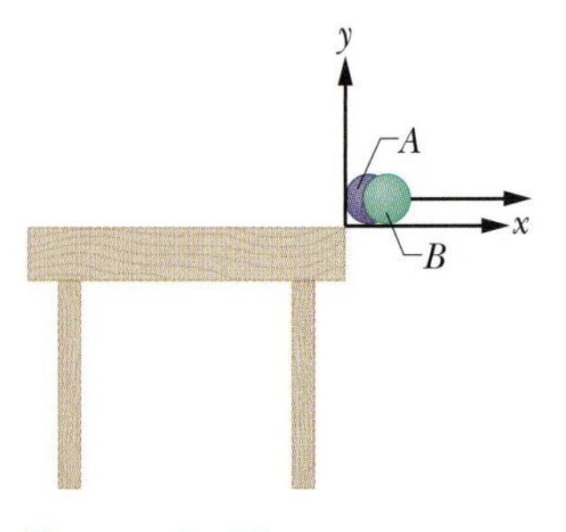


FIGURE 5-45$a$ ■ Problem 65.

(a) Which ball do you think will hit the ground first? Explain your reasons for thinking so.

Figure 5-45$b$ shows a number of graphs of a quantity vs. time. In each case, the horizontal axis is the time axis. The vertical axis is unspecified. For each of the items below, select which graph could be a plot of that quantity vs. time. If none of the graphs are possible, write N. The time axes are taken to have $t = 0$ at the instant both balls leave the table. Use the $x$ and $y$ axes shown in the figure.
(b) the $x$-component of the velocity of ball $B$? (c) the $y$-component of the velocity of ball $A$? (d) the $y$-component of the acceleration of ball $A$? (e) the $y$-component of the force on ball $B$? (f) the $y$-component of the force on ball $A$? (g) the $x$-component of the velocity of ball $A$? (h) the $y$-component of the acceleration of ball $B$?

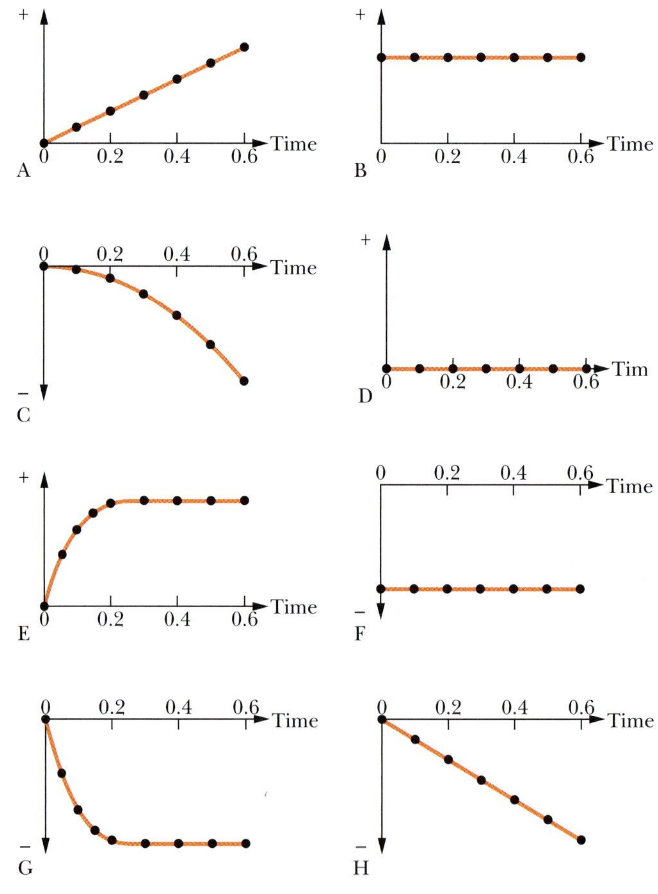


FIGURE 5-45$b$ ■ Problem 65.

**66. Properties of a Projectile** A heavy projectile is thrown and follows a path something like the one shown in Fig. 5-46. For each of the quantities in the list (a)–(d) below, select a direction from the list (A–G) that describes it. If you think that none of the choices apply, write N.

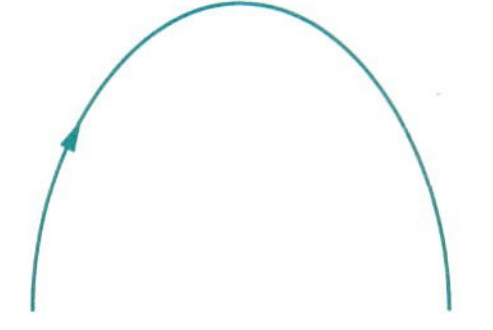

**FIGURE 5-46** ■ Problem 66.

*Quantities*:

**(a)** The projectile's velocity when it is at the highest point
**(b)** The force on the projectile when it is part way up
**(c)** The force on the projectile when it is at the highest point
**(d)** The projectile's acceleration when it is part way down

*Choices*:

A. Points straight up
B. Points straight down
C. Points directly to the left
D. Points directly to the right
E. Is equal to zero
F. Points somewhat upward and to the right
G. Points somewhat upward and to the left
N. None of the above

**67. Passing by the Spanish Guns** In C. S. Forster's novel *Lieutenant Hornblower* (set in the early 1800s), a British naval vessel tries to sneak by a Spanish garrison. The ship passes as far away from the Spanish guns as it can—a distance $s$. The Spanish gunner knows that his gun has a muzzle velocity whose magnitude is equal to $v_1$. (a) Once the gun is fired, what controls the motion of the cannonball? Write the equations that determine the vector position of the cannonball after it leaves the cannon. You may ignore air resistance. (b) Suppose the gunner inclines his gun upward at an angle $\theta$ to the horizontal. Solve the equations you have written in part (a) to obtain expressions that can be evaluated to give the position of the cannonball at any time, $t$. (c) If the gunner wants the cannonball to hit the ship, he must choose his angle correctly. Explain how he can calculate the correct angle. (Again, you may ignore air resistance.) (d) If the muzzle velocity of the cannonball has a magnitude of 100 m/s and the ship is a distance of half a kilometer away, find the angle the gunner should use. (Take g to be 10 m/s$^2$.)

**FIGURE 5-47** ■ Problem 67.

*You may need one or more of the following trigonometric identities (i.e., these are true for all angles, $\theta$):*

$$\sin^2\theta + \cos^2\theta = 1 \qquad \cos^2\theta - \sin^2\theta = \cos 2\theta$$
$$\tan\theta = (\sin\theta)/(\cos\theta) \qquad 2\sin\theta\cos\theta = \sin 2\theta$$

**68. Who Killed Adam Able?** A person shoved out of a window makes just as good a projectile as a golf ball rolling off a table. Read the murder mystery entitled A Damnable Man that follows. In order to solve the crime, read the section on projectile motion in your text carefully and reason out for yourself what variables might be important in solving the crime. In fact, not all of the information given in the mystery is relevant and some information, which you can find for yourself by observation and experiment, is missing. Solve the crime by presenting a clear explanation of the equations and calculations you used. (*Hint*: If you are in the physics lab and shove your lab partner fairly hard, you will likely find that your partner ends up with a speed of about 2 m/s.)

*A Damnable Man*
by Kevin Laws

It is a warm, quiet, humid night in the city—the traffic has died away and there isn't even a cooling breeze. There is a busy hotel that is so well built that sounds don't carry through the windows. The hotel has impressively large rooms. This is obvious from the outside because there is more space between floors and rooms than normal. The rooms appear to have 14-foot-high ceilings, nice plate glass windows that slide open, and fully two-foot-thick floors for ducting and sound insulation. This is the type of hotel that people like to stay at when someone else is paying the bill.

Outside the hotel, a man is speaking quietly with the doorman, then begins to measure the plush runway carpet for replacement. He is reeling out the tape measure between the hotel and the curb when a scream breaks the quiet. Looking up, he sees a man falling toward him. Stunned, he drops the tape measure and runs for the safety of the hotel. The doorman stands, horrified, as the man completes his fall with a sickening sound, ensuring that the carpet must be replaced. At intense times, people can think of the strangest things, and the carpet-man finds this to be true . . . all he can think of are the bloodstains left on his tape measure. Even if they are cleaned off, he doesn't think he can use it again without thinking of tonight. Even measuring with another will be hard, and 18 feet will be indelibly marked in his memory—that's where the blood stains are.

The police arrive and quickly conclude that it is not a suicide—among the victim's personal effects, they find pictures and records that indicate he has been blackmailing four other occupants of the hotel. He also has bruises on his shins where the ledge at the bottom of the tall hotel window would have hit them; he must have been pushed pretty hard. Adam Able is the dead man's name, as it appears on the driver's license in his wallet. His license indicates that Adam was 5' 11" tall and weighed 160 lb. He has been blackmailing Adrianna Myers, a frail widow in Room 356; Steven Caine, a newspaper reporter in Room 852; Mark Johnson, a body builder in Room 1956; and Stanley Michaels, an actor in Room 2754. All of the suspects admit they were in their rooms at the time of the murder. **WHO KILLED ADAM ABLE?**

**69. Digital Projectile One** In this problem and the one that follows you will be asked to use VideoPoint, VideoGraph, or some other video analysis program and a spreadsheet to explore and analyze the nature of a projectile launch depicted in a digital movie. If you use VideoPoint, one appropriate movie has filename PASCO106. In this movie a small ball of mass 9.5 g is launched at an angle, $\theta$, with respect to the horizontal. Your instructor may suggest an alternative file for your use.

Open the movie PASCO106. For simplicity you might want to set the origin in the video analysis at the location of the ball at time $t$ = 0. Also, for immediate visual feedback on your results you should use the *View Window* to set up graphs of $x$ vs. $t$ and $y$ vs. $t$ before you begin the analysis.

**(a)** What is the approximate launch angle $\theta$? Measure this angle with respect to the horizontal. Explain how you found the angle.
**(b)** Explain in which direction, $x$ or $y$, the ball has a constant velocity and cite the real evidence (not just theoretical) for this constant velocity. (*Hint*: Use markers of various sorts on the digital movie to demonstrate that the ball is moving at a constant velocity in one of the directions and not in the other.)
**(c)** Explain in which direction, $x$ or $y$, the ball is accelerating. Cite real evidence (not just theoretical) for this acceleration. (*Hint*: Use markers of various sorts on the digital movie.)
**(d)** Theoretically, what is the net vertical force on the 9.5 g ball when it is rising? Falling? Turning around? What is the observational basis for this theoretical assumption?
**(e)** Theoretically, what is the net horizontal force on the 9.5 g ball when it is rising? Falling? Turning around? What is the observational basis for this theoretical assumption?
**(f)** What do you predict will happen to the shapes of the $x$ vs. $t$ and $y$ vs. $t$ graphs if you rotate your coordinate system so that the $x$ axis points in the vertical direction and the $y$ axis points in the horizontal direction?
**(g)** Rotate your coordinate system so that the $x$ axis points in the vertical direction and the $y$ axis points in the horizontal direction. What happens to the shapes of the graphs? Is this what you predicted?

**70. Digital Projectile Two** In this problem you will use VideoPoint, VideoGraph, or some other video analysis program and a spreadsheet to explore and analyze the nature of a projectile launch depicted in a digital movie. If you use VideoPoint, one appropriate movie has filename PASCO106. In this movie a small ball of mass 9.5 g is launched at an angle, $\theta$, with respect to the horizontal. Your instructor may suggest an alternative file for your use.

Open the movie PASCO106. Use the VideoPoint software and spreadsheet modeling to find the equation that describes: the horizontal motion $x$ vs. $t$ and the equations that describe the vertical motion $y$ vs. $t$.

**(a)** Hand in the printout of your two models. Place your name, date and section # on it, and answer questions (b) through (d) at the bottom of the page.
**(b)** According to your horizontal model, what is the equation that describes the horizontal position of the ball, $x$, as a function of time? What is its horizontal acceleration, $a_x$? What is its initial horizontal velocity, $v_{1x}$?
**(c)** According to your vertical model, what is the equation that describes the vertical position, $y$, of the ball as a function of time? What is the value of the ball's vertical acceleration, $a_y$? What is its initial vertical velocity, $v_{1y}$?
**(d)** Use the components $v_{1x}$ and $v_{1y}$ to compute the initial speed of the ball. What is the launch angle with respect to the horizontal?
**(e)** Compare your answer to part (d) to your approximation from part (a) of the previous problem.

**71. Curtain of Death** A large metallic asteroid strikes Earth and quickly digs a crater into the rocky material below ground level by launching rocks upward and outward. The following table gives five pairs of launch speeds and angles (from the horizontal) for such rocks, based on a model of crater formation. (Other rocks, with intermediate speeds and angles, are also launched.) Suppose that you are at $x = 20$ km when the asteroid strikes the ground at time $t_1 = 0$ and position $x = 0$ (Fig. 5-48). (a) At $t_2 = 20$ s, what are the $x$ and $y$ coordinates of the rocks headed in your direction from launches $A$ through $E$? (b) Plot these coordinates and then sketch a curve through the points to include rocks with intermediate launch speeds and angles. The curve should give you an idea of what you would see as you look up into the approaching rocks and what dinosaurs must have seen during asteroid strikes long ago.

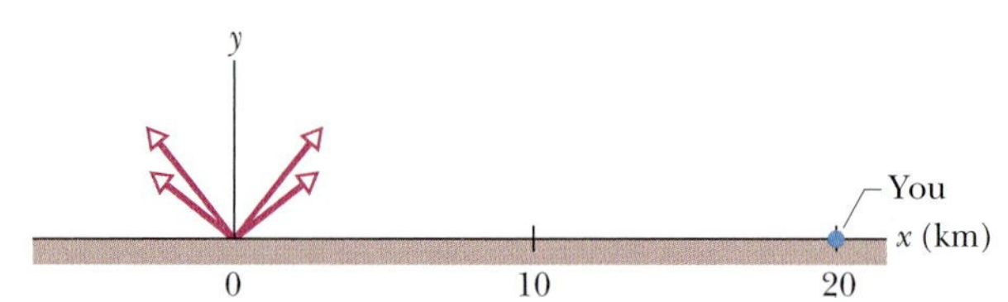


FIGURE 5-48 ■ Problem 71.

| Launch | Speed (m/s) | Angle (degrees) |
|---|---|---|
| *A* | 520 | 14.0 |
| *B* | 630 | 16.0 |
| *C* | 750 | 18.0 |
| *D* | 870 | 20.0 |
| *E* | 1000 | 22.0 |

# 6 Identifying and Using Forces

Cats, who enjoy sleeping on window sills, are often kept in apartment buildings. When a cat accidentally falls out of a window and onto a sidewalk, the extent of injury (such as the number of fractured bones or the certainty of death) *decreases* with height if the fall is more than seven or eight floors. (There is even a record of a cat who fell 32 floors and suffered only slight damage to its thorax and one tooth.)

**How can the damage possibly decrease with height?**

*The answer is in this chapter.*

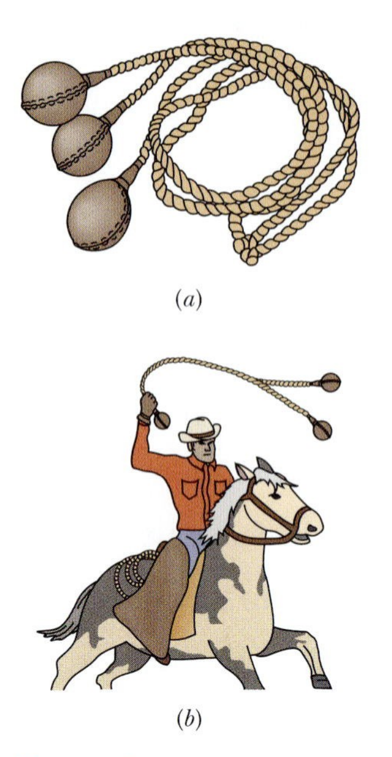

FIGURE 6-1 ■ (*a*) A modern Inuit bola. (*b*) A sketch of a gaucho using a bola.

## 6-1 Combining Everyday Forces

It is common for objects to experience multiple forces that do not act along the same line. We saw examples of this in Section 5-2 in our brief consideration of the motion of a ball falling under the influence of both a gravitational force and air drag forces. In Section 5-7 we discussed the motion involved in the hammer throw and that of a rock rotating on a string. The hammer and the rock experience both a vertical gravitational force and changing centripetal forces that are almost horizontal.

The bola shown in Fig. 6-1 is another example of a system that experiences multiple forces acting in more than one dimension. The bola is a prehistoric weapon devised for capturing relatively large animals. The analysis of the bola's motion as it is whirled about, released, and encounters an animal is very complex. At any given moment the spherical end of a flying bola experiences a gravitational force, the pull of the rope, and an air drag force.

In this chapter, you will learn more about the characteristics of these everyday forces and how they can be superimposed using vector addition to find net forces. In addition, we will consider how to apply Newton's laws to predict motion and to identify hidden forces. As you will see, the ability to identify forces and use them along with Newton's laws to predict motion is extremely useful for two reasons. First, engineers can use their knowledge of the forces on a system to predict the motion of system components. This ability is vital in the design of a range of devices from bridges to aircraft. Second, the belief physicists have in the validity of Newton's laws of motion leads them to combine acceleration measurements with Newtonian analysis techniques to identify and characterize invisible forces. This approach to the discovery of forces was introduced in Section 3-9.

## 6-2 Net Force as a Vector Sum

In Chapter 3 we presented experiments that demonstrate that when two or more forces act on an object that moves in a straight line, it is the *net* force that determines how the object's motion will change. For one-dimensional motion the net force turns out to be the vector sum of the forces acting on the object. We call this the **principle of superposition for forces.** If we use the rules of two-dimensional vector addition that we learned about in Chapter 4, can we apply the principle of superposition in cases where the forces do not lie along a single line?

Countless experiments have demonstrated that the principle of superposition also works in two (and three) dimensions. For example, consider the rotating rock discussed in Fig. 5-24. As the rock rotates, it experiences both a gravitational and a string force as shown in Fig. 6-2. We already know that $\vec{F}^{\text{grav}} = -mg\hat{j}$ where $m$ is the mass of the rock. If we attach a spring scale between the rock and the string, we can measure the string force $\vec{F}^{\text{string}}$. If the rock is rotating in a circle in a horizontal plane and we measure its centripetal acceleration, we find that it is related to the forces on the rock by

$$\vec{F}^{\text{net}} = \vec{F}^{\text{grav}} + \vec{F}^{\text{string}} = m\vec{a}. \tag{6-1}$$

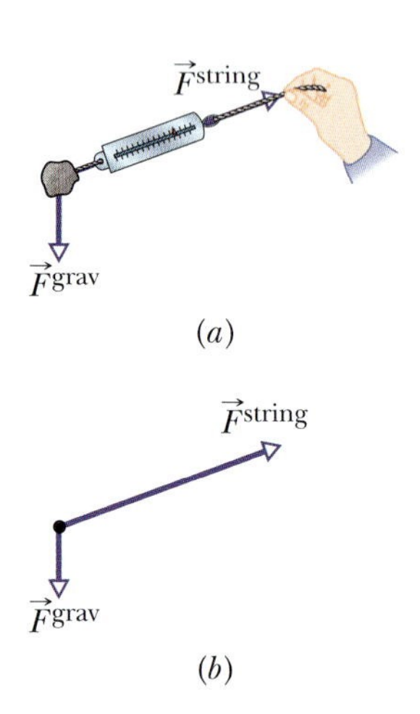


FIGURE 6-2 ■ (*a*) At any particular moment there are two forces on a rock twirled on the end of a string—a gravitational force and a string force. Here the directions of the forces are indicated for the case where the rock rotates in a horizontal plane. (*b*) A free-body diagram showing the tails of the two force vectors at a point that represents the rock on which they act.

Here the net force that leads to the measured acceleration turns out indeed to be the two-dimensional vector sum (or superposition) of the two forces acting on the rock. We can find the vector sum of two or more force vectors by using the graphical method explained in Section 4-3, or we can resolve the vectors into components using the method presented in Section 4-4.

Another way to verify experimentally that the superposition of force vectors in two dimensions is a vector sum is to set up a situation in which the net force in a plane is zero. For example, we can pull on a ring with three spring scales in such a way that the ring is stationary. In this case, we know the acceleration of the ring, and hence the net force on the ring is zero. Every time we do this, we find that the vector sum of three forces is zero. An example is shown in Fig. 6-3. Here the sum of $F_{Ax}\hat{\text{i}}$ and $F_{Bx}\hat{\text{i}}$ gives us a vector component that has the same magnitude as $F_{Cx}\hat{\text{i}}$ but has the opposite sign so as to cancel it. Thus, the net force is zero as shown in Fig. 6-3*d*. After many such experiments, we become convinced that the net force on an object is the vector sum of the individual forces acting on the object, even if those forces do not act along a single line.

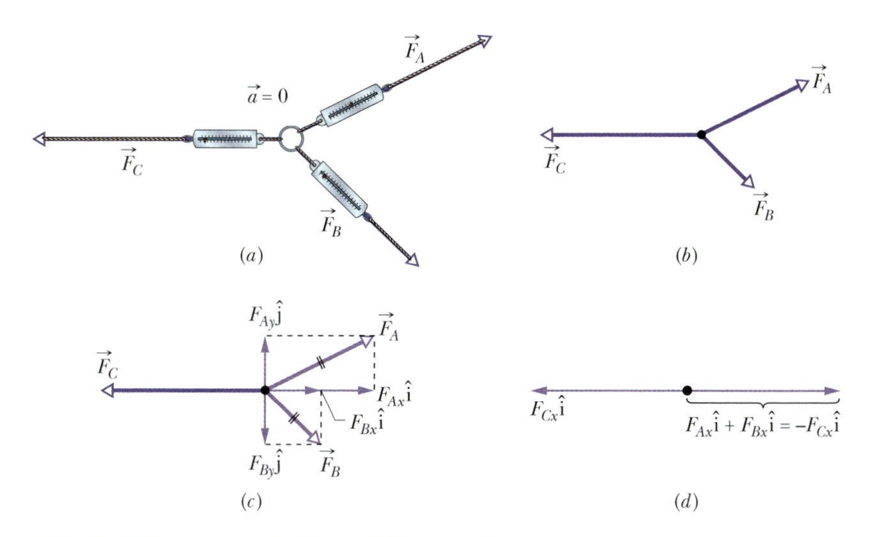


**FIGURE 6-3** ■ (*a*) If the ring does not accelerate under the influence of the three forces, we conclude that the net force on it is zero, hence the vector sum of $\vec{F}_A$, $\vec{F}_B$, and $\vec{F}_C$ is zero. (*b*) A free-body diagram showing the tails of the three vectors at a point that represents the center of the ring on which they act. (*c*) Using the component method of resolving the vectors $\vec{F}_A$, $\vec{F}_B$, and $\vec{F}_C$ verifies that their sum is zero. (*d*) The sum of the *x*-components of $F_{Ax}$ and $F_{Bx}$ is $-F_{Cx}$.

## Free-Body Diagrams in Two Dimensions

In Chapter 3 we found that it was important to keep track of the magnitudes and directions of the forces acting on an object if we wanted to use Newton's Second Law ($\vec{F}^{\text{net}} = m\vec{a}$) to determine the object's acceleration. The same is true for cases in which the forces do not lie along a single line. We introduced the idea of using a *free-body diagram* for this purpose in Section 3-7. The procedures for drawing free-body diagrams for two- and three-dimensional forces are similar to those used for one-dimensional forces: (1) Identify the object for which the motion is to be analyzed and represent it as a point. (2) Identify all the forces acting on the object and represent each force vector with an arrow. The tail of each force vector should be on the point. Draw the arrow in the direction of the force. Represent the relative magnitudes of the forces through the relative lengths of the arrows. (3) Label each force vector so that it is clear which force it represents.

Figures 6-2*b* and 6-3*b* are free-body diagrams for the situations depicted in the first part of those figures.

## Newton's Second Law in Multiple Dimensions

The preceding example hints at another important point regarding multiple forces acting along different lines. Namely, forces (or components of forces) in perpendicular dimensions are independent and separable. That is, Newton's Second Law $\vec{F}^{\text{net}} = m\vec{a}$ can be written as two (or three) component equations:

$$F_x^{\text{net}} = ma_x, \quad F_y^{\text{net}} = ma_y, \quad \text{and} \quad F_z^{\text{net}} = ma_z.$$

We will focus on two-dimensional examples in this chapter.

This statement regarding the separable nature of forces and components of forces should not be especially surprising. Recall from Chapter 5 that horizontal and vertical motions are independent and separable. That is, an acceleration in one dimension only affects the motion in that dimension. Therefore, we could treat two-dimensional motions as two separate one-dimensional cases. Since net force and acceleration are directly related, the independent and separable nature of acceleration is a direct hint that forces behave this way.

If three forces act on an object, then we can expand $F_x^{\text{net}} = ma_x$ to get

$$F_{A\,x} + F_{B\,x} + F_{C\,x} = ma_x,$$

where $F_{A\,x}$ is the $x$-component of force $A$, $F_{B\,x}$ is the $x$-component of force $B$, and so on. The $x$-component of the acceleration is $a_x$. These components are signed quantities. This means that although the components are not vectors, they can still be either positive or negative. So, we need to be careful when we begin substituting in actual values for the components that we include the correct sign.

We can use a similar expansion to find $ma_y$ and so on.

## A Word about Notation

Recall from earlier chapters that $\vec{F}$ represents a vector. The magnitude (that is, size) of the vector is represented by $|\vec{F}|$ when we want to stress that the value is always positive. More commonly, the magnitude is simply represented as $F$. That is, a vector quantity represented without the arrow over it is the magnitude of the vector, which is always positive. $F_x$ and $F_y$ represent vector components and may be positive or negative depending on what direction $\vec{F}$ points in relation to the chosen coordinate system.

We have already introduced several different forces including gravitational, tension, and friction forces. These are important, everyday forces. In the rest of this chapter, we will add to our list of common forces and discuss those we have already introduced in more detail.

**READING EXERCISE 6-1:** A helicopter is moving to the right at a constant horizontal velocity due to the force on it caused by its rotor. It also experiences a downward gravitational force and a horizontal drag force as shown in the diagram below. Which of the following diagrams is a correct free-body diagram representing the forces on the helicopter?

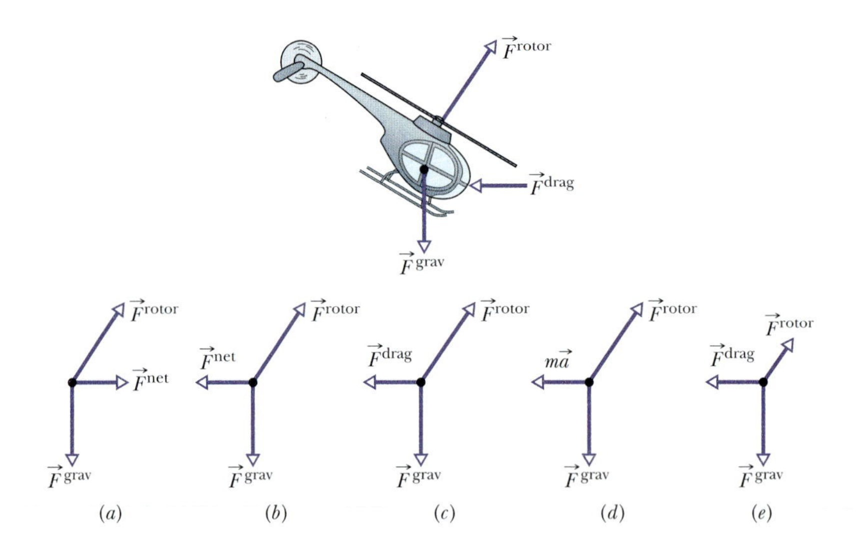

## TOUCHSTONE EXAMPLE 6-1: Tug-of-War

In a two-dimensional tug-of-war, Alex, Betty, and Charles pull horizontally on an automobile tire at the angles shown in the overhead view of Fig. 6-4*a*. The tire remains stationary in spite of the three pulls. Alex pulls with a force $\vec{F}_A$ of magnitude 220 N, and Charles pulls with force $\vec{F}_C$ of magnitude 170 N. The direction of $\vec{F}_C$ is not given. What is the magnitude of Betty's force $\vec{F}_B$?

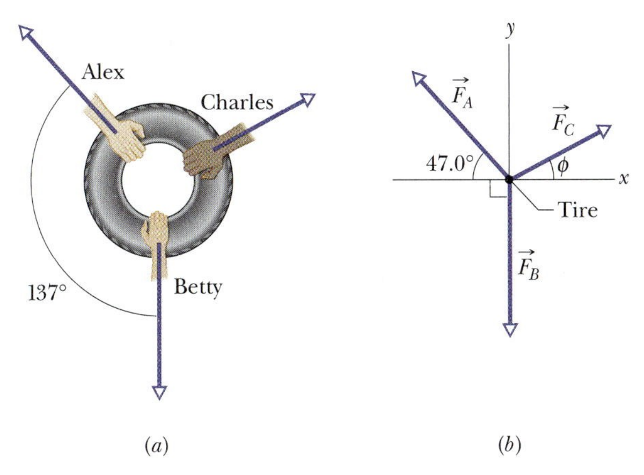


**FIGURE 6-4** ■ (*a*) An overhead view of three people pulling on a tire. (*b*) A free-body diagram for the tire.

**SOLUTION** ■ Because the three forces pulling on the tire do not accelerate the tire, the tire's acceleration is $\vec{a} = 0$ (that is, the forces are in equilibrium). The **Key Idea** here is that we can relate that acceleration to the net force $\vec{F}^{\text{net}}$ on the tire with Newton's Second Law ($\vec{F}^{\text{net}} = m\vec{a}$), which we can write as

$$\vec{F}_A + \vec{F}_B + \vec{F}_C = m(0) = 0,$$

or

$$\vec{F}_B = -\vec{F}_A - \vec{F}_C. \tag{6-2}$$

The free-body diagram for the tire is shown in Fig. 6-4*b*, where we have conveniently centered a coordinate system on the tire and assigned $\phi$ to the angle between the $x$ axis and $\vec{F}_C$.

We want to solve for the magnitude of $\vec{F}_B$. Although we know both magnitude and direction for $\vec{F}_A$, we know only the magnitude of $\vec{F}_C$ and not its direction. Thus, with unknowns on both sides of Eq. 6-2, we cannot directly solve it on a vector-capable calculator. Instead we must rewrite Eq. 6-2 in terms of components for either the $x$ or the $y$ axis. If the sum of the forces is zero, it must also be that the sum of the $x$-components of the forces is zero *and* the sum of the $y$-components is zero. Since $\vec{F}_B$ is directed along the $y$ axis, we choose that axis and write

$$F_{By} = -F_{Ay} - F_{Cy}.$$

Note that we have dropped the arrows over our symbols and added a subscript "$y$" here. We did this because we are now dealing with components of the vectors as opposed to the vectors themselves. Evaluating these components with their angles and using the angle 133° ( = 180° − 47.0°) for $\vec{F}_A$, we obtain

$$F_B \sin(-90°) = -F_A \sin 133° - F_C \sin \phi,$$

where $F_A$, $F_B$, and $F_C$ denote vector magnitudes (not components). Using the given data for the magnitudes, yields

$$-F_B = -(220 \text{ N})(\sin 133°) - (170 \text{ N}) \sin \phi. \tag{6-3}$$

However, we do not know $\phi$.

We can find $\phi$ by rewriting Eq. 6-2 for the $x$ axis as

$$F_{Bx} = -F_{Ax} - F_{Cx}$$

and then as

$$F_B \cos(-90°) = -F_A \cos 133° - F_C \cos \phi,$$

which gives us

$$0 = -(220 \text{ N})(\cos 133°) - (170 \text{ N}) \cos \phi$$

and

$$\phi = \cos^{-1} - \frac{(220 \text{ N})(\cos 133°)}{170 \text{ N}} = 28.04°.$$

Inserting this into Eq. 6-3, we find

$$F_B = 241 \text{ N}. \qquad \text{(Answer)}$$

# 6-3 Gravitational Force and Weight

## Gravitational Force

As we discussed in Section 3-9, gravitational forces result from interactions between masses and can act over long distances. Although gravitational interactions between any two masses are always present, they are only noticeable when at least one of the masses is very large. We have already presented experimental evidence in Section 3-9 that the gravitational pull of the Earth on an object is directly proportional to the object's mass. We use the constant of proportionality, denoted $g$, to relate the gravitational force to mass from Eq. 3-9:

$$\vec{F}^{grav} = -mg\hat{j}, \tag{6-4}$$

where the constant $g$, known as the **local gravitational strength,** is a positive scalar and $\hat{j}$ is a unit vector that points up. The minus sign tells us that the gravitational force points down. Close to the Earth's surface, the value of $g$ is 9.8 N/kg.

## Weight

Weight is a commonly used synonym for the magnitude of the gravitational force acting on an object.

> The weight $W$ of a body is a scalar quantity that equals the magnitude $|\vec{F}^{grav}|$ of the local gravitational force exerted by the Earth or some other massive astronomical object (such as the moon) on the body.

$$W = |\vec{F}^{grav}| = mg \quad \text{(weight)} \tag{6-5}$$

To *weigh* a body means to measure its weight. As we mentioned in Section 3-9, we can measure gravitational force and hence weight, using a balance, a spring scale, or an electronic scale. Sometimes scales are marked in mass units. Since the value of $g$ changes as we move away from the Earth, scales are only accurate for measuring mass when the value of $g$ is the same as it is where the scale was calibrated.

Weight must be measured when the body is not accelerating vertically relative to the astronomical object attracting it. For example, you can accurately measure your weight on a scale in your bathroom or on a fast train moving horizontally. However, if you repeat the measurement with the scale in an accelerating elevator, the reading on the scale differs from your weight because of the vertical acceleration. This was first discussed in Section 2-4.

Note that the weight $W$, which has SI units of newtons, and the local gravitational strength $g$, which has SI units of newtons per kilogram, are not components of vectors, which can be positive or negative. Instead they are both magnitudes and *are always positive*.

## Mass Versus Weight

Unfortunately, everyday speech sometimes leads us to believe that the terms "weight" and "mass" are interchangeable. Although the weight of a body (given by $W = mg$) is proportional to its mass, *weight and mass are not the same thing*. Mass has a standard unit of kilograms whereas weight is the magnitude of a force, with a standard unit of newtons. If you move a body to a location such as the surface of the Moon where the value of the local gravitational strength $g$ is different, the body's mass (how much "stuff" the object is made up of) is *not* different, but its weight is. For example, the weight of a bowling ball with a mass of 7.2 kg is 71 N on Earth. On the Moon, this same bowling ball would have the same mass, but a weight of only 12 N. This is because the local gravitational strength is only about one-sixth of its value on Earth.

**READING EXERCISE 6-2:** Suppose you are given two different objects, a balance like the one shown in Fig. 3-9, and a spring scale like the one shown in Fig. 3-23. Describe how you could determine whether the two objects have the same mass. What might you do to determine the weight of one of the objects? Is the weight of each object the same as the mass of the object? Is the ratio of the masses the same as the ratio of the weights? ■

**READING EXERCISE 6-3:** Comment on the accuracy of the statement the patient is making in the *Frank & Ernest* cartoon.

## 6-4 Contact Forces

As we have mentioned, the gravitational force can act over large distances and exists even if the two interacting objects are not touching. Hence, we sometimes refer to the gravitational force as an "action at a distance" force. In contrast, forces such as tension and frictional forces only exist when there is contact between interacting objects. We call forces of this kind "contact" forces. In order to understand the nature of contact forces between solid objects, it is helpful to learn more about the atomic nature of solids.

### An Idealized Model of a Solid

Modern scientists have strong evidence that solids in our everyday world are made of atoms. It is very hard to compress a solid object or pull it apart. The forces between atoms seem to behave like springs. When you push on a spring that is at its natural or equilibrium length, it resists compression by pushing back on you. But when you pull on a spring, it also resists stretching by pulling back on you. This has led physicists to create an idealized model for a solid as an array of atoms held together by forces that behave like very stiff springs, each having an equilibrium length of about $10^{-10}$ m. A three-dimensional model of a possible array of atoms in a simple solid is shown in Fig. 6-5*a*. This model is explained in more detail in Section 13-5. (As we will see in Chapter 22, the force between atoms in a solid can be understood in terms of the electromagnetic forces between the charged particles in atoms.)

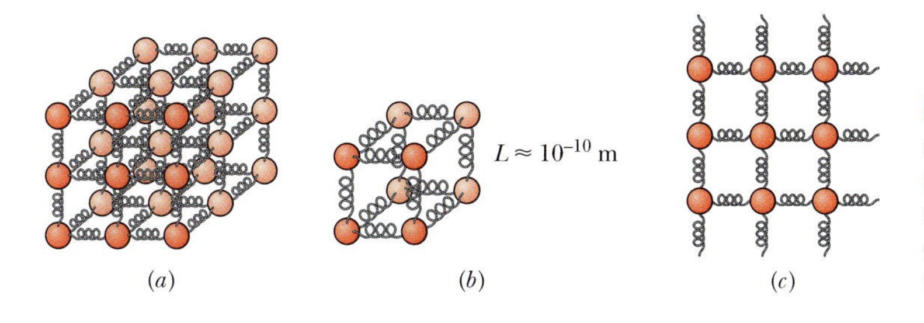


**FIGURE 6-5** ■ An idealized model of a solid consisting of atoms separated by tiny springs. (*a*) A model consisting of stiff springs ("atomic bonds") holding balls ("atoms") together. (*b*) Eight atoms at the corner of a cube show the three-dimensional nature of a small hunk of the idealized solid. (*c*) A depiction of a few of the atoms that lie in the plane of the paper.

### Using the Model to Understand Contact Forces

How can we use this simplified model to help us understand contact forces? Let's consider what happens when you push on an innerspring mattress (Fig. 6-6). As you push, the springs in the mattress become compressed under your finger and push back

**FIGURE 6-6** ■ This physical model of a solid as a matrix of atoms separated by tiny springs behaves rather like an innerspring mattress. Our "solid" is compressed just slightly by the force exerted on it by a finger. According to Newton's Third Law the "solid" then exerts an equal and opposite upward force back on the finger.

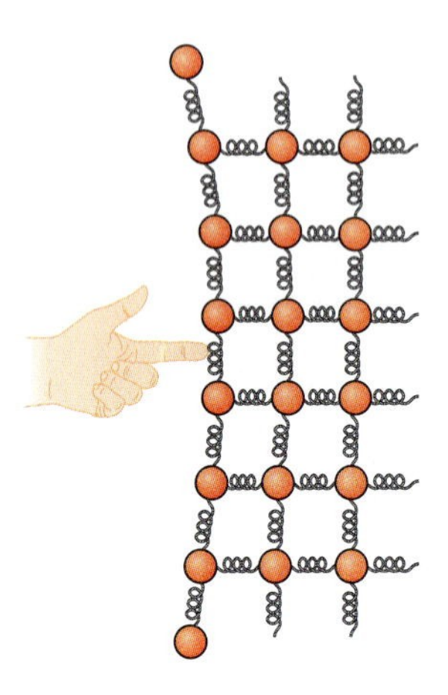

**FIGURE 6-7** ■ Compressing an idealized solid wall with a force exerted by a finger. The deformation of the wall is exaggerated. The wall exerts an oppositely directed force with the same magnitude back on the finger.

on your finger. According to Newton's Third Law, the force you exert on the mattress springs is equal in magnitude and opposite in direction to the force the mattress springs exert on your finger.

Similarly, if you push on a wall it compresses (it is deformed, bent, or buckled ever so slightly), and it pushes back on you (Fig. 6-7). The compression of the wall is hard to see because its billions and billions of tiny atomic springs are much stiffer than the mattress springs. But the harder you push, the more compressed the wall becomes and the larger the force the wall exerts on your finger. When you push harder, it hurts, because in accord with Newton's Third Law, the surface is also pushing back on your finger with a larger force. You can feel (and see) your finger becoming more and more compressed due to the force exerted on it by the wall. Ouch! Try it!

We call a force exerted perpendicular to a surface a **normal force** and denote it as $\vec{N}$. Note that in this context *normal* is a technical term that derives from a Latin term *norma* meaning "carpenter's square." It is a synonym for perpendicular and does not mean "ordinary."

When one object, such as a wall, exerts a contact force on another object, such as your finger, the force is not necessarily perpendicular to the surfaces in contact. However, you can decompose the force vector into a parallel component and a perpendicular component as shown in Fig. 6-8*b*. We call the component vector perpendicular to the surfaces in contact the **normal force.** We call the component vector parallel to the surfaces the **friction force** and denote it as $\vec{f}$.

In mathematical terms the decomposition of the contact force on your finger, $\vec{F}^{\text{contact}}_{\text{on finger}}$, is given by the sum of the two perpendicular force vectors,

$$\vec{F}^{\text{contact}}_{\text{on finger}} = \vec{N} + \vec{f}. \tag{6-6}$$

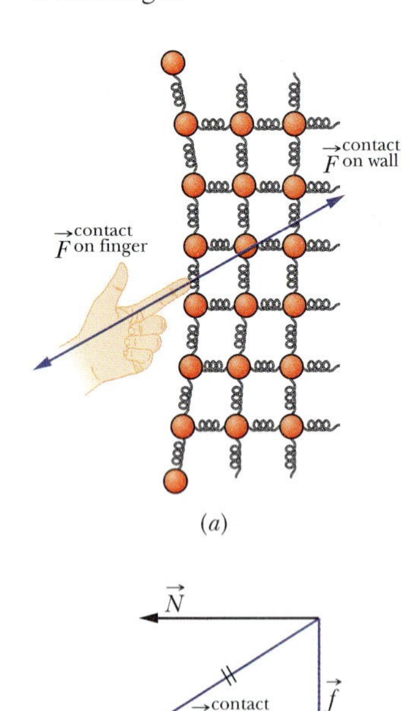


**FIGURE 6-8** ■ (*a*) Compressing an idealized solid surface with a contact force that is neither purely perpendicular nor purely parallel to the surface. (*b*) This force exerted by the wall on the finger can be decomposed into parallel and perpendicular components.

## The Normal Force

Let's consider a couple of situations in which normal forces are exerted on stationary blocks as shown in Fig. 6-9. The normal force exerted on one object by another object is always directed perpendicular to the surfaces that are in contact and away from the surface of the object exerting the force. We can use our idealized atomic model to explain this. The atoms at the surfaces of the objects that are in close contact interact so as to oppose being pushed closer together. As a result of Newton's Third Law, we can see that:

> When one body exerts a force with a component that is perpendicular to the surface of another body, the other body (even one with a seemingly rigid surface) deforms and pushes back on the first body with an opposing normal force $\vec{N}$ that is also perpendicular to the surfaces that are in contact.

**1. A Vertical Wall:** A block that is pushed against a wall experiences a normal force from the wall. An example of this is shown in Fig. 6-9*a*. Since the block is not moving,

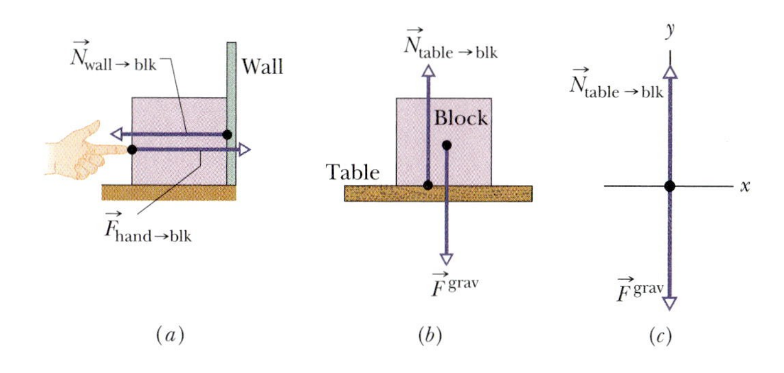


**FIGURE 6-9** ■ (*a*) A hand pushes a block into a wall with a force $\vec{F}_{hand \to blk}$. Since the block can't move, it compresses the wall, which pushes back on it with a normal force $\vec{N}_{wall \to blk}$. (*b*) A block resting on a tabletop experiences a normal force $\vec{N}$ perpendicular to the tabletop. (*c*) The corresponding free-body diagram for the block.

the net force on it must be zero. For now we will just consider the horizontal forces on the block given by

$$\vec{F}_{hand \to blk} - \vec{N}_{wall \to blk} = 0 \quad \text{or} \quad \vec{N}_{wall \to blk} = -\vec{F}_{hand \to blk} \qquad \text{(special case 1).} \qquad (6\text{-}7)$$

**2. A Horizontal Table:** Likewise, any object that rests on a table, shelf, or the ground near the Earth's surface experiences a normal force. Figure 6-9*b* shows an example. A block of mass $m$ lies on a table's horizontal surface. It is not moving in spite of the fact that it has a gravitational force $\vec{F}^{grav}$ on it due to the Earth. In other words, the block should fall but the table is in the way! We must conclude that if the block does not accelerate, the net force on the block must be zero,

$$\vec{F}^{net} = \vec{F}^{grav} + \vec{N}_{table \to blk} = 0 \quad \text{or} \quad \vec{N}_{table \to blk} = -\vec{F}^{grav}_{blk} \qquad \text{(special case 2).}$$

So the table must be pushing up on the block with normal force $\vec{N}_{table \to blk}$ that is equal to $-\vec{F}^{grav}_{blk}$. A free-body diagram for the block is shown in Fig. 6-9*c*. Forces $\vec{F}^{grav}$ and $\vec{N}_{table \to blk}$ are the only two forces on the block, and they are both vertical. We can write Newton's Second Law in terms of components along a positive upward $y$ axis.

The component, $F_y^{grav}$, of the gravitational force is $-mg$. So, if there is no vertical acceleration and no other vertical forces act on the object, the magnitude of the normal force on an object resting on a horizontal surface is $mg$. Since its direction is up, if we use the coordinates shown in Fig. 6-9*c*, then

$$\vec{N}_{table \to blk} = +mg\,\hat{j} \qquad \text{(special case 2).} \qquad (6\text{-}8)$$

**Single Normal Force as an Idealization:** The normal force exerted by the surface of the table on the block is actually the sum of billions of contact interactions between surface atoms in the table and block. However, the use of a single force vector to summarize external forces that act in the same direction as shown in Fig. 6-9 is a useful simplification. It is conventional to draw a single upward arrow at the point where the middle of the bottom surface of the block touches the table, as shown in Fig. 6-10.

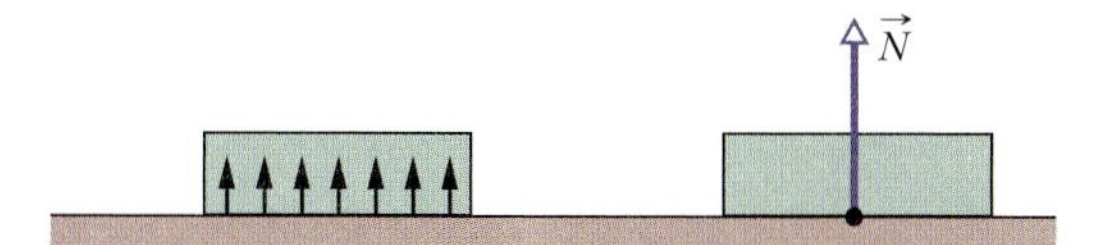


**FIGURE 6-10** ■ For simplification, many small force vectors supporting the bottom of the block are replaced by a single large force vector acting through the center of the block.

**Normal Force in an Elevator:** Suppose a block is placed in an elevator that is accelerating in an upward direction. How would that change the normal force it experiences? In Chapter 2, we discussed how a person riding in such an elevator would feel heavy while accelerating upward and feel light while accelerating downward (see Fig. 2-12). This brings us to the idea of *apparent weight*. A common bathroom scale reading is a measurement of the normal force exerted by the scale on your feet. In normal usage of the scale (in other words, you are standing still on the scale in a space that is not accelerating vertically), the scale will measure your weight. This is because the scale reading (normal force from the scale on your feet) is related to your weight through Newton's $F_y^{net} = ma_y$ relation.

**READING EXERCISE 6-4:** In Figure 6-9*b*, is the magnitude of the normal force $\vec{N}$ greater than, less than, or equal to $mg$ if the body and table are in an elevator that is moving upward (a) at constant speed, (b) at increasing speed, and (c) at decreasing speed? ■

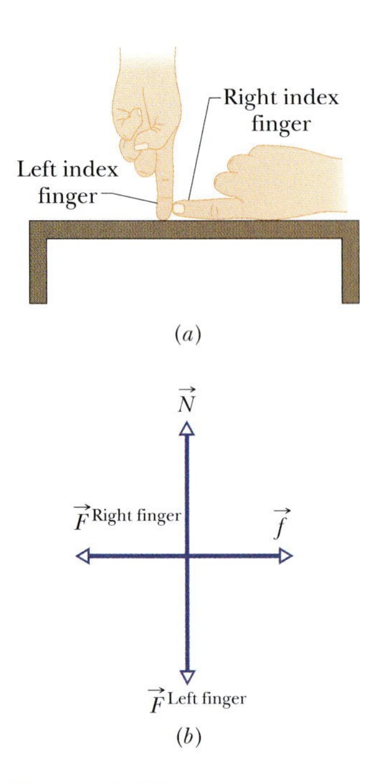


**FIGURE 6-11** ■ (*a*) Applying a horizontal force toward the left to an object (such as a fingertip) that is in contact with a surface. (*b*) A free-body diagram showing the forces on the left fingertip. If the object is not accelerating, there must be a friction force on it toward the right that is equal in magnitude and opposite in direction to the force applied on the left fingertip by the right index finger.

## The Friction Force Component

Let's consider the friction component of a general contact force. As we discussed earlier, this is the component of the contact force that is parallel to the surface. Suppose the tip of your finger is the object of interest. You would like to study the friction component of the contact force that a fairly smooth table can exert on your fingertip and how it might be related to the normal force. Tilt your left finger so it is vertical (is at an angle of about 90° with the horizontal). Try the following activities while maintaining the 90° angle with respect to the surface of the table:

**Activity 1:** Press on the table with your left index finger, first with a small force and then a larger force, and feel the increase in the normal force the table exerts on your fingertip.

**Activity 2:** Now take the index finger of your right hand and apply enough horizontal force to your left index finger so that it glides along at a constant velocity. (See Figure 6-11.) Is there a horizontal friction force acting? If so, why?

**NOTE:** In order to make your fingertip slide across the table at constant velocity, you must continually push on it in a horizontal direction. Can your applied force be the only horizontal force on your fingertip? No, because if it were, then your fingertip would accelerate. Thus, if we are not willing to give up on Newton's Second Law, we must assume that there is a second force, directed opposite to the applied force but with the same magnitude, so that the two forces balance out. This idea that a second force exists is represented in both Fig. 6-11 and by the following $x$-component equations:

$$F_x^{net} = F_x^{l\,finger} + f_x^{stat} = ma_x = 0 \quad \text{so} \quad F_x^{l\,finger} = -f_x.$$

Since both forces are purely horizontal, this gives us

$$\vec{F}^{l\,finger} = -\vec{f}. \tag{6-9}$$

**Activity 3:** What happens to the friction force when the normal force on your fingertip increases? Once again, adjust your constant applied force so that your left fingertip is moving at a constant velocity. Next, increase the normal force on your left fingertip just enough so your fingertip stops moving. Then get your left fingertip moving at a constant velocity again by applying more horizontal force with your right finger.

If you do Activity 3 carefully, you should conclude that the friction force on an object opposes the direction of its slipping over the surface and that it is greater when the normal force on the object becomes larger.

Contact friction forces are unavoidable in our daily lives. They are literally everywhere. If we were not able to counteract them, they would stop every moving object and bring to a halt every rotating shaft. On the other hand, if friction were totally absent, we could not walk, travel in a car, or ride a bicycle. In some cases, the effects of friction are very small compared to other forces and can be ignored. In other cases, to simplify a situation, friction is assumed to be negligible even though it may not really be. In either case, if the intention is to ignore the effects of friction, the interface between the object and the surface is called *frictionless*.

Contact friction depends on many factors, and it turns out that friction forces can behave very differently depending on the normal force, the nature of the surfaces that are in contact, and other factors. It is not always obvious when looking at surfaces whether the friction forces will be large or small. Sometimes smooth surfaces have greater friction forces than rough ones.

Understanding the relationship between friction forces and atomic and molecular interactions is a very active field of research in both physics and the engineering sciences. These relationships are not completely understood. Unlike Newton's laws of motion, which scientists believe hold to a high degree of accuracy when applied to everyday objects in our surroundings, some of the characteristics of friction that we describe here are only valid for certain common types of interacting surfaces. *Thus, the friction equations that we present here are sometimes useful approximations, but they do not always apply.*

In the next two subsections, we will explore some common characteristics of *kinetic friction*, in which one surface moves relative to another, and of *static friction*, in which the surfaces in contact are stationary relative to one another.

## Kinetic Friction Forces

Imagine that you give a book a quick push and send it sliding across a long horizontal countertop. As you expect, the book slows and then stops. We showed data on this behavior in Section 3-2. What does this observation tell us about the nature of the interaction between the book and the countertop? Based on our definition of velocity and on the data shown in Fig. 3-3, we suspect that the book has a *constant* acceleration. This acceleration is parallel to the surface, and in the direction opposite the book's velocity. Once again, we have no reason to believe that Newton's Second Law is not valid in this situation. Hence, from $F_x^{net} = ma_x$, we must assume that a contact friction force that is constant acts on the book in the same direction as the acceleration (parallel to the counter surface, in the direction opposite the book's velocity relative to the table) as is shown in Fig. 6-12.

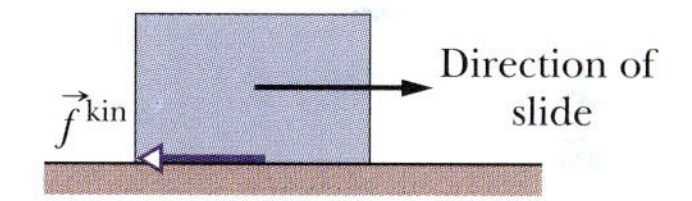


**FIGURE 6-12** ■ A friction force $\vec{f}^{kin}$ opposes the slide of a body over a surface.

In both the example of keeping your fingertip moving at a constant velocity and the example of watching a book with an initial velocity slide to a stop with a constant acceleration, an object is experiencing a **kinetic friction force** $\vec{f}^{kin}$. The word "kinetic" indicates that the object is moving relative to a surface. The phenomenon of "contact friction" can be explained by assuming that there is an attractive force between the atoms at the surfaces of the two objects. The attraction between two very smooth surfaces such as glass panes is consistent with this assumption and is known as **adhesion.**

What might the kinetic friction force depend on? Imagine sending an object sliding across a countertop as we discussed above. Would the book slow down more or less quickly if we slide the book across a carpeted floor instead of the smooth countertop? Would it slow down more or less quickly if we slide it across ice instead? Does the rate at which an object slows down seem to depend on its velocity? Would the book slow down more or less quickly if it has more mass or an additional applied downward force on it so the normal force between the surfaces is larger?

We can answer some of these questions for several situations by looking at the graph presented in Fig. 3-3. This graph shows the velocity as a function of time for three situations where objects slide to a stop on surfaces. You will likely find it helpful to refer back to that figure now (page 59). We can also draw inferences from the fingertip motions earlier in this section. Here are some observations and conclusions about kinetic friction.

**KINETIC FRICTION—SOME OBSERVATIONS AND CONCLUSIONS**

**OBSERVATION 1 ON THE INFLUENCE OF THE RELATIVE VELOCITY BETWEEN SURFACES:** The graphs in Fig. 3-3 tell us that in three situations involving different combinations of objects and surfaces, the objects all slow to a stop with constant acceleration and hence experience a constant kinetic friction force. *Conclusion: Kinetic friction forces appear to be independent of the magnitude of the velocity of the object relative to the surface over which the object is sliding, but act in a direction opposite to the direction of the velocity.*

**OBSERVATION 2 ON THE NATURE OF THE SLIDING SURFACES:** The graphs in Fig. 3-3 tell us that in three situations involving different combinations of object and surfaces, the rate of the stopping acceleration is different. *Conclusion: Kinetic friction forces appear to depend on the nature of the surfaces that are in contact with one another.*

**OBSERVATION 3 ON THE INFLUENCE OF THE NORMAL FORCE:** When you completed Activity 3 earlier in this section, you observed that the applied force needed to keep your fingertip moving at a constant velocity increases when the normal force on your fingertip becomes larger. *Conclusion: Kinetic friction forces appear to increase when the normal force on a sliding object increases and thus depend on how hard the objects are being pushed together.*

Is there a mathematical relationship between the magnitude of the kinetic friction force on an object and the magnitude of the normal force the object experiences? A plausible relationship would be that these two force magnitudes are proportional to each other. Let's look at the results of a simple experiment in which we can measure the kinetic friction force as a function of the normal force on a sliding block. In this experiment, we use a spring scale to measure how much horizontal force we need to apply to pull a wooden block along at a constant velocity. (See Fig. 6-13.) We can determine the magnitude of the friction force by using the fact that it must be equal to the magnitude of the applied force if the moving block doesn't accelerate (Eq. 6-9). If the table surface is horizontal, and the only other vertical force on the block is the gravitational force, then the normal force is given by $\vec{N} = mg\hat{j}$ (Eq. 6-8). That is, $N_y = mg$. The normal force can be changed by piling more mass on the block. We can then measure the kinetic friction force again.

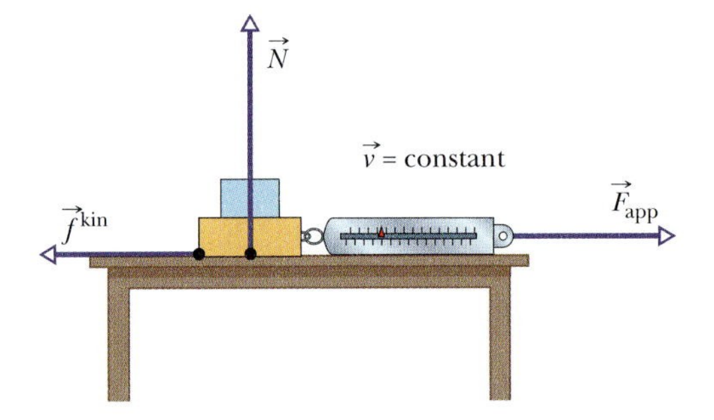


**FIGURE 6-13** ■ A block is pulled along at a constant velocity with a horizontal applied force, measured by a spring scale. This force is countered by a kinetic friction force of the same magnitude. Since the tabletop is horizontal, the magnitude of the normal force on the block is equal to the product of its mass $m$ and the gravitational acceleration constant $g$.

The data shown in Fig. 6-14 reveal that for a Velcro-covered wood block sliding on a Formica table surface, the magnitude of the friction force is proportional to that of the normal force with a constant of proportionality given by $\mu^{kin} = 0.21$. Turning the block on its side to reduce the area in contact does not affect this constant of proportionality.

Results similar to those shown in Fig. 6-14 for many situations reveal that the magnitude of the friction force for dry sliding is usually proportional to the magnitude of the normal forces pressing surfaces together and does not depend on other factors. Thus, for the purposes of the systems we will deal with in this book, the magnitude of the kinetic friction force, $\vec{f}^{kin}$, can be expressed as

$$f^{kin} = \mu^{kin} N, \tag{6-10}$$

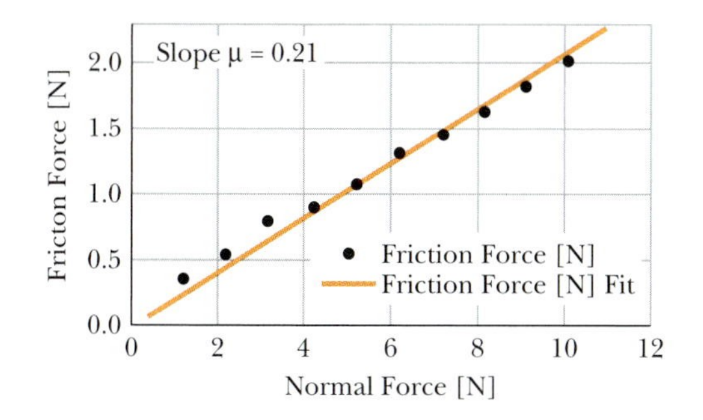


**FIGURE 6-14** ■ A graph of data showing that when a block is pulled along at a constant velocity, the magnitude of the kinetic friction force is directly proportional to the normal force exerted by the surface it slides over.

where $\mu^{\text{kin}}$ is the slope of the linear graph that relates the magnitude of the kinetic friction force $f^{\text{kin}}$ and the magnitude of the normal force $N$. The slope $\mu^{\text{kin}}$ is called the **coefficient of kinetic friction.**

The coefficient $\mu^{\text{kin}}$ is a dimensionless scalar that must be determined experimentally. Its value depends on certain properties of both the body and the surface. Hence, the coefficients are usually referred to with the preposition "between," as in "the value of $\mu^{\text{kin}}$ *between* a book and countertop is 0.04, but the value *between* rock-climbing shoes and rock is as much as 0.9." Based on our observations, we assume that the value of $\mu^{\text{kin}}$ does not depend on the speed at which the body slides along the surface. Note that $f^{\text{kin}} = \mu^{\text{kin}}N$ is *not* a vector equation. The direction of $\vec{f}^{\text{kin}}$ is always parallel to the surface and *opposes* the sliding motion.

## Static Friction Forces

Do friction forces continue to act on an object once it stops sliding? The answer to this question is more complicated than simply "yes" or "no." Start out by imagining a large, heavy box sitting on a horizontal, carpeted floor. You push on the box, but the box does not move. Unless we are to believe that Newton's Second Law ($\vec{F}^{\text{net}} = m\vec{a}$) is not valid in this situation, we must assume that there is some other force acting on the box that is counteracting the application of the push force. That is, there must be a force acting in the opposite direction that is exactly equal in magnitude to the push force. We will call this opposing force a **static friction force.** The word "static" is used to signify that the object is not moving relative to the surface as shown in Fig. 6-15*b-d*.

Now imagine that you push even harder on the box as shown in Fig. 6-15*c* and *d*. The box still does not move. Apparently the friction force can change in magnitude, otherwise it would no longer balance your applied force. In other words, if you push on an object in an attempt to slide it across a surface and the object does not slide, then we know that there is a static friction force. This force acts in the direction opposite the push with a matching magnitude, regardless of how hard you push. If you stop pushing on the box, that oppositely directed force must disappear. How do we know this? Because if you removed the push force, and the static friction force *did not* disappear as well, then the box would accelerate in the direction of the friction force. We know from everyday observation that this does not happen. So the static friction force appears to be a very strange force that changes magnitude in response to other forces.

This situation is in no way specific to the example of the box on carpet. At the interface between any two solids prior to slipping, the static friction force starts at zero when no applied force is present and increases as the force that tends to produce slipping increases. The static friction force adjusts in magnitude to exactly counteract the applied force (usually a push or pull) at every instant. The static friction force mirrors the applied force. If the applied force is zero, then the static friction force is zero. If the applied force has a horizontal component that is 10 N, the static friction force has a horizontal component that is 10 N. We call forces that behave like the static friction force **passive forces.** Passive forces are forces that change in magnitude in response to other forces.

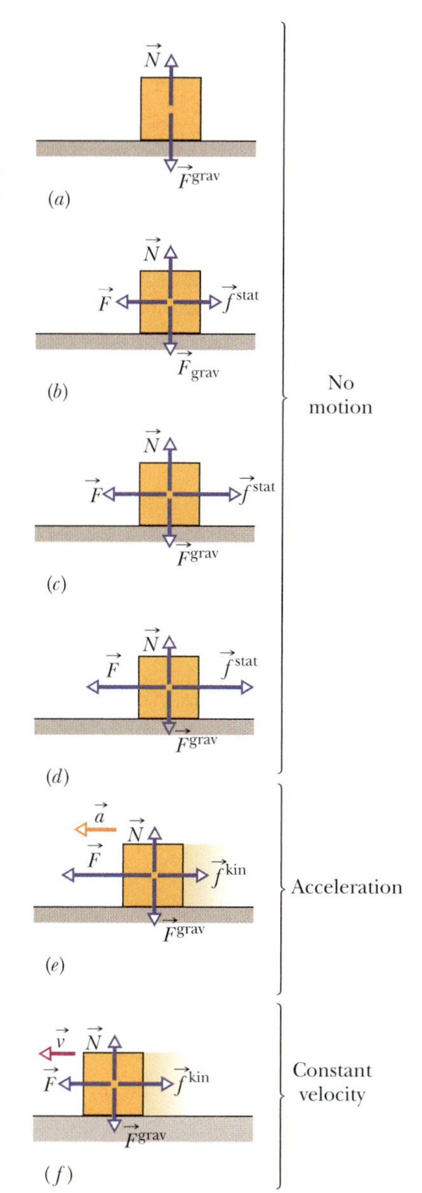


**FIGURE 6-15** ■ (*a*) There are no horizontal forces on a stationary block. (*b–d*) An external force $\vec{F}$ applied to the block is balanced by a static friction force $\vec{f}^{\text{stat}}$. As $\vec{F}$ is increased, $\vec{f}^{\text{stat}}$ also increases, until $\vec{f}^{\text{stat}}$ reaches a certain maximum value. (*e*) The block then "breaks away," accelerating suddenly in the direction of $\vec{F}$. (*f*) If the block is now to move with constant velocity, the magnitude $|\vec{F}|$ of the applied force must be reduced from the maximum value it had just before the block broke away.

Now imagine that you push on the box with all your strength. Finally, the box begins to slide. Evidently, there is a maximum magnitude of the static friction force. When you exceed that maximum magnitude, your push force is larger than the opposing static friction force and the box accelerates in the direction of your push. A typical sequence of static frictional force responses to applied forces is shown in Fig. 6-15. This sequence is consistent with the experimental results obtained when an electronic force sensor is used to monitor the force on a block as a function of time. The experimental setup is shown in Fig. 6-16 and a graph of the results is shown in Fig. 6-17.

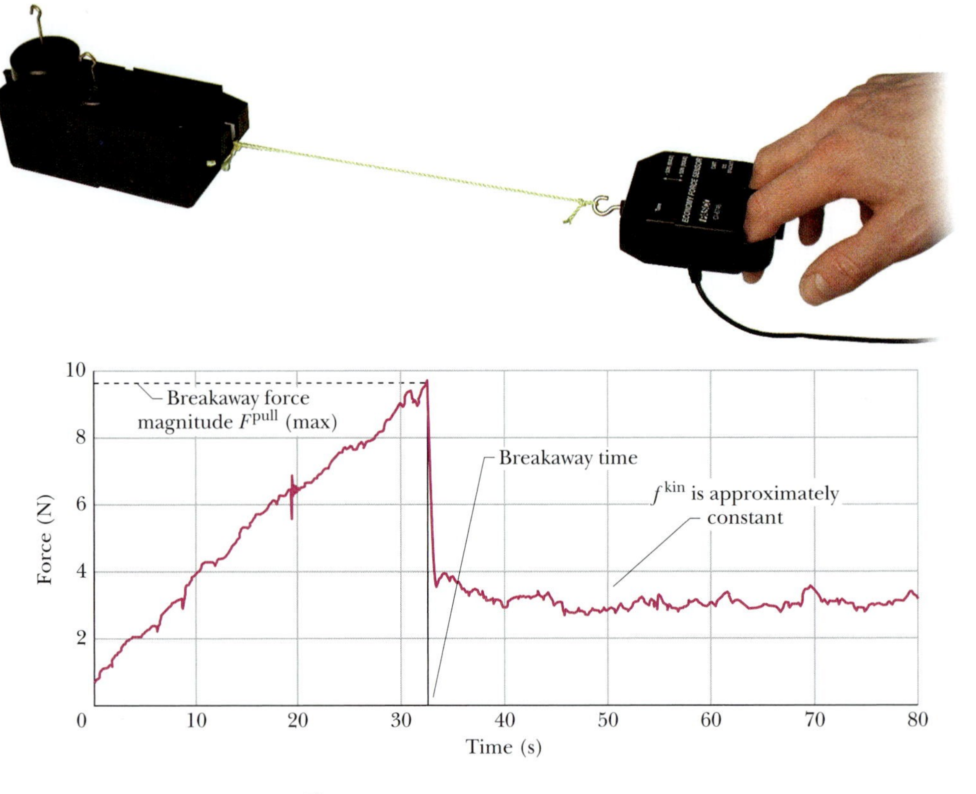


**FIGURE 6-16** ■ The apparatus used for the static friction experiment includes an electronic force sensor attached to a computer data acquisition system (not pictured).

**FIGURE 6-17** ■ Graph of the magnitude of the static friction force on a wooden block as a function of time. This force opposes a steadily increasing applied force between 0.0 s and 32 s. At 32 s the block suddenly "breaks away" and starts moving. At about 40 s, it starts moving at a steady velocity as a kinetic friction force with a magnitude that is less than the static force starts acting.

As the pulling force, $\vec{F}^{\text{pull}}$, increases, the block remains at rest. Then, when a "breakaway" force is reached, it moves very suddenly. That is, the magnitude of the friction force, $\vec{f}^{\text{stat}}$, keeps increasing to oppose the pulling force in accordance with Newton's Second Law until the object "breaks free" and starts to move. Hence, we express the magnitude of the static friction force as

$$f^{\text{stat}} = |\vec{f}^{\text{stat}}| \leq \mu^{\text{stat}} N, \tag{6-11}$$

where $\mu^{\text{stat}}$ is known as the **coefficient of static friction** and $N$ is the magnitude of the normal force on the body from the surface. Just as for kinetic friction, the coefficient $\mu^{\text{stat}}$ is dimensionless and determined experimentally. Its value depends on certain properties of both the body and the surface, and so is referred to with the preposition "between."

Usually, the magnitude of the kinetic friction force, which acts when there is motion, is less than the maximum magnitude of the static friction force, which acts when there is no motion. We see this in the data shown in Fig. 6-17. Thus, if you wish the block to move across the surface with a constant speed, you must usually decrease the magnitude of the applied force once the block begins to move, as in Fig. 6-15*f*. Another common behavior for a certain range of applied forces is to see slip-and-stick behavior in which an object breaks away, slides to a stop, breaks away again, and so on. We will not deal with the slip-stick phenomenon in this book.

**READING EXERCISE 6-5:** Figure 6-17 shows the result of the experiment in which a 295.6 g block with a 500 g mass on it is pulled along a table with a steadily increasing force until it breaks away at $t = 32$ s. (a) What is the coefficient of static friction, $\mu^{\text{stat}}$, between the table and the mass? (b) What is the coefficient of kinetic friction $\mu^{\text{kin}}$? ■

**READING EXERCISE 6-6:** A block lies on a floor. (a) What is the magnitude of the friction force exerted on it by the floor if the block is not being pushed? (b) If a horizontal force of 5 N is now applied to the block, but the block does not move, what is the magnitude of the friction force on it? (c) If the maximum value $\vec{f}^{\,\text{max}}$ of the static friction force on the block is 10 N, will the block move if the magnitude of the horizontally applied force is 8 N? (d) If the magnitude is 12 N? (e) What is the magnitude of the friction force in part (c)? ■

**READING EXERCISE 6-7:** Discuss and explain the following statement using the terms related to friction forces that are presented above: "If we were not able to counteract them, frictional forces would stop every moving object and bring to a halt every rotating shaft. On the other hand, if friction were totally absent, we could not walk or ride a bicycle." ■

## Tension

So far we considered contact forces between objects that are not attached and that can be pulled apart fairly easily. Let's consider one more type of contact force—a force that occurs when a long thin object such as a rod or string is attached to other objects at each of its ends. For example, consider a leash with a dog straining at one end and the dog's owner pulling the other end, a handle bolted to a pot that is too massive to move and is being pulled by a cook, or a string with one end attached to a ceiling and the other end attached to a hanging mass. In all three cases, a long narrow object that is stretched is transmitting forces from an object at one of its ends to an object at its other end. We say that a long narrow object that is being pulled taut by opposing forces is under **tension.** In order to use Newton's laws of motion to analyze the forces and motions of the objects that are attached to the ends of strings or rods, we need to understand more about the phenomenon of tension.

What do we observe about tension? Let's consider a stationary rubber band that connects two force probes like that shown in Fig. 6-18. We observe that the forces the rubber band exerts on the force probes at each end have the same magnitude but act in opposite directions.

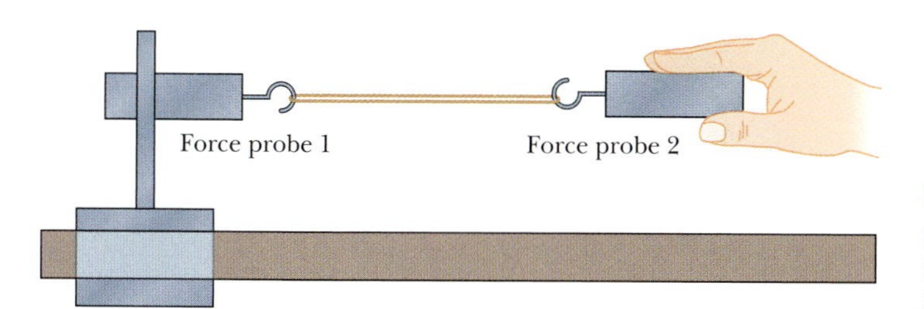


**FIGURE 6-18** ■ A rubber band is connecting two force probes. Each probe detects the same magnitude of force, but the force on probe 1 is in the opposite direction of the force on probe 2. This observation is not surprising as it is entirely consistent with Newton's laws.

We also observe that the tension force is present everywhere along the rubber band. Although it is not readily observable, the tension everywhere along the rubber band is in fact equal in magnitude to the applied forces at the ends that caused the rubber band or string to stretch. Thus, when a taut rubber band (that is not accelerating) is attached to an object, it exerts a tension force on the object that is directed along the rubber band and away from the object. This tension expresses itself as a pulling force, but only at the ends of the rubber band. These same observations hold for most long thin connectors including strings, cords, and ropes.

**An Atomic Model for Tension** Our simple model of solid matter, as consisting of atoms connected by springs, is very helpful in understanding how objects that are under tension can transmit forces. Suppose a very, very thin string, having only one strand of atoms, is connected by small interatomic springs. Figure 6-19*a* shows the natural length of the string. Figure 6-19*b* shows the string when it is extended by equal and opposite forces applied to its ends so that it is not accelerating.

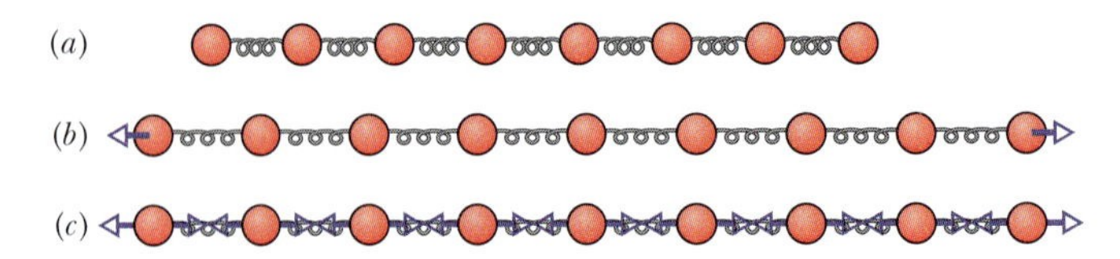


**FIGURE 6-19** ■ A string is idealized as a line of atoms with springs representing the mutual interaction forces between them.

In our idealization, we have assumed only one strand of atoms. Obviously, real strings, cords, and ropes have many strands of molecules consisting of complex arrays of atoms. Although many strands will make a string or rope stronger, it will not change the ideas presented in our simple model.

Assuming that our ideal string is not accelerating, each atom must have zero net force on it. The atom on the left end of the string must be experiencing an attractive force from the neighboring atom to its right that is equal in magnitude and opposite in direction to the applied force on the end. However, each stretched spring represents a force of interaction between neighboring atoms that must obey Newton's Third Law. Thus, the leftmost atom must be exerting an attractive force on its neighboring atom that is "equal and opposite" to the force that atom exerts on it. These pairs of mutual interaction forces exist throughout the string, as shown in Fig. 6-19*c*. The magnitude of these interaction forces each atom experiences has been given a special name. It is called the *tension in the string*, which we denote as $T$. In contrast to the tension force, tension, which we often denote with a $T$, is a scalar quantity that is always positive with no inherent direction associated with it. Hence, we will often denote a tension force $\vec{T}$ that points (for example) in the positive $y$ direction as $\vec{T} = +T\hat{j}$ and one that points in a negative direction as $\vec{T} = -T\hat{j}$.

Next, let's use Newton's laws to examine the effect of tension associated with the motion of a skier being towed by a snowmobile by means of a nylon cord as depicted in Fig. 6-20. We consider two situations—one in which the system is not accelerating and the other in which it is. The snowmobile moves forward when its treads, which are turning, dig into the snow and push against it. However, assume for now that the runners on the skis and those at the front of the snowmobile experience no friction forces.

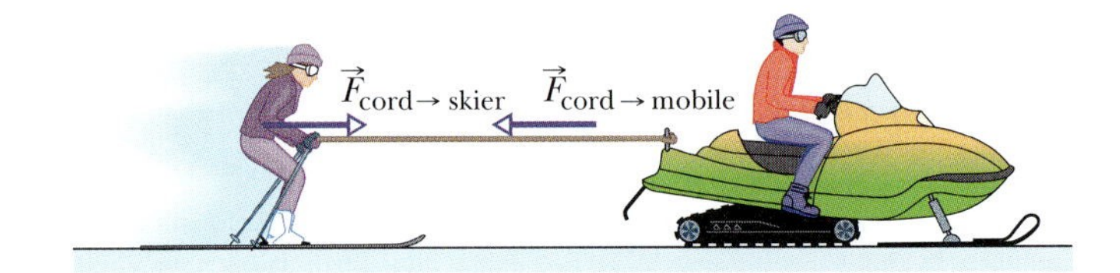


**FIGURE 6-20** ■ The cord connecting a skier and a snowmobile exerts oppositely directed forces on the skier and the snowmobile.

**Tension for a Nonaccelerating System** Remember that if the system is not accelerating, then it moves at a constant velocity. Furthermore, Newton's Third Law tells us that the force between any two objects in the system that are in contact is equal and opposite. For example, at the left end of the cord, the skier feels a pulling force acting along the direction of the cord, which we denote as $\vec{F}_{\text{cord}\rightarrow\text{skier}}$, and the cord experiences an oppositely directed force from the skier, denoted by $\vec{F}_{\text{skier}\rightarrow\text{cord}}$. A similar situation applies to the interaction forces at the right end of the cord, so that

$$\vec{F}_{\text{cord}\rightarrow\text{skier}} = -\vec{F}_{\text{skier}\rightarrow\text{cord}} \quad \text{and} \quad \vec{F}_{\text{cord}\rightarrow\text{mobile}} = -\vec{F}_{\text{mobile}\rightarrow\text{cord}}. \tag{6-12}$$

But we already know from Newton's Second Law for $\vec{a} = 0$ that the net force on the cord must be zero. Since the net force is zero, $F_x^{net}$ (the sum of the $x$-components of the forces) must be zero. Hence,

$$F_{\text{cord } x}^{\text{net}} = F_{\text{skier}\rightarrow\text{cord } x} + F_{\text{mobile}\rightarrow\text{cord } x} = 0 \quad \text{or} \quad F_{\text{skier}\rightarrow\text{cord } x} = -F_{\text{mobile}\rightarrow\text{cord } x}.$$

The forces the skier and snowmobile exert on the cord are purely horizontal. So we have

$$\vec{F}_{\text{skier}\rightarrow\text{cord}} = -\vec{F}_{\text{mobile}\rightarrow\text{cord}}. \tag{6-13}$$

This result agrees with the observation reported in Fig. 6-18. Namely, forces exerted by the ends of a taut cord have the same magnitude. Even more significantly, we can combine Eqs. 6-12 and 6-13 to show that

$$\vec{F}_{\text{skier}\rightarrow\text{mobile}} = -\vec{F}_{\text{mobile}\rightarrow\text{skier}}.$$

When nonaccelerating objects are connected by a string, cord, or rope, they interact in accordance with Newton's Third Law *through the connector* as if they are in direct contact.

**An Accelerating System** Suppose the snowmobile driver pushes in his throttle and increases his velocity at a constant rate. Now the system has an acceleration $\vec{a}$, and the cord connecting the skier to the snowmobile must experience the same acceleration. In terms of the $x$-components we get

$$F_{\text{cord } x}^{\text{net}} = F_{\text{skier}\rightarrow\text{cord } x} + F_{\text{mobile}\rightarrow\text{cord } x} = m_{\text{cord}}\, a_x. \tag{6-14}$$

This tells us that if the cord has a nonzero mass, then the force of the snowmobile on the right end of the cord must be greater than the force on the left end to maintain an acceleration. However, in many situations, including this one showing the snowmobile pulling a skier, the mass of the cord is so much less than the mass of the entire system that it can be taken to be zero.

Taking the direction of motion to be along the positive $x$ axis, we can write the tension forces in terms of the positive scalar $T$ representing the tension,

$$\vec{F}_{\text{skier}\rightarrow\text{cord}} = -T_L\hat{\text{i}} \quad \text{and} \quad \vec{F}_{\text{mobile}\rightarrow\text{cord}} = +T_R\hat{\text{i}}, \tag{6-15}$$

where $T_L$ is the tension on the left side of the cord and $T_R$ is the tension on the right side of the cord. Then we can rewrite Eq. 6-14 in terms of the tension difference and the $x$-component of acceleration to get $T_R - T_L = m_{\text{cord}}\, a_x$. But the snowmobile force, which serves to accelerate the entire system, is given by $F_{\text{mobile}\rightarrow\text{sys } x} = m^{\text{tot}} a_x$. Solving the last two equations for $a_x$ and rearranging terms gives us the ratio

$$\frac{T_L - T_R}{F_{\text{mobile}\rightarrow\text{sys } x}} = \frac{m_{\text{cord}}}{m^{\text{tot}}}. \tag{6-16}$$

Let's consider the implications of this equation. In most situations, the mass of the cord is much less than the mass of the system. Whenever that is true, the difference in tension at the ends of the cord is much smaller than the force that accelerates the system. For example, assume the skier's mass and the snowmobile's mass together total 200 kg, and the mass of the cord she grips is 1 kg. The ratio of these masses gives us only a 0.5% difference in tension forces at the ends of the cord. For most everyday purposes, this force difference at the ends of an accelerating cord is negligible. In laboratory experiments, masses of between 100 g and 2 kg are typically connected by

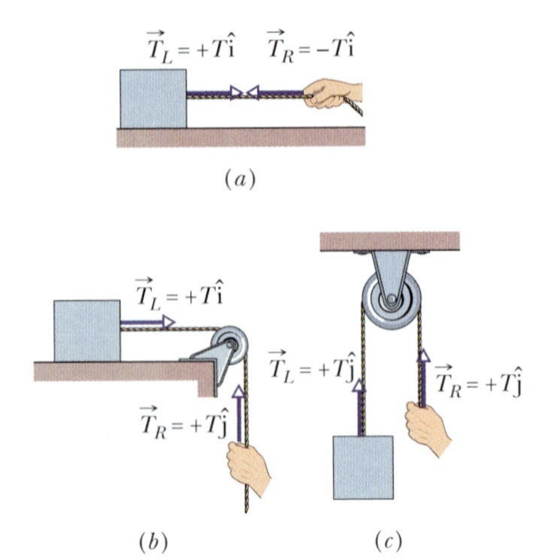


**FIGURE 6-21** ■ (*a*) The cord, pulled taut, is under tension. If its mass is negligible, it pulls on the body and the hand with force of magnitude $T = |\vec{T}|$, even if it runs around a massless, frictionless pulley as in (*b*) and (*c*).

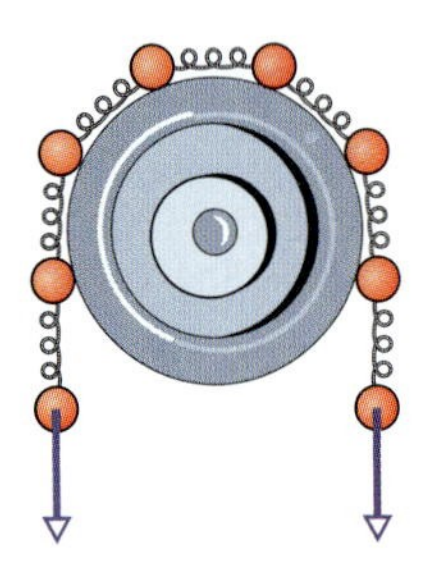

**FIGURE 6-22** ■ Tension in a taut string still exists even when it undergoes direction changes.

fishing line capable of sustaining tensions of well over 100 N. A 1 m length of this type of fishing line has a mass of about 0.25 g, so the force differences are usually less than 1%. For cases where the mass of a connecting cord is very small compared to the masses of the objects attached to its ends, we can assume that the tension is *essentially* the same at all points along the cord. When we can legitimately make this simplifying approximation, we say that we have a **massless string.**

**Pulleys and Direction Change** What happens if a "massless" cord stretched over a "massless" pulley changes direction as shown in Fig. 6-21*b* and *c*? Is the tension still the same everywhere in the cord? Let's examine our atomic model. If a string is wrapped around a pulley, its direction is different at one end than at the other. However, each tiny segment of the string only changes direction ever so slightly. The direction change is less than it is in Fig. 6-22 where we have only placed eight atoms in the chain. We conclude that the magnitude of the tension forces that are spread throughout the string also do not change significantly when the string bends around other objects. This conclusion is supported by experiments in which spring scales are inserted in various places along a string that bends while it is under tension.

Any solid object that is attached at two ends and pulled can transmit tension forces from one end to another. Some objects are quite elastic, such as rubber bands or weak springs, others are more rigid, such as strings and rods. Small rubber bands, light-duty springs, and strings cannot stand compressive forces. They are so long and narrow that they buckle under compression. Alternatively, rods and heavy springs do not buckle under compression. In some of the analyses that follow, you will be dealing with "massless" strings and springs that buckle under compression forces.

**READING EXERCISE 6-8:** Consider Figure 6-21*c* and assume that the pulley is massless but the cord is *not*. Is the magnitude of the pull force on the cord exerted by the hand equal to (=), less than (<), or greater than (>) the magnitude of the pull force exerted by the block when the block is moving upward (a) at constant speed, (b) at increasing speed, and (c) at decreasing speed? Explain. ■

## TOUCHSTONE EXAMPLE 6-2: Einstein's Elevator

In Fig. 6-23*a*, a passenger of mass $m = 72.2$ kg stands on a platform scale in an elevator cab. We are concerned with the scale readings when the cab is stationary and when it is moving up or down.

(a) Find a general solution for the scale reading, whatever the vertical motion of the cab happens to be.

**SOLUTION** ■ One **Key Idea** here is that the scale reading is equal to the magnitude of the normal force $\vec{N}$ the scale exerts on the passenger. The only other force acting on the passenger is the gravitational force $\vec{F}^{\text{grav}}$, as shown in the free-body diagram of the passenger in Fig. 6-23*b*.

A second **Key Idea** is that we can relate the forces on the passenger to the acceleration $\vec{a}$ of the passenger with Newton's Second Law ($\vec{F}^{\text{net}} = m\vec{a}$). However, recall that we can use this law only in an inertial frame. If the cab accelerates, then it is *not* an inertial frame. So we choose the ground to be our inertial frame and make any measure of the passenger's acceleration relative to it.

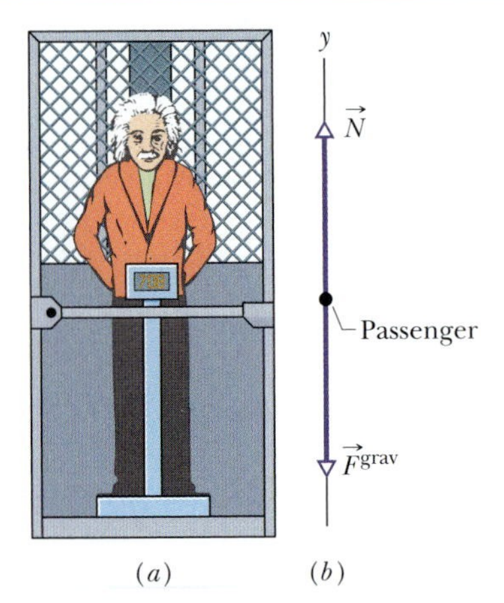


**FIGURE 6-23** ■ (*a*) A passenger stands on a platform scale that indicates his weight or apparent weight. (*b*) The free-body diagram for the passenger, showing the normal force $\vec{N}$ on him from the scale and the gravitational force $\vec{F}^{\text{grav}}$.

Because the two forces on the passenger and the passenger's acceleration are all directed vertically, along the $y$ axis shown in Fig.

6-23*b*, we can use Newton's Second Law written for *y*-components ($F_y^{net} = ma_y$) to get

$$N_y + F_y^{grav} = ma_y$$

or

$$N_y = -F_y^{grav} + ma_y. \quad (6\text{-}17)$$

This tells us that the scale reading, which is equal to $N_y$ (provided $N_y \geq 0$), depends on the vertical acceleration $a_y$ of the cab. Since $\vec{F}^{grav} = -mg\hat{j}$, the *y*-component of the gravitational force, $F_y^{grav} = -mg$. This gives us

$$N_y = m(g + a_y). \quad \text{(Answer)} \quad (6\text{-}18)$$

This tells us that the scale reading is larger than the passenger's static weight, *mg*, when the elevator accelerates upward, since then $a_y > 0$. But if the elevator is accelerating *downward*, then $a_y$ is negative and the scale reads less than the passenger's static weight. This is true, as long as the downward acceleration is smaller than *g*. If the downward acceleration is greater than *g*, $(g + a_y)$ in Eq. 6-18 is a negative value. In that case, $N_y = 0$, since $N_y$ can never be negative. (Why not?)

(b) What does the scale read if the cab is stationary or moving upward at a constant 0.50 m/s?

**SOLUTION** ■ The **Key Idea** here is that for any constant velocity (zero or otherwise), the acceleration $a_y$ of the passenger is zero. Substituting this and other known values into Eq. 6-18, we find

$$N_y = (72.2 \text{ kg})(9.8 \text{ m/s}^2 + 0) = 708 \text{ N}. \quad \text{(Answer)}$$

This is just the weight of the passenger and is equal to the magnitude $F^{grav}$ of the gravitational force on him.

(c) What does the scale read if the cab accelerates upward at 3.20 m/s² and downward at 3.20 m/s²?

**SOLUTION** ■ For $a_y = +3.20 \text{ m/s}^2$, Eq. 6-18 gives

$$N_y = (72.2 \text{ kg})(9.8 \text{ m/s}^2 + 3.20 \text{ m/s}^2)$$
$$= 939 \text{ N}, \quad \text{(Answer)}$$

and for $a_y = -3.20 \text{ m/s}^2$, it gives

$$N_y = (72.2 \text{ kg})(9.8 \text{ m/s}^2 - 3.20 \text{ m/s}^2)$$
$$= 477 \text{ N}. \quad \text{(Answer)}$$

So for an upward acceleration (either the cab's upward speed is increasing or its downward speed is decreasing), the scale reading is greater than the passenger's weight. Similarly, for a downward acceleration (either the cab's upward speed is decreasing or its downward speed is increasing), the scale reading is less than the passenger's weight.

(d) During the upward acceleration in part (c), what is the magnitude $F^{net}$ of the net force on the passenger, and what is the magnitude $a_{p,cab}$ of the passenger's acceleration as measured in the frame of the cab? Does $\vec{F}^{net} = m\vec{a}_{p,cab}$?

**SOLUTION** ■ One **Key Idea** here is that the magnitude $F^{grav}$ of the gravitational force on the passenger does not depend on the motion of the passenger or the cab, so from part (b), $F^{grav}$ is 708 N. From part (c), the magnitude *N* of the normal force on the passenger during the upward acceleration is the 939 N reading on the scale. Thus, the net force on the passenger is

$$F_y^{net} = N_y + F_y^{grav} = N - F^{grav} = 939 \text{ N} - 708 \text{ N} = 231 \text{ N}, \quad \text{(Answer)}$$

during the upward acceleration. However, the acceleration $\vec{a}_{p,cab}$ of the passenger relative to the frame of the cab is zero. Thus, in the noninertial frame of the accelerating cab, $\vec{F}^{net}$ is not equal to $m\vec{a}_{p,cab}$. This is an example of the fact that Newton's Second Law does not hold in noninertial (that is, accelerating) frames of reference.

## TOUCHSTONE EXAMPLE 6-3: Pulling a Block

In Fig. 6-24*a*, a hand *H* pulls on a taut horizontal rope *R* (of mass $m = 0.200$ kg) that is attached to a block *B* (of mass $M = 5.00$ kg). The resulting acceleration $\vec{a}$ of the rope and block across the frictionless surface has constant magnitude 0.300 m/s² and is directed to the right. We will call this the positive direction for the *x* axis. Note that this rope is not "massless;" we return to this feature in part (d).

(a) Identify all the third-law force pairs for the horizontal forces in Fig. 6-24*a* and show how the vectors in each pair are related.

**SOLUTION** ■ The **Key Idea** here is that a third-law force pair arises when two bodies interact; the forces of the pair are equal in magnitude and opposite in direction, and the force on each body is due to the other body. The "exploded view" of Fig. 6-24*b* shows

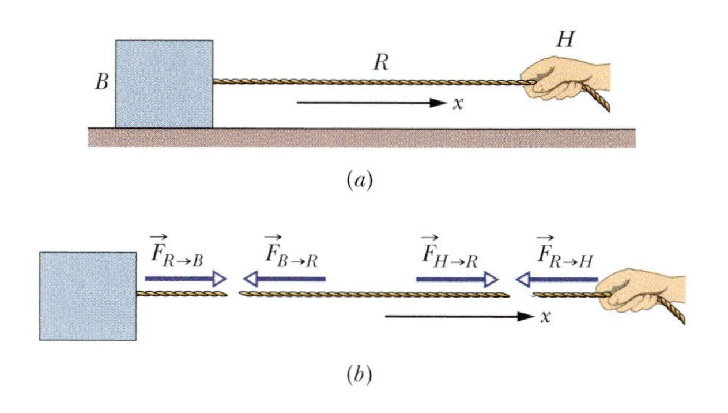


**FIGURE 6-24** ■ (*a*) Hand *H* pulls on rope *R*, which is attached to block *B*. (b) An exploded view of block, rope, and hand, with the forces between block and rope and between rope and hand.

that here there are two such force pairs for the horizontal forces. At the hand-rope boundary, we have the force $\vec{F}_{H\to R}$ exerted by the hand on the rope and the force $\vec{F}_{R\to H}$ exerted by the rope on the hand. These forces are a Newton's Third Law force pair and so are equal in magnitude and opposite in direction. They are related by

$$\vec{F}_{H\to R} = -\vec{F}_{R\to H}. \quad \text{(Answer)}$$

Similarly, at the rope–block boundary we have

$$\vec{F}_{R\to B} = -\vec{F}_{B\to R}. \quad \text{(Answer)}$$

(b) What is the magnitude of the force $\vec{F}_{R\to B}$ that the rope exerts on the block?

**SOLUTION** ■ We know that the block has an acceleration $\vec{a}$ in the positive direction of the $x$ axis. The only force acting on the block along that axis is $\vec{F}_{R\to B}$. The **Key Idea** here is that we can relate force $\vec{F}_{R\to B}$ to acceleration $\vec{a}$ by Newton's Second Law. Because both vectors are along the $x$ axis, we use the $x$ component version of the law ($F_x^{\text{net}} = ma_x$), writing

$$F_{R\to B\,x} = Ma_x.$$

Substituting known values, we find that the magnitude of $\vec{F}_{R\to B}$, which we denote $F_{R\to B}$ and is equal to $F_{R\to B\,x}$, is

$$F_{R\to B} = (5.00 \text{ kg})(0.300 \text{ m/s}^2) = 1.50 \text{ N}. \quad \text{(Answer)}$$

(c) What is the magnitude of the force $\vec{F}_{B\to R}$ that the block exerts on the rope?

**SOLUTION** ■ From (a), we know that $\vec{F}_{B\to R} = -\vec{F}_{R\to B}$, so $\vec{F}_{B\to R}$ has the magnitude

$$F_{B\to R} = F_{R\to B} = 1.50 \text{ N}. \quad \text{(Answer)}$$

(d) What is the magnitude of the force $\vec{F}_{H\to R}$ that the hand exerts on the rope?

**SOLUTION** ■ A **Key Idea** here is that, with the rope taut, the rope and block form a system on which $\vec{F}_{H\to R}$ acts. The mass of the system is $m + M$. For this system, Newton's Second Law for $x$-components gives us

$$\begin{aligned} F_{H\to R\,x} &= (m + M)a_x \\ &= (0.200 \text{ kg} + 5.00 \text{ kg})(0.300 \text{ m/s}^2) \\ &= 1.56 \text{ N} \quad \text{(Answer)} \quad (6\text{-}19) \end{aligned}$$

Now note that the magnitude of the force $\vec{F}_{H\to R}$ on the rope from the hand (1.56 N) is greater than the magnitude of the force $\vec{F}_{R\to B}$ on the block from the rope [1.50 N, from part (b) above]. The reason is that $\vec{F}_{R\to B}$ must accelerate only the block but $\vec{F}_{H\to R}$ must accelerate both the block and the rope, and the rope's mass $m$ is not negligible. If we let $m \to 0$ in Eq. 6-19, then we find 1.50 N, the same magnitude as at the other end. We often assume that an interconnecting rope is massless so that we can approximate the forces at its two ends as having the same magnitude.

## TOUCHSTONE EXAMPLE 6-4: Three Cords

In Fig. 6-25*a*, a block $B$ of mass $M = 15$ kg hangs by a cord from a knot $K$ of mass $m_K$, which hangs from a ceiling by means of two other cords. The cords have negligible mass, and the magnitude of the gravitational force on the knot is negligible compared to the gravitational force on the block. What are the tensions in the three cords?

**SOLUTION** ■ Let's start with the block because it has only one attached cord. The free-body diagram in Fig. 6-25*b* shows the forces on the block: gravitational force $\vec{F}^{\text{grav}}$ (with a magnitude of $Mg$) and force $\vec{T}_C$ from the attached cord. A **Key Idea** is that we can relate these forces to the acceleration of the block via Newton's Second Law ($\vec{F}^{\text{net}} = m\vec{a}$). Because the forces are both vertical, we choose the vertical component version of the law, $F_y^{\text{net}} = ma_y$, and write

$$F_y^{\text{net}} = T_{C\,y} + F_y^{\text{grav}} = T_C - Mg = Ma_y.$$

Substituting 0 for the block's acceleration $a_y$, we find

$$T_{C\,y} - Mg = M(0) = 0.$$

This means that the two forces on the block are equal in magnitude. Substituting for $M$ (= 15.0 kg) and $g$ and solving for $T_{C\,y}$ yields

$$T_{C\,y} = 147 \text{ N}. \quad \text{(Answer)}$$

*Note*: Although $\vec{T}_C$ and $\vec{F}^{\text{grav}}$ are equal in magnitude and opposite in direction, they are *not* a Newton's Third Law force pair. Why?

We next consider the knot in the free-body diagram of Fig. 6-25*c*, where the negligible gravitational force on the knot is not included. The **Key Idea** here is that we can relate the three other forces acting on the knot to the acceleration of the knot via Newton's Second Law ($\vec{F}^{\text{net}} = m\vec{a}$) by writing

$$\vec{T}_A + \vec{T}_B + \vec{T}_C = m_K\vec{a}_K.$$

Substituting 0 for the knot's acceleration $\vec{a}_K$ yields

$$\vec{T}_A + \vec{T}_B + \vec{T}_C = 0, \quad (6\text{-}20)$$

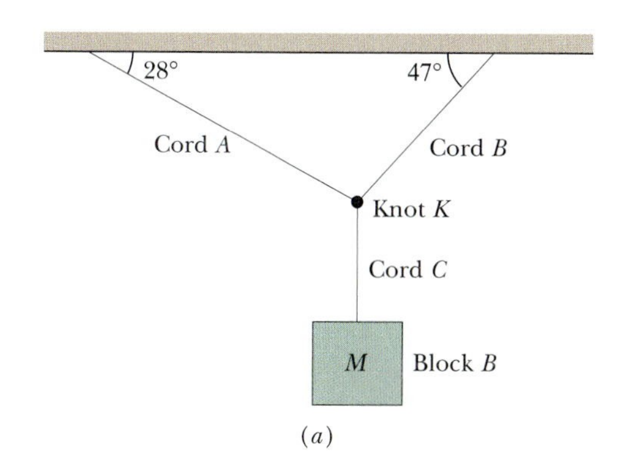


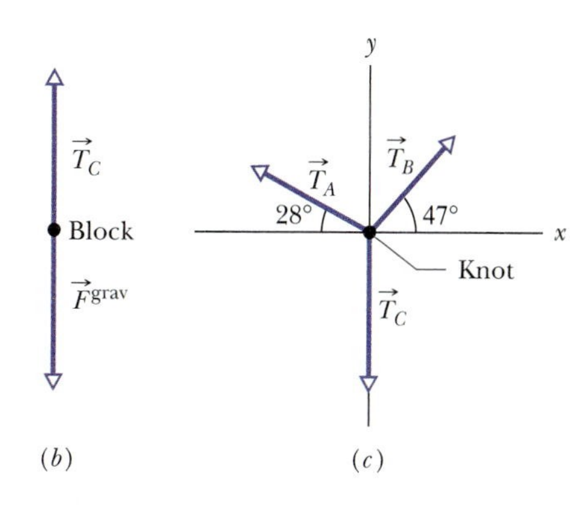


**FIGURE 6-25** ■ (*a*) A block of mass *m* hangs from three cords. (*b*) A free-body diagram for the block. (*c*) A free-body diagram for the knot at the intersection of the three cords.

which means that the three forces on the knot are in equilibrium. Although we know both magnitude and angle for $\vec{T}_C$, we know only the angles and not the magnitudes for $\vec{T}_A$ and $\vec{T}_B$. With unknowns in two vectors, we cannot solve Eq. 6-20 for $\vec{T}_A$ or $\vec{T}_B$ directly on a vector-capable calculator. Instead, we rewrite Eq. 6-20 in terms of components along the *x* and *y* axes. For the *x* axis, we have

$$T_{Ax} + T_{Bx} + T_{Cx} = 0,$$

which, using the given data, yields

$$-|\vec{T}_A|\cos 28° + |\vec{T}_B|\cos 47° + 0 = 0, \qquad (6\text{-}21)$$

or alternatively

$$|\vec{T}_A|\cos 152° + |\vec{T}_B|\cos 47° + 0 = 0.$$

Similarly, for the *y* axis we rewrite Eq. 6-20 as

$$T_{Ay} + T_{By} + T_{Cy} = 0$$

or

$$|\vec{T}_A|\sin 28° + |\vec{T}_B|\sin 47° - |\vec{T}_C| = 0.$$

Substituting our previous result for $T_C$ then gives us

$$|\vec{T}_A|\sin 28° + |\vec{T}_B|\sin 47° - 147\text{ N} = 0. \qquad (6\text{-}22)$$

We cannot solve Eq. 6-21 or Eq. 6-22 separately because each contains two unknowns, but we can solve them simultaneously because they contain the same two unknowns. Doing so (either by substitution, by adding or subtracting the equations appropriately, or by using the equation-solving capability of a calculator), we discover

$$|\vec{T}_A| = 104\text{ N} \quad \text{and} \quad |\vec{T}_B| = 134\text{ N}. \qquad \text{(Answer)}$$

Thus, the magnitudes of the tensions in the cords are 104 N in cord *A*, 134 N in cord *B*, and 147 N in cord *C*.

## 6-5 Drag Force and Terminal Speed

If you are riding in a car and put your hand out the window, you feel nothing when the car is first starting up. But as you speed up, the forces on your hand become larger and larger. The force you feel on your hand is called **air drag.** The magnitude of the air drag increases as the velocity of your hand relative to the air increases. Air drag is another common force, but it is only important when an object is moving relatively rapidly.

Air is a fluid. A **fluid** is anything that can flow—generally either a gas or a liquid. When there is a relative velocity between a fluid and a body (either because the body moves through the fluid or because the fluid moves past the body), the body experiences a **drag force** $\vec{D}$ that opposes the relative motion and points in the direction in which the fluid flows relative to the body. Like contact forces, air drag forces are ultimately the result of billions of tiny electromagnetic forces between air molecules and another object.

Here we examine only cases in which air is the fluid, the body is blunt (like your hand or a baseball) rather than slender (like a javelin), and the relative motion is fast enough so that the air becomes turbulent (breaks up into swirls) behind the body. In such cases, experiments reveal that the magnitude $D = |\vec{D}|$ of the drag force is re-

**FIGURE 6-26** ■ This skier crouches in an "egg position" to minimize her effective cross-sectional area and thus the air drag acting on her.

lated to the relative speed $v = |\vec{v}|$ by an experimentally determined **drag coefficient** $C$ according to

$$D = \tfrac{1}{2}C\rho A v^2, \tag{6-23}$$

where $\rho$ is the air density (mass per volume) and $A$ is the **effective cross-sectional area** of the body (the area of a cross section taken perpendicular to the velocity $\vec{v}$). The drag coefficient $C$ (typical values range from 0.4 to 1.0) is not truly a constant for a given body, because if $v$ varies significantly, the value of $C$ can vary as well. Here, we ignore such complications.

Downhill speed skiers know well that drag depends on the cross-sectional area ($A$) and speed squared ($v^2$). To reach high speeds a skier must reduce the drag force as much as possible by, for example, riding the skis in the "egg position" (Fig. 6-26) to minimize cross-sectional area $A$.

When a blunt body falls from rest through air, the drag force $\vec{D}$ is directed upward; its magnitude gradually increases from zero as the speed of the body increases. This upward force $\vec{D}$ opposes the downward gravitational force, $\vec{F}^{\text{grav}} = -mg\,\hat{j}$, on the body. We can relate these forces to the body's acceleration by writing Newton's Second Law in terms of vector components for a vertical $y$ axis ($F_y^{\text{net}} = ma_y$),

$$F_y^{\text{net}} = (D_y + F_y^{\text{grav}}) = (+D - mg) = ma_y, \tag{6-24}$$

where $m$ is the mass of the body. Experience tells us that $D$ increases as the velocity of the falling object relative to the air increases. As suggested in Fig. 6-27, if the body falls long enough the force magnitudes, $D$ and $F^{\text{grav}}$, eventually equal each other as shown in Fig. 6-27*c*. According to Eq. 6-24, when this happens $a_y = 0$, and the body's speed no longer increases. The body then falls at a constant speed, called the *terminal speed* $v_t$. To find the terminal speed, we set $a_y = 0$ in Eq. 6-24 and use that relation for the magnitude of the drag force given by $D = \tfrac{1}{2}C\rho A v^2$ (Eq. 6-23). Then the terminal speed is given by

$$v_t = \sqrt{\frac{2mg}{C\rho A}}. \tag{6-25}$$

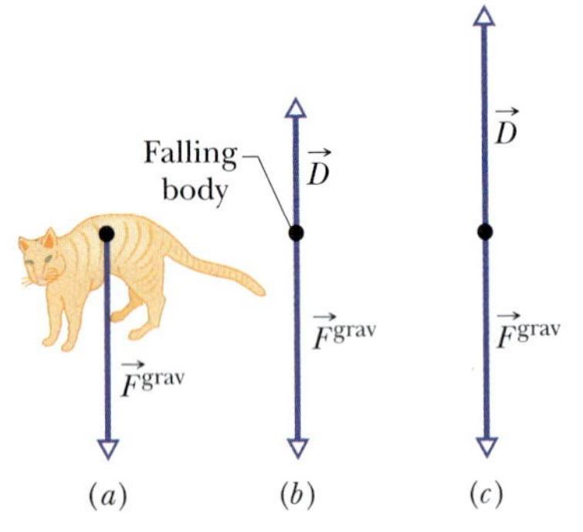


**FIGURE 6-27** ■ The forces that act on a body falling through air: (*a*) the body when it has just begun to fall and (*b*) the free-body diagram a little later, after a drag force has developed. (*c*) The drag force has increased until it balances the gravitational force on the body. The body now falls at its constant terminal speed.

Table 6-1 gives values of the terminal speed for some common objects.

According to calculations* based on the assumption that $D = \tfrac{1}{2}CA\rho v^2$, a cat must fall about six floors to reach terminal speed. Until it does so, $mg > D_y$ and the cat ac-

**TABLE 6-1**
**Some Terminal Speeds in Air**

| Object | Terminal Speed (m/s) | 95% Distance[a] (m) |
|---|---|---|
| Shot (from shot put) | 145 | 2500 |
| Sky diver (typical) | 60 | 430 |
| Baseball | 42 | 210 |
| Tennis ball | 31 | 115 |
| Basketball | 20 | 47 |
| Ping-Pong ball | 9 | 10 |
| Raindrop (radius = 1.5 mm) | 7 | 6 |
| Parachutist (typical) | 5 | 3 |

[a]This is the distance through which the body must fall from rest to reach 95% of its terminal speed.

*Source:* Adapted from Peter J. Brancazio, *Sport Science* New York: Simon & Schuster (1984).

celerates downward because of the net downward force. Recall from Chapter 2 that your body is an accelerometer, not a speedometer. Because the cat also senses the acceleration, it is frightened and keeps its feet underneath its body, its head tucked in, and its spine bent upward, making its cross-sectional area ($A$) small, so its terminal speed $v_t$ becomes relatively large. If the cat maintains this position, it could be injured on landing. However, if the cat shown at the top of the chapter opening photo reaches $v_t$, its acceleration vanishes so it relaxes, stretching its legs and neck horizontally outward and straightening its spine (it then resembles a flying squirrel). These actions increase its area $A$ and hence the magnitude of the drag force $D_y$ acting on it. The cat begins to slow its descent because now the magnitude of its upward drag force is greater than the downward gravitational force. Eventually, a new, smaller terminal velocity is reached. The decrease in terminal velocity reduces the possibility of serious injury on landing. Just before hitting the ground, the cat pulls its legs back beneath its body to prepare for the landing.

Humans often fall from great heights for fun when sky diving. However, in April 1987, during a jump, sky diver Gregory Robertson noticed that fellow sky diver Debbie Williams had been knocked unconscious in a collision with a third sky diver and was unable to open her parachute. Robertson, who was well above Williams at the time and who had not yet opened his parachute for the 4 km plunge, reoriented his body head-down to minimize his cross-sectional area and maximize his downward speed. Reaching an estimated terminal velocity of 320 km/h, he caught up with Williams and then went into a horizontal "spread eagle" (as shown in Fig. 6-28) to increase his drag force. He could then grab her. He opened her parachute and then, after releasing her, his own, a scant 10 s before impact. Williams received extensive internal injuries due to her lack of control on landing but survived.

**FIGURE 6-28** ■ A sky diver in a horizontal "spread eagle" maximizes the air drag.

**READING EXERCISE 6-9:** Near the ground, is the speed of large raindrops greater than ($>$), less than ($<$), or equal to ($=$) the speed of small raindrops? Assume that all raindrops are spherical and have the same drag coefficient $C$. **Beware!** More than one factor is involved. ■

## 6-6 Applying Newton's Laws

Now that you have learned about several types of forces that can act on an object, you have the basic knowledge needed to analyze the accelerations and forces experienced by bodies in an interacting system. However, you will need to use your knowledge in an organized fashion to predict how a system will move or to identify unknown forces based on observations of system motions.

There are several key steps that we suggest you use in performing an analysis. These steps are an extension of those presented in Sections 3-7 and 6-2. The steps are outlined in more detail in Touchstone Example 6-5:

1. Construct a diagram of the system you wish to analyze.
2. Isolate the bodies of interest in the system on your diagram. Identify the types, directions, and approximate magnitudes of the forces acting on each body. Label the forces to indicate the type of force ($\vec{F}^{\text{grav}}$, $\vec{N}$, $\vec{f}$, $\vec{T}$).
3. Construct a free-body diagram representing each body as a point. Place the tails of the labeled force vectors for that body at its point. If possible, show the angles these vectors make with respect to each other as well as the relative magnitudes of the vectors.

*W. O. Whitney and C. J. Mehlhaff, "High-Rise Syndrome in Cats," *The Journal of the American Veterinary Medical Association*, 1987, Vol. 191, pp. 1399–1403.

4. Predict the direction of the acceleration and draw a special acceleration vector in that direction and label it with $\vec{a}$. Then choose a coordinate system so that one axis lies parallel to the direction of the predicted acceleration.

5. Write down Newton's Second Law in vector form for each body in the system. Then decompose the vectors into a pair of one-dimensional equations for each body,

$$\vec{a} = \frac{1}{m}\vec{F}^{\,net} \Rightarrow \qquad a_x = \frac{1}{m}F_x^{\,net} \quad \text{and} \quad a_y = \frac{1}{m}F_y^{\,net}.$$

Remember that we drop the vector notation (arrows) when we write the one-dimensional equations. These equations associate the components of vectors.

6. Solve the set of equations for each dimension ($x$ and $y$) separately to find the unknown vector components.

In Touchstone Example 6-5 we show how these six steps can be used to find the forces that act on a block of known mass as it slides up an incline.

## TOUCHSTONE EXAMPLE 6-5: Sliding Up a Ramp

Figure 6-29 shows a 300 g block on a 30° incline. The block is moving up the incline at a constant velocity because a string that passes over a pulley is attached to a falling mass. We assume that the mass of the string and pulley are negligible and that there is no friction in the pulley and no friction between the incline and the block. (a) What force is the string exerting on the block during the time that the block is moving up the plane at constant velocity? (b) What is the normal force that the incline is exerting on the block?

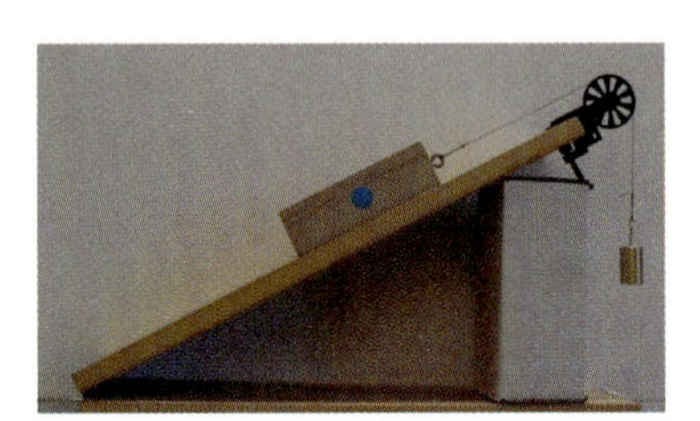

**FIGURE 6-29** ■ Photograph of a block on an inclined plane that is moving at a constant velocity.

**SOLUTION** ■ The **Key Idea** is that because the block is not accelerating, the net force on it must be zero (according to Newton's Second Law). If we follow the steps outlined in Section 6-6 we can identify the forces on the block, choose a coordinate system, and decompose the vectors into components. Since the components along each axis must add up to zero, we can solve our equations for the magnitude and direction of the force of the string on the block.

**Step One: Construct a Diagram of the System** Figure 6-30*a* shows the essential features of the system of interest needed to answer question (a) including the incline, the block, and the string pulling on the block. The figure is more abstract than the photograph of Fig. 6-29.

**Step Two: Isolate the Objects of Interest and Identify the Forces** There is only one object of interest in this problem—the block.

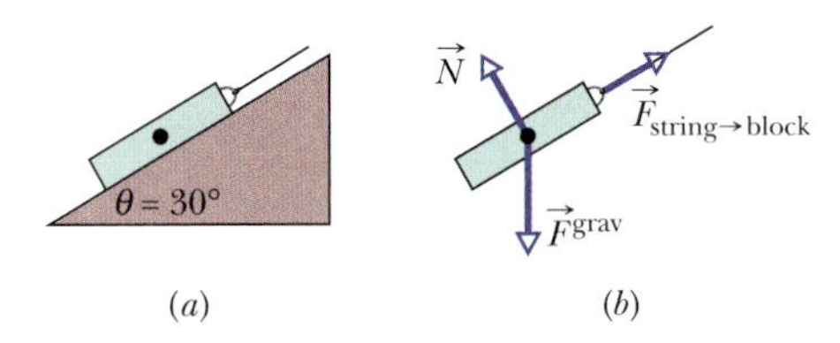


**FIGURE 6-30** ■ (*a*) Step one sketch of just those parts of the system of interest for solving the problem. (*b*) Step two sketch of just the block and the forces acting on it with labels.

Thus, we only need to diagram and identify the forces on it. There are three forces acting on the block. First, there is the gravitational force that the Earth exerts on the block that acts vertically downward. Next, there is the normal force that is at right angles (normal) to the surface of the incline. Finally, there is the tension force along the direction of the string that is exerted on the upper end of the block. These forces are shown in Fig. 6-30*b*. *Note*: Although each bit of mass on the block is being pulled downward by the Earth, we can idealize this force and assume it acts at the center of the block. Likewise we assume that the normal force exerted on the block by the inclined plane surface acts like a single force at the middle of the surface of the block that is in contact with the incline. We realize it is the vector sum of billions of smaller normal vectors acting at all points along the surface of contact of the block.

**Step Three: Construct a Free-Body Diagram** To analyze a system using Newton's Second Law, we draw a free-body diagram for each object in our system. Usually the object experiencing forces is represented by a dot. Then, a vector representing each force that acts on that object is drawn with its tail on the dot. Each vector should be pointing in the direction of the particular force being represented. Also, if the relative magnitudes of the forces are known, the

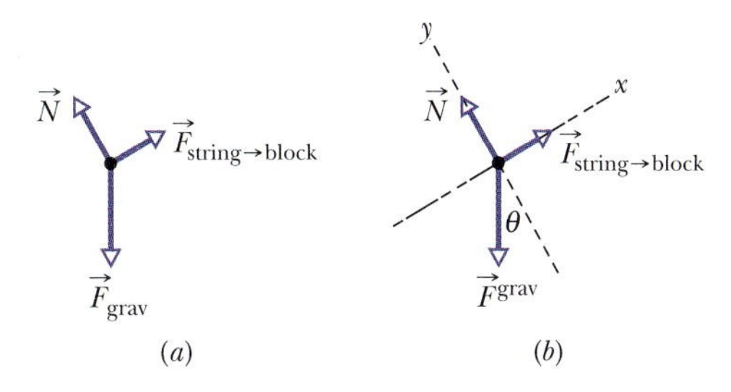


**FIGURE 6-31** ■ Steps three and four free-body diagrams for the forces on the block: (*a*) without a coordinate system and (*b*) with a coordinate system.

lengths of the vectors should represent those magnitudes. In this example, we only need a free-body diagram for one object—the block. A clearly labeled arrow showing the predicted direction of the acceleration of the object should also be included. We have no acceleration in this case, so no acceleration vector is included. The free-body diagram for the block is shown in Fig. 6-31*a*.

**Step Four: Predict the Direction of the Acceleration and Choose a Coordinate System** In choosing the coordinate system for this particular situation, it is useful to break away from our standard practice of having the $y$ axis be a vertical axis and the $x$ axis be horizontal. In general, it is helpful to have one of the axes chosen so it is in the direction of either the acceleration of the object of interest or the forces we are trying to find. One force on the block points up the incline (the string force). Another force is perpendicular to the incline (the normal force). Let's choose "up the incline" as the direction of the positive $x$ axis and a $y$ axis that is perpendicular to the incline (shown in Fig. 6-31*b*). In this coordinate system, only the gravitational force vector will need to be decomposed. Note that using a standard coordinate system would not be incorrect, just less convenient.

We can use some basic geometry to convince ourselves that the gravitational force vector makes an angle of 30° with respect to the negative $y$ axis.

**Step Five: Apply Newton's Second Law and Decompose the Force Vectors** Recall that the block is moving with constant velocity so the vector sum of the forces acting on it must be zero. Thus we can write

$$\vec{F}^{\,\text{net}} = m\vec{a} = 0 \quad \text{so} \quad \vec{F}^{\,\text{grav}} + \vec{N} + \vec{F}_{\text{string}\to\text{block}} = 0.$$

But in order for $\vec{F}^{\,\text{net}} = 0$ we must have $F_x^{\,\text{net}} = 0$ and $F_y^{\,\text{net}} = 0$. Therefore,

$$F_x^{\,\text{grav}} + N_x + F_{\text{string}\to\text{block}\,x} = 0 \quad \text{and}$$

$$F_y^{\,\text{grav}} + N_y + F_{\text{string}\to\text{block}\,y} = 0.$$

Recall that here $F_x^{\,\text{grav}}$ denotes the $x$-component of the gravitational force, $N_x$ denotes the $x$-component of the normal force, and so on. These components are not vectors and so do not have arrows above them. The only vector that needs decomposition is the gravitational force vector. This decomposition is shown in Fig. 6-32. Since we know by inspection that the gravitational force components are negative, they are expressed with explicit signs as

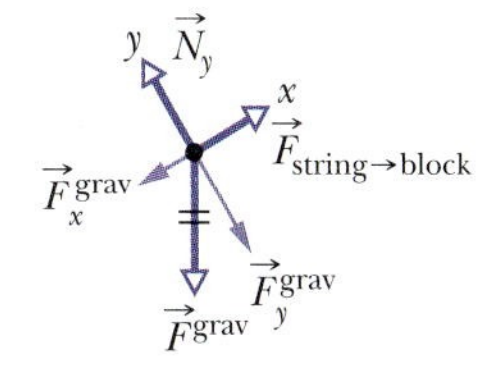


**FIGURE 6-32** ■ Decomposition of the gravitational force vector into components along the chosen $x$ and $y$ axes.

$$F_x^{\,\text{grav}} = -F^{\text{grav}} \sin\theta \quad \text{and} \quad F_y^{\,\text{grav}} = -F^{\text{grav}} \cos\theta.$$

The angle $\theta$ between the downward-pointing force vector and the $y$ axis is 30°. By inspecting the diagram, we see that the normal force vector points along the positive $y$ axis and the tension force vector points along the positive $x$ axis. So, these vectors can be written as

$$\vec{N} = +N\hat{\text{j}} \quad \text{and} \quad \vec{F}_{\text{string}\to\text{block}} = +T\hat{\text{i}},$$

where $T$ is a positive scalar representing the tension in the string and $N$ is a positive scalar representing the magnitude of the normal force. Our expression for $F_x^{\,\text{net}} = 0$ then becomes $+T - F^{\text{grav}} \sin\theta = 0$. Our expression for $F_y^{\,\text{net}} = 0$ then becomes $N - F^{\text{grav}} \cos\theta = 0$. Thus,

$$T = mg\sin\theta,$$

and

$$N = mg\cos\theta.$$

We know how to find the values of the gravitational force components in terms of the mass of the block $m$, the local gravitational strength constant $g$, and the angle $\theta$:

$$T = mg\sin\theta = 0.300\text{ kg} \times 9.8\text{ m/s}^2 \times \sin 30° = 1.47\text{ N}$$

$$N = mg\cos\theta = 0.300\text{ kg} \times 9.8\text{ m/s}^2 \times \sin 30° = 2.55\text{ N}.$$

Finally, rounding to 2 significant figures gives us

$$\vec{F}_{\text{string}\to\text{block}} = +T\hat{\text{i}} = +(1.5\text{ N})\hat{\text{i}} \qquad \text{(Answer)}$$

$$\vec{N} = +(2.5\text{ N})\hat{\text{j}}. \qquad \text{(Answer)}$$

*A final note*: This example shows the basics for a relatively simple analysis. If we had taken friction into account and picked a part of the motion that is accelerated, the problem would have been more complicated. However, the basic steps would be exactly the same. To master the techniques of analysis for more complex situations, you will also need to study the rest of the of touchstone examples in this chapter.

## TOUCHSTONE EXAMPLE 6-6: Breaking Loose

Figure 6-33*a* shows a coin of mass $m$ at rest on a book that has been tilted at an angle $\theta$ with the horizontal. By experimenting, you find that when $\theta$ is increased to 13°, the coin is on the *verge* of sliding down the book, which means that even a slight increase beyond 13° produces sliding. What is the coefficient of static friction $\mu^{\text{stat}}$ between the coin and the book?

**SOLUTION** ■ If the book were frictionless, the coin would surely slide down it for any tilt of the book because of the gravitational force on the coin. Thus, one **Key Idea** here is that a frictional force $f^{\text{stat}}$ must be holding the coin in place. A second **Key Idea** is that, because the coin is *on the verge* of sliding *down* the book, that force is at its *maximum* magnitude $f^{\max}$ and is directed *up* the book. Also, from Eq. 6-11, we know that $f^{\max} = \mu^{\text{stat}} N$, where $N$ is the magnitude of the normal force $\vec{N}$ on the coin from the book. Thus,

$$f^{\max} = \mu^{\text{stat}} N,$$

from which

$$\mu^{\text{stat}} = \frac{f^{\text{stat}}}{N}. \tag{6-26}$$

To evaluate this equation, we need to find the force magnitudes $f^{\text{stat}}$ and $N$. To do that, we use another **Key Idea:** When the coin is on the verge of sliding, it is stationary and thus its acceleration $\vec{a}$ is zero. We can relate this acceleration to the forces on the coin with Newton's Second Law ($\vec{F}^{\text{net}} = m\vec{a}$). As shown in the free-body diagram of the coin in Fig. 6-33*b*, these forces are (1) the frictional force $\vec{f}^{\text{stat}}$, (2) the normal force $\vec{N}$, and (3) the gravitational force $\vec{F}^{\text{grav}}$ on the coin, with magnitude equal to $mg$. Then, from Newton's Second Law with $\vec{a} = 0$, we have

$$\vec{f}^{\text{stat}} + \vec{N} + \vec{F}^{\text{grav}} = 0. \tag{6-27}$$

To find $f^{\text{stat}}$ and $N$, we rewrite Eq. 6-27 for components along the $x$ and $y$ axes of the tilted coordinate system in Fig. 6-33*b*. For the $x$ axis and with $mg$ substituted for $|\vec{F}^{\text{grav}}|$, we have

$$f_x^{\text{stat}} + N_x + F_x^{\text{grav}} = f_x^{\text{stat}} + 0 - mg \sin\theta = 0,$$

so

$$f_x^{\text{stat}} = f^{\text{stat}} = +mg \sin\theta. \tag{6-28}$$

Similarly, for the $y$ axis we have

$$f_y^{\text{stat}} + N_y + F_y^{\text{grav}} = 0 + N - mg\cos\theta = 0,$$

so

$$N = +mg\cos\theta. \tag{6-29}$$

Substituting Eqs. 6-28 and 6-29 into Eq. 6-26 produces

$$\mu^{\text{stat}} = \frac{mg\sin\theta}{mg\cos\theta} = \tan\theta, \tag{6-30}$$

which here means

$$\mu^{\text{stat}} = \tan 13° = 0.23. \quad \text{(Answer)}$$

Actually, you do not need to measure $\theta$ to get $\mu^{\text{stat}}$. Instead, measure the two lengths shown in Fig. 6-33*a* and then substitute $h/d$ for $\tan\theta$ in Eq. 6-30.

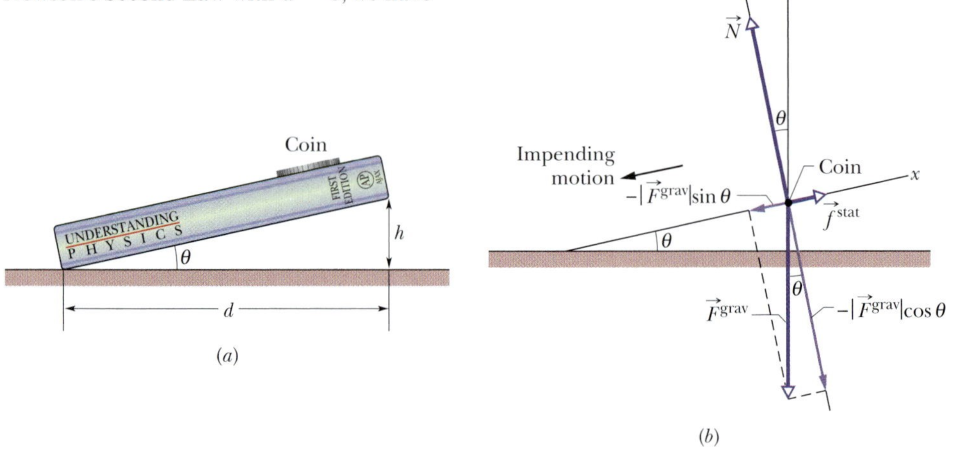


**FIGURE 6-33** ■ (*a*) A coin on the verge of sliding down a book. (*b*) A free-body diagram for the coin, showing the three forces (drawn to scale) that act on it. The gravitational force $\vec{F}^{\text{grav}}$ is shown resolved into its components along the $x$ and the $y$ axes, whose orientations are chosen to simplify the problem. Component $F_x^{\text{grav}} = -|\vec{F}^{\text{grav}}|\sin\theta$ tends to slide the coin down the book. Component $F_y^{\text{grav}} = -|\vec{F}^{\text{grav}}|\cos\theta$ presses the coin onto the book.

## TOUCHSTONE EXAMPLE 6-7: Accelerated by Friction

A 40 kg slab rests on a frictionless floor. A 10 kg block rests on top of the slab (Fig. 6-34). The coefficient of static friction $\mu^{\text{stat}}$ between the block and the slab is 0.60, whereas their kinetic friction coefficient $\mu^{\text{kin}}$ is 0.40. The 10 kg block is pulled by a horizontal force with a magnitude of 100 N. What are the resulting accelerations of (a) the slab and (b) the block?

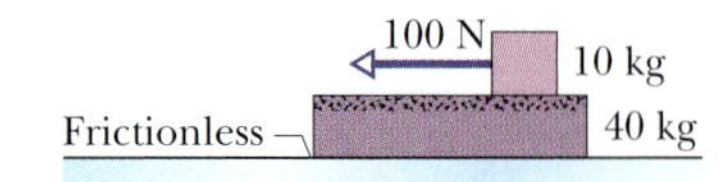


**FIGURE 6-34**

**SOLUTION** ■ The first **Key Idea** here is that we should apply Newton's Second Law *separately* to the slab ($m_{slab}$ = 40 kg) and to the block ($m_{block}$ = 10 kg) to obtain the acceleration of each:

$$\vec{a}_{slab} = \frac{\vec{F}^{net}_{slab}}{m_{slab}} \quad \text{and} \quad \vec{a}_{block} = \frac{\vec{F}^{net}_{block}}{m_{block}}.$$

To find the net force on each of these objects, we can draw a free-body diagram for each, as shown in Fig. 6-35. The direction of the frictional force from the slab on the block, $\vec{f}_{slab \to block}$, is determined by considering the direction of the block's impending motion. The direction of the frictional force from the block on the slab, $\vec{f}_{block \to slab}$, is inferred from Newton's Third Law.

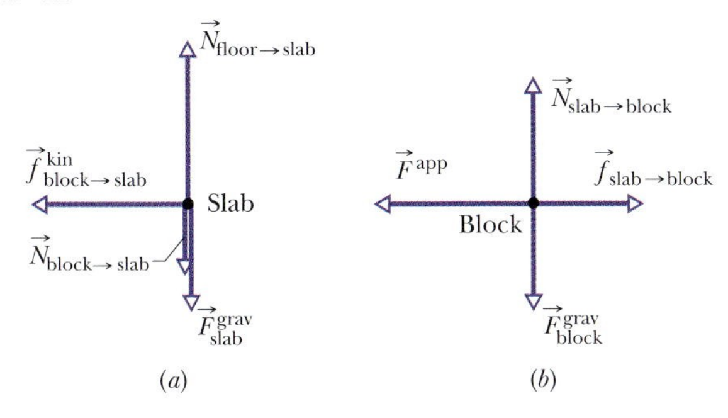


**FIGURE 6-35** ■ The free-body diagram showing all the forces acting on (*a*) the 40 kg slab and (*b*) the 10 kg block. These forces are not drawn to scale.

Since we expect the block and the slab to accelerate to the left, let's decide that the positive $x$ axis is pointing to the left and the $y$ axis is pointing straight up. Then

$$a_{slab\,y} = 0 \quad \text{so} \quad F^{net}_{slab\,y} = 0, \tag{6-31}$$

and

$$a_{block\,y} = 0 \quad \text{so} \quad F^{net}_{block\,y} = 0. \tag{6-32}$$

To calculate the horizontal accelerations of the block and the slab, we must first determine whether the frictional force of interaction between them is static or kinetic. The **Key Idea** here is that the maximum static frictional force's magnitude is limited to be no larger than $f^{max} = \mu^{stat} N$. Applying this to the slab, we find that

$$f^{max}_{block \to slab} = \mu^{stat} N_{block \to slab}. \tag{6-33}$$

Newton's Third Law tells us that

$$N_{block \to slab} = N_{slab \to block}, \tag{6-34}$$

and Fig. 6-35*b* and Eq. 6-31 tell us that

$$N_{slab \to block} = F^{grav}_{block} = m_{block}\, g. \tag{6-35}$$

Combining these ideas yields

$$f^{max}_{block \to slab} = \mu^{stat} m_{block}\, g$$
$$= (0.60)(10\text{ kg})(9.8\text{ N/kg}) = 59\text{ N}.$$

If this limit is not exceeded, then static friction would keep the block and the slab locked together, accelerating with a common acceleration

$$a_x = \frac{F_x^{net}}{(m_{slab} + m_{block})}$$
$$= 100\text{ N}/(40\text{ kg} + 10\text{ kg}) = +2.00\text{ m/s}^2.$$

Newton's Second Law (written in terms of $x$-components) tells us that this acceleration requires $F^{net}_{slab\,x} = m_{slab}\, a_x = (40\text{ kg})(2.00\text{ m/s}^2) = 80.0$ N. But Fig. 6-35*a* shows that the only horizontal force acting on the slab is the frictional force from the block. Since 80 N are required to accelerate the slab but we found the static frictional force is limited to 59 N, we can conclude that the block and the slab *cannot* be locked together by static friction. The block must be sliding to the left on top of the slab. This means that

$$F^{net}_{block\,x} = F_x^{app} + f^{kin}_{slab \to block}, \tag{6-36}$$

and

$$F^{net}_{slab\,x} = f^{kin}_{block \to slab} = \mu^{kin} N_{block \to slab}. \tag{6-37}$$

Combining Eqs. 6-34, Eq. 6-35, and Eq. 6-37 yields $F^{net}_{slab\,x} = \mu^{kin} m_{block}\, g$ or

$$\vec{F}^{net}_{slab} = (\mu^{kin} m_{block}\, g)\hat{\imath}, \tag{6-38}$$

so

$$\vec{a}_{slab} = \frac{\vec{F}^{net}_{slab}}{m_{slab}}$$
$$= (0.40)(10\text{ kg})(9.8\text{ m/s}^2)/(40\text{ kg})\hat{\imath}$$
$$= (0.98\text{ m/s}^2)\hat{\imath}. \qquad \text{(Answer)}$$

It's interesting to note that a frictional force causes the slab to speed up, not slow down. The same is true when you start running from rest. To accelerate, you push backwards on the ground with your shoes. The ground, courtesy of Newton's Third Law, pushes forward on you, accelerating you forward. It is actually the static frictional force that the ground exerts on you that accelerates you.

Finally, to calculate the acceleration of the block, we note that the net force on the block in the $y$ direction is zero and that (by Newton's Third Law) $\vec{f}^{kin}_{slab \to block} = -\vec{f}^{kin}_{block \to slab}$. Therefore,

$$\vec{F}^{net}_{block} = \vec{F}^{app} + \vec{f}^{kin}_{slab \to block}$$
$$= \vec{F}^{app} - \vec{f}^{kin}_{block \to slab}$$
$$= \vec{F}^{app} - \mu^{kin} m_{block}\, g\hat{\imath}.$$

So

$$\vec{a}_{block} = \frac{\vec{F}^{net}_{block}}{m_{block}}$$
$$= \frac{(100\text{ N})\hat{\imath} - (0.40)(10\text{ kg})(9.8\text{ m/s}^2)\hat{\imath}}{10\text{ kg}}$$
$$= [10\text{ m/s}^2 - (0.4)(9.8\text{ m/s}^2)]\hat{\imath}$$
$$= (+6.1\text{ m/s}^2)\hat{\imath}. \qquad \text{(Answer)}$$

### TOUCHSTONE EXAMPLE 6-8: Banked Curve

You cannot always count on friction to get your car around a curve, especially if the road is icy or wet. That is why highway curves are banked. Suppose that a car of mass $m$ moves at a constant speed $v$ of 20 m/s around a curve, now banked, whose radius $R$ is 190 m (Fig. 6-36*a*). What bank angle $\theta$ makes reliance on friction unnecessary?

**SOLUTION** ■ A centripetal force must act on the car if the car is to move along the circular path. A **Key Idea** is that the track is banked so as to tilt the normal force $\vec{N}$ on the car toward the center of the circle (Fig. 6-36*b*). Thus, $\vec{N}$ now has a centripetal component $N_r$, directed inward along a radial axis $r$. We want to find the value of the bank angle $\theta$ such that this centripetal component keeps the car on the circular track without need of friction.

A second **Key Idea** is to keep the $y$ axis vertical and the $x$ axis horizontal rather than in the direction of the incline. This enables us to find the radial component of the normal force more easily.

As Fig. 6-36*b* shows (and as you should verify), the angle that $\vec{N}$ makes with the vertical is equal to the bank angle $\theta$ of the track. Thus, the radial component $N_r$ is equal to $+N \sin\theta$ where $N$ is the magnitude of the normal force. We can now write Newton's Second Law for components along the $r$ axis ($F_r^{\text{net}} = ma_r$) as

$$+N\sin\theta = m\left(+\frac{v^2}{R}\right). \qquad (6\text{-}39)$$

We cannot solve this equation for the value of $\theta$ because it also contains the unknowns $N$ and $m$.

We next consider the forces and acceleration along the $y$ axis in Fig. 6-36*b*. The vertical component of the normal force is $N_y = N\cos\theta$, the gravitational force $\vec{F}^{\text{grav}}$ on the car is $(-mg)\hat{j}$, and the acceleration of the car along the $y$ axis is zero. Thus, we can write Newton's Second Law for components along the $y$ axis ($F_y^{\text{net}} = ma_y$) as

$$+N\cos\theta - mg = m(0),$$

from which

$$N\cos\theta = mg. \qquad (6\text{-}40)$$

This too contains the unknowns $N$ and $m$, but note that dividing Eq. 6-39 by Eq. 6-40 neatly eliminates both those unknowns. Doing so, replacing $\sin\theta/\cos\theta$ with $\tan\theta$ and solving for $\theta$, then yield

$$\theta = \tan^{-1}\left(\frac{v^2}{gR}\right)$$

$$= \tan^{-1}\left(\frac{(20\text{ m/s})^2}{(9.8\text{ m/s}^2)(190\text{ m})}\right) = 12°. \qquad \text{(Answer)}$$

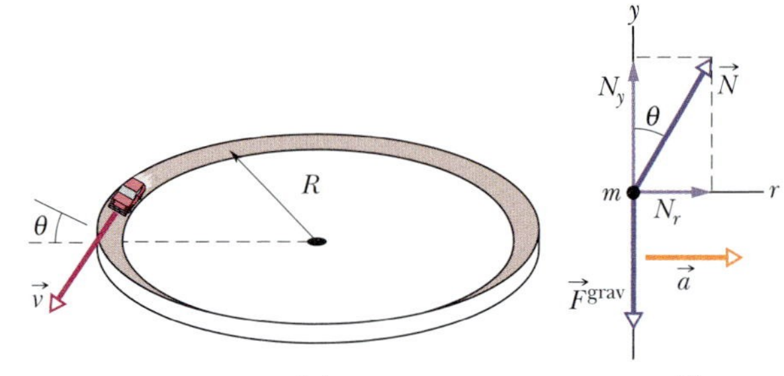

**FIGURE 6-36** ■ (*a*) A car moves around a curved banked road at constant speed. The bank angle is exaggerated for clarity. (*b*) A free-body diagram for the car, assuming that friction between tires and road is zero. The radially inward component of the normal force provides the necessary centripetal force. The resulting acceleration is also radially inward.

## 6-7 The Fundamental Forces of Nature

According to Newton's Third Law, forces between two objects always act in pairs. In the study of the structure of matter, physicists have used a belief in mutual interactions to study the nature of forces. As we learn more about matter and how it behaves, we explore the nature of forces by observing changes in the motion of objects that interact. Using these observations, scientists have identified only four types of forces.

The most familiar of these forces are the **gravitational force,** of which falling and weight are our most familiar examples, and the **electromagnetic force,** which, at a fundamental level, is the basis of all the other forces we considered in this chapter. The electromagnetic force is the combination of electrical forces and magnetic forces. Electromagnetic forces enable an electrically charged balloon to stick to a wall and a magnet to pick up an iron nail. In fact, aside from the gravitational force, *any* force that we can experience directly as a push or pull is electromagnetic in nature. That is, all such forces, including friction forces, normal forces, contact forces, and tension forces arise from electromagnetic forces exerted by one atom on another. For example, the tension in a taut cord exists only because its atoms attract one another. When pulled apart a bit, while normal forces result from atoms repelling each other when being pushed together.

Only two other fundamental forces are known, and they both act over such short distances that we cannot experience them directly through our senses. They are the **weak force,** which is involved in certain kinds of radioactive decay, and the **strong force,** which binds together the quarks that make up protons and neutrons and is the "glue" that holds together an atomic nucleus.

Physicists have long believed that nature has an underlying simplicity and that the number of fundamental forces can be reduced. Einstein spent most of his working life trying to interpret these forces as different aspects of a single *superforce.* He failed, but in the 1960s and 1970s, other physicists showed that the weak force and the electromagnetic force are different aspects of a single **electroweak force.** The quest for further reduction continues today, at the very forefront of physics. Table 6-2 lists the progress that has been made toward **unification** (as the goal is called) and gives some hints about the future.

**TABLE 6-2**
**The Quest for the Superforce—A Progress Report**

| Date | Researcher | Achievement |
|---|---|---|
| 1687 | Newton | Showed that the same laws apply to astronomical bodies and to objects on Earth. Unified celestial and terrestrial mechanics. |
| 1820<br>1830s | Oersted<br>Faraday | Showed, by brilliant experiments, that the then separate sciences of electricity and magnetism are intimately linked. |
| 1873 | Maxwell | Unified the sciences of electricity, magnetism, and optics into the single subject of electromagnetism. |
| 1979 | Glashow, Salam, Weinberg | Received the Nobel Prize for showing that the weak force and the electromagnetic force could be different aspects of a single *electroweak force.* This combination of forces reduced the number of forces viewed as fundamental forces from four to three. |
| 1984 | Rubbia, van der Meer | Received the Nobel Prize for verifying experimentally the predictions of the theory of the electroweak force. |
| | | **Work in Progress** |
| *Grand unification theories* (*GUTs*): Seek to unify the electroweak force and the strong force. | | |
| *Supersymmetry theories:* Seek to unify all forces, including the gravitational force, within a single framework. | | |
| *Superstring theories*: Interpret point-like particles, such as electrons, as being unimaginably tiny, closed loops. Strangely, extra dimensions beyond the familiar four dimensions of space-time appear to be required. | | |

# Problems

## SEC. 6-2 ■ NET FORCE AS A VECTOR SUM

**1. Standard Body** If the 1 kg standard body has an acceleration of 2.00 m/s$^2$ at 20° to the positive direction of the $x$ axis, then what are (a) the $x$-component and (b) the $y$-component of the net force on it, and (c) what is the net force in unit-vector notation?

**2. Chopping Block** Two horizontal forces act on a 2.0 kg chopping block that can slide over a frictionless kitchen counter, which lies in an $xy$ plane. One force is $\vec{F}_A = (3.0\text{ N})\hat{\text{i}} + (4.0\text{ N})\hat{\text{j}}$. Find the acceleration of the chopping block in unit-vector notation when the other force is (a) $\vec{F}_B = (-3.0\text{ N})\hat{\text{i}} + (-4.0\text{ N})\hat{\text{j}}$, (b) $\vec{F}_B = (-3.0\text{ N})\hat{\text{i}} + (4.0\text{ N})\hat{\text{j}}$, and (c) $\vec{F}_B = (3.0\text{ N})\hat{\text{i}} + (-4.0\text{ N})\hat{\text{j}}$.

**3. Two Horizontal Forces** Only two horizontal forces act on a 3.0 kg body. One force is 9.0 N, acting due east, and the other is 8.0 N, acting 62° north of west. What is the magnitude of the body's acceleration?

**4. Two Forces** While two forces act on it, a particle is to move at the constant velocity $\vec{v} = (3\text{ m/s})\hat{\text{i}} - (4\text{ m/s})\hat{\text{j}}$. One of the forces is $\vec{F}_A = (2\text{ N})\hat{\text{i}} + (-6\text{ N})\hat{\text{j}}$. What is the other force?

**5. Three Forces** Three forces act on a particle that moves with unchanging velocity $\vec{v} = (2\text{ m/s})\hat{\text{i}} - (7\text{ m/s})\hat{\text{j}}$. Two of the forces are $\vec{F}_A = (2\text{ N})\hat{\text{i}} + (3\text{ N})\hat{\text{j}}$ and $\vec{F}_B = (-5\text{ N})\hat{\text{i}} + (8\text{ N})\hat{\text{j}}$. What is the third force?

**6. Three Astronauts** Three astronauts, propelled by jet backpacks, push and guide a 120 kg asteroid toward a processing dock, exerting the forces shown in Fig. 6-37. What is the asteroid's acceleration (a) in unit-vector notation and as (b) a magnitude and (c) a direction?

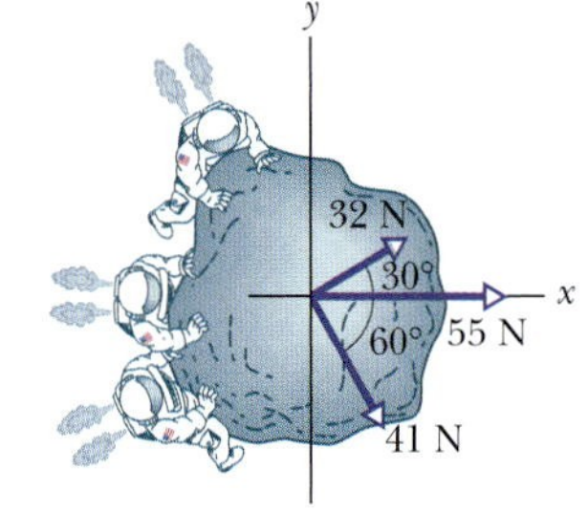


FIGURE 6-37 ■ Problem 6.

**7. The Box** There are two forces on the 2.0 kg box in the overhead view of Fig. 6-38 but only one is shown. The figure also shows the acceleration of the box. Find the second force (a) in unit-vector notation and as (b) a magnitude and (c) a direction.

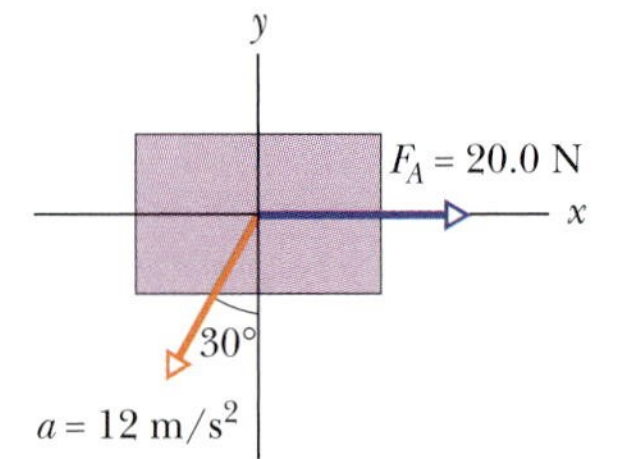


FIGURE 6-38 ■ Problem 7.

**8. A Tire** Figure 6-39 is an overhead view of a 12 kg tire that is to be pulled by three ropes. One force ($\vec{F}_A$, with magnitude 50 N) is indicated. Orient the other two forces $\vec{F}_B$ and $\vec{F}_C$ so that the magnitude of the resulting acceleration of the tire is least, and find that magnitude if (a) $F_B = 30$ N, $F_C = 20$ N; (b) $F_B = 30$ N, $F_C = 10$ N; and (c) $F_B = F_C = 30$ N.

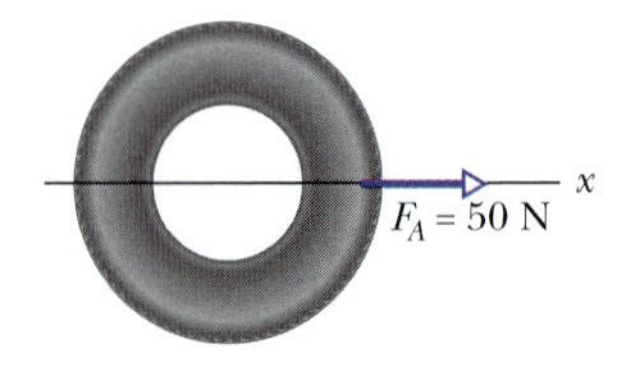


FIGURE 6-39 ■ Problem 8.

## SEC. 6-3 ■ GRAVITATIONAL FORCE AND WEIGHT

**9. Salami on a Cord** (a) An 11.0 kg salami is supported by a cord that runs to a spring scale, which is supported by another cord from the ceiling (Fig. 6-40*a*). What is the reading on the scale, which is marked in weight units? (b) In Fig. 6-40*b* the salami is supported by a cord that runs around a pulley and to a scale. The opposite end of the scale is attached by a cord to a wall. What is the reading on the scale? (c) In Fig. 6-40*c* the wall has been replaced with a second 11.0 kg salami on the left, and the assembly is stationary. What is the reading on the scale now?

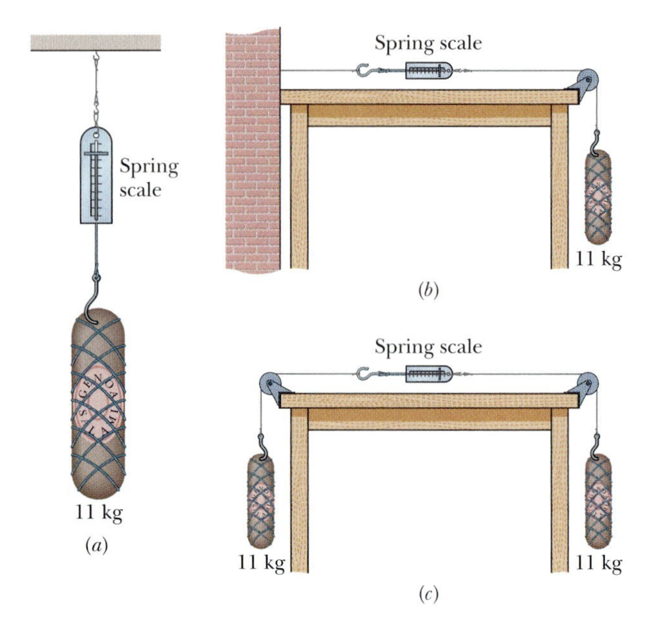


FIGURE 6-40 ■ Problem 9.

**10. Spaceship on the Moon** A spaceship lifts off vertically from the Moon, where the freefall acceleration is 1.6 m/s$^2$. If the spaceship has an upward acceleration of 1.0 m/s$^2$ as it lifts off, what is the magnitude of the force of the spaceship on its pilot, who weighs 735 N on Earth?

## SEC. ■ 6-4 CONTACT FORCES

**11. A Bureau** A bedroom bureau with a mass of 45 kg. including drawers and clothing, rests on the floor. (a) If the coefficient of static friction between the bureau and the floor is 0.45, what is the magnitude of the minimum horizontal force that a person must apply to start the bureau moving? (b) If the drawers and clothing, with 17 kg mass, are removed before the bureau is pushed, what is the new minimum magnitude?

**12. Scrambled Eggs** The coefficient of static friction between Teflon and scrambled eggs is about 0.04. What is the smallest angle from the horizontal that will cause the eggs to slide across the bottom of a Teflon-coated skillet?

**13. Baseball Player** A baseball player with mass $m = 79$ kg, sliding into second base, is retarded by a frictional force of magnitude 470 N. What is the coefficient of kinetic friction $\mu^{\text{kin}}$ between the player and the ground?

**14. The Mysterious Sliding Stones** Along the remote Racetrack Playa in Death Valley. California, stones sometimes gouge out prominent trails in the desert floor, as if they had been migrating (Fig 6-41). For years curiosity mounted about why the stones moved. One explanation was that strong winds during the occasional rainstorms would drag the rough stones over ground softened by rain. When the desert dried out, the trails behind the stones were hard-baked in place. According to measurements, the coefficient of kinetic friction between the stones and the wet playa ground is about 0.80. What horizontal force is needed on a stone of typical mass 20 kg to maintain the stone's motion once a gust has started it moving? (Story continues with Problem 42.)

FIGURE 6-41 ■ Problem 14.

**15. A Crate** A person pushes horizontally with a force of 220 N on a 55 kg crate to move it across a level floor. The coefficient of kinetic friction is 0.35. (a) What is the magnitude of the frictional force? (b) What is the magnitude of the crate's acceleration?

**16. A House on a Hill** A house is built on the top of a hill with a nearby 45° slope (Fig. 6-42). An engineering study indicates that the slope angle should be reduced because the top layers of soil along the slope might slip past the lower layers. If the static coefficient of friction between two such layers is 0.5, what is the least angle $\phi$ through which the present slope should be reduced to prevent slippage?

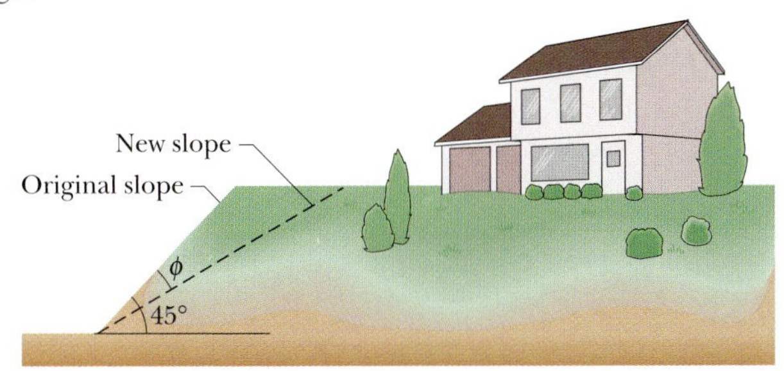


FIGURE 6-42 ■ Problem 16.

**17. Hockey Puck** A 110 g hockey puck sent sliding over ice is stopped in 15 m by the frictional force on it from the ice. (a) If its initial speed is 6.0 m/s, what is the magnitude of the frictional force? (b) What is the coefficient of friction between the puck and the ice?

**18. Rock Climber** In Fig. 6-43 a 49 kg rock climber is climbing a "chimney" between two rock slabs. The static coefficient of friction between her shoes and the rock is 1.2; between her back and the rock it is 0.80. She has reduced her push against the rock until her back and her shoes are on the verge of slipping. (a) Draw a free-body diagram of the climber. (b) What is her push against the rock? (c) What fraction of her weight is supported by the frictional force on her shoes?

FIGURE 6-43 ■ Problem 18.

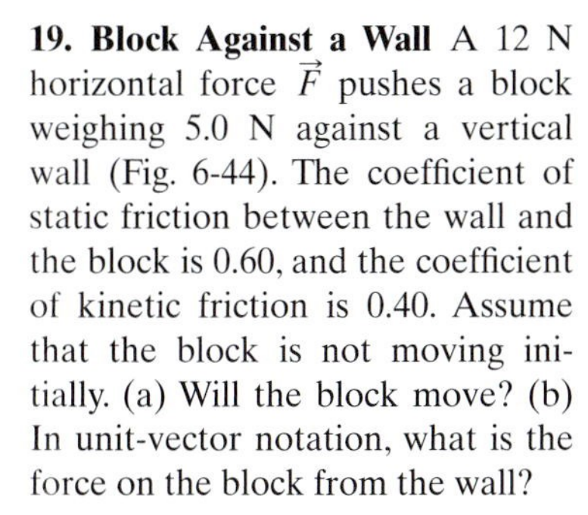

**19. Block Against a Wall** A 12 N horizontal force $\vec{F}$ pushes a block weighing 5.0 N against a vertical wall (Fig. 6-44). The coefficient of static friction between the wall and the block is 0.60, and the coefficient of kinetic friction is 0.40. Assume that the block is not moving initially. (a) Will the block move? (b) In unit-vector notation, what is the force on the block from the wall?

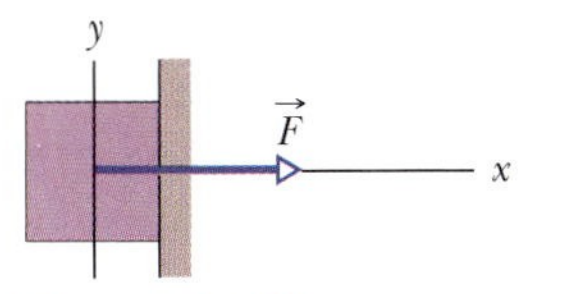


FIGURE 6-44 ■ Problem 19.

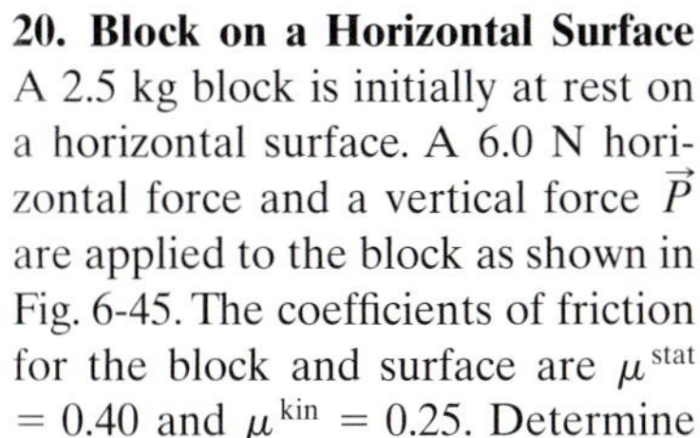

**20. Block on a Horizontal Surface** A 2.5 kg block is initially at rest on a horizontal surface. A 6.0 N horizontal force and a vertical force $\vec{P}$ are applied to the block as shown in Fig. 6-45. The coefficients of friction for the block and surface are $\mu^{\text{stat}} = 0.40$ and $\mu^{\text{kin}} = 0.25$. Determine the magnitude and direction of the frictional force acting on the block if the magnitude of $\vec{P}$ is (a) 8.0 N, (b) 10 N, and (c) 12 N.

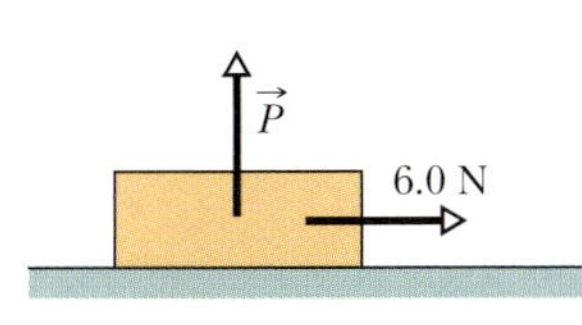


FIGURE 6-45 ■ Problem 20.

**21. Pile of Sand** A worker wishes to pile a cone of sand onto a circular area in his yard. The radius of the circle is $R$, and no sand is to spill onto the surrounding area (Fig. 6-46). If $\mu^{\text{stat}}$ is the static coefficient of friction between each layer of sand along the slope and the sand beneath it (along which is might slip), show that the greatest volume of sand that can be stored in this manner is $\pi\mu^{\text{stat}}R^3/3$. (The volume of a cone is $Ah/3$, where $A$ is the base area and $h$ is the cone's height.)

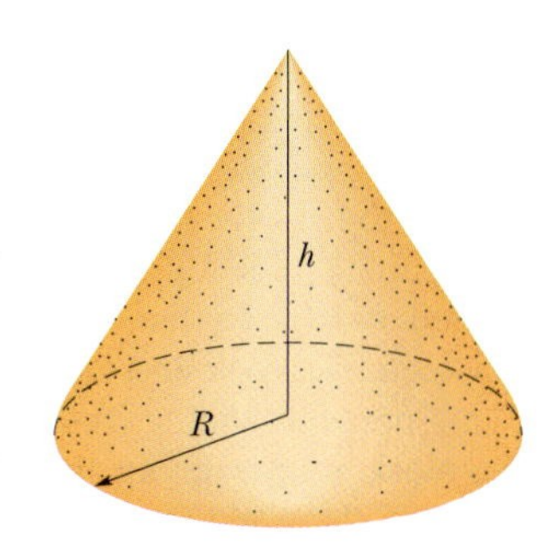


FIGURE 6-46 ■ Problem 21.

**22. Worker and Crate** A worker pushes horizontally on a 35 kg crate with a force of magnitude 110 N. The coefficient of static friction between the crate and the floor is 0.37. (a) What is the frictional force on the crate from the floor? (b) What is the maximum magnitude $f^{\text{stat}}_{\text{max}}$ of the static frictional force under the circumstances? (c) Does the crate move? (d) Suppose, next, that a second worker pulls directly upward on the crate to help out. What is the least vertical pull that will allow the first worker's 110 N push to move the crate? (e) If, instead, the second worker pulls horizontally to help out, what is the least pull that will get the crate moving?

**23. A Crate is Dragged** A 68 kg crate is dragged across a floor by pulling on a rope attached to the crate and inclined 15° above the horizontal. (a) If the coefficient of static friction is 0.50, what minimum force magnitude is required from the rope to start the crate moving? (b) If $\mu^{\text{kin}} = 0.35$, what is the magnitude of the initial acceleration of the crate?

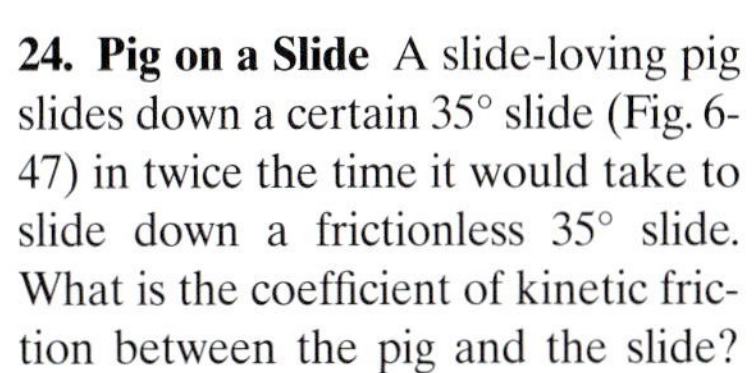

**24. Pig on a Slide** A slide-loving pig slides down a certain 35° slide (Fig. 6-47) in twice the time it would take to slide down a frictionless 35° slide. What is the coefficient of kinetic friction between the pig and the slide?

FIGURE 6-47 ■ Problem 24.

**25. Blocks $A$ and $B$** In Fig. 6-48 blocks $A$ and $B$ have weights of 44 N and 22 N, respectively. (a) Determine the minimum weight of block $C$ to keep $A$ from sliding if $\mu^{\text{stat}}$ between $A$ and the table is 0.20. (b) Block $C$ suddenly is lifted off $A$. What is the acceleration of block $A$ if $\mu^{\text{kin}}$ between $A$ and the table is 0.15?

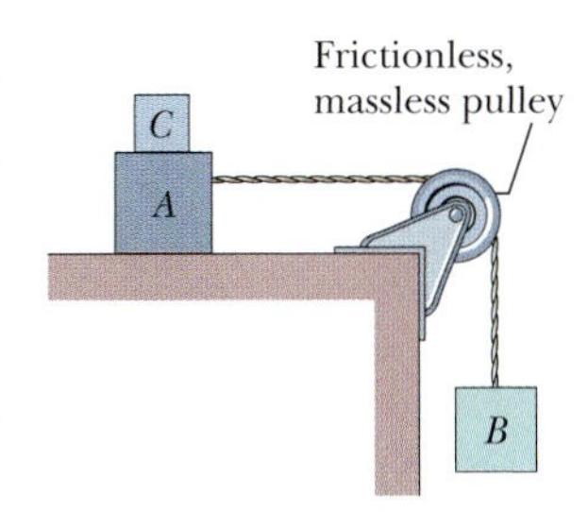


FIGURE 6-48 ■ Problem 25.

**26. Block Pushed at an Angle** A 3.5 kg block is pushed along a horizontal floor by a force $\vec{F}$ of magnitude 15 N at an angle $\theta = 40°$ with the horizontal (Fig. 6-49). The coefficient of kinetic friction between the block and the floor is 0.25. Calculate

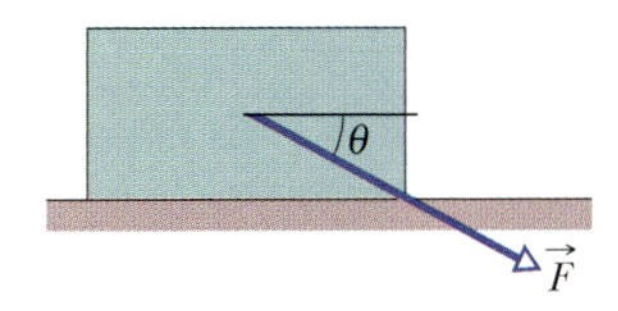


FIGURE 6-49 ■ Problem 26.

the magnitudes of (a) the frictional force on the block from the floor and (b) the acceleration of the block.

**27. Mountain Side** Figure 6-50 shows the cross section of a road cut into the side of a mountain. The solid line $AA'$ represents a weak bedding plane along which sliding is possible. Block $B$ directly above the highway is separated from uphill rock by a large crack (called a *joint*), so that only friction between the block and the bedding plane prevents sliding. The mass of the block is $1.8 \times 10^7$ kg, the *dip angle* $\theta$ of the bedding plane is 24°, and the coefficient of static friction between block and plane is 0.63. (a) Show that the block will not slide. (b) Water seeps into the joint and expands upon freezing, exerting on the block a force $\vec{F}$ parallel to $AA'$. What minimum value of $F$ will trigger a slide?

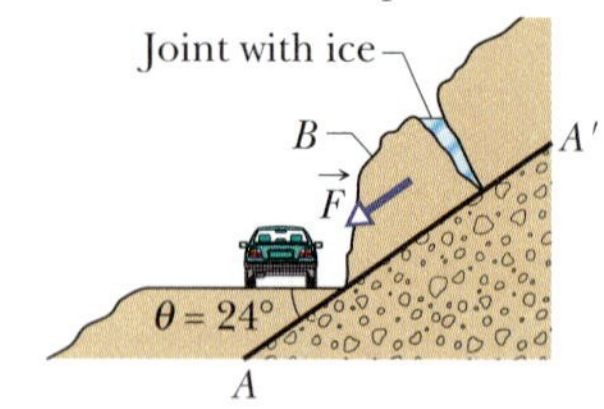


FIGURE 6-50 ■ Problem 27.

**28. Penguin Sled** A loaded penguin sled weighing 80 N rests on a plane inclined at 20° to the horizontal (Fig. 6-51). Between the sled and the plane, the coefficient of static friction is 0.25, and the coefficient of kinetic friction is 0.15. (a) What is the minimum magnitude of the force $\vec{F}$, parallel to the plane, that will prevent the sled from slipping down the plane? (b) What is the minimum magnitude $F$ that will start the sled moving up the plane? (c) What value of $F$ is required to move the sled up the plane at constant velocity?

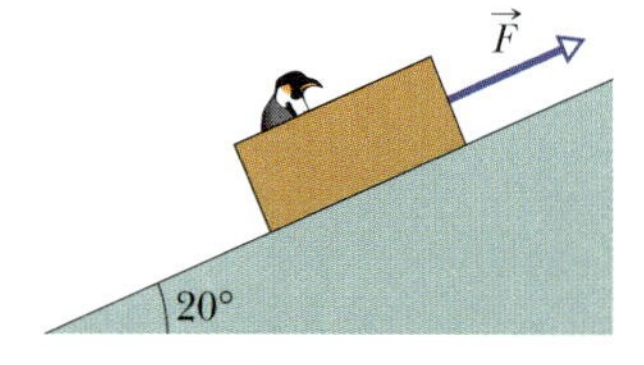


FIGURE 6-51 ■ Problem 28.

**29. Block on a Table** Block $B$ in Fig. 6-52 weighs 711 N. The coefficient of static friction between block and table is 0.25; assume that the cord between $B$ and the knot is horizontal. Find the maximum weight of block $A$ for which the system will be stationary.

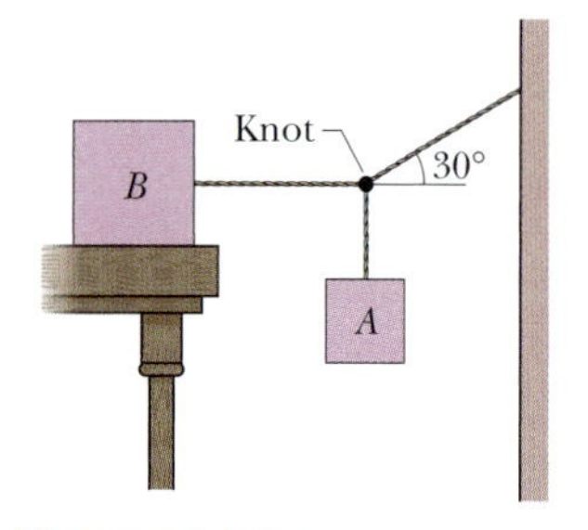


FIGURE 6-52 ■ Problem 29.

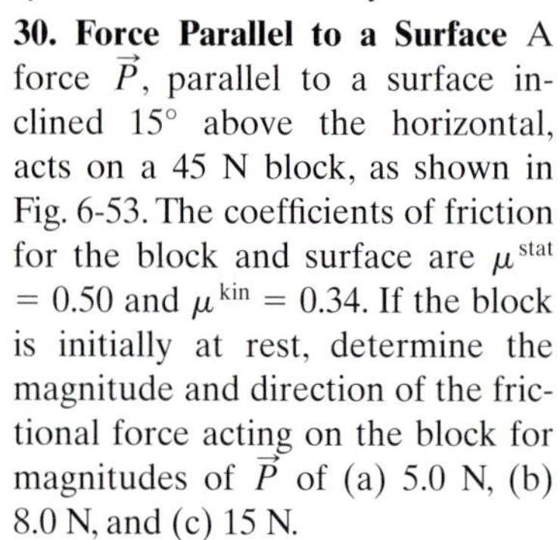

**30. Force Parallel to a Surface** A force $\vec{P}$, parallel to a surface inclined 15° above the horizontal, acts on a 45 N block, as shown in Fig. 6-53. The coefficients of friction for the block and surface are $\mu^{\text{stat}} = 0.50$ and $\mu^{\text{kin}} = 0.34$. If the block is initially at rest, determine the magnitude and direction of the frictional force acting on the block for magnitudes of $\vec{P}$ of (a) 5.0 N, (b) 8.0 N, and (c) 15 N.

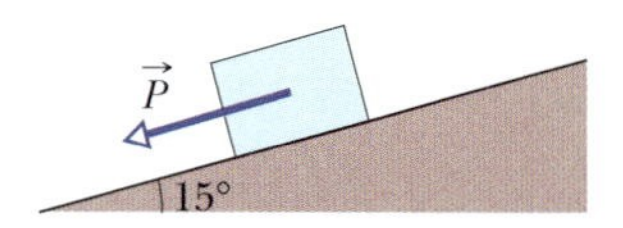


FIGURE 6-53 ■ Problem 30.

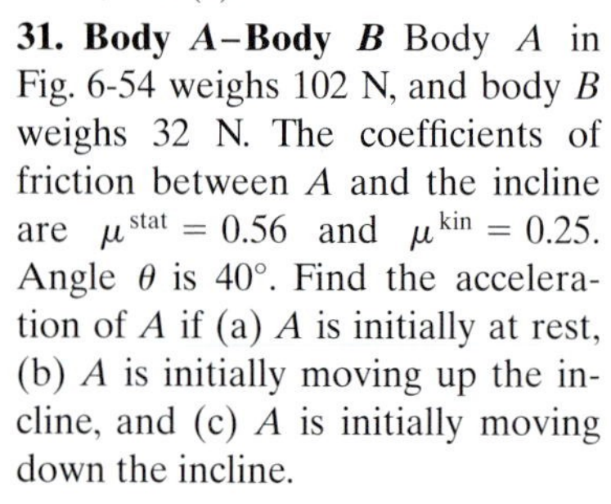

**31. Body A–Body B** Body $A$ in Fig. 6-54 weighs 102 N, and body $B$ weighs 32 N. The coefficients of friction between $A$ and the incline are $\mu^{\text{stat}} = 0.56$ and $\mu^{\text{kin}} = 0.25$. Angle $\theta$ is 40°. Find the acceleration of $A$ if (a) $A$ is initially at rest, (b) $A$ is initially moving up the incline, and (c) $A$ is initially moving down the incline.

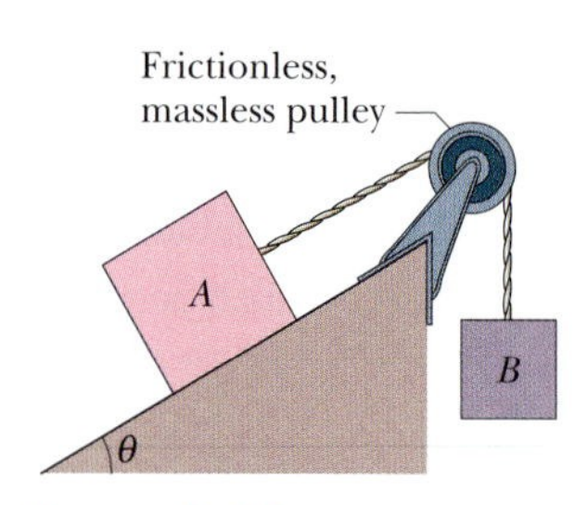


FIGURE 6-54 ■ Problem 31 and 32.

**32. Two Blocks and a Pulley** In Fig. 6-54, two blocks are connected over a pulley. The mass of block $A$ is 10 kg and the coefficient of kinetic friction between $A$ and the incline is 0.20. Angle $\theta$ of the incline is 30°. Block $A$ slides down the incline at constant speed. What is the mass of block $B$?

**33. Two Blocks Massless String** Two blocks of weights 3.6 N and 7.2 N are connected by a massless string and slide down a 30° inclined plane. The coefficient of kinetic friction between the lighter block and the plane is 0.10; that between the heavier block and the plane is 0.20. Assuming that the lighter block leads, find (a) the magnitude of the acceleration of the blocks and (b) the tension in the string. (c) Describe the motion if, instead, the heavier block leads.

**34. Box of Cheerios®** In Fig. 6-55, a box of Cheerios® and a box of Wheaties® are accelerated across a horizontal surface by a horizontal force $\vec{F}$ applied to the Cheerios® box. The magnitude of the frictional force on the Cheerios® box is 2.0 N, and the magnitude of the frictional force on the Wheaties® box is 4.0 N. If the magnitude of $\vec{F}$ is 12 N, what is the magnitude of the force on the Wheaties® box from the Cheerios® box?

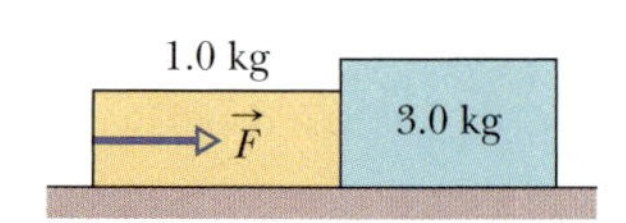


FIGURE 6-55 ■ Problem 34.

**35. Blocks Not Attached** The two blocks (with $m = 16$ kg and $M = 88$ kg) shown in Fig. 6-56 are not attached. The coefficient of static friction between the blocks is $\mu^{\text{stat}} = 0.38$, but the surface beneath the larger block is frictionless. What is the minimum magnitude of the horizontal force $\vec{F}$ required to keep the smaller block from slipping down the larger block?

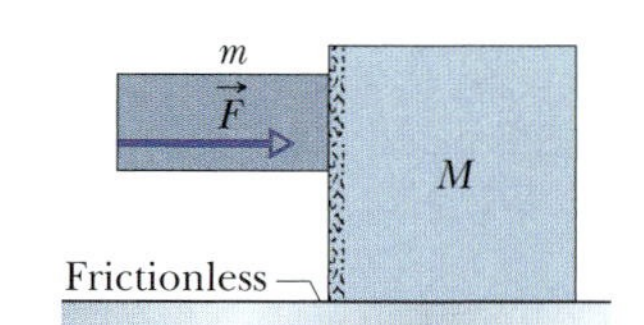


FIGURE 6-56 ■ Problem 35.

**36. Aunts and Uncles** In Fig. 6-57, a box of ant aunts (total mass $m_A = 1.65$ kg) and a box of ant uncles (total mass $m_B = 3.30$ kg) slide down an inclined plane while attached by a massless rod parallel to the plane. The angle of incline is $\theta = 30°$. The coefficient of kinetic friction between the aunt box and the incline is $\mu_A^{\text{kin}} = 0.226$; that between the uncle box and the incline is $\mu_B^{\text{kin}} = 0.113$. Compute (a) the tension in the rod and (b) the common acceleration of the two boxes. (c) How would the answers to (a) and (b) change if the uncles trailed the aunts?

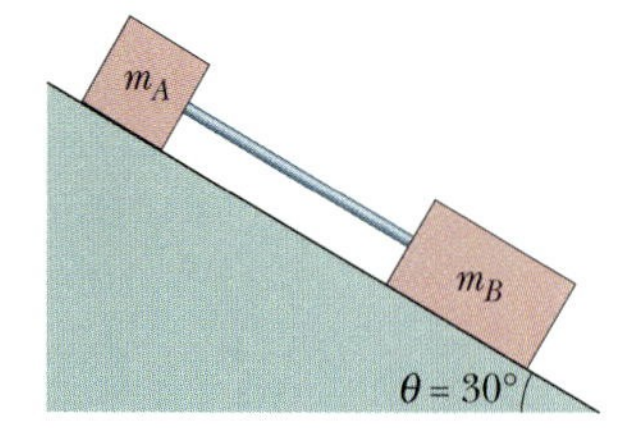


FIGURE 6-57 ■ Problem 36.

**37. Block on a Slab** A 40 kg slab rests on a frictionless floor. A 10 kg block rests on top of the slab (Fig. 6-58). The coefficient of static friction $\mu^{\text{stat}}$ between the block and the slab is 0.60, whereas their kinetic friction coefficient $\mu^{\text{kin}}$ is 0.40. The 10 kg block is pulled by a horizontal force with a magnitude of 100 N. What are the resulting accelerations of (a) the block and (b) the slab?

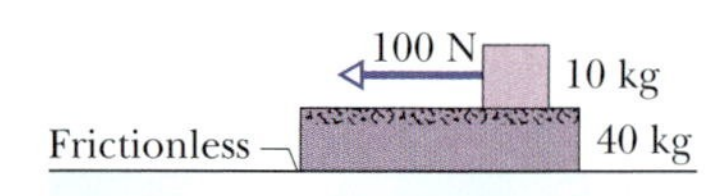


FIGURE 6-58 ■ Problem 37.

**38. A Locomotive** A locomotive accelerates a 25-car train along a level track. Every car has a mass of $5.0 \times 10^4$ kg and is subject to a frictional force $f^{kin} = (250\ \text{N}\cdot\text{s/m})\vec{v}$. At the instant when the speed of the train is 30 km/h, the magnitude of its acceleration is 0.20 m/s$^2$. (a) What is the tension in the coupling between the first car and the locomotive? (b) If this tension is equal to the maximum force the locomotive can exert on the train, what is the steepest grade up which the locomotive can pull the train at 30 km/h?

**39. Crate in a Trough** In Fig. 6-59, a crate slides down an inclined right-angled trough. The coefficient of kinetic friction between the crate and the trough is $\mu^{kin}$. What is the acceleration of the crate in terms of $\mu^{kin}$, $\theta$, and $g$?

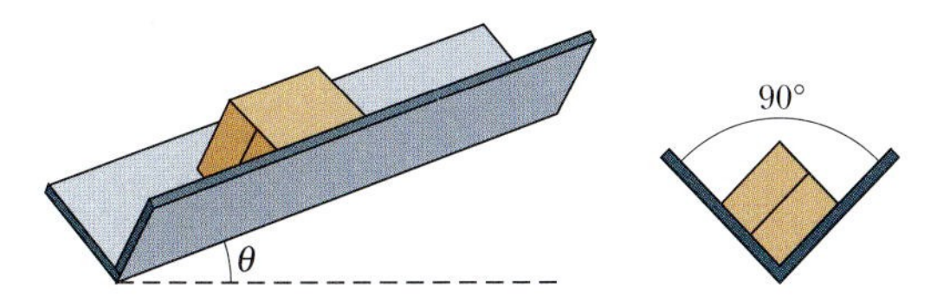


FIGURE 6-59 ■ Problem 39.

**40. Box of Sand** An initially stationary box of sand is to be pulled across a floor by means of a cable in which the tension should not exceed 1100 N. The coefficient of static friction between the box and the floor is 0.35. (a) What should be the angle between the cable and the horizontal in order to pull the greatest possible amount of sand, and (b) what is the weight of the sand and box in that situation?

**41. Boat with Engine Off** A 1000 kg boat is traveling at 90 km/h when its engine is shut off. The magnitude of the frictional force $\vec{f}^{kin}$ between boat and water is proportional to the speed $v$ of the boat; $\vec{f}^{kin} = (70\ \text{N}\cdot\text{s/m})\vec{v}$. Find the time required for the boat to slow to 45 km/h.

## SEC. 6-5 ■ DRAG FORCE AND TERMINAL SPEED

**42. Continuation of Problem 14** First reread the explanation of how the wind might drag desert stones across the playa. Now assume that Eq. 6-23 gives the magnitude of the air drag force on the typical 20 kg stone, which presents a vertical cross-sectional area to the wind of 0.040 m$^2$ and has a drag coefficient $C$ of 0.80. Take the air density to be 1.21 kg/m$^3$, and the coefficient of kinetic friction to be 0.80. (a) In kilometers per hour, what wind speed $V$ along the ground is needed to maintain the stone's motion once it has started moving? Because winds along the ground are retarded by the ground, the wind speeds reported for storms are often measured at a height of 10 m. Assume wind speeds are 2.00 times those along the ground, (b) For your answer to (a), what wind speed would be reported for the storm and is that value reasonable for a high-speed wind in a storm?

**43. Missile** Calculate the drag force on a missile 53 cm in diameter cruising with a speed of 250 m/s at low altitude, where the density of air is 1.2 kg/m$^3$. Assume $C = 0.75$.

**44. Sky Diver** The terminal speed of a sky diver is 160 km/h in the spread-eagle position and 310 km/h in the nosedive position. Assuming that the diver's drag coefficient $C$ does not change from one position to the other, find the ratio of the effective cross-sectional area $A$ in the slower position to that in the faster position.

**45. Jet Vs. Prop-Driven Transport** Calculate the ratio of the drag force on a passenger jet flying with a speed of 1000 km/h at an altitude of 10 km to the drag force on a prop-driven transport flying at half the speed and half the altitude of the jet. At 10 km the density of air is 0.38 kg/m$^3$, and at 5.0 km it is 0.67 kg/m$^3$. Assume that the airplanes have the same effective cross-sectional area and the same drag coefficient $C$.

## SEC. 6-6 ■ APPLYING NEWTON'S LAWS

**46. Block on an Incline** Refer to Fig. 6-29. Let the mass of the block be 8.5 kg and the angle $\theta$ be 30°. The block moves at constant velocity. Find (a) the tension in the cord and (b) the normal force acting on the block. (c) If the cord is cut, find the magnitude of the block's acceleration.

**47. Electron Moving Horizontally** An electron with a speed of $1.2 \times 10^7$ m/s moves horizontally into a region where a constant vertical force of $4.5 \times 10^{-16}$ N acts on it. The mass of the electron is $9.11 \times 10^{-31}$ kg. Determine the vertical distance the electron is deflected during the time it has moved 30 mm horizontally.

**48. Tarzan** Tarzan, who weighs 820 N, swings from a cliff at the end of a 20 m vine that hangs from a high tree limb and initially makes an angle of 22° with the vertical. Immediately after Tarzan steps off the cliff, the tension in the vine is 760 N. Choose a coordinate system for which the $x$ axis points horizontally away from the edge of the cliff and the $y$ axis points upward. (a) What is the force of the vine on Tarzan in unit-vector notation? (b) What is the net force acting on Tarzan in unit-vector notation? What are the (c) magnitude and (d) direction of the net force acting on Tarzan? What are the (e) magnitude and (f) direction of Tarzan's acceleration?

**49. Skier on a Rope Tow** A 50 kg skier is pulled up a frictionless ski slope that makes an angle of 8.0° with the horizontal by holding onto a tow rope that moves parallel to the slope. Determine the magnitude of the force of the rope on the skier at an instant when (a) the rope is moving with a constant speed of 2.0 m/s and (b) the rope is moving with a speed of 2.0 m/s but that speed is increasing at a rate of 0.10 m/s$^2$.

**50. Running Armadillo** For sport, a 12 kg armadillo runs onto a large pond of level, frictionless ice with an initial velocity of 5.0 m/s along the positive direction of an $x$ axis. Take its initial position on the ice as being the origin. It slips over the ice while being pushed by a wind with a force of 17 N in the positive direction of the $y$ axis. In unit-vector notation, what are the animal's (a) velocity and (b) position vector when it has slid for 3.0 s?

**51. Sphere Suspended from a Cord** A sphere of mass $3.0 \times 10^{-4}$ kg is suspended from a cord. A steady horizontal breeze pushes the sphere so that the cord makes a constant angle of 37° with the vertical. Find (a) the magnitude of that push and (b) the tension in the cord.

**52. Skier in the Wind** A 40 kg skier comes directly down a frictionless ski slope that is inclined at an angle of 10° with the horizontal while a strong wind blows parallel to the slope. Determine the magnitude and direction of the force of the wind on the skier if (a) the magnitude of the skier's velocity is constant, (b) the magnitude of the skier's velocity is increasing at a rate of 1.0 m/s$^2$. and (c) the magnitude of the skier's velocity is increasing at a rate of 2.0 m/s$^2$.

**53. Jet Engine** A 1400 kg jet engine is fastened to the fuselage of a passenger jet by just three bolts (this is the usual practice). Assume that each bolt supports one-third of the load. (a) Calculate the force on each bolt as the plane waits in line for clearance to take off. (b) During flight, the plane encounters turbulence, which suddenly imparts an upward vertical acceleration of 2.6 m/s$^2$ to the plane. Calculate the force on each bolt now.

**54. Pulling a Crate** A worker drags a crate across a factory floor by pulling on a rope tied to the crate (Fig. 6-60). The worker exerts a force of 450 N on the rope, which is inclined at 38° to the horizontal, and the floor exerts a horizontal force of 125 N that opposes the motion. Calculate the magnitude of the acceleration of the crate if (a) its mass is 310 kg or (b) its weight is 310 N.

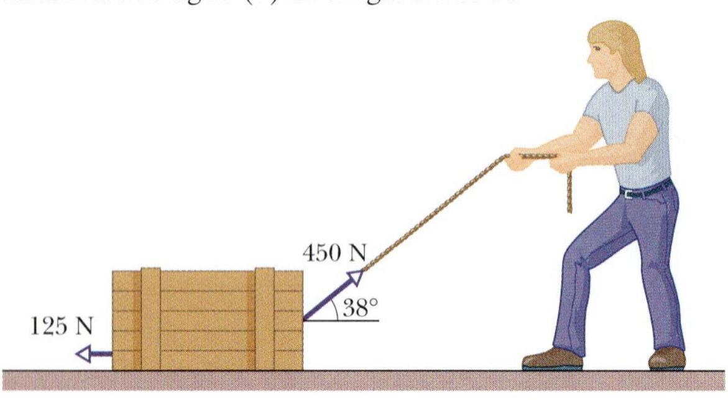


FIGURE 6-60 ■ Problem 54.

**55. Motorcycle Rider** A motorcycle and 60.0 kg rider accelerate at 3.0 m/s² up a ramp inclined 10° above the horizontal. (a) What is the magnitude of the net force acting on the rider? (b) What is the magnitude of the force on the rider from the motorcycle?

**56. One on an Incline—One Hanging** A block of mass $m_A$ = 3.70 kg on a frictionless inclined plane of angle 30.0° is connected by a cord over a massless, frictionless pulley to a second block of mass $m_B$ = 2.30 kg hanging vertically (Fig. 6-61). What are (a) the magnitude of the acceleration of each block and (b) the direction of the acceleration of the hanging block? (c) What is the tension in the cord?

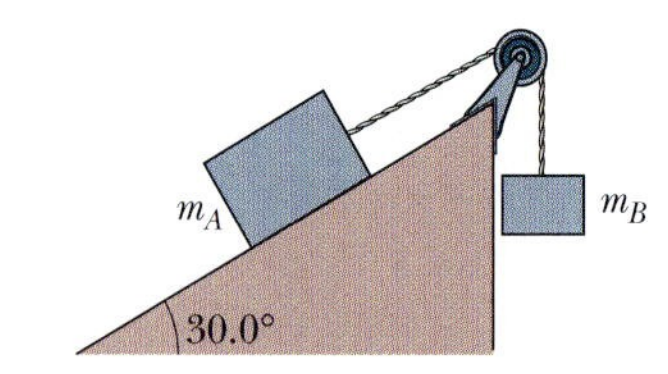


FIGURE 6-61 ■ Problem 56.

**57. Pencil Box** In Fig. 6-62, a 1.0 kg pencil box on a 30° frictionless incline is connected to a 3.0 kg pen box on a horizontal frictionless surface. The pulley is frictionless and massless. (a) If the magnitude of the applied force $\vec{F}$ is 2.3 N, what is the tension in the connecting cord? (b) What is the largest value that the magnitude of $\vec{F}$ may have without the connecting cord becoming slack?

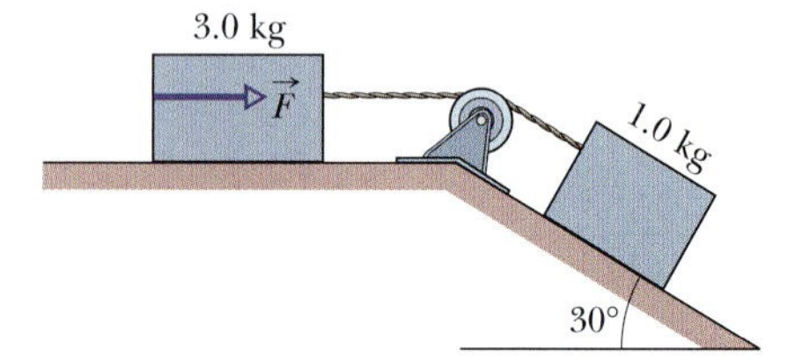


FIGURE 6-62 ■ Problem 57.

**58. Projected Up an Incline** A block is projected up a frictionless inclined plane with initial speed $v_1$ = 3.50 m/s. The angle of incline is $\theta$ = 32.0°. (a) How far up the plane does it go? (b) How long does it take to get there? (c) What is its speed when it gets back to the bottom?

**59. Horse-Drawn Barge** In earlier days, horses pulled barges down canals in the manner shown in Fig. 6-63. Suppose the horse pulls on the rope with a force of 7900 N at an angle of 18° to the direction of motion of the barge, which is headed straight along the canal. The mass of the barge is 9500 kg, and its acceleration is 0.12 m/s². What are the (a) magnitude and (b) direction of the force on the barge from the water?

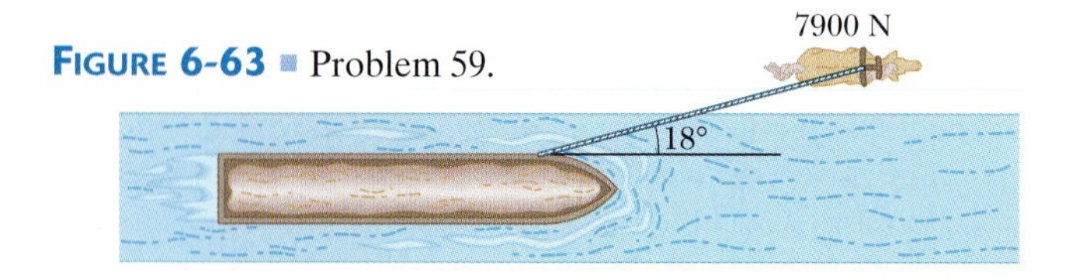


FIGURE 6-63 ■ Problem 59.

**60. Lifting a Block** In Fig. 6-64, a 5.00 kg block is pulled along a horizontal frictionless floor by a cord that exerts a force of magnitude $F$ = 12.0 N at an angle $\theta$ = 25.0° above the horizontal. (a) What is the magnitude of the block's acceleration? (b) The force magnitude $F$ is slowly increased. What is its value just before the block is lifted (completely) off the floor? (c) What is the magnitude of the block's acceleration just before it is lifted (completely) off the floor?

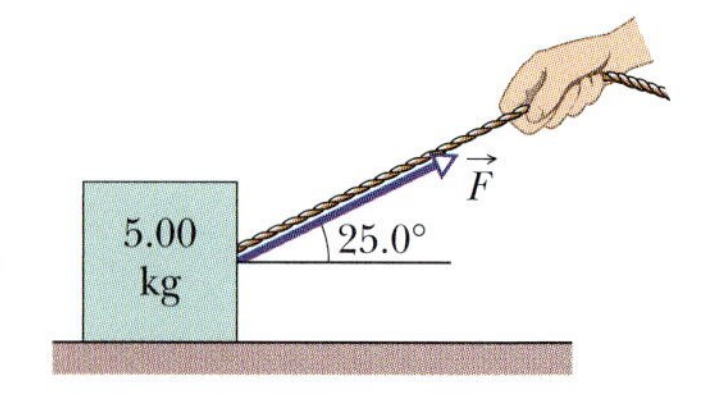


FIGURE 6-64 ■ Problem 60.

**61. A Rope Must Sag** A block of mass $M$ is pulled along a horizontal frictionless surface by a rope of mass $m$, as shown in Fig. 6-65. A horizontal force $\vec{F}$ is applied to one end of the rope. (a) Show that the rope *must* sag, even if only by an imperceptible amount. Then, assuming the sag is negligible, find (b) the acceleration of rope and block, (c) the force on the block from the rope, and (d) the tension in the rope at its midpoint.

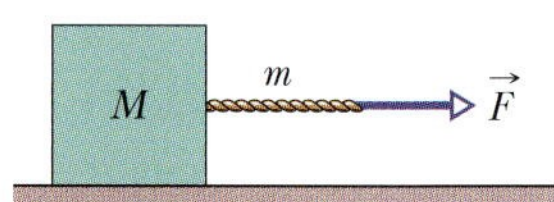


FIGURE 6-65 ■ Problem 61.

**62. Crate at Constant Speed** In Fig. 6-66, a 100 kg crate is pushed at constant speed up the frictionless 30.0° ramp by a horizontal force $\vec{F}$. What are the magnitudes of (a) $\vec{F}$ and (b) the force on the crate from the ramp?

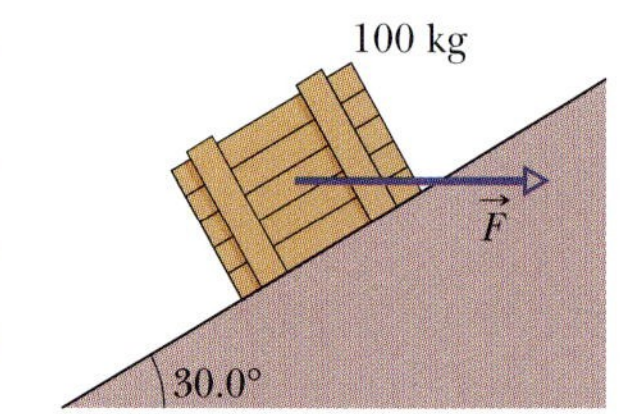


FIGURE 6-66 ■ Problem 62.

**63. Alpine Cable Car** Figure 6-67 shows a section of an alpine cable-car system. The maximum permissible mass of each car with occupants is 2800 kg. The cars, riding on a support cable, are pulled by a second cable attached to each pylon (support tower); assume the cables are straight. What is the difference in tension between adjacent sections of pull cable if the cars are at the maximum permissible mass and are being accelerated up the 35° incline at 0.81 m/s²?

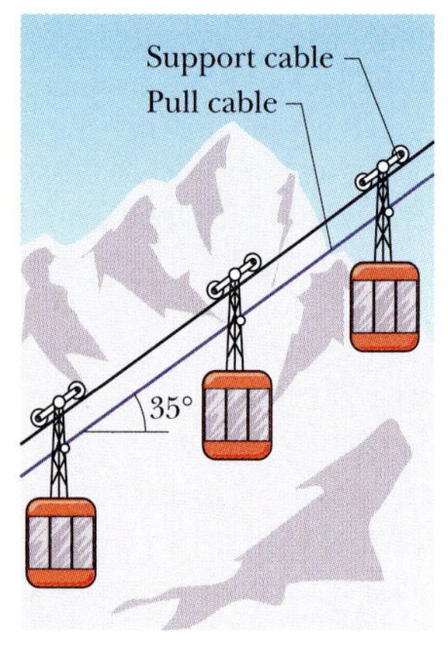


FIGURE 6-67 ■ Problem 63.

**64. Bobsled Run** During an Olympic bobsled run, the Jamaican team makes a turn of radius 7.6 m at a speed of 96.6 km/h. What is their acceleration in $g$-units? (1 $g$-unit = 9.8 m/s².)

**65. Grand Prix** Suppose the coefficient of static friction between the road and the tires on a Formula One car is 0.6 during a Grand Prix auto race. What speed will put the car on the verge of sliding as it rounds a level curve of 30.5 m radius?

**66. Roller Coaster** A roller-coaster car has a mass of 1200 kg when fully loaded with passengers. As the car passes over the top of a circular hill of radius 18 m, its speed is not changing. What are the magnitude and direction of the force of the track on the car at the top of the hill if the car's speed is (a) 11 m/s and (b) 14 m/s?

**67. Flat Track** What is the smallest radius of an unbanked (flat) track around which a bicyclist can travel if her speed is 29 km/h and the coefficient of static friction between tires and track is 0.32?

**68. Amusement Park Ride** An amusement park ride consists of a car moving in a vertical circle on the end of a rigid boom of negligible mass. The combined weight of the car and riders is 5.0 kN, and the radius of the circle is 10 m. What are the magnitude and direction of the force of the boom on the car at the top of the circle if the car's speed there is (a) 5.0 m/s and (b) 12 m/s?

**69. Puck on a Table** A puck of mass $m$ slides on a frictionless table while attached to a hanging cylinder of mass $M$ by a cord through a hole in the table (Fig. 6-68). What speed keeps the cylinder at rest?

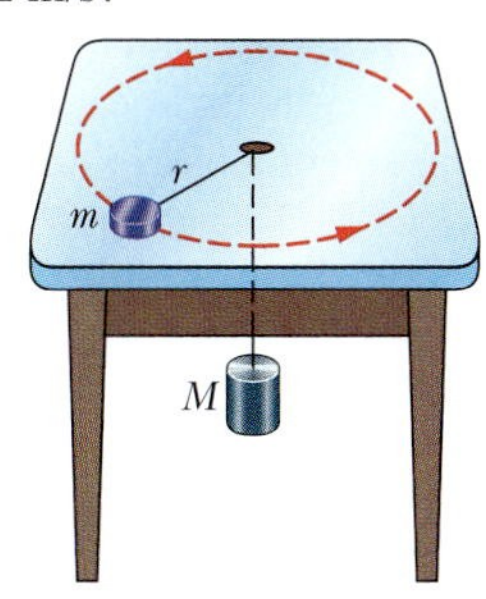

FIGURE 6-68 ■ Problem 69.

**70. Bicyclist** A bicyclist travels in a circle of radius 25.0 m at a constant speed of 9.00 m/s. The bicycle–rider mass is 85.0 kg. Calculate the magnitudes of (a) the force of friction on the bicycle from the road and (b) the *net* force on the bicycle from the road.

**71. Student on Ferris Wheel** A student of weight 667 N rides a steadily rotating Ferris wheel (the student sits upright). At the highest point, the magnitude of the normal force $\vec{N}$ on the student from the seat is 556 N. (a) Does the student feel "light" or "heavy" there? (b) What is the magnitude of $\vec{N}$ at the lowest point? (c) What is the magnitude $N$ if the wheel's speed is doubled?

**72. Old Streetcar** An old streetcar rounds a flat corner of radius 9.1 m, at 16 km/h. What angle with the vertical will be made by the loosely hanging hand straps?

**73. Flying in a Circle** An airplane is flying in a horizontal circle at a speed of 480 km/h. If its wings are tilted 40° to the horizontal, what is the radius of the circle in which the plane is flying? (See Fig. 6-69.) Assume that the required force is provided entirely by an "aerodynamic lift" that is perpendicular to the wing surface.

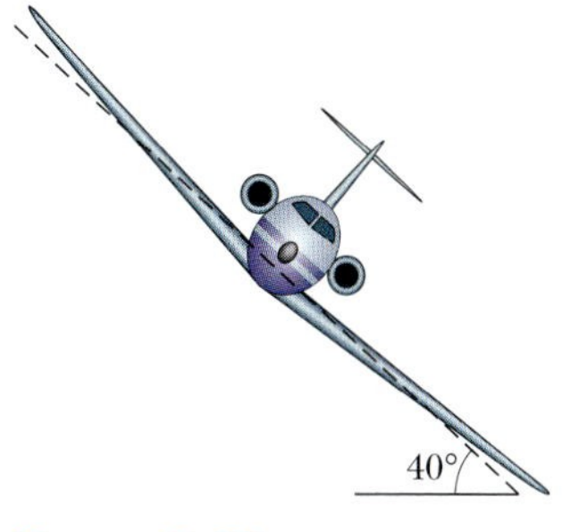

FIGURE 6-69 ■ Problem 73.

**74. High-Speed Railway** A high-speed railway car goes around a flat, horizontal circle of radius 470 m at a constant speed. The magnitudes of the horizontal and vertical components of the force of the car on a 51.0 kg passenger are 210 N and 500 N, respectively. (a) What is the magnitude of the net force (of *all* the forces) on the passenger? (b) What is the speed of the car?

**75. Ball Connected to a Rod** As shown in Fig. 6-70, a 1.34 kg ball is connected by means of two massless strings to a vertical, rotating rod. The strings are tied to the rod and are taut. The tension in the upper string is 35 N. (a) Draw the free-body diagram for the ball. What are (b) the tension in the lower string, (c) the net force on the ball, and (d) the speed of the ball?

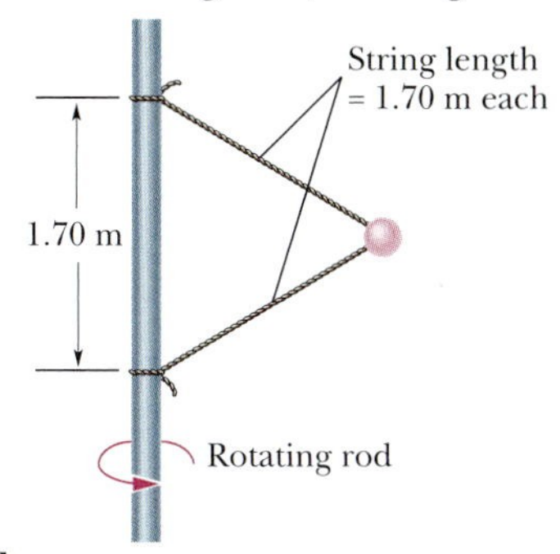

FIGURE 6-70 ■ Problem 75.

**76. Pushing the Second Block** A 2.0 kg block and a 1.0 kg block are connected by a string and are pushed across a horizontal surface by a force applied to the 1.0 kg block as shown in Fig. 6-71. The coefficient of kinetic friction between the blocks and the horizontal surface is 0.20. If the magnitude of $\vec{F}$ is 20 N, what is the tension in the string that connects the blocks?

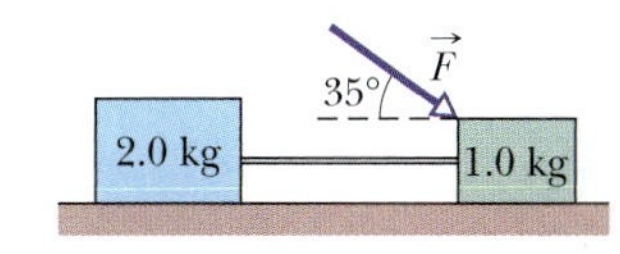

FIGURE 6-71 ■ Problem 76.

# Additional Problems

**77. Engineering a Highway Curve** If a car goes through a curve too fast, the car tends to slide out of the curve, as discussed in Touchstone Example 6-8. For a banked curve with friction, a frictional force acts on a fast car to oppose the tendency to slide out of the curve; the force is directed down the bank (in the direction in which water would drain). Consider a circular curve of radius $R = 200$ m and bank angle $\theta$, where the coefficient of static friction between tires and pavement is $\mu^{\text{stat}}$. A car is driven around the curve as shown in Fig. 6-72. (a) Find an expression for the car speed $v^{\text{max}}$ that puts the car on the verge of sliding out. (b) On the same graph, plot $v^{\text{max}}$ versus angle $\theta$ for the range 0° to 50°, first for $\mu^{\text{stat}} = 0.60$ (dry pavement) and then for $\mu^{\text{stat}} = 0.050$ (wet or icy pavement). In kilometers per hour, evaluate $v^{\text{max}}$ for a bank angle of $\theta = 10°$ and for (c) $\mu^{\text{stat}} = 0.60$ and (d) $\mu^{\text{stat}} = 0.050$. (Now you can see why accidents occur in highway curves when wet or icy conditions are not obvious to drivers, who tend to drive at normal speeds.)

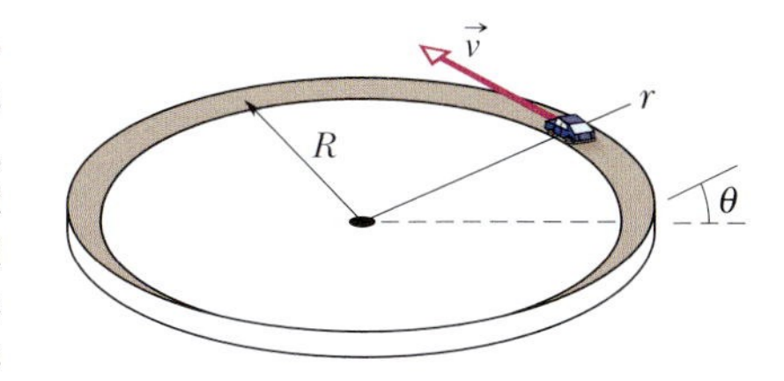

FIGURE 6-72 ■ Problem 77.

**78. Change in Conditions** In the early afternoon, a car is parked on a street that runs down a steep hill, at an angle of 35.0° relative to the horizontal. Just then the coefficient of static friction between the tires and the street surface is 0.725. Later, after nightfall, a sleet storm hits the area, and the coefficient decreases due to both the ice and a chemical change in the road surface because of the temperature decrease. By what percentage must the coefficient decrease if the car is to be in danger of sliding down the street?

**79. Moving People at the Airport** While traveling, I passed through Charles de Gaulle Airport in Paris, France. The airport has some interesting devices, including a "people mover"—a moving strip of rubber like a horizontal escalator without steps. It became interesting when the mover entered a plastic tube bent up at an angle to take me to the next terminal. I managed to get a photograph of it (Fig. 6-73). If you were building this people mover for the architect, what material would you choose for the surface of the moving strip? (*Hint*: You want to be sure that people standing on the strip do not tend to slide down it. Figure out what coefficient of friction you need to keep from sliding down and then look up coefficients of friction in tables in reference books to get a material appropriate for the slipperiest shoes.)

FIGURE 6-73 ■ Problem 79.

**80. Expert Witness** You testify as an *expert witness* in a case involving an accident in which car $A$ slid into the rear of car $B$, which was stopped at a red light along a road headed down a hill (Fig. 6-74). You find that the slope of the hill is $\theta = 12.0°$, that the cars were separated by distance $d = 24.0$ m when the driver of car $A$ put the car into a slide (it lacked any automatic anti-brake-lock system), and that the speed of car $A$ at the onset of braking was $v_1 = 18.0$ m/s. With what speed did car $A$ hit car $B$ if the coefficient of kinetic friction was (a) 0.60 (dry road surface) and (b) 0.10 (road surface covered with wet leaves)?

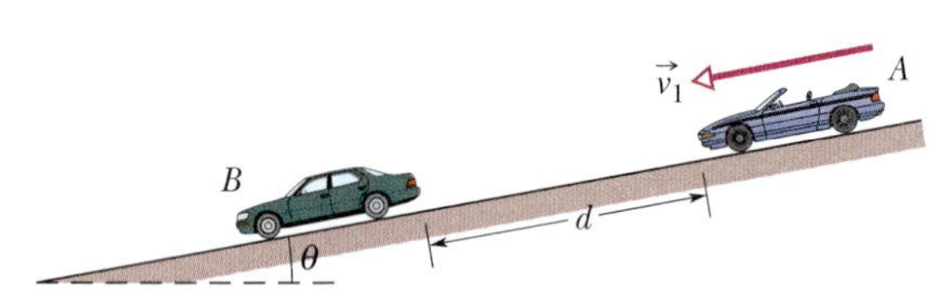


FIGURE 6-74 ■ Problem 80.

**81. Luggage Transport** Luggage is transported from one location to another in an airport by a conveyor belt. At a certain location, the belt moves down an incline that makes an angle of 2.5° with the horizontal. Assume that with such a slight angle there is no slipping of the luggage. Determine the magnitude and direction of the frictional force by the belt on a box weighing 69 N when the box is on the inclined portion of the belt for the following situations: (a) The belt is stationary. (b) The belt has a speed of 0.65 m/s that is constant. (c) The belt has a speed of 0.65 m/s that is increasing at a rate of 0.20 m/s$^2$. (d) The belt has a speed of 0.65 m/s that is decreasing at a rate of 0.20 m/s$^2$. (e) The belt has a speed of 0.65 m/s that is increasing at a rate of 0.57 m/s$^2$.

**82. Bolt on a Rod** A bolt is threaded onto one end of a thin horizontal rod, and the rod is then rotated horizontally about its other end. An engineer monitors the motion by flashing a strobe lamp onto the rod and bolt, adjusting the strobe rate until the bolt appears to be in the same eight places during each full rotation of the rod (Fig. 6-75). The strobe rate is 2000 flashes per second; the bolt has mass 30 g and is at radius 3.5 cm. What is the magnitude of the force on the bolt from the rod?

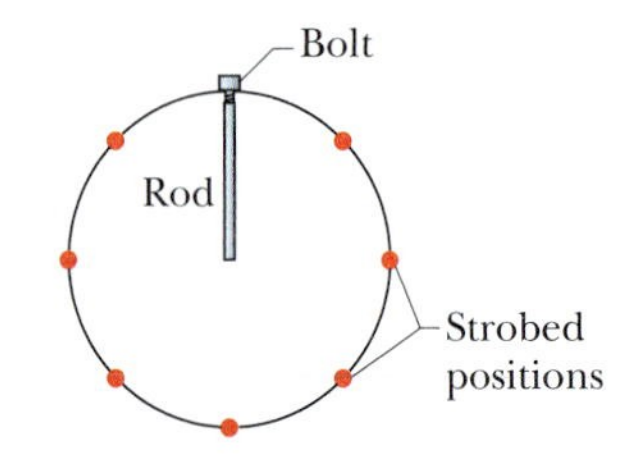


FIGURE 6-75 ■ Problem 82.

**83. From the Graph** A 4.10 kg block is pushed along a floor by a constant applied force that is horizontal and has a magnitude of 40.0 N. Figure 6-76 gives the block's speed $v$ versus time $t$ as the block moves along an $x$ axis on the floor. What is the coefficient of kinetic friction between the block and the floor?

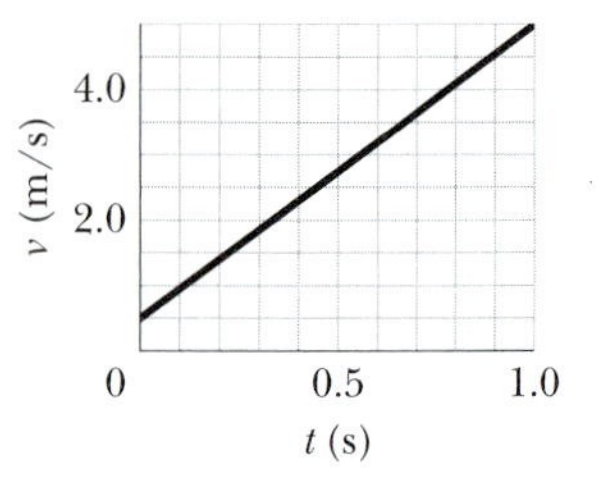


FIGURE 6-76 ■ Problem 83.

**84. Tapping a Rolling Ball** Figure 6-77 shows a multiple exposure strobe photograph of a ball rolling on a horizontal table. The image marked with a heavy arrow occurs at time $t = 0$ and the ball moves to the right at that instant. Each image of the ball occurs 1/30 s later than the one immediately to its left. Using the coordinate system shown in Fig. 6-77, sketch qualitatively accurate (i.e., we don't care about the values but we do care about the shape) graphs of each of the following variables as a function of time: $x$ coordinate, $y$ coordinate, $x$-component of velocity, $y$-component of velocity, $x$-component of the net force on the ball, and $y$-component of the net force on the ball. The time at which the "kink" in the path occurs is $t = t_1$. Be sure to note this important time on your graphs.

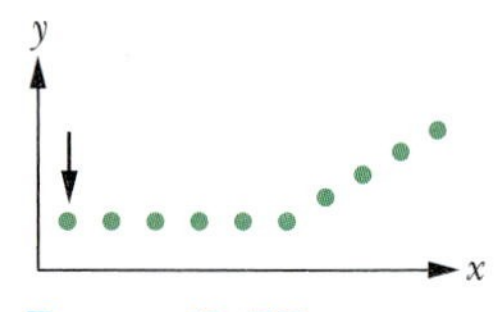


FIGURE 6-77 ■ Problem 84.

**85. Ball on a Ramp** Figure 6-78 shows a multiple-exposure photograph of a ball rolling up an inclined plane. (The ball is rolling in the dark, the camera lens is held open, and a brief flash occurs every 3/4 sec, four times in total.) The leftmost ball corresponds to an instant just after the ball was released. The rightmost ball is at the highest point the ball reaches.

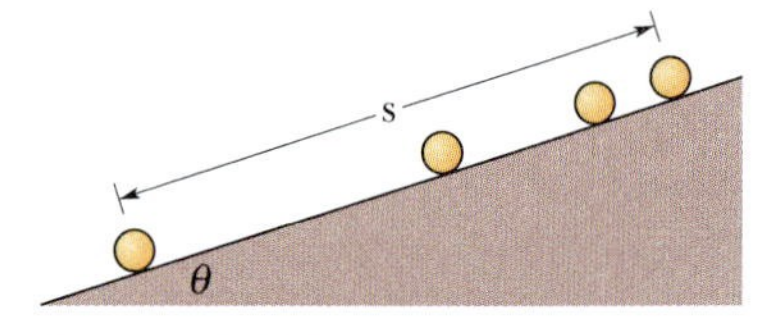


FIGURE 6-78 ■ Problem 85.

**(a)** Copy this picture on your paper. Draw an arrow at each of the four ball locations to indicate the velocity of the ball at that instant. Make the relative lengths of the arrows indicate the relative magnitudes of the velocities. Explain what is happening ("tell the story" of the picture).

**(b)** For the instant of time when the ball is at the second position shown from the left, draw a free-body diagram for the ball and indicate all forces acting on it.

**(c)** If your force diagram doesn't include an arrow pointing up the ramp, explain why the ball keeps rolling up the ramp.

**(d)** If the mass of the ball is $m$, what is its acceleration?

**(e)** If the angle $\theta$ is equal to 30°, how long is the distance $s$?

**86. Motion Graphs (a)** Suppose you were to push on a bowling ball on a smooth floor at a 45° angle as shown in Fig. 6-79*a* and then leave it alone to roll. Sketch a graph frame like that shown in Fig. 6-79*a*, and then sketch a prediction of the ball's motion both before and after you stop pushing. Note on your graph the point at which you stop pushing and explain the basis for your prediction.

**(b)** If the initial speed of the ball is 3.5 m/s, what is the magnitude of the $x$-component of velocity, $v_{1x}$? Is it positive or negative? What is $v_{1y}$? Is it positive or negative?

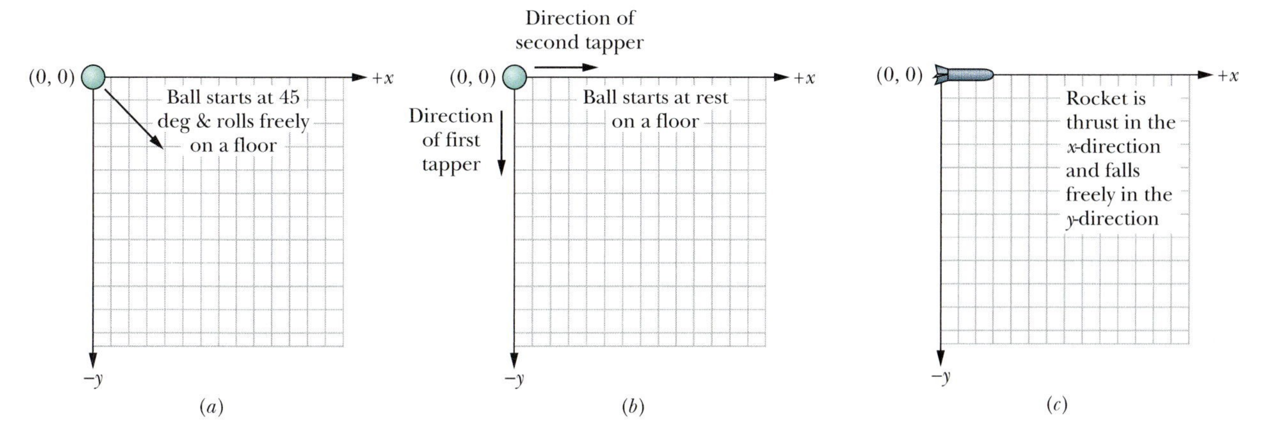


**FIGURE 6-79** ■ Problem 86.

**(c)** Suppose you and your partner were to tap the ball *very* rapidly. Each set of taps is at right angles to the other as shown in Fig. 6-79*b*. Sketch a graph frame like that shown in Fig. 6-79*b*, and sketch a prediction of the ball's motion on your graph. Explain the basis for your prediction.

**(d)** Suppose a rocket ship is thrust from a tower at a constant acceleration that has a magnitude of about 9.8 m/s$^2$ in the $x$ direction and is allowed to fall freely toward Earth in the $y$ direction. Sketch a graph frame like that shown in Fig. 6-79*c*, and sketch a prediction of the rocket's motion on your graph. Explain the basis for your prediction.

**87. Wanda Lifts Weights** Wanda is working out with weights and manages to lift a light rope with a 10 kg mass hanging from it. When she is through lifting the right side of the rope and the left side of the rope each make an angle of $\theta = 15°$ with respect to the horizontal. See Fig. 6-80.

**FIGURE 6-80** ■ Problem 87.

**(a)** Draw a free-body diagram showing the forces on the midpoint of the rope (where it is the lowest).

**(b)** What are the magnitudes of each of her pulling forces $\vec{F}_A$ and $\vec{F}_B$?

**(c)** How hard would Wanda have to pull with each hand to raise the 10 kg mass so that the rope becomes perfectly horizontal?

**88. Constant Speed on a Race Track** The race track shown in Fig. 6-81 has two straight sections connected by semicircular ends. A car is traveling in a clockwise direction around the track at a constant speed. Assume that air resistance is negligible. Draw three sketches of the race track.

**FIGURE 6-81** ■ Problem 88.

**(a)** On the first sketch show the velocity vector at each of the numbered points 1–4. Make the relative lengths of the vectors consistent with the relative magnitudes of the velocity at the four points.

**(b)** On the second sketch show the acceleration vectors at each of the numbered points 1–4. Make the relative lengths of the vectors consistent with the relative magnitudes of the acceleration at the four points. *Hint*: Use the techniques developed in Chapter 5 to draw vectors representing the acceleration or change in velocity.

**(c)** Horizontal forces are needed to maintain the car's motion around the track. These are provided by road friction and by road forces where the track is banked at the curves. On the third sketch show the vectors representing the required horizontal forces at each of the numbered points 1–4. Make the relative lengths of the vectors consistent with the relative magnitudes of the force at the four points.

*Note*: This exercise is adapted from A. Arons, *Homework and Test Questions for Introductory Physics Teaching* (New York: Wiley, 1994), Chapter 3.

**89. Pulling on the Ceiling** Suppose a person exerts a force of 50 N on one end of a rope as shown in Fig. 6-82.

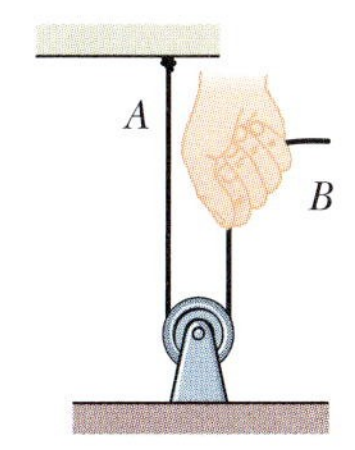


**FIGURE 6-82** ■ Problem 89.

**(a)** What are the magnitude and direction of the force at point $A$ exerted on the rope by the ceiling?

**(b)** What are the magnitude and direction of the force exerted on the ceiling by the rope? How does the force get transmitted from one end of the rope to the other? What does the stretching of the rope have to do with this?

**(c)** What are the magnitude and direction of the force the rope exerts on the person's hand at point $B$?

**(d)** Draw a diagram with vector arrows indicating the *relative magnitudes* and *directions* of the forces the rope exerts on the ceiling at point $A$ and the force the rope exerts on the person's hand at point $B$.

**90. Thinking About Normal Forces** Suppose you push on a flexible piece of stretched fabric with a force of 5.0 N as shown in Fig. 6-83*a*. The fabric assembly is fixed and does not move.

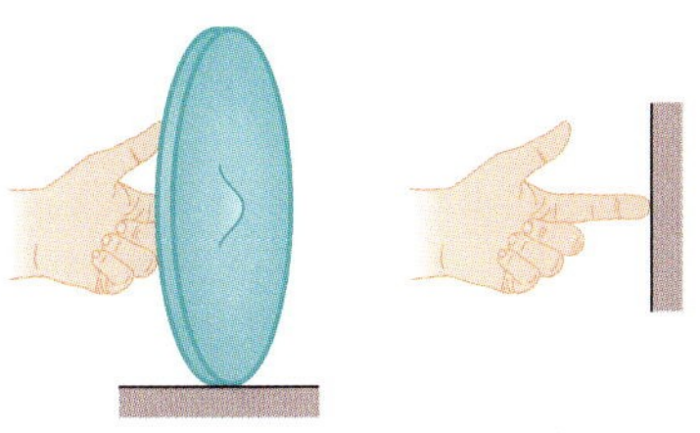

**FIGURE 6-83*a*** ■ Problem 90.

**(a)** What are the direction and magnitude of the normal force exerted back on the finger by the sheet? Is this normal force zero? If not, is it larger, smaller, or the same as the normal force would be if the fabric did not stretch?
**(b)** Discuss the role the stretching of the fabric plays in regard to this normal force.
**(c)** Suppose you push in the same way on a wall as shown in Fig. 6-83*b*. What are the direction and magnitude of the normal force exerted back on the finger by the wall?
**(d)** Does the wall stretch noticeably? What causes the wall to be able to exert a force on the finger? How does the wall "know" what force to exert back on the hand?

**91. Forces in a Car** Suppose you are sitting in a car that is speeding up. Assume the car has rear-wheel drive.

**(a)** Draw free-body diagrams for your own body, the seat in which you are sitting (apart from the car), the car (apart from the seat), and the road surface where the tires and the road interact.
**(b)** Describe each force in words; show larger forces with longer arrows.
**(c)** Identify the third-law pairs of forces.
**(d)** Explain carefully in your own words the origin of the force imparting acceleration to the car.

**92. The Sliding Pizza** One day I was coming home late from work and stopped to pick up a pizza for dinner. I put the pizza box on the dashboard of my car and pushed it forward against the windshield and left against the steering wheel to prevent it from falling. (See Fig. 6-84.) Before I started driving, I realized that the box could still slide to the right or back toward the seat. When driving, do I have to worry more about it sliding when I turn left or when I turn right? Do I have to worry more when I speed up or when I slow down? Explain your answer in terms of the physics you have learned.

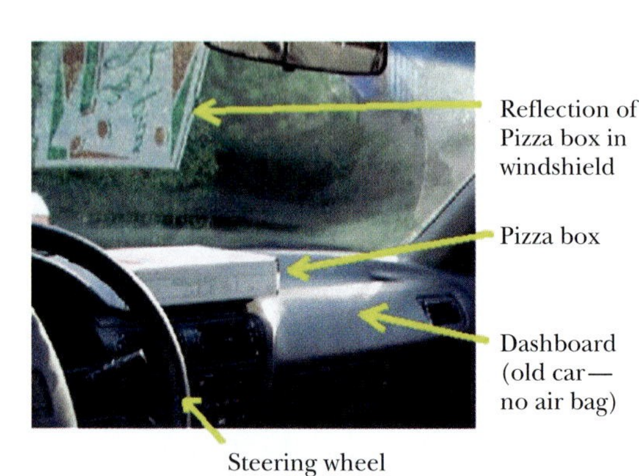


**FIGURE 6-84** ■ Problem 92.

**93. The Farmer and the Donkey** An old Yiddish joke is told about a farmer in Chelm, a town famous for the lack of wisdom of its inhabitants. One day the farmer was going to the mill to have a bag of wheat ground into flour. He was riding to the mill on his donkey, with the sack of wheat thrown over the donkey's back behind him. On his way, he met a friend. His friend chastised him. "Look at you! You must weigh 200 pounds and that sack of flour must weigh 100. That's a very small donkey! Together, you're too much weight for him to carry!" On his way to the mill the farmer thought about what his friend had said. On his way home, he passed his friend again, confident that this time the friend would be satisfied. The farmer still rode the donkey, but this time he carried the 100 pound bag of flour on his own shoulder!

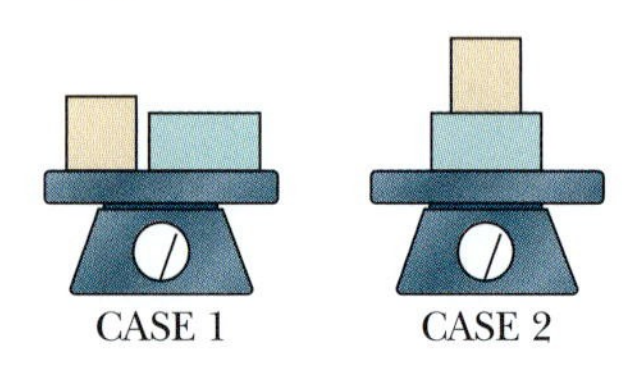


**FIGURE 6-85** ■ Problem 93.

Our common sense and intuitions seem to suggest that it doesn't matter how you arrange things; they'll weigh the same. Let's be certain that the Newtonian framework we are developing yields our intuitive result. Analyze the problem by considering the simplified picture shown in Fig. 6-85. Two blocks rest on a scale. One block weighs 10 N, the other 25 N. In case 1 the blocks are arranged on the scale as shown in the figure on the left. In case 2 the blocks are arranged as shown on the right. Each system has come to rest. Analyze the forces on the blocks and on the scale in the two cases by isolating the objects—each block and the scale—and show that according to the principles of Newton's laws, the total force exerted on the scale by both blocks together must be the same in both cases. (*Note*: It's not enough to say: "They have to be the same." That's just restating your intuition. We need to see that *reasoning using only the principles of our Newtonian framework* leads to the same conclusion.)

**94. Pulling Two Boxes (a)** A worker is trying to pull a pair of heavy crates along the floor with a rope. The rope is attached to the lower crate, which has a mass $M$. The upper crate has a mass $m$ and the coefficient of static friction between the crate and the floor is $\mu^{\text{stat}}$. If the rope is held at an angle $\theta$ as shown in Fig. 6-86, what is the magnitude of the maximum force the worker can exert without the lower crate beginning to slide?

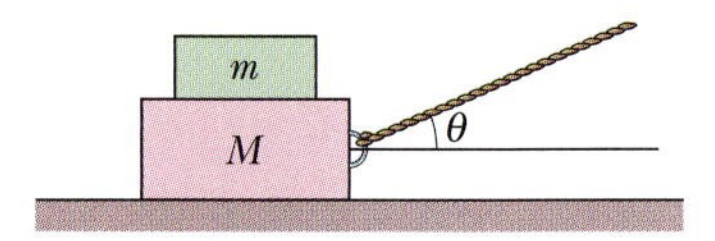


**FIGURE 6-86** ■ Problem 94.

**(b)** The worker knows that the lower crate has a mass of 50 kg and the upper crate has a mass of 10 kg. She finds that if she pulls with a force of 120 N at an angle of 60° she can keep the crates sliding at a constant speed. Can you use this information to find the coefficient of kinetic friction $\mu^{\text{kin}}$ between the lower crate and the floor? If you can, do it. If you can't, explain why not.
**(c)** In a different situation, she finds that she can pull a lower crate of mass 30 kg and an upper crate of mass 7.5 kg with a constant velocity of 50 cm/s pulling at an angle of 45°. Can you use this information to find the coefficient of kinetic friction $\mu^{\text{kin}}$ between the lower crate and the floor? If you can, do it. If you can't, explain why not.

**95. Tricking Bill** A student, whom we will call Bill, was about to go out on a date when his roommate, Bob, asked him to hold a pail against the ceiling with a broom for a moment. After Bill complied, the roommate mentioned that the pail was filled with water and left. See Fig. 6-87.

**(a)** Draw a free-body diagram showing all the forces acting on the pail. For each force, be sure you identify the kind of force and the object whose interaction with the pail is responsible for the force.
**(b)** Suppose Bill wants to slide the pail a few feet to one side so he can get to a chair in the room. Are there any other forces not specified in your answer to part (a) that become relevant?

**FIGURE 6-87** ■ Problem 95.

**(c)** Suppose the pail weighs 1 pound, it has 6 pounds of water in it, the maximum coefficient of static friction $\mu^{\text{stat}}_{\text{broom}}$ between the broom and pail is 0.3, and the maximum coefficient of static friction $\mu^{\text{stat}}_{\text{ceiling}}$ between the pail and the ceiling is 0.5. Can Bill slide the pail? Explain.

**96. Friction is Doing *What?*** A large block is resting on the table. On top of that block rests another, smaller block, as shown in Fig. 6-88. You press on the larger block to start it moving. After about 0.25 s, it is moving at a constant speed and the block on the top is not slipping.

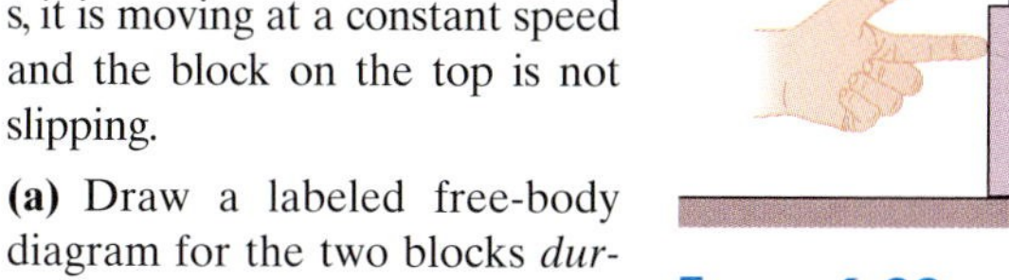


FIGURE 6-88 ■ Problem 96.

**(a)** Draw a labeled free-body diagram for the two blocks *during the time when they are accelerating*, specifying all the forces acting on the blocks. (Be sure to specify the type of force and the object causing each force.) Wherever you can, compare the magnitudes of forces.

**(b)** Draw a labeled free-body diagram for the two blocks *during the time when they are moving at a constant speed*, specifying all the forces acting on the blocks. (Be sure to specify the type of force and the object causing each force.) Wherever you can, compare the magnitudes of forces.

**(c)** Suppose the bottom block has a mass of 0.4 kg and the coefficient of friction between the block and the table is 0.3. The top block has a mass of 0.1 kg and the coefficient of friction between the two blocks is 0.2. What force do you need to exert to keep the blocks moving at a constant speed of 10 cm/s? (You may use $g = 10$ N/kg and you may treat kinetic and static friction as the same.)

**97. Al and George Pushing the Truck** George left the lights on in his truck while at a truck stop in Kansas and his battery went dead. Fortunately, his friend Al is there, although Al is driving his Geo Metro. Since the road is very flat, George is able to convince Al to give his truck a long, slow push to get it up to 20 miles/hour. At this speed, George can engage the truck's clutch and the truck's engine should start up. (See Fig. 6-89.)

FIGURE 6-89 ■ Problem 97.

**(a)** Al begins to push the truck. It takes him 5 minutes to get the truck up to a speed of 20 miles/hour. Draw separate free-body diagrams for the Geo and for the truck during the time that Al's Geo is pushing the truck. List all the horizontal forces in order by magnitude from largest to smallest. If any are equal, state that explicitly. Explain your reasoning.

**(b)** If the truck is accelerating uniformly over the 5 minutes, how far does Al have to push the truck before George can engage the clutch?

**(c)** Suppose the mass of the truck is 4000 kg, the mass of the car is 800 kg, and the coefficient of static friction between the vehicles and the road is 0.1. At one instant when they are trying to get the truck moving, the car is pushing the truck and exerting a force of 1000 N, but neither vehicle moves. What is the static frictional force between the truck and the road? Explain your reasoning.

**98. Pushing a Carriage** A young man is pushing a baby carriage at a constant velocity along a level street. A friend comes by to chat and the young man lets go of the carriage. It rolls on for a bit, slows, and comes to a stop. At time $t = 0$ the young man is walking with a constant velocity. At time $t_1$ he releases the carriage. At time $t_2$ the carriage comes to rest. Sketch qualitatively accurate (i.e., we don't care about the values but we do care about the shape) graphs of each of the following variables versus time:

(a) position of the carriage, (b) velocity of the carriage, (c) acceleration of the carriage, (d) net force on the carriage, (e) force the man exerts on the carriage, (f) force of friction on the carriage. Be sure to note the important times $t = 0$, $t_1$, and $t_2$ on the time axes of your graphs. Take the positive direction to be the direction in which the man was initially walking.

**99. A Two-Stage Rocket** Students in a school rocketry club have prepared a two-stage rocket. The rocket has two small engines. The first will fire for a time, getting the rocket up partway. Then the first-stage engine drops off, revealing a second engine. After a little time, that engine will fire and take the rocket up even higher.

The rocket starts firing its engines at a time $t = 0$. From that instant, it begins to move upward with a constant acceleration. This continues until time $t_1$. The rocket drops the first stage and continues upward briefly until time $t_2$, at which point the second stage begins to fire and the rocket again accelerates upward, this time with a larger (but again constant) acceleration. Sometime during this second period of acceleration, our recording apparatus stops.

Sketch qualitatively accurate (i.e., we don't care about the values but we do care about the shape) graphs of the height of the rocket, $y$, its velocity, $v_y$, its acceleration, $a_y$, the force on the rocket that results from the firing of the engine, $\vec{F}_y$, and the net force on the rocket, $\vec{F}_y^{\text{net}}$. Take the positive direction as upward. Be sure to note times $t = 0$, $t_1$, and $t_2$ on the time axes of your graphs.

**100. Pushing a Box** A worker is pushing a cart along the floor. At first, the worker has to push hard in order to get the cart moving. After a while, it is easier to push. Finally, the worker has to pull back on the cart in order to bring it to a stop before it hits the wall. The force exerted by the worker on the cart is purely horizontal. Take the direction the worker is going as positive.

Figure 6-90 shows graphs of some of the physical variables of the problem. Match the graphs with the variables in the list at the

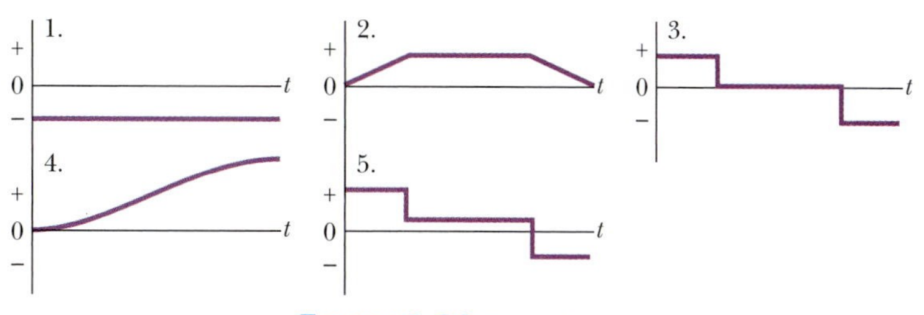


FIGURE 6-90 ■ Problem 100.

left below. You may use a graph more than once or not at all. *Note*: The time axes are to the same scale, but the ordinates $y$ axes are not.

**(a)** Friction force
**(b)** Force exerted by the worker
**(c)** Net force
**(d)** Acceleration
**(e)** Velocity

**101. Comparing a Light and Heavy Object** Consider a metal sphere two inches in diameter and a feather. For each quantity in the list below, indicate the relation between the quantity for the sphere and feather. Is it the same, greater, or lesser? Explain in each case why you gave the answer you did.

**(a)** The gravitational force
**(b)** The time it will take to fall a given distance in air
**(c)** The time it will take to fall a given distance in vacuum
**(d)** The total force on the object when falling in vacuum
**(e)** The total force on the object when falling in air

**102. Hitting the Green** A golfer is trying to hit a golf ball onto the green as shown in Fig. 6-91. The green is a horizontal distance $s$ from his tee and it is up on the side of a hill a height $h$ above his tee. When he strikes the ball it leaves the tee at an angle $\theta$ to the horizontal. He wants to know with what speed, $v_1$, the ball must leave the tee in order to reach the height $h$ at the distance $s$.

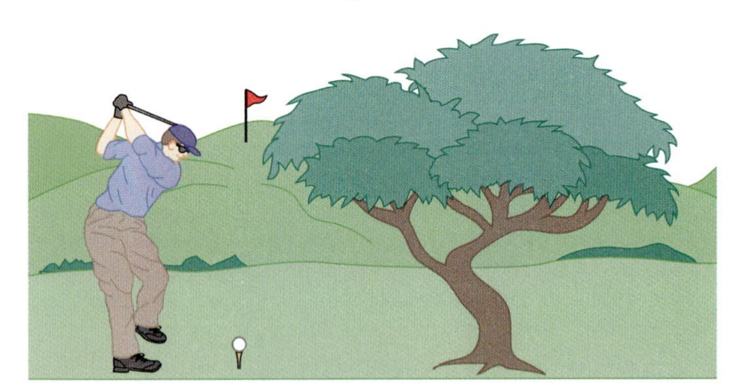

FIGURE 6-91 ■ Problem 102.

**(a)** Once he has struck the ball, what controls its motion? Write the equations that determine the vector acceleration of the golf ball afler it leaves the tee. Be sure to specify your coordinate system. For this part of the problem you may ignore air resistance.
**(b)** Solve the equations you have written in (a) to obtain expressions that can be evaluated to give the position of the ball at any time, $t$.
**(c)** If the golfer wants his ball to land in the right place, he must hit it so that it leaves the tee with the right speed. Explain how he can calculate it. (Again, you may ignore air resistance.) Find an equation for the initial speed in terms of the problem's givens.
**(d)** If the ball leaves the tee at an angle of 30°, $s = 100$ m, and $h = 10$ m, find the speed with which the ball leaves the tee.
**(e)** Now consider the effect of air resistance. Suppose that a good model for the force of air resistance is Newton's drag law,

$$\vec{F} = -b\,|\vec{v}|\,\vec{v}$$

where $|\vec{v}|$ is the speed and $b$ is a constant. Consider three points on the ball's trajectory: halfway up, at its highest point, and halfway down. Discuss the direction of the resistance force at each point. Qualitatively (do not attempt a calculation!), what will the effect of air resistance be on the ball's motion?

**103. Air Resistance 1: Dimensional Analysis** We know that as an object passes through the air, the air exerts a resistive force on it. Suppose we have a spherical object of radius $R$ and mass $m$. What might the force plausibly depend on?

- It might depend on the properties of the object. The only ones that seem relevant are $m$ and $R$.
- It might depend on the object's coordinate and its derivatives: $\vec{r}, \vec{v}, \vec{a}, \ldots$.
- It might depend on the properties of the air, such as the density, $\rho$.

**(a)** Explain why it is plausible that the force the air exerts on a sphere depends on $R$ but implausible that it depends on $m$.
**(b)** Explain why it is plausible that the force the air exerts depends on the object's speed through it, $|\vec{v}|$, but not on its position, $\vec{r}$, or acceleration, $\vec{a}$.
**(c)** Dimensional analysis is the use of units (e.g., meters, seconds, or newtons) associated with quantities to reason about the relationship between the quantities. Using dimensional analysis, construct a plausible form for the force that air exerts on a spherical body moving through it.

**104. Counterweights** The use of counterweights to help devices move up and down with a minimum of effort is common in engineering. For example, counterweights are used to help people open and close old-fashioned windows and to move up and down in elevators. Imagine that an engineer working for the Disney Epcot Center is asked to design a ride that allows people to travel up and down a sloped hill to get a view of the entire Epcot Center while other tourists move straight up and down an artificial cliff on the other side of the incline. Our engineer builds a small prototype of his device using a low-friction cart on an inclined track attached to a falling mass. His goal is to see whether he can actually apply Newton's laws to this situation and if it is okay to neglect the effects of friction.

In this exercise you will analyze data collected from a digital movie of the situation discussed above and shown in Fig. 6-92a. If you have access to VideoPoint you can view the digital movie yourself. It is entitled PASCO098. Your instructor may provide you with a different but similar movie. The cart in PASCO098 has a mass $m_c = .510$ kg and is accelerated up a ramp that has a 21° incline. A string attached to the cart exerts a force on it. The string transmits a force to the cart because its other end is attached by means of a pulley to a falling mass of $m_f = .184$ kg.

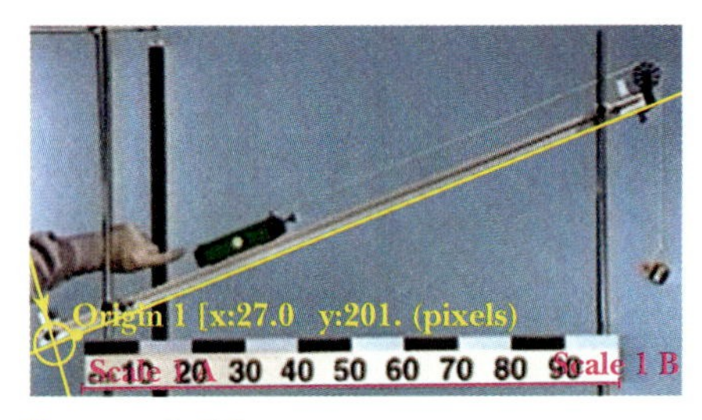


FIGURE 6-92a ■ Problem 104.

Table 6-3 contains position vs. time measurements for the cart in PASCO098 along an $x$ axis. The $x$ axis is rotated from the horizontal direction so that it lies along the ramp. Using these data you can determine the acceleration, if any, of the cart. (It is best to enter the data into a spreadsheet for analysis.) Finally, you will use Newton's laws along with the information on the angle of the incline and the masses of the cart and the falling mass to determine (theoretically) what the acceleration of the cart is. Our goal is to determine whether the theoretically calculated motion and the actual motion (as described by the data in Table 6-3) agree.

**(a)** Enter the data in Table 6-3 into a spreadsheet program. Determine what kind of motion the cart experiences. Is it a constant velocity? If so what are the magnitude and direction of the velocity? Is the motion a constant acceleration? If so, what are the magnitude and direction of the acceleration? (You may want to use equation-fitting software in answering this question). Cite the evidence that leads you to give the answers you did.

**(b)** What is the value of the net force on the cart in the $x$ direction (along the incline)?

**(c)** Sketch a diagram of the cart like that shown in Fig. 6-92*b*. Draw a free-body diagram showing the directions of *all* the forces on the cart including the gravitational force, $\vec{F}^{\text{grav}}$, the normal force, $\vec{N}$, and the string force due to its tension, $T$.

**(d)** Consider the situation in which the cart and falling mass move with a constant velocity. Choose a coordinate system in which the positive $x$ axis is directed up along the ramp (rotated from the horizontal). *Assume that there is no friction in the pulley or cart bearings*! Show that by taking components of these forces along the $x$ axis the magnitude, $F_x^{\text{net}}$, of the net force on the cart in the $x$ direction, can be calculated using the equation

$$F_x^{\text{net}} = T - m_c\, g \sin\theta = 0,$$

where the gravitational constant, $g$, is $+9.8$ N/kg.

**(e)** Assume that since the cart and falling mass are connected by the string they have the same magnitude of velocity. Also assume that the tension in the string is the same at all points along the string so that the magnitude of the string force at point $A$ on the cart is the same as the magnitude of the string force at point $B$ on the falling mass. Show that if the net force on the falling mass is zero, then $T - F^{\text{grav}} = 0$, where $F^{\text{grav}} = m_f g$.

**(f)** Use the equations you derived in parts (d) and (e) to show that if the velocity of the cart and falling mass system are constant, then theoretically $m_f g$ ought to equal $m_c g \sin\theta$.

**(g)** Use the given values of $m_c$ and $m_f$ (also available on the title screen of the PASCO098 movie) along with the angle of the incline to verify that $m_f g$ and $m_c g \sin\theta$ have the same values to two significant digits. This equality, if it exists, confirms the agreement between theory and experiment.

**(h)** Also discuss why the answers should only be good to two significant figures.

**TABLE 6-3**
**Problem 104**

| Time (sec) | $x$(m) |
|---|---|
| 0.000 | 0.002929 |
| 0.2050 | 0.03956 |
| 0.4100 | 0.08465 |
| 0.6150 | 0.1221 |
| 0.8200 | 0.1659 |
| 1.025 | 0.2038 |
| 1.230 | 0.2463 |
| 1.435 | 0.2885 |
| 1.640 | 0.3301 |
| 1.845 | 0.3676 |
| 2.050 | 0.4114 |
| 2.255 | 0.4472 |
| 2.460 | 0.4931 |
| 2.665 | 0.5297 |
| 2.870 | 0.5748 |
| 3.075 | 0.6165 |
| 3.280 | 0.6624 |

**FIGURE 6-92*b*** Problem 104.

# 7 Translational Momentum

A karate master undergoes extensive training to thicken the bones and strengthen the muscles in his or her hands. This enables him or her to break stacks of concrete patio blocks with a single blow. Although novices cannot perform this feat, they are able to break 3/4-inch-thick pine boards quite easily. For example, Tom Casiani, an introductory physics student at Dickinson College, broke a stack of nine pine boards in spite of the fact that he had never done any karate before taking physics.

**How can novices break pine boards, but not concrete slabs, without sustaining injuries?**

---

*The answer is in this chapter.*

## 7-1 Collisions and Explosions

In the last few chapters, we have focused on understanding how forces affect the motion of an object. We have specifically discussed several common forces and practiced using Newton's Second Law, $\vec{F}^{\text{net}} = m\vec{a}$, to determine the acceleration of an object that experiences steady forces. Although there are many situations in which one can determine the acceleration of an object by summing the forces acting on it, there are other situations in which using the equation $\vec{F}^{\text{net}} = m\vec{a}$ is not possible. For example, when objects collide or a large object explodes into smaller fragments, the event can happen so rapidly that it is impossible to keep track of the interaction forces.

Collisions and explosions range from the microscopic scale of subatomic particles (Fig. 7-1*b*) to the astronomic scale of colliding stars and galaxies, so an understanding of these processes is of great interest to physicists. In fact, many physicists today spend their time playing "the collision game." The goal of this game is to find out as much as possible about the forces that act during rapid interactions between particles or during the explosion of a particle into fragments. We have developed techniques for learning about rapid interactions by determining the state of the particles before and after they interact. Indeed, most of our understanding of the subatomic world—electrons, protons, neutrons, muons, quarks, and the like—comes from experiments involving collisions and explosions.

In this chapter we will define a new quantity known as linear or translational momentum to help us study collision processes. Since explosions are actually collision processes in reverse, it turns out that we can use the same methods in studying both phenomena. We shall use the following formal definition of a collision.

> A **COLLISION** or **EXPLOSION** is an isolated event in which two or more bodies exert relatively strong forces on each other over a short time compared to the period over which their motions take place.

By *relatively strong forces* we mean that the collision or explosion forces are considerably larger than other forces that might be acting on the system. Similarly, a *relatively short time* means that the weaker forces (other than the collision or explosion forces) have not had enough time to accelerate the system elements noticeably.

In order to analyze collisions or explosions, we distinguish between times that are *before, during*, and *after* an event, as suggested in Fig. 7-2. Figure 7-2 shows two colliding bodies and indicates that the forces associated with the collision are forces that the bodies exert *on each other*.

**READING EXERCISE 7-1:** According to our definition of collision, which, if any, of the following events qualify as collisions? Explain. (*a*) Suppose it took the ocean liner *Titanic* 60 seconds to plow into an iceberg and come to a stop. (*b*) During a volley, a tennis ball usually is in the air for less than a second. Suppose a tennis racket is in contact with a ball for 2 seconds. ■

(*a*)

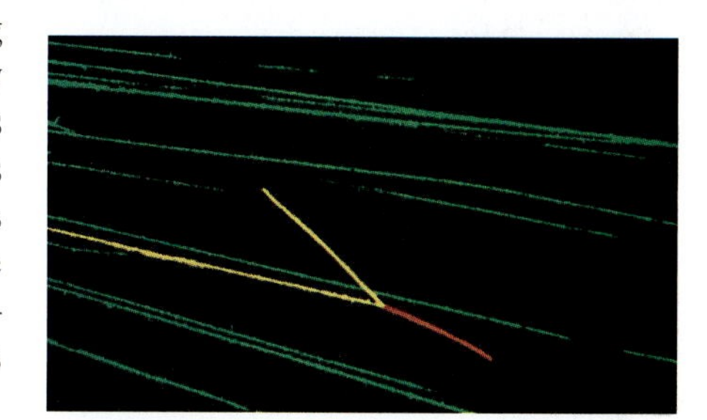

(*b*)

(*c*)

**FIGURE 7-1** ■ Collisions range widely in scale. (*a*) Meteor Crater in Arizona is about 1200 m wide and 200 m deep. (*b*) An alpha particle coming in from the left bounces off a nitrogen nucleus that had been stationary and that now moves toward the bottom right. (*c*) In a tennis match, the ball is in contact with the racquet for about 4 ms in each collision (for a cumulative time of approximately 1 s in a set).

## 7-2 Translational Momentum of a Particle

Consider a winter accident on a narrow icy road in which a compact car skids into a loaded pickup truck that is moving toward it. If you want to predict the motions of the vehicles after that crash, what do you need to know about the vehicles? Many people would guess that both the mass and the velocity of each vehicle make a difference. It turns out that the product of these two quantities, which we will soon begin calling by the name momentum, is a very useful concept in predicting the outcome of collisions.

In fact, Newton did not use the ideas of acceleration and velocity in his original descriptions of the laws of motion. He developed his laws, in part, by studying colli-

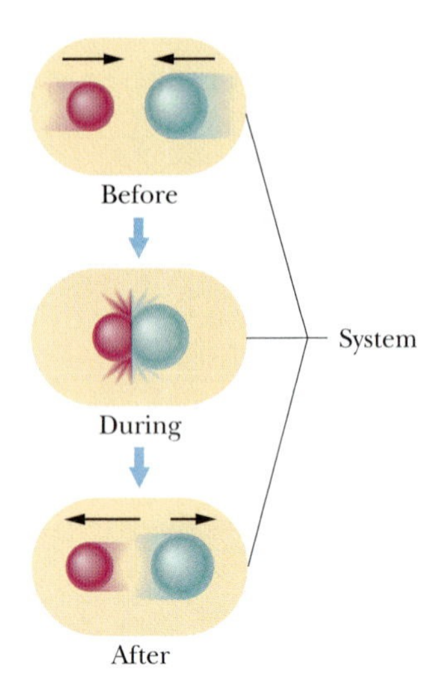


**FIGURE 7-2** ■ Stages in a collision between two bodies.

sions, and this led him to introduce the concept of momentum. As is the case for "acceleration," momentum is a word that has several meanings in everyday language but only a single precise meaning in physics. The **translational momentum** of a particle is a vector $\vec{p}$, defined as

$$\vec{p} \equiv m\vec{v} \quad \text{(definition of translational momentum)}, \tag{7-1}$$

where $m$ is the mass of the particle and $\vec{v}$ is its instantaneous velocity. Since $m$ is always a positive scalar quantity, the momentum vector and the velocity vector are always in the same direction. This relation also tells us that the SI unit for momentum is the kilogram-meter per second (the unit for mass multiplied by the unit for velocity).

Some people use the phrase "linear momentum" rather than the phase "translational momentum" when discussing the product $m\vec{v}$. However, this momentum is associated with the movement of an object from one position to another, regardless of whether the overall motion of the object occurs along a line. For example, the equation $\vec{p} = m\vec{v}$ serves to define the momentum of a projectile following a parabolic path or a small rock rotating in a circle. Hence, the term "translational" is a better adjective than "linear."

The adjective "translational" is often dropped, leaving us with just the term "momentum." However, it serves to distinguish this type of momentum from *rotational momentum*, which is introduced in Chapter 12.

Newton expressed his second law of motion in terms of momentum as follows:

> The rate of change of the momentum of a particle is proportional to the net force acting on the particle and is in the direction of that force.

In equation form this statement is

$$\vec{F}^{\text{net}} = \frac{d\vec{p}}{dt} \quad \text{(single particle)}. \tag{7-2}$$

We can relate this statement of the second law to the familiar $\vec{F}^{\text{net}} = m\vec{a}$ by substituting $m\vec{v}$ for $\vec{p}$ and pulling the mass, which is a constant, out of the derivative so that

$$\vec{F}^{\text{net}} = \frac{d\vec{p}}{dt} = \frac{d}{dt}(m\vec{v}) = m\frac{d\vec{v}}{dt} = m\vec{a}.$$

Thus, the equations $\vec{F}^{\text{net}} = d\vec{p}/dt$ and $\vec{F}^{\text{net}} = m\vec{a}$ are equivalent expressions of Newton's Second Law of Motion as it applies to the motion of a particle whose mass remains constant.

What these relations are telling us is that a nonzero net force on a body causes it to undergo a momentum change. This should not come as a surprise. A nonzero net force results in an acceleration of the object on which the force acts. That acceleration produces a change in velocity, and the change in velocity is associated with a change in momentum. Another way to think of this is that a nonzero net force—for example, the push on a cart—is what "gives" the cart its change in momentum.

**READING EXERCISE 7-2:** The figure to the right gives the translational momentum versus time for a particle moving along an axis. A force directed along the axis acts on the particle, causing its momentum to change. (a) Rank the four regions indicated according to the magnitude of the force, greatest first. (b) In which region is the particle slowing down?

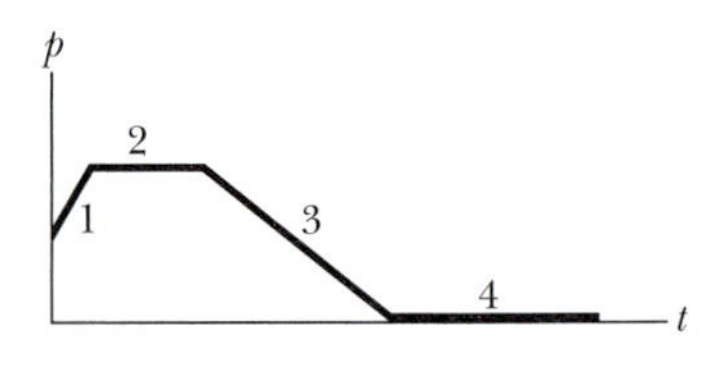

## 7-3 Isolated Systems of Particles

We are often interested in the behavior of a collection of particles that interact only with each other. We can draw an imaginary boundary around the particles, but complications can arise if some of the particles experience net forces that originate outside the "system" boundary. In order to effectively study interactions between particles, we must limit our focus to **isolated systems** of particles. (See Fig. 7-3.)

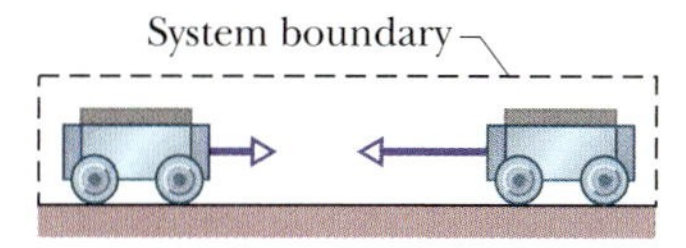


**FIGURE 7-3** ■ The vector sum of the normal force on each cart and the gravitational force on it is zero. If the track is also essentially frictionless, the carts form an *isolated system*. The collision forces they exert on each other are inside the system and can be studied fairly easily.

> An **ISOLATED SYSTEM** is defined as a collection of particles that can interact with each other but whose interactions with the environment outside the collection have a negligible effect on their motions.

Basically the particles in an isolated system experience no significant net external forces. Two carts that are about to collide on a frictionless track form an isolated system. Why? Because even though each cart experiences a normal force from the track and a gravitational force from the Earth, the net force on each cart that originates outside the system boundary is zero.

Let's consider another example. Suppose two galaxies collide in outer space and exert a complex series of gravitational and electromagnetic forces on each other. Let's take these two galaxies to be our system. If these galaxies are far from other astronomical bodies, their gravitational interactions with entities outside the system will be small—especially in comparison to the internal forces they exert on each other. These galaxies can be considered to be an isolated system. If the particles in our two galaxies are strongly attracted to neighboring galaxies, they do not form an isolated system. Additional examples of isolated systems are shown in Fig. 7-4.

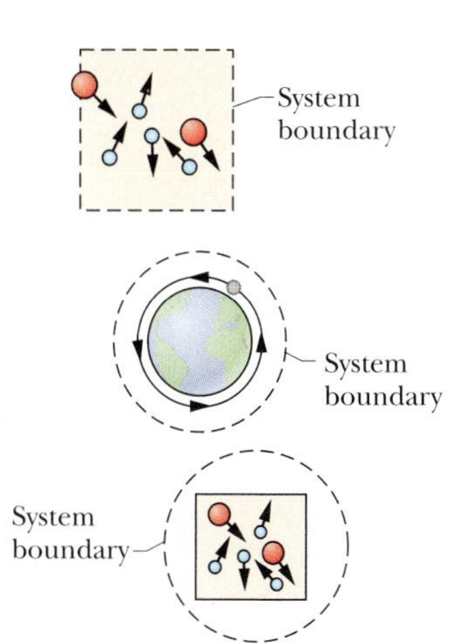

System boundary
Pucks riding on a cushion of air on an air table interact with each other before hitting the walls of the table. Friction forces with the surface of the table are negligible. The system is temporarily isolated—until a puck hits an air table wall.
System boundary
An orbiting satellite and the Earth interact. Forces between these objects and others such as the sun and moon are considered to have negligible effect on their motions.
System boundary
Gas molecules interact with each other and with the walls of their container. Other forces, such as those of the table holding up the container and the gravitational force, are considered to have a negligible effect on the motions of the molecules and container.


**FIGURE 7-4** ■ Examples of isolated systems.

### Momentum for a System of Particles

If we apply Newton's Second and Third Laws to an isolated system of particles, we can learn about changes in the total momentum of the system. Let's consider a system of $n$ particles, each with its own mass, velocity, and translational momentum. The particles in the system may interact with each other. The system as a whole has a total translational momentum $\vec{p}_{\text{sys}}$, which is defined to be the vector sum of the individual particles' translational momenta. Thus,

$$\begin{aligned}\vec{p}_{\text{sys}} &= \vec{p}_A + \vec{p}_B + \vec{p}_C + \cdots + \vec{p}_n \\ &= m_A\vec{v}_A + m_B\vec{v}_B + m_C\vec{v}_C + \cdots + m_n\vec{v}_n.\end{aligned} \quad (7\text{-}3)$$

> The translational momentum of a system is the vector sum of the momenta of the individual particles.

If the system is not isolated, then external forces are also acting on the system. Recall that for a single particle such as particle $A$,

$$\vec{F}_A^{\text{net}} = \frac{d\vec{p}_A}{dt}.$$

Hence, we have a separate Newton's Second Law equation for each of the $n$ particles, telling how that particle will respond to the forces it feels:

$$\vec{F}_A^{\text{net}} = \frac{d\vec{p}_A}{dt}, \qquad \vec{F}_B^{\text{net}} = \frac{d\vec{p}_B}{dt}, \qquad \vec{F}_C^{\text{net}} = \frac{d\vec{p}_C}{dt}, \ldots .$$

But the total momentum of the system, $\vec{p}_{\text{sys}}$, is given by the sum of the momenta of the particles in the system, so that

$$\vec{p}_{\text{sys}} = \vec{p}_A + \vec{p}_B + \vec{p}_C + \cdots .$$

Since the derivative of a sum is the same as the sum of the derivatives, the rate of change of the system momentum is given by

$$\frac{d\vec{p}_{\text{sys}}}{dt} = \frac{d(\vec{p}_A + \vec{p}_B + \vec{p}_C \ldots)}{dt} = \frac{d\vec{p}_A}{dt} + \frac{d\vec{p}_B}{dt} + \frac{d\vec{p}_C}{dt} + \cdots . \tag{7-4}$$

However, according to Eq. 7-2, the net force on any one of the particles is given by $\vec{F}^{\text{net}} = d\vec{p}/dt$, so the rate of change of the total momentum of the system is equal to the sum of the forces felt by each of the $n$ particles:

$$\frac{d\vec{p}_{\text{sys}}}{dt} = \vec{F}_A^{\text{net}} + \vec{F}_B^{\text{net}} + \vec{F}_C^{\text{net}} + \cdots = \vec{F}_{\text{sys}}^{\text{net}}.$$

In words, the sum of all forces acting on all the particles in the system is equal to the time rate of change of the total momentum of the system. That leaves us with the general statement:

$$\vec{F}_{\text{sys}}^{\text{net}} = \frac{d\vec{p}_{\text{sys}}}{dt} \quad \text{(system of particles)}. \tag{7-5}$$

In principle, $\vec{F}^{\text{net}}$ is the sum of all forces on particles in the system. This includes forces from particles within the system acting on other particles within the system (called **internal forces**). It also includes forces from objects outside the system acting on objects within the system (called **external forces**). In practice, all the internal forces occur as third-law pairs that cancel. Thus, the contribution of the internal forces to the overall net force is zero. Hence, $\vec{F}^{\text{net}}$ is always just the sum of all *external* forces acting on the system. This equation is the generalization of the single-particle equation $\vec{F}^{\text{net}} = d\vec{p}/dt$ (Eq. 7-2) to a system of many particles.

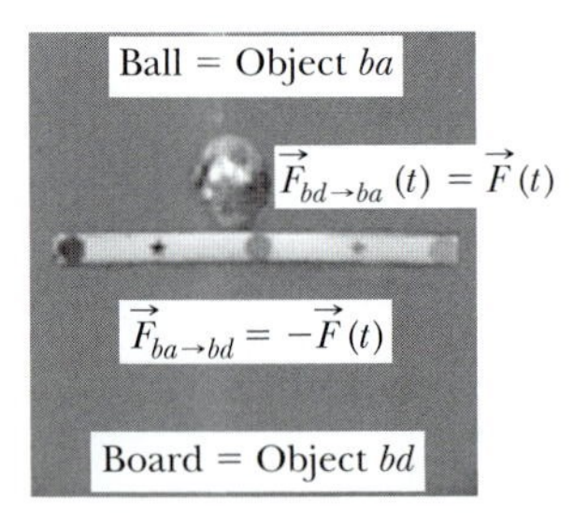


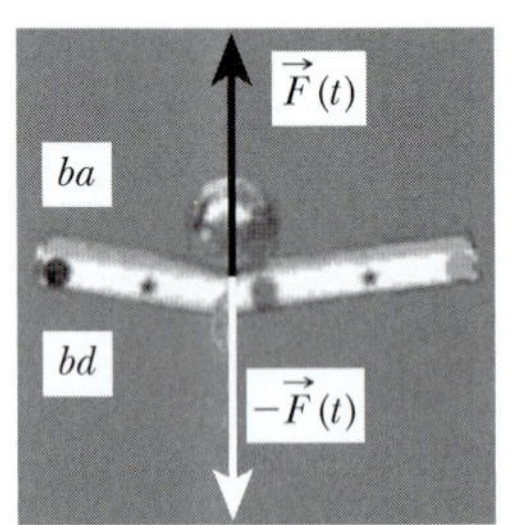


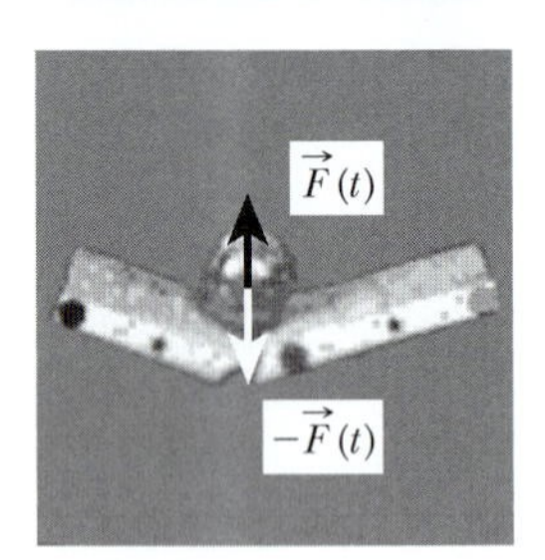


**FIGURE 7-5** ▪ A lead ball ($ba$) of mass 0.850 kg collides with a pine board ($bd$). During the collision, the lead ball exerts a force of $\vec{F}_{ba\to bd} = -\vec{F}(t)$ on the board and the board exerts force $\vec{F}_{bd\to ba} = +F(t)$ on the ball. Forces $\vec{F}(t)$ and $-\vec{F}(t)$ are a third-law force pair. Their magnitudes vary with time during the collision, but at any given instant those magnitudes are equal. A digital video clip of the collision was recorded at 250 frames/second. The ball and the board are in contact for only about 0.012 s. (Courtesy of Robert Teese.)

## 7-4 Impulse and Momentum Change

Although a pine board is not really very particle-like, it is instructive to consider a two-body system consisting of a falling lead ball ($ba$) that collides with a pine board ($bd$) (shown in Fig. 7-5). At any given moment, the force that the board exerts on the ball can be denoted as $\vec{F}_{bd \to ba}$. It is obvious that before the falling ball makes contact with the board, the interaction forces are negligible. When the ball first makes contact with the board, the magnitudes of the interaction forces are relatively small. As the force magnitudes reach a maximum, the ball causes the board to flex and begin to break. After the board breaks, the forces are zero again. Thus, we expect a graph of the magnitude of the force exerted on the ball by the board to look something like that shown in Fig. 7-6*a*.

A video analysis of the acceleration of the ball shows that the peak force exerted on it by the board is about 800 N. This is also the peak force that a karate expert breaking a board would experience. However, the fact that it takes less than one-hundredth of a second (two frames in Fig. 7-5) helps prevent injury. Note that the gravitational force is less than 10 N, so we can neglect it.

As the forces act over time they change the translational momentum of both objects. The amount of change will depend on how the forces vary over time. To see this quantitatively, let us apply Newton's Second Law in the form $\vec{F}_{bd \to ba} = d\vec{p}_{ba}/dt$ to the lead ball depicted in Fig. 7-5. If we denote the net force on the ball as $\vec{F}^{\,\text{net}}(t) = \vec{F}_{bd \to ba} + \vec{F}^{\,\text{grav}}_{ba} \approx \vec{F}_{bd \to ba}$ and $\vec{p}_{ba}$ as $\vec{p}$, then

$$d\vec{p} = \vec{F}^{\,\text{net}}(t)\,dt, \tag{7-6}$$

in which $\vec{F}^{\,\text{net}}(t)$ is a time-varying force on the ball with magnitude given by the curve in Fig. 7-6*a*. Let us integrate $d\vec{p} = \vec{F}^{\,\text{net}}(t)\,dt$ over the collision interval $\Delta t$ from an initial time $t_1$ (just before the collision) to a final time $t_2$ (just after the collision). We obtain

$$\int_{\vec{p}_1}^{\vec{p}_2} d\vec{p} = \int_{t_1}^{t_2} \vec{F}^{\,\text{net}}(t)\,dt, \tag{7-7}$$

where $\vec{p}_1$ represents the momentum of the ball at time $t_1$ just before the collision, and $\vec{p}_2$ represents the momentum at time $t_2$ just after the collision. The left side of this equation is $\vec{p}_2 - \vec{p}_1$, which is the change in translational momentum of the lead ball. The right side of the equation is a measure of both the strength and the duration of the collision force exerted on the ball by the board. It is defined as a vector quantity called the **impulse** $\vec{J}$. In general, the impulse an object experiences due to a collision force $\vec{F}^{\,\text{net}}(t)$ is defined as

$$\vec{J} \equiv \int_{t_1}^{t_2} \vec{F}^{\,\text{net}}(t)\,dt \quad \text{(impulse defined).} \tag{7-8}$$

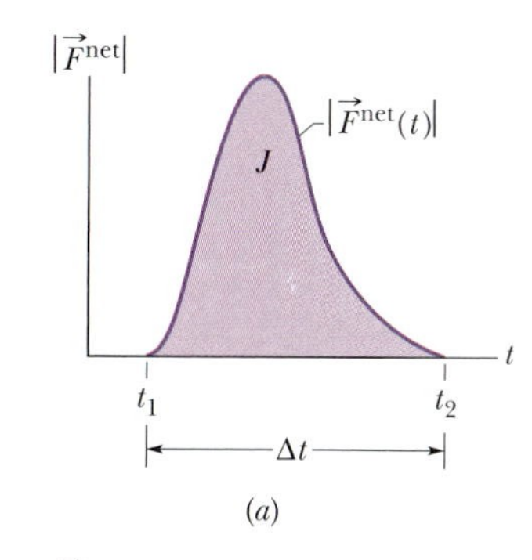


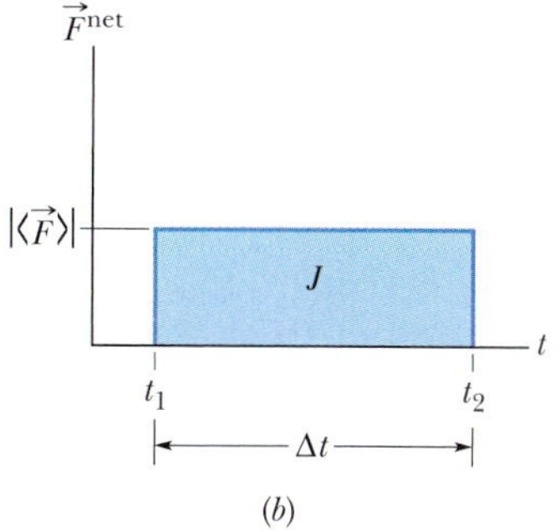


**FIGURE 7-6** ■ (*a*) The sketched graph is an idealization of what the magnitude of the time-varying net force that acts on an object during a collision shown in Fig. 7-5 might look like. The time interval $\Delta t$ for the collision is only about 1/100th of a second for the ball falling on a board as shown in Fig. 7-5. (*b*) The height of the rectangle represents the magnitude of the average net force acting on an object during the same time interval $\Delta t$. The area within the rectangle is equal to the area under the curve (or integral) in (*a*).

When $\vec{F}^{\,\text{net}}(t)$ does not change direction during the collision, this relation tells us that the magnitude of the impulse is equal to the area under the $|\vec{F}^{\,\text{net}}(t)|$ curve of Fig. 7-6*a*.

When an object undergoes a collision it is considerably easier to measure its momentum change than it is to determine its impulse curve. Thus it is quite common to consider the time interval over which the colliding objects are in contact with each other. Then, if we assume that the force is constant during that time interval, we have a feel for the magnitude of the average force that an object experiences. If we denote the average net force during a collision as $\langle \vec{F}^{\,\text{net}} \rangle$, then we can relate it to momentum change using the expression

$$\langle \vec{F}^{\,\text{net}} \rangle = \frac{\vec{J}}{\Delta t} = \frac{\vec{p}_2 - \vec{p}_1}{\Delta t} = \frac{\Delta \vec{p}}{\Delta t} \quad \text{(average net force during a collision).} \tag{7-9}$$

Graphically we can represent the magnitude of the average force on an object $|\vec{F}^{\,\text{net}}|$ as the area within the rectangle of Fig. 7-6*b* that is equal to the area under the $|\vec{F}^{\,\text{net}}(t)|$ curve of Fig. 7-6*a* over the same time interval.

In the case of the example in Fig. 7-5 of the ball (object *ba*) breaking the board (object *bd*), the time period in which the ball is in contact with the board turns out to be only $\Delta t \approx 0.01$ s while the magnitude of the momentum change is $|\Delta\vec{p}_{ba}| =$ 0.53 kg · m/s. This gives us an average net collision force on the ball of magnitude

$$|\langle\vec{F}^{\,\text{net}}_{ba}\rangle| = \frac{\Delta\vec{p}_{ba}}{\Delta t} = \frac{.53\text{ kg}\cdot\text{m/s}}{0.01\text{ s}} \approx 50\text{ N}.$$

The general relation

$$\int_{\vec{p}_1}^{\vec{p}_2} d\vec{p} = \int_{t_1}^{t_2} \vec{F}^{\,\text{net}}(t)\, dt$$

tells us that the change in the translational momentum of any object is equal to the impulse that acts on that object. Thus, for an object in a collision:

$$\vec{J} = \vec{p}_2 - \vec{p}_1 \quad \text{(impulse-momentum theorem).} \tag{7-10}$$

This relation is called the **impulse-momentum theorem;** it tells us that impulse and translational momentum are both vectors and have the same units and dimensions. The impulse-momentum theorem can also be written in component form as

$$p_{2x} - p_{1x} = \Delta p_x = J_x, \tag{7-11}$$

$$p_{2y} - p_{1y} = \Delta p_y = J_y, \tag{7-12}$$

and

$$p_{2z} - p_{1z} = \Delta p_z = J_z. \tag{7-13}$$

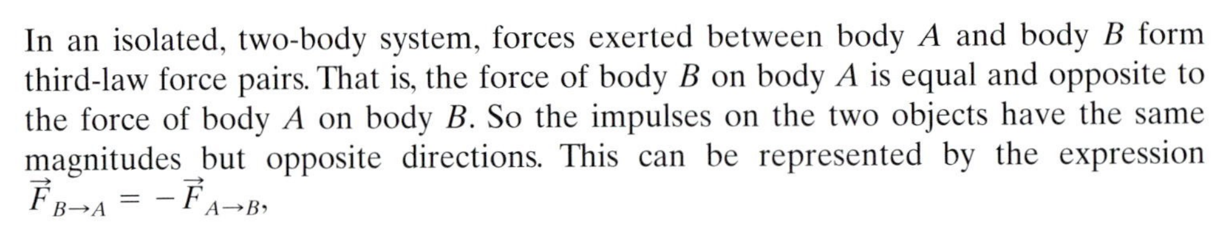

In an isolated, two-body system, forces exerted between body $A$ and body $B$ form third-law force pairs. That is, the force of body $B$ on body $A$ is equal and opposite to the force of body $A$ on body $B$. So the impulses on the two objects have the same magnitudes but opposite directions. This can be represented by the expression $\vec{F}_{B\to A} = -\vec{F}_{A\to B}$,

so that $\quad \vec{J}_{B\to A} = -\vec{J}_{A\to B} \quad$ (isolated 2-body system).

**READING EXERCISE 7-3:** Have you ever been in an egg-tossing contest? The idea of this adventure is to work as a team of two people tossing a raw egg back and forth. After each successful toss (success = unbroken egg), each team member must take a step back. Pretty soon, you have to throw the egg quite hard to get it across to your partner. If you catch an egg of mass $m$ that is coming toward you with velocity $\vec{v}$, what is the magnitude of the change in momentum that the egg undergoes? Would this value change if you catch the egg more quickly or more slowly? Explain. Suppose that the time it takes you to bring the egg to a stop in your hand is $\Delta t$. Are you more likely to have a "successful" catch if $\Delta t$ is large or small? Why? How do you physically react in order to make $\Delta t$ larger? ■

**READING EXERCISE 7-4:** The figure to the right shows an overhead view of a ball bouncing from a vertical wall without any change in its speed. Consider the change $\Delta\vec{p}$ in the ball's translational momentum. (a) Is $\Delta p_x$ positive, negative, or zero? (b) Is $\Delta p_y$ positive, negative, or zero? (c) What is the direction of $\Delta\vec{p}$?

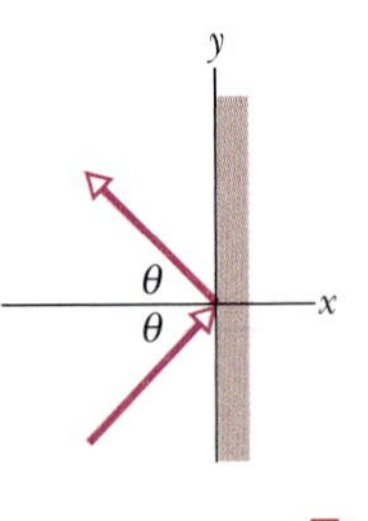


■

## TOUCHSTONE EXAMPLE 7-1: Ball and Bat

A pitched 140 g baseball, in horizontal flight with a speed $v_1$ of 39.0 m/s, is struck by a bat. After leaving the bat, the ball travels in the opposite direction with speed $v_2$, also 39.0 m/s.

(a) What impulse $\vec{J}$ acts on the ball while it is in contact with the bat during the collision?

**SOLUTION** ■ The **Key Idea** here is that momentum is a vector quantity, so even though the magnitude of the momentum does not change, there is a significant change in momentum due to the direction change of the ball. We must calculate the impulse from the change in the ball's translational momentum, using Eq. 7-10 for one-dimensional motion. Let us choose the direction in which the ball is initially moving to be the negative direction. From Eq. 7-10 we have

$$\begin{aligned} J_x &= p_{2x} - p_{1x} = mv_{2x} - mv_{1x} \\ &= (0.140\ \text{kg})(39.0\ \text{m/s}) - (0.140\ \text{kg})(-39.0\ \text{m/s}) \\ &= 10.9\ \text{kg}\cdot\text{m/s}. \qquad \text{(Answer)} \end{aligned}$$

With our sign convention, the initial velocity of the ball is negative and the final velocity is positive. The impulse turns out to be positive, which tells us that the direction of the impulse vector acting on the ball is the direction in which the bat is swinging.

(b) The impact time $\Delta t$ for the baseballl–bat collision is 1.20 ms. What average net force acts on the baseball?

**SOLUTION** ■ The **Key Idea** here is that the average net force is the ratio of the impulse $\vec{J}$ to the duration $\Delta t$ of the collision (see Eq. 7-9). Thus,

$$\begin{aligned} \langle F_x^{\text{net}} \rangle &= \frac{J_x}{\Delta t} = \frac{10.9\ \text{kg}\cdot\text{m/s}}{0.00120\ \text{s}} \\ &= 9080\ \text{N}. \qquad \text{(Answer)} \end{aligned}$$

Note that this is the *average* net force. The *maximum* net force is larger. The sign of the average force on the ball from the bat is positive, which means that the direction of the force vector is the same as that of the impulse vector.

In defining a collision, we assumed that no significant external force acts on the colliding bodies. The gravitational force always acts on the ball, whether the ball is in flight or in contact with the bat. However, this force, with a magnitude of $mg = 1.37$ N, is negligible compared to the average force exerted by the bat, which has a magnitude of 9080 N. We are quite safe in treating the collision as "isolated during the short collision time period."

(c) Now suppose the collision is not head-on, and the ball leaves the bat with a speed $v_2$ of 45.0 m/s at an upward angle of 30.0° (Fig. 7-7). What now is the impulse on the ball?

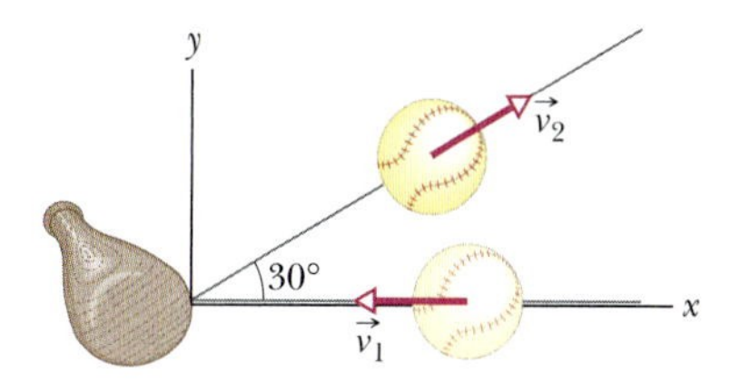


**FIGURE 7-7** ■ A bat collides with a pitched baseball, sending the ball off at an angle of 30° from the horizontal.

**SOLUTION** ■ The **Key Idea** here is that now the collision is two-dimensional because the ball's outward path is not along the same axis as its incoming path. Thus, we must use vectors to find the impulse $\vec{J}$. From Eq. 7-10, we can write

$$\vec{J} = \vec{p}_2 - \vec{p}_1 = m\vec{v}_2 - m\vec{v}_1.$$

Thus,

$$\vec{J} = m(\vec{v}_2 - \vec{v}_1). \qquad (7\text{-}14)$$

We can evaluate the right side of this equation directly on a vector-capable calculator, since we know that the mass $m$ is 0.140 kg, the final velocity $\vec{v}_2$ is 45.0 m/s at 30.0°, and the initial velocity $\vec{v}_1$ is 39.0 m/s at 180°.

Instead, we can evaluate Eq. 7-14 in component form. To do so, we first place an $xy$ coordinate system as shown in Fig. 7-7. Then along the $x$ axis we have

$$\begin{aligned} J_x &= p_{2x} - p_{1x} = m(v_{2x} - v_{1x}) \\ &= (0.140\ \text{kg})[(45.0\ \text{m/s})(\cos 30.0°) - (-39.0\ \text{m/s})] \\ &= 10.92\ \text{kg}\cdot\text{m/s}. \end{aligned}$$

Along the $y$ axis,

$$\begin{aligned} J_y &= p_{2y} - p_{1y} = m(v_{2y} - v_{1y}) \\ &= (0.140\ \text{kg})[(45.0\ \text{m/s})(\sin 30.0°) - 0] \\ &= 3.150\ \text{kg}\cdot\text{m/s}. \end{aligned}$$

The impulse is then

$$\vec{J} = (10.9\hat{\text{i}} + 3.15\hat{\text{j}})\ \text{kg}\cdot\text{m/s}, \qquad \text{(Answer)}$$

and the magnitude and direction of $\vec{J}$ are

$$J = |\vec{J}| = \sqrt{J_x^2 + J_y^2} = 11.4\ \text{kg}\cdot\text{m/s}$$

and

$$\theta = \tan^{-1}\frac{J_y}{J_x} = 16°. \qquad \text{(Answer)}$$

## TOUCHSTONE EXAMPLE 7-2: Carts Colliding

A moving cart coming from the left has a mass of 1.8 kg and an initial velocity component of +0.3 m/s. It then collides with a stationary cart to its right with a mass of 0.8 kg. After the collision the 1.8 kg cart slows down and the 0.8 kg cart moves away from it at a brisk velocity as shown in Fig. 7-8.

Let's consider two collisions for which the right cart is given the *same* momentum after the collision. In one case the right cart has a deformable rubber stopper attached to its force sensor. In the other case the rubber stopper is replaced with a more rigid metal hook. *What effect does the deformability of the surfaces in contact during the collision have on the collision process? In particular, what information do measured impulse curves give us about the duration of the collision and the maximum force experienced by the right cart?* How can we use the impulse curve to estimate the momentum transferred to the right cart during the collision?

(a) Use the measured impulse curves shown in Fig. 7-9 to find the approximate collision times when the collision involves contact between a metal hook and a deformable rubber stopper (shown in case $a$). Compare that to the collision time when the contact is between two metal hooks (shown in case $b$).

**SOLUTION** ■ The **Key Idea** here is that during the time that the collision force is significantly above zero, the two colliding objects are in contact. It is clear from the graph (case $a$) that the collision time when the deformable rubber stopper is the point of contact is about 22 ms or $22 \times 10^{-3}$ s. When the rubber stopper is replaced with a more rigid metal hook, the collision time, as shown on the graph (case $b$), is reduced to about 15 ms or $15 \times 10^{-3}$ s. Another **Key Idea** is that the collision times for highly deformable objects are greater than they are for less deformable objects.

(b) Also use the measured impulse curves to compare the maximum forces experienced by the initially stationary cart for the two types of collisions (metal–rubber and metal–metal). Use the impulse-momentum theorem to explain why one maximum force is greater than the other.

**SOLUTION** ■ It is clear from the graph (case $a$) that the peak force when the deformable rubber stopper is the point of contact is about 40 N (case $a$), while the peak force when the rubber stopper is replaced with a more rigid metal hook is greater at approximately 49 N (case $b$). The **Key Idea** here is that if an object experiences a certain momentum change, the impulse-momentum theorem can be used to relate the momentum change to the impulse curve by the equation

$$p_{2x} - p_{1x} = J_x = \int_{t_1}^{t_2} F_x \, dt = \langle F_x \rangle \Delta t.$$

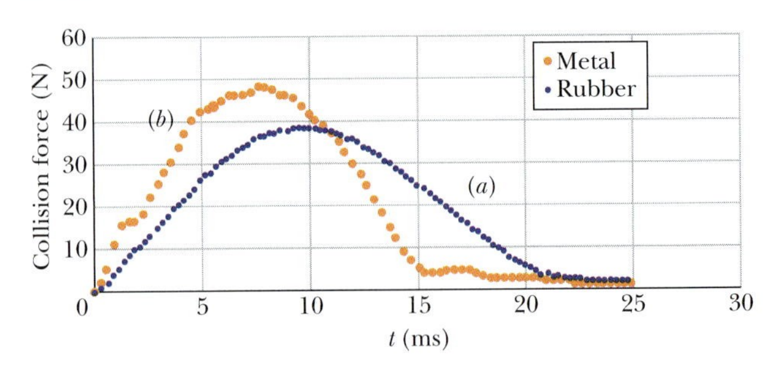


**FIGURE 7-9** ■ Impulse curves when ($a$) the point of contact on the force sensor is a deformable rubber stopper, and ($b$) the point of contact on the force sensor is a piece of hard metal.

Thus, since the duration of the contact, $\Delta t$, is longer in a slow collision than in a fast one, the average force and hence the peak force must be smaller in a slow collision. Conversely the peak force during a rapid collision is greater than it would be in a slow collision.

(c) Use the measured impulse curves to estimate the magnitude of the momentum transferred to the right cart during each type of collision. You can approximate the impulse "curves" as triangles with the base being the contact time and the height equal to the peak force. Verify that both curves predict that approximately (in this case, to one significant figure) the same momentum was imparted to the cart in each case in spite of the fact that the collision times and peak forces are different.

**SOLUTION** ■ The **Key Ideas** here are that the momentum change of the right cart is equal to the impulse imparted to it and that this impulse is an integral that can be calculated by finding the area under the impulse or force vs. time curve. In the special case where the right cart is initially at rest, this momentum change is also the final momentum of the right cart. If we approximate this curve as a triangle, this area can be computed using the familiar equation Area = $(\frac{1}{2})$ base × height.

Rubber stopper:

$$\begin{aligned} J_x &= \text{Area} = \tfrac{1}{2} bh = \tfrac{1}{2} \Delta t F_x^{\text{peak}} \\ &= \tfrac{1}{2}(22 \text{ ms} \times 40 \text{ N}) \\ &= \tfrac{1}{2}(22 \times 10^{-3} \text{ s} \times 40 \text{ N}) \\ &= 0.4 \text{ N} \cdot \text{s}. \end{aligned} \quad \text{(Answer)}$$

$$\begin{aligned} p_{2x} - p_{1x} &= p_{2x} - 0 = \Delta p_x = J_x \\ p_{2x} &= 0.4 \text{ N} \cdot \text{s} = 0.4 \text{ kg} \cdot \text{m/s}. \end{aligned}$$

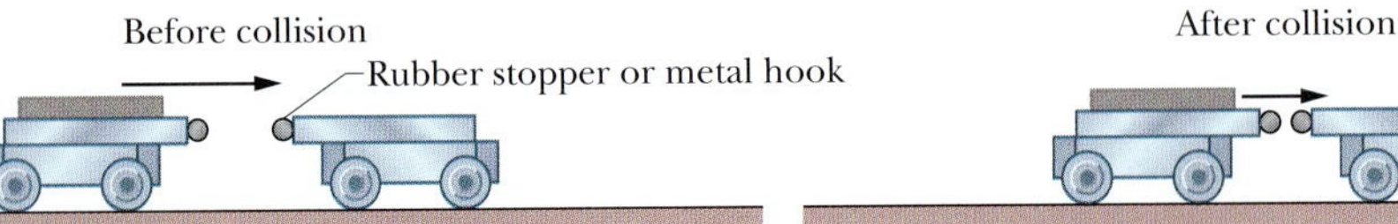


**FIGURE 7-8** ■ A depiction of the motion of two carts before and after a collision.

Metal hook:

$$\begin{aligned} J_x &= \text{Area} = \tfrac{1}{2}bh = \tfrac{1}{2}\Delta t F_x^{\text{peak}} \\ &= \tfrac{1}{2}(15\text{ ms} \times 49\text{ N}) \\ &= \tfrac{1}{2}(15 \times 10^{-3}\text{ s} \times 49\text{ N}) \\ &= 0.4\text{ N}\cdot\text{s}. \qquad \text{(Answer)} \end{aligned}$$

$$\begin{aligned} p_{2x} - p_{1x} = p_{2x} - 0 &= \Delta p_x = J_x \\ p_{2x} &= 0.4\text{ N}\cdot\text{s} = 0.4\text{ kg}\cdot\text{m/s}. \end{aligned}$$

**Note on Impulse and Karate Injuries:** We can use the differences in the shapes of the two impulse curves to explain how it is possible for beginners who are not trained in the art of karate to break pine boards, but not patio blocks, without sustaining injuries. Breaking a board is a complex process that must obey the law of conservation of momentum that will be introduced in Section 7-5, and work and energy relationships that will be introduced in Chapter 9. It turns out that the concept of impulse is one of the critical factors in karate, so that in order to break a concrete block or a board, a certain impulse must be imparted to it.

When struck, the board or block bends, storing energy like a stretched spring does, until a critical deformation needed to break it is reached. In fact, a clean, knot-free pine board that is hit along its grain is relatively easy to break. One of several factors that make breaking a pine board less injurious is the fact that a board deforms much more than a concrete block before breaking. Thus, for a given impulse, the duration of the collision is significantly longer when a pine board breaks than when a concrete block breaks. As we saw in part (b) of this touchstone example, this means that for a given impulse much less peak force will be exerted on the board by the hand. Since Newton's Third Law holds, then it also means that the hand experiences a much lower peak force than it would striking a concrete block hard enough to break it. A lower peak force on the hand reduces the chance that the fifth metacarpal bone in the hand will break.

## 7-5 Newton's Laws and Momentum Conservation

What happens to the momentum, $\vec{p}_{\text{sys}}$, of a system of particles that is *isolated* so there is no net force acting? Assume that the particles are interacting with each other and undergoing all sorts of collisions that obey Newton's Third Law. What happens to the momentum of the overall system if $\vec{F}^{\text{net}} = 0$ from all sources both external and internal? We know from Newton's Second Law that

$$\vec{F}^{\text{net}} = \frac{d\vec{p}_{\text{sys}}}{dt} = 0,$$

and so

$$\vec{p}_{\text{sys}} = \text{constant} \quad \text{(for an isolated system).} \qquad (7\text{-}15)$$

> If no net external force acts on a system of particles, the total translational momentum $\vec{p}_{\text{sys}}$ of the system cannot change.

This result is called the **law of conservation of translational momentum.** It is a natural consequence of Newton's laws. This law can also be written in equation form as

$$\vec{p}_{\text{sys}\,1} = \vec{p}_{\text{sys}\,2} \quad \text{(isolated system),} \qquad (7\text{-}16)$$

where $\vec{p}_{\text{sys}\,1}$ is the total momentum of all the particles in a system at time $t_1$ and $\vec{p}_{\text{sys}\,2}$ is the system momentum at time $t_2$. In words, this equation says that, for an isolated system, the total translational momentum at any initial time $t_1$ is equal to the total translational momentum at any later time $t_2$. This is not to say that the momenta of individual particles within the system do not change. Particles inside a system can undergo changes in momentum. However, they must do so by exchanging momentum with other particles in the system so that the total system momentum remains constant.

In the next section we will consider two colliding carts that form an isolated system and look at how they exchange their momenta in a way that conserves the total momentum of the system.

## 7-6 Simple Collisions and Conservation of Momentum

Suppose that two very low friction carts roll along a smooth, level track. What happens to them before, during, and after they collide? We know by analysis with Newton's Second Law that the external forces on the carts (the gravitational force pulling downward and the normal forces of the track holding them up) cancel each other out, so $\vec{F}^{\text{net}} = 0$. Thus, the system is isolated, so we predict that the total momentum of the two-cart system will be conserved. In other words, each cart should change its momentum in such a way that the total change in system momentum is zero.

In this section, we will examine two different types of collisions for simple systems that are isolated: (1) a collision in which the hard rubber end of a more massive cart hits the hard rubber end of a less massive cart and the two carts bounce off each other, and (2) a collision in which the rubber ends are replaced with Velcro or clay so that the carts stick together after the collision. Is it possible for momentum to be conserved in these two very different situations?

### A Bouncy Collision

Our first case, the bouncy collision, is depicted in Fig. 7-10. Two bodies having almost the same speed but different masses are just about to have a *one-dimensional collision* (meaning that the motions before and after the collision are along the same straight line). Imagine that these two objects bounce off one another immediately following their collision. What happens during the collision? Does the cart on the right with more mass on it exert more force on the cart on the left? Less force? The same force?

**FIGURE 7-10** ■ Two carts of different masses undergo a "bouncy" collision. The collision forces can be measured 4000 times a second using electronic force sensors attached to a computer data acquisition system.

These carts are outfitted with electronic force sensors, so we can measure the collision forces. The impulse curves indicating the changes in forces on each of the carts during the time of impact are shown in Fig. 7-11.

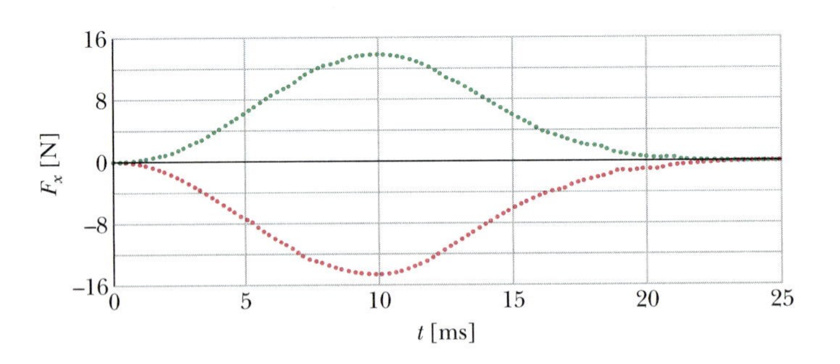


**FIGURE 7-11** ■ The top graph displays the $x$-component of force the left cart exerts on the right cart. The bottom graph shows the force the right cart exerts on the left cart. The collision forces for the two carts are equal and opposite on a moment-by-moment basis. The time of contact is less 25 ms or about 1/40th of a second.

The fact that the interaction forces have equal magnitudes and are oppositely directed at every moment of contact is yet another experimental verification of Newton's Third Law. It shows that there is no net internal force in this two "particle" system. If the total momentum of the system is to be considered, we expect that the change in momentum of the left cart will be equal and opposite to the change in momentum of the right cart. However, since the mass of the right cart is greater and momentum is the product of mass and velocity, the right cart must have a smaller change in velocity than the less massive cart on the left. You are familiar with this fact. When a massive bowling ball hits a bowling pin, the magnitude of the pin's velocity is much larger than that of the ball. An observation of the two carts bouncing off each other confirms the prediction that the more massive cart on the right undergoes less velocity change than the cart on the left.

This conclusion can be expressed mathematically. Using Eq. 7-16,

Total momentum $\vec{p}_{\text{sys }1}$ (before the collision) = total momentum $\vec{p}_{\text{sys }2}$ (after the collision).

We can also express this mathematically in terms of the momentum of each cart as

$$\vec{p}_{A\,1} + \vec{p}_{B\,1} = \vec{p}_{A\,2} + \vec{p}_{B\,2} \quad \text{(conservation of translational momentum).} \tag{7-17}$$

Because the motion is one-dimensional, we can drop the vector arrows and use only components along the direction of the motion. Thus, from $\vec{p} = m\vec{v}$, we can rewrite this expression in terms of the masses and velocity components of the particles. For example, if we choose an $x$ axis along the line of motion, then

$$m_A v_{A\,x}(t_1) + m_B v_{B\,x}(t_1) = m_A v_{A\,x}(t_2) + m_B v_{B\,x}(t_2) \quad (x\text{-component}), \tag{7-18}$$

where $v_{A\,x}(t_1)$ is the $x$-component of object $A$'s velocity at time $t_1$. As we discussed while treating one-dimensional motions in previous chapters, it is essential when substituting actual values for the components into an equation that we use the correct sign (+ or −) to denote the direction of motion of each object along the chosen axis.

Here we have used an experimental verification of Newton's Third Law and a belief that Newton's Second Law is valid to assert that momentum ought to be conserved for an isolated system. Are we correct? Indeed, if we measure masses and use a computer data acquisition system or video analysis software to find velocity components before and after a collision, it is possible to verify momentum conservation experimentally for bouncy collisions. In the next subsection we will describe the details of this type of experimental verification for a sticky collision.

## A Sticky Collision

To discuss a collision in which the particles stick together, we can replace the rubber cart bumpers with Velcro or gooey clay blobs. Another way to explore a sticky collision is to drop a stationary mass onto our low-friction cart. We can gently place the stationary mass on top of the moving cart and record what happens to the cart velocity with a video camera. We will describe how a video analysis of the cart position on video frames (1) enables us to confirm that momentum is conserved and (2) enables us to use our knowledge of momentum conservation to predict the final velocity of any sticky collision between two particle-like objects that form an isolated system. (See Fig. 7-12.)

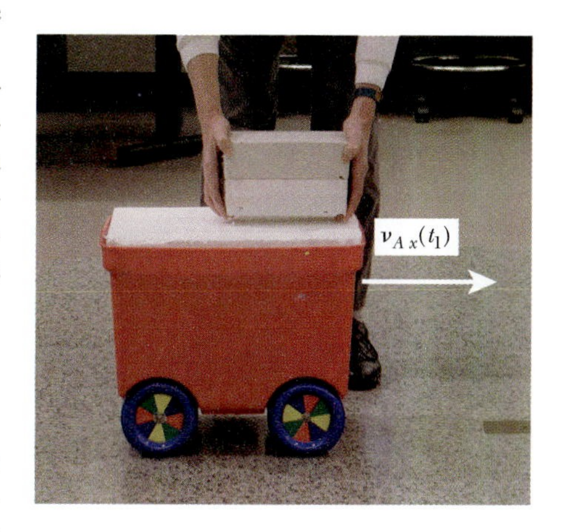


**FIGURE 7-12** ■ A single frame of a video clip shows two bricks being placed on top of a cart as it moves toward the right with an initial velocity component of 1.78 m/s. The cart slows down noticeably once the bricks are placed on top of it.

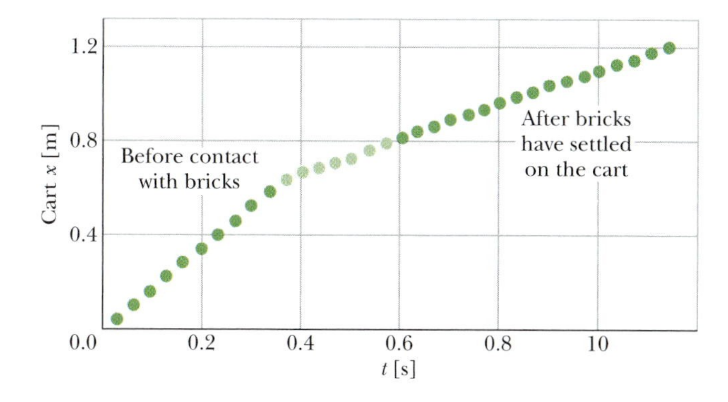


**FIGURE 7-13** ■ A graph based on video analysis shows how the position of the moving cart in Fig. 7-12 changes before, during, and after bricks are placed gently on top of it. The slope of the graph before the mass touches the cart (0.00 to 0.33 s) gives the cart's initial velocity. The slope of the graph after the mass has fully settled on the cart (0.60 to 1.13 s) gives the final velocity of the cart–mass system.

**The Video Analysis:** As we view the video frames and locate the cart position in each frame, we see that in this case the two objects "stick" together following their gentle "collision." By determining the slope of the position vs. time graph for the first few frames we find that the cart of mass $m_A = 2.84$ kg is moving from left to right with an initial $x$-component of velocity of $v_{A\,x}(t_1) = +1.78$ m/s. (See Fig. 7-13.) After two bricks of total mass $m_B = 4.26$ kg are placed gently on the cart, the combined masses continue to move more slowly from left to right with a system velocity component given by $v_{\text{sys}\,x}(t_2) = 0.712$ m/s.

**(1) Confirmation of Momentum Conservation:** Let's check to see that $\vec{p}_{\text{sys}\,2} = \vec{p}_{\text{sys}\,1}$ (that is, Eq. 7-16 holds). We use our data to find the initial momentum of the system. The bricks (denoted as $B$) have no initial velocity, so

$$\vec{p}_{\text{sys}\,x}(t_1) = p_{A\,x}(t_1)\hat{\text{i}} + p_{B\,x}(t_1)\hat{\text{i}} = m_A v_{A\,x}(t_1)\hat{\text{i}} + m_B v_{B\,x}(t_1)\hat{\text{i}}$$
$$= (2.84 \text{ kg})(1.78 \text{ m/s})\hat{\text{i}} + 0\,\hat{\text{i}} \tag{7-19}$$
$$\vec{p}_{\text{sys}\,x}(t_1) = \vec{p}_{\text{sys}\,1} = (5.06 \text{ kg}\cdot\text{m/s})\hat{\text{i}}.$$

To find the final momentum of the system we note that after their collision, the cart and the bricks move together with the *same* velocity. Thus

$$\vec{p}_{\text{sys}\,x}(t_2) = p_{A\,x}(t_2)\hat{\text{i}} + p_{B\,x}(t_2)\hat{\text{i}} = m_A v_{A\,x}(t_2)\hat{\text{i}} + m_B v_{B\,x}(t_2)\hat{\text{i}} \tag{7-20}$$

with

$$v_{A\,x}(t_2) = v_{B\,x}(t_2) = v_{\text{sys}\,x}(t_2),$$

so

$$\vec{p}_{\text{sys}\,x}(t_2) = \vec{p}_{\text{sys}\,2} = (m_A + m_B)\vec{v}_{\text{sys}\,x}(t_2)$$
$$= (2.84 \text{ kg} + 4.26 \text{ kg})(0.712 \text{ m/s})\hat{\text{i}}$$
$$= (5.06 \text{ kg}\cdot\text{m/s})\hat{\text{i}}.$$

There is uncertainty associated with any experimental measurements. Even though video analysis is a very fine tool for motion analysis, we were quite fortunate to have our initial and final momentum values agree to three significant figures. That doesn't usually happen in momentum conservation experiments.

**(2) Predicting the Final Velocity:** If you can correctly identify an isolated system and apply momentum conservation, a knowledge of the initial velocities of a two-particle system can enable you to predict the velocities after a sticky collision. We merely need to equate the last terms in Eqs. 7-19 and 7-20 and solve for the final velocity. For example, with $\vec{v}_{B\,x}(t_1) = 0$, this gives

$$v_{\text{sys}\,x}(t_2)\hat{\text{i}} = \frac{m_A}{m_A + m_B} v_{A\,x}(t_1)\hat{\text{i}}. \tag{7-21}$$

For our cart–brick situation this would give us a predicted final velocity of

$$v_{\text{sys}\,x}(t_2) = \frac{2.84 \text{ kg}}{2.84 \text{ kg} + 4.26 \text{ kg}}(1.78 \text{ m/s}) = 0.712 \text{ m/s} \quad \text{(predicted final speed).}$$

Note that the speed $|\vec{v}_{\text{sys}\,x}(t_2)|$ of the combined masses after the collision must be less than the speed $|\vec{v}_{A\,x}(t_1)|$ of the mass that was moving before the collision, because the mass ratio $m_A/(m_A + m_B)$ is always less than one.

Remember that regardless of whether the objects involved in the collision bounce off one another or stick together, the total translational momentum of a system is conserved so long as there is no net external force acting on it. Friction is an external force that often renders a system nonisolated and hence interferes with momentum conservation.

Our consideration of bouncy and sticky collisions is enough to get us started analyzing collisions. However, many collisions are not completely bouncy or completely sticky. In Chapter 10, we will use the concept of mechanical energy conservation to refine our understanding of collisions.

**READING EXERCISE 7-5:** Consider two small frictionless carts of equal mass that are resting on a level track with a firecracker wedged between them. When the firecracker explodes, the carts fly apart. Is translational momentum conserved in this case? (State any assumptions you made in formulating your answer.) Explain in detail why momentum is conserved or why it isn't. ■

### TOUCHSTONE EXAMPLE 7-3: Exploding Box

A fireworks box with mass $m = 6.0$ kg slides with speed $v = 4.0$ m/s across a frictionless floor in the positive direction along an $x$ axis. It suddenly explodes into two pieces. One piece, with mass $m_A = 2.0$ kg, moves in the positive direction along the $x$ axis with speed $v_A = 8.0$ m/s. What is the velocity of the second piece, with mass $m_B$?

**SOLUTION** ■ There are two **Key Ideas** here. First, we could get the velocity of the second piece if we knew its momentum, because we already know its mass is $m_B = m - m_A = 4.0$ kg. Second, we can relate the momenta of the two pieces to the original momentum of the box if momentum is conserved. Let's check.

Our reference frame will be that of the floor. Our system consists initially of the box and then of the two pieces. The box and pieces each experience a normal force from the floor and a gravitational force. However, those forces are both vertical and cancel out (sum to zero). The forces produced by the explosion are internal to the system. Thus, the horizontal component of the momentum of the system is conserved, and we can apply momentum conservation (Eq. 7-16) along the $x$ axis.

The initial momentum of the system is that of the box:

$$\vec{p}_{\text{sys}\,1} = m\vec{v}.$$

Similarly, we can write the final momenta of the two pieces as

$$\vec{p}_{A\,2} = m_A\vec{v}_A \quad \text{and} \quad \vec{p}_{B\,2} = m_B\vec{v}_B.$$

The final total momentum $\vec{p}_{\text{sys}\,2}$ of the system is the vector sum of the momenta of the two pieces:

$$\vec{p}_{\text{sys}\,2} = \vec{p}_{A\,2} + \vec{p}_{B\,2} = m_A\vec{v}_A + m_B\vec{v}_B.$$

Since all the velocities and momenta in this problem are vectors along the $x$ axis, we can write them in terms of their $x$-components. Doing so, we now obtain

$$p_{\text{sys}\,x}(t_1) = p_{\text{sys}\,x}(t_2)$$

or

$$mv_x(t_1) = m_A v_{A\,x}(t_2) + m_B v_{B\,x}(t_2).$$

Inserting known data, we find

$$(6.0\text{ kg})(4.0\text{ m/s}) = (2.0\text{ kg})(8.0\text{ m/s}) + (4.0\text{ kg})v_{B\,x}(t_2)$$

and thus

$$v_{B\,x}(t_2) = 2.0\text{ m/s}.$$

Since all the momenta and velocities in the vertical direction are zero, our final result is

$$\vec{v}_B = v_{B\,x}\hat{\text{i}} = (2.0\text{ m/s})\hat{\text{i}}, \qquad \text{(Answer)}$$

and the second piece also moves in the positive direction along the $x$ axis.

## 7-7 Conservation of Momentum in Two Dimensions

What happens when one object strikes another with a glancing blow? As shown in Fig. 7-14, the objects can come off at an angle with respect to each other. Can we still apply the law of conservation of momentum?

The principle of conservation of momentum is applicable to collisions in two or three dimensions just as it is in one dimension, as long as the net force on the system is zero in each of the dimensions. If the net force is not zero in one of the dimensions, momentum is not conserved in that dimension in accordance with Eqs. 7-11, 7-12, and 7-13. For convenience, we choose a two-dimensional coordinate system. Then we can

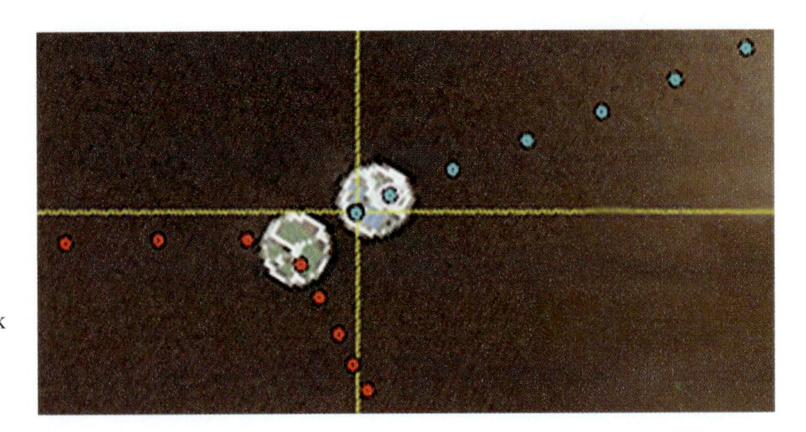

FIGURE 7-14 ■ A video analysis of a collision between two pucks. A puck traveling on an air table hits another stationary puck with a glancing blow. They both travel off in different velocities in such a way that momentum is conserved.

decompose the momentum conservation equation $\vec{p}_{A1} + \vec{p}_{B1} = \vec{p}_{A2} + \vec{p}_{B2}$ (Eq. 7-17) into components. When conservation of momentum is applied to multidimensional motion, it is applied in each direction separately. In other words, the single expression $\vec{p}_{A1} + \vec{p}_{B1} = \vec{p}_{A2} + \vec{p}_{B2}$ can be replaced with up to three expressions that involve unit vectors associated with three orthogonal coordinate axes directions. In the two-dimensional case, these are

$$p_{A\,x}(t_1)\hat{\mathrm{i}} + p_{B\,x}(t_1)\hat{\mathrm{i}} = p_{A\,x}(t_2)\hat{\mathrm{i}} + p_{B\,x}(t_2)\hat{\mathrm{i}}, \quad (7\text{-}22)$$

and

$$p_{A\,y}(t_1)\hat{\mathrm{j}} + p_{B\,y}(t_1)\hat{\mathrm{j}} = p_{A\,y}(t_2)\hat{\mathrm{j}} + p_{B\,y}(t_2)\hat{\mathrm{j}}, \quad (7\text{-}23)$$

where the subscripts denote the momenta for particles $A$ and $B$ along each of the coordinate axes $x$ and $y$.

This set of equations describes the relationships that have to be satisfied by the initial and final momenta of the particles as a result of momentum conservation in two dimensions. In terms of the object's masses and velocities, the equations above can be expressed in terms of components:

$x$-components: $$m_A v_{A\,x}(t_1) + m_B v_{B\,x}(t_1) = m_A v_{A\,x}(t_2) + m_B v_{B\,x}(t_2), \quad (7\text{-}24)$$

and $y$-components: $$m_A v_{A\,y}(t_1) = +m_B v_{B\,y}(t_1) = m_A v_{A\,y}(t_2) + m_B v_{B\,y}(t_2). \quad (7\text{-}25)$$

We can use either set of two equations above (7-22 and 7-23 or 7-24 and 7-25) to analyze a collision. We will choose which set to use based on the information available to us.

If we determine the angles that the objects make with respect to various axes before and after a collision, we can often calculate the $x$- and $y$-components of the momenta or velocities using trigonometry. This is shown in Fig. 7-15 for two pucks that have different masses. *We also must take special care to associate the correct sign (to denote direction) with each term in the expressions above.* For example, Fig. 7-15 shows a collision between a projectile body and a target body initially at rest. The impulses between the bodies have sent the bodies off at angles $\theta_A$ and $\theta_B$ measured relative to the $x$ axis, along which object $A$ initially traveled. In this situation, we would rewrite $\vec{p}_{A\,1} + \vec{p}_{B\,1} = \vec{p}_{A\,2} + \vec{p}_{B\,2}$ for components along the $x$ axis as

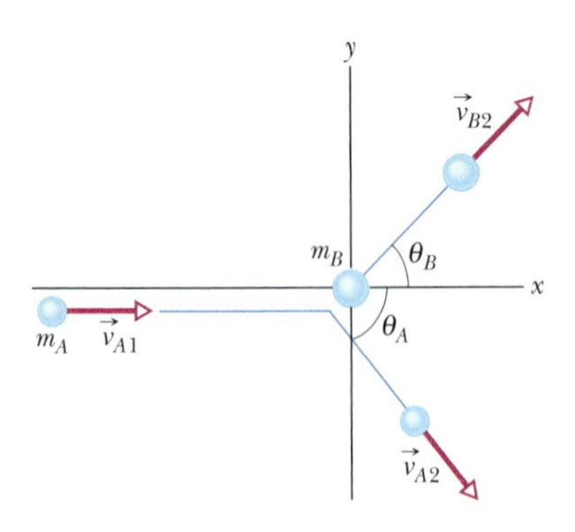


FIGURE 7-15 ■ An object of mass $m_A$ hits a second object of mass $m_B$ at a glancing blow, and each object moves off at an angle with respect to the original line of motion (defined here as the positive $x$ axis).

$$m_A v_{A\,x}(t_1) + m_B v_{B\,x}(t_1) = m_A v_{A\,x}(t_2) + m_B v_{B\,x}(t_2),$$

or

$$m_A|\vec{v}_{A\,1}| + 0 = m_A|\vec{v}_{A\,2}|\cos\theta_A + m_B|\vec{v}_{B\,2}|\cos\theta_B,$$

and along the $y$ axis as

$$m_A v_{A\,y}(t_1) + m_B v_{B\,y}(t_1) = m_A v_{A\,y}(t_2) + m_B v_{B\,y}(t_2),$$

or $$0 + 0 = -m_A|\vec{v}_{A\,2}|\sin\theta_A + m_B|\vec{v}_{B\,2}|\sin\theta_B.$$

The minus sign in the first term to the right of the equal sign above is very important. It indicates that the $y$-component of velocity for $m_A$ is downward.

**READING EXERCISE 7-6:** An initially stationary device lying on a frictionless floor explodes into two pieces, which then slide across the floor. One piece slides in the positive direction along an $x$ axis. (a) What is the sum of the momenta of the two pieces after the explosion? (b) Can the second piece move at an angle to the $x$ axis? Why or why not? (c) What is the direction of the momentum of the second piece? ■

**READING EXERCISE 7-7:** Consider a system that contains the Earth and a grapefruit. The grapefruit starts off at rest and falls a certain distance, at which point its velocity has increased to 2 m/s. What is the change in momentum of the grapefruit? What is the change in momentum of the Earth? What is the approximate change in speed of the Earth associated with this change in momentum? State any estimates you made in answering the question. ■

## TOUCHSTONE EXAMPLE 7-4: Skaters Embrace

Two skaters collide and embrace, "sticking" together after impact, as suggested by Fig. 7-16, where the origin is placed at the point of collision. Alfred, whose mass $m_A$ is 83 kg, is originally moving east with speed $v_A = 6.2$ km/h. Barbara, whose mass $m_B$ is 55 kg, is originally moving north with speed $v_B = 7.8$ km/h.

(a) What is the velocity $\vec{v}_{\text{sys}\,2}$ of the couple after they collide?

**SOLUTION** ■ One **Key Idea** here is the assumption that the two skaters form an isolated system. That is, during the collision we assume no *net* external force acts on them. In particular, we neglect any frictional force on their skates from the ice because the peak collision forces are much larger than the friction forces. With that assumption, we can apply conservation of the total translational momentum $\vec{p}_{\text{sys}}$ by writing $\vec{p}_{\text{sys}\,1} = \vec{p}_{\text{sys}\,2}$ as

$$m_A\vec{v}_{A\,1} + m_B\vec{v}_{B\,1} = (m_A + m_B)\vec{v}_{\text{sys}\,2}. \tag{7-26}$$

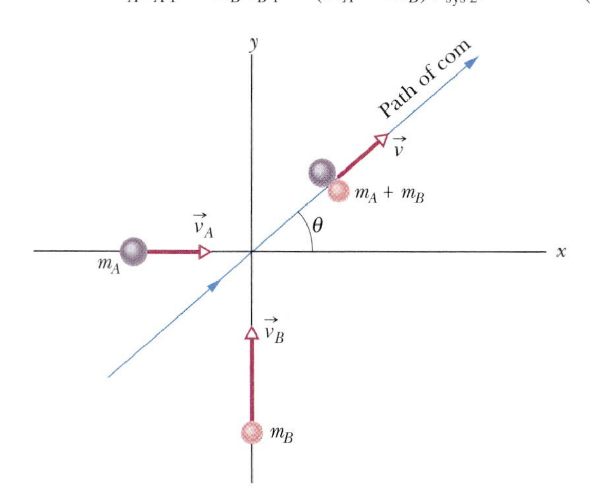


**FIGURE 7-16** ■ Two skaters, Alfred ($A$) and Barbara ($B$), represented by spheres in this simplified overhead view, have a "sticky" collision. Afterward, they move off together at angle $\theta$, with speed $v$.

Solving for the system velocity $\vec{v}_{\text{sys}\,2} = \vec{v}$ after collision gives us

$$\vec{v} = \frac{m_A\vec{v}_A + m_B\vec{v}_B}{m_A + m_B}.$$

We can solve this directly on a vector-capable calculator by substituting given data for the symbols on the right side. We can also solve it by applying a second **Key Idea** (one we have used before) and then some algebra: The idea is that the total translational momentum of the system is conserved separately for components along the $x$ axis and $y$ axis shown in Fig. 7-16. Writing Eq. 7-26 in component form for the $x$ axis and noting that $\vec{v}_A = v_A\hat{\text{i}} + 0\hat{\text{j}}$ yields

$$m_Av_A + m_B(0) = (m_A + m_B)|\vec{v}|\cos\theta, \tag{7-27}$$

and for the $y$ axis, since $\vec{v}_B = 0\hat{\text{i}} + v_B\hat{\text{j}}$,

$$m_A(0) + m_Bv_B = (m_A + m_B)|\vec{v}|\sin\theta. \tag{7-28}$$

We cannot solve either of these equations separately because they both contain two unknowns ($|\vec{v}|$ and $\theta$), but we can solve them simultaneously by dividing Eq. 7-28 by Eq. 7-27. We get

$$\tan\theta = \frac{m_Bv_B}{m_Av_A} = \frac{(55\text{ kg})(7.8\text{ km/h})}{(83\text{ kg})(6.2\text{ km/h})} = 0.834.$$

Thus,

$$\theta = \tan^{-1} 0.834 = 39.8° \approx 40°. \quad \text{(Answer)}$$

From Eq. 7-28, with $m_A + m_B = 138$ kg, we then have a final system speed of

$$v = |\vec{v}| = \frac{m_Bv_B}{(m_A + m_B)\sin\theta} = \frac{(55\text{ kg})(7.8\text{ km/h})}{(138\text{ kg})(\sin 39.8°)}$$
$$= 4.86\text{ km/h} \approx 4.9\text{ km/h}. \quad \text{(Answer)}$$

FIGURE 7-17 ■ Liftoff of Project Mercury spacecraft.

## 7-8 A System with Mass Exchange—A Rocket and Its Ejected Fuel

In the systems we have dealt with so far, we have assumed that the total mass of the system remains constant; no mass is added or removed from the system. Such systems are called **closed.** Sometimes, as in a rocket (Fig. 7-17), the mass does not stay constant. Most of the mass of a rocket on its launching pad is fuel, all of which will eventually be burned and ejected from the nozzle of the rocket engine. A rocket accelerates by ejecting some of its own mass in the form of exhaust gases. It turns out that both the rate at which the fuel burns and the velocity of the ejected fuel particles relative to the rocket are constant.

We handle the variation of the mass of the rocket as the rocket accelerates by applying Newton's Second Law, not to the rocket alone but to the rocket and its ejected combustion products taken together. The mass of *this* system does *not* change as the rocket accelerates.

### Finding the Acceleration

Let's consider the acceleration of this rocket in deep space with no gravitational or atmospheric drag forces acting on it. To simplify our observation of what happens, suppose that at an arbitrary time $t_1$ when the rocket has a total mass $M$, we happen to be in an inertial reference frame that moves at a constant velocity that is exactly the same as the rocket's velocity. What do we observe in a short time interval $dt$?

At time $t_1$ the rocket is not moving relative to us (see Fig. 7-18*a*). After a time interval $dt$, the rocket has ejected a small amount of burned fuel of mass $dm$ at a velocity relative to the rocket, which we call $\vec{v}^{\,\text{rel}}$.

Our system consists of the rocket and the exhaust products released during interval $dt$. The system is closed and isolated, so the translational momentum of the system must be conserved during $dt$; that is,

$$\vec{p}_{\text{sys }1} = \vec{p}_{\text{sys }2}. \tag{7-29}$$

However, at time $t_1$ when the rocket is not moving relative to us, we observe that the initial momentum of the system is zero. Thus, at a later time $dt$ the total momentum of the system must still be zero. As the mass $dm$ of burned fuel flies off at a velocity $\vec{v}^{\,\text{rel}}$ the rocket that now has a very slightly smaller mass of $M - dm$ must recoil in the opposite direction with a small increase in its velocity of $d\vec{v}$ as shown in Fig. 7-18*b*. In order to keep the total momentum of the rocket–fuel system zero we must have

$$\vec{p}_{\text{sys }1} = 0 = \vec{p}_{\text{sys }2} = dm(\vec{v}^{\,\text{rel}}) + (M - dm)\,d\vec{v}. \tag{7-30}$$

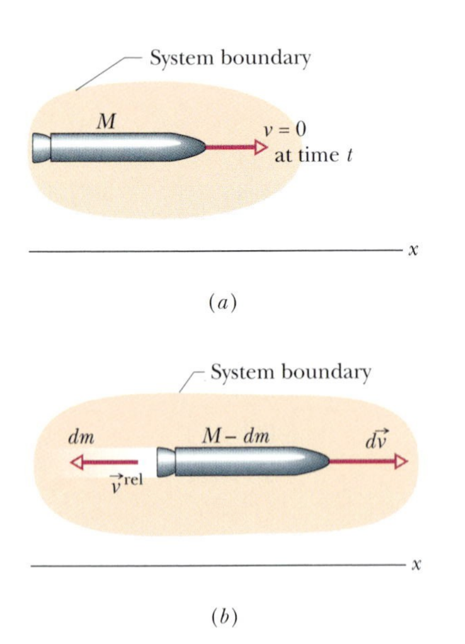


FIGURE 7-18 ■ (*a*) An accelerating rocket of mass $M$ at time $t_1$, as seen from an inertial reference frame. (*b*) The same rocket, but at time $t_1 + dt$. The exhaust products released during interval $dt$ are shown.

Since the rocket mass $M \gg dm$, the total rocket mass $M$ is always much greater than the mass of fuel ejected in a short time, so we can rewrite the momentum conservation equation as

$$dm(\vec{v}^{\,\text{rel}}) + M\,d\vec{v} \approx 0. \tag{7-31}$$

Dividing each term by $dt$ and rearranging terms gives us

$$-\frac{dm}{dt}\vec{v}^{\,\text{rel}} = M\frac{d\vec{v}}{dt}. \tag{7-32}$$

If we note that the change in the rocket mass due to the loss of the ejected fuel during the time interval $dt$ is given by $dM = -dm$, we can replace $-dm/dt$ with $dM/dt$. Since

$dv/dt$ is the acceleration of the rocket relative to the inertial reference, the expression above becomes

$$\frac{dM}{dt}\vec{v}^{\,\text{rel}} = M\vec{a} \qquad \text{(first rocket equation).} \tag{7-33}$$

This equation holds at any instant, with the mass $M$, the *fuel consumption rate* $R = -dM/dt$, and the acceleration $\vec{a}$ evaluated at that instant. Note that $\vec{v}^{\,\text{rel}}$ and $\vec{a}$ point in opposite directions because we chose $\vec{v}^{\,\text{rel}}$ to be the velocity of the *ejected gas relative to the rocket* rather than the other way around. This is not at first apparent in Eq. 7-33 until you remember that $dM/dt$ is negative. The left side of this equation has the dimensions of a force ($\text{kg}\cdot\text{m/s}^2 = \text{N}$) and depends only on design characteristics of the rocket engine—namely, the rate $R$ at which it consumes fuel mass and the speed $\vec{v}^{\,\text{rel}}$ with which that mass is ejected relative to the rocket.

We call the term $-R\vec{v}^{\,\text{rel}}$ the **thrust** of the rocket engine and represent it with $\vec{F}^{\,\text{thrust}}$. Newton's Second Law emerges clearly if we write $-R\vec{v}^{\,\text{rel}} = M\vec{a}$ as $\vec{F}^{\,\text{thrust}} = M\vec{a}$, in which $\vec{a}$ is the acceleration of the rocket at the time that its mass is $M$. Notice that $\vec{F}^{\,\text{thrust}}$ points in the same direction that the rocket is accelerating, even though $\vec{v}^{\,\text{rel}}$ points in the opposite direction. Since $dM/dt$ is intrinsically negative, $R = -dM/dt$ is positive.

## Finding the Velocity Change

How will the velocity of a rocket change as it consumes its fuel? Recall that the change in the rocket mass due to the loss of the ejected fuel during the time interval $dt$ is given by $dM = -dm$. Then we can rewrite Eq. 7-30, which is $M d\vec{v} = -(dm)\vec{v}^{\,\text{rel}}$, and rearrange the terms to get

$$d\vec{v} = \vec{v}^{\,\text{rel}}\frac{dM}{M}$$

where integrating gives us

$$\int_{\vec{v}_1}^{\vec{v}_2} d\vec{v} = \vec{v}^{\,\text{rel}}\int_{M_1}^{M_2}\frac{dM}{M},$$

in which $M_1 = M(t_1)$ represents the initial mass of the rocket at time $t_1$ and $M_2 = M(t_2)$ is the mass of the rocket at some later ("final") time $t_2$. Evaluating the integrals then gives

$$\vec{v}_2 - \vec{v}_1 = \vec{v}^{\,\text{rel}}\ln\frac{M_2}{M_1} = -\vec{v}^{\,\text{rel}}\ln\frac{M_1}{M_2} \qquad \text{(second rocket equation),} \tag{7-34}$$

for the increase in the speed of the rocket during the change in mass from $M_1$ to $M_2$. (The symbol "ln" in this equation means the *natural logarithm.*) The final mass is always less than the initial mass so the natural log will always be positive. But the velocity of the ejected fuel relative to the rocket is also in the opposite direction as the velocity change of the rocket. This always gives us a velocity change in a direction opposite that of mass ejection.

We see here the advantage of multistage rockets, in which $M_2$ is reduced by discarding successive stages when their fuel is depleted. Discarding rocket stages means there is less mass to accelerate. An ideal rocket would reach its destination with only its payload remaining.

FIGURE 7-19a ■ Liftoff of the Mercury-Redstone rocket that sent the first American astronaut, Alan Shepard, into space in 1961.

## Thrust Forces at Liftoff

In the first few seconds of liftoff, the fuel consumption rate is not large enough to change the overall mass $M$ of a typical modern rocket by a noticeable amount. Thus, its mass $M$ is approximately constant. We can use this fact along with Eq. 7-33 in the analysis of video images of a NASA rocket to find the thrust forces of the rocket. As an example, we will do an analysis of the Mercury-Redstone rocket that lifted Alan Shepard into space in 1961. An image of the rocket during liftoff is shown in Fig. 7-19*a*. However, at liftoff we are not in deep space, so the net force on the rocket is the vector sum of the thrust force of the rocket acting in an upward direction and the downward force of the gravitational attraction of the Earth. Therefore,

$$\vec{F}^{\text{thrust}} + \vec{F}^{\text{grav}} = M\vec{a} \quad \text{(at liftoff from the Earth's surface).}$$

Taking the positive $y$ direction to be vertically upward, this simplifies to

$$F_y^{\text{thrust}}\hat{\mathrm{j}} - Mg\hat{\mathrm{j}} = Ma_y\hat{\mathrm{j}}.$$

The $y$ position as a function of time of the Mercury-Redstone rocket liftoff is shown in Fig. 7-19*b*.

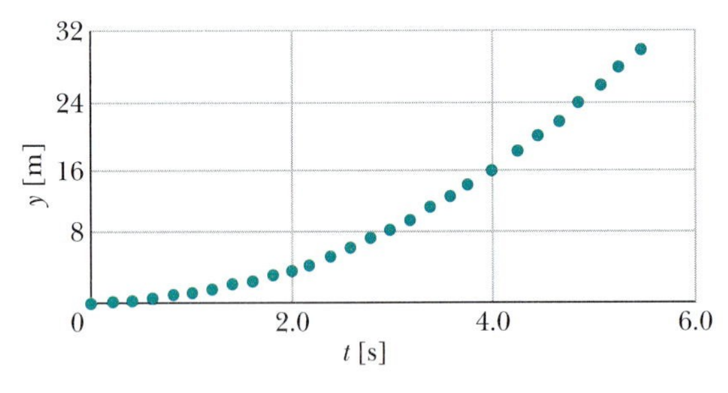


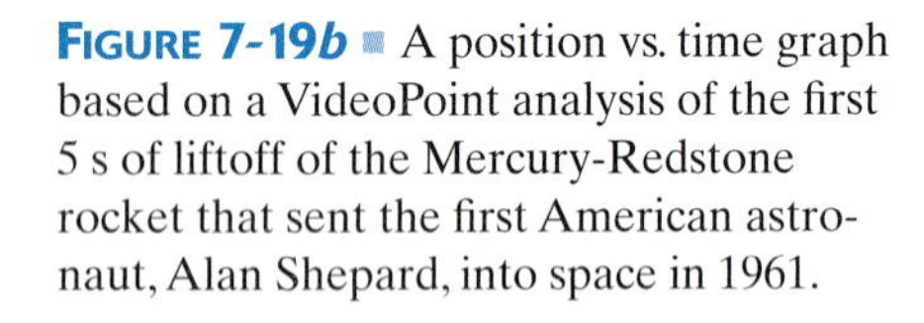

FIGURE 7-19b ■ A position vs. time graph based on a VideoPoint analysis of the first 5 s of liftoff of the Mercury-Redstone rocket that sent the first American astronaut, Alan Shepard, into space in 1961.

Fitting the curve with a quadratic function gives an upward acceleration of magnitude 1.1 m/s$^2$. The mass of the Mercury-Redstone rocket with full fuel and payload is $M = 3.0 \times 10^4$ kg. Thus, the $y$-component of the thrust force is given by

$$F_y^{\text{thrust}} = M(a_y + g) = (3.0 \times 10^4\,\text{kg})(1.1 + 9.8)\,\text{m/s}^2 = 33 \times 10^5\,\text{N}.$$

If we know the fuel consumption rate we can also find the relative velocity with which fuel is ejected from the rocket using the first rocket equation (Eq. 7-33) given by $R\vec{v}^{\,\text{rel}} = M\vec{a}$.

## TOUCHSTONE EXAMPLE 7-5: Rocket Thrust

A rocket whose initial mass $M_1$ is 850 kg consumes fuel at the rate $R = 2.3$ kg/s. The speed $v^{\text{rel}}$ of the exhaust gases relative to the rocket engine is 2800 m/s.

(a) What thrust does the rocket engine provide?

**SOLUTION** ■ The **Key Idea** here is that the magnitude of the thrust $F^{\text{thrust}}$ is equal to the product of the fuel consumption rate $R$ and the relative speed $v_{\text{rel}}$ at which exhaust gases are expelled:

$$F^{\text{thrust}} = Rv^{\text{rel}} = (2.3 \text{ kg/s})(2800 \text{ m/s})$$
$$= 6440 \text{ N} \approx 6400 \text{ N}.$$

(b) What is the initial acceleration of the rocket launched from a spacecraft?

**SOLUTION** ■ We can relate the thrust $\vec{F}^{\text{thrust}}$ of a rocket to the resulting acceleration $\vec{a}$ with $\vec{F}^{\text{thrust}} = M\vec{a}$, where $M$ is the rocket's mass. The **Key Idea**, however, is that $M$ decreases and the magnitude of the acceleration $a$ increases as fuel is consumed. Because we want the initial value of the acceleration here, we must use the initial value $M_1$ of the mass, finding that

$$\vec{a} = \frac{\vec{F}^{\text{thrust}}}{M_1} = \frac{6440 \text{ N}\,\hat{\text{i}}}{850 \text{ kg}} = (7.6 \text{ m/s}^2)\,\hat{\text{i}}. \quad \text{(Answer)}$$

(c) Suppose that the mass $M_2$ of the rocket when its fuel is exhausted is 180 kg. What is its speed relative to the spacecraft at that time? Assume that the spacecraft is so massive that the launch does not alter its speed.

**SOLUTION** ■ The **Key Idea** here is that the rocket's final speed $v_2$ (when the fuel is exhausted) depends on the ratio $M_1/M_2$ of its initial mass to its final mass, as given by Eq. 7-34. With the initial speed $v_1 = 0$, we have

$$\vec{v}_2 = -\vec{v}^{\text{rel}} \ln\left(\frac{M_1}{M_2}\right)$$
$$= -(-2800 \text{ m/s}\,\hat{\text{i}}) \ln\left(\frac{850 \text{ kg}}{180 \text{ kg}}\right)$$
$$= (2800 \text{ m/s}) \ln(4.72)\,\hat{\text{i}} \approx 4300 \text{ m/s}\,\hat{\text{i}}. \quad \text{(Answer)}$$

Note that the ultimate speed of the rocket can exceed the exhaust speed $v^{\text{rel}}$.

# Problems

## SEC. 7-2 ■ TRANSLATIONAL MOMENTUM OF A PARTICLE

**1. Same Momentum** Suppose that your mass is 80 kg. How fast would you have to run to have the same translational momentum as a 1600 kg car moving at 1.2 km/h?

**2. VW Beetle** How fast must an 816 kg VW Beetle travel to have the same translational momentum as a 2650 kg Cadillac going 16 km/h?

**3. Radar** An object is tracked by a radar station and found to have a position vector given by $\vec{r} = [(3500 \text{ m}) - (160 \text{ m/s})t]\hat{\text{i}} + (2700 \text{ m})\hat{\text{j}}$ with $\vec{r}$ in meters and $t$ in seconds. The radar station's $x$ axis points east, its $y$ axis north, and its $z$ axis vertically up. If the object is a 250 kg meteorological missile, what are (a) its translational momentum and (b) its direction of motion?

## SEC. 7-4 ■ IMPULSE AND MOMENTUM CHANGE

**4. Ball Moving Horizontally** A 0.70 kg ball is moving horizontally with a speed of 5.0 m/s when it strikes a vertical wall. The ball rebounds with a speed of 2.0 m/s. What is the magnitude of the change in translational momentum of the ball?

**5. Cue Ball** A 0.165 kg cue ball with an initial speed of 2.00 m/s bounces off the rail in a game of pool, as shown from an overhead view in Fig. 7-20. For $x$ and $y$ axes located as shown, the bounce reverses the $y$-component of the ball's velocity but does not alter the $x$-component. (a) What is $\theta$ in Fig 7-20? (b) What is the change in the ball's momentum in unit-vector notation? (The fact that the ball rolls is not relevant to either question.)

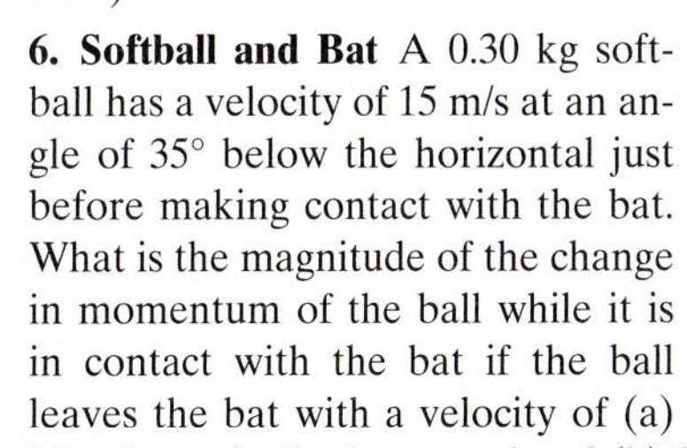

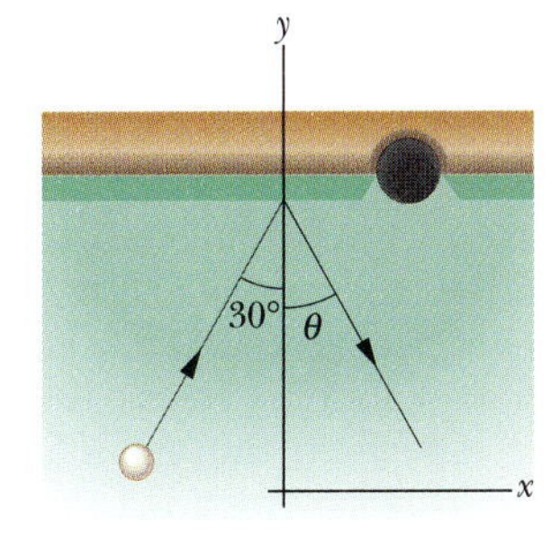


FIGURE 7-20 ■ Problem 5.

**6. Softball and Bat** A 0.30 kg softball has a velocity of 15 m/s at an angle of 35° below the horizontal just before making contact with the bat. What is the magnitude of the change in momentum of the ball while it is in contact with the bat if the ball leaves the bat with a velocity of (a) 20 m/s, vertically downward and (b) 20 m/s, horizontally away from the batter and back toward the pitcher?

**7. Stationary Ball-Impulse** A cue stick strikes a stationary pool ball, with an average force of 50 N over a time of 10 ms. If the ball has mass 0.20 kg, what speed does it have just after impact?

**8. Average Force During Crash** The National Transportation Safety Board is testing the crash-worthiness of a new car. The 2300 kg vehicle, moving at 15 m/s, is allowed to collide with a bridge abutment, which stops it in 0.56 s. What is the magnitude of the average force that acts on the car during the impact?

**9. Average Force of Bat** A 150 g baseball pitched at a speed of 40 m/s is hit straight back to the pitcher at a speed of 60 m/s. What is the magnitude of the average force on the ball from the bat if the bat is in contact with the ball for 5.0 ms?

**10. Henri LaMothe** Until he was in his seventies, Henri LaMothe excited audiences by belly-flopping from a height of 12 m into 30 cm of water (Fig. 7-21). Assuming that he stops just as he reaches the bottom of the water and estimating his mass, find the magnitudes of (a) the average force and (b) the average impulse on him from the water.

FIGURE 7-21 ■ Problem 10.

**11. Steel Ball** A force that averages 1200 N is applied to a 0.40 kg steel ball moving at 14 m/s in a collision lasting 27 ms. If the force is in a direction opposite the initial velocity of the ball, find the final speed and direction of the ball.

**12. Chute Failure** In February 1955, a paratrooper fell 370 m from an airplane without being able to open his chute but happened to land in snow, suffering only minor injuries. Assume that his speed at impact was 56 m/s (terminal speed), that his mass (including gear) was 85 kg, and that the magnitude of the force on him from the snow was at the survivable limit of $1.2 \times 10^5$ N. What are (a) the minimum depth of snow that would have stopped him safely and (b) the magnitude of the impulse on him from the snow?

**13. Rebounding Ball** A 1.2 kg ball drops vertically onto a floor, hitting with a speed of 25 m/s. It rebounds with an initial speed of 10 m/s. (a) What impulse acts on the ball during the contact? (b) If the ball is in contact with the floor for 0.020 s, what is the magnitude of the average force on the floor from the ball?

**14. Superman** It is well known that bullets and other missiles fired at Superman simply bounce off his chest (Fig.7-22). Suppose that a gangster sprays Superman's chest with 3 g bullets at the rate of 100 bullets/min, and the speed of each bullet is 500 m/s. Suppose too that the bullets rebound straight back with no change in speed. What is the magnitude of the average force on Superman's chest from the stream of bullets?

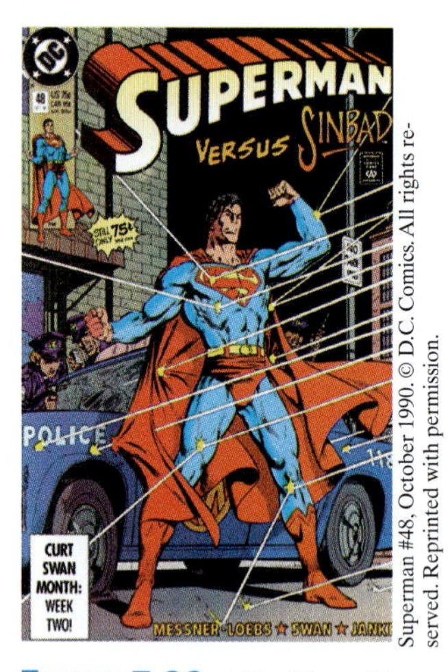




FIGURE 7-22 ■ Problem 14.

**15. Inattentive Driver** A 1400 kg car moving at 5.3 m/s is initially traveling north in the positive $y$ direction. After completing a 90° right-hand turn to the positive $x$ direction in 4.6 s, the inattentive operator drives into a tree, which stops the car in 350 ms. In unit-vector notation, what is the impulse on the car (a) due to the turn and (b) due to the collision? What is the magnitude of the average force that acts on the car (c) during the turn and (d) during the collision? (e) What is the angle between the average force in (c) and the positive $x$ direction?

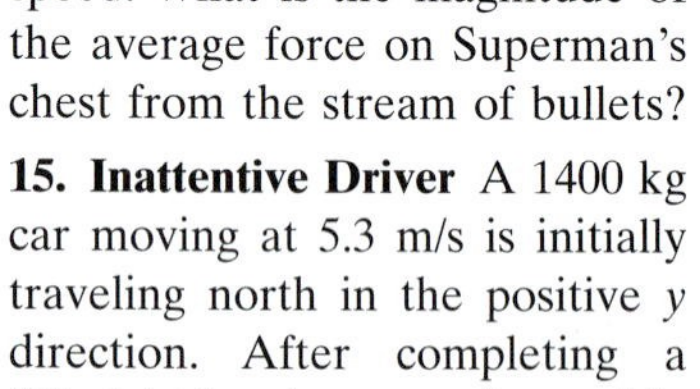

**16. Softball** A 0.30 kg softball has a velocity of 12 m/s at an angle of 35° below the horizontal just before making contact with a bat. The ball leaves the bat 2.0 ms later with a vertical velocity of magnitude 10 m/s as shown in Fig. 7-23. What is the magnitude of the average force of the bat on the ball during the ball–bat contact?

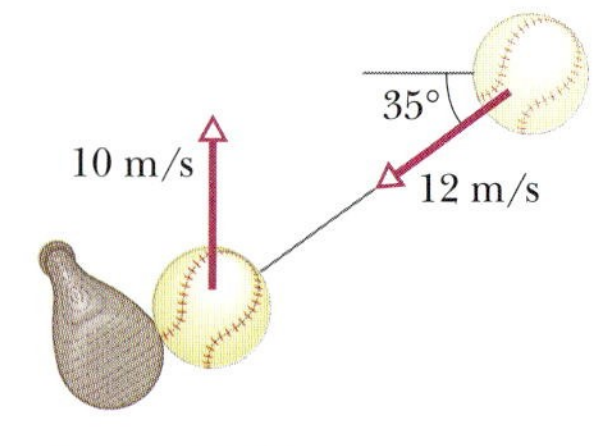


FIGURE 7-23 ■ Problem 16.

**17. Force and Impulse** The magnitude of an unbalanced force on a 10 kg object increases at a constant rate from zero to 50 N in 4.0 s, causing the initially stationary object to move. What is the object's speed at end of the 4.0 s?

**18. Thunderstorm** During a violent thunderstorm, hail of diameter 1.0 cm falls directly downward at a speed of 25 m/s. There are estimated to be 120 hailstones per cubic meter of air. (a) What is the mass of each hailstone (density = 0.92 g/cm$^3$)? (b) Assuming that the hail does not bounce, find the magnitude of the average force on a flat roof measuring 10 m $\times$ 20 m due to the impact of the hail. (*Hint*: During impact, the force on a hailstone from the roof is approximately equal to the net force on the hailstone, because the gravitational force on it is small.)

**19. Pellet Gun** A pellet gun fires ten 2.0 g pellets per second with a speed of 500 m/s. The pellets are stopped by a rigid wall. What are (a) the momentum of each pellet and (b) the magnitude of the average force on the wall from the stream of pellets? (c) If each pellet is in contact with the wall for 0.6 ms, what is the magnitude of the average force on the wall from each pellet during contact? (d) Why is this average force so different from the average force calculated in (b)?

**20. Superball Hits Wall** Figure 7-24 shows an approximate plot of force magnitude versus time during the collision of a 58 g Superball with a wall. The initial velocity of the ball is 34 m/s perpendicular to the wall; it rebounds directly back with approximately the same speed, also perpendicular to the wall. What is $F^{max}$, the maximum magnitude of the force on the ball from the wall during the collision?

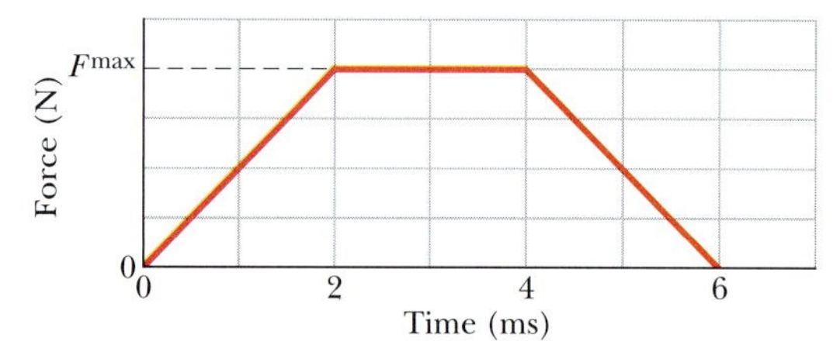


FIGURE 7-24 ■ Problem 20.

**21. Spacecraft** A spacecraft is separated into two parts by detonating the explosive bolts that hold them together. The masses of the parts are 1200 kg and 1800 kg; the magnitude of the impulse on each part from the bolts is 300 N · s. With what relative speed do the two parts separate because of the detonation?

**22. Ball Strikes Wall** In the overhead of Fig. 7-25, a 300 g ball with a speed $v$ of 6.0 m/s strikes a wall at an angle $\theta$ of 30° and then rebounds with the same speed and angle. It is in contact with the wall for 10 ms. (a) What is the impulse on the ball from the wall? (b) What is the average force on the wall from the ball?

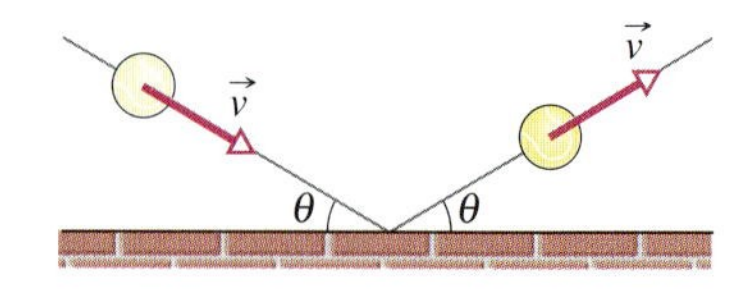


FIGURE 7-25 ■ Problem 22.

**23. Two Barges** In Fig. 7-26, two long barges are moving in the same direction in still water, one with a speed of 10 km/h and the other with a speed of 20 km/h. While they are passing each other, coal is shoveled from the slower to the faster one at a rate of 1000 kg/min. How much additional force must be provided by the driving engines of (a) the fast barge and (b) the slow barge if neither is to change speed? Assume that the shoveling is always perfectly sideways and that the frictional forces between the barges and the water do not depend on the mass of the barges.

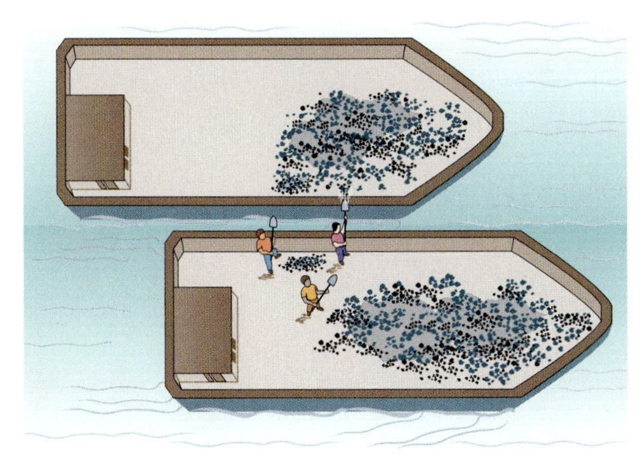

FIGURE 7-26 ■ Problem 23.

## SEC. 7-6 ■ SIMPLE COLLISIONS AND CONSERVATION OF MOMENTUM

**24. Two Blocks** Two blocks of masses 1.0 kg and 3.0 kg are connected by a spring and rest on a frictionless surface. They are given velocities toward each other such that the 1.0 kg block travels initially at 1.7 m/s toward the other. What is the initial velocity of the other block if the system has no initial momentum?

**25. Meteor Impact** Meteor Crater in Arizona (Fig 7-1*a*) is thought to have been formed by the impact of a meteor with Earth some 20,000 years ago. The mass of the meteor is estimated at $5 \times 10^{10}$ kg, and its speed at 7200 m/s. What speed would such a meteor give Earth in a head-on collision?

**26. Bullet Strikes Wooden Block** A 5.20 g bullet moving at 672 m/s strikes a 700 g wooden block at rest on a frictionless surface. The bullet emerges, traveling in the same direction with its speed reduced to 428 m/s. What is the resulting speed of the block?

**27. Man Throws Stone** A 91 kg man lying on a surface of negligible friction shoves a 68 g stone away from him, giving it a speed of 4.0 m/s. What velocity does the man acquire as a result?

**28. Mechanical Toys** A mechanical toy slides along an $x$ axis on a frictionless surface with a velocity of $(-0.40 \text{ m/s})\hat{i}$ when two internal springs separate the toy into three parts, as given in the table. What is the velocity of part $A$?

| Part | Mass (kg) | Velocity (m/s) |
|---|---|---|
| $A$ | 0.50 | ? |
| $B$ | 0.60 | $0.20\hat{i}$ |
| $C$ | 0.20 | $0.30\hat{i}$ |

**29. Icy Road** Two cars $A$ and $B$ slide on an icy road as they attempt to stop at a traffic light. The mass of $A$ is 1100 kg, and the mass of $B$ is 1400 kg. The coefficient of kinetic friction between the locked wheels of either car and the road is 0.13. Car $A$ succeeds in stopping at the light, but car $B$ cannot stop and rear-ends car $A$. After the collision, $A$ stops 8.2 m ahead of its position at impact, and $B$ 6.1 m ahead; see Fig. 7-27. Both drivers had their brakes locked throughout the incident. Using the material in Chapters 2 and 6, find the speed of (a) car $A$ and (b) car $B$ immediately after impact. (c) Use conservation of translational momentum to find the speed at which car $B$ struck car $A$. On what grounds can the use of momentum conservation be criticized here?

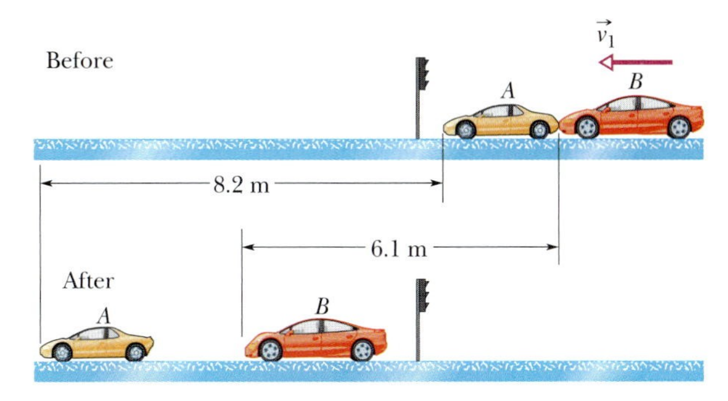


FIGURE 7-27 ■ Problem 29.

**30. Bullet and Two Blocks** In Fig. 7-28*a*, a 3.50 g bullet is fired horizontally at two blocks at rest on a frictionless tabletop. The bullet passes through the first block, with mass 1.20 kg, and embeds itself in the second, with mass 1.80 kg. Speeds of 0.630 m/s and 1.40 m/s, respectively, are thereby given to the blocks (Fig. 7-28*b*). Neglecting the mass removed from the first block by the bullet, find (a) the speed of the bullet immediately after it emerges from the first block and (b) the bullet's original speed.

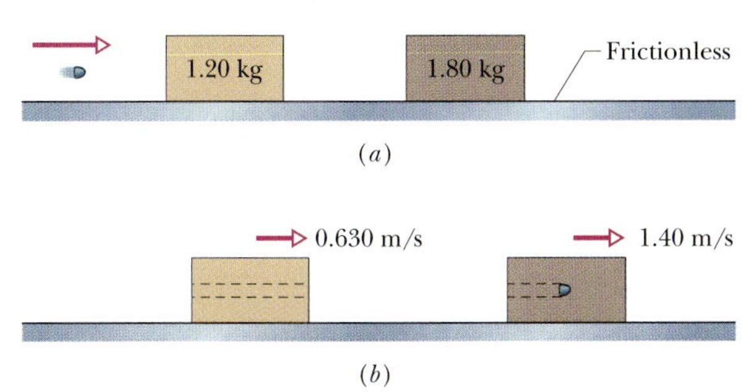


FIGURE 7-28 ■ Problem 30.

**31. Man on a Cart** A 75 kg man is riding on a 39 kg cart traveling at a speed of 2.3 m/s. He jumps off with zero horizontal speed relative to the ground. What is the resulting change in the speed of the cart?

**32. Block and Bullet** A bullet of mass 4.5 g is fired horizontally into a 2.4 kg wooden block at rest on a horizontal surface. The bullet stops in the block. The speed of the block immediately after the bullet stops relative to it is 2.7 m/s. At what speed is the bullet fired?

**33. Water in a Rocket Sled** A rocket sled with a mass of 2900 kg moves at 250 m/s on a set of rails. At a certain point, a scoop on the sled dips into a trough of water located between the tracks and scoops water into an empty tank on the sled. By applying the principle of conservation of translational momentum, determine the speed of the sled after 920 kg of water has been scooped up. Ignore any retarding force on the scoop.

**34. Bullet Fired Upward** A 10 g bullet moving directly upward at 1000 m/s strikes and passes through the center of a 5.0 kg block initially at rest (Fig. 7-29). The bullet emerges from the block moving directly upward at 400 m/s. To what maximum height does the block then rise above its initial position? (*Hint*: Use free-fall equations from Chapter 3.)

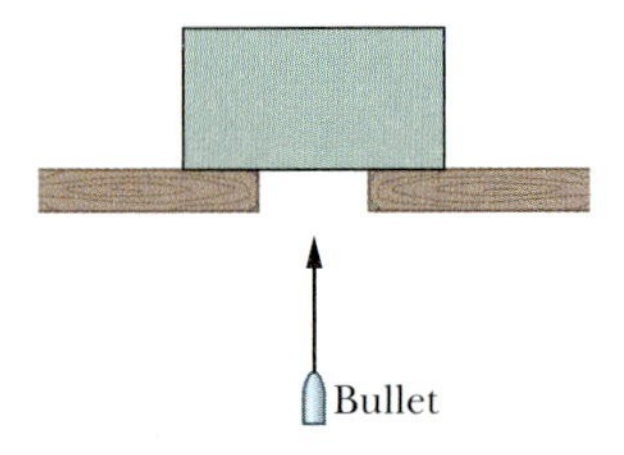


FIGURE 7-29 ■ Problem 34.

**35. Projectile Body** A projectile body of mass $m_A$ and initial velocity $v_{A\,1}$ collides with an initially stationary target body of mass $m_B$ in a one-dimensional collision. What are the velocities of the bodies after the collision if they stick together?

**36. Two Blocks Collide** A 5.0 kg block with a speed of 3.0 m/s collides with a 10 kg block that has a speed of 2.0 m/s in the same direction. After the collision, the 10 kg block is observed to be traveling in the original direction with a speed of 2.5 m/s. What is the velocity of the 5.0 kg block immediately after the collision?

**37. Last Stage of a Rocket** The last stage of a rocket, which is traveling at a speed of 7600 m/s, consists of two parts that are clamped together: a rocket case with a mass of 290.0 kg and a payload capsule with a mass of 150.0 kg. When the clamp is released, a compressed spring causes the two parts to separate with a relative speed of 910.0 m/s. What are the speeds of (a) the rocket case and (b) the payload after they have separated? Assume that all velocities are along the same line.

**38. Man on a Flatcar** A railroad flatcar of weight $W$ can roll without friction along a straight horizontal track. Initially, a man of weight $w$ is standing on the car, which is moving to the right with speed $v_1$ (see Fig. 7-30). What is the change in velocity of the car if the man runs to the left (in the figure) so that his speed relative to the car is $v^{\text{rel}}$?

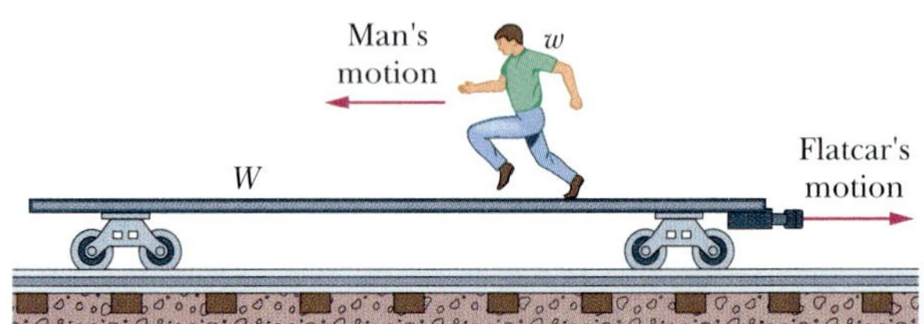


FIGURE 7-30 ■ Problem 38.

**39. Space Vehicle** A space vehicle is traveling at 4300 km/h relative to Earth when the exhausted rocket motor is disengaged and sent backward with a speed of 82 km/h relative to the command module. The mass of the motor is four times the mass of the module. What is the speed of the command module relative to Earth just after the separation?

**40. Projectile Body Two** A projectile body of mass $m_A$ and initial velocity $v_{A\,1} = 10.0$ m/s collides with an initially stationary target body of mass $m_B = 2.00 m_A$ in a one-dimensional collision. What is the velocity of $m_B$ following the collision if the two masses stick together?

## SEC. 7-7 ■ CONSERVATION OF MOMENTUM IN TWO DIMENSIONS

**41. Ice-Skating Man** A 60 kg man is ice-skating due north with a velocity of 6.0 m/s when he collides with a 38 kg child. The man and child stay together and have a velocity of 3.0 m/s at an angle of 35° north of east immediately after the collision. What are the magnitude and direction of the velocity of the child just before the collision?

**42. Barge Collision** A barge with mass $1.50 \times 10^5$ kg is proceeding downriver at 6.2 m/s in heavy fog when it collides with a barge heading directly across the river (see Fig. 7-31). The second barge has mass $2.78 \times 10^5$ kg and before the collision is moving at 4.3 m/s. Immediately after impact, the second barge finds its course deflected by 18° in the downriver direction and its speed increased to 5.1 m/s. The river current is approximately zero at the time of the accident. What are the speed and direction of motion of the first barge immediately after the collision?

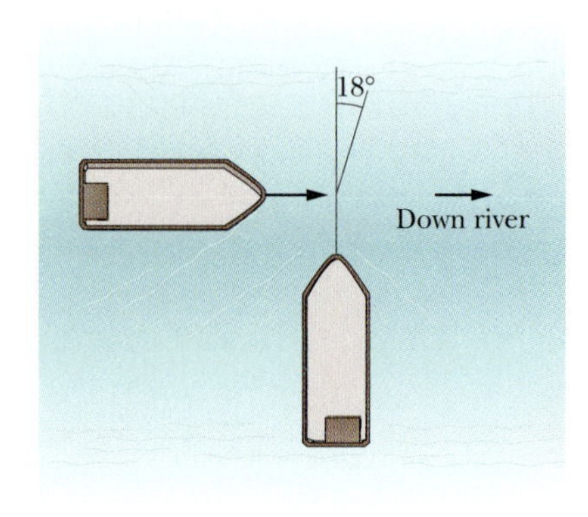


FIGURE 7-31 ■ Problem 42.

**43. Package Explodes** A 2.65 kg stationary package explodes into three parts that then slide across a frictionless floor. The package had been at the origin of a coordinate system. Part $A$ has mass $m_A = 0.500$ kg and velocity $(10.0 \text{ m/s}\, \hat{\text{i}} + 12.0 \text{ m/s}\, \hat{\text{j}})$. Part $B$ has mass $m_B = 0.750$ kg, a speed of 14.0 m/s, and travels at an angle 110° counterclockwise from the positive direction of the $x$ axis. (a) What is the speed of part $C$? (b) In what direction does it travel?

**44. Particle Collision** A 2.00 kg "particle" traveling with velocity $\vec{v} = (4.0 \text{ m/s})\hat{\text{i}}$ collides with a 4.00 kg "particle" traveling with velocity $\vec{v} = (2.0 \text{ m/s})\hat{\text{j}}$. The collision connects the two particles. What then is their velocity in (a) unit-vector notation and (b) magnitude-angle notation?

**45. Two Vehicles** Two vehicles $A$ and $B$ are traveling west and south, respectively, toward the same intersection, where they collide and lock together. Before the collision, $A$ (total weight 12.0 kN) has a speed of 64.4 km/h, and $B$ (total weight 16.0 kN) has a speed of 96.6 km/h. Find the (a) magnitude and (b) direction of the velocity of the (interlocked) vehicles immediately after the collision, assuming the collision is isolated.

**46. Tin Cookie** A 2.0 kg tin cookie, with an initial velocity of 8.0 m/s to the east, collides with a stationary 4.0 kg cookie tin. Just after the collision, the cookie has a velocity of 4.0 m/s at an angle of 37° north of east. Just then, what are (a) the magnitude and (b) the direction of the velocity of the cookie tin?

**47. Colliding Balls** A 5.0 kg ball moving due east at 4.0 m/s collides with a 4.0 kg ball moving due west at 3.0 m/s. Just after the collision, the 5.0 kg ball has a velocity of 1.2 m/s, due south. What is the magnitude of the velocity of the 4.0 kg ball just after the collision?

**48. Particle Collision Two** A collision occurs between a 2.00 kg particle traveling with velocity $\vec{v} = (-4.00 \text{ m/s})\hat{\text{i}} + (-5.00 \text{ m/s})\hat{\text{j}}$ and a 4.00 kg particle traveling with velocity $\vec{v} = (6.00 \text{ m/s})\hat{\text{i}} + (-2.00 \text{ m/s})\hat{\text{j}}$. The collision connects the two particles. What then is their velocity in (a) unit-vector notation and (b) magnitude-angle notation?

**49. Suspicious Package** A suspicious package is sliding on frictionless surface when it explodes into three pieces of equal masses and with the velocities (1) 7.0 m/s, north, (2) 4.0 m/s, 30° south of west, and (3) 4.0 m/s, 30° south of east. (a) What is the velocity (magnitude and direction) of the package before it exploded?

**50. Mess Kit** A 4.0 kg mess kit sliding on a frictionless surface explodes into two 2.0 kg parts, one moving at 3.0 m/s, due north, and the other at 5.0 m/s, 30° north of east. What is the original speed of the mess kit?

**51. Radioactive Nucleus** A certain radioactive nucleus can transform to another nucleus by emitting an electron and a neutrino. (The *neutrino* is one of the fundamental particles of physics.) Suppose that in such a transformation, the initial nucleus is stationary, the electron and neutrino are emitted along perpendicular paths, and the magnitudes of the translational momenta are $1.2 \times 10^{-22}$ kg · m/s for the electron and $6.4 \times 10^{-23}$ kg · m/s for the neutrino. As a result of the emissions, the new nucleus moves (recoils). (a) What is the magnitude of its translational momentum? What is the angle between its path and the path of (b) the electron (c) the neutrino?

**52. Internal Explosion** A 20.0 kg body is moving in the positive $x$ direction with a speed of 200 m/s when, due to an internal explosion, it breaks into three parts. One part, with a mass of 10.0 kg, moves away from the point of explosion with a speed of 100 m/s in the positive $y$ direction. A second fragment, with a mass of 4.00 kg, moves in the negative $x$ direction with a speed of 500 m/s. What is the velocity of the third (6.00 kg) fragment?

**53. Vessel at Rest Explodes** A vessel at rest explodes, breaking into three pieces. Two pieces, having equal mass, fly off perpendicular to one another with the same speed of 30 m/s. The third piece has three times the mass of each other piece. What are the magnitude and direction of its velocity immediately after the explosion?

**54. Proton–Proton Collision** A proton with a speed of 500 m/s collides with another proton initially at rest. The projectile and target protons then move along perpendicular paths, with the projectile path at 60° from the original direction. After the collision, what are the speeds of (a) the target proton and (b) the projectile proton?

**55. Box Sled** A 6.0 kg box sled is coasting across frictionless ice at a speed of 9.0 m/s when a 12 kg package is dropped into it from above. What is the new speed of the sled?

**56. Two Balls** Two balls $A$ and $B$, having different but unknown masses, collide. Initially, $A$ is at rest and $B$ has speed $v$. After the collision, $B$ has speed $v/2$ and moves perpendicularly to its original motion. (a) Find the direction in which ball $A$ moves after the collision. (b) Show that you cannot determine the speed of $A$ from the information given.

**57. Two Objects, Same Mass** After a collision, two objects of the same mass and same initial speed are found to move away together at $\frac{1}{2}$ their initial speed. Find the angle between the initial velocities of the objects.

**58. Sliding on Ice** Two 30 kg children, each with a speed of 4.0 m/s, are sliding on a frictionless frozen pond when they collide and stick together because they have Velcro straps on their jackets. The two children then collide and stick to a 75 kg man who was sliding at 2.0 m/s. After this collision, the three-person composite is stationary. What is the angle between the initial velocity vectors of the two children?

**59. Alpha Particle and Oxygen** An alpha particle collides with an oxygen nucleus that is initially at rest. The alpha particle is scattered at an angle of 64.0° from its initial direction of motion, and the oxygen nucleus recoils at an angle of 51.0° on the opposite side of that initial direction. The final speed of the nucleus is $1.20 \times 10^5$ m/s. Find (a) the final speed and (b) the initial speed of the alpha particle. (In atomic mass units, the mass of an alpha particle is 4.0 u, and the mass of an oxygen nucleus is 16 u.)

**60. Two Bodies Collide** Two 2.0 kg bodies, $A$ and $B$, collide. The velocities before the collision are $\vec{v}_{A1} = (15\text{ m/s})\hat{\text{i}} + (30\text{ m/s})\hat{\text{j}}$ and $\vec{v}_{B1} = (-10\text{ m/s})\hat{\text{i}} + (5.0\text{ m/s})\hat{\text{j}}$. After the collision, $\vec{v}_{A2} = (-5.0\text{ m/s})\hat{\text{i}} + (20\text{ m/s})\hat{\text{j}}$. What is the final velocity of $B$?

**61. Game of Pool** In a game of pool, the cue ball strikes another ball of the same mass and initially at rest. After the collision, the cue ball moves at 3.50 m/s along a line making an angle of 22.0° with its original direction of motion, and the second ball has a speed of 2.00 m/s. Find (a) the angle between the direction of motion of the second ball and the original direction of motion of the cue ball and (b) the original speed of the cue ball.

**62. Billiard Ball** A billiard ball moving at a speed of 2.2 m/s strikes an identical stationary ball with a glancing blow. After the collision, one ball is found to be moving at a speed of 1.1 m/s in a direction making a 60° angle with the original line of motion. Find the velocity of the other ball.

**63. Three Balls** In Fig. 7-32, ball $A$ with an initial speed of 10 m/s collides with stationary balls $B$ and $C$, whose centers are on a line perpendicular to the initial velocity of ball $A$ and that are initially in contact with each other. The three balls are identical. Ball $A$ is aimed directly at the contact point, and all motion is frictionless. After the collision, balls $B$ and $C$ have the same speed 6.93 m/s, but ball $B$ moves at an angle of 30° above the horizontal and ball $C$ moves at an angle of 30° below the horizontal. What is the velocity of ball $A$ after the collision?

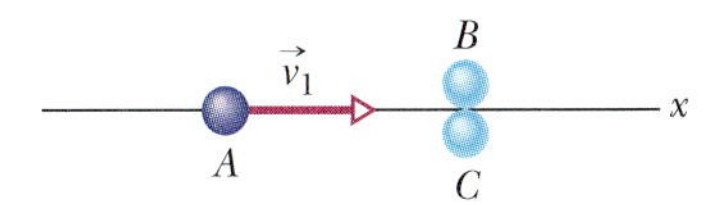

FIGURE 7-32 ■ Problem 63.

## SEC. 7-8 ■ A SYSTEM WITH MASS EXCHANGE—A ROCKET AND ITS EJECTED FUEL

**64. Railroad Car with Grain** A railroad car moves at a constant speed of 3.20 m/s under a grain elevator. Grain drops into it at the rate of 540 kg/min. What is the magnitude of the force needed to keep the car moving at constant speed if friction is negligible?

**65. Space Probe** A 6090 kg space probe, moving nose-first toward Jupiter at 105 m/s relative to the Sun, fires its rocket engine, ejecting 80.0 kg of exhaust at a speed of 253 m/s relative to the space probe. What is the final velocity of the probe?

**66. Moving Away From Solar System** A rocket is moving away from the solar system at a speed of $6.0 \times 10^3$ m/s. It fires its engine, which ejects exhaust with a speed of $3.0 \times 10^3$ m/s relative to the rocket. The mass of the rocket at this time is $4.0 \times 10^4$ kg, and its acceleration is 2.0 m/s². (a) What is the thrust of the engine? (b) At what rate, in kilograms per second is exhaust ejected during the firing?

**67. Deep Space** A rocket, which is in deep space and initially at rest relative to an inertial reference frame, has a mass of $2.55 \times 10^5$ kg, of which $1.81 \times 10^5$ kg is fuel. The rocket engine is then fired for 250 s, during which fuel is consumed at the rate of 480 kg/s. The speed of the exhaust products relative to the rocket is 3.27 km/s. (a) What is the rocket's thrust? After the 250 s firing, what are the (b) mass and (c) speed of the rocket?

**68. Mass Ratio** Consider a rocket that is in deep space and at rest relative to an inertial reference frame. The rocket's engine is to be fired for a certain interval. What must be the rocket's *mass ratio* (ratio of initial to final mass) over that interval if the rocket's original speed relative to the inertial frame is to be equal to (a) the exhaust speed (speed of the exhaust products relative to the rocket) and (b) 2.0 times the exhaust speed?

**69. Lunar Mission** During a lunar mission, it is necessary to increase the speed of a spacecraft by 2.2 m/s when it is moving at 400 m/s relative to the Moon. The speed of the exhaust products from the rocket engine is 1000 m/s relative to the spacecraft. What fraction of the initial mass of the spacecraft must be burned and ejected to accomplish the speed increase?

**70. Set for Vertical Firing** A 6100 kg rocket is set for vertical firing from the ground. If the exhaust speed is 1200 m/s, how much gas must be ejected each second if the thrust (a) is to equal the magnitude of the gravitational force on the rocket and (b) is to give the rocket an initial upward acceleration of 21 m/s$^2$?

# Additional Problems

**71. Break a Leg (Not!)** When jumping straight down, you can be seriously injured if you land stiff-legged. One way to avoid injury is to bend your knees upon landing to reduce the force of the impact. Suppose you have a mass $m$ and you jump off a wall of height $h$.

**(a)** Use what you learned about constant acceleration motion to find the speed with which you hit the ground. Assume you simply step off the wall, so your initial $y$ velocity is zero. Ignore air resistance. (Express your answer in terms of the symbols given.)
**(b)** Suppose that the time interval starting when your feet first touch the ground until you stop is $\Delta t$. Calculate the (average) net force acting on you during that interval. (Again, express your answer in terms of the symbols given.)
**(c)** Suppose $h = 1$ m. If you land stiff-legged, the time it takes you to stop may be as short as 2 ms, whereas if you bend your knees, it might be as long as 0.1 s. Calculate the average net force that would act on you in the two cases.
**(d)** The net force on you while you are stopping includes both the force of gravity and the force of the ground pushing up. Which of these forces do you think does you the injury? Explain your reasoning.
**(e)** For the two cases in part (c), calculate the upward force the ground exerts on you.

**72. Finding Momentum Change and Impulse** Consider the graphs shown in Fig. 7-33. These graphs depict two force magnitude vs. time curves and several related momentum vs. time graphs. They describe a low-friction cart traveling along an $x$ axis with a force sensor attached to it. The cart–force sensor system has a mass of 0.50 kg. The cart undergoes a series of collisions. It collides with a hard wall and with a wall that is padded with soft foam. Sometimes there is a small clay blob on the wall causing the cart–force sensor system to stick to the wall after the collision.

**(a)** What is the approximate momentum change associated with graph $a$? With graph $d$? Determine this change by taking approximate readings from the graphs. Show your calculations!
**(b)** Which of the two impulse curves, $A$ or $B$, might lead to the momentum change depicted in graph $a$? In graph $d$? Explain the reasons for your answer.
**(c)** Suppose the forces on the cart–force sensor system were described by graph $A$. What would its velocity change be?

**73. Relating Impulse Curves to Collisions** Suppose you collected $F_x$ vs. $t$ and $p_x$ vs. $t$ data for a series of collisions for an important project report and then you lost your notes. Fortunately you still

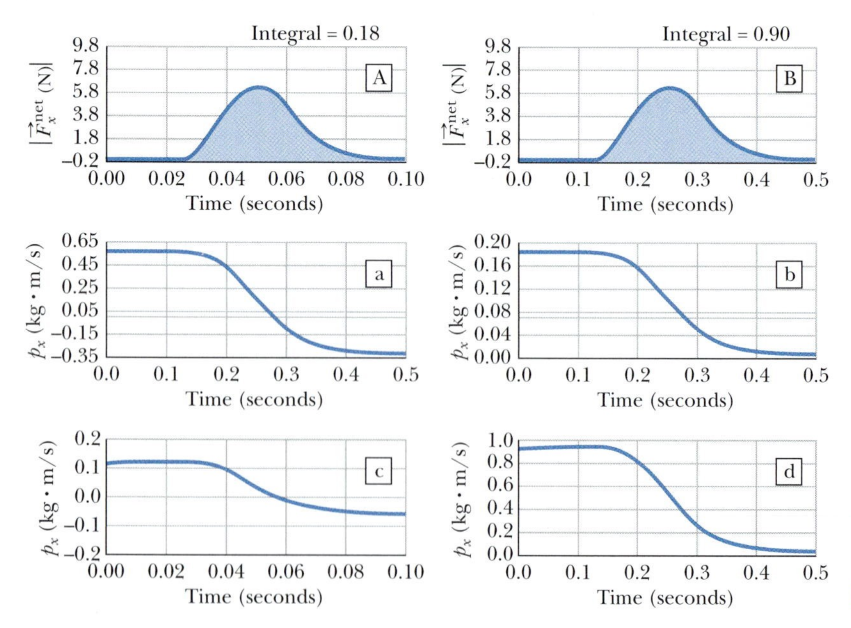


FIGURE 7-33 ■ Problems 72 and 73.

have your data on a computer disk. You open up the files and find the graphs shown in Fig. 7-33. You don't know which graph corresponds to which collision, but you are able to reconstruct some of your work by asking and answering the following questions:

**(a)** Which $F_x^{net}$ vs. $t$ graph, $A$ or $B$, probably resulted from collisions between the cart–force sensor system and a soft, padded wall? Which one probably resulted from collisions between the force sensor and a hard wall? Explain in words the reasons for your answer.
**(b)** Which $p_x$ vs. $t$ graphs probably resulted from collisions between the cart–force sensor system and a padded wall? Which ones probably resulted from collisions between the cart–force sensor system and a hard wall? Explain the reasons for your answers. (*Hint*: There may be more than one graph for each type of collision.)
**(c)** Which $p_x$ vs. $t$ graphs correspond to a situation in which the cart bounces back? Which $p_x$ vs. $t$ graphs correspond to a situation in which you placed a small clay blob on the force sensor hook so the cart sticks to the wall that it collides with? Explain the reasons for your answers. (*Hint*: There may be more than one graph for each type of collision.)

**74. Carts and Graphs** Two carts on an air track are pushed toward each other. Initially, cart $A$ moves in the positive $x$ direction and cart $B$ moves in the negative $x$ direction. The carts bounce off each other. The graphs in Fig. 7-34 describe some of the variables associated with the motion as a function of time. For each item in the list below, identify which graph is a possible display of that variable as a function of time. If none apply, write N (for none).

**(a)** the momentum of cart $A$
**(b)** the force on cart $B$
**(c)** the force on cart $A$
**(d)** the position of cart $A$
**(e)** the position of cart $B$

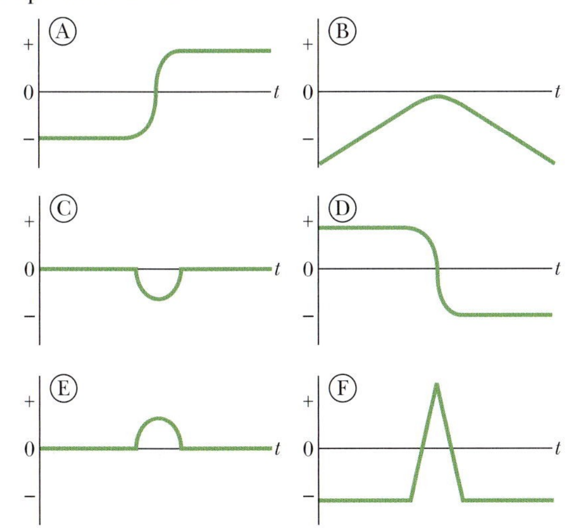


FIGURE 7-34 ■ Problem 74.

**75. Colliding Carts** Two carts are riding on an air track as shown in Fig. 7-35$a$. At clock time $t = 0$, cart B is at the origin traveling in the negative $x$ direction with a velocity $\vec{v}_{B1}$. At that time, cart A is at the position shown and is at rest. Cart B has twice the mass of cart A. The carts "bump" each other, but don't stick.

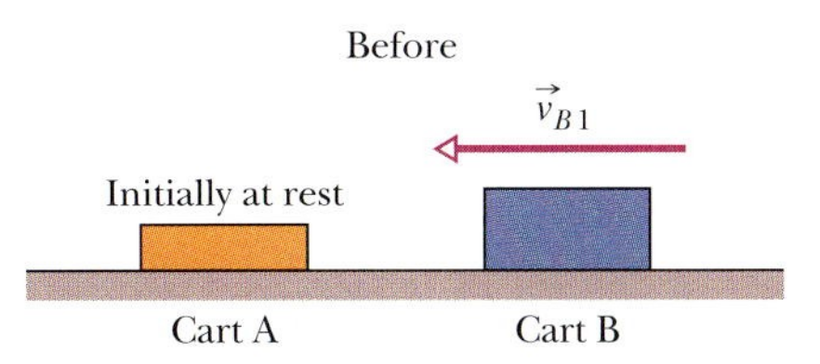


FIGURE 7-35$a$ ■ Problem 75.

The graphs shown in Fig. 7-35$b$ are a number of possible plots for the various physical parameters associated with the two carts. Each graph has two curves, one for each cart and labeled with the cart's letter. For each property (a)–(e), select the number 1, 2, etc., of the graphs that could be a plot of the property.

**(a)** The forces *exerted by* the carts
**(b)** The position of the carts
**(c)** The velocity of the carts
**(d)** The acceleration of the carts
**(e)** The momentum of the carts

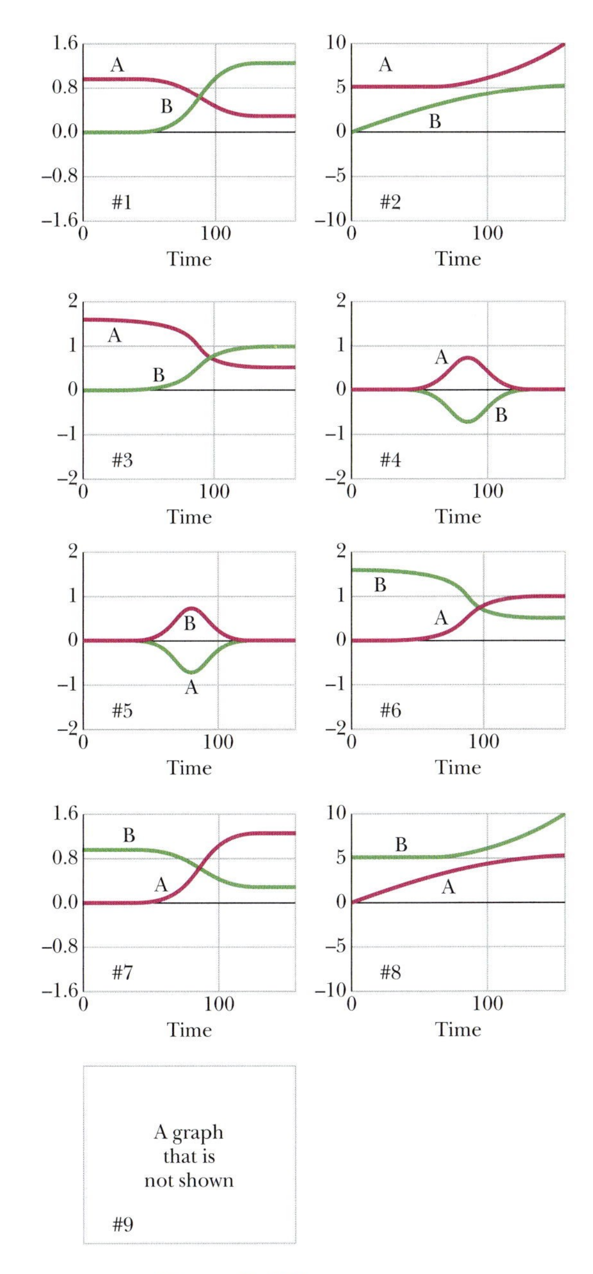


FIGURE 7-35$b$ ■ Problem 75.

**76. Could Newton Predict the "Third Law"?** Isaac Newton studied many types of collisions and invented the definition of momentum about twenty years before he developed his three laws of motion. As a result of his observations of collision processes, he formulated the law of conservation of momentum as a statement of experimental fact.

Let's assume for the sake of argument that Newton had already defined the concepts of force and momentum but had not yet formulated his laws of motion. Also assume that he had an electronic force sensor and was able to verify the impulse-momentum theorem. Explain in words how Newton could use the impulse-momentum theorem and the law of conservation of momentum to predict the existence of the third law of motion and to explain the nature of the interaction forces between two colliding objects.

**77. Taking Cyrano to the Moon** In Edmund Rostand's famous play, *Cyrano de Bergerac*, Cyrano, in an attempt to distract a suitor from visiting Roxanne, claims to have descended to Earth from the Moon and proclaims to have invented six novel and fantastical methods for traveling to the Moon. One is as follows.

*Sitting on an iron platform—thence*
*To throw a magnet in the air. This is*
*A method well conceived—the magnet flown,*
*Infallibly the iron will pursue:*
*Then quick! relaunch your magnet, and you thus*
*Can mount and mount unmeasured distances!**

In an old cartoon, there is another version of this method. A character in the old West is on a hand-pumped, two-person rail car. After getting tired of pumping the handle up and down to make the car move along the rails, he takes out a magnet, hangs it from a fishing pole, and holds it in front of the cart. The magnet pulls the cart toward it, which pushes the magnet forward, and so on, so the cart moves forward continually. What do you think of these methods? Can some version of them work? Discuss in terms of the physics you have learned.

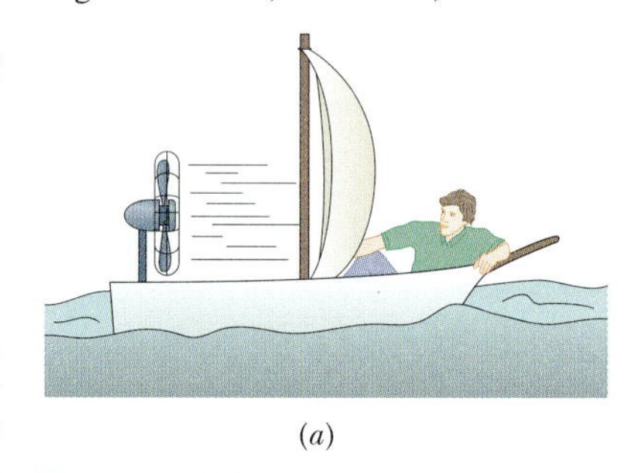
(a)
FIGURE 7-36a ■ Problem 78.

**78. Self Propulsion** People have forever been cooking up schemes for low-energy propulsion. Of course, we believe that whatever is designed had better be compatible with the laws of physics. Several schemes are shown below. Which ones do you think will work? Answer the questions detailed in (a) through (d) by referring to Fig. 7-36.

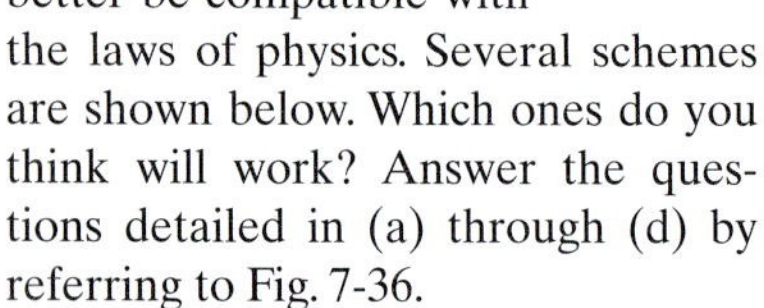

**(a)** In Fig. 7-36*a*, a lazy fisherman turns on a battery-operated fan and blows air onto the sail of his boat. Will he go anywhere? If he moves, what will his direction be? Explain.

**(b)** In Fig. 7-36*b*, a clever child is dangling a large magnet out in front of her wagon. It attracts a smaller magnet that she has attached to the front of her cart. Will she go anywhere? If she moves, what will her direction be? Explain.

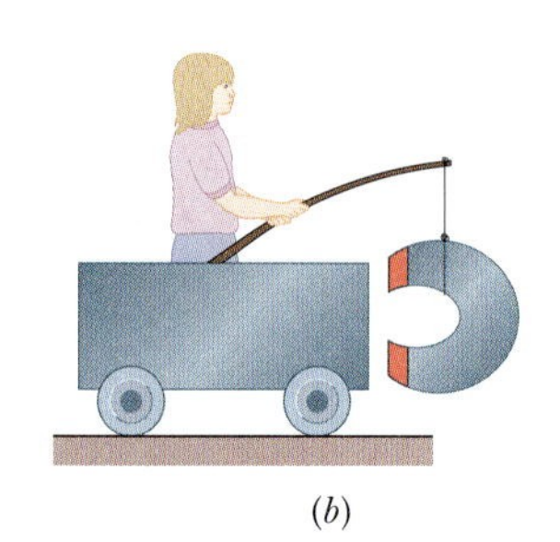
(b)
FIGURE 7-36b ■ Problem 78.

**(c)** In Fig. 7-36*c*, an astronaut is floating in outer space and wants to move backward. She tosses a ball out in front of her. Will she go anywhere? If she moves, what will her direction be? Explain.

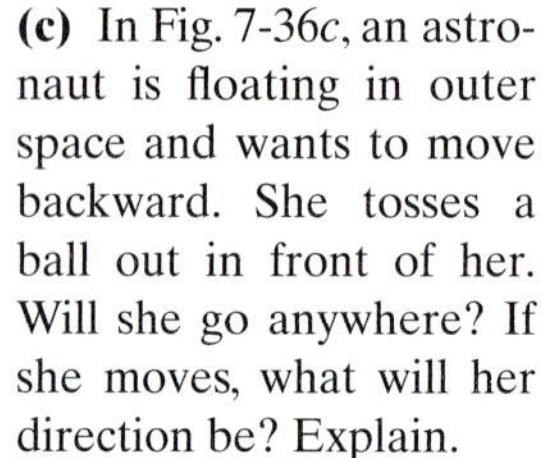

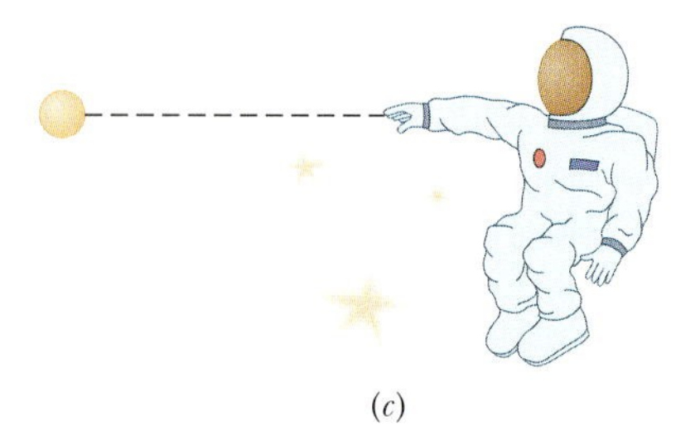
(c)
FIGURE 7-36c ■ Problem 78.

**(d)** In Fig. 7-36*d*, a college student on roller blades has a carbon dioxide container strapped to her back. The carbon dioxide jets out behind her as shown. Will she go anywhere? If she moves, what will her direction be? Explain.

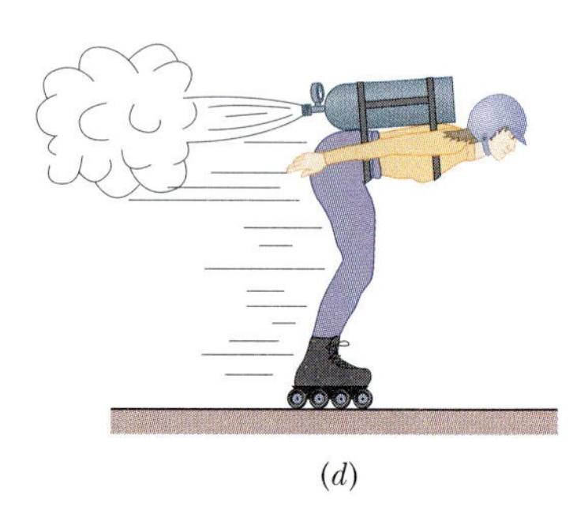
(d)
FIGURE 7-36d ■ Problem 78.

**79. The Ice-Skating Professor** A professor of physics is going ice skating for the first time. He has gotten himself into the middle of an ice rink and cannot figure out how to make the skates work. Every motion he makes simply causes his feet to slip on the ice and leaves him in the same place he started. He decides that he can get off the ice by throwing his gloves in the opposite direction.

**(a)** Suppose he has a mass $M$ and his gloves have a mass $m$. If he throws the gloves as hard as he can away from him, they leave his hand with a velocity $\vec{v}_{\text{glove}}$. Explain whether or not he will move. If he does move, calculate his velocity, $\vec{v}_{\text{prof}}$.

**(b)** Discuss his motion from the point of view of the forces acting on him.

**(c)** If the ice rink is 10 m in diameter and the skater starts in the center, estimate how long it will take him to reach the edge, assuming there is no friction at all.

**80. When Can You Conserve Momentum?** The principle of conservation of momentum is useful in some situations and not in others. Describe how you obtain the impulse-momentum theorem from Newton's Second Law and what situations lead to momentum conservation. How would you decide whether conservation of momentum could be useful in a particular problem?

**81. Momentum Conservation in Subsystems** Can a system whose momentum is conserved be made up of smaller systems whose individual momenta are not conserved? Explain why or why not and give an example.

**82. The Rabbit and the Eagle** You are working for the Defenders of Wildlife on the protection of the bald eagle, an endangered species. Walt Disney Productions, Inc. has agreed to help your cause by producing an animated movie about the bald eagle. You have set up a dramatic scene in which a young rabbit is frightened by the shadow of the eagle and starts bounding toward the east at 30 m/s as the eagle swoops down vertically at a speed of 15 m/s. A moment before the eagle contacts it, the rabbit bounds off a cliff and is captured in mid-air. (See Fig. 7-37.) The animators want to know how

*Translated from the French by Gladys Thomas and Mary F. Guillemard, e-text prepared by Sue Asscher, distributed by Project Gutenberg.

to portray what happens just after the capture. If the eagle has a mass of 2.5 kg and the rabbit has a mass of 0.8 kg, what is the *velocity* of the eagle with the rabbit in its talons just after the capture? (Include a diagram of the situation before and after capture with vectors showing the initial and final velocities.)

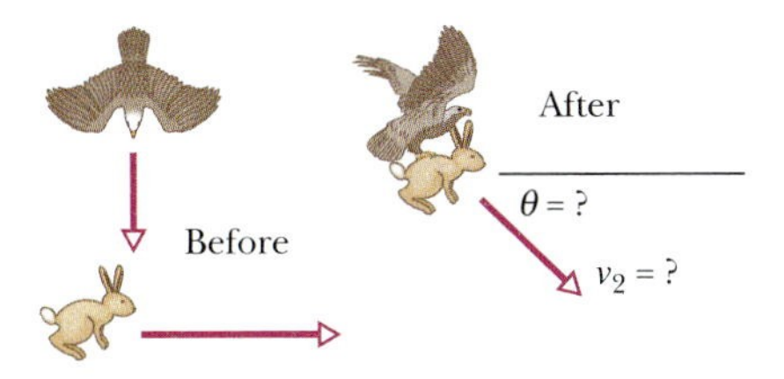


FIGURE 7-37 ■ Problem 82.

**83. Air Resistance 1: Estimating the Effect** The force of air resistance on a sphere of radius $R$ can plausibly be argued to have the form

$$\vec{F}^{\text{ drag}} = -\tfrac{1}{2}C\rho R^2|v|\vec{v} = -b|v|\vec{v},$$

where $\vec{v}$ is the vector velocity and $|\vec{v}|$ is its magnitude (the speed). The density of the air, $\rho$, is about 1 kg/m$^3$—1/1000 that of water. The parameter $C$ is a dimensionless constant.

If we drop a steel ball and a styrofoam ball from a height of $s$, the steel ball reaches the ground when the styrofoam ball is still a bit above the ground. Call this distance $h$. Estimate the air resistance coefficient $C$ as follows:

**(a)** Assume the effect of air resistance on the steel sphere is negligible. Calculate approximately how long the steel sphere takes to fall to the ground ($\Delta t_{\text{ste}}$) and how fast it is traveling just before it hits ($v_{\text{ste}}$). Express your answers in terms of $s$, $g$, and $m$.
**(b)** Since the steel and styrofoam were not very different, use $\langle\vec{v}_{\text{ste}}\rangle$, the average velocity of the steel ball during its fall to calculate an average air resistance force, $\langle\vec{F}^{\text{drag}}\rangle = -b\langle\vec{v}\rangle^2$ acting on the styrofoam sphere during its fall. Express this force in terms of $b$, $m$ (the mass of the styrofoam sphere), $g$, $s$, and $h$.
**(c)** The average velocity of the steel ball is $\langle\vec{v}_{\text{ste}}\rangle = s/\Delta t_{\text{ste}}$. The average velocity of the styrofoam sphere was $\langle\vec{v}_{\text{sty}}\rangle = (s - h)/\Delta t_{\text{ste}}$. Assume this difference, $\Delta\langle\vec{v}\rangle$, is caused by the average air resistance force acting over the time $\Delta t_{\text{ste}}$ with our basic Newton's law formula:

$$\langle\vec{F}^{\text{drag}}\rangle\Delta t_{\text{ste}} = m\,\Delta\langle\vec{v}\rangle.$$

Use this to show that

$$b \cong \frac{mh}{s^2}.$$

**(d)** A styrofoam ball of radius $R = 5$ cm and mass $m = 50$ g is dropped with a steel ball from a height of $s = 2$ m. When the steel ball hits, the styrofoam is about $h = 10$ cm above the ground. Calculate $b$ (for the styrofoam sphere) and $C$ (for any sphere).

**84. Air Resistance 2: Deriving the Equation** In this problem, you will derive an explicit form of Newton's drag law for air resistance, whose structure we derived by dimensional analysis in Problem 6-103. The derivation below will provide the dimensionless coefficient that we were unable to find by dimensional analysis.

**(a)** Consider a small particle of mass $m$ that is initially at rest. (Ignore gravity.) The particle is approached by a very massive wall moving toward it along an $x$ axis with a speed $v$. After the wall hits it, what speed will the small particle have? (*Hint*: Consider first the case of the small particle moving toward a stationary wall with a velocity $-v$. Analyze what happens.)
**(b)** Suppose the moving wall is a disk of radius $R$ moving at a speed $v$ in a direction perpendicular to the plane of the disk. If there are $N$ small particles per unit volume in the region of space the disk is sweeping through, how many of them will the disk encounter in a small time $\Delta t$?
**(c)** Calculate the total momentum transferred to the air in the time $\Delta t$ by the disk, assuming that there are $N$ air particles per unit volume and they each have mass $m$.
**(d)** Find the force the disk exerts on the air and the force the air exerts on the disk. How do you know?
**(e)** Show that the force you calculated has the form

$$\vec{F}^{\text{ drag}} = -\tfrac{1}{2}C\rho R^2|\vec{v}|\vec{v}$$

and find the dimensionless constant, $C$.

**85. Juggler** This problem is based on the analysis of a digital movie depicting a juggler. If you are using VideoPoint, view the movie entitled DSON007. Your instructor may provide you with a different movie to analyze or ask you to use the data presented in Fig. 7-38*b*. We track the motion of the white baseball of mass 0.138 kg in Fig. 7-38*a*, which is being caught and thrown in a smooth motion. The figure shows alternate frames depicting the catch and throw from just before to just after the juggler's hand is in contact with the ball. The data presented in Fig. 7-38*b* include a least-squares fit for frames 33–39 of the digital video shown in Fig. 7-38*a*. During all of these frames the ball is in contact with the juggler's hand. (Although the time codes are correct, the digital capture system missed recording a few frames between $t = 1.567$ s and $t = 1.700$ s.)

The goal of this problem is to consider the catch–throw process as a slow collision between the juggler's hand and the ball. In particular we would like you to verify that the impulse-momentum theorem holds for this situation. You should assume that the data and analysis presented here are correct and that Newton's Second Law is valid.

**(a)** Examine the $y$ position of the ball as a function of time for a time period during which the ball is in the juggler's hand (frames 33–39 in Fig. 7-38*a*). Express each fit coefficient and its uncertainty (that is, the standard deviation of the mean) to the correct number of significant figures. Write down the equation that allows you to calculate $y$ as a function of $t$.
**(b)** What is the nature of the vertical motion of the ball during the time it is being caught and thrown? Is its vertical velocity component zero, a constant, constantly changing, or is something else going on? Cite the reasons for your answer. What are the magnitude and direction of the vertical acceleration, $a_y$, of the ball?
**(c)** Calculate the *instantaneous* vertical velocity of the ball just as it's being caught (frame 33). Calculate the *instantaneous* vertical velocity of the ball just as it's being released (frame 39). (*Hints*: Use three significant figures in your coefficients. You can either interpret the physical meaning of the fit coefficient $a_1$ and then use the kinematic equation relating velocity to acceleration, initial velocity (at $t$ = 0.000 s), and time, or you can take the derivative with respect to time of the $y$ vs. $t$ equation you just wrote down in part (a).)
**(d)** Assuming the vertical acceleration of the ball is constant while it is in the juggler's hand, what is the *net* vertical force on the ball during the entire catch–throw process? Draw a free-body diagram showing the magnitudes and directions of the forces on the ball. What are the magnitude and direction of the gravitational force on the ball? What are the magnitude and direction of the vertical force the juggler exerts on the ball?

FIGURE 7-38a ■ Problem 85.

**DSON007: Juggling Data**

| Frame | $t$(s) | $y$(m) |
|---|---|---|
| 33 | 1.500 | 0.416 |
| 34 | 1.533 | 0.330 |
| 35 | 1.567 | 0.249 |
| 36 | 1.700 | 0.152 |
| 37 | 1.733 | 0.213 |
| 38 | 1.767 | 0.305 |
| 39 | 1.800 | 0.421 |

$a_0 = 34.7656$ m

$a_1 = -41.9051$ m/s

$a_2 = 12.6780$ m/s$^2$

The fit graph is given by $y_{\text{fit}}(m) = a_0 + a_1 t + a_2 t^2$.

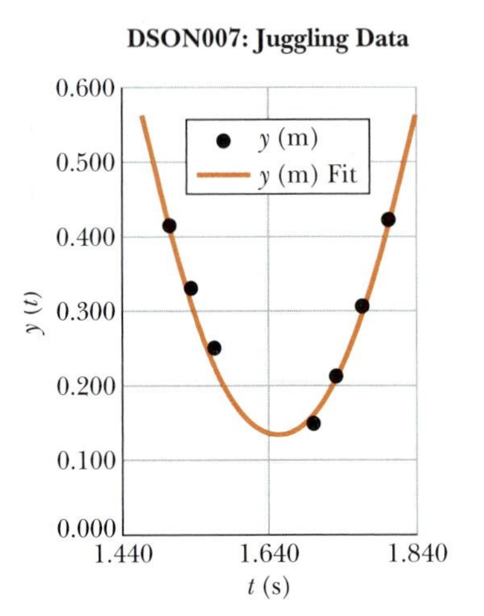


FIGURE 7-38b ■ Problem 85.

**(e)** Identify any Newton's Third Law pairs for this situation. Identify what object is exerting the gravitational force on the ball. According to Newton's Third Law, how is the ball interacting with that object?
**(f)** Find the vertical momentum of the ball when it first falls into the juggler's hand (as in frame 33). Also find the vertical momentum of the ball when it is just about to leave the juggler's hand (as in frame 39). What are the magnitude and direction of the *momentum change*, $\Delta p_y$, in the vertical direction that the ball undergoes during this time period? *Beware*: Momentum is a vector quantity. Do not fall into the trap of simply subtracting the *magnitudes* of the two momenta.
**(g)** How much time, $\Delta t$, does the ball spend in the hand of the juggler? Calculate the *impulse* transmitted to the ball by the net force on it during the catch–throw "collision."
**(h)** Compare the change in momentum to the impulse imparted to the ball. Does the impulse-momentum theorem seem to hold to the appropriate number of significant figures?

**86. Momentum Before and After a Sticky Collision** This problem is based on the analysis of a digital movie depicting a collision between two carts. Before the collision, one cart is moving and one cart is stationary. Following the collision, the two carts stick together. If you are using VideoPoint, view the movie entitled PASCO028. It depicts a cart of mass 2 kg colliding with a stationary cart of mass 1 kg. Your instructor may provide you with a different movie to analyze.
**(a)** Use video analysis software and a spreadsheet to find the initial momentum of the two-cart system before collision. Explain the method you used and show all your data and calculations.
**(b)** Use video analysis software and a spreadsheet to find the final momentum of the two-cart system after collision. Explain the method you used and show all your data and calculations.
**(c)** What is the percent difference between the momentum of the system before the collision and after the collision? Within the limits of experimental uncertainty, is the total momentum of the two-cart system conserved? Why or why not?
**(d)** If you found that the total momentum after collision is less than that before the collision, you can either conclude that: (1) momentum is still conserved but some of it is transferred to the track (that is the whole Earth) or (2) the law of conservation of momentum has failed. Assuming that the law of conservation of momentum still holds, how much momentum is transferred to the track and Earth? Remember that momentum is a vector quantity, and you must specify both the magnitude and direction of this momentum.
**(e)** Why don't you see the track move just after the collision?

# 8 Extended Systems

If you leap forward, chances are that your head and torso will follow a parabolic path, like a baseball thrown in from the outfield. However, when a skilled ballet dancer leaps across the stage in a *grand jeté*, the path taken by her head and torso is nearly horizontal during much of the jump. She seems to be floating across the stage. The audience may not know much about projectile motion, but they still sense that something unusual has happened.

**How does the ballerina seemingly "turn off" the gravitational force?**

*The answer is in this chapter.*

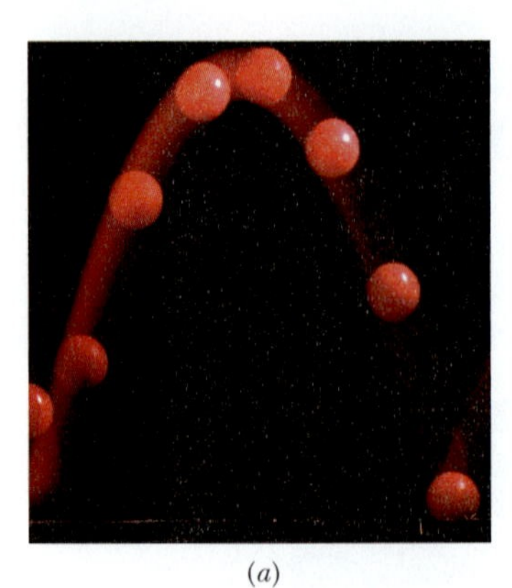

(a)

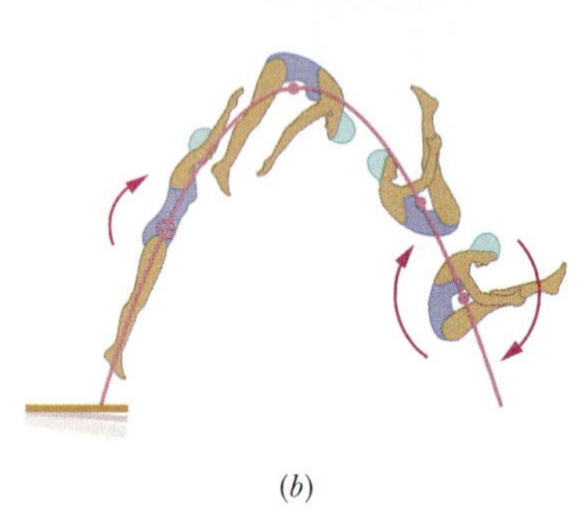

(b)

**FIGURE 8-1** ■ (a) A bouncing ball follows a parabolic path. (b) A diver bounces off a board. Even though many points on her body that are not marked follow complex paths, a special point that also follows a parabolic path (shown by the dots) can be calculated based on the positions of the diver's body parts.

## 8-1 The Motion of Complex Objects

Up to this point we have focused our discussions on objects that can be considered to move as particles. In order to treat an object as a particle, every point on the object must be moving with the same velocity and acceleration. Although this requirement simplifies the analysis of motion, it is not commonly the case with everyday objects.

Here is an example. The motion of a rotating diver shown in Fig. 8-1*b* is clearly more complicated than that of the bouncing ball in Fig. 8-1*a*. Every part of the diver moves in a different manner than every other part, so we cannot describe her as a tossed particle. Instead, we must consider her as a system of particles. In large, complicated systems, it is often difficult to keep track of all the parts, and we cannot make predictions about the motion of the parts using the physics we have learned for particles. In fact, even a baseball, which seems to move as a particle, is usually spinning as it moves through the air.

So why is it that we have been able to treat objects like baseballs as particles in the previous chapters? And how do we handle the analysis of more complex systems, like divers and rotating baseball bats? We answer these questions in this chapter.

## 8-2 Defining the Position of a Complex Object

Even if we only have two objects in a system, their motions can be quite complex. Suppose two stars attract each other gravitationally so they are moving relative to one another. At the same time that the stars are exerting forces on each other, external forces could cause this two-star system to accelerate. But what is it that accelerates? Where is this two-star system actually located? At the location of the first star? The second star? Somewhere else? In this section we will show that we can define a position that can be used to describe accurately where a system is located and how the system accelerates.

Let's start by considering two particles, $A$ and $B$, that attract one another as shown in Fig. 8-2. Suppose they have external forces $\vec{F}_A^{\text{ext}}$ and $\vec{F}_B^{\text{ext}}$ acting on them. Applying Newton's Second Law ($\vec{F}^{\text{net}} = m\vec{a}$) to each particle in this system gives us

$$\vec{F}_A^{\text{ext}} + \vec{F}_{B\to A} = m_A\vec{a}_A, \tag{8-1}$$

and

$$\vec{F}_B^{\text{ext}} + \vec{F}_{A\to B} = m_B\vec{a}_B, \tag{8-2}$$

where $\vec{F}_{B\to A}$ and $\vec{F}_{A\to B}$ are the internal forces that the two particles in the system exert on each other. In order to get the net force acting on the "system," we must add Eqs. 8-1 and 8-2 together. Since $\vec{F}_{B\to A}$ and $\vec{F}_{A\to B}$ are equal and opposite forces (by Newton's Third Law), they cancel each other and we are left with an expression for the net force on the system of

$$\vec{F}_{\text{sys}}^{\text{net}} = \vec{F}_A^{\text{ext}} + \vec{F}_B^{\text{ext}} = m_A\vec{a}_A + m_B\vec{a}_B. \tag{8-3}$$

However, applying Newton's Second Law directly to the entire system also gives us

$$\vec{F}_{\text{sys}}^{\text{net}} = M_{\text{sys}}\vec{a}_{\text{sys}}, \tag{8-4}$$

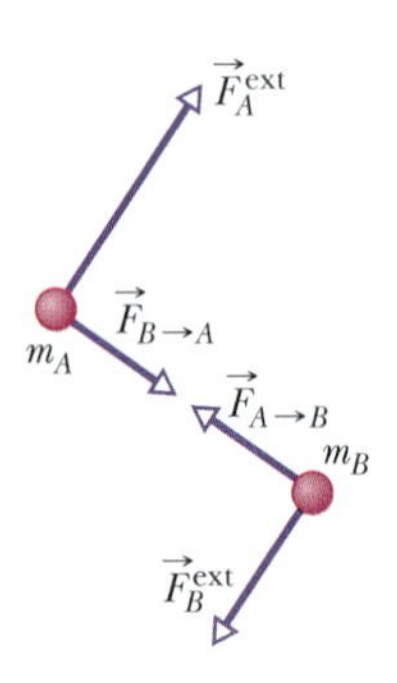


**FIGURE 8-2** ■ Two particles connected by a "massless" rod (not shown) can experience different external forces while exerting equal and opposite internal forces on each other.

where $M_{\text{sys}} = m_A + m_B$ is the total mass and $\vec{a}_{\text{sys}}$ is the acceleration of the system taken as a whole. A look at Fig. 8-2 tells us that the particles could be in orbit about each other or moving together while the system is rotating and accelerating along a line. So what do we mean by the "acceleration of the system as a whole?" If we combine Eqs. 8-3 and 8-4 we can use the result as a basis for defining a point in space that represents the system's acceleration. This result is given by

$$\vec{F}^{\,\text{net}}_{\text{sys}} = M_{\text{sys}}\vec{a}_{\text{sys}} = m_A\vec{a}_A + m_B\vec{a}_B. \tag{8-5}$$

Solving for the system acceleration gives us

$$\vec{a}_{\text{sys}} = \frac{1}{M_{\text{sys}}}(m_A\vec{a}_A + m_B\vec{a}_B). \tag{8-6}$$

This expression indicates that the acceleration of the system can be viewed as a "weighted average" of the particle accelerations.

If we choose a coordinate system, we can locate the particles in the system in terms of their position vectors $\vec{r}_A$ and $\vec{r}_B$ as shown in Fig. 8-3. Remember that acceleration is related to position by $\vec{a} = d^2\vec{r}/dt^2$—our expression for the acceleration suggests that we can define the effective "position" of the system as

$$\vec{R}^{\,\text{eff}} = \frac{1}{M_{\text{sys}}}(m_A\vec{r}_A + m_B\vec{r}_B) \quad \text{(two particle system).} \tag{8-7}$$

We can verify that this expression for the position of the object makes sense by taking its derivative with respect to time twice. When we do that, we find that we get back the equation for the system acceleration that we derived in Eq. 8-6.

If we had considered a more complex system of $N$ particles we would have come to a similar expression for the effective position, $\vec{R}^{\,\text{eff}}$, of the system in terms of the system mass and the sum of the products of the individual masses and position vectors,

$$\vec{R}^{\,\text{eff}} = \frac{1}{M_{\text{sys}}}(m_A\vec{r}_A + m_B\vec{r}_B + m_C\vec{r}_C + \cdots + m_N\vec{r}_N) \quad (N \text{ particle system}). \tag{8-8}$$

In the next section we explore the properties of this expression and compare $\vec{R}^{\,\text{eff}}$ to the location of the balancing point for a system of objects.

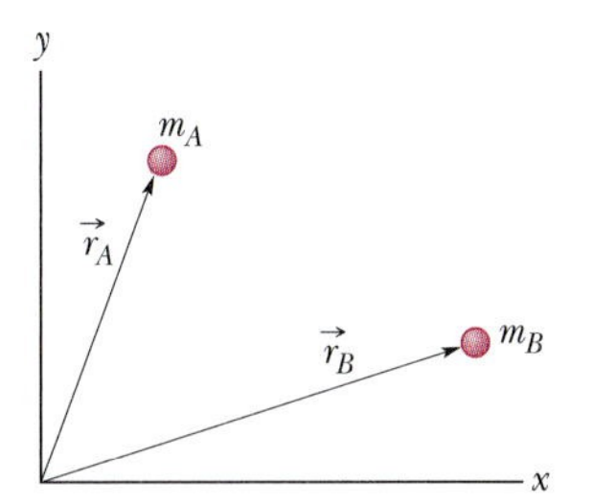


**FIGURE 8-3** ▪ If we choose a coordinate system to describe our two particles mathematically, the position vectors describing the location of each of the particles are $\vec{r}_A$ and $\vec{r}_B$, respectively.

## 8-3 The Effective Position—Center of Mass

Consider the system shown in Fig. 8-4. If the two masses are equal, $m_A = m_B$, then from

$$\vec{R}^{\,\text{eff}} = \frac{1}{M_{\text{sys}}}(m_A\vec{r}_A + m_B\vec{r}_B),$$

we get

$$\vec{R}^{\,\text{eff}} = \tfrac{1}{2}(\vec{r}_A + \vec{r}_B) \quad \text{(equal masses).}$$

If we want to consider the two-particle system's effective position quantitatively, then we must pick a coordinate system to determine $\vec{r}_A$ and $\vec{r}_B$. If we choose one of the axes of the coordinate system to lie on a line connecting the particles, it is easy to see that when the masses are equal, then $\vec{R}^{\,\text{eff}}$, the effective position of the system, is midway between the two objects on the line connecting them (Fig. 8-4*a*). If we imagined that the system particles are connected by a massless rod and tried to balance such an object on our finger, we would find that the balancing point is also halfway between the two masses whenever $m_A$ is equal to $m_B$ (Fig. 8-4*b*). For this reason we define the effective position of a system that is calculated using Eq. 8-8 as the **center**

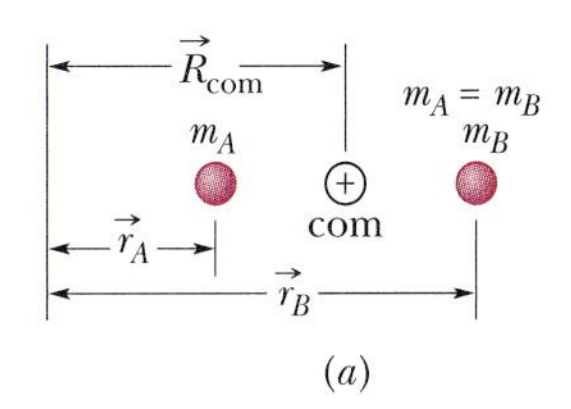


(*a*)

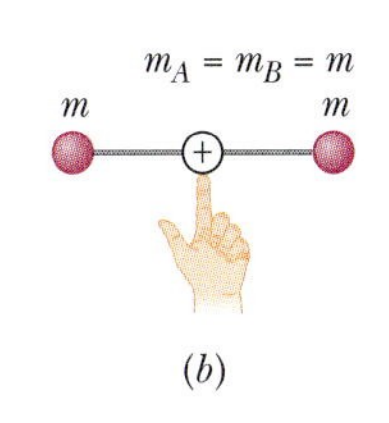


(*b*)

**FIGURE 8-4** ▪ (*a*) If the masses in a two-particle system are equal, then the center of mass of the system is always on a line halfway between the two particles. (*b*) If we connect the two particles having identical masses with a massless rod, the balance point of the system also turns out to be halfway between the particles.

**of mass (com)** of the system. We denote the location of the center of mass as $\vec{R}_{\text{com}} \equiv \vec{R}^{\text{eff}}$. With experiment and careful observation, we can determine that special balancing point or the center of mass of almost any system. In general:

> The center of mass (com) of a body or a system of bodies is its balancing point. It is the point that moves as though all of the mass were concentrated there and the system behaves as if all the external forces are applied there.

What happens to the center of mass of a two-particle system if the masses are not equal? If we let particle $B$ be twice the mass of particle $A$ so $m_B = 2m_A$, we find that

$$\vec{R}_{\text{com}} = \frac{1}{M_{\text{sys}}}(m_A\vec{r}_A + m_B\vec{r}_B) = \frac{1}{m_A + m_B}(m_A\vec{r}_A + 2m_A\vec{r}_B)$$

$$= \frac{m_A}{m_A + 2m_A}(\vec{r}_A + 2\vec{r}_B),$$

or

$$\vec{R}_{\text{com}} = \tfrac{1}{3}(\vec{r}_A + 2\vec{r}_B) \quad \text{(special case for } m_B = 2m_A\text{)}.$$

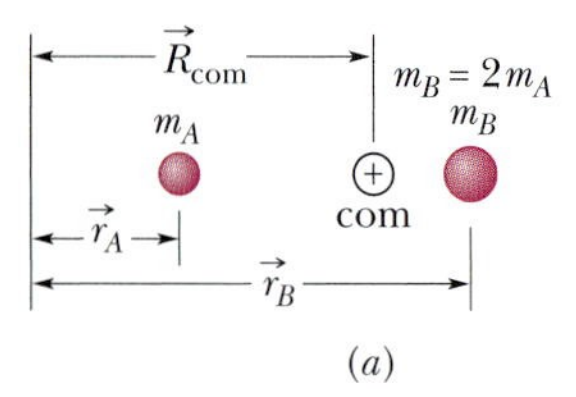


(*a*)

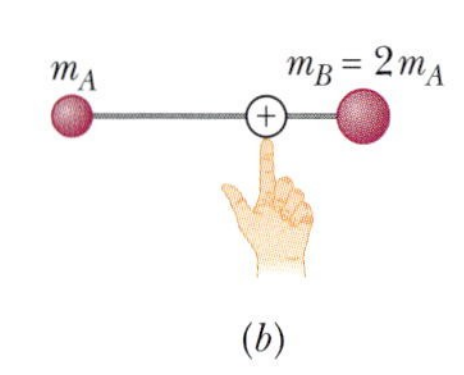


(*b*)

**FIGURE 8-5** ■ (*a*) If the masses in a two-particle system are not equal, then the center of mass is always on a line between the two particles but is closer to the more massive particle. (*b*) If we connect the two particles having unequal masses with a massless rod, the balance point of the system turns out to be at the same location as the calculated center of mass of the system.

In words, the center of mass or "effective position" of this system is located along the line joining the centers of the two masses, two-thirds of the way from the less massive object $m_A$ and one-third of the way from the more massive object $m_B$ (Fig. 8-5*a*). If we connected the two particles in the system with a massless rod and tried to balance it on our finger, we would find that the center of mass is also the same as the balancing point (Fig. 8-5*b*).

Physicists love to look at something complicated and find something simple and familiar in it. Fortunately, this turns out to be the case with the complicated motions of particle systems. For example, recall the diver who is rotating as she falls through the air as shown in Fig. 8-1*b*. We can consider her body to be a system made up of many individual particles that can exert internal forces on each other. If we neglect air drag, then the only significant external force on her is a constant gravitational force that acts downward. If we were to calculate the location of her center of mass at each moment during her fall, we would find that the calculated center of mass of the diver moves in a very simple parabolic path.

Actually calculating the center of mass of an athlete or dancer who is constantly changing her configuration seems like an impossible task. However, we can think of an athlete as a series of particles connected by massless rods. We locate each particle near the center of a linear body segment (such as ankle to toe, knee to ankle, hip to knee, and so on) and assign it the mass of the body part it represents. We find that we can use the techniques described in the next section to perform computer-aided calculations of an athlete's center of mass. Such an analysis performed on a series of video frames always gives a parabolic path when the athlete or dancer is jumping. An example of this is presented at the end of Section 8-5 for a ballerina performing a *grand jeté*. This analysis and countless others provide us with experimental verification that the concept of center of mass is useful when tracking the motion of complex systems that experience external forces.

## 8-4 Locating a System's Center of Mass

Let's consider how to calculate the center of mass (com) for a system consisting of a few particles that lie along a chosen $x$ axis. Figure 8-5 shows two particles of masses $m_A$ and $m_B$ separated by a distance $d$. Here, $x_A$ is the $x$ coordinate of $m_A$'s position and

$x_B$ is the $x$ coordinate of $m_B$'s position. We can write the expression for the $x$ coordinate of the center of mass of this system as

$$x_{\text{com}} = \frac{m_A x_A + m_B x_B}{M_{\text{sys}}}, \tag{8-9}$$

in which $M_{\text{sys}}$ is the total mass of the system. (Here, $M_{\text{sys}} = m_A + m_B$.) We can extend this equation to a more general situation in which $N$ particles are strung out along the $x$ axis. Then the total mass is $M_{\text{sys}} = m_A + m_B + \cdots + m_N$, and the location of the center of mass is

$$x_{\text{com}} = \frac{m_A x_A + m_B x_B + m_C x_C + \cdots + m_N x_N}{M_{\text{sys}}}. \tag{8-10}$$

If the particles are distributed in three dimensions, then we can start with our expression for $\vec{R}_{\text{com}}$ (Eq. 8-8) and express each position vector in terms of its $x$-, $y$-, and $z$-components. For example, the $i$th position vector is given by

$$\vec{r}_i = x_i\hat{\text{i}} + y_i\hat{\text{j}} + z_i\hat{\text{k}}.$$

It is not difficult to show that when all the position vectors in a system of $N$ particles are expressed in their rectangular coordinates using the equation above, then

$$\vec{R}_{\text{com}} = X_{\text{com}}\hat{\text{i}} + Y_{\text{com}}\hat{\text{j}} + Z_{\text{com}}\hat{\text{k}}.$$

The components of the center of mass of a system of particles are

$$X_{\text{com}} = \frac{1}{M_{\text{sys}}}(m_A x_A + m_B x_B + m_C x_C + \cdots),$$

$$Y_{\text{com}} = \frac{1}{M_{\text{sys}}}(m_A y_A + m_B y_B + m_C y_C + \cdots), \quad \text{(center of mass vector components-particle system)}. \tag{8-11}$$

$$Z_{\text{com}} = \frac{1}{M_{\text{sys}}}(m_A z_A + m_B z_B + m_C z_C + \cdots)$$

We can use the equations for $X_{\text{com}}$ and $Y_{\text{com}}$ to calculate the center of mass of a system of three pucks on an air table that have masses of 100, 200, and 300 g. The location of the center of each puck is shown in the data table in Fig. 8-6. The diagram shows the locations of the pucks in a rectangular coordinate system along with the calculated location of the center of mass of the system.

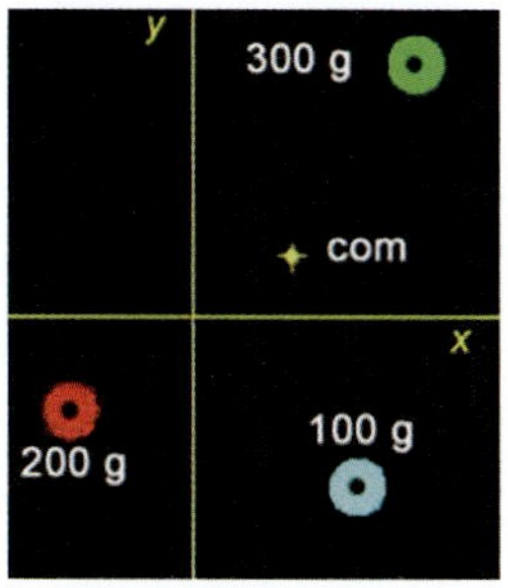

| $m$(g) | $x$(cm) | $y$(cm) |
|---|---|---|
| 100 | 16.8 | –16.8 |
| 200 | –12.2 | –9.4 |
| 300 | 22.7 | 24.1 |
| $M_{\text{sys}}$(g) | $X_{\text{com}}$(cm) | $Y_{\text{com}}$(cm) |
| 600 | 10.1 | 6.1 |

**FIGURE 8-6** ■ Three pucks gliding on an air table form a system. Equation 8-11 can be used to locate the center of mass of the system at $x = 10.1$ cm and $y = 6.1$ cm.

## Solid Bodies

Some systems have too many particles to keep track of individually. It would be an enormous task to calculate the location of the center of mass using the summation technique described above. A solid object can be treated as a "continuous distribution" of matter. The term *continuous* implies that the "particles" that make up the object are no longer clearly separable. The particles then become differential mass elements, $dm$, the sums (shown in Eq. 8-11) become integrals, and the coordinates of the center of mass vector components are defined as

$$X_{\text{com}} = \frac{1}{M_{\text{sys}}} \int x\,dm, \quad Y_{\text{com}} = \frac{1}{M_{\text{sys}}} \int y\,dm, \quad Z_{\text{com}} = \frac{1}{M_{\text{sys}}} \int z\,dm \tag{8-12}$$

(center of mass vector components-continuous system),

where $M_{\text{sys}}$ is the mass of the system.

If you are clever and don't enjoy doing unnecessary integrations, you can bypass one or more of the integrals above if an object has a point, a line, or a plane of symmetry. In these cases, the center of mass of such an object then lies at that point, on that line, or in that plane. For example, the center of mass of a uniform sphere (which has a point of symmetry) is at the center of the sphere (which is the point of symmetry). The center of mass of a uniform cone (whose axis is a line of symmetry) lies on the axis of the cone. The task required to determine the location of the center of mass of the cone is then reduced to determining where along this axis the center of mass is located. For example, the center of mass of a banana (which has a plane of symmetry that splits it into two equal parts) lies somewhere in that plane.

The center of mass of an object need not lie within the object. There is no dough at the center of mass of a doughnut, and no iron at the center of mass of a horseshoe.

Evaluating these integrals for most common objects (like a television set) would be difficult, so here we shall consider only *uniform* solid objects. Such an object has *uniform density,* or mass per unit volume. That is, the density $\rho$ (Greek letter rho) is the same for any given segment of the object as for the whole object:

$$\rho = \frac{dm}{dV} = \frac{M_{\text{sys}}}{V} \quad \text{(uniform object density)}, \tag{8-13}$$

where $dV$ is the volume occupied by a mass element $dm$, and $V$ is the total volume of the object. If we substitute $dm = (M_{\text{sys}}/V)\,dV$ into Eq. 8-12 we find that

$$X_{\text{com}} = \frac{1}{V} \int x\,dV, \quad Y_{\text{com}} = \frac{1}{V} \int y\,dV, \quad Z_{\text{com}} = \frac{1}{V} \int z\,dV. \tag{8-14}$$

**READING EXERCISE 8-1:** The figure shows a uniform square plate from which four identical squares at the corners will be removed. (a) Where is the center of mass of the plate originally? Where is it after the removal of (b) square 1, (c) squares 1 and 2, (d) squares 1 and 3, (e) squares 1, 2, and 3, (f) all four squares? Answer in terms of quadrants, axes, or points (without calculation, of course).

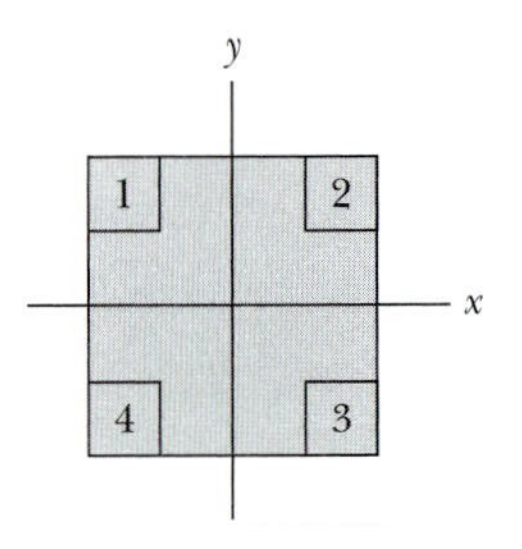

## TOUCHSTONE EXAMPLE 8-1: Three Masses

Three particles of masses $m_A = 1.2$ kg, $m_B = 2.5$ kg, and $m_C = 3.4$ kg form an equilateral triangle of edge length $a = 140$ cm. Where is the center of mass of this three-particle system?

**SOLUTION** ■ A **Key Idea** to get us started is that we are dealing with particles instead of an extended solid body, so we can use Eq. 8-11 to locate their center of mass. The particles are in the plane of the equilateral triangle, so we need only the first two equations. A second **Key Idea** is that we can simplify the calculations by choosing the $x$ and $y$ axes so that one of the particles is located at the origin and the $x$ axis coincides with one of the triangle's sides (Fig. 8-7). The three particles then have the following coordinates:

| Particle | Mass (kg) | $X$ (cm) | $Y$ (cm) |
|---|---|---|---|
| $A$ | 1.2 | 0 | 0 |
| $B$ | 2.5 | 140 | 0 |
| $C$ | 3.4 | 70 | 121 |

The total mass $M_{sys}$ of the system is 7.1 kg.

From Eq. 8-11, the coordinates of the center of mass are

$$X_{com} = \frac{m_A X_A + m_B X_B + m_C X_C}{M_{sys}}$$

$$= \frac{(1.2\text{ kg})(0) + (2.5\text{ kg})(140\text{ cm}) + (3.4\text{ kg})(70\text{ cm})}{7.1\text{ kg}}$$

$$= 83\text{ cm}, \qquad \text{(Answer)}$$

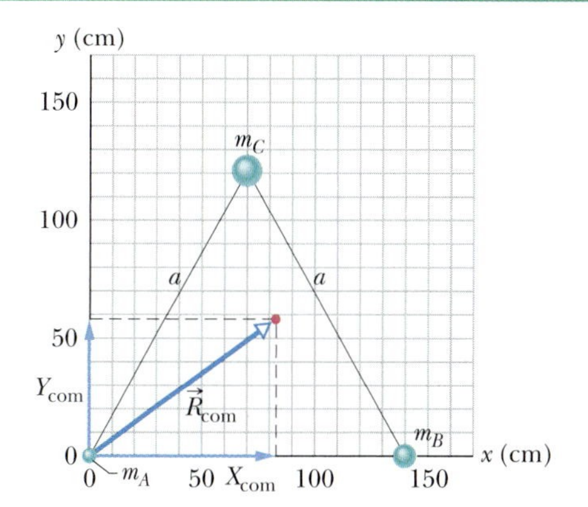


**FIGURE 8-7** ■ Three particles form an equilateral triangle of edge length $a$. The center of mass is located by the position vector $\vec{R}_{com}$.

and

$$Y_{com} = \frac{m_A Y_A + m_B Y_B + m_C Y_C}{M_{sys}}$$

$$= \frac{(1.2\text{ kg})(0) + (2.5\text{ kg})(0) + (3.4\text{ kg})(121\text{ cm})}{7.1\text{ kg}}$$

$$= 58\text{ cm}. \qquad \text{(Answer)}$$

In Fig. 8-7, the center of mass is located by the position vector $\vec{R}_{com}$, which has components $X_{com}$ and $Y_{com}$.

## TOUCHSTONE EXAMPLE 8-2: U-Shaped Object

The U-shaped object pictured in Fig. 8-8 has outside dimensions of 100 mm on each side, and each of its three sides is 20 mm wide. It was cut from a uniform sheet of plastic 6.0 mm thick. Locate the center of mass of this object.

**SOLUTION** ■ A **Key Idea** here is to break the U-shaped object up into pieces, each having an easily located center of mass. We can then replace each piece by a point mass located at the center of mass of that piece, and then use Eq. 8-11 to locate the center of mass of the whole object.

As shown in Fig. 8-8, we can think of the U-shaped object as made up of two vertical bars, each 100 mm long by 20 mm wide, joined together by one horizontal bar 60 mm long and 20 mm wide. Let's place the origin of our coordinate system at the lower-left rear corner of the U, with the $x$ axis across its base and the $y$ axis along its left edge.

To locate the center of mass of each of the bars, we will use the **Key Idea** that the center of mass of a symmetric object of uniform density is located at its geometric center. This means that the center of mass of each bar is exactly halfway from either end, halfway from either side, and halfway between the top and bottom surfaces of the plastic sheet. Putting a dot at the center of each of

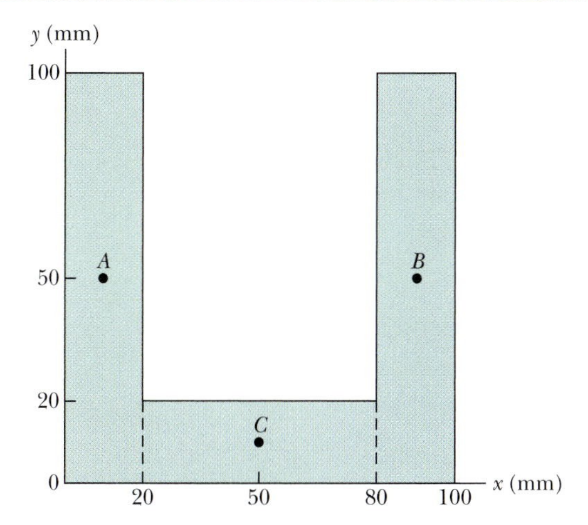


**FIGURE 8-8** ■ The U-shaped object can be broken up into three uniform rectangular bars, $A$, $B$, and $C$, as shown in the figure. For convenience we'll take the corners of the U to be square and not rounded. The dot on each bar shows the location of its center of mass. We will use the coordinate system shown to locate the center of mass of each bar and then to locate the center of mass of the whole object.

the bars, $A$, $B$, and $C$ in Fig. 8-8, we can now write down their locations in the coordinate system pictured there:

| Object | Mass | $X$ | $Y$ | $Z$ |
|---|---|---|---|---|
| Left bar | $M_A$ | 10 mm | 50 mm | 3 mm |
| Right bar | $M_B$ | 90 mm | 50 mm | 3 mm |
| Bottom bar | $M_C$ | 50 mm | 10 mm | 3 mm |

To learn more about the relative masses of the three bars, we can use the **Key Idea** that the mass of each bar is proportional to its volume since the bars have a uniform common density. In this case the relationship is even simpler: since the three bars are each the same width and thickness, each one's mass is directly proportional to its length, so

$$M_A = M_B = M \quad \text{and} \quad M_C = M(60\text{ mm})/(100\text{ mm}) = 0.6M,$$

and $$M_{\text{sys}} = M_A + M_B + M_C = M + M + 0.6M = 2.6M.$$

Replacing each bar by a point mass at its center of mass, Eq. 8-11 gives us the location of the center of mass of the entire U-shaped object:

$$X_{\text{com}} = \frac{M_A X_A + M_B X_B + M_C X_C}{M_{\text{sys}}}$$

$$= \frac{\cancel{M}(10\text{ mm}) + \cancel{M}(90\text{ mm}) + 0.6\cancel{M}(50\text{ mm})}{2.6\cancel{M}}$$

$$= 50\text{ mm}, \qquad \text{(Answer)}$$

$$Y_{\text{com}} = \frac{M_A Y_A + M_B Y_B + M_C Y_C}{M_{\text{sys}}}$$

$$= \frac{\cancel{M}(50\text{ mm}) + \cancel{M}(50\text{ mm}) + 0.6\cancel{M}(10\text{ mm})}{2.6\cancel{M}}$$

$$= 40.769\text{ mm} \cong 41\text{ mm}, \qquad \text{(Answer)}$$

and $$Z_{\text{com}} = \frac{M_A Z_A + M_B Z_B + M_C Z_C}{M_{\text{sys}}}$$

$$= \frac{\cancel{M}(3\text{ mm}) + \cancel{M}(3\text{ mm}) + 0.6\cancel{M}(3\text{ mm})}{2.6\cancel{M}}$$

$$= 3\text{ mm}, \qquad \text{(Answer)}$$

so $$\vec{R}_{\text{com}} = (50\text{ mm})\hat{\text{i}} + (41\text{ mm})\hat{\text{j}} + (3\text{ mm})\hat{\text{k}}. \qquad \text{(Answer)}$$

## TOUCHSTONE EXAMPLE 8-3: Crescent-Shaped Object

Figure 8-9*a* shows a uniform metal plate $P$ of radius $2R$ from which a disk of radius $R$ has been stamped out (removed) in an assembly line. Using the $xy$ coordinate system shown, locate the center of mass, com$_P$, of the plate.

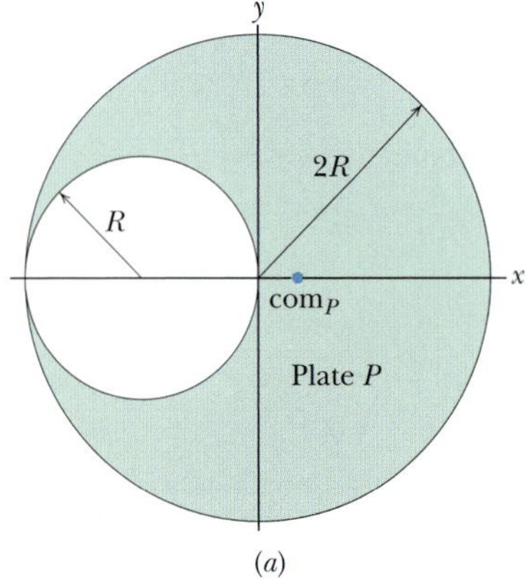


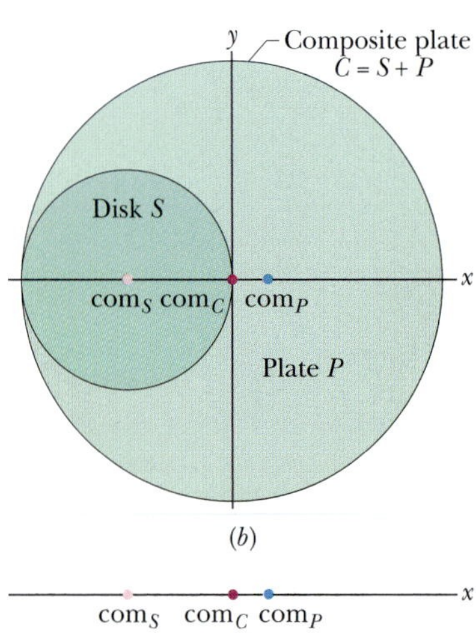


**FIGURE 8-9** (*a*) Plate $P$ is a metal plate of radius $2R$, with a circular hole of radius $R$. The center of mass of $P$ is at point com$_P$. (*b*) Disk $S$ has been put back into place to form a composite plate $C$. The center of mass, com$_S$, of disk $S$ and the center of mass, com$_C$, of plate $C$ are shown. (*c*) The center of mass, com$_{S+P}$, of the combination of $S$ and $P$ coincides with com$_C$, which is at $x = 0$.

**SOLUTION** First, let us roughly locate the center of mass of plate $P$ by using the **Key Idea** of symmetry. We note that the plate is symmetric about the $x$ axis (we get the portion below that axis by rotating the upper portion about the axis). Thus, com$_P$ must be on the $x$ axis. The plate is not symmetric about the $y$ axis. However, because there is somewhat more mass on the right of the $y$ axis, com$_P$ must be somewhat to the right of that axis. Thus, the location of com$_P$ should be roughly as indicated in Fig. 8-9*a*.

Another **Key Idea** here is that plate $P$ is an extended solid body, so we can use Eq. 8-14 to find the actual coordinates of com$_P$. However, that procedure is difficult. A much easier way is to use this **Key Idea**: In working with centers of mass, we can take the mass of any *uniform* object to be concentrated in a particle located at the object's center of mass. Here is how we do so:

First, put the stamped-out disk (call it disk $S$) back into place (Fig. 8-9*b*) to form the original composite plate (call it plate $C$). Because of its circular symmetry, the center of mass, com$_S$, for disk $S$ is at the center of $S$, at $x = -R$ (as shown). Similarly, the center of mass, com$_C$, for composite plate $C$ is at the center of $C$, at the origin (as shown). We then have the following:

| Plate | Center of Mass | Location of of com | Mass |
|---|---|---|---|
| $P$ | com$_P$ | $X_P = ?$ | $m_P$ |
| $S$ | com$_S$ | $X_S = -R$ | $m_S$ |
| $C$ | com$_C$ | $X_C = 0$ | $m_C = m_S + m_P$ |

Now we use the **Key Idea** of concentrated mass: Assume that mass $m_S$ of disk $S$ is concentrated in a particle at $X_S = -R$, and mass $m_P$ is concentrated in a particle at $X_P$ (Fig. 8-9*c*). Next treat these two particles as a two-particle system, using Eq. 8-9 to find their center of mass $X_{S+P}$. We get

$$X_{S+P} = \frac{m_S X_S + m_P X_P}{m_S + m_P}. \qquad (8\text{-}15)$$

Next note that the combination of disk $S$ and plate $P$ is composite plate $C$. Thus, the position $X_{S+P}$ of com$_{S+P}$ must coincide with the position $X_C$ of com$_C$, which is at the origin, so $X_{S+P} = X_C = 0$. Substituting this into Eq. 8-15 and solving for $X_P$, we get

$$X_P = -X_S \frac{m_S}{m_P}. \qquad (8\text{-}16)$$

Now we seem to have a problem, because we do not know the masses in Eq. 8-16. However, we can relate the masses to the face areas of $S$ and $P$ by noting that

$$\begin{aligned} \text{Mass} &= \text{density} \times \text{volume} \\ &= \text{density} \times \text{thickness} \times \text{area}. \end{aligned}$$

Then $$\frac{m_S}{m_P} = \frac{\text{density}_S}{\text{density}_P} \times \frac{\text{thickness}_S}{\text{thickness}_P} \times \frac{\text{area}_S}{\text{area}_P}.$$

Because the plate is uniform, the densities and thicknesses are equal; we are left with

$$\frac{m_S}{m_P} = \frac{\text{area}_S}{\text{area}_P} = \frac{\text{area}_S}{\text{area}_C - \text{area}_S} = \frac{\pi R^2}{\pi(2R)^2 - \pi R^2} = \frac{1}{3}.$$

Substituting this and $X_S = -R$ into Eq. 8-16, we have

$$X_P = \tfrac{1}{3}R. \qquad \text{(Answer)}$$

## 8-5 Newton's Laws for a System of Particles

If you roll a billiard ball at a second billiard ball that is at rest, you expect that the two-ball system will continue to have some forward motion after impact. You would be surprised to see both balls come back toward you. But what do we actually observe when one billiard ball rolling at a constant velocity hits another resting ball that has the same mass?

What we observe is that the center of mass of the two-ball system continues to move forward, its motion completely unaffected by the collision. If you focus on the center of mass (which is always halfway between two particles that have the same mass) you can easily convince yourself that this is so. No matter whether the collision is glancing, head-on, or somewhere in between, the center of mass continues to move forward, just as if the collision had never occurred. This is depicted in Fig. 8-10 for a head-on collision.

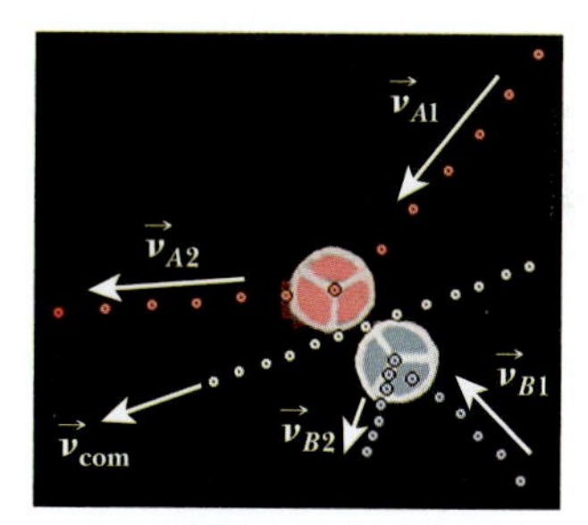


**FIGURE 8-10** ■ Two pucks of equal mass glide along an air table. They strike each other a glancing blow. A video analysis shows a point halfway between them in each frame moving at a constant velocity.

Let's consider another simple situation. Two pucks with the same mass are moving and collide with a glancing blow. Using a digital video clip of this collision, we can track the locations of the two pucks frame by frame and mark a point halfway between the two puck centers. These halfway points, shown as white dots in Fig. 8-10, represent the center of mass of the two-puck system. It is clear that the center of mass of the system is moving in a straight line at constant speed.

Let us look into this center of mass motion theoretically. Why should we expect the center of mass of the billiard balls or a collection of pucks on an air table to move with a constant velocity? Let's start with an assemblage of $N$ particles of different masses and shapes like those shown in Fig. 8-11. These objects are floating just above a level air table. We are interested not in the individual motions of these particles, but *only* in the motion of the center of mass of the system. We use balancing points to find the center of mass of each of the oddly shaped objects in the system. We can then use these locations for each object to calculate the center of mass of the system as a whole. As we discussed in Section 8-2,

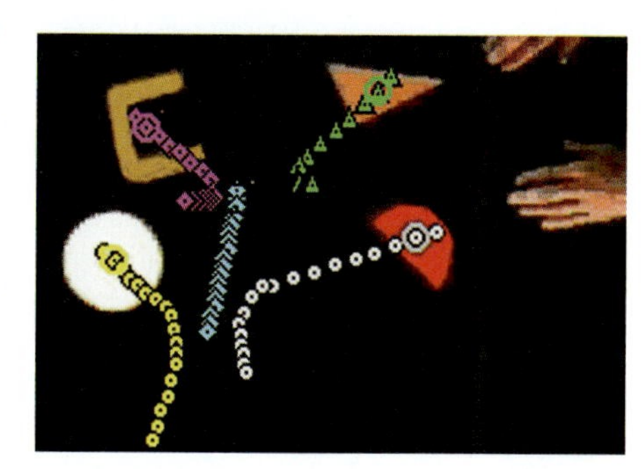

**FIGURE 8-11** ■ Four shapes, each having a different mass, collide in the center of an air table. The path of the center of mass of each shape is found using video analysis. The calculated path of the center of mass of the system is shown as diamonds. Note that this calculated system center of mass moves along a straight line at a constant velocity before, during, and after the collisions that take place between the various objects.

$$\vec{F}^{\text{ net}}_{\text{sys}} = M_{\text{sys}}\vec{a}_{\text{sys}} \qquad \text{(system of particles)}, \qquad (8\text{-}17)$$

where $M_{\text{sys}}$ is the total mass of the system and as we now know, $\vec{a}_{\text{sys}} = \vec{a}_{\text{com}}$ is the acceleration of the system's center of mass. This equation is Newton's Second Law for the motion of the center of mass of a system of particles. However, the meaning of the three quantities that appear in $\vec{F}^{\text{ net}}_{\text{sys}} = M_{\text{sys}}\vec{a}_{\text{sys}}$ must be carefully interpreted.

1. $\vec{F}_{\text{sys}}^{\text{net}}$ is the sum of *all external forces* that act on the system. Forces on one part of the system from another (*internal forces*) do not matter. (By Newton's Third Law, we know that the internal forces cancel each other out when the system is considered as a whole).
2. $M_{\text{sys}}$ is the *total mass* of the objects in the system. We assume that no mass enters or leaves the system as it moves, so that $M_{\text{sys}}$ remains constant. Such a system is said to be **closed.**
3. Although Newton's Second Law allows us to determine the acceleration of the *center of mass*, $\vec{a}_{\text{com}}$ of the system from the net force on it, in some situations we may have no information about the acceleration of any other point in the system.

$\vec{F}_{\text{sys}}^{\text{net}} = M_{\text{sys}}\vec{a}_{\text{com}}$ is equivalent to three equations involving the components of $\vec{F}^{\text{net}}$ and $\vec{a}_{\text{com}}$ along three coordinate axes that can be chosen. These equations are

$$F_{\text{sys}\,x}^{\text{net}} = M_{\text{sys}}a_{\text{com}\,x}, \quad F_{\text{sys}\,y}^{\text{net}} = M_{\text{sys}}a_{\text{com}\,y}, \quad F_{\text{sys}\,z}^{\text{net}} = M_{\text{sys}}a_{\text{com}\,z}. \tag{8-18}$$

**Application to the Air Table Objects** Once the objects on the air table are set into motion, no net external force acts on the system. This is because the external forces consist of a downward gravitational force and the upward normal force on each object. These forces cancel each other out. Thus there is no net force on the system. Since $\vec{F}_{\text{sys}}^{\text{net}} = 0$, we know that $\vec{a}_{\text{com}} = 0$ also. Because acceleration is the rate of change of velocity, we conclude that the velocity of the center of mass of the system of four objects does not change. When various objects collide, the forces that come into play are *internal* forces on one object from another. Such forces do not contribute to the net force, which remains zero. Thus, even though the velocities of the four objects change individually as a result of the forces the objects feel from within the system, the center of mass of the system continues to move with unchanged velocity (Fig. 8-11).

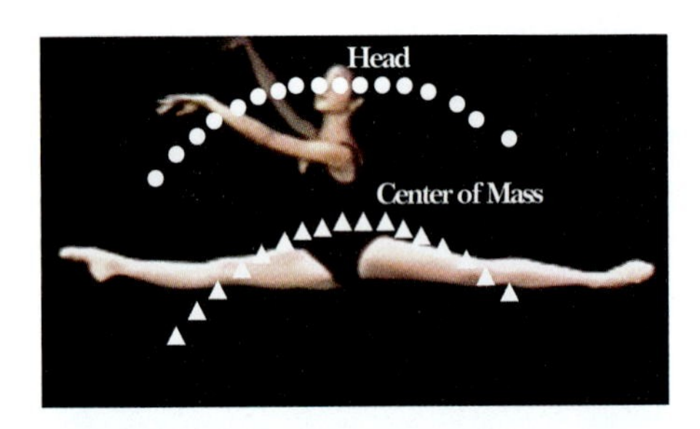


FIGURE 8-12 ■ A video analysis of the *grand jeté* shows that the center of mass of the dancer moves in a parabolic path while her head moves horizontally at the peak of her jump. (See Fig. 1-14 in Ch. 1 for more details).

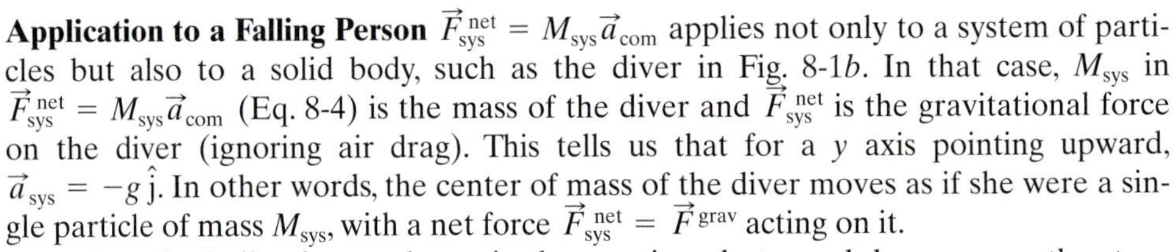

**Application to a Falling Person** $\vec{F}_{\text{sys}}^{\text{net}} = M_{\text{sys}}\vec{a}_{\text{com}}$ applies not only to a system of particles but also to a solid body, such as the diver in Fig. 8-1*b*. In that case, $M_{\text{sys}}$ in $\vec{F}_{\text{sys}}^{\text{net}} = M_{\text{sys}}\vec{a}_{\text{com}}$ (Eq. 8-4) is the mass of the diver and $\vec{F}_{\text{sys}}^{\text{net}}$ is the gravitational force on the diver (ignoring air drag). This tells us that for a $y$ axis pointing upward, $\vec{a}_{\text{sys}} = -g\,\hat{\text{j}}$. In other words, the center of mass of the diver moves as if she were a single particle of mass $M_{\text{sys}}$, with a net force $\vec{F}_{\text{sys}}^{\text{net}} = \vec{F}^{\text{grav}}$ acting on it.

When the ballet dancer shown in the opening photograph leaps across the stage in a *grand jeté*, she raises her arms and stretches her legs out horizontally as soon as her feet leave the stage (Fig. 8-12). These actions shift her center of mass upward through her body. Although the shifting center of mass faithfully follows a parabolic path across the stage, its movement, relative to the body, decreases the height that is attained by her head and torso, relative to that of a normal jump. The result is that the head and torso follow a nearly horizontal path, giving an illusion that the dancer is floating as shown in the Fig. 8-12 video analysis.

**READING EXERCISE 8-2:** The halfway point between the two pucks in Fig. 8-10 is moving in a straight line. If each frame is exactly 1/15th of a second later than the previous frame, (a) what evidence is there that the speed is constant? (b) If the distance between the first location of the center of mass and the last location is 0.41 m, what is the speed of the center of mass? ■

**READING EXERCISE 8-3:** Two skaters on frictionless ice hold opposite ends of a pole of negligible mass. An axis runs along the pole, and the origin of the axis is at the center of mass of the two-skater system. One skater, Fred, weighs twice as much as the other skater, Ethel. Where do the skaters meet if (a) Fred pulls hand over hand along the pole so as to draw himself to Ethel, (b) Ethel pulls hand over hand to draw herself to Fred, and (c) both skaters pull hand over hand? ■

## TOUCHSTONE EXAMPLE 8-4: Center-of-Mass Acceleration

The three particles in Fig. 8-13*a* are initially at rest. Each experiences an *external* force due to bodies outside the three-particle system. The directions are indicated, and the magnitudes are $F_A = 6.0$ N, $F_B = 12$ N, and $F_C = 14$ N. What is the magnitude of the acceleration of the center of mass of the system, and in what direction does it move?

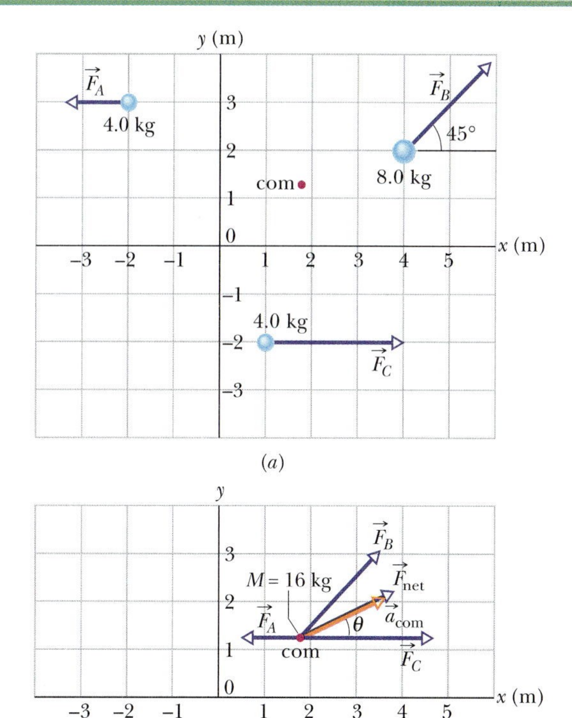


**FIGURE 8-13** (*a*) Three particles, initially at rest in the positions shown, are acted on by the external forces shown. The center of mass, com, of the system is marked. (*b*) The forces are now transferred to the center of mass of the system, which behaves like a particle with a mass $M$ equal to the total mass of the system. The net external force $\vec{F}^{\text{net}}$ and the acceleration $\vec{a}_{\text{com}}$ of the center of mass are shown.

**SOLUTION** The position of the center of mass, calculated by the method of Touchstone Example 8-1, is marked by a dot in Fig. 8-13. One **Key Idea** here is that we can treat the center of mass as if it were a real particle, with a mass equal to the system's total mass $M_{\text{sys}} = 16$ kg. We can also treat the three external forces as if they act at the center of mass (Fig. 8-13*b*).

A second **Key Idea** is that we can now apply Newton's Second Law ($\vec{F}^{\text{net}} = m\vec{a}$) to the center of mass, writing

$$\vec{F}^{\text{net}}_{\text{sys}} = M_{\text{sys}}\vec{a}_{\text{com}}, \tag{8-19}$$

or

$$\vec{F}_A + \vec{F}_B + \vec{F}_C = M_{\text{sys}}\vec{a}_{\text{com}},$$

so

$$\vec{a}_{\text{com}} = \frac{\vec{F}_A + \vec{F}_B + \vec{F}_C}{M_{\text{sys}}}. \tag{8-20}$$

Equation 8-19 tells us that the acceleration $\vec{a}_{\text{com}}$ of the center of mass is in the same direction as the net external force $\vec{F}^{\text{net}}_{\text{sys}}$ on the system (Fig. 8-13*b*). Because the particles are initially at rest, the center of mass must also be at rest. As the center of mass then begins to accelerate, it must move off in the common direction of $\vec{a}_{\text{com}}$ and $\vec{F}^{\text{net}}_{\text{sys}}$.

We can evaluate the right side of Eq. 8-20 directly on a vector-capable calculator, or we can rewrite Eq. 8-20 in component form, find the components of $\vec{a}_{\text{com}}$, and then find $\vec{a}_{\text{com}}$. Along the $x$ axis, we have

$$a_{\text{com}\,x} = \frac{F_{Ax} + F_{Bx} + F_{Cx}}{M_{\text{sys}}} = \frac{-6.0\text{ N} + (12\text{ N})\cos 45° + 14\text{ N}}{16\text{ kg}} = 1.03\text{ m/s}^2.$$

Along the $y$ axis, we have

$$a_{\text{com}\,y} = \frac{F_{Ay} + F_{By} + F_{Cy}}{M_{\text{sys}}} = \frac{0 + (12\text{ N})\sin 45° + 0}{16\text{ kg}} = 0.530\text{ m/s}^2.$$

From these components, we find that $\vec{a}_{\text{com}}$ has the magnitude

$$a_{\text{com}} = \sqrt{(a_{\text{com}\,x})^2 + (a_{\text{com}\,y})^2} = 1.16\text{ m/s}^2 \approx 1.2\text{ m/s}^2, \quad \text{(Answer)}$$

and the angle (from the positive direction of the $x$ axis)

$$\theta = \tan^{-1}\frac{a_{\text{com}\,y}}{a_{\text{com}\,x}} = 27°. \quad \text{(Answer)}$$

## 8-6 The Momentum of a Particle System

Since the effective position vector describing a system of $N$ particles is the same as its center of mass, we can express Eq. 8-8 as

$$\vec{R}_{\text{com}} = \vec{R}^{\text{eff}} = \frac{1}{M_{\text{sys}}}(m_A\vec{r}_A + m_B\vec{r}_B + m_C\vec{r}_C + \cdots), \tag{8-21}$$

in which $M_{sys}$ is the system's total mass and $m_A\vec{r}_A$, $m_B\vec{r}_B$, and so on represent the product of the mass and position vector of each of the particles in the system. This expression can be rewritten as

$$M_{sys}\vec{R}_{com} = m_A\vec{r}_A + m_B\vec{r}_B + m_C\vec{r}_C + \cdots. \tag{8-22}$$

Differentiating the expression above with respect to time gives

$$M_{sys}\vec{v}_{com} = m_A\vec{v}_A + m_B\vec{v}_B + m_C\vec{v}_C + \cdots. \tag{8-23}$$

Here $\vec{v}_A(=d\vec{r}_A/dt)$ is the velocity of particle $A$ and $\vec{v}_{com}(=d\vec{R}_{com}/dt)$ is the velocity of the center of mass.

Now consider the translational momentum of this same system. The system as a whole has a total translational momentum $\vec{p}_{sys}$, which is defined to be the vector sum of translational momenta of the particles in the system. Thus,

$$\begin{aligned}\vec{p}_{sys} &= \vec{p}_A + \vec{p}_B + \vec{p}_C + \cdots \\ &= m_A\vec{v}_A + m_B\vec{v}_B + m_C\vec{v}_C + \cdots.\end{aligned} \tag{8-24}$$

If we compare this equation with $M_{sys}\vec{v}_{com} = m_A\vec{v}_A + m_B\vec{v}_B + m_C\vec{v}_C + \cdots$ (Eq. 8-23), we see that

$$\vec{p}_{sys} = M_{sys}\vec{v}_{com} \quad \text{(translational momentum, system of particles)}, \tag{8-25}$$

which gives us another way to define the translational momentum of a system of particles:

> The translational momentum of a system of particles is equal to the product of the total mass $M_{sys}$ of the system and the velocity $\vec{v}_{com}$ of the center of mass.

So we can determine the total momentum of a system either by determining the vector sum of the individual momenta of parts or by taking the total mass of the system and multiplying by the velocity of the center of mass of the system. Either path leads us to the same value.

If we take the time derivative of $\vec{p}_{sys} = M_{sys}\vec{v}_{com}$, we find

$$\frac{d\vec{p}_{sys}}{dt} = M_{sys}\frac{d\vec{v}_{com}}{dt}, \tag{8-26}$$

or

$$\frac{d\vec{p}_{sys}}{dt} = M_{sys}\vec{a}_{com}. \tag{8-27}$$

Comparing $\vec{F}^{net}_{sys} = M\vec{a}_{com}$ (Eq. 8-17) with Eq. 8-27 allows us to write Newton's Second Law for a system of particles in the equivalent form

$$\vec{F}^{net}_{sys} = \frac{d\vec{p}_{sys}}{dt} \quad \text{(system of particles)}, \tag{8-28}$$

where $\vec{F}^{net}_{sys}$ is the net external force acting on the particles in the system. As we discussed in Chapter 7, this equation is the generalization of the single-particle equation $\vec{F}^{net} = d\vec{p}/dt$ to a system of many particles. In its new form, $d\vec{p}_{sys}/dt = M_{sys}\vec{a}_{com}$, the introduction of the concept of the center of mass of a system gives us an additional technique for determining the rate at which the total momentum of a system changes.

# Problems

## SEC. 8-4 ■ LOCATING A SYSTEM'S CENTER OF MASS

**1. Particle-Like Object** A 4.0 kg particle-like object is located at $x = 0$, $y = 2.0$ m; a 3.0 kg particle-like object is located at $x = 3.0$ m, $y = 1.0$ m. At what (a) $x$ and (b) $y$ coordinates must a 2.0 kg particle-like object be placed for the center of mass of the three-particle system to be located at the origin?

**2. 2D Center of Mass of Three Objects** Consider Fig. 8-14. Three masses located in the $x$-$y$ plane have the following coordinates; a 5 kg mass has coordinates given by $(2, -3)$ m; a 4 kg mass has coordinates $(-4, 2)$ m; a 2 kg mass has coordinates $(3, 3)$ m. Find the coordinates of the center of mass to two significant figures.

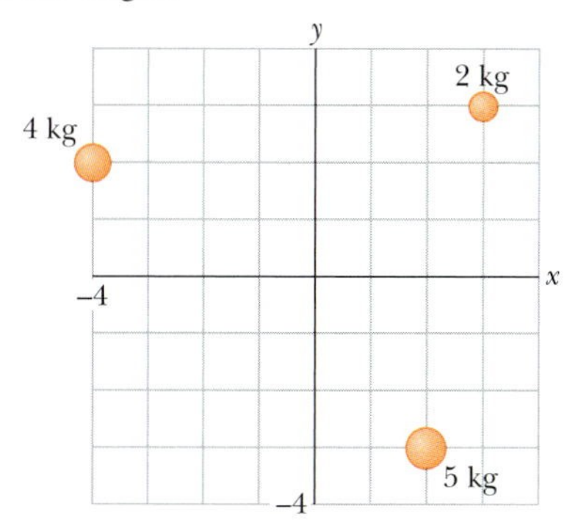


FIGURE 8-14 ■ Problem 2.

**3. Earth–Moon System** (a) How far is the center of mass of the Earth–Moon system from the center of Earth? (Appendix C gives the masses of Earth and the Moon and the distance between the two.) (b) Express the answer to (a) as a fraction of Earth's radius $R_e$.

**4. Carbon Monoxide** A distance of $1.131 \times 10^{-10}$ m lies between the centers of the carbon and oxygen atoms in a carbon monoxide (CO) gas molecule. Locate the center of mass of a CO molecule relative to the carbon atom. (Find the masses of C and O in Appendix F.)

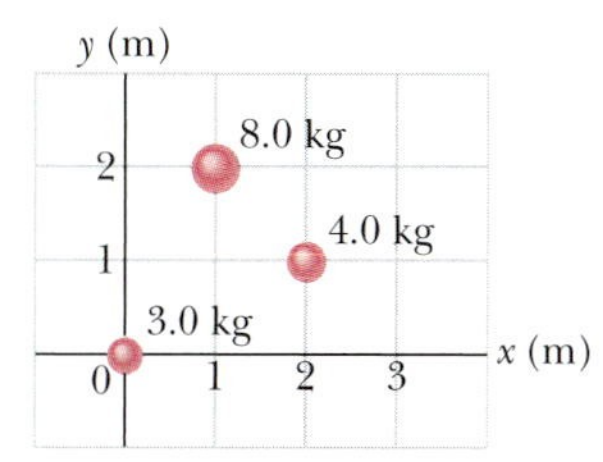


FIGURE 8-15 ■ Problem 5.

**5. Three-Particle System** What are (a) the $x$ coordinate and (b) the $y$ coordinate of the center of mass of the three-particle system shown in Fig. 8-15? (c) What happens to the center of mass as the mass of the topmost particle is gradually increased?

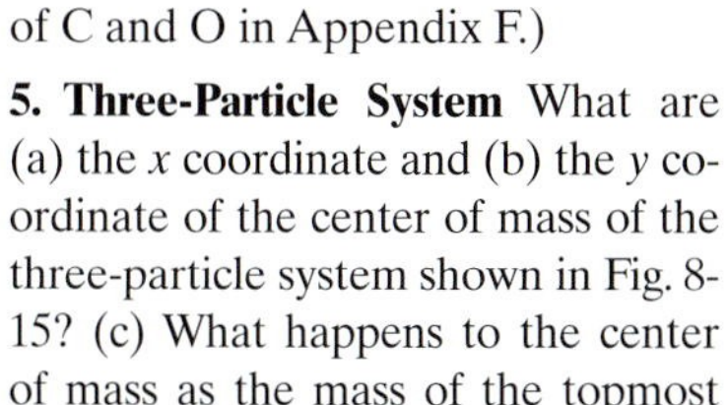

**6. Three Thin Rods** Three thin rods, each of length $L$, are arranged in an inverted U, as shown in Fig. 8-16. The two rods on the arms of the U each have mass $M$; the third rod has mass $3M$. Where is the center of mass of the assembly?

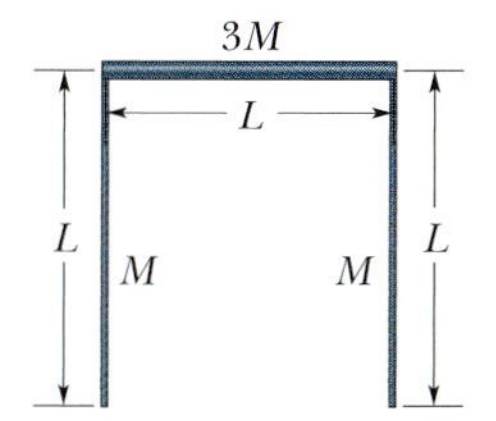


FIGURE 8-16 ■ Problem 6.

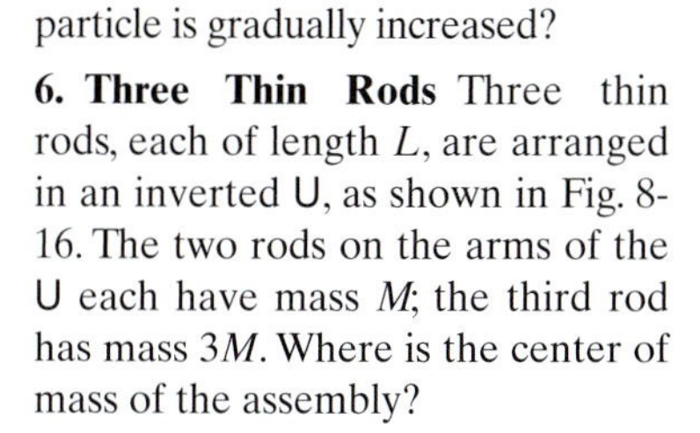

**7. Uniform Square** A uniform square plate 6 m on a side has had a square piece 2 m on a side cut out of it (Fig. 8-17). The center of that piece is at $x = 2$ m, $y = 0$. The center of the square plate is at $x = y = 0$. Find (a) the $x$ coordinate and (b) the $y$ coordinate of the center of mass of the remaining piece.

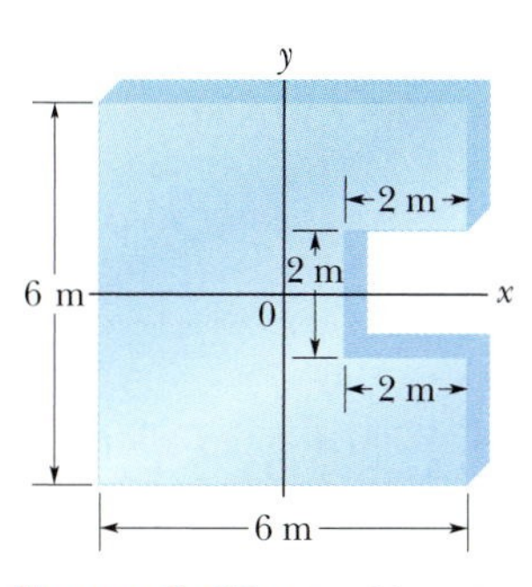


FIGURE 8-17 ■ Problem 7.

**8. Composite Slab** Figure 8-18 shows the dimensions of a composite slab; half the slab is made of aluminum (density = 2.70 g/cm$^3$) and half is made of iron (density = 7.85 g/cm$^3$). Where is the center of mass of the slab?

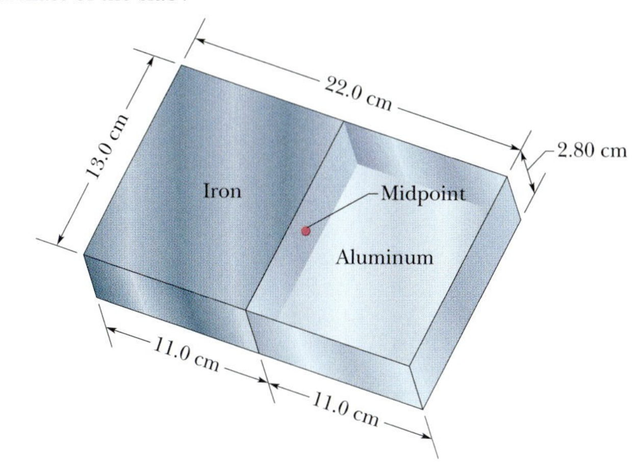


FIGURE 8-18 ■ Problem 8.

**9. Ammonia** In the ammonia ($NH_3$) molecule (see Fig. 8-19), the three hydrogen (H) atoms form an equilateral triangle; the center of the triangle is $9.40 \times 10^{-11}$ m from each hydrogen atom. The nitrogen (N) atom is at the apex of a pyramid, with the three hydrogen atoms forming the base. The nitrogen-to-hydrogen atomic mass ratio is 13.9, and the nitrogen-to-hydrogen distance is $10.14 \times 10^{-11}$ m. Locate the center of mass of the molecule relative to the nitrogen atom.

FIGURE 8-19 ■ Problem 9.

**10. Metal Cube** Figure 8-20 shows a cubical box that has been constructed from a metal plate of uniform density and negligible thickness. The box is open at the top and has edge length 40 cm. Find (a) the $x$ coordinate, (b) the $y$ coordinate, and (c) the $z$ coordinate of the center of mass of the box.

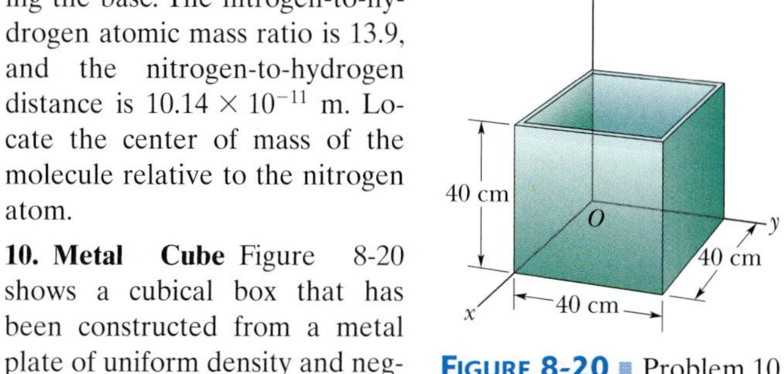


FIGURE 8-20 ■ Problem 10.

**11. Cylindrical Can** A right cylindrical can with mass $M$, height $H$, and uniform density is initially filled with soda of mass $m$ (Fig. 8-21). We punch small holes in the top and bottom to drain the soda; we then consider the height $h$ of the center of mass of the can and any soda within it. What is $h$ (a) initially and (b) when all the soda has drained? (c) What happens to $h$ during the

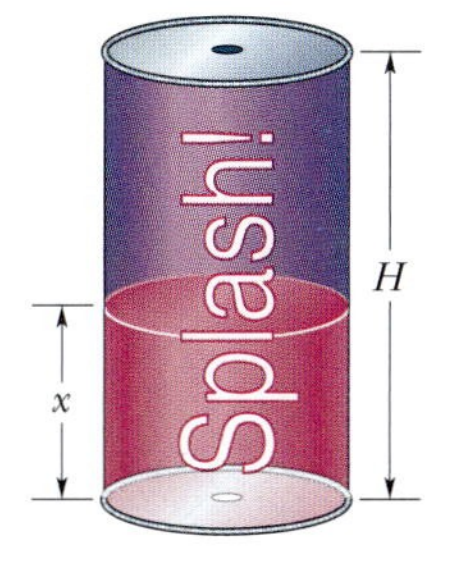


FIGURE 8-21 ■ Problem 11.

draining of the soda? (d) If $x$ is the height of the remaining soda at any given instant, find $x$ (in terms of $M$, $H$, and $m$) when the center of mass reaches its lowest point.

**12. Clustered Problem 1** In Fig. 8-22*a*, a uniform wire forms an isosceles triangle of base $B$ and height $H$. (a) Find the $x$ and $y$ coordinates of the figure's center of mass by assuming that each side can be replaced with a particle of the same mass as that side and positioned at the center of the side. (*Be careful*: Note that the base and, say, the left-hand side do not have the same mass.) (b) Use Eq. 8-12 to find the $x$ and $y$ coordinates of the center of mass of the left-hand side.

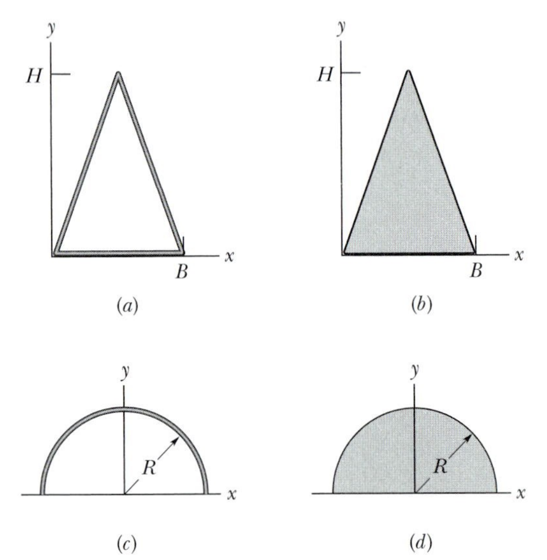


FIGURE 8-22 ■ Problems 12 through 15.

**13. Clustered Problem 2** Figure 8-22*b* shows a uniform, solid plate in the shape of an isosceles triangle with base $B$ and height $H$. What are the $x$ and $y$ coordinates of the plate's center of mass?

**14. Clustered Problem 3** In Fig. 8-22*c*, a uniform wire forms a semicircle of radius $R$. What are the $x$ and $y$ coordinates of the figure's center of mass?

**15. Clustered Problem 4** Figure 8-22*d* shows a uniform, solid plate in the shape of a semicircle with radius $R$. What are the $x$ and $y$ coordinates of the plate's center of mass?

**16. Great Pyramid** The Great Pyramid of Cheops at El Gizeh, Egypt (Fig. 8-23*a*), had height $H = 147$ m before its topmost stone fell. Its base is a square with edge length $L = 230$ m (see Fig. 8-23*b*). Its volume $V$ is equal to $L^2H/3$. Assuming $\rho = 1.8 \times 10^3$ kg/m$^3$ is its uniform density, find the original height of its center of mass above the base.

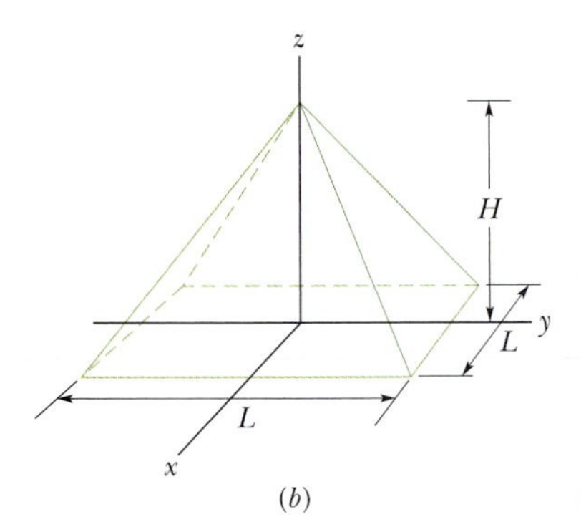

FIGURE 8-23 ■ Problem 16.

**17. Four Particles** At a certain instant, four particles have the $xy$ coordinates and velocities given in the following table. At that instant, what are (a) the coordinates of their center of mass and (b) the velocity of their center of mass?

| Particle | Mass (kg) | Position (m) | Velocity (m/s) |
|---|---|---|---|
| 1 | 2.0 | 0, 3.0 | $-9.0 \text{m/s}\, \hat{j}$ |
| 2 | 4.0 | 3.0, 0 | $6.0 \text{m/s}\, \hat{i}$ |
| 3 | 3.0 | 0, −2.0 | $6.0 \text{m/s}\, \hat{j}$ |
| 4 | 12 | −1.0, 0 | $-2.0 \text{m/s}\, \hat{i}$ |

**18. Inverse Ratios** Show that the ratio of the distances of two particles from their center of mass is the inverse ratio of their masses.

**19. *xy* Coordinates** A 2.00 kg particle has the $xy$ coordinates ($-1.20$ m, 0.500 m) and a 4.00 kg particle has the $xy$ coordinates (0.600 m, $-0.750$ m). Both lie on a horizontal plane. At what $xy$ coordinates must you place a 3.00 kg particle such that the center of mass of the three-particle system has the coordinates ($-0.500$ m, $-0.700$ m)?

**20. Uniform Plate** What are (a) the $x$ coordinate and (b) the $y$ coordinate of the center of mass for the uniform plate shown in Fig. 8-24?

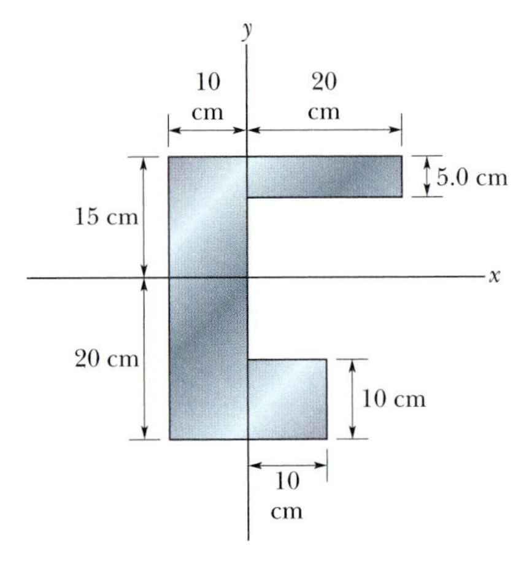


FIGURE 8-24 ■ Problem 20.

## SEC. 8-6 ■ MOMENTUM OF A PARTICLE SYSTEM

**21. Peanut Butter and Jelly** At $t_1 = 0$, a 1.0 kg jelly jar is projected vertically upward from the base of a 50-m-tall building with an initial velocity of 40 m/s. At the same instant and directly overhead, a 2.0 kg peanut butter jar is dropped from rest from the top of the building. How far above ground level is the center of mass of the two-jar system at $t_2 = 3.0$ s?

**22. Two Skaters With Pole** Two skaters, one with mass 65 kg and the other with mass 40 kg, stand on an ice rink holding a pole of length 10 m and negligible mass. Starting from the ends of the pole, the skaters pull themselves along the pole until they meet. How far does the 40 kg skater move?

**23. Old Chrysler** An old Chrysler with mass 2400 kg is moving along a straight stretch of road at 80 km/h. It is followed by a Ford with mass 1600 kg moving at 60 km/h. How fast is the center of mass of the two cars moving?

**24. Ladder on a Balloon** A man of mass $m$ clings to a rope ladder suspended below a balloon of mass $M$; see Fig. 8-25. The balloon is stationary with respect to the

FIGURE 8-25 ■ Problem 24.

ground. (a) If the man begins to climb the ladder at speed $v$ (with respect to the ladder), in what direction and with what speed (with respect to the ground) will the balloon move? (b) What is the state of the motion after the man stops climbing?

**25. A Stone is Dropped** A stone is dropped at $t_1 = 0$. A second stone, with twice the mass of the first, is dropped from the same point at $t_2 = 100$ ms. (a) How far below the release point is the center of mass of the two stones at $t_3 = 300$ ms? (Neither stone has yet reached the ground.) (b) How fast is the center of mass of the two-stone system moving at that time?

**26. Traffic Signal** A 1000 kg automobile is at rest at a traffic signal. At the instant the light turns green, the automobile starts to move with a constant acceleration of 4.0 m/s$^2$. At the same instant a 2000 kg truck, traveling at a constant speed of 8.0 m/s, overtakes and passes the automobile. (a) How far is the center of mass of the automobile–truck system from the traffic light at $t_2 = 3.0$ s? (b) What is the speed of the center of mass of the automobile–truck system then?

**27. Shell Explodes** A shell is shot with an initial velocity $\vec{v}_1$ of 20 m/s, at an angle of 60° with the horizontal. At the top of the trajectory, the shell explodes into two fragments of equal mass (Fig. 8-26). One fragment, whose speed immediately after the explosion is zero, falls vertically. How far from the gun does the other fragment land, assuming that the terrain is level and that air drag is negligible?

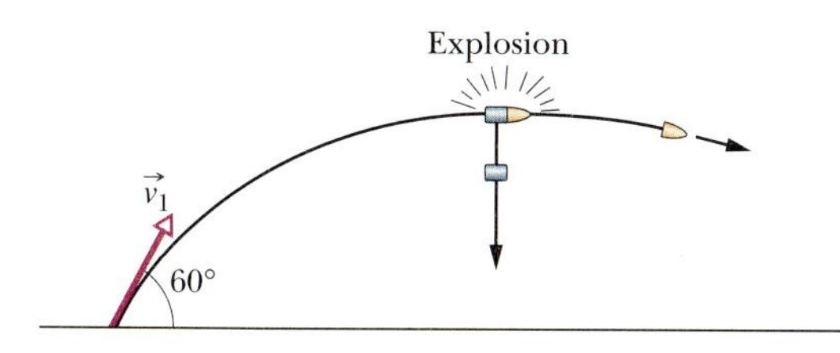


FIGURE 8-26 ■ Problem 27.

**28. Big Olive** A big olive ($m = 0.50$ kg) lies at the origin and a big Brazil nut ($M = 1.5$ kg) lies at the point (1.0, 2.0) m in an $xy$ plane. At $t = 0$, a force $\vec{F}_o = (2\text{N})\hat{\text{i}} + (3\text{N})\hat{\text{j}}$ begins to act on the olive, and a force $\vec{F}_n = (-3\text{N})\hat{\text{i}} + (-2\text{N})\hat{\text{j}}$ begins to act on the nut. In unit-vector notation, what is the displacement of the center of mass of the olive–nut system at $t_2 = 4.0$ s, with respect to its position at $t_1 = 0$?

**29. Sugar Containers** Two identical containers of sugar are connected by a massless cord that passes over a massless, frictionless pulley with a diameter of 50 mm (Fig. 8-27). The two containers are at the same level. Each originally has a mass of 500 g. (a) What is the horizontal position of their center of mass? (b) Now 20 g of sugar is transferred from one container to the other, but the containers are prevented from moving. What is the new horizontal position of their center of mass, relative to the central axis through the lighter container? (c) The two containers are now released. In what direction does the center of mass move? (d) What is its acceleration?

FIGURE 8-27 ■ Problem 29.

**30. Ricardo and Carmelita** Ricardo, of mass 80 kg, and Carmelita, who is lighter, are enjoying Lake Merced at dusk in a 30 kg canoe. When the canoe is at rest in the placid water, they exchange seats, which are 3.0 m apart and symmetrically located with respect to the canoe's center. Ricardo notices that the canoe moves 40 cm relative to a submerged log during the exchange and calculates Carmelita's mass, which she has not told him. What is it?

**31. Dog in a Boat** In Fig. 8-28*a*, a 4.5 kg dog stands on an 18 kg flatboat and is 6.1 m from the shore. He walks 2.4 m along the boat toward shore and then stops. Assuming there is no friction between the boat and the water, find how far the dog is then from the shore. (*Hint*: See Fig. 8-28*b*. The dog moves leftward and the boat moves rightward, but does the center of mass of the *boat* + *dog* system move?)

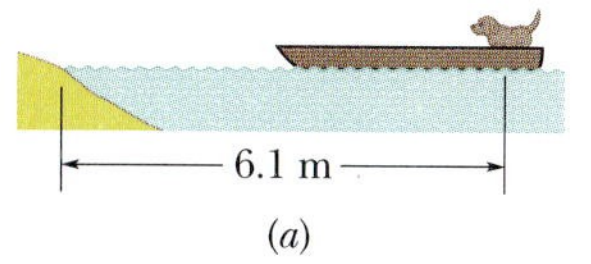


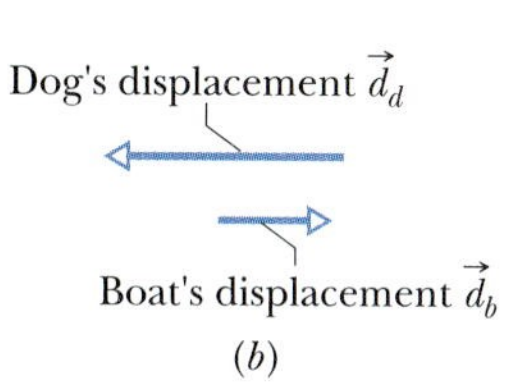


FIGURE 8-28 ■ Problem 31.

**32. A Certain Nucleus** A certain nucleus, at rest, transforms into three particles. Two of them are detected; their masses and velocities are as shown in Fig. 8-29. (a) In unit-vector notation, what is the translational momentum of the third particle, with a mass of $11.7 \times 10^{-27}$ kg?

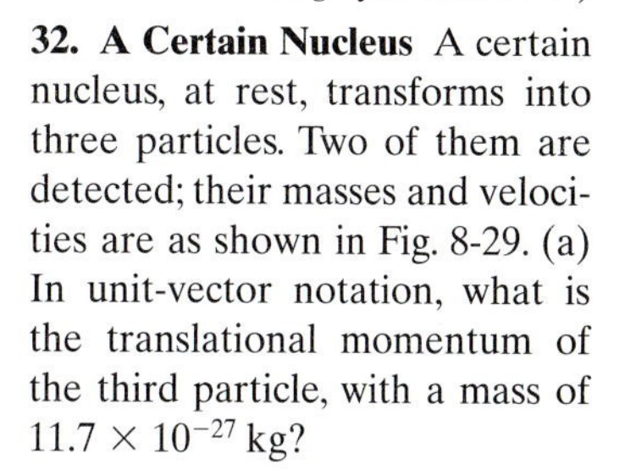

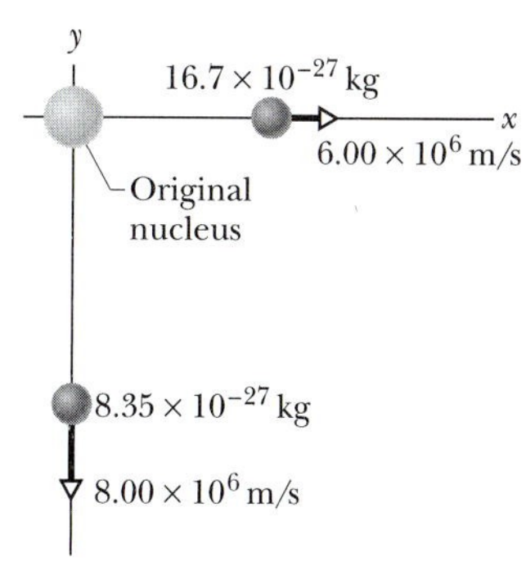


FIGURE 8-29 ■ Problem 32.

**33. Father and Child** A 40 kg child and her 75 kg father simultaneously dive from a 100 kg boat that is initially motionless. The child dives horizontally toward the east with a speed of 2.0 m/s, and the father dives toward the south with a speed of 1.5 m/s at an angle of 37° above the horizontal. (Assume the boat's vertical motion due to the father's dive does not alter its horizontal motion.) Determine the magnitude and direction of the velocity of the boat along the water's surface immediately after their dives.

**34. Sumo Wrestler** A 2140 kg railroad flatcar, which can move with negligible friction, is motionless next to a platform. A 242 kg sumo wrestler runs at 5.3 m/s along the platform (parallel to the track) and then jumps onto the flatcar. What is the speed of the flatcar if he then (a) stands on it, (b) runs at 5.3 m/s relative to the flatcar in his original direction, and (c) turns and runs at 5.3 m/s relative to the flatcar opposite his original direction?

**35. Block Released from Rest** A 2.00 kg block is released from rest over the side of a very tall building at time $t_1 = 0$. At time $t_2 = 1.00$ s, a 3.00 kg block is released from rest at the same point. The first block hits the ground at $t_3 = 5.00$ s. Plot, for the time interval $t_1 = 0$ to $t_4 = 6.00$ s, (a) the position and (b) the speed of the center of mass of the two-block system. Take $y = 0$ at the release point.

**36. Speed of COM** At the instant a 3.0 kg particle has a velocity of 6.0 m/s in the negative $y$ direction, a 4.0 kg particle has a velocity of 7.0 m/s in the positive $x$ direction. What is the speed of the center of mass of the two-particle system?

**37. Car and Truck** A 1500 kg car and a 4000 kg truck are moving north and east, respectively, with constant velocities. The center of mass of the car–truck system has a velocity of 11 m/s in a direction 55° north of east. (a) What is the magnitude of the car's velocity? (b) What is the magnitude of the truck's velocity?

**38. Cannon in a Flatcar** A cannon and a supply of cannonballs are inside a sealed railroad car of length $L$, as in Fig.8-30. The cannon fires to the right, the car recoils to the left. Fired cannonballs travel a horizontal distance $L$ and remain in the car after hitting the far wall and landing on the floor there. (a) After all the cannonballs have been fired, what is the greatest distance the car could have moved from its original position? (b) What is the speed of the car just after the last cannonball has completed its motion?

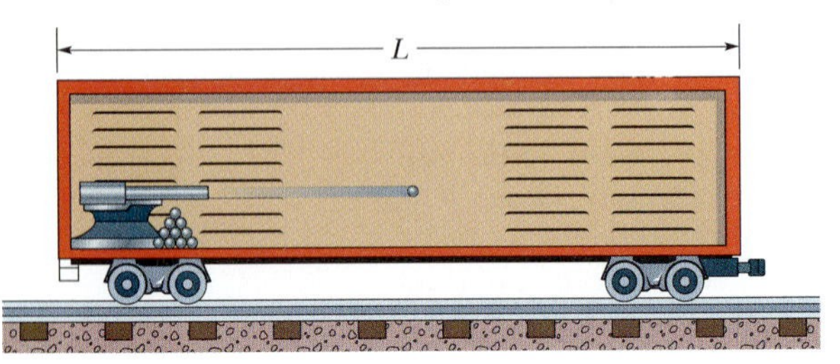


FIGURE 8-30 ■ Problem 38.

**39. Cannon Tilted Up** A 1400 kg cannon, which fires a 70.0 kg shell with a speed of 556 m/s relative to the muzzle, is set at an elevation angle of 39.0° above the horizontal. The cannon is mounted on frictionless rails so that it can recoil freely. (a) At what speed relative to the ground is the shell fired? (b) At what angle with the ground is the shell fired? (*Hint*: The horizontal component of the momentum of the system remains unchanged as the cannon is fired.)

**40. Table of Three** The following table gives the masses of three objects and, at a certain instant, the coordinates $(x, y)$ and the velocities of the objects. At that instant, what are the (a) position and (b) velocity of the center of mass of the three-particle system, and (c) what is the net translational momentum of the system?

| Object | Mass (kg) | Coordinates (m) | Velocity (m/s) |
|---|---|---|---|
| 1 | 4.00 | (0.00, 0.00) | $(1.50\text{ m/s})\hat{i} - (2.50\text{m/s})\hat{j}$ |
| 2 | 3.00 | (7.00, 3.00) | 0.00 |
| 3 | 5.00 | (3.00, 2.00) | $(2.00\text{ m/s})\hat{i} - (1.00\text{m/s})\hat{j}$ |

# Additional Problems

**41. Iceboat** You are on an iceboat on frictionless, flat ice; you and the boat have a combined mass $M$. Along with you are two stones with masses $m_A$ and $m_B$ such that $M = 6.00m_A = 12.0m_B$. To get the boat moving, you throw the stones rearward, either in succession or together, but in each case with a certain speed $v^{\text{rel}}$ relative to the boat after the stone is thrown. What is the resulting speed of the boat if you throw the stones (a) simultaneously, (b) in the order $m_A$ and then $m_B$, and (c) in the order $m_B$ and then $m_A$?

**42. *P* and *Q*** Two particles $P$ and $Q$ are initially at rest 1.0 m apart. $P$ has a mass of 0.10 kg and $Q$ a mass of 0.30 kg. $P$ and $Q$ attract each other with a constant force of $1.0 \times 10^{-2}$ N. No external forces act on the system. (a) Describe the motion of the center of mass. (b) At what distance from $P$'s original position do the particles collide?

**43. Suspicious Package** A suspicious package is sliding on a frictionless surface when it explodes into three pieces of equal masses and with the velocities (1) 7.0 m/s, north, (2) 4.0 m/s, 30° south of west, and (3) 4.0 m/s, 30° south of east. (a) What is the velocity (magnitude and direction) of the package before it explodes? (b) What is the displacement of the center of mass of the three-piece system (with respect to the point where the explosion occurs) 3.0 s after the explosion?

**44. Mass on an Air Track** Figure 8-31 shows an arrangement with an air track, in which a cart is connected by a cord to a hanging block. The cart has mass $m_A = 0.600$ kg and its center is initially at $xy$ coordinates $(-0.500\text{ m}, 0.000\text{ m})$; the block has mass $m_B = 0.400$ kg and its center is initially at $xy$ coordinates $(0, -0.100\text{ m})$. The mass of the cord and pulley are negligible. The cart is released from rest, and both cart and block move until the cart hits the pulley. The friction between the cart and the air track and between the pulley and its axle is negligible. (a) In unit-vector notation, what is the acceleration of the center of mass of the cart–block system? (b) What is the velocity of the center of mass as a function of time $t$? (c) Sketch the path taken by the system's center of mass. (d) If the path is curved, does it bulge upward to the right or downward to the left? If, instead, it is straight, give the angle between it and the $x$ axis.

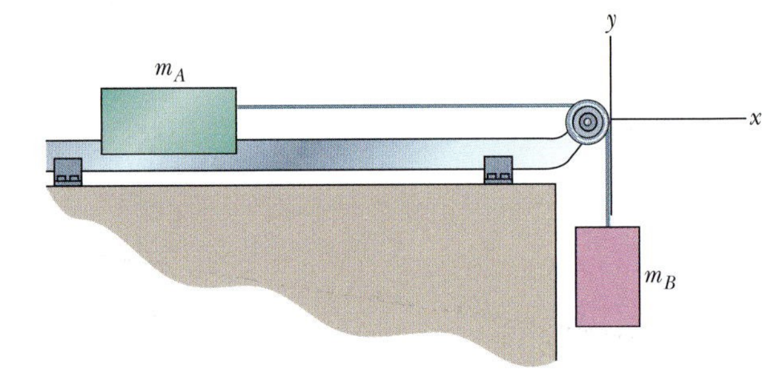


FIGURE 8-31 ■ Problem 44.

**45. *Left Alone, Write Your Own*** For one or more of the following situations, write a problem involving physics in this chapter, using the style of the Touchstone Examples and providing realistic data, graphs of the variables, and explained solutions: (a) determining the center of mass of a large object, (b) a system separated into parts by an internal explosion, (c) someone climbing or descending a structure, (d) track and field events.

**46. Car on a Boat** The script for an action movie calls for a small race car (of mass 1500 kg and length 3.0 m) to accelerate along a flat-top boat (of mass 4000 kg and length 14 m), from one end to the other. The car will then jump the gap between the boat and a somewhat lower dock. You are the technical advisor for the movie. The boat will initially touch the dock as shown in Fig. 8-32. Assume the boat can slide through the water without significant resistance, and that both the car and the boat can be approximated as uniform in their mass distribution. Determine what the width of the gap will be just as the car is about to make the jump.

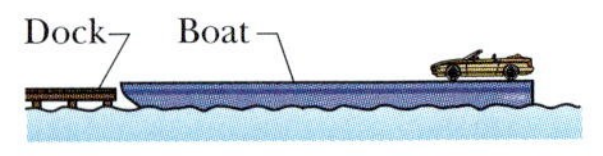


FIGURE 8-32 ■ Problem 46.

**47. Two Carts with Unequal Masses** Suppose you examine a digital movie of two carts with different masses that undergo a collision (for example, PASCO020 in VideoPoint). You will find there is a point between the two carts that moves at the same constant velocity both before and after the collision. We call this special point the center of mass of the two-cart system. In the PASCO020 movie, where one cart has twice the mass of the other, analysis of the video indicates that the center of mass is one-third of the distance between the two carts (measured relative to the more massive cart).

A similar situation is depicted in Fig. 8-33. The figure shows a moment in time when the cart *centers* just happen to be 1.000 m apart. For the situation in Fig. 8-33, show that the equation

$$x_{\text{com}} = \frac{m_A x_A + m_B x_B}{m_A + m_B}$$

gives a center of mass for these two carts that is one-third of the distance between them (measured from the more massive cart).

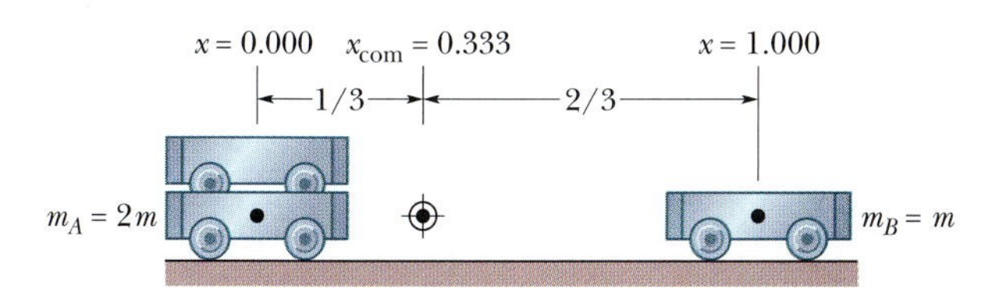


**FIGURE 8-33** ■ Problem 47.

# 9 Kinetic Energy and Work

In the weight-lifting competition of the 1996 Olympics, Andrey Chemerkin lifted a record-breaking 260.0 kg from the floor to over his head (about 2 m). In 1957, Paul Anderson stooped beneath a reinforced wood platform, placed his hands on a short stool to brace himself, and then pushed upward on the platform with his back, lifting the platform and its load about a centimeter. On the platform were auto parts and a safe filled with lead. The composite weight of the load was 27 900 N (6270 lb)!

**Who did more work on the objects he lifted—Chemerkin or Anderson?**

*The answer is in this chapter.*

## 9-1 Introduction

A ski jumper who wants to understand her motion along a curved track is presented with a special challenge. If she tries to use Newton's laws of motion to predict her speed at each location along that track, she has to account for the fact that the net force and her acceleration keep changing as the slope of the track changes. The goal of this chapter is to devise a way to simplify the analysis of motions like those of our ski jumper shown in Fig. 9-1.

**FIGURE 9-1** ■ It is difficult to use Newton's Second Law to analyze the motion of a ski jumper traveling down a curved ramp and predict her velocity at the bottom of the ramp.

We can begin by drawing on ideas presented in Chapter 7. There we introduce two new concepts—*momentum* and *impulse*—and use them to derive an alternate form of Newton's Second Law known as the *impulse–momentum theorem*. This theorem, expressed in Eq. 7-10, tells us that the impulse $\vec{J}$ on a moving particle is equal to its momentum change.

$$\vec{J} = \int_{t_1}^{t_2} \vec{F}^{\,\text{net}}(t)\,dt = m\vec{v}_2 - m\vec{v}_1 \quad \text{(impulse–momentum theorem).} \tag{7-10}$$

One of the most useful aspects of the impulse–momentum theorem is that we can use it without having to keep track of the particle's position.

Can we derive another alternate form of Newton's law to relate a particle's velocity and position changes without keeping track of time? In this chapter we simplify the analysis of complex motions like that of the skier by proving an analogous theorem called the *net work–kinetic energy theorem*. But, in order to "derive" our new theorem we introduce two additional concepts—*work* and *kinetic energy*.

We will start our development of the new theorem by introducing the concept of work, $W$, in analogy to the impulse represented by the integral in Eq. 7-10. Initially we consider a very simple situation in which a net force acts along the line of motion of a particle. In this case the concept of work as a one-dimensional analogy to impulse involving position changes rather than time changes would be

$$W = \int_{x_1}^{x_2} F_x^{\text{net}}\,dx \quad \text{(one-dimensional position analogy to impulse).} \tag{9-1}$$

Here $F_x^{\text{net}}$ and $dx$ are components of force and infinitesimal position change vectors along an $x$ axis, $x_1$ is the initial position of the particle and $x_2$ is its position at a later time. In order for the integral to be unique and well defined we will also add the requirement that the component of the force is either constant or only varies with $x$. That is, we will consider the integral

$$\int_{x_1}^{x_2} F_x^{\text{net}}(x)\,dx$$

where the $x$ in parentheses signifies the force on the particle varies with location along the $x$ axis. After developing the concept of work we introduce the concept of kinetic energy and then derive the *net work–kinetic energy theorem* for one-dimensional motions. Next we apply the concept of work and the new theorem to the analysis of some motions that result from the actions of common one-dimensional forces.

Although we begin with one-dimensional situations, we will extend the equations we derive to two (and three) dimensions. As part of this process we will also introduce a method for finding a scalar product of two vectors. Then in Section 9-9 toward the end of the chapter, we demonstrate how the two-dimensional form of our new net work–kinetic energy theorem enables us to determine the speed of the skier as a function of her location along a frictionless ramp in a very simple manner.

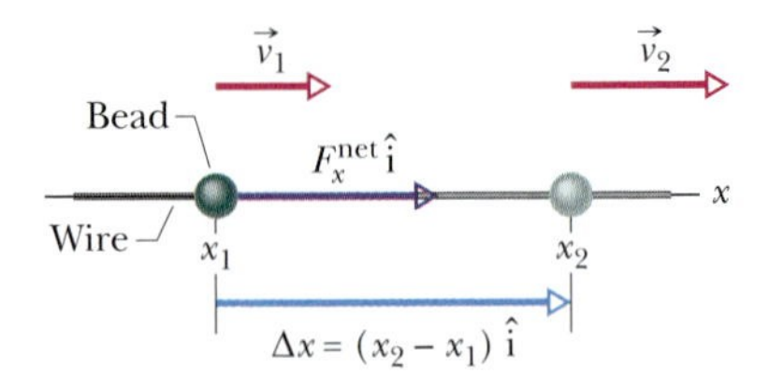


**FIGURE 9-2** ■ A simple situation showing a bead on a wire that experiences a net force that is directed along the wire. The bead moves in the same direction as the force.

# 9-2 Introduction to Work and Kinetic Energy

## One-Dimensional Relationship for a Net Force

In order to find an alternative form of Newton's Second Law that relates position and velocity, suppose our particle is a bead moving along a wire, as in Fig. 9-2. Suppose the net force on the bead is a combination of an applied force and a friction force that acts as the bead slides along. We know that if a net force acts on the bead along its direction of motion, its position will change in the direction of the force and its speed will increase. If the direction of the net force is opposite to that of the bead's motion, the bead's speed will decrease.

In our example, the force, $\vec{F}^{net} = F_x^{net}\hat{i}$, is directed along the wire as shown in Fig. 9-2. This force causes the bead to accelerate in the same direction as the force. We can use Newton's Second Law to relate the force and acceleration components as

$$F_x^{net}(x) = ma_x, \tag{9-2}$$

where $m$ is the bead's mass. As the bead moves through a displacement $\Delta\vec{x} = (x_2 - x_1)\hat{i}$, the force changes the bead's velocity from an initial value $\vec{v}_1$ to another value we will call $\vec{v}_2$. Using the definition of acceleration as the rate of velocity change over a short time interval $dt$, this gives us

$$F_x^{net}(x) = ma_x = m\frac{dv_x}{dt}.$$

In order to relate the velocity and position, we perform two mathematical operations: First we multiply each term in the equation above by the $x$-component of velocity, $v_x$. Second, we use the definition of $v_x$ as $dx/dt$ to substitute for $v_x$ on the left side of our new equation. This gives us

$$F_x^{net}(x)\frac{dx}{dt} = mv_x\frac{dv_x}{dt}. \tag{9-3}$$

Since the bead's mass $m$ is constant, we can use the chain rule of differentiation to see that the term on the right can be rewritten as

$$mv_x\frac{dv_x}{dt} = \frac{d(\frac{1}{2}mv_x^2)}{dt}. \tag{9-4}$$

If we substitute the expression on the right side of Eq. 9-4 for the term $mv_x\,dv_x/dt$ in Eq. 9-3, we get

$$F_x^{net}(x)\frac{dx}{dt} = \frac{d(\frac{1}{2}mv_x^2)}{dt}.$$

Now we can eliminate $dt$ from Eq. 9-3 by realizing that during the same infinitesimal time interval $dt$ the $x$-component of force times the infinitesimal change in $x$ is equal to the change in the expression $\frac{1}{2}mv_x^2$. This gives us the following equality between differentials

$$F_x^{net}(x)\,dx = d(\tfrac{1}{2}mv_x^2). \tag{9-5}$$

Because the net force on the bead and its velocity are not necessarily constant over the full displacement shown in Fig. 9-2, we must integrate both sides of Eq. 9-5 to determine the relationship between position change and velocity change due to our variable force.

$$\int_{x_1}^{x_2} F_x^{\text{net}}(x)\,dx = \int_{v_{1x}}^{v_{2x}} d(\tfrac{1}{2}mv_x^2) = \tfrac{1}{2}mv_{2x}^2 - \tfrac{1}{2}mv_{1x}^2. \quad (9\text{-}6)$$

In summary, by using Newton's Second Law, the definitions of velocity and acceleration, and the rules of calculus, we have derived a new form of the second law that relates how a variable force acting over a distance will change the speed of a particle of mass $m$.

Note that the expression on the left side of this equation,

$$\int_{x_1}^{x_2} F_x^{\text{net}}(x)\,dx,$$

is identical to the expression we developed (in Eq. 9-1) as the one-dimensional (1D) integral of force over position that is analogous to the integral of force over time used in the impulse-momentum theorem.

## Initial Definitions of Work and Kinetic Energy

The left and right sides of Eq. 9-6 are the basis for two new and very important definitions. We define the integral as the **net work,** $W^{\text{net}}$, done on a particle as it moves from an initial to a final position, so that

$$W^{\text{net}} \equiv \int_{x_1}^{x_2} F_x^{\text{net}}(x)\,dx \quad \text{(**net work** definition—1D net force and displacement).} \quad (9\text{-}7)$$

If the net force component $F_x^{\text{net}}(x)$ along a line is made up of the sum of several force components $F_{Ax}(x) + F_{Bx}(x) + F_{Cx}(x) + \cdots$, we see that the contribution of each force to the net work is

$$W^{\text{net}} \equiv \int_{x_1}^{x_2} F_x^{\text{net}}(x)\,dx = \int_{x_1}^{x_2} (F_{Ax}(x) + F_{Bx}(x) + F_{Cx}(x) + \cdots)\,dx$$
$$= \int_{x_1}^{x_2} F_{Ax}(x)\,dx + \int_{x_1}^{x_2} F_{Bx}(x)\,dx + \int_{x_1}^{x_2} F_{Cx}(x)\,dx + \cdots.$$

If we define the work associated with a single force component $F_{Ax}(x)$ as

$$W_A \equiv \int_{x_1}^{x_2} F_{Ax}(x)\,dx \quad \text{(definition of work—1D **single** force and displacement),} \quad (9\text{-}8)$$

we see that the net work is given by

$$W^{\text{net}} = W_A + W_B + W_C + \cdots. \quad (9\text{-}9)$$

So there are two ways to calculate the net work. One is to sum the force components before the net work is calculated. The other is to calculate the work associated with each of the components separately. The net work is then determined by adding up the work done by each force component.

The right side of Eq. 9-6 tells us how the speed of a particle of mass $m$ is changed by the net work done on it. This $\frac{1}{2}mv_x^2$ is the factor that is changed by the work done on the bead. Because we are talking about translational motion (as distinct from rotational motion), we call this factor **translational kinetic energy** (or often just "kinetic energy"). In the most general sense, kinetic energy is a quantity associated with motion. However, in this chapter, we limit our discussion to the motion of a single particle, or the center of mass of systems of particles and extended objects. When we turn our attention to the study of thermodynamics later in the book, we will have to revisit the concept of kinetic energy to take into account the fact that particles within objects may well be moving, even if the center of mass is not. So we define the translational kinetic energy, $K$, of a particle-like object in terms of its mass and the square of the speed of its center of mass as

$$K \equiv \tfrac{1}{2}mv_x^2 \quad \text{(definition of kinetic energy for 1D motion).} \tag{9-10}$$

According to this definition the term on the right side of Eq. 9-6 given by $\frac{1}{2}mv_{2x}^2 - \frac{1}{2}mv_{1x}^2$ represents the *change* in the object's translational kinetic energy.

Using our new definitions for net work and translational kinetic energy, Eq. 9-6 can be rewritten in streamlined form as

$$W^{\text{net}} = K_2 - K_1 = \Delta K \quad \text{(the net work-energy theorem).} \tag{9-11}$$

Equation 9-11 is known as the **net work-kinetic energy theorem.** This theorem tells us that when a net force acts along the direction of motion of a particle that moves from one position to another, the particle's kinetic energy changes by $\frac{1}{2}mv_{2x}^2 - \frac{1}{2}mv_{1x}^2$. Equation 9-11 is known as a theorem because Eqs. 9-6 and 9-9 were derived mathematically from Newton's Second Law. It is analogous in many ways to the *impulse-momentum theorem* (also derived from Newton's Second Law) that relates the action of a net force over time to the change in momentum of a particle.

## Units of Work and Energy

The SI unit for both work and kinetic energy (and every other type of energy) is the **joule** (J). It is defined directly from $K = \frac{1}{2}mv^2$ (Eq. 9-10) in terms of the units for mass and velocity as $\text{kg} \cdot \text{m}^2/\text{s}^2$. It is easy to show that the units for work, which is the product of a force in newtons (N) and a distance in meters (m), are $\text{N} \cdot \text{m}$, which also reduce to $\text{kg} \cdot \text{m}^2/\text{s}^2$. In summary,

$$1 \text{ joule} = 1 \text{ J} = 1 \text{ kg} \cdot \text{m}^2/\text{s}^2 = 1 \text{ N} \cdot \text{m} \quad \text{(SI unit for energy).}$$

Other units of energy which you may encounter are the erg (or $\text{g} \cdot \text{cm}^2/\text{s}^2$) and the foot-pound.

## Generalizing Work and Kinetic Energy Concepts

So far we have only related position and velocity for the very special case of a net force acting along a line of motion. What if the forces on a particle that make up the net force have components that do not lie along the direction of motion of our body of interest? If forces do not act parallel to a particle's displacement, how can we multiply and then integrate two vectors such as force and displacement to calculate work?

The remainder of this chapter is devoted to understanding how to calculate the work done by forces in some common situations. This will enable us to apply the net work-kinetic energy theorem to relate how changes in a particle's position due to a net force are related to changes in its speed.

**READING EXERCISE 9-1:** A particle moves along an $x$ axis. Does the kinetic energy of the particle increase, decrease, or remain the same if the particle's velocity changes (a) from $-3$ m/s to $-2$ m/s and (b) from $-2$ m/s to 2 m/s? (c) In each situation, is the net work done on the particle positive, negative, or zero? ■

## 9-3 The Concept of Physical Work

So far we have only discussed the work done on small particles with no internal structure. Now we would like to apply the concept of work to changes in motion of familiar extended objects. If we push on an object that deforms and changes its shape as it moves, then it's impossible to describe its motion in terms of a single displacement. For this reason, when we apply the concept of work to extended objects, we are assuming that these objects are particle-like as defined in Section 2-1. Thus, when we refer to doing work on an object, we assume the object is rigid enough that the work done distorting it is negligible compared to the work that displaces its center of mass.

Now let's get back to work. In casual conversation, most of us think of work as an expenditure of effort. It takes effort to push a rigid box down the hallway or to lift it. But you also expend effort to hold a heavy object steady in midair or to shove against a massive object that won't budge. If we examine our expression for work (Eq. 9-7), we see that at least for one-dimensional motion, work is given by the product of the components of a force along the line of motion of a particle and the particle's displacement along that line. This means that even though shoving really hard on a massive object that is at rest takes a lot of effort, *no physical work is done on it* unless it starts to move in the direction of the force. As we examine ways to define work in more general situations, we will find that the definition of work in physics requires that

> No work is done on a rigid object by a force unless there is a component of the force along the object's line of motion.

We have defined work in such a way that it requires a force and a displacement. How do we know how much work is done? Let's consider how much effort it takes to push a heavy, very rigid box down a hallway with a steady force. In this special case the force and the displacement of the box are in the same direction. In the next section we will consider what happens when the force acting on an object is in the opposite direction as its displacement.

Suppose you push the box to the right so the $x$-component of its displacement is $\Delta x = x_2 - x_1$. Now imagine that you use the same steady force to push the same box through twice the distance so its displacement component is $2\Delta x$. How much more effort did this take? How much effort would it take to just watch the box? The answers to these questions give us important insights into the nature of work. Namely, for a given force the magnitude of physical work *should* be proportional to the distance that an object is moved. This is consistent with the way we have defined work—so that it is proportional to the distance an object moves under the influence of forces.

Imagine pushing another larger box through a rightward displacement of $\Delta x$ using twice the force. How much more effort would you guess it takes to push the larger box than it takes to push the smaller box? You can get a feel for this by pushing one and then two larger textbooks of identical mass across a tabletop as shown in Fig. 9-3.

If you took a moment to do this experiment, you found that it takes about twice the effort to push two books through a displacement of $\Delta x$ using a force of $2\vec{F}_x$ as it did to push one book through the same displacement with a force of $\vec{F}_x$. We can conclude that the amount (that is, the absolute value) of work done is not just proportional to the distance an object is moved—it is also proportional to the magnitude of the force acting on the object. The concepts of proportionality between displacement force and the amount of physical work done can be summarized by the equation

$$|W| = |F_x \Delta x| \quad \text{(amount of work done by a steady force along a line of motion).}$$

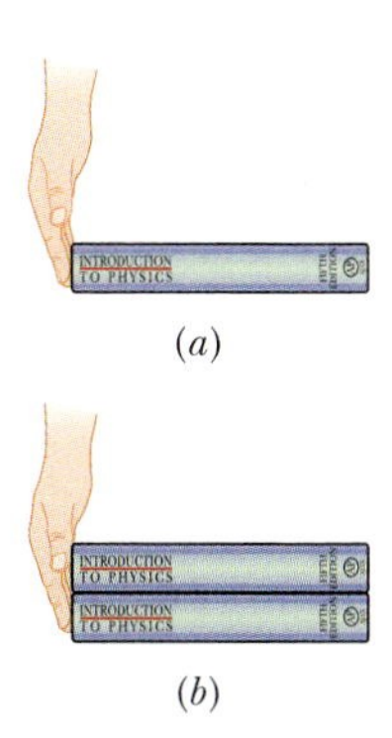


**FIGURE 9-3** ■ Pushing one and then two textbooks across a tabletop with a small but steady force.

In the next section we will use the mathematical definition of work to consider how much work is done under the influence of a steady force and also the circumstances under which work is positive or negative.

**READING EXERCISE 9-2:** The figure shows four situations in which a force acts on a box while the box either slides to the right with a displacement $\Delta\vec{x}$ or doesn't budge as indicated. The force on each box is shown. Rank the situations according to the amount of the work done by the force on the box during its displacement from greatest to least.

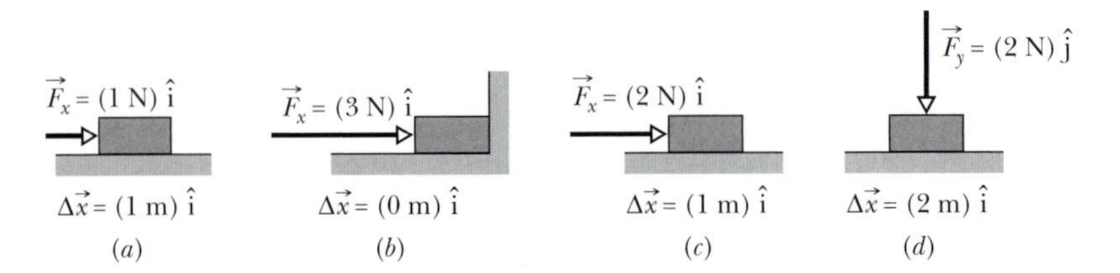


## 9-4 Calculating Work for Constant Forces

### One-Dimensional Forces and Motions Along the Same Line

Let's start by reconsidering the formal definition of the work associated with one-dimensional motion (presented in Eq. 9-8),

$$W \equiv \int_{x_1}^{x_2} F_x(x)\, dx \quad \text{(definition of work for a 1D motion and force).} \tag{9-12}$$

As we established in Section 9-1, $F_x(x)$ denotes the $x$-component of a force that can vary with $x$. However, if the force does not vary with $x$, then we can simply denote $F_x(x)$ as $F_x$ and take it out of the integral. This allows us to write

$$W \equiv \int_{x_1}^{x_2} F_x\, dx = F_x(x_2 - x_1) = F_x \Delta x,$$

so that

$$W = F_x\, \Delta x \quad \text{(work done by a constant force along a line of motion).} \tag{9-13}$$

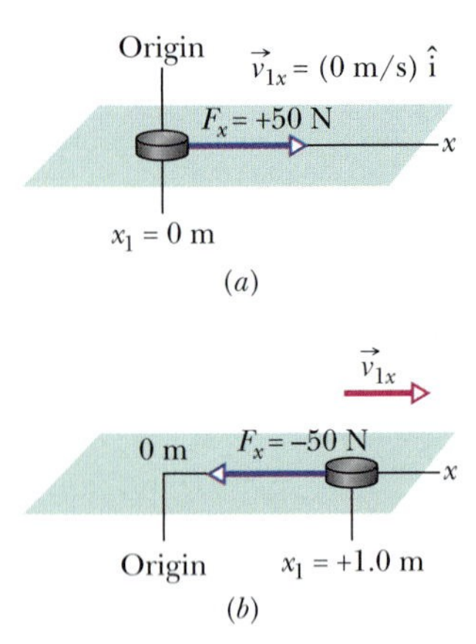


FIGURE 9-4 ■ (*a*) A puck is pushed from rest with a positive force component. (*b*) A puck that is already moving in a positive direction is pushed with a force that has a negative force component along a chosen $x$ axis.

**Positive vs. Negative Work** At the end of the last section we presented the equation $|W| = |F_x \Delta x|$ to represent the amount of work done on a rigid object. This is an informal expression we developed by imagining the effort needed to slide rigid objects on a table or down a hall. Equation 9-13 that we just derived is very similar except it has no absolute value signs. Since both $F_x$ and $\Delta x$ represent components of vectors along an axis, either component can be positive or negative. If this is the case, then the sign of the work, $W$, calculated as the products of these components can also be positive and negative. This raises some questions. How can we tell when the work done by a force will be positive? Negative? Does this mean that work is a vector component? To answer these questions let's consider the work done on a puck that is free to move along a line on a sheet of ice with no friction forces on it.

Imagine that the puck is initially at rest at your chosen origin. When you push it to the right with a steady horizontal force component of $F_x = +50$ N, it speeds up until its $x$ position is $+1$ m (Fig. 9-4*a*). The work you do on the puck is positive since it is given by

$$W = F_x \Delta x = F_x(x_2 - x_1) = (+50\ \text{N})(+1\ \text{m} - 0\ \text{m}) = +50\ \text{J} \quad \text{(speeding up).}$$

Suppose that now as the puck is moving away from the origin along your chosen positive $x$ axis, you suddenly take your other hand and push on it in the opposite direction with a steady horizontal force component of $F_x = -50$ N. Since you are pushing the puck in a direction opposite to its motion, it starts slowing down and reaches a zero velocity at a distance of 2.0 m from the origin (Fig. 9-4*b*). In this case the work you do on the puck while slowing it down is negative since it is given by

$$W = F_x \Delta x = F_x(x_2 - x_1) = (-50\text{ N})(+2\text{ m} - 1\text{ m}) = -50\text{ J} \quad \text{(slowing down).}$$

If we consider many similar situations with different types of forces acting, we can make the following general statement about the sign of the work done by a single force or net force acting on the center of mass of a rigid object:

**POSITIVE VS. NEGATIVE WORK:** If a single (or net) force has a component that acts in the direction of the object's displacement, the work that the force does is *positive*. If a single (or net) force has a component that acts in a direction opposite to the object's displacement, the work it does is *negative*.

**Work Is a Scalar Quantity** Recall that in Section 2-2 we defined a scalar (unlike the component of a vector) to be a quantity that is independent of coordinate systems. Based on this definition, we can see that work is a scalar quantity even though it can be positive or negative. If we rotate our $x$ axis by 180° so that all the components of force and displacement in our puck example above change sign, the sign of the work (which is the product of components) would not change sign. So even though there are directions associated with both force and displacement, there is no direction associated with the positive or negative work done by a force on an object. Therefore, *work is a scalar quantity.*

## Work Done by a Gravitational Force

We next examine the work done on an object by a particular type of constant force—namely, the gravitational force. Suppose a particle-like object of mass $m$, such as a tomato, is thrown upward with initial speed of $v_{1\,y}$ as in Fig. 9–5*a*. As it rises, it is slowed by a gravitational force $\vec{F}^{\text{grav}}$ that acts downward in the direction opposite the tomato's motion. We expect that $\vec{F}^{\text{grav}}$ does negative work on the tomato as it rises because the force is in the direction opposite the motion.

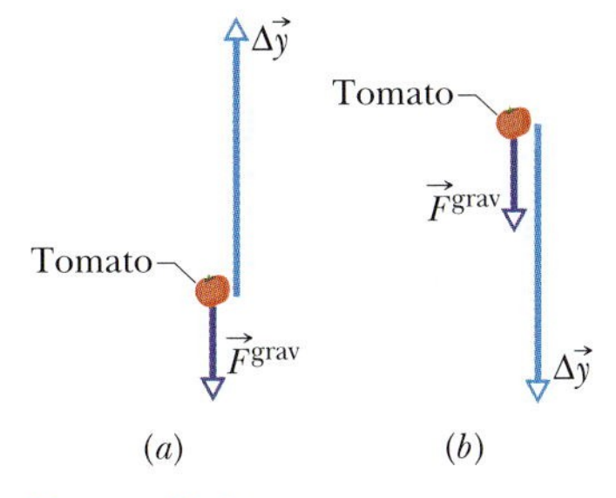


**FIGURE 9-5** ■ If the only force acting on a tomato is gravitational: (*a*) As the tomato rises, the gravitational force does negative work on the object. (*b*) As the tomato falls downward, the gravitational force does positive work on it.

To verify this, let's choose our $y$ axis to be positive in the upward direction, so that the $y$-component of the gravitational force (acts downward) is given by $F_y^{\text{grav}} = -mg$. We calculate the work done on the tomato as $W^{\text{grav}} = F_y^{\text{grav}} \Delta y$ where the $y$-component of displacement is given by $\Delta y = y_2 - y_1$.

Since during the rise $y_2$ is greater than $y_1$, $\Delta y$ is positive. The gravitational force component is negative and so we can write

$$W^{\text{grav}} = -mg(y_2 - y_1) = -mg\,|\Delta y| \quad \text{(rising object).} \tag{9-14}$$

After the object has reached its maximum height it begins falling back down. We expect the work done by the gravitational force to be positive in this case because the force and motion are in the same direction. Here $y_2$ is less than $y_1$ (Fig. 9-5*b*). Hence both the $y$-component of the gravitational force $F_y^{\text{grav}}$ and displacement $\Delta y = y_2 - y_1$ are negative. This gives us an expression for positive work of

$$W^{\text{grav}} = mg\,|\Delta y| \quad \text{(falling object).} \tag{9-15}$$

Thus, as we saw for the puck on the ice, the work done by the gravitational force is positive when the force and the displacement of the tomato are in the same direction and the work done by the force is negative when the force and displacement are in opposite directions.

### TOUCHSTONE EXAMPLE 9-1: Crepe Crate

During a storm, a crate of crepe is sliding across a slick, oily parking lot through a displacement $\Delta\vec{x} = (-3.0\text{ m})\hat{\imath}$ while a steady wind pushes against the crate with a force $\vec{F} = (2.0\text{ N})\hat{\imath}$. The situation and coordinate axes are shown in Fig. 9-6.

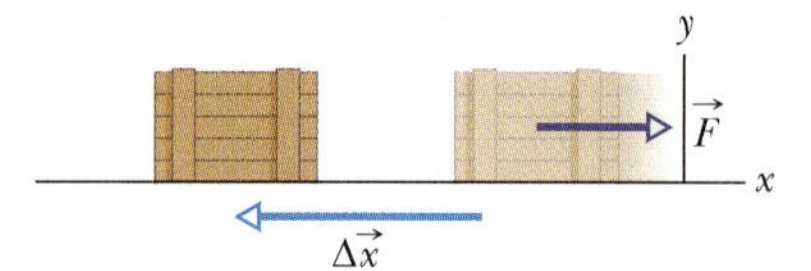


**FIGURE 9-6** ■ A constant force $\vec{F}$ created by the wind slows a crate down as it undergoes a displacement $\Delta\vec{x}$.

(a) How much work does this force from the wind do on the crate during the displacement?

**SOLUTION** ■ The **Key Idea** here is that, because we can treat the crate as a particle and because the wind force is constant ("steady") in both magnitude and direction during the displacement, we can use Eq. 9-13 ($W = F_x\,\Delta x$) to calculate the work,

$$W = F_x\,\Delta x = (2.0\text{ N})(-3.0\text{ m}) = -6.0\text{ J}. \quad \text{(Answer)}$$

So, the wind's force does negative 6.0 J of work on the crate.

(b) If the crate has a kinetic energy of 10 J at the beginning of displacement $\Delta\vec{x}$, what is its kinetic energy at the end of $\Delta\vec{x}$ assuming $\vec{F} = \vec{F}^{\text{net}}$?

**SOLUTION** ■ The **Key Idea** here is that, because the force does negative work on the crate, it reduces the crate's kinetic energy. Using the work-kinetic energy theorem in the form of Eq. 9-11, we have

$$K_2 = K_1 + W^{\text{net}} = 10\text{ J} + (-6.0\text{ J}) = 4.0\text{ J}. \quad \text{(Answer)}$$

Because the kinetic energy is decreased to 4.0 J, the crate has been slowed.

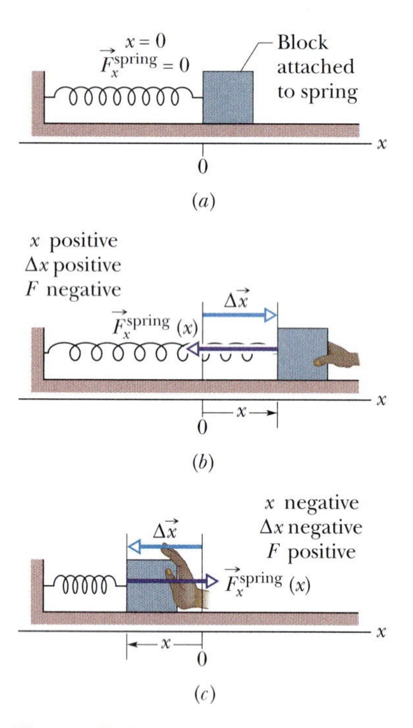


**FIGURE 9-7** ■ One end of a spring is attached to a fixed wall and the other end to a block that is free to slide. (*a*) The origin of an $x$ axis is located at the point where the relaxed spring is connected to the block. (*b*) The block and spring are given a positive displacement. Note the direction of the restoring force $\vec{F}_x^{\text{spring}}(x)$ exerted by the spring. (*c*) The spring is compressed by a negative amount $x$. Again, note the direction of the restoring force.

## 9-5 Work Done by a Spring Force

So far, we have limited our discussion to the work done by *constant* forces that do not change with position or time. Our goal in this section and the next is to explore how to calculate the work done by variable forces. A very common one-dimensional variable force is the spring force. The spring force is of great interest because many forces in our natural and man-made surroundings have the same mathematical form as the spring force. Examples include the interaction between atoms bound in a solid, the flexing of a bridge under the weight of vehicles, and the sway of a building during an earthquake—as long as displacements remain small. Thus, by examining this one idealized force, you can gain an understanding of many phenomena.

As you may have experienced, the magnitude of the force exerted by a spring increases when it is stretched or compressed more. Figure 9-7*a* shows a spring in its **relaxed state**—that is, neither compressed nor extended. One end is fixed, and a rigid block is attached to the other, free end. If we stretch the spring by pulling the block to the right, as in Fig. 9-7*b*, the spring pulls back on the block toward the left. (Because a spring's force acts to restore the relaxed state, it is sometimes said to be a *restoring force.*) If we compress the spring by pushing the block to the left, as in Fig. 9-7*c*, the spring now pushes on the block back toward the right.

To a good approximation for many springs, the force $\vec{F}_x^{\text{spring}}$ exerted by it is proportional to the displacement $\Delta\vec{x}$ of the free end from its relaxed position. As usual, the fact that $\vec{F}_x^{\text{spring}}$ depends on $x$ is symbolized by writing it as a function of $x$, $\vec{F}_x^{\text{spring}}(x)$. If its displacement is not too large, many spring-like objects have a *spring force* given by

$$\vec{F}_x^{\text{spring}}(x) = -k\,\Delta\vec{x} \quad \text{(Hooke's law for a 1D ideal spring).} \tag{9-16}$$

This "law" is named after Robert Hooke, an English scientist of the late 1600s. Since Hooke's law is based on the measured behavior of specific objects, it does not have the same status as Newton's laws.

The minus sign in $\vec{F}_x^{\text{spring}}(x) = -k\,\Delta\vec{x}$ (Eq. 9-16) indicates that the spring force is always opposite in direction from the displacement of the free end so the force is "restoring." The constant of proportionality $k$ is called the **spring constant.** It is always positive and is a measure of the stiffness of the spring. The larger $k$ is, the stiffer the spring—that is, the stronger will be its pull or push for a given displacement. The SI unit for $k$ is the N/m.

In Fig. 9-7, an $x$ axis has been placed parallel to the length of a spring, with the origin ($x = 0$) at the position of the free end when the spring is in its relaxed state. For this coordinate system and arrangement, we can write $\vec{F}_x^{\text{spring}}(x) = -k\Delta\vec{x}$ in component form as

$$F_x^{\text{spring}}(x) = -kx \qquad \text{(Hooke's law for } x = 0 \text{ in the relaxed state).} \tag{9-17}$$

The equation correctly describes ideal spring behavior. It tells us that if $x$ is positive (the spring is stretched toward the right on the $x$ axis), then the component $F_x^{\text{spring}}(x)$ is negative (it is a pull toward the left). If $x$ is negative (the spring is compressed toward the left), then the component $F_x^{\text{spring}}(x)$ is positive (it is a push toward the right). Also note that Hooke's law gives us a *linear* relationship between $F_x$ and $x$.

## Work Done by a Spring Force

In the situation shown in Fig. 9-7, the spring force components and displacements lie along the same line, so we can substitute the spring force in the more general expression presented in Eq. 9-8 to determine the work done by a one-dimensional variable force. We get

$$W^{\text{spring}} \equiv \int_{x_1}^{x_2} F_x^{\text{spring}}(x)\,dx. \tag{Eq. 9-8}$$

To apply this equation to the work done by the spring force as the block in Fig. 9-7*a* moves, let us make two simplifying assumptions about the spring and block. (1) The spring is *massless*; that is, its mass is negligible compared to the block's mass. (2) The spring is *ideal* so it obeys Hooke's law exactly. Making these simplifying assumptions might seem to make the results unreal. But for many interesting cases, these simplifications give us results that agree fairly well with experimental findings.

Back to the integral. We use Hooke's law (Eq. 9-16) to substitute $-kx$ for the component $F_x^{\text{spring}}(x)$. We also pull $k$ out of the integral since it is constant. Thus, we get

$$\begin{aligned} W^{\text{spring}} &= \int_{x_1}^{x_2} (-kx)\,dx = -k\int_{x_1}^{x_2} x\,dx \\ &= (-\tfrac{1}{2}k)[x^2]_{x_1}^{x_2} = -\tfrac{1}{2}k(x_2^2 - x_1^2), \end{aligned} \tag{9-18}$$

so the work done on the block by the spring force as the block moves is

$$W^{\text{spring}} = +\tfrac{1}{2}kx_1^2 - \tfrac{1}{2}kx_2^2 \qquad \text{(work by a spring force).} \tag{9-19}$$

This work $W^{\text{spring}}$, done by the spring force, can have a positive or negative value, depending on whether the block is moving toward or away from its zero position. This is quite similar to the way the gravitational work done on the tomato in the previous section changes as the tomato rises and falls.

Note that the final position $x_2$ appears in the *second* term on the right side of Eq. 9-19. Therefore,

The work done by the spring force on the block $W^{\text{spring}}$ is positive if the block moves closer to the relaxed position ($x = 0$). The work done by the spring force on the block is negative if the block moves farther away from $x = 0$. It is zero if the block ends up at the same distance from $x = 0$.

**READING EXERCISE 9-3:** For three situations, the initial and final positions, along the $x$ axis for the block in Fig. 9-7 are, respectively, (a) $-3$ cm, 2 cm; (b) 2 cm, 3 cm; and (c) $-2$ cm, 2 cm. In each situation, is the work done by the spring force on the block positive, negative, or zero?

## TOUCHSTONE EXAMPLE 9-2: Pralines and a Spring

A package of spicy Cajun pralines lies on a frictionless floor, attached to the free end of a spring in the arrangement of Fig. 9-7*a*. An applied force of magnitude $F^{\text{app}} = 4.9$ N would be needed to hold the package stationary at $x_2 = 12$ mm (Fig. 9-7*b*).

(a) How much work does the spring force do on the package if the package is pulled rightward from $x_1 = 0$ to $x_3 = 17$ mm?

**SOLUTION** ■ A **Key Idea** here is that as the package moves from one position to another, the spring force does work on it as given by Eq. 9-19. We know that the initial position $x_1$ is 0 and the final position $x_3$ is 17 mm, but we do not know the spring constant $k$.

We can probably find $k$ with Eq. 9-16 (Hooke's law), but we need a second **Key Idea** to use it: if the package were held stationary at $x_2 = 12$ mm, the spring force would have to balance the applied force (by Newton's Second Law). Thus, the $x$-component of the spring force $F_x^{\text{spring}}$ would have to be $-4.9$ N (toward the left in Fig. 9-7*b*), so Eq. 9-16 gives us

$$k = -\left(\frac{F_x^{\text{spring}}}{x_2}\right) = -\left(\frac{-4.9\ \text{N}}{12 \times 10^{-3}\ \text{m}}\right) = +408\ \text{N/m}.$$

Now, with the package at $x_3 = 17$ mm, Eq. 9-19 yields

$$W^{\text{spring}} = -\tfrac{1}{2}kx_3^2 = -\tfrac{1}{2}(408\ \text{N/m})(17 \times 10^{-3}\ \text{m})^2$$
$$= -0.059\ \text{J} = -59\ \text{mJ}. \qquad \text{(Answer)}$$

(b) Next, the package is moved leftward from $x_3 = 17$ mm to $x_4 = -12$ mm. How much work does the spring force do on the package during this displacement? Explain the sign of this work.

**SOLUTION** ■ The **Key Idea** here is the first one we noted in part (a). Now $x_3 = +17$ mm and $x_4 = -12$ mm where $x_3$ and $x_4$ are two positions of the spring relative to its equilibrium position. They do not represent displacements. Eq. 9-19 yields

$$W^{\text{spring}} = \tfrac{1}{2}kx_3^2 - \tfrac{1}{2}kx_4^2 = \tfrac{1}{2}k(x_3^2 - x_4^2)$$
$$= \tfrac{1}{2}(408\ \text{N/m})[(17 \times 10^{-3}\ \text{m})^2 - (-12 \times 10^{-3}\ \text{m})^2]$$
$$= 0.030\ \text{J} = 30\ \text{mJ}. \qquad \text{(Answer)}$$

This work done on the block by the spring force is positive because the block ends up closer to the spring's relaxed position.

## TOUCHSTONE EXAMPLE 9-3: Cumin Canister

In Fig. 9-8, a cumin canister of mass $m = 0.40$ kg slides across a horizontal frictionless counter with velocity $\vec{v} = v_x\hat{\text{i}}(-0.50\ \text{m/s})\hat{\text{i}}$. It then runs into and compresses a spring of spring constant $k = 750$ N/m. When the canister is momentarily stopped by the spring, by what amount $\Delta x$ is the spring compressed?

**SOLUTION** ■ There are three **Key Ideas** here:

1. The work $W^{\text{spring}}$ done on the canister by the spring force is related to the requested displacement $\Delta x = x_2 - x_1$ by Eq. 9-19 ($W^{\text{spring}} = \tfrac{1}{2}kx_1^2 - \tfrac{1}{2}kx_2^2$).
2. Since $\vec{F}^{\text{spring}} = \vec{F}^{\text{net}}$, the work $W^{\text{spring}}$ is also related to the kinetic energy of the canister by Eq. 9-11 ($W^{\text{net}} = K_2 - K_1$).

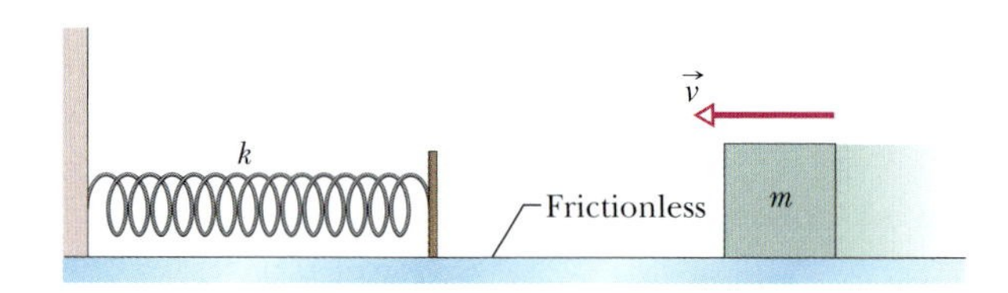


**FIGURE 9-8** ■ A canister of mass $m$ moves at velocity $\vec{v}$ toward a spring with spring constant $k$.

3. The canister's kinetic energy has an initial value of $K_1 = \tfrac{1}{2}mv_x^2$ and a value of zero when the canister is momentarily at rest.

Putting the first two of these ideas together, and noting that $x_1 = 0$ here since the spring is initially uncompressed, we write the net work-kinetic energy theorem for the canister as

$$K_2 - K_1 = -\tfrac{1}{2}kx_2^2.$$

Substituting according to the third idea makes this

$$0 - \tfrac{1}{2}mv_x^2 = -\tfrac{1}{2}kx_2^2.$$

Simplifying, solving for $x$, and substituting known data then give us

$$x_2 = \pm\sqrt{\frac{mv_x^2}{k}} = \pm\sqrt{\frac{(0.40\text{ kg})(-0.50\text{ m/s})^2}{750\text{ N/m}}}$$
$$= \pm 1.2 \times 10^{-2}\text{ m}$$
$$= \pm 1.2\text{ cm}.$$

We reject $x_2 = +1.2$ cm as a solution, since clearly the mass moves to the left as it compresses the spring. So

$$\Delta x = x_2 - x_1$$
$$= -1.2\text{ cm} - 0\text{ cm}$$
$$= -1.2\text{ cm}. \quad \text{(Answer)}$$

## 9-6 Work for a One-Dimensional Variable Force—General Considerations

### Calculating Work for Well-Behaved Forces

In the previous section, we were able to find the work done by our spring force using calculus to perform the integration called for in Eq. 9-8. This is because the spring force is a "well-behaved," continuous mathematical function that can be integrated. If you know the function $\vec{F}_x(x)$, you can substitute it into Eq. 9-8, introduce the proper limits of integration, carry out the integration, and thus find the work. (Appendix E contains a list of common integrals.) In summary, whenever a one-dimensional variable force is a function that can be integrated using the rules of calculus, the use of Eq. 9-8 is the preferred way to find the work done by the force on an object that moves along that same line.

### Calculating Work Using Numerical Integration

Suppose that instead of a spring force, our variable force on an object is caused by someone pushing and pulling erratically on the bead sliding along a wire depicted in Fig. 9-2. In that case, the force will probably not vary with $x$ the way a familiar mathematical function does, so we cannot use the rules of calculus to perform our integration. Whenever this is the case, we can use numerical methods to examine the variable force during small displacements where the force is approximately constant. We can then calculate the work done during each small displacement, and we can add each contribution to the work together to determine the total work. In this situation we are doing a **numerical integration.**

Let's start our exploration of numerical integration by considering the $x$-component of a one-dimensional force that varies as a particle moves. A general plot of such a *one-dimensional variable force* is shown in Fig. 9-9*a*. One method for finding the work done on the particle is to divide the distance between the initial location of a particle, $x_1$, and its final location, $x_N$, into $N$ small steps of width $\Delta x$. We can choose a large $N$ so that the values of $\Delta x$ are small enough so the force component along the $x$ axis $\vec{F}_x(x)$ is reasonably constant over that interval. Let $\langle F_{x\,n}(x)\rangle$ be the component representing the average value of $F_x(x)$ within the $n$th interval. As shown in Fig. 9-9*b* or *c*, $\langle F_{x\,n}(x)\rangle$ is the height of the $n$th strip. The value of $x$ for the $n$th strip is given by $x_n = (n - \tfrac{1}{2})\Delta x$, where $\Delta x = (x_N - x_0)/N$.

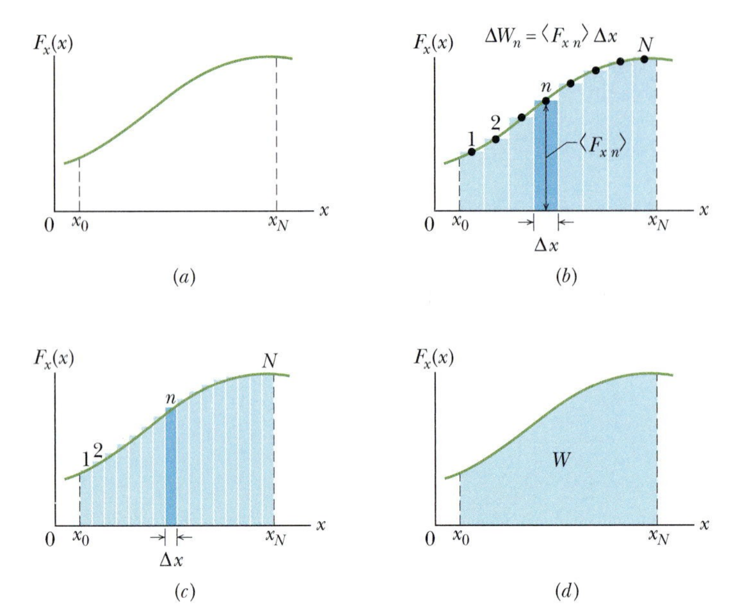


**FIGURE 9-9** ■ A particle only moves in one dimension. (*a*) A one-dimensional force component $F_x(x)$ is plotted against the displacement $x_N - x_0$ of the particle. (*b*) Same as (*a*) but with the area under the curve divided into narrow strips. (*c*) Same as (*b*) but with the area divided into narrower strips. (*d*) The limiting case. The work done by the force is given in Eq. 9-8 and is represented by the shaded area between the curve and the $x$ axis and between $x_0$ and $x_N$.

With $\langle F_{x\,n} \rangle$ taken to be constant, the small increment of work $\Delta W_n$ done by the force in the $n$th interval is approximately given by Eq. 9-13 as

$$\Delta W_n \cong \langle F_{x\,n} \rangle \Delta x. \tag{9-20}$$

Referring to the most darkly shaded region in Fig. 9-9*b* or *c*, we see that $\Delta W_n$ is then equal to the area of the $n$th rectangular strip.

To approximate the total work $W$ done by the force as the particle moves from $x_0$ to $x_N$, we add the areas of all the strips between $x_0$ and $x_N$ in Fig. 9-9*c*,

$$W \cong \sum_{n=1}^{N} \Delta W_n = \sum_{n=1}^{N} \langle F_{x\,n}(x) \rangle \Delta x. \tag{9-21}$$

This is not an exact calculation of the actual work done because the broken "skyline" formed by the tops of the rectangular strips in Fig. 9-9*b* (representing the values of $\langle F_{x\,n} \rangle$ as constants) only approximates the actual curve of $\vec{F}_x(x)$.

If needed in a particular situation we can make the approximation better by reducing the strip width $\Delta x$ and using more strips, as in Fig. 9-9*c*. Once the strip width is sufficiently small, Eq. 9-21 can be used to compute the total work done by the variable force.

## Defining the Integral

It is interesting to note that in the limit where the strip width approaches zero, the number of strips then becomes infinitely large and we approach an exact result,

$$W = \lim_{\Delta x \to 0} \sum_{n=1}^{N} \langle F_{n\,x}(x) \rangle \Delta x. \tag{9-22}$$

This limit is precisely what we mean by the integral of the function $F_x(x)$ between the limits $x_0$ and $x_N$. Thus, Eq. 9-22 becomes

$$W = \int_{x_0}^{x_N} F_x(x)\,dx \quad \text{(work done by a variable force in one dimension).} \qquad \text{(Eq. 9-8)}$$

Geometrically, the work is equal to the area between the $\vec{F}_x(x)$ curve and the $x$ axis, taken between the limits $x_0$ and $x_N$ (shaded in Fig. 9-9*d*). Remember that whenever $F_x$ is negative, the area between the graph of $F_x$ and the $x$ axis is also negative.

## TOUCHSTONE EXAMPLE 9-4: Work on a Stone

A 2.0 kg stone moves along an $x$ axis on a horizontal frictionless surface, acted on by only a force $F_x(x)$ that varies with the stone's position as shown in Fig. 9-10.

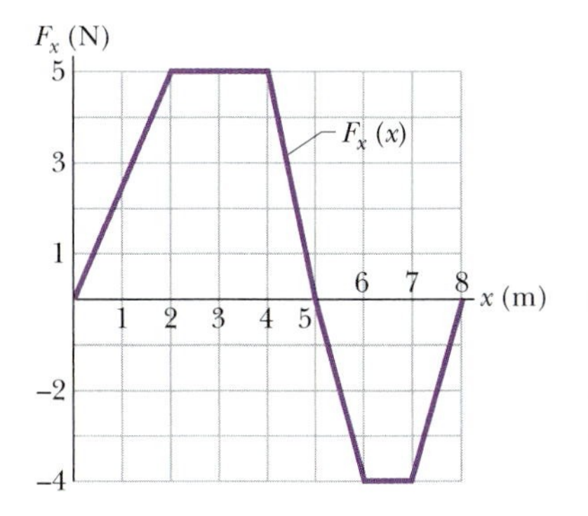


**FIGURE 9-10** ■ A graph showing the variation of a one-dimensional force component with a stone's position.

(a) How much work is done on the stone by the force as the stone moves from its initial point at $x_1 = 0$ to $x_2 = 5$ m?

**SOLUTION** ■ A **Key Idea** is that the work done by a single one-dimensional force is given by Eq. 9-8:

$$W = \int_{x_1}^{x_2} F_x(x)\, dx.$$

Here the limits are $x_1 = 0$ m and $x_2 = 5$ m, and $F_x(x)$ is given by Fig. 9-10. A second **Key Idea** is that we can easily evaluate the integral graphically from Fig. 9-10. To do so, we find the area between the plot of $F_x(x)$ and the $x$ axis, between the limits $x_1 = 0$ m and $x_2 = 5$ m. Note that we can split that area into three parts: a right triangle at the left (from $x = 0$ m to $x = 2$ m), a central rectangle (from $x = 2$ m to $x = 4$ m), and a triangle at the right (from $x = 4$ m to $x = 5$ m).

Recall that the area of a triangle is $\frac{1}{2}$(base)(height). The work $W_{0\to5}$ that was done on the stone from $x_1 = 0$ and $x_2 = 5$ m is then

$$W_{0\to5} = \tfrac{1}{2}(2\text{ m})(5\text{ N}) + (2\text{ m})(5\text{ N}) + \tfrac{1}{2}(1\text{ m})(5\text{ N})$$
$$= 17.5\text{ J}. \qquad \text{(Answer)}$$

(b) The stone starts from rest at $x_1 = 0$ m. What is its speed at $x = 8$ m?

**SOLUTION** ■ A **Key Idea** here is that the stone's speed is related to its kinetic energy, and its kinetic energy is changed because of the net work done on the stone by the force. Because the stone is initially at rest, its initial kinetic energy $K_1$ is 0. If we write its final kinetic energy at $x_3 = 8$ m as $K_3 = \frac{1}{2}mv_3^2$, then we can write the work-kinetic energy theorem of Eq. 9-11 ($K_3 = K_1 + W^{\text{net}}$) as

$$\tfrac{1}{2}mv_3^2 = 0 + W_{0\to8}, \qquad (9\text{-}23)$$

where $W_{0\to8}$ is the work done on the stone from $x_1 = 0$ m to $x_3 = 8$ m.

A second **Key Idea** is that, as in part (a), we can find the work graphically from Fig. 9-10 by finding the area between the plotted curve and the $x$ axis. However, we must be careful about signs. We must take an area to be positive when the plotted curve is above the $x$ axis and negative when it is below the $x$ axis. We already know that work $W_{0\to5} = 17.5$ J, so completing the calculation of the area gives us

$$W_{0\to8} = W_{0\to5} + W_{5\to8}$$
$$= 17.5\text{ J} - \tfrac{1}{2}(1\text{ m})(4\text{ N}) - (1\text{ m})(4\text{ N}) - \tfrac{1}{2}(1\text{ m})(4\text{ N})$$
$$= 9.5\text{ J}.$$

Substituting this and $m = 2.0$ kg into Eq. 9-23 and solving for $v_3$, we find

$$v_3 = 3.1\text{ m/s}. \qquad \text{(Answer)}$$

## 9-7 Force and Displacement in More Than One Dimension

In this section we will explore a quite general situation in which a particle moves in a curved three-dimensional path while acted upon by a three-dimensional force that could vary with the position of the particle and might not act in the same direction as the particle's displacement. How can we calculate the work done on the particle by the force in this much more complex situation?

Before undertaking this more general treatment of work, we will actually start our exploration with a simple example of the work done by a constant two-dimensional force acting on a sled that is moving in only one dimension. Our simple example will lead us to conclude that we need to devise a general method for finding work as the product of two vectors.

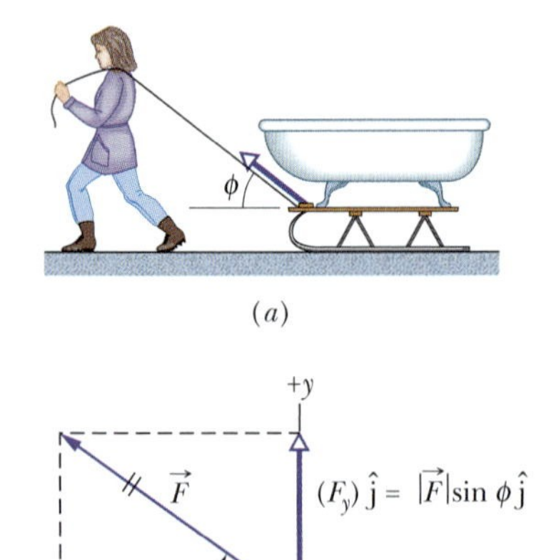


**FIGURE 9-11** ■ (*a*) A sled is pulled by a rope that makes an angle $\phi$ with the ground as it moves toward the left. (*b*) The components of the pulling force $\vec{F}$ along a positive $x$ axis and perpendicular to it.

## Work Done by a Force Applied at an Angle

Imagine that you are pulling a loaded sled with no friction present (Fig. 9-11*a*). You hold the rope handle of the sled at some angle $\phi$ relative to the ground. You pull as hard as you can and the sled starts to move. However, you find that you are getting tired quickly and still have a significant distance to go. What would you do? Is the situation hopeless? One thing that you could try is to change the angle at which you pull on the handle of the sled. Should you make the angle $\phi$ larger or smaller?

As you probably know from your everyday experiences, you must pull more or less horizontally on a heavy object to pull it along. If you make the angle $\phi$ smaller, then you will pull the sled more efficiently. As discussed in Chapter 6, this is because the perpendicular force component can only change the direction of the motion (and in this situation we assume the sled glides on top of packed snow that prevents it from moving down). Only the component of a force along the line of motion is effective in changing an object's speed. So, by the work-kinetic energy theorem, it must be that only the component of a force along the line of motion contributes to the work done by the force. Saying this more formally:

> To calculate the work done on an object by a force during a displacement, we use only the component of force along the line of the object's displacement. The component of force perpendicular to the displacement does zero work.

From Fig. 9-11*b*, we see that we can write the $x$-component of the force $F_x$ in terms of the magnitude of the force and the angle $\phi$ between the force and the positive $x$ axis. That is,

$$F_x = |\vec{F}|\cos\phi. \tag{9-24}$$

To find the work done, we can use Eq. 9-13 ($W = F_x\Delta x$) for a constant force to get

$$W = F_x\,\Delta x = (|\vec{F}|\cos\phi)\,\Delta x \quad \text{(work for displacement parallel to an } x \text{ axis)}, \tag{9-25}$$

where $\Delta x$ is the sled's displacement. Since the sled is moving to the left in the direction of the pull, both $F_x$ and $\Delta x$ are positive in our coordinate system and so the work done by the force is positive.

We can derive a similar but more general expression for the work done by a two-dimensional force along any line of displacement $\Delta\vec{r}$ (that does not necessarily lie along a chosen axis). To do this we must always take the angle $\phi$ *between the force and the direction of the displacement* (rather than the direction of a positive axis). In this case the expression for the work becomes

$$W = |\vec{F}||\Delta\vec{r}|\cos\phi \quad \text{(work in terms of angle between } \Delta\vec{r} \text{ and } \vec{F}\text{)}. \tag{9-26}$$

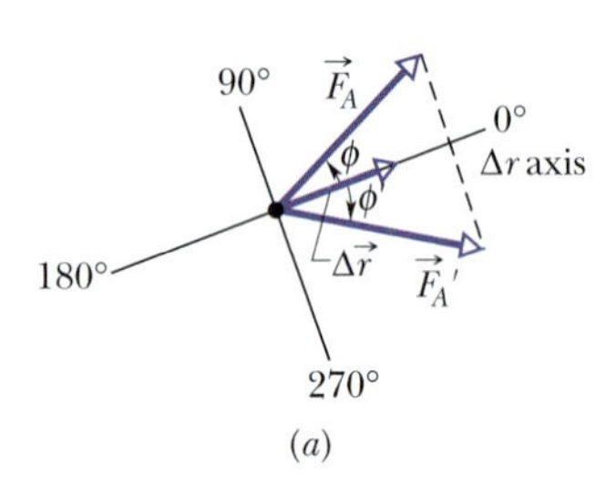


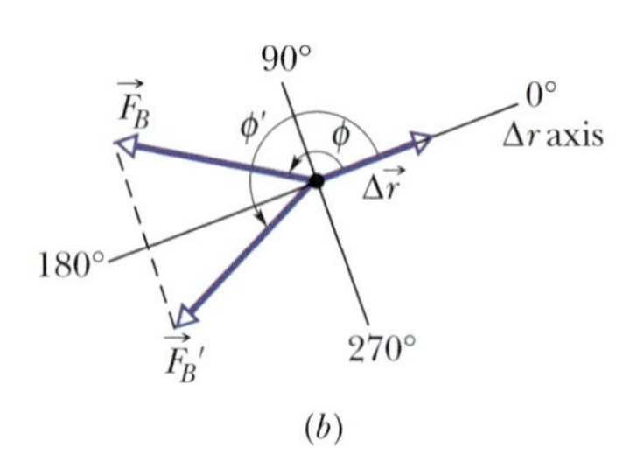


**FIGURE 9-12** ■ (*a*) If force $\vec{F}_A$ or $\vec{F}_A'$ has an angle $\phi < 90°$ or $\phi > 270°$ with respect to the displacement vector $\Delta\vec{r}$ of an object, its component relative to the displacement will be positive. (*b*) If a force $\vec{F}_B$ or $\vec{F}'_B$ has an angle $90° < \phi < 270°$ with respect to the displacement $\Delta\vec{r}$ of an object, its components relative to the displacement will be negative.

As shown in Fig. 9-12, using the angle between the force and displacement and *absolute values* for both, the sign of the work comes out correctly.

**Why the Sign of the Work Is Correct in Eq. 9-26** As shown in Fig. 9-12, if we set the angle $\phi$ in $W = |\vec{F}||\Delta\vec{r}|\cos\phi$ (Eq. 9-26) to any value less than 90°, then $\cos\phi$ is positive and so is the work. If $\phi$ is greater than 90° (up to 180°), then $\cos\phi$ is negative and so is the work. Referring to Fig. 9-11, we see that this way of determining the sign of the work done by an applied force is equivalent to determining the sign based on whether there is a component of force in the same or opposite direction as the motion. (Can you see why the work is zero when $\phi = 90°$?)

You can use similar considerations to determine the sign of work for $180° < \phi < 270°$ and for $270° < \phi < 360°$. Notice that once again it is the *relative* directions of the force and displacement vectors that determine the work done. As we already stated, no matter which way you choose to have the coordinate system pointing, the work for a particular process (including its sign) stays the same. Thus, work is indeed a scalar quantity.

*Cautions*: There are two restrictions to using the equations above to calculate work done on an object by a force. First, the force must be a *constant force*; that is, it must not change in magnitude or direction as the object moves through its displacement $\Delta\vec{r}$. Second, the object must be *particle-like*. This means that the object must be *rigid* and not change shape as its center of mass moves.

**Net Work Done by Several Forces** Suppose the net force on a rigid object is given by $\vec{F}^{\,\text{net}} = \vec{F}_A + \vec{F}_B + \vec{F}_C \cdots$, and we want to calculate the net work done by these forces. As we discussed earlier, it is simple to prove mathematically that the **net work** done on the object is the sum of the work done by the individual forces. We can calculate the net work in two ways: (1) We can use $W_A = |\vec{F}_A||\Delta\vec{r}|\cos\phi$ (Eq. 9-26) where $\phi$ is the angle between the direction of $\vec{F}_A$ and the object's displacement to find the work done by each force and then sum those works. Work is a scalar quantity, so summing the work done by the forces is as simple as adding up positive and negative numbers. (2) Alternatively, we can first find the net force $\vec{F}^{\,\text{net}}$ by finding the vector sum of the individual forces. Then we can use $W = |\vec{F}||\Delta\vec{r}|\cos\phi$ (Eq. 9-26), substituting the magnitude of $\vec{F}^{\,\text{net}}$ for the magnitude of $\vec{F}$, and the angle between the directions of the net force and the displacement for $\phi$.

## Work Done by a Three-Dimensional Variable Force

In general, even if a force varies with position, a particle can move through an infinitesimal displacement $d\vec{r}$ while being acted on by a three-dimensional force $\vec{F}(\vec{r})$. The displacement can be expressed in rectangular coordinates as

$$d\vec{r} = dx\hat{\text{i}} + dy\hat{\text{j}} + dz\hat{\text{k}}. \tag{9-27}$$

If we restrict ourselves to considering forces with rectangular components that depend only on the position component of the particle along a given axis, then

$$\vec{F}(\vec{r}) = F_x(x)\hat{\text{i}} + F_y(y)\hat{\text{j}} + F_z(z)\hat{\text{k}}. \tag{9-28}$$

Given the fact that no work is done unless there is a force component along the line of displacement, we can write the infinitesimal amount of work $dW$ done on the particle by the force $\vec{F}(\vec{r})$ as

$$dW = F_x(x)\,dx + F_y(y)\,dy + F_z(z)\,dz. \tag{9-29}$$

The work $W$ done by $\vec{F}$ while the particle moves from an initial position $\vec{r}_1$ with coordinates $(x_1, y_1, z_1)$ to a final position $\vec{r}_2$ with coordinates $(x_2, y_2, z_2)$ is then

$$W = \int_{\vec{r}_1}^{\vec{r}_2} dW = \int_{x_1}^{x_2} F_x(x)\,dx + \int_{y_1}^{y_2} F_y(y)\,dy + \int_{z_1}^{z_2} F_z(z)\,dz. \tag{9-30}$$

Note that if $\vec{F}(\vec{r})$ has only an $x$-component, then the $y$ and $z$ terms in the equation above are zero, so this equation reduces to

$$W = \int_{x_1}^{x_2} F_x(x)\,dx, \qquad \text{(Eq. 9-8)}$$

which is known as a **line integral.**

**READING EXERCISE 9-4:** The figure shows four situations in which a force acts on a box while the box slides rightward a distance $|\Delta\vec{x}|$ across a frictionless floor. The magnitudes of the forces are identical; their orientations are as shown. Rank the situations according to the work done on the box by the force during the displacement, from most positive to most negative.

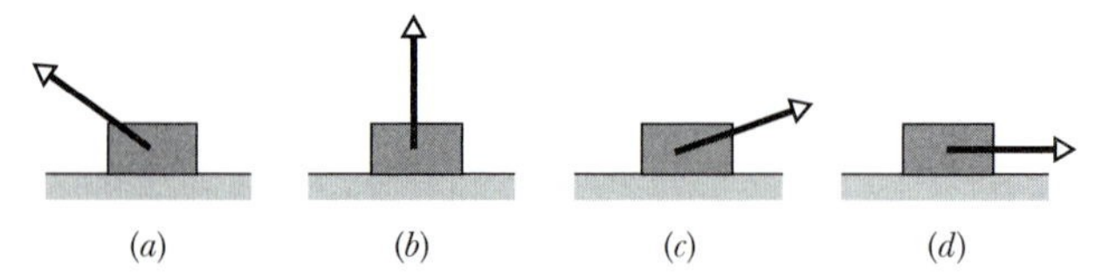


## TOUCHSTONE EXAMPLE 9-5: Sliding a Safe

Figure 9-13*a* shows two industrial spies sliding an initially stationary 225 kg floor safe a displacement $\Delta\vec{r}$ of magnitude 8.50 m, along a straight line toward their truck. The push $\vec{F}_1$ of Spy 001 is 12.0 N, directed at an angle of 30° downward from the horizontal; the pull $\vec{F}_2$ of Spy 002 is 10.0 N, directed at 40° above the horizontal. The magnitudes and directions of these forces do not change as the safe moves, and the floor and safe make frictionless contact.

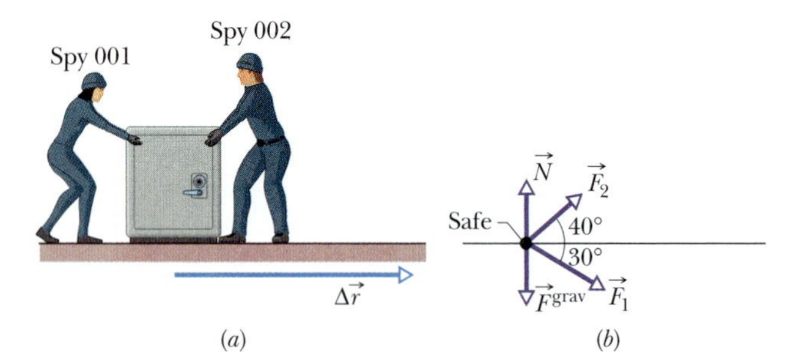


**FIGURE 9-13** (*a*) Two spies move a floor safe through displacement $\Delta\vec{r}$. (*b*) A free-body diagram showing the forces on the safe.

(a) What is the work done on the safe by forces $\vec{F}_1$ and $\vec{F}_2$ during the displacement $\Delta\vec{r}$?

**SOLUTION** We use two **Key Ideas** here. First, the work $W$ done on the safe by the two forces is the sum of the works they do individually. Second, because we can treat the safe as a particle and the forces are constant in both magnitude and direction, we can use Eq. 9-26,

$$(W = |\vec{F}||\Delta\vec{r}|\cos\phi),$$

to calculate those works. Note: $\cos(+\phi) = \cos(-\phi)$. From this and the free-body diagram for the safe in Fig. 9-13*b*, the work done by $\vec{F}_1$ is

$$W_1 = |\vec{F}_1||\Delta\vec{r}|\cos\phi_1 = (12.0\text{ N})(8.50\text{ m})(\cos 30°)$$
$$= 88.33\text{ J},$$

and the work done by $\vec{F}_2$ is

$$W_2 = |\vec{F}_2||\Delta\vec{r}|\cos\phi_2 = (10.0\text{ N})(8.50\text{ m})(\cos 40°)$$
$$= 65.11\text{ J}.$$

Thus, the work done by both forces is

$$W = W_1 + W_2 = 88.33\text{ J} + 65.11\text{ J}$$
$$= 153.4\text{ J} \approx 153\text{ J}. \quad \text{(Answer)}$$

During the 8.50 m displacement, therefore, the spies transfer 153 J of energy to the kinetic energy of the safe.

(b) During the displacement, what is the work $W^{grav}$ done on the safe by the gravitational force $\vec{F}^{grav}$ and what is the work $W^{Normal}$ done on the safe by the normal force $\vec{N}$ from the floor?

**SOLUTION** The **Key Idea** is that, because these forces are constant in both magnitude and direction, we can find the work they do with Eq. 9-26. Thus, with $mg$ as the magnitude of the gravitational force, we write

$$W^{grav} = mg|\Delta\vec{r}|\cos 90° = mg|\Delta\vec{r}|(0) = 0, \quad \text{(Answer)}$$

and

$$W^{Normal} = N|\Delta\vec{r}|\cos 90° = N|\Delta\vec{r}|(0) = 0. \quad \text{(Answer)}$$

We should have known this result. Because these forces are perpendicular to the displacement of the safe, they do zero work on the safe and do not transfer any energy to or from it.

# 9-8 Multiplying a Vector by a Vector: The Dot Product

In the previous section, we discussed how to calculate the work done by a force that acts at some angle to the direction of an object's motion. We saw that in one dimension work is defined as the scalar product of two vector components (force and displacement). This may seem strange, but it is because only the *component* of the force along a line relative to the direction of displacement contributes to the work done by the force. This type of relationship between two vector quantities is so common that mathematicians have defined an operation to represent it. That operation is called the *dot* (or *scalar*) *product*. Learning about how to represent and calculate this product will lead us to a more general mathematical definition of work for three-dimensional situations. Application of the dot product will make the key equations we derived in the previous section easier to represent.

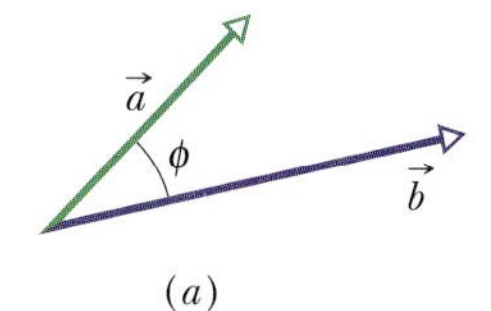

(*a*)

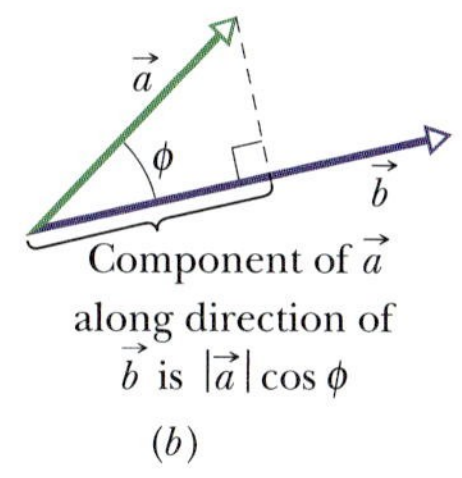

(*b*)

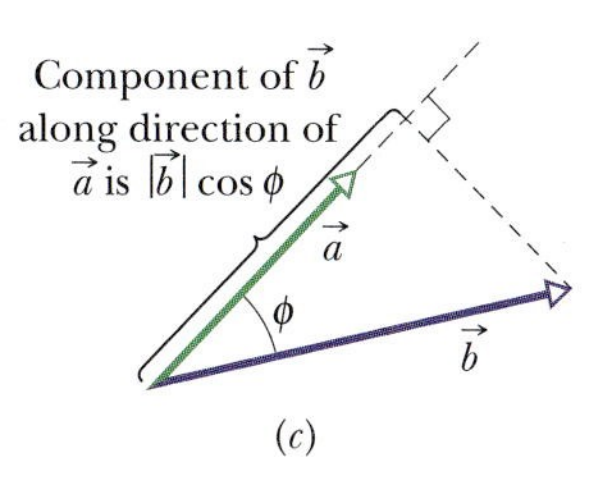

(*c*)

**FIGURE 9-14** ■ (*a*) Two vectors $\vec{a}$ and $\vec{b}$ with an angle $\phi$ between them. Since each vector has a component along the direction of the other vector, the same dot product results from: (*b*) multiplying the component of $\vec{a}$ on $\vec{b}$ by $|\vec{b}|$ or (*c*) multiplying the component of $\vec{b}$ on $\vec{a}$ by $|\vec{a}|$.

## The Dot Product of Two Vectors

The **scalar or dot product** of the vectors $\vec{a}$ and $\vec{b}$ in Fig. 9-14*a* is written as $\vec{a}\cdot\vec{b}$ and defined to be

$$\vec{a}\cdot\vec{b} \equiv |\vec{a}||\vec{b}|\cos\phi \quad \text{(definition of scalar product)}, \tag{9-31}$$

where $|\vec{a}|$ is the magnitude of $\vec{a}$, $|\vec{b}|$ is the magnitude of $\vec{b}$, and $\phi$ is the angle between $\vec{a}$ and $\vec{b}$ (or, more properly, between the directions of $\vec{a}$ and $\vec{b}$). There are actually two such angles, $\phi$ and $360° - \phi$. Either can be used in $\vec{a}\cdot\vec{b} = |\vec{a}||\vec{b}|\cos\phi$, because their cosines are the same.

Note that there are only scalars on the right side of $\vec{a}\cdot\vec{b} = |\vec{a}||\vec{b}|\cos\phi$ (including the value of $\cos\phi$). Thus $\vec{a}\cdot\vec{b}$ on the left side represents a scalar quantity. Being scalars, the values of these quantities do not change, no matter how we choose to define our coordinate system. Because of the dot placed between the two vectors to denote this product, the name usually used for it is "dot product" and $\vec{a}\cdot\vec{b}$ is spoken as "a dot b."

As in the case of work, the dot product can be regarded as the product of two quantities: (1) the magnitude of one of the vectors and (2) the component of the second vector along the direction of the first vector. For example, in Fig. 9-14*b*, $\vec{a}$ has a component ($|\vec{a}|\cos\phi$) along the direction of $\vec{b}$. Note that a perpendicular dropped from the head of $\vec{a}$ to $\vec{b}$ determines that component. Alternatively, $\vec{b}$ has a component $|\vec{b}|\cos\phi$ along the direction of $\vec{a}$.

> If the angle $\phi$ between two vectors is 0°, the component of one vector along the other is maximum, and so also is the dot product of the vectors. If the angle $\phi$ between two vectors is 180°, the component of one vector along the other is a minimum. If, instead, $\phi$ is 90°or 270°, the component of one vector along the other is zero, and so is the dot product.

Equation 9-31 ($\vec{a}\cdot\vec{b} = |\vec{a}||\vec{b}|\cos\phi$) is sometimes rewritten as follows to emphasize the components:

$$\vec{a}\cdot\vec{b} = |\vec{a}|(|\vec{b}|\cos\phi) = (|\vec{a}|\cos\phi)|\vec{b}|. \tag{9-32}$$

Here, ($|\vec{b}|\cos\phi$) is the component of $\vec{b}$ along $\vec{a}$, and ($|\vec{a}|\cos\phi$) is the component of $\vec{a}$ along $\vec{b}$. The commutative law applies to a scalar product, so we can write

$$\vec{a}\cdot\vec{b} = \vec{b}\cdot\vec{a}.$$

When two vectors are in unit-vector notation in one, two, or three dimensions, it can be shown mathematically that we will get the same result shown in Eq. 9-32 by writing the dot product as

$$\vec{a} \cdot \vec{b} = (a_x\hat{\text{i}} + a_y\hat{\text{j}} + a_z\hat{\text{k}}) \cdot (b_x\hat{\text{i}} + b_y\hat{\text{j}} + b_z\hat{\text{k}}), \tag{9-33}$$

which we can expand according to the distributive law: Each component of the first vector is to be "dotted" with each component of the second vector. For example, the first step is

$$a_x\hat{\text{i}} \cdot (b_x\hat{\text{i}} + b_y\hat{\text{j}} + b_z\hat{\text{k}}) = a_xb_x(\hat{\text{i}} \cdot \hat{\text{i}}) + a_xb_y(\hat{\text{i}} \cdot \hat{\text{j}}) + a_xb_z(\hat{\text{i}} \cdot \hat{\text{k}}).$$

Since $\hat{\text{i}}$ is perpendicular to both $\hat{\text{j}}$ and $\hat{\text{k}}$, there is no component of $\hat{\text{i}}$ along the other two unit vectors, the angle between them is 90°, and so $\hat{\text{i}} \cdot \hat{\text{j}} = \hat{\text{i}} \cdot \hat{\text{k}} = 0$. On the other hand, $\hat{\text{i}}$ is completely along $\hat{\text{i}}$, the angle here is 0°, and so $\hat{\text{i}} \cdot \hat{\text{i}} = 1$. Therefore,

$$a_x\hat{\text{i}} \cdot (b_x\hat{\text{i}} + b_y\hat{\text{j}} + b_z\hat{\text{k}}) = a_xb_x.$$

If we continue along these lines, we find that

$$\vec{a} \cdot \vec{b} = a_xb_x + a_yb_y + a_zb_z. \tag{9-34}$$

### Defining the Work Done as a Dot Product

If the force is constant over a displacement $\Delta\vec{r}$, we can use the definition of a dot product above and the relationship $W = |\vec{F}||\Delta\vec{r}| \cos \phi$ in Eq. 9-26 to produce an alternative mathematical definition for work,

$$W \equiv \vec{F} \cdot \Delta\vec{r} \quad \text{(definition of work done by a constant force).} \tag{9-35}$$

If the force is variable we can still use the definition of a dot product above along with the relationship presented in Eq. 9-30, where we integrated over infinitesimal displacements:

$$W = \int_{\vec{r}_1}^{\vec{r}_2} dW = \int_{x_1}^{x_2} F_x(x)\, dx + \int_{y_1}^{y_2} F_y(y)\, dy + \int_{z_1}^{z_2} F_z(z)\, dz,$$

to produce a more general alternative mathematical definition for work:

$$W \equiv \int_{\vec{r}_1}^{\vec{r}_2} \vec{F}(\vec{r}) \cdot d\vec{r} \quad \text{(definition of work done by a variable force).} \tag{9-36}$$

This dot product representation of work has some advantages. For one, the notation is more compact. It is also especially useful for calculating work when $\vec{F}$ and $d\vec{r}$ or $\Delta\vec{r}$ are given in unit-vector notation because we can exploit the fact that $\vec{a} \cdot \vec{b} = a_xb_x + a_yb_y + a_zb_z$ (Eq. 9-34).

## 9-9 Net Work and Translational Kinetic Energy

### Generalizing the Net Work-Kinetic Energy Theorem

We know from Newton's laws that if you apply a force to an object in the same direction as the object's motion, the object's speed will increase. From our discussion of work, we also know that the force does positive work on the object. If you apply the force in the direction opposite the direction of the object's motion, the object's speed will decrease, and we know that in that case the force will do negative work on the object. This suggests that work done by forces correlates with changes in speed. We used these considerations to relate work and kinetic energy for the special case of a bead on a wire that experiences a single force in the direction of the wire. In doing so, we developed a net work-kinetic energy theorem for one-dimensional forces and motions given by

$$W^{\text{net}} = K_2 - K_1 \quad \text{(the 1D net work-kinetic energy theorem)}, \qquad \text{(Eq. 9-11)}$$

where

$$W^{\text{net}} \equiv \int_{x_1}^{x_2} F_x^{\text{net}}(x)\, dx \text{ (Eq. 9-7)} \quad \text{and} \quad K = \tfrac{1}{2}mv_x^2. \qquad \text{(Eq. 9-10)}$$

Can we extend this to our more general three-dimensional situation? Fortunately we can combine

$$\int_{x_1}^{x_2} F_x^{\text{net}}(x)\, dx = \tfrac{1}{2}mv_{2x}^2 - \tfrac{1}{2}mv_{1x}^2, \qquad \text{(Eq. 9-6)}$$

and our general expression for work in three dimensions (Eq. 9-30) and rearrange terms to get

$$\begin{aligned} W^{\text{net}} &= \int_{x_1}^{x_2} F_x^{\text{net}}(x)\, dx + \int_{y_1}^{y_2} F_y^{\text{net}}(y)\, dy + \int_{z_1}^{z_2} F_z^{\text{net}}(z)\, dz \\ &= \tfrac{1}{2}m(v_{2x}^2 + v_{2y}^2 + v_{2z}^2) - \tfrac{1}{2}m(v_{1x}^2 + v_{1y}^2 + v_{1z}^2). \end{aligned} \qquad \text{(9-37)}$$

Since the speed of a particle moving in three dimensions is given by $v^2 = v_x^2 + v_y^2 + v_z^2$ and

$$W \equiv \int_{\vec{r}_1}^{\vec{r}_2} \vec{F}(\vec{r}) \cdot d\vec{r} = \int_{x_1}^{x_2} F_x(x)\, dx + \int_{y_1}^{y_2} F_y(y)\, dy + \int_{z_1}^{z_2} F_z(z)\, dz,$$

this reduces to

$$W^{\text{net}} = K_2 - K_1 \quad \text{(3D net work-kinetic energy theorem for variable force)}, \qquad \text{(9-38)}$$

where $K \equiv \frac{1}{2}mv^2$ represents a more general definition of the kinetic energy of a particle of mass $m$ with its center of mass moving with speed $v$. In words, Eq. 9-38 tells us that

Net work done on the particle = Change in its translational kinetic energy.

This relationship is valid in one, two, or three dimensions.

## Experimental Verification of the Net Work-Kinetic Energy Theorem

Experimental verification of the one-dimensional net work-kinetic energy theorem is shown in Figs. 9-15 and 9-16. A low-friction cart with a force sensor attached to it is pulled along a smooth track from $x_1 = 0.6$ m to $x_2 = 1.2$ m with a variable applied force. The applied force is measured with a force sensor. The distance along the track is measured with a motion detector. Both measurements are fed to a computer for

**FIGURE 9-15** ■ A variable force is applied to a force sensor attached to a cart on a horizontal track. The cart's position and velocity are recorded using a motion sensor. The friction force is negligible.

display. If we ignore friction, then $W^{\text{net}} = W^{\text{app}}$. The net work is given by the area under the curve obtained when data for the net force vs. distance is graphed. This area (determined by numerical integration as in Fig. 9-9) gives us

$$W^{\text{net}} \equiv \int_{x_1}^{x_2} F_x^{\text{net}}(x)\, dx = 1.3\text{ J}.$$

The distance and time data are used to determine the velocity of the cart at each location. The cart mass ($m = 1.5$ kg) and velocity are then used to determine the translational kinetic energy of the cart as a function of its location along the track. The change in kinetic energy between $x_1 = 0.6$ m and $x_2 = 1.6$ m is

$$\Delta K = K_2 - K_1 = 1.4\text{ J} - 0.1\text{ J} = 1.3\text{ J},$$

as expected according to the net work-kinetic energy theorem.

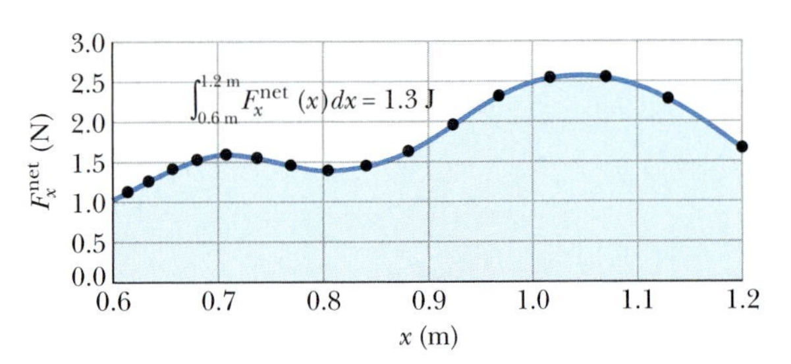


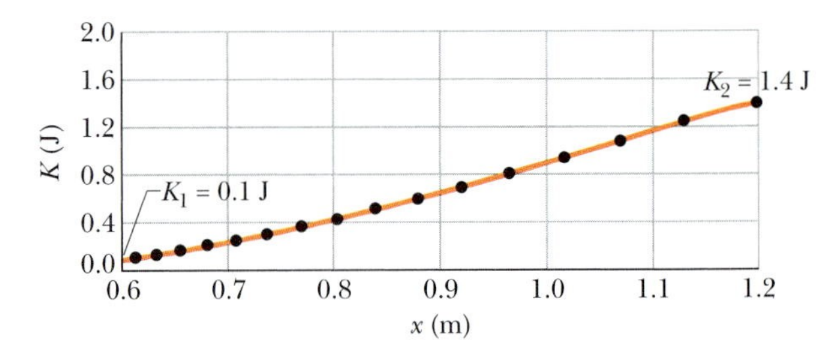


**FIGURE 9-16** ■ Experimental verification of the net work-kinetic energy theorem for a cart undergoing one-dimensional horizontal motion under the influence of a variable applied force and a negligible friction force.

## Lifting and Lowering—Net Work and Kinetic Energy

Suppose we *lift* a particle-like object by applying a vertical force $\vec{F}_y$ to it as shown in Fig. 9-17. During the upward displacement, our applied force does *positive* work $W^{\text{app}}$ on the object while the gravitational force does *negative* work $W^{\text{grav}}$ on it. Our force adds energy to (or transfers energy *to*) the object while the gravitational force removes energy from (or transfers energy *from*) it. By $\Delta K = K_2 - K_1$ (Eq. 9-38), the change $\Delta K$ in the translational kinetic energy of the object due to these two energy transfers is

$$\Delta K = K_2 - K_1 = W^{\text{net}} = W^{\text{app}} + W^{\text{grav}}. \tag{9-39}$$

This equation also applies if we lower the object. However, then the gravitational force tends to transfer energy *to* the object whereas our force tends to transfer energy *from* it.

A common situation involves an object that is stationary before and after being lifted. For example, suppose you lift a book from the floor to a shelf. Then $K_2$ and $K_1$ are both zero, and $\Delta K = K_2 - K_1 = W^{\text{net}} = W^{\text{app}} + W^{\text{grav}}$ reduces to

$$W^{\text{net}} = W^{\text{app}} + W^{\text{grav}} = 0\text{ N},$$

or

$$W^{\text{app}} = -W^{\text{grav}} \quad \text{(if object starts and ends at rest).} \tag{9-40}$$

**FIGURE 9-17** ■ An upward force is applied to an object in the presence of a downward gravitational force: (*a*) As the object rises, the applied force does positive work while the gravitational force does negative work. (*b*) As the object is lowered the applied force does negative work on the object while the gravitational force does positive work.

Note that we get the same result if $K_2$ and $K_1$ are not zero but are still equal. This result means that the work done by the applied force is the negative of the work done by the gravitational force. That is, the applied force transfers the same amount of energy to the object as the gravitational force takes away from it (whenever the initial and final speeds of an object are the same).

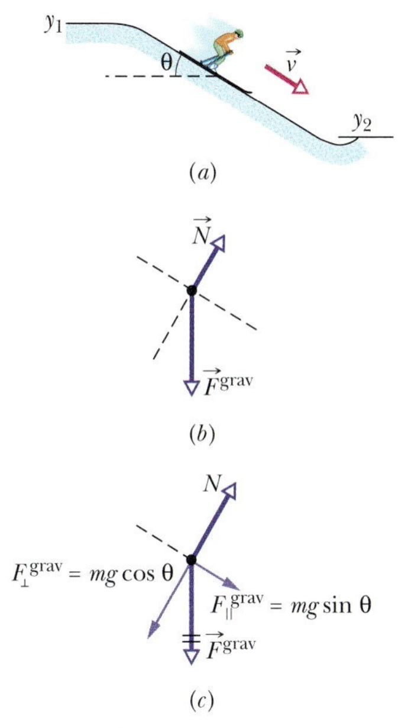


**FIGURE 9-18** ■ (*a*) A curved ramp makes an angle $\theta$ with respect to the horizontal at the location of a skier. (*b*) A free-body diagram showing the forces on the skier. (*c*) A diagram showing the resolution of $\vec{F}^{grav}$ into the components parallel and perpendicular to the ramp.

## Falling on an Incline—The Skier on a Curved Ramp

Finally, we are ready to return to the question of how to find the speed of a skier (shown in Fig. 9-1) as a function of how far she has descended on a frictionless ramp. Suppose we would like to know the speed of the skier shown in Fig. 9-1 as she leaves the end of a long curved ramp in order to predict how far she can jump. Let us revisit our initial claim that the net work-kinetic energy theorem is much more useful than Newton's Second Law for this calculation. The net work-kinetic energy theorem is only useful in this particular case if we make the simplifying assumptions that: (1) we can neglect frictional forces; (2) the skier doesn't push with her poles as she slides down the ramp; and (3) she holds her body rigid.

Given these assumptions, we can determine the net force on the skier's center of mass when she is at an arbitrary location on the ramp (Fig. 9-18*a*). This net force is the sum of forces shown in the free-body diagram in Fig. 9-18*b*. Figure 9-18*c* shows the components of the normal force and the gravitational force parallel and perpendicular to the ramp. Since there is no motion perpendicular to the ramp at a given location, the force components perpendicular to the ramp cancel out. So the net force acts parallel to the ramp. Its component down the ramp is given by $F_{\|}^{net} = mg\sin\theta$ where $\theta$ is the angle between the horizontal and the ramp.

Our problem now is to take into account the fact that $\theta$ keeps changing along the curved ramp. To do this we can divide the ramp into a whole series of tiny ramps having sides $dx$, $dy$, and a length $dr$ as shown in Fig. 9-19*a*. A greatly enlarged picture of one of these infinitesimal ramps is shown in Fig. 9-19*b*. The infinitesimal work done in traveling a distance $dr$ down any one of the tiny ramps is given by

$$dW = F_{\|}dr = (-mg\sin\theta\, dr), \tag{9-41}$$

but since $\sin\theta = dy/dr$, we see that $dW$ becomes simply $(-mg)\,dy$. This is a very profound result because it tells us that the work done by a rigid object as it falls down a frictionless ramp does not depend on the angle of the ramp but only on the constant factor $-mg$ and the vertical distance through which the object's center of mass falls. We will return to this idea in Chapter 10.

If we integrate the net force over the collection of tiny ramps we get

$$\begin{aligned} W^{net} &= \int_{\vec{r}_1}^{\vec{r}_2} \vec{F}^{net}(\vec{r}) \cdot d\vec{r} = \int_{r_1}^{r_2} F_{\|}\, dr \\ &= \int_{y_1}^{y_2} (-mg)\, dy = -mg(y_2 - y_1) = -mg\,\Delta y = F_y^{grav}\,\Delta y, \end{aligned} \tag{9-42}$$

where $\Delta y$ (like $F_y^{grav}$) is a negative quantity because $y_2 < y_1$.

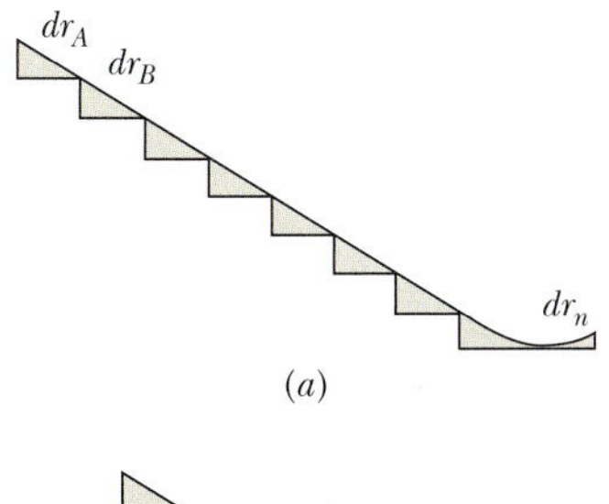


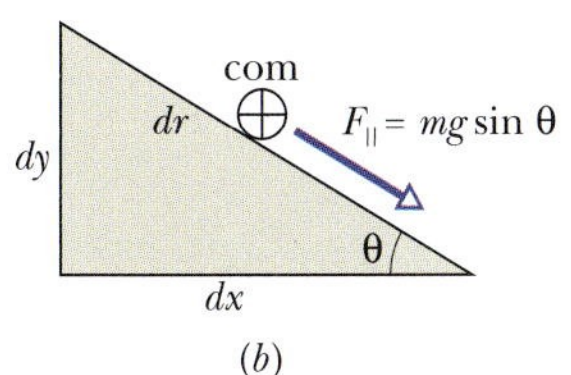


**FIGURE 9-19** ■ (*a*) The ramp can be divided into many smaller ramps, each possibly having a different $\theta$. (*b*) A ramp of infinitesimal length $dr$ with a vertical component $dy$ and a horizontal component $dx$.

Now that we have obtained a simple expression for the net work done by the gravitational force as the skier goes down the ramp, we can use the net work-kinetic energy theorem (Eq. 9-38) to find the skier's speed at the bottom of the ramp. If the skier starts from rest so that her initial speed is $v_1 = 0$ m/s, then

$$W^{net} = K_2 - K_1 \text{ so that } W^{net} = mg\Delta y = \tfrac{1}{2}m(v_2^2). \tag{9-43}$$

Solving for the final speed gives us

$$v_2 = \sqrt{2g\,\Delta y}. \tag{9-44}$$

Note that using Newton's Second Law to find this speed would be extremely difficult because it requires us to keep track of the angle of the ramp at each location. We will further explore the advantages of the net work-kinetic energy theorem for other situations in the next chapter.

## TOUCHSTONE EXAMPLE 9-6: Weight Lifting

Let us return to the lifting feats of Andrey Chemerkin shown on the opening page of this chapter.

(a) Chemerkin made his record-breaking lift with rigidly connected objects (a barbell and disk weights) having a total mass $m = 260.0$ kg. He lifted them a distance of 2.0 m. During the lift, how much work was done on the objects by the gravitational force $\vec{F}^{grav}$ acting on them?

**SOLUTION** The **Key Idea** here is that we can treat the rigidly connected objects as a single particle and thus use Eq. 9-14,

$$W^{grav} = -mg\Delta y,$$

to find the work $W^{grav}$ done on them by $\vec{F}^{grav}$. The total weight $mg$ was 2548 N, and $\Delta y = +2.0$ m. Thus,

$$W^{grav} = -mg\Delta y = -(2548 \text{ N})(2.0 \text{ m}) = -5100 \text{ J}. \quad \text{(Answer)}$$

(b) How much work was done on the objects by Chemerkin's force during the lift?

**SOLUTION** We do not have an expression for Chemerkin's force on the object, and even if we did, his force was certainly not constant. Thus, one **Key Idea** here is that we *cannot* just substitute his force into Eq. 9-12 to find his work. However, we know that the objects were stationary at the start and end of the lift, so that $K_2 - K_1 = 0$. Therefore, as a second **Key Idea**, we know by the net work-kinetic energy theorem that the work $W^{app}$ done by Chemerkin's applied force was the negative of the work $W^{grav}$ done by the gravitational force $\vec{F}^{grav}$. Equation 9-40 expresses this fact and gives us

$$W^{app} = -W^{grav} = +5100 \text{ J}. \quad \text{(Answer)}$$

(c) While Chemerkin held the objects stationary above his head, how much work was done on them by his force?

**SOLUTION** The **Key Idea** is that when he supported the objects, they were stationary. Thus, their displacement $\Delta\vec{r} = 0$ and, by Eq. 9-36, the work done on them was zero (even though supporting them was a very tiring task).

(d) How much work was done by the force Paul Anderson applied to lift objects with a total weight of 27 900 N a distance of 1.0 cm?

**SOLUTION** Following the argument of parts (a) and (b) but now with $mg = 27\,900$ N and $\Delta y = 1.0$ cm, we find

$$W^{app} = -W^{grav} = -(-mg\,\Delta y) = +mg\,\Delta y$$
$$= (27\,900 \text{ N})(0.010 \text{ m}) = 280 \text{ J}. \quad \text{(Answer)}$$

Anderson's lift required a tremendous upward force but only a small energy transfer of 280 J because of the short displacement involved.

## TOUCHSTONE EXAMPLE 9-7: Crate on a Ramp

An initially stationary 15.0 kg crate of cheese wheels is pulled, via a cable, a distance $d = 5.70$ m up a frictionless ramp to a height $h$ of 2.50 m, where it stops (Fig. 9-20*a*).

(a) How much work $W^{grav}$ is done on the crate by the gravitational force $\vec{F}^{grav}$ during the lift?

**SOLUTION** A **Key Idea** is that we can treat the crate as a particle and thus use Eq. 9-26 ($W = |\vec{F}||\Delta\vec{r}|\cos\phi$) to find the work $W^{grav}$ done by $\vec{F}^{grav}$. However, we do not know the angle $\phi$ between the directions of $\vec{F}^{grav}$ and displacement $\Delta\vec{r}$. From the crate's free-body diagram in Fig. 9-20*b*, we find that $\phi$ is $\theta + 90°$, where $\theta$ is the (unknown) angle of the ramp. Equation 9-26 then gives us

$$W^{grav} = mgd\cos(\theta + 90°) = -mgd\sin\theta, \quad (9\text{-}45)$$

where we have used a trigonometric identity to simplify the expression. The result seems to be useless because $\theta$ is unknown. But (continuing with physics courage) we see from Fig. 9-20*a* that $\sin\theta = h/d$, where $h$ is a known quantity. With this substitution, Eq. 9-45 becomes

$$W^{grav} = -mgh$$
$$= -(15.0 \text{ kg})(9.8 \text{ N/kg})(2.50 \text{ m}) \quad (9\text{-}46)$$
$$= -368 \text{ J}. \quad \text{(Answer)}$$

Note that Eq. 9-46 tells us that the work $W^{grav}$ done by the gravitational force depends on the vertical displacement but (perhaps surprisingly) not on the horizontal displacement. (Again, we return to this point in Chapter 10.)

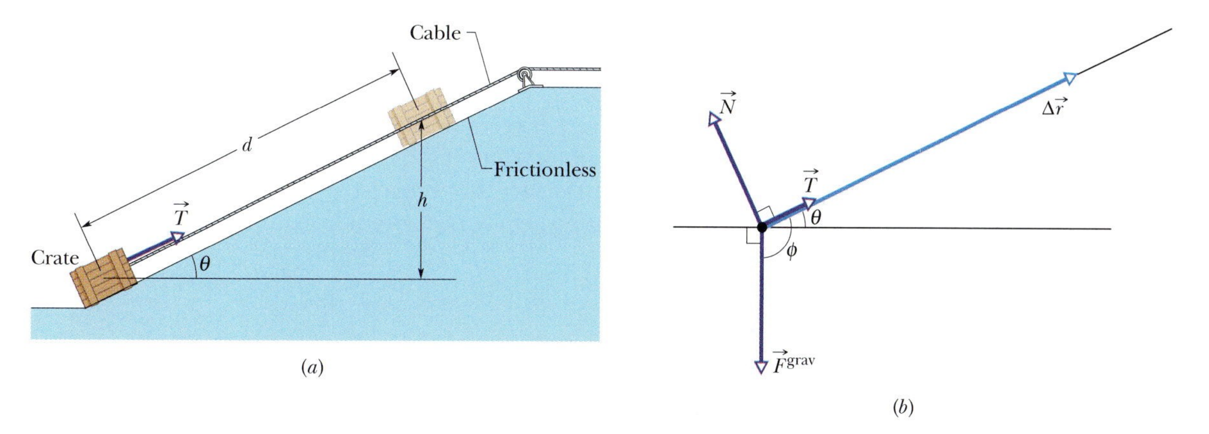


**FIGURE 9-20** ■ (*a*) A crate is pulled up a frictionless ramp by a force $\vec{T}$ parallel to the ramp. (*b*) A free-body diagram for the crate, showing all the forces on it. Its displacement $\Delta\vec{r}$ is also shown.

(b) How much work $W^{\text{rope}}$ is done on the crate by the force $\vec{T}$ from the cable during the lift?

**SOLUTION** ■ We cannot just substitute the force magnitude $T$ for $|\vec{F}|$ in Eq. 9-26 ($W = |\vec{F}||\Delta\vec{r}|\cos\phi$) because we do not know the value of $T$. However, a **Key Idea** to get us going is that we can treat the crate as a particle and then apply the work-kinetic energy theorem ($W^{\text{net}} = \Delta K$) to it. Because the crate is stationary before and after the lift, the change $\Delta K$ in its kinetic energy is zero. For the net work $W^{\text{net}}$ done on the crate, we must sum the works done by all three forces acting on the crate. From (a), the work $W^{\text{grav}}$ done by the gravitational force $\vec{F}^{\text{grav}}$ is $-368$ J. The work $W^{\text{Normal}}$ done by the normal force $\vec{N}$ on the crate from the ramp is zero because $\vec{N}$ is perpendicular to the displacement. We want the work $W^{\text{rope}}$ done by $\vec{T}$. Thus, the work-kinetic energy theorem gives us

$$\Delta K = W^{\text{rope}} + W^{\text{grav}} + W^{\text{Normal}},$$

or

$$0 = W^{\text{rope}} - 368\text{ J} + 0,$$

and so

$$W^{\text{rope}} = 368\text{ J}. \qquad \text{(Answer)}$$

## 9-10 Power

A contractor wishes to lift a load of bricks from the sidewalk to the top of a building using a winch. We can now calculate how much work the force applied by the winch must do on the load to make the lift. The contractor, however, is much more interested in the *rate* at which that work is done. Will the job take 5 minutes (acceptable) or a week (unacceptable)?

The rate at which work is done by a force is called the **power.** If an amount of work $W$ is done in an amount of time $\Delta t$ by a force, we define the **average power** due to the work done by a force during that time interval as

$$\langle P\rangle \equiv \frac{W}{\Delta t} \quad \text{(definition of average power).} \qquad (9\text{-}47)$$

We define the **instantaneous power** $P$ as the instantaneous rate of doing work, so that

$$P \equiv \frac{dW}{dt} \quad \text{(definition of instantaneous power),} \qquad (9\text{-}48)$$

where $dW$ is the infinitesimal amount of work done in an infinitesimal time interval $dt$. Suppose we know the work $W(t)$ done by a force as a continuous well-behaved function of time. Then to get the instantaneous power $P$ at, say, time $t = 3.0$ s during the work, we would first take the time derivative of $W(t)$, and then evaluate the result for $t = 3.0$ s.

The SI unit of power is the joule per second. This unit is used so often that it has a special name, the **watt** (W), after James Watt (who greatly improved the rate at which steam engines could do work). In the British system, the unit of power is the foot-pound per second. Often the horsepower is used. Some relations among these units are

$$1 \text{ watt} = 1 \text{ W} = 1 \text{ J/s} = 0.738 \text{ ft} \cdot \text{lb/s} \tag{9-49}$$

and

$$1 \text{ horsepower} = 1 \text{ hp} = 550 \text{ ft} \cdot \text{lb/s} = 746 \text{ W}. \tag{9-50}$$

Inspection of Eq. 9-47 shows that we can express work as power multiplied by time, $W = \langle P \rangle \Delta t$. When we do this, we commonly use the unit of kilowatt-hour. Thus,

$$\begin{aligned} 1 \text{ kilowatt-hour} &= 1 \text{ kW} \cdot \text{h} = (10^3 \text{ W})(3600 \text{ s}) \\ &= 3.6 \times 10^6 \text{ J} = 3.6 \text{ MJ}. \end{aligned} \tag{9-51}$$

Perhaps because the unit of kilowatt-hour appears on our utility bills, it has become identified as an electrical unit. However, the kilowatt-hour can be used equally well as a unit for other examples of work (or energy). Thus, if you pick up this book from the floor and put it on a tabletop, you are free to report the work that you have done as $4 \times 10^{-6}$ kW · h (or alternatively converting to milliwatts to get as 4 mW · h).

We can also express the rate at which a force does work on a particle (or particle-like object) in terms of that force and the body's velocity. For a particle that is moving along a straight line (say, the $x$ axis) and acted on by a constant force $\vec{F}$ directed at some angle $\phi$ to that line, $P = dW/dt$ (Eq. 9-48) becomes

$$P = \frac{dW}{dt} = \frac{(|\vec{F}|\cos\phi)|dx|}{dt} = |\vec{F}|\cos\phi\left|\frac{dx}{dt}\right|,$$

but since $v_x = dx/dt$, we get

$$P = |\vec{F}||v_x|\cos\phi. \tag{9-52}$$

Reorganizing the right side of this equation as the dot product $\vec{F} \cdot \vec{v}$ we may rewrite Eq. 9-52 as

$$P = \vec{F} \cdot \vec{v} \quad \text{(instantaneous power)}. \tag{9-53}$$

For example, the truck in Fig. 9-21 exerts a force $\vec{F}$ on the trailing load, which has velocity $\vec{v}$ at some instant. The instantaneous power due to $\vec{F}$ is the rate at which $\vec{F}$ does work on the load at that instant and is given by Eq. 9-52 and $P = \vec{F} \cdot \vec{v}$ (Eq. 9-53). Saying that this power is "the power of the truck" is often acceptable, but we should keep in mind what is meant: Power is the rate at which the applied *force* does work.

**FIGURE 9-21** ■ The power due to the truck's applied force on the trailing load is the rate at which that force does work on the load.

**READING EXERCISE 9-5:** A block moves with uniform circular motion because a cord tied to the block is anchored at the center of a circle. Is the power due to the force on the block from the cord positive, negative, or zero?

## TOUCHSTONE EXAMPLE 9-8: Average and Instantaneous Power

A horizontal cable accelerates a suspicious package across a frictionless horizontal floor. The amount of work that has been done by the cable's force on the package is given by $W(t) = (0.20\ \text{J/s}^2)t^2$.

(a) What is the average power $\langle P \rangle$ due to the cable's force in the time interval $t_1 = 0$ s to $t_2 = 10$ s?

**SOLUTION** ■ The **Key Idea** here is that the average power $\langle P \rangle$ is the ratio of the amount of work $W$ done in the given time interval to that time interval (Eq. 9-47). To find the work $W$, we evaluate the amount of work that has been done, $W(t)$, at $t = 0$ s and $t = 10$ s. At those times, the cable has done work $W_1$ and $W_2$, respectively:

$$W_1 = (0.20\ \text{J/s}^2)(0\ \text{s})^2 = 0\ \text{J} \quad \text{and} \quad W_2 = (0.20\ \text{J/s}^2)(10\ \text{s})^2 = 20\ \text{J}.$$

Therefore, in the 10 s interval, the work done is $W_2 - W_1 = 20$ J. Equation 9-47 then gives us

$$\langle P \rangle = \frac{W}{\Delta t} = \frac{20\ \text{J}}{10\ \text{s}} = 2.0\ \text{W}. \quad \text{(Answer)}$$

Thus, during the 10 s interval, the cable does work at the average rate of 2.0 joules per second.

(b) What is the instantaneous power $P$ due to the cable's force at $t = 3.0$ s, and is $P$ then increasing or decreasing?

**SOLUTION** ■ The **Key Idea** here is that the instantaneous power $P$ at $t = 3.0$ s is the time derivative of the work $dW/dt$ evaluated at $t = 3.0$ s (Eq. 9-48). Taking the derivative of $W(t)$ gives us

$$P = \frac{dW}{dt} = \frac{d}{dt}[(0.20\ \text{J/s}^2)t^2] = (0.40\ \text{J/s}^2)t.$$

This result tells us that as time $t$ increases, so does $P$. Evaluating $P$ for $t = 3.0$ s, we find

$$P = (0.40\ \text{J/s}^2)(3.0\ \text{s}) = 1.20\ \text{W}. \quad \text{(Answer)}$$

Thus, at $t = 3.0$s, the cable is doing work at the rate of 1.20 joules per second, and that rate is increasing.

# Problems

### SEC. 9-2 ■ INTRODUCTION TO WORK AND KINETIC ENERGY

**1. Electron in Copper** If an electron (mass $m = 9.11 \times 10^{-31}$ kg) in copper near the lowest possible temperature has a kinetic energy of $6.7 \times 10^{-19}$ J, what is the speed of the electron?

**2. Large Meteorite vs. TNT** On August 10, 1972, a large meteorite skipped across the atmosphere above western United States and Canada, much like a stone skipped across water. The accompanying fireball was so bright that it could be seen in the daytime sky (see Fig. 9-22 for a similar event). The meteorite's mass was about $4 \times 10^6$ kg; its speed was about 15 km/s. Had it entered the atmosphere vertically, it would have hit Earth's surface with about the same speed. (a) Calculate the meteorite's loss of kinetic energy (in joules) that would have been associated with the vertical impact. (b) Express the energy as a multiple of the explosive energy of 1 megaton of TNT, which is $4.2 \times 10^{15}$ J. (c) The energy associated with the atomic bomb explosion over Hiroshima was equivalent to 13 kilotons of TNT. To how many Hiroshima bombs would the meteorite impact have been equivalent?

FIGURE 9-22 ■ Problem 2. A large meteorite skips across the atmosphere in the sky above the Ottawa region.

**3. Calculate Kinetic Energy** Calculate the kinetic energies of the following objects moving at the given speeds: (a) a 110 kg football linebacker running at 8.1 m/s; (b) a 4.2 g bullet at 950 m/s; (c) the aircraft carrier *Nimitz*, $40.2 \times 10^8$ kg at 32 knots.

**4. Father Racing Son** A father racing his son has half the kinetic energy of the son, who has half the mass of the father. The father speeds up by 1.0 m/s and then has the same kinetic energy as the son. What are the original speeds of (a) the father and (b) the son?

**5. A Proton is Accelerated** A proton (mass $m = 1.67 \times 10^{-27}$ kg) is being accelerated along a straight line at $3.6 \times 10^{13}$ m/s$^2$ in a machine. If the proton has an initial speed of $2.4 \times 10^7$ m/s and travels 3.5 cm, what then is (a) its speed and (b) the increase in its kinetic energy?

**6. Vehicle's Kinetic Energy** If a vehicle with a mass of 1200 kg has a speed of 120 km/h, what is the vehicle's kinetic energy as determined by someone at rest alongside the vehicle's road?

**7. Truck Traveling North** A 2100 kg truck traveling north at 41 km/h turns east and accelerates to 51 km/h. (a) What is the change in the kinetic energy of the truck? What are the (b) magnitude and (c) direction of the change in the translational momentum of the truck?

**8. Two Pieces from One** An object, with mass $m$ and speed $v$ relative to an observer, explodes into two pieces, one three times as massive as the other; the explosion takes place in deep space. The less massive piece stops relative to the observer. How much kinetic energy is added to the system in the explosion, as measured in the observer's reference frame? *Hint*: Translational momentum is conserved.

**9. Freight Car** A railroad freight car of mass $3.18 \times 10^4$ kg collides with a stationary caboose car. They couple together, and 27.0% of the initial kinetic energy is transferred to nonconservative forms of energy (thermal, sound, vibrational, and so on). Find the mass of the caboose. *Hint*: Translational momentum is conserved.

**10. Two Chunks** An 8.0 kg body is traveling at 2.0 m/s with no external force acting on it. At a certain instant an internal explosion occurs, splitting the body into two chunks of 4.0 kg mass each. The explosion gives the chunks an additional 16 J of kinetic energy. Neither chunk leaves the line of original motion. Determine the speed and direction of motion of each of the chunks after the explosion. *Hint*: Translational momentum is conserved.

**11. Kinetic Energy and Impulse** A ball having a mass of 150 g strikes a wall with a speed of 5.2 m/s and rebounds with only 50% of its initial kinetic energy. (a) What is the speed of the ball immediately after rebounding? (b) What is the magnitude of the impulse on the wall from the ball? (c) If the ball was in contact with the wall for 7.6 ms, what was the magnitude of the average force on the ball from the wall during this time interval?

**12. Unmanned Space Probe** A 2500 kg unmanned space probe is moving in a straight line at a constant speed of 300 m/s. Control rockets on the space probe execute a burn in which a thrust of 3000 N acts for 65.0 s. (a) What is the change in the magnitude of the probe's translational momentum if the thrust is backward, forward, or directly sideways? (b) What is the change in kinetic energy under the same three conditions? Assume that the mass of the ejected burn products is negligible compared to the mass of the space probe.

## SEC. 9-6 ■ WORK FOR A ONE-DIMENSIONAL VARIABLE FORCE

**13. Graph of Acceleration** Figure 9-23 gives the acceleration of a 2.00 kg particle as it moves from rest along an $x$ axis while a constant applied force $\vec{F}^{app}$ acts on it from $x = 0$ m to $x = 9$ m. How much work has the force done on the particle when the particle reaches (a) $x = 4$ m, (b) $x = 7$ m, and (c) $x = 9$ m? What is the particle's speed and direction of travel when it reaches (d) $x = 4$ m, (e) $x = 7$ m, and (f) $x = 9$ m?

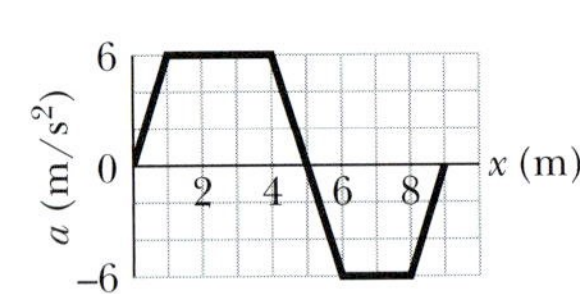


FIGURE 9-23 ■ Problem 13.

**14. Can of Nuts and Bolts** A can of nuts and bolts is pushed 2.00 m along an $x$ axis by a broom along the greasy (frictionless) floor of a car repair shop in a version of shuffleboard. Figure 9-24 gives the work $W$ done on the can by the constant horizontal force from the broom, versus the can's position $x$. (a) What is the magnitude of that force? (b) If the can had an initial kinetic energy of 3.00 J, moving in the positive direction of the $x$ axis, what is its kinetic energy at the end of the 2.00 m displacement?

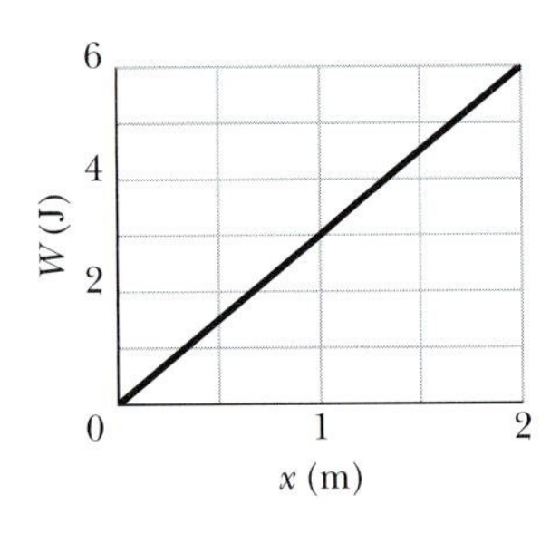


FIGURE 9-24 ■ Problem 14.

**15. Single Force** A single force acts on a body that moves along an $x$ axis. Figure 9-25 shows the velocity component $v_x$ versus time $t$ for the body. For each of the intervals $AB$, $BC$, $CD$, and $DE$, give the sign (plus or minus) of the work done by the force on the body or state that the work is zero.

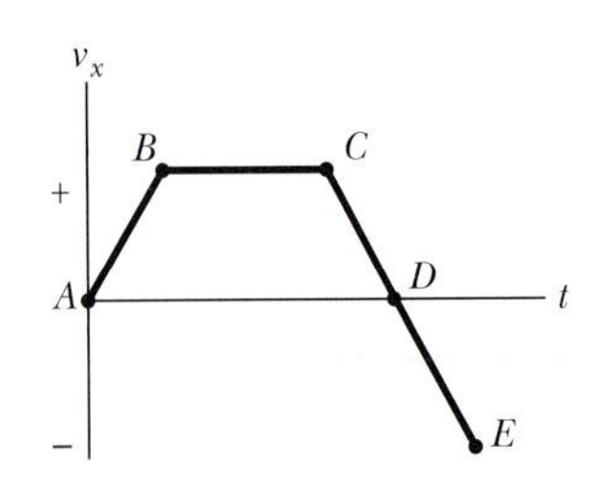


FIGURE 9-25 ■ Problem 15.

**16. Block Attached to a Spring** The block in Fig. 9-7 lies on a horizontal frictionless surface and is attached to the free end of the spring, with a spring constant of 50 N/m. Initially, the spring is at its relaxed length and the block is stationary at position $x = 0$ m. Then an applied force with a constant magnitude of 3.0 N pulls the block in the positive direction of the $x$ axis, stretching the spring until the block stops. When that stopping point is reached, what are (a) the position of the block, (b) the work that has been done on the block by the applied force, and (c) the work that has been done on the block by the spring force? During the block's displacement, what are (d) the block's position when its kinetic energy is maximum and (e) the value of that maximum kinetic energy?

**17. Luge Rider** A luge and rider, with a total mass of 85 kg, emerge from a downhill track onto a horizontal straight track with an initial speed of 37 m/s. If they slow at a constant rate of 2.0 m/s², (a) what magnitude $F$ is required for the slowing force, (b) what distance $d$ do they travel while slowing, and (c) what work $W$ is done on them by the slowing force? What are (d) $F$, (e) $d$, and (f) $W$ if the luge and the rider slow at a rate of 4.0 m/s²?

**18. Work from Graph** A 5.0 kg block moves in a straight line on a horizontal frictionless surface under the influence of a force that varies with position as shown in Fig. 9-26. How much work is done by the force as the block moves from the origin to $x = 8.0$ m?

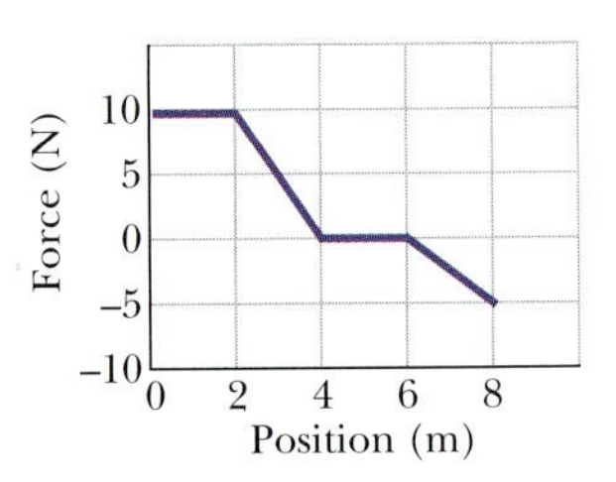


FIGURE 9-26 ■ Problem 18.

**19.** A 10 kg brick moves along an $x$ axis. Its acceleration as a function of its position is shown in Fig. 9-27. What is the net work done on the brick by the force causing the acceleration as the brick moves from $x = 0$ m to $x = 8.0$ m?

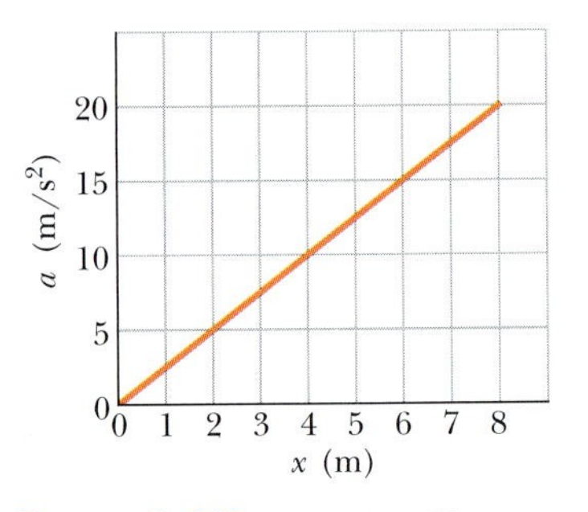


FIGURE 9-27 ■ Problem 19.

**20. Velodrome** (a) In 1975 the roof of Montreal's Velodrome, with a weight of 360 kN, was lifted by 10 cm so that it could be centered. How much work was done on the roof by the forces making the lift? (b) In 1960, Mrs. Maxwell Rogers of Tampa, Florida, reportedly raised one end of a car that had fallen onto her son when a jack failed. If her panic lift effectively raised 4000 N (about $\frac{1}{4}$ of the car's weight) by 5.0 cm, how much work did her force do on the car?

**21. Two Pulleys and a Canister** In Fig. 9-28, a cord runs around two massless, frictionless pulleys; a canister with mass $m = 20$ kg hangs from one pulley; and you exert a force $\vec{F}$ on the free end of the cord. (a) What must be the magnitude of $\vec{F}$ if you are to lift the canister at a constant speed? (b) To lift the canister by 2.0 cm, how far must you pull the free end of the cord? During that lift, what is the work done on the canister by (c) your force (via the cord) and (d) the gravitational force on the canister? (*Hint*: When a cord loops around a pulley as shown, it pulls on the pulley with a net force that is twice the tension in the cord.)

FIGURE 9-28 ■ Problem 21.

**22. Spring at MIT** During spring semester at MIT, residents of the parallel buildings of the East Campus dorms battle one another with large catapults that are made with surgical hose mounted on a window frame. A balloon filled with dyed water is placed in a pouch attached to the hose, which is then stretched through the width of the room. Assume that the stretching of the hose obeys Hooke's law with a spring constant of 100 N/m. If the hose is stretched by 5.00 m and then released, how much work does the force from the hose do on the balloon in the pouch by the time the hose reaches its relaxed length?

**23. Plot $F(x)$** The force on a particle is directed along an $x$ axis and given by $\vec{F} = F_0(x/x_0 - 1)$. Find the work done by the force in moving the particle from $x = 0$ to $x = 2x_0$ by (a) plotting $\vec{F}(x)$ and measuring the work from the graph and (b) integrating $\vec{F}(x)$.

**24. Block Dropped on a Spring** A 250 g block is dropped onto a relaxed vertical spring that has a spring constant of $k$ = 2.5 N/cm (Fig. 9-29). The block becomes attached to the spring and compresses the spring 12 cm before turning around. While the spring is being compressed, what work is done on the block by (a) the gravitational force on it and (b) the spring force? (c) What is the speed of the block just before it hits the spring? (Assume that friction is negligible.) (d) If the speed at impact is doubled, what is the maximum compression of the spring?

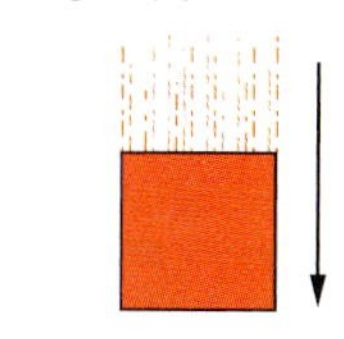

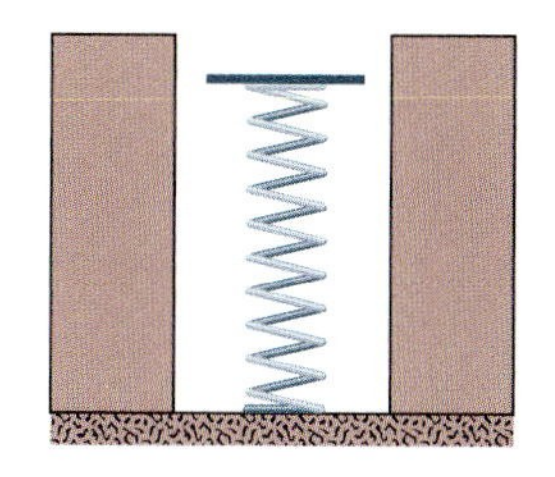

FIGURE 9-29 ■ Problem 24.

**25. Bird Cage** A spring with a spring constant of 15 N/cm has a cage attached to one end (Fig. 9-30). (a) How much work does the spring force do on the cage when the spring is stretched from its relaxed length by 7.6 mm? (b) How much additional work is done by the spring force when the spring is stretched by an additional 7.6 mm?

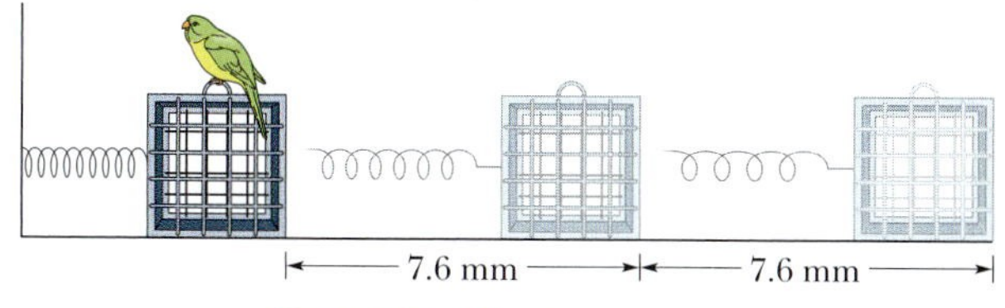


FIGURE 9-30 ■ Problem 25.

## SEC. 9-7 ■ FORCE AND DISPLACEMENT IN MORE THAN ONE DIMENSION

**26. Constant Force** A constant force of magnitude 10 N makes an angle of 150° (measured counterclockwise) with the positive $x$ direction as it acts on a 2.0 kg object moving in the $xy$ plane. How much work is done on the object by the force as the object moves from the origin to the point with position vector $(2.0\text{ m})\hat{i} - (4.0\text{ m})\hat{j}$?

**27. Force on a Particle** A force $\vec{F} = (4.0\text{ N})\hat{i} + (c\text{ N})\hat{j}$ acts on a particle as the particle goes through displacement $\vec{d} = (3.0\text{ m})\hat{i} - (2.0\text{ m})\hat{j}$. (Other forces also act on the particle.) What is the value of $c$ if the work done on the particle by force $\vec{F}$ is (a) zero, (b) 17 J, and (c) −18 J?

**28. Crate on an Incline** To push a 25.0 kg crate up a frictionless incline, angled at 25.0° to the horizontal, a worker exerts a force of magnitude 209 N parallel to the incline. As the crate slides 1.50 m, how much work is done on the crate by (a) the worker's applied force, (b) the gravitational force on the crate, and (c) the normal force exerted by the incline on the crate? (d) What is the total work done on the crate?

**29. Cargo Canister** Figure 9-31 shows an overhead view of three horizontal forces acting on a cargo canister that was initially stationary but that now moves across a frictionless floor. The force magnitudes are $F_A$ = 3.00 N, $F_B$ = 4.00 N, and $F_C$ = 10.0 N. What is the net work done on the canister by the three forces during the first 4.00 m of displacement?

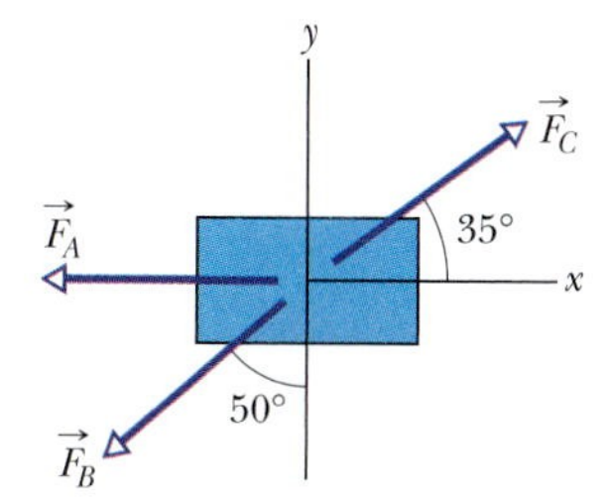


FIGURE 9-31 ■ Problem 29.

**30. A Particle Moves** A particle moves along a straight path through displacement $\vec{d} = (8\text{ m})\hat{i} - (c\text{ m})\hat{j}$ while force $\vec{F} = (2\text{ N})\hat{i} - (4\text{ N})\hat{j}$ acts on it. (Other forces also act on the particle.) What is the value of $c$ if the work done by $\vec{F}$ on the particle is (a) zero, (b) positive, and (c) negative?

**31. Worker Pulling Crate** To pull a 50 kg crate across a horizontal frictionless floor, a worker applies a force of 210 N, directed 20° above the horizontal. As the crate moves 3.0 m, what work is done on the crate by (a) the worker's force, (b) the gravitational force on the crate, and (c) the normal force on the crate from the floor? (d) What is the total work done on the crate?

**32. Floating Ice Blocks** A floating ice block is pushed through a displacement $\vec{d} = (15\text{ m})\hat{i} - (12\text{ m})\hat{j}$ along a straight embankment by rushing water, which exerts a force $\vec{F} = (210\text{ N})\hat{i} - (150\text{ N})\hat{j}$ on the block. How much work does the force do on the block during the displacement?

**33. Coin on a Frictionless Plane** A coin slides over a frictionless plane and across an $xy$ coordinate system from the origin to a point with $xy$ coordinates (3.0 m, 4.0 m) while a constant force acts on it. The force has magnitude 2.0 N and is directed at a counterclockwise angle of 100° from the positive direction of the $x$ axis. How much work is done by the force on the coin during the displacement?

**34. Work Done by 2-D Force** What work is done by a force $\vec{F} = ((2\text{ N/m})x)\hat{i} + (3\text{ N})\hat{j}$, with $x$ in meters, that moves a particle from a position $\vec{r}_1 = (2\text{ m})\hat{i} + (3\text{ m})\hat{j}$ to a position $\vec{r}_2 = -(4\text{ m})\hat{i} - (3\text{ m})\hat{j}$?

## SEC. 9-9 ■ NET WORK AND TRANSLATIONAL KINETIC ENERGY

**35. Cold Hot Dogs** Figure 9-32 shows a cold package of hot dogs sliding rightward across a frictionless floor through a distance $d$ = 20.0 cm while three forces are applied to it. Two of the forces are horizontal and have the magnitudes $F_A$ = 5.00 N and $F_B$ = 1.00 N;

the third force is angled down by $\theta = -60.0°$ and has the magnitude $F_C = 4.00$ N. (a) For the 20.0 cm displacement, what is the *net* work done on the package by the three applied forces, the gravitational force on the package, and the normal force on the package? (b) If the package has a mass of 2.0 kg and an initial kinetic energy of 0 J, what is its speed at the end of the displacement?

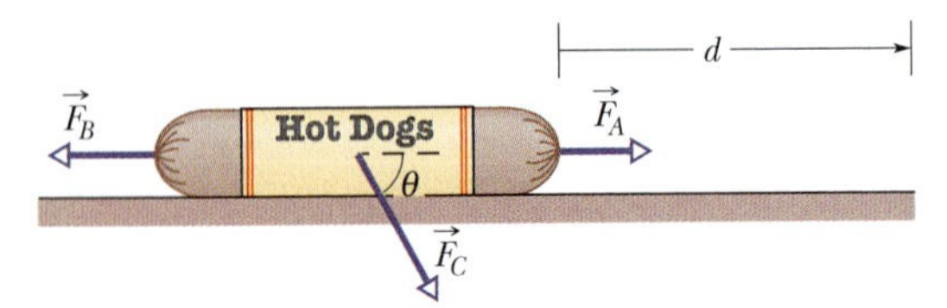


FIGURE 9-32 ■ Problem 35.

**36. Air Track** A 1.0 kg standard body is at rest on a frictionless horizontal air track when a constant horizontal force $\vec{F}$ acting in the positive direction of an $x$ axis along the track is applied to the body. A stroboscopic graph of the position of the body as it slides to the right is shown in Fig. 9-33. The force $\vec{F}$ is applied to the body at $t_1 = 0.0$ s, and the graph records the position of the body at 0.50 s intervals. How much work is done on the body by the applied force $\vec{F}$ between $t_1 = 0.0$ s and $t_2 = 2.0$ s?

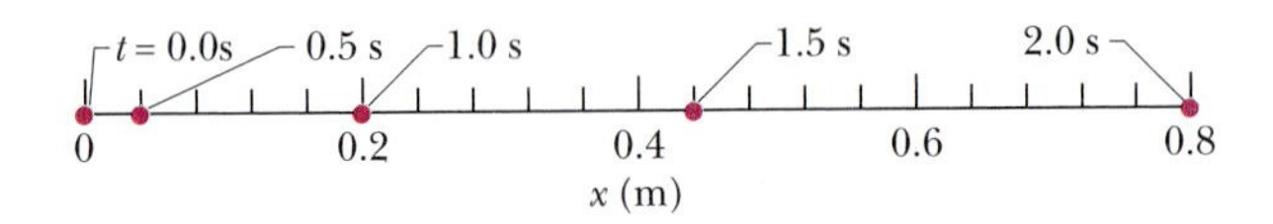


FIGURE 9-33 ■ Problem 36.

**37. Three Forces** Figure 9-34 shows three forces applied to a trunk that moves leftward by 3.00 m over a frictionless floor. The force magnitudes are $F_A = 5.00$ N, $F_B = 9.00$ N, and $F_C = 3.00$ N. During the displacement, (a) what is the net work done on the trunk by the three forces and (b) does the kinetic energy of the trunk increase or decrease?

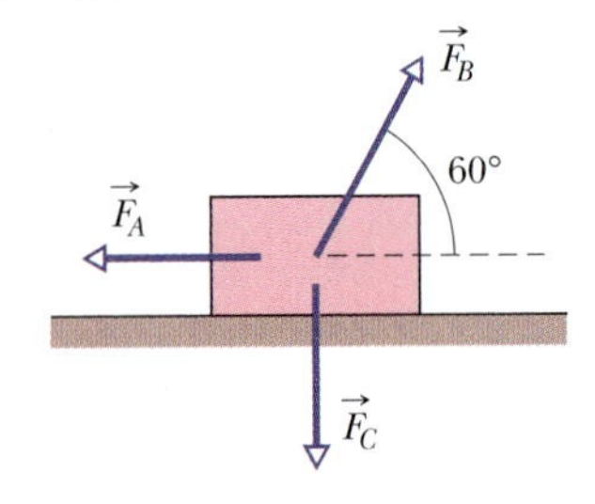


FIGURE 9-34 ■ Problem 37.

**38. Block of Ice** In Fig. 9-35, a block of ice slides down a frictionless ramp at angle $\theta = 50°$, while an ice worker pulls up the ramp (via a rope) with a force of magnitude $F_r = 50$ N. As the block slides through distance $d = 0.50$ m along the ramp, its kinetic energy increases by 80 J. How much greater would its kinetic energy have been if the rope had not been attached to the block?

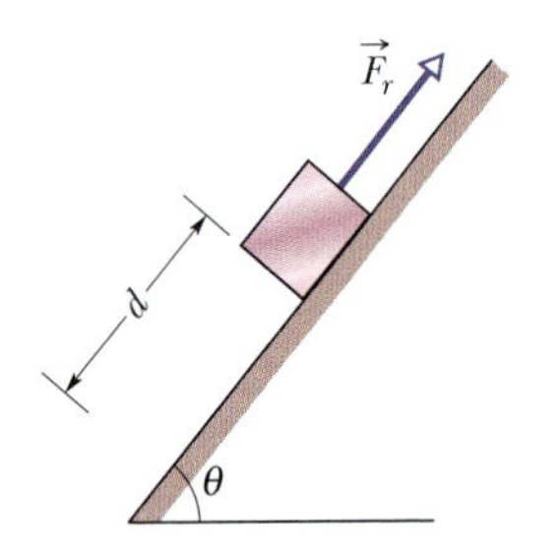


FIGURE 9-35 ■ Problem 38.

**39. Helicopter** A helicopter hoists a 72 kg astronaut 15 m vertically from the ocean by means of a cable. The acceleration of the astronaut is $g/10$. How much work is done on the astronaut by (a) the force from the helicopter and (b) the gravitational force on her? What are the (c) kinetic energy and (d) speed of the astronaut just before she reaches the helicopter?

**40. Given $x(t)$** A force acts on a 3.0 kg particle-like object in such a way that the position of the object as a function of time is given by $x = (3 \text{ m/s})t - (4 \text{ m/s}^2)t^2 + (1 \text{ m/s}^3)t^3$ with $x$ in meters and $t$ in seconds. Find the work done on the object by the force from $t_1 = 0.0$ s to $t_2 = 4.0$ s. (*Hint*: What are the speeds at those times?)

**41. Lowering a Block** A cord is used to vertically lower an initially stationary block of mass $M$ at a constant downward acceleration of $g/4$. When the block has fallen a distance $d$, find (a) the work done by the cord's force on the block, (b) the work done by the gravitational force on the block, (c) the kinetic energy of the block, and (d) the speed of the block.

**42. Force Applied Downward** In Fig. 9-36*a*, a 2.0 N force is applied to a 4.0 kg block at a downward angle $\theta$ as the block moves rightward through 1.0 m across a frictionless floor. Find an expression for the speed $v_2$ of the block at the end of that distance if the block's initial velocity is (a) 0.0 m/s and (b) 1.0 m/s to the right. (c) The situation in Fig. 9-36*b* is similar in that the block is initially moving at 1.0 m/s to the right, but now the 2.0 N force is directed downward to the left. Find an expression for the speed $v_2$ of the block at the end of the 1.0 m distance. (d) Graph all three expressions for $v_2$ versus downward angle $\theta$, for $\theta = 0°$ to $\theta = -90°$. Interpret the graphs.

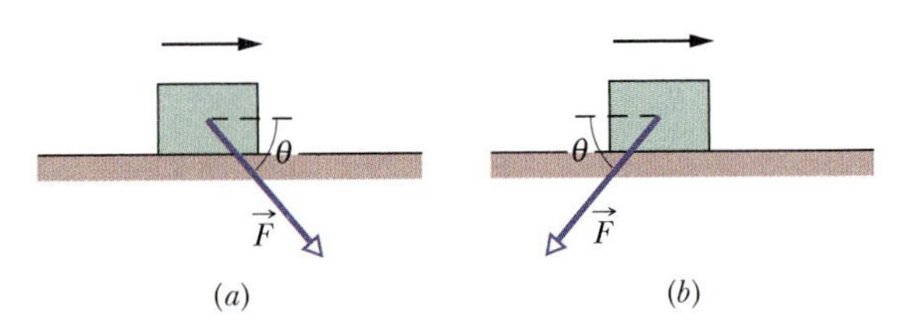


FIGURE 9-36 ■ Problem 42.

**43. Canister and One Force** The only force acting on a 2.0 kg canister that is moving in an $xy$ plane has a magnitude of 5.0 N. The canister initially has a velocity of 4.0 m/s in the positive $x$ direction, and some time later has a velocity of 6.0 m/s in the positive $y$ direction. How much work is done on the canister by the 5.0 N force during this time?

**44. Block of Ice Slides** A 45 kg block of ice slides down a frictionless incline 1.5 m long and 0.91 m high. A worker pushes up against the ice, parallel to the incline, so that the block slides down at constant speed. (a) Find the magnitude of the worker's force. How much work is done on the block by (b) the worker's force, (c) the gravitational force on the block, (d) the normal force on the block from the surface of the incline, and (e) the net force on the block?

**45. Cave Rescue** A cave rescue team lifts an injured spelunker directly upward and out of a sinkhole by means of a motor-driven cable. The lift is performed in three stages, each requiring a vertical distance of 10.0 m: (a) the initially stationary spelunker is accelerated to a speed of 5.00 m/s; (b) he is then lifted at the constant speed of 5.00 m/s; (c) finally he is slowed to zero speed. How much work is done on the 80.0 kg rescuee by the force lifting him during each stage?

**46. Work-Kinetic Energy** The only force acting on a 2.0 kg body as the body moves along the $x$ axis varies as shown in Fig. 9-37. The velocity of the body at $x = 0.0$ m is 4.0 m/s. (a) What is the kinetic energy of the body at $x = 3.0$ m? (b) At what value of $x$ will the body have a ki-

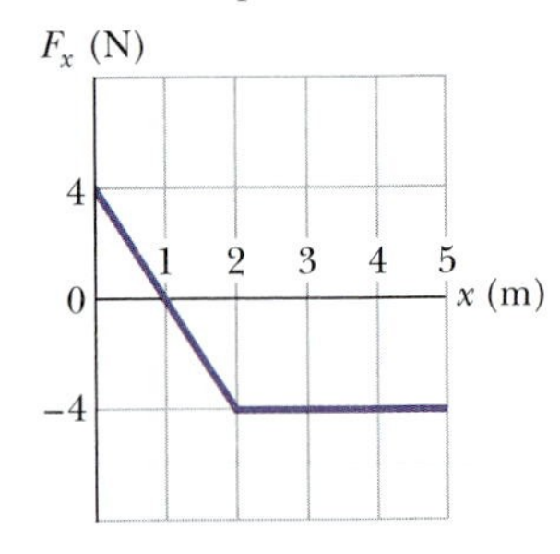


FIGURE 9-37 ■ Problem 46.

netic energy of 8.0 J? (c) What is the maximum kinetic energy attained by the body between $x = 0.0$ m and $x = 5.0$ m?

**47. Block at Rest** A 1.5 kg block is initially at rest on a horizontal frictionless surface when a horizontal force in the positive direction of an $x$ axis is applied to the block. The force is given by $\vec{F}(x) = (2.5\text{ N} - x^2\text{ N/m}^2)\hat{\text{i}}$, where $x$ is in meters and the initial position of the block is $x = 0.0$ m. (a) What is the kinetic energy of the block as it passes through $x = 2.0$ m? (b) What is the maximum kinetic energy of the block between $x = 0.0$ m and $x = 2.0$ m?

## SEC. 9-10 ■ POWER

**48. Average Rate of Work** The loaded cab of an elevator has a mass of $3.0 \times 10^3$ kg and moves 210 m up the shaft in 23 s at constant speed. At what average rate does the force from the cable do work on the cab?

**49. Block Pulled at Constant Speed** A 100 kg block is pulled at a constant speed of 5.0 m/s across a horizontal floor by an applied force of 122 N directed 37° above the horizontal. What is the rate at which the force does work on the block?

**50. Resistance to Motion** Resistance to the motion of an automobile consists of road friction, which is almost independent of speed, and air drag, which is proportional to speed-squared. For a certain car with a weight of 12,000 N, the net resistant force $\vec{F}$ is given by $\vec{F} = [300\text{ N} + (1.8\text{ N}\cdot\text{s}^2/\text{m}^2)v_x^2]\hat{\text{i}}$, where $\vec{F}$ is in newtons and $v_x$ is in meters per second. Calculate the power (in horsepower) required to accelerate the car at 0.92 m/s$^2$ when the speed is 80 km/h.

**51. A Force Acts on a Body** A force of 5.0 N acts on a 15 kg body initially at rest. Compute the work done by the force in (a) the first, (b) the second, and (c) the third seconds and (d) the instantaneous power due to the force at the end of the third second.

**52. Rope Tow** A skier is pulled by a tow rope up a frictionless ski slope that makes an angle of 12° with the horizontal. The rope moves parallel to the slope with a constant speed of 1.0 m/s. The force of the rope does 900 J of work on the skier as the skier moves a distance of 8.0 m up the incline. (a) If the rope moved with a constant speed of 2.0 m/s, how much work would the force of the rope do on the skier as the skier moved a distance of 8.0 m up the incline? At what rate is the force of the rope doing work on the skier when the rope moves with a speed of (b) 1.0 m/s and (c) 2.0 m/s?

**53. Freight Elevator** A fully loaded, slow-moving freight elevator has a cab with a total mass of 1200 kg, which is required to travel upward 54 m in 3.0 min, starting and ending at rest. The elevator's counterweight has a mass of only 950 kg, so the elevator motor must help pull the cab upward. What average power is required of the force the motor exerts on the cab via the cable?

**54. Ladle Attached to Spring** A 0.30 kg ladle sliding on a horizontal frictionless surface is attached to one end of a horizontal spring (with $k = 500$ N/m) whose other end is fixed. The ladle has a kinetic energy of 10 J as it passes through its equilibrium position (the point at which the spring force is zero). (a) At what rate is the spring doing work on the ladle as the ladle passes through its equilibrium position? (b) At what rate is the spring doing work on the ladle when the spring is compressed 0.10 m and the ladle is moving away from the equilibrium position?

**55. Towing a Boat** The force (but not the power) required to tow a boat at constant velocity is proportional to the speed. If a speed of 4.0 km/h requires 7.5 kW, how much power does a speed of 12 km/h require?

**56. Transporting Boxes** Boxes are transported from one location to another in a warehouse by means of a conveyor belt that moves with a constant speed of 0.50 m/s. At a certain location the conveyor belt moves for 2.0 m up an incline that makes an angle of 10° with the horizontal, then for 2.0 m horizontally, and finally for 2.0 m down an incline that makes an angle of 10° with the horizontal. Assume that a 2.0 kg box rides on the belt without slipping. At what rate is the force of the conveyor belt doing work on the box (a) as the box moves up the 10° incline, (b) as the box moves horizontally, and (c) as the box moves down the 10° incline?

**57. Horse Pulls Cart** A horse pulls a cart with a force of 40 lb at an angle of 30° above the horizontal and moves along at a speed of 6.0 mi/h. (a) How much work does the force do in 10 min? (b) What is the average power (in horsepower) of the force?

**58. Object Accelerates Horizontally** An initially stationary 2.0 kg object accelerates horizontally and uniformly to a speed of 10 m/s in 3.0 s. (a) In that 3.0 s interval, how much work is done on the object by the force accelerating it? What is the instantaneous power due to that force (b) at the end of the interval and (c) at the end of the first half of the interval?

**59. A Sprinter** A sprinter who weighs 670 N runs the first 7.0 m of a race in 1.6 s, starting from rest and accelerating uniformly. What are the sprinter's (a) speed and (b) kinetic energy at the end of the 1.6 s? (c) What average power does the sprinter generate during the 1.6 s interval?

**60. The *Queen Elizabeth 2*** The luxury liner *Queen Elizabeth 2* has a diesel-electric powerplant with a maximum power of 92 MW at a cruising speed of 32.5 knots. What forward force is exerted on the ship at this speed? (1 knot = 1.852 km/h.)

**61. Swimmer** A swimmer moves through the water at an average speed of 0.22 m/s. The average drag force opposing this motion is 110 N. What average power is required of the swimmer?

**62. Auto Starts from Rest** A 1500 kg automobile starts from rest on a horizontal road and gains a speed of 72 km/h in 30 s. (a) What is the kinetic energy of the auto at the end of the 30 s? (b) What is the average power required of the car during the 30 s interval? (c) What is the instantaneous power at the end of the 30 s interval, assuming that the acceleration is constant?

**63. A Locomotive** A locomotive with a power capability of 1.5 MW can accelerate a train from a speed of 10 m/s to 25 m/s in 6.0 min. (a) Calculate the mass of the train. Find (b) the speed of the train and (c) the force accelerating the train as functions of time (in seconds) during the 6.0 min interval. (d) Find the distance moved by the train during the interval.

# Additional Problems

**64. Estimate, Then Integrate** (a) Estimate the work done by the force represented by the graph of Fig. 9-38 in displacing a particle from $x_1 = 1$ m to $x_2 = 3$ m. (b) The curve is given by $F_x = a/x^2$, with $a = 9\ \text{N}\cdot\text{m}^2$. Calculate the work using integration.

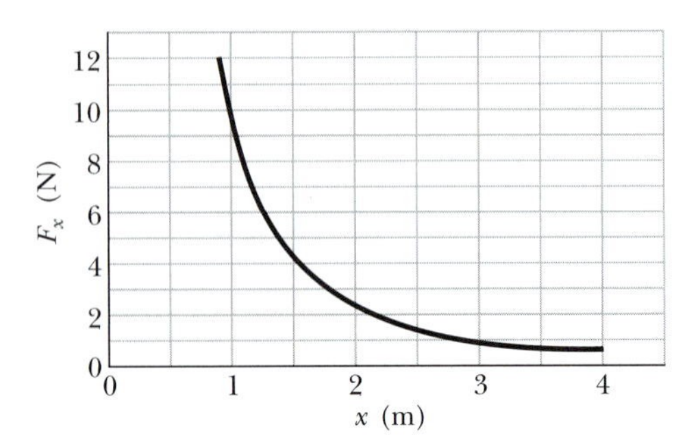


FIGURE 9-38 ■ Problem 64.

**65. Explosion at Ground Level** An explosion at ground level leaves a crater with a diameter that is proportional to the energy of the explosion raised to the $\frac{1}{3}$ power; an explosion of 1 megaton of TNT leaves a crater with a 1 km diameter. Below Lake Huron in Michigan there appears to be an ancient impact crater with a 50 km diameter. What was the kinetic energy associated with that impact, in terms of (a) megatons of TNT (1 megaton yields $4.2 \times 10^{15}$ J) and (b) Hiroshima bomb equivalents (13 kilotons of TNT each)? (Ancient meteorite or comet impacts may have significantly altered Earth's climate and contributed to the extinction of the dinosaurs and other life-forms.)

**66. Pushing a Block** A hand pushes a 3 kg block along a table from point $A$ to point $C$ as shown in Fig. 9-39. The table has been prepared so that the left half of the table (from $A$ to $B$) is frictionless. The right half (from $B$ to $C$) has a nonzero coefficient of friction equal to $\mu^{\text{kin}}$. The hand pushes the block from $A$ to $C$ using a constant force of 5 N. The block starts off at rest at point $A$ and comes to a stop when it reaches point $C$. The distance from $A$ to $B$ is $\frac{1}{2}$ meter and the distance from $B$ to $C$ is also $\frac{1}{2}$ meter.

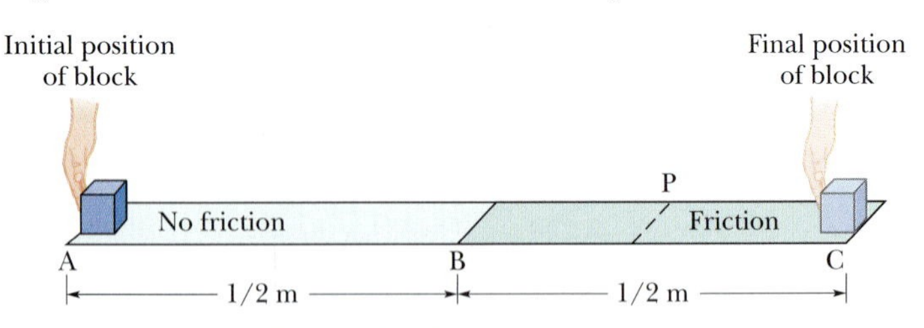


FIGURE 9-39 ■ Problem 66.

**(a)** Describe in words the motion of the block as it moves from $A$ to $C$.

**(b)** Draw a free-body diagram for the block when it is at point $P$.

**(c)** What is the direction of the acceleration of the block at point $P$? If it is 0, state that explicitly. Explain your reasoning.

**(d)** Does the magnitude of the acceleration increase, decrease, or remain the same as the block moves from $B$ to $C$? Explain your reasoning.

**(e)** What is the net work done on the object as it moves from $A$ to $B$? From $B$ to $C$?

**(f)** Calculate the coefficient of friction $\mu^{\text{kin}}$.

**67. Continental Drift** According to some recent highly accurate measurements made from satellites, the continent of North America is drifting at a rate of about 1 cm per year. Assuming a continent is about 50 km thick, estimate the kinetic energy the continental United States has as a result of this motion.

**68. Fan Carts p&E** Two fan carts labeled $A$ and $B$ are placed on opposite sides of a table with their fans pointed in the same direction as shown in Fig. 9-40. Cart $A$ is weighted with iron bars so it is twice as massive as cart $B$. When each fan is turned on, it provides the same constant force on the cart independent of its mass. Assume that friction is small enough to be neglected. The fans are set with a timer so that after they are switched on, *they stay on for a fixed length of time*, $\Delta t$, *and then turn off.*

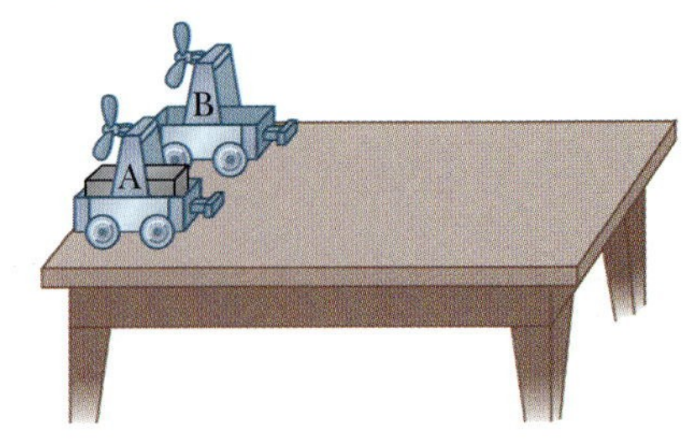


FIGURE 9-40 ■ Problem 68.

**(a)** *Just after the fans turn off*, which of the following statements is true about the magnitude of the momenta of the two carts?

**(i)** $p_A > p_B$
**(ii)** $p_A = p_B$
**(iii)** $p_A < p_B$

**(b)** *Just after the fans turn off*, which of the following statements is true about the kinetic energies of the two carts?

**(i)** $K_A > K_B$
**(ii)** $K_A = K_B$
**(iii)** $K_A < K_B$

**(c)** Which of the following statements are true? You may choose as many as you like, or none. If you choose none, write N.

**(i)** After the fans are turned on, each cart moves at a constant velocity, but the two velocities are different from each other.
**(ii)** The kinetic energy of each cart is conserved.
**(iii)** The momentum of each cart is conserved.

**69. Sticky Carts** Two identical carts labeled $A$ and $B$ are initially resting on a smooth track. The coordinate system is shown in Fig. 9-41$a$. The cart on the right, cart $B$, is given a push to the left and is released. The clock is then started. At $t_1 = 0$, cart $B$ moves in the direction shown with a speed $v_1$. The carts hit and stick to each other. The graphs in Fig. 9-41$b$ describe some of the variables associated with the motion as a function of time, but without labels on the vertical axis. For the experiment described and for each item in the list below, identify which graph (or graphs) is a possible display of that variable as a function of time, assuming a proper scale and units. "The system" refers to carts $A$ and $B$ together. Friction is so small that it can be ignored. If none apply, write N.

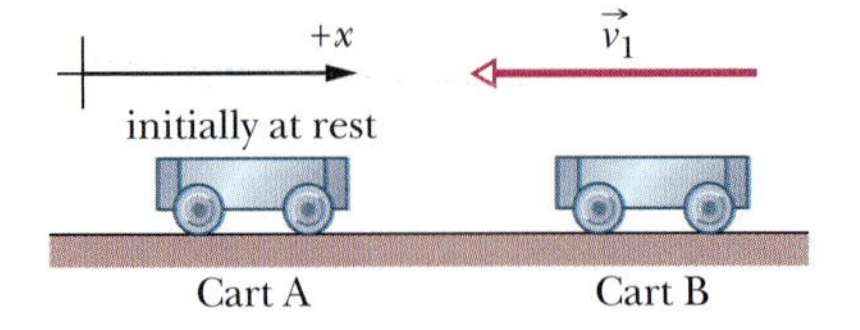


FIGURE 9-41$a$ ■ Problem 69.

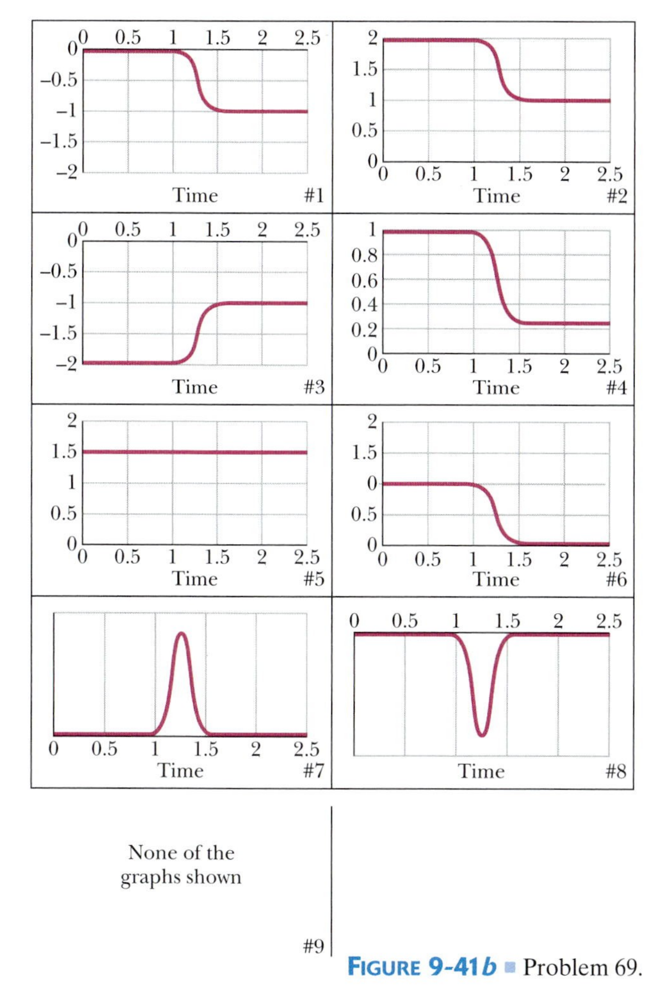


FIGURE 9-41*b* ■ Problem 69.

**(a)** The $x$-component of momentum of cart $B$
**(b)** The $x$-component of force on cart $A$
**(c)** The $x$-component of total momentum of the system
**(d)** The kinetic energy of cart $B$
**(e)** The total kinetic energy of the system

**70. Graphs and Carts** Two *identical* carts are riding on an air track. Cart $A$ is given a quick push in the positive $x$ direction toward cart $B$. When the carts hit, they stick to each other. The graphs shown in Fig. 9-42 describe some of the variables associated with the motion as a function of time beginning just *after* the push is completed. For the experiment described and for each item in the list below, identify which graph (or graphs) is a possible display of that variable as a function of time.

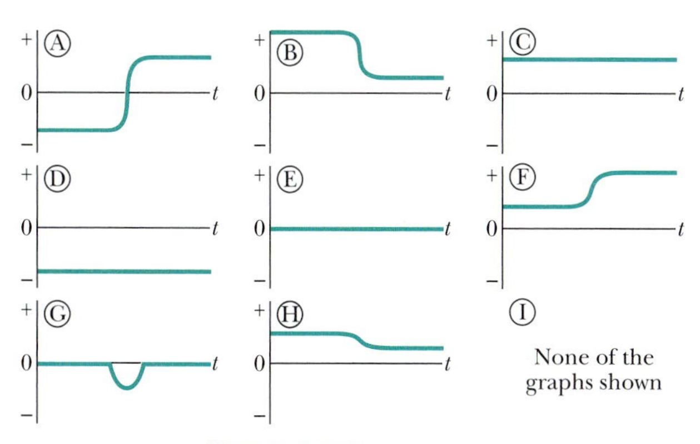


FIGURE 9-42 ■ Problem 70.

**(a)** The momentum of cart $A$
**(b)** The total momentum of the two carts
**(c)** The kinetic energy of cart $A$
**(d)** The force on cart $A$
**(e)** The force on cart $B$

**71. Rebound to the Left** A 5.0 kg block travels to the right on a rough, horizontal surface and collides with a spring. The speed of the block *just before* the collision is 3.0 m/s. The block continues to move to the right, compressing the spring to some maximum extent. The spring then forces the block to begin moving to the left. As the block rebounds to the left, it leaves the now uncompressed spring at 2.2 m/s. If the coefficient of kinetic friction between the block and surface is 0.30, determine (a) the work done by friction while the block is in contact with the spring and (b) the maximum distance the spring is compressed.

**72. Rescue** A helicopter lifts a stretcher with a 74 kg accident victim in it out of a canyon by applying a vertical force on the stretcher. The stretcher is attached to a guide rope, which is 50 meters long and makes an angle of 37° with respect to the horizontal. See Fig. 9-43. What is the work done by the helicopter on the injured person and stretcher?

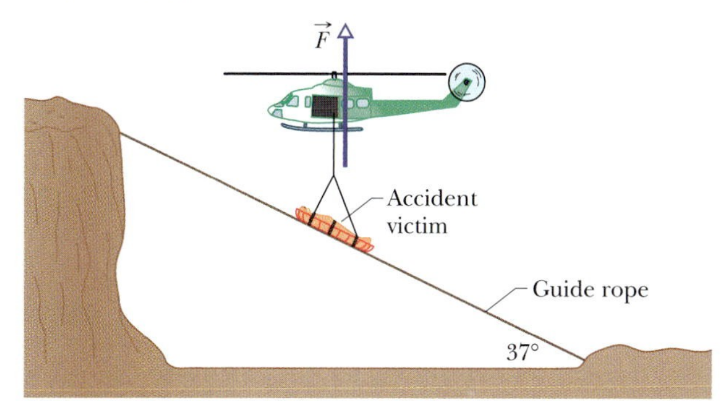


FIGURE 9-43 ■ Problem 72.

**73. A Spring** Idealized data for a spring's displacement $\Delta\vec{x}$ from its equilibrium position as a function of an external force, $\vec{F}^{\text{ext}}$, are shown in Fig. 9-44.

| $\vec{F}^{\text{ext}}$ [N] | $\Delta\vec{x}$ [cm] |
|---|---|
| 0.0 | 0 |
| 1.0 | 5 |
| 2.0 | 10 |
| 3.0 | 15 |
| 4.0 | 20 |

FIGURE 9-44 ■ Problem 73.

**(a)** Draw a properly scaled and carefully labeled graph of $\vec{F}^{\text{ext}}$ vs. $\Delta\vec{x}$ for these data.
**(b)** Does this spring obey Hooke's law? Why or why not?
**(c)** What is the value of its spring constant $k$?
**(d)** Shade the area on your graph that represents the amount of work done in stretching the spring from a displacement or extension of 0 cm to one of 5 cm. Also shade the area on the graph that represents the amount of work done in stretching the spring from a displacement or extension of 15 cm to one of 20 cm. Are the shaded areas approximately the same size? What does the size of the shaded area indicate about the work done in these two cases?
**(e)** Explain why the amount of work done in the second case is different from the amount done in the first case, even though the change in length of the spring is the same in both cases.

**74. Variable Force** The center of mass of a cart having a mass of 0.62 kg starts with a velocity of −2.5 m/s along an $x$ axis. The cart starts from a position of 5.0 m and moves without any noticeable friction acting on it to a position of 0.0 meters. During this motion, a fan assembly exerts a force on the cart in a positive $x$ direction.

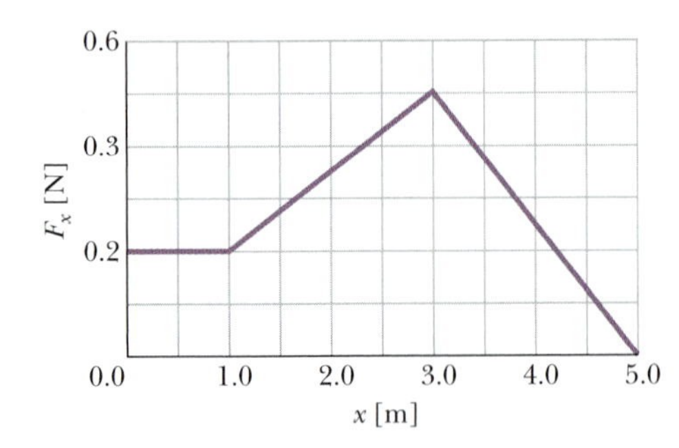


FIGURE 9-45 ■ Problem 74.

However, instead of being powered by batteries, the fan is driven by a voltage source that is programmed to change with its distance from a motion detector. This program leads to a variable force as shown in Fig. 9-45.

(a) What is the work done on the cart by the fan as the cart moves from 5.0 m to 0.0 m? (b) What is the change in kinetic energy of the cart between 5.0 m and 0.0 m? (c) What is the final velocity of the cart when it is at 0.0 m?

**75. Given a Shove** An ice skater of mass $m$ is given a shove on a frozen pond. After the shove she has a speed of $v_1 = 2$ m/s. Assume that the only horizontal force that acts on her is a slight frictional force between the blades of the skates and the ice.

**(a)** Draw a free-body diagram showing the horizontal force and the two vertical forces that act on the skater. Identify these forces. **(b)** Use the net work-kinetic energy theorem to find the distance the skater moves before coming to rest. Assume that the coefficient of kinetic friction between the blades of the skates and the ice is $\mu^{\text{kin}} = 0.12$.

**76. Karate Board Tester** The karate board tester shown in Fig. 9-46$a$ is a destructive testing device that allows one to determine the deformation of the center of a pine karate board as a function of the forces applied to it.
The displacement component, $\Delta x$, of the center of a pine board from its equilibrium position increases as a function of the $x$-component of an external force, $F_x^{\text{ext}}$, applied to it as shown in the data table of Fig. 9-46$b$.

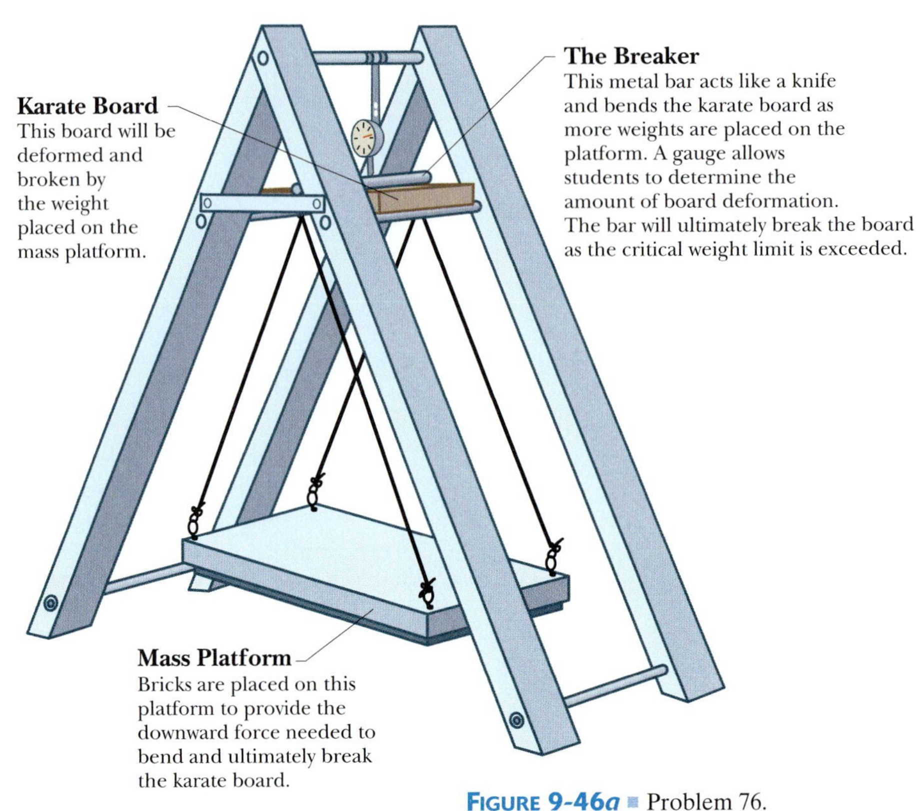

Karate Board
This board will be deformed and broken by the weight placed on the mass platform.
The Breaker
This metal bar acts like a knife and bends the karate board as more weights are placed on the platform. A gauge allows students to determine the amount of board deformation.
The bar will ultimately break the board as the critical weight limit is exceeded.
Mass Platform
Bricks are placed on this platform to provide the downward force needed to bend and ultimately break the karate board.


FIGURE 9-46$a$ ■ Problem 76.

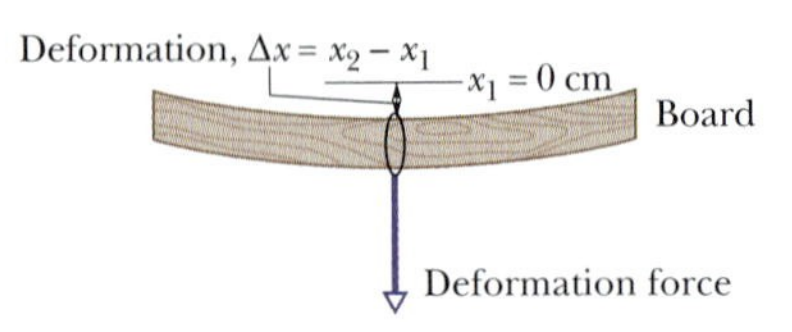


| $F_x^{\text{ext}}$[N] | $\Delta x$[cm] |
|---|---|
| 0 | 0.000 |
| 167 | 0.156 |
| 326 | 0.312 |
| 461 | 0.446 |
| 567 | 0.602 |

FIGURE 9-46$b$ ■ Problem 76.

**(a)** Draw a properly scaled and carefully labeled graph of $F^{\text{ext}}$ vs. $\Delta x$ for these data.
**(b)** Does this pine board obey Hooke's law? Why or why not?
**(c)** What is the value of the effective spring constant $k$ for the board?
**(d)** Shade the area on the graph that represents the amount of work done in stretching the center of the board from a displacement or extension of 0.000 cm to one of 0.156 cm. Also shade the area on the graph that represents the amount of work done in stretching the board from a displacement or extension of 0.446 cm to one of 0.602 cm. Are the shaded areas approximately the same size? What does the size of the shaded area indicate about the work done in these two cases?
**(e)** Explain why the amount of work done in the second case is different from the amount done in the first case, even though the change in displacement of the board is the same in both cases.

**77. Karate Chop Movie** In the movie DSON012 (available in VideoPoint or from your instructor) a physics student breaks a stack of eight pine boards. The thickness of the stack of boards with spacers is 0.34 m. In answering the following questions, treat any work done by gravitational forces on the student's hand as negligible.

**(a)** Use video analysis software to analyze the motion of the student's hand in the vertical or $y$ direction. By using data from frames 3–5, find the velocity of the student's hand just before he hits the boards. By using data from frames 7–9, find the velocity of the student's hand just after he breaks all the boards.
**(b)** Assume that the effective mass of the student's hand is 1.0 kg. Use the net work-kinetic energy theorem to find the work done on the student's hand by the boards.

# 10 Potential Energy and Energy Conservation

The prehistoric people of Easter Island carved hundreds of giant stone statues in their quarry, then moved them to sites all over the island. How they managed to move them by as much as 10 km without the use of sophisticated machines has been a hotly debated subject, with many fanciful theories about the source of the required energy.

**How could this have been accomplished using only primitive means?**

*The answer is in this chapter.*

**FIGURE 10-1** ■ While lifting a massive barbell, a powerlifter increases the separation between the barbell and Earth and rearranges the Earth–barbell system.

## 10-1 Introduction

In Chapter 9 we introduced the concepts of work and kinetic energy. We then derived a net work-kinetic energy theorem to describe what happens to the kinetic energy of a single rigid object when work is done on it. In this chapter we will consider systems composed of several objects that interact with one another. We are interested in what happens when forces from objects outside the system (*external* forces) change the arrangement of the interacting parts.

Let's consider two systems that can be reconfigured by external forces. The first system consists of an Earth–barbell system that has its arrangement changed when a weight lifter (outside of the system) pulls the barbell and the Earth apart by pulling up on the barbell with his arms and pushing down on the Earth with his feet (Fig. 10-1). The second system consists of two crates and a floor. This system is rearranged by a person (again, outside the system) who pushes the crates apart by pushing on one crate with her back and the other with her feet (Fig. 10-2). Although the external forces changing each system's configuration are exerted by a person pushing in opposite directions on two objects, there is an obvious difference between these two situations. Namely, as soon as the weight lifter stops pushing in both directions, the system's parts (barbell and Earth) fall back together. When the person stops pushing in opposite directions on the crates, the crates do not snap back together.

The internal interaction forces between the parts of these two systems differ. (Recall that we can call forces between objects within a system *internal forces*.) The lifter's forces are opposed by gravitational forces, but the crate-separator's forces are opposed by sliding friction forces. The lifter has to do a considerable amount of work to raise the barbell. However, when the barbell is dropped, we know that it picks up speed and gains kinetic energy as the Earth and the barbell move toward each other. In what sense can we say that the work the weight lifter did has been stored in the new configuration of the Earth–barbell system? And why does the work done by the woman separating the crates seem to be lost rather than stored away?

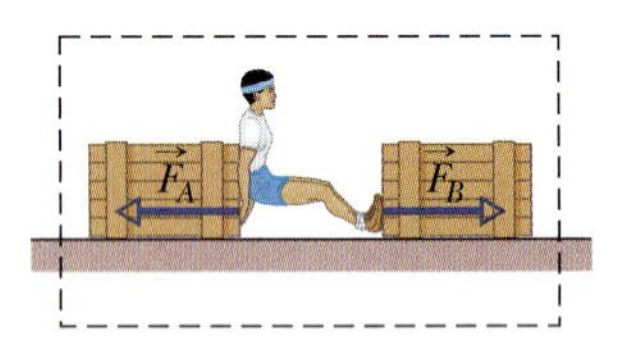


**FIGURE 10-2** ■ A woman exerts equal and opposite forces on two crates. The work she does on the crates causes the crate–floor system to be rearranged.

## 10-2 Work and Path Dependence

There are many types of internal forces that can do work on a system of interacting objects. Examples include gravitational forces, sliding friction forces, spring forces, and air drag forces. How can we tell whether the work done by a certain type of internal force is "stored" or "used up" when the arrangement of a system changes?

A test has been devised for determining whether the work done by a particular type of force is "stored" or "used up." This test involves considering the work done by an internal interaction force when one part of a system moves. Consider a preliminary description of this test, which we will refine later:

> **TEST OF A SYSTEM'S ABILITY TO "STORE" WORK DONE BY INTERNAL FORCES (PRELIMINARY STATEMENT):** If the work done by a force between two objects within a system as some object in the system moves does not depend on the path taken, then the work done by this (internal) force can be stored in the system.

This test doesn't seem so strange when we apply it to some simple situations we have already discussed involving gravitational and friction forces.

### The Path Independence Test for a Gravitational Force

Consider the skier traveling down a curved frictionless ramp as shown in Fig. 10-3. We showed in Section 9-9 that the net work done on the skier as she travels down the ramp is given by

$$W^{\text{net}} = -mg(y_2 - y_1) = F_y^{\text{grav}}\Delta y \qquad \text{(Eq. 9-42)}$$

and so does not depend on the shape of the ramp but only on the vertical component of the gravitational force and the vertical displacement of her center of mass. This result derives from that fact that whenever the skier has a component of motion in a horizontal direction the horizontal displacement is perpendicular to the Earth's gravitational force. These horizontal "detours" do not contribute to the work done by the Earth's gravitational force on the skier's center of mass.

Thus the work done on a particle by a gravitational force seems to be independent of the path taken to get from $y_1$ to $y_2$ as shown for the skier in Fig. 10-3. The gravitational force passes the test! If we do work on the system of skier and Earth to raise the skier to the top of the ramp, she can fall down again gaining kinetic energy, just as the barbell a weight lifter raises can fall down again. In both cases, we have to overcome the opposing internal gravitational force. In both cases, our work seems to be stored within the system.

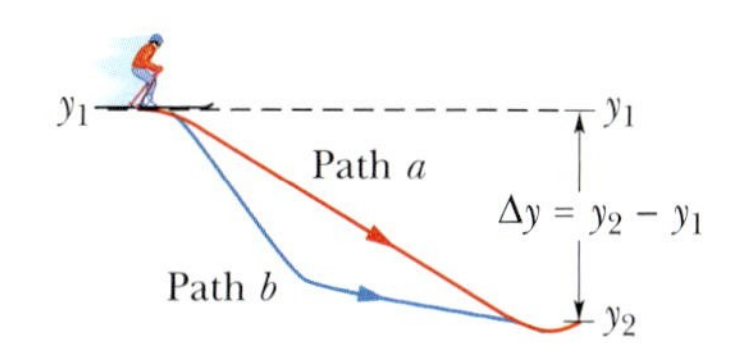


**FIGURE 10-3** ■ The gravitational work done on a skier descending on a frictionless ramp depends only on the gravitational force and her vertical displacement $\Delta y$ and not on the shape of the ramp.

## Path Dependence of Work Done by a Friction Force

Our consideration of the skier indicates that the gravitational force does work that is path independent. But what about an object such as a crate that is displaced in the presence of a friction force? Is the work that the opposing friction force does on the crate path independent?

Consider pushing one of the crates shown in Fig. 10-2 along a level floor. Suppose the surface of the floor is quite uniform so that when you push with a constant magnitude of force, the crate moves at a constant speed. According to Newton's Second Law, if the acceleration of the crate is zero the net force on it is zero. So the external force you apply and the internal friction force must be equal and opposite. In this special case the friction force is steady. It always acts in a direction that opposes the displacement. If you push the crate directly from point 1 to point 2, you are taking it along path $a$ as shown in Fig. 10-4. The work done by friction along that path is always negative (since the force and displacement are in opposite directions) and is given by

$$W^{\text{fric}}_{1\to 2} = -|f^{\text{kin}}|d,$$

where $d$ is the distance between points 1 and 2. Suppose instead we push the crate along path $b$ from points 1 to 4, then points 4 to 3 and then points 3 to 2, where the distance on each leg of the path is also $d$. The work done by the friction force is still negative and is given by

$$W^{\text{fric}}_{1\to 4\to 3\to 2} = (-|f^{\text{kin}}|d) + (-|f^{\text{kin}}|d) + (-|f^{\text{kin}}|d) = -3|f^{\text{kin}}|d.$$

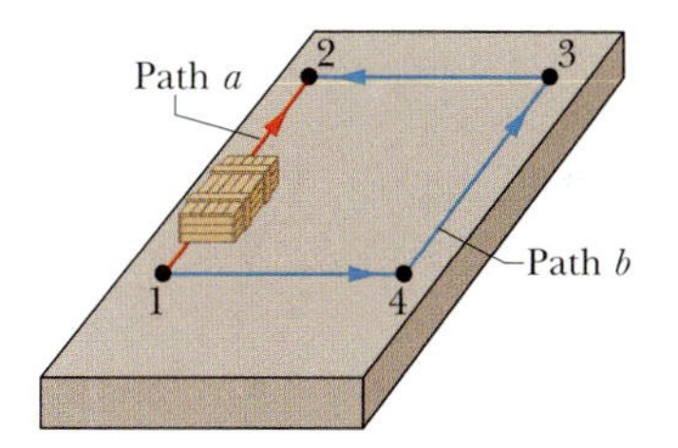


**FIGURE 10-4** ■ The work done on a crate by the friction forces acting on it is much greater when it is pushed from point 1 to 2 along path $b$ than along path $a$.

We see that the kinetic friction force does not pass the path independence test! The negative work done on the crate by the friction force is three times greater for path $b$ than for path $a$. In general, if a friction force of constant magnitude is the only internal force in a system, then the work needed to get from point 1 to point 2 is proportional to the length of the path taken. Thus kinetic friction is a path-dependent force. This suggests that path dependence is an indicator of whether or not external work done on a system can be stored. When you do external work on the crate that is part of a crate–floor system, the system cannot then use the external work you do on it to rearrange itself after you stop pushing.

## Conservative Forces and Path Independence

So far we have seen that the work done on a system that has gravitational forces acting between its parts seems to be path independent and seems to "store external work." Alternatively, if friction forces act between system parts, the opposite seems to hold. The system cannot store external work, and the friction forces do work that is not path independent. It is customary to define gravitational and other forces that do path independent work as **conservative forces** and forces that do not as **nonconservative forces.** The term "conservative" implies that something related to work is "stored" or conserved when the parts of a system are rearranged. In the next few sections of this chapter we will explore the concept of "conserved" in more detail.

## General Statements about Conservative Forces

There is an alternative way to apply the path independence test to the work done by a force. It has to do with the net work associated with motion along a closed path—that is, motion in which an object makes a round trip through space, returning to its original location. Figure 10-5*b* shows an arbitrary round trip for a particle that has work done on it by a single force during its trip. The particle moves from an initial point 1 to point 2 along path *a* and then back to point 1 along path *b*. The internal force does work on the particle as the particle moves along each path. Without worrying about where positive work is done and where negative work is done, let us just represent the work between points 1 and 2 as the particle moves along path *a* as $W_{1\rightarrow 2}^{\text{path } a}$. Then we can denote the work done between points 1 and 2 if the particle moves along path *b* as $W_{1\rightarrow 2}^{\text{path } b}$ (Fig 10-5*a*). If the force is conservative, then the net work done is the same for either path,

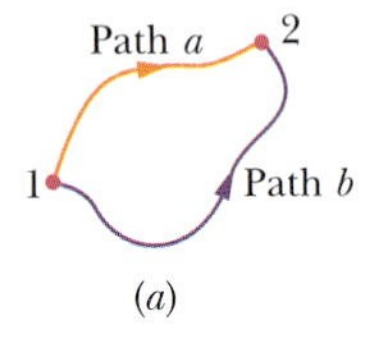


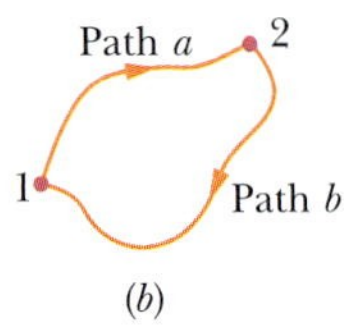


**FIGURE 10-5** ■ When a particle is acted on by a conservative force, the work done by the force is: (*a*) *independent* of whether the particle moves from point 1 to point 2 by following either path *a* or path *b*; or (*b*) *zero* if the particle makes any possible round trip from 1 back to point 1. One possible round trip includes moving to point 2 along path *a* and then back to point 1 along path *b*.

$$W_{1\rightarrow 2}^{\text{path } a} = W_{1\rightarrow 2}^{\text{path } b} \quad \text{so that} \quad W_{1\rightarrow 2}^{\text{path } a} - W_{1\rightarrow 2}^{\text{path } b} = 0.$$

However, if we move in the opposite direction and go along path *b* from point 2 to point 1, then all the increments of displacement change sign, and work done in one direction is the negative of work done in the other direction. This is given by

$$W_{1\rightarrow 2}^{\text{path } b} = -W_{2\rightarrow 1}^{\text{path } b}.$$

Thus we get the following expression for the work done on a particle as it makes a round trip along a closed path traveling from point 1 to point 2 along path *a* and then back from point 2 to point 1 along path *b*,

$$W_{1\rightarrow 2}^{\text{path } a} + W_{2\rightarrow 1}^{\text{path } b} = 0 \qquad \text{(conservative force only).} \qquad (10\text{-}1)$$

This equation tells us that the work done by a conservative force along any closed path is zero.

> **CONSERVATIVE FORCE TEST:** The work done by a conservative force on a particle moving between two points does not depend on the path taken by the particle. An alternative statement of this test is that the net work done by a conservative force on a particle moving around any closed path is zero.

The path independence of conservative forces has another useful aspect. If you need to calculate the work done by a conservative force along a given path between two points and the calculation is difficult, you can find the work by using another path between those two points for which the calculation is easier.

## The Conservative Force Test for a Spring Force

So far the only systems we have introduced that have conservative internal forces are those in which the gravitational force acts alone. Let's consider another system consisting of a wall, a rigid block, and an ideal table that does not exert friction forces on the block. We assume that the wall and the block interact because they are connected by an ideal spring (see Fig. 10-6). The end of the spring that is free to move exerts a force on the block in a direction opposite to the displacement from the spring's relaxed position. According to Hooke's law, the component of the spring force on the block is given by $F_x^{spring}(x) = -kx$, where $x$ is the displacement from a relaxed state at $x = 0$ (Eq. 9-17). We used this force and the definition of work to show that the work done on the block by the spring force is

$$W^{spring} = +\tfrac{1}{2}kx_1^2 - \tfrac{1}{2}kx_2^2, \qquad \text{(Eq. 9-19)}$$

whenever the spring is stretched or compressed from position $x_1$ to position $x_2$, along one-dimensional paths.

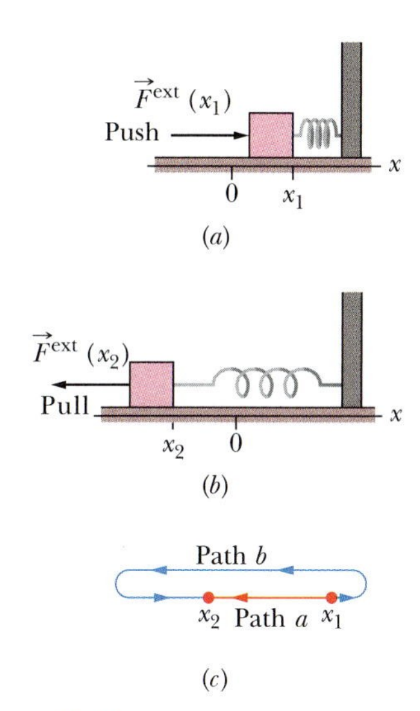


**FIGURE 10-6** ■ A block is attached to a spring that is anchored to a wall. (*a*) An external force is used to push the block so the position component of the end of the spring is $x_1$. (*b*) Then, the block is pulled out so the position component of the end of the spring is $x_2$. (*c*) Two of many possible paths the spring end can take to get from $x_1$ to $x_2$.

Before discussing the application of the conservative force test, we need to point out that a single spring exerts a force that is inherently one-dimensional. So thinking about paths that the block might take under the influence of net external forces does not make sense unless we restrict ourselves to situations for which the net external force acts along the line of the spring. For the sake of discussion we choose the line of the spring to be the $x$ axis.

Let's apply the test that says that if the spring force is conservative, then the work done along any one-dimensional path is the same. Figure 10-6 shows a block that is pushed inward and then pulled outward by an external force. Eq. 9-19 indicates the work done by the spring. The equation describing the spring's work depends only on the two locations $x_1$ and $x_2$ and not on how the spring got from one location to the other. For example, you could start the spring end at location $x_1$ and push it in further, then pull it out past $x_2$ and finally back to $x_2$. The work will be the same no matter what one-dimensional path you take—that is, as long as you don't impose very large displacements on the spring that cause Hooke's law to break down.

Since the work done by ideal spring forces is path independent, we can conclude that

> The ideal spring force is a conservative force, as is the gravitational force.

The alternate test that requires the work done by the spring force to be zero on a round trip is also true, since for a round trip $x_2 = x_1$ so $W^{spring} = \frac{1}{2}kx_1^2 - \frac{1}{2}kx_2^2 = 0$. The fact that the spring force is zero in a round trip makes sense. Suppose the spring starts out in a compressed position. When you push on a spring and compress it further, the spring force opposes its displacement and the work done by the spring is negative. If you then pull the spring back to its original but still compressed position, the spring force and the displacement are in the same direction and the work done by the spring is positive. So the negative work done by the spring while it is being pushed in and the positive work it does while being pulled out add up to zero.

When a spring attached to a wall is stretched and released it naturally heads toward its equilibrium position. This is not unlike the weight lifter's mass naturally falling back toward the Earth. The external work done on the block in opposition to the spring force seems to be stored in the wall–spring–block system.

## The Conservative Force Test for a Car on a Hot Wheels® Track

Let's see how our two conservative force tests are applied to a fairly complex system consisting of a low-friction toy car, a Hot Wheels® track, and the Earth. Are the internal forces the system exerts on the car conservative?

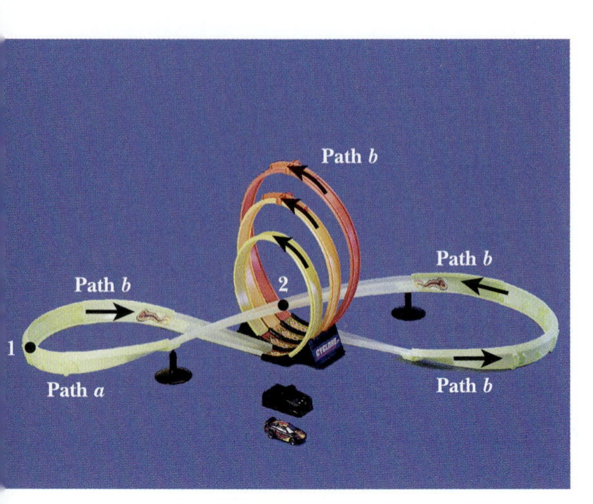


**FIGURE 10-7** ■ This Hot Wheels® race track allows toy cars to travel around a closed path. The point labeled 2 is an overpass.

**Using the Path Independence Test** There are two possible paths that a "low-friction" toy car could take traveling between points 1 and 2 on the track shown in Fig. 10-7. The car could travel uphill directly from point 1 to point 2 along path $a$. Alternatively it could travel many times further by taking path $b$. On that path it proceeds downhill, passes under the top ramp, goes through three loop-the-loops, and traverses the figure-eight loop on the left before returning to point 2.

We can use the net work-kinetic energy theorem ($W^{\text{net}} = \Delta K$) developed in Chapter 9 to determine whether or not the internal forces do work on the car along certain paths. The net work is given by the sum of the works done by the internal forces that the rest of the system exerts on the car. In equation form this is $W^{\text{net}} = W^{\text{int}} = W^{\text{norm}} + W^{\text{grav}} + W^{\text{fric}}$.

The normal forces do not contribute to the work done on the car. This is because these forces are always perpendicular to the direction of the car's displacement at any point along the track. So the internal work is done by only a combination of the gravitational and friction forces so that $W^{\text{net}} = W^{\text{grav}} + W^{\text{fric}}$.

If we start the car along path $a$ at point 1 with a certain amount of kinetic energy, then we can measure its kinetic energy at point 2 to determine the net internal work done on it by the rest of the system. We can make the same observation for the car as it goes from point 1 to point 2 along path $b$. If the kinetic energy change was the same along both paths, then the net work done along each path would be the same and we would conclude that the combination of the gravitational and friction forces is conservative. However, measurements tell us that a different amount of kinetic energy is lost along path $b$ than along the more direct path $a$. We conclude that the combination of friction and gravitational forces acting on the car is not conservative. Since we believe that gravitational forces are conservative, we suspect that friction is the problem.

**Closed-Path Test** It turns out that the closed-path test is a lot easier to apply in this case. All we have to do is ask the question: Is the net work done on the car in going around a closed loop (say, from point 1 to point 1) zero? If the answer is yes, then according to the net work-kinetic energy theorem given by $W^{\text{net}} = W^{\text{grav}} + W^{\text{fric}} = \Delta K = 0$, the car would lose no kinetic energy in making a complete loop. However, for this Hot Wheels® track we observe that the car does slow down, so there must be a loss in kinetic energy. As expected, therefore, the combination of gravitational and friction forces does not pass this logically equivalent conservative force test.

What if friction were not present? If we could devise a magic car with no friction in its wheel bearings, then the only type of force capable of doing work on the car as it traveled would be the conservative gravitational force on the car due to the Earth. In this case, the net work done on the car would be zero around a closed loop, and the car would lose no kinetic energy.

**READING EXERCISE 10-1:** The figure shows three paths connecting points 1 and 2. A single force $\vec{F}$ does the indicated work on a particle moving along each path in the indicated direction. On the basis of this information, is force $\vec{F}$ conservative?

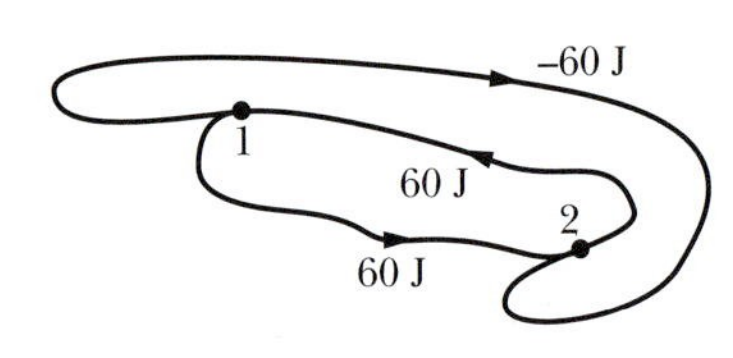


■

**READING EXERCISE 10-2:** In applying the path independence test for conservative forces to the car traveling on the Hot Wheels® track, we made the statement: "Measurements tell us that a different amount of kinetic energy is lost along path $b$ than along the more direct path $a$." Which path do you think will have the most kinetic energy loss associated with it? Explain the reasons for your answer. ■

### TOUCHSTONE EXAMPLE 10-1: Cheese on a Track

Figure 10-8*a* shows a 2.0 kg block of slippery cheese that slides along a frictionless track from point 1 to point 2. The cheese travels through a total distance of 2.0 m along the track, and a net vertical distance of 0.80 m. How much work is done on the cheese by the gravitational force during the slide?

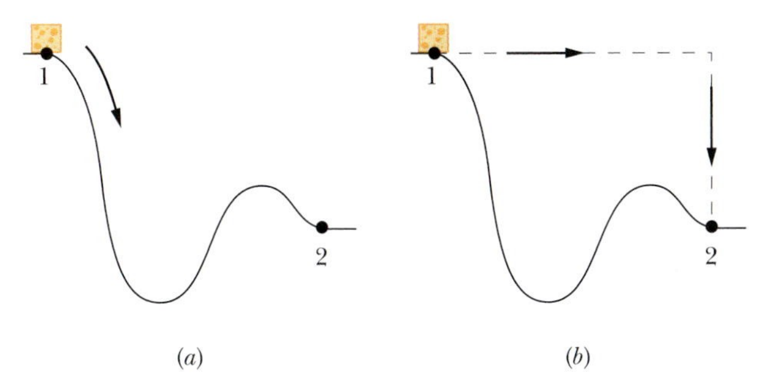


**FIGURE 10-8** ■ (*a*) A block of cheese slides along a frictionless track from point 1 to point 2. (*b*) Finding the work done on the cheese by the gravitational force is easier along the dashed path than along the actual path taken by the cheese; the result is the same for both paths.

**SOLUTION** ■ A **Key Idea** here is that we *cannot* use Eq. 9-26 ($W^{grav} = |\vec{F}^{grav}||\Delta\vec{r}|\cos\phi$) to calculate the work done by the gravitational force $\vec{F}^{grav}$ as the cheese moves along the track. The reason is that the angle $\phi$ between the directions of $\vec{F}^{grav}$ and the displacement $\Delta\vec{r}$ varies along the track in an unknown way. (Even if we did know the shape of the track and could calculate $\phi$ along it, the calculation could be very difficult.)

A second **Key Idea** is that because $\vec{F}^{grav}$ is a conservative force, we can find the work by choosing some other path between 1 and 2—one that makes the calculation easy. Let us choose the dashed path in Fig. 10-8*b*; it consists of two straight segments. Along the horizontal segment, the angle $\phi$ is a constant 90°. Even though we do not know the displacement along that horizontal segment, Eq. 9-26 tells us that the work $W^{horiz}$ done there is

$$W^{horiz} = mg|\Delta\vec{r}|\cos 90^\circ = 0.$$

Along the vertical segment, the magnitude of the displacement $|\Delta\vec{r}|$ is 0.80 m and, with $\vec{F}^{grav}$ and $\Delta\vec{r}$ both downward, the angle $\phi$ is a constant 0°. Thus, Eq. 9-26 gives us, for the work $W^{vert}$ done along the vertical part of the dashed path,

$$\begin{aligned} W^{vert} &= mg|\Delta\vec{r}|\cos 0^\circ \\ &= (2.0\text{ kg})(9.8\text{ m/s}^2)(0.80\text{ m})(1) = 15.7\text{ J}. \end{aligned}$$

The total work done on the cheese by $\vec{F}^{grav}$ as the cheese moves from point *a* to point *b* along the dashed path is then

$$W = W^{horiz} + W^{vert} = 0 + 15.7\text{ J} \approx 16\text{ J}. \quad \text{(Answer)}$$

This is also the work done as the cheese moves along the track from 1 to 2.

## 10-3 Potential Energy as "Stored Work"

If external work can be stored when a system of objects is rearranged, we refer to the system as a "conservative system." In this section we will define a new quantity called potential energy as a measure of stored work in a conservative system.

### Rearranging a Gravitational System

Consider the external work that weightlifter Sun Ruiping does separating the Earth and the 118.5 kg barbell shown in Fig 10-9. Ruiping acts as an external agent that does work on *both* the Earth and the barbell in opposing the gravitational force as she pushes up on the barbell with her hands and down on the Earth with her feet. The *net* work done on the Earth–barbell system during the time the barbell is raised is the sum of the external work, $W_{sys}^{ext}$, done on the system by our weightlifter and the internal gravitational work, $W_{sys}^{int} = W_{sys}^{grav}$, that the two objects in the system exert on each other. This can be summarized by the equation

$$W_{sys}^{net} = W_{sys}^{ext} + W_{sys}^{int} \quad \text{(net work on a system).} \tag{10-2}$$

Since the velocities of both the barbell and the Earth are zero before and after the lift, there is no change in the kinetic energy of the Earth–barbell system as a

(a)

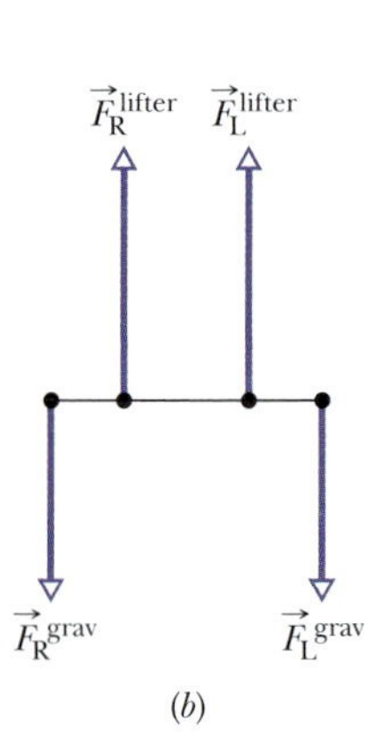


(b)

**FIGURE 10-9** ■ (*a*) China's Sun Ruiping broke the world record for the snatch lift in October 2002. In the middle part of her lift the barbell has an upward acceleration. Thus the positive force she exerted on the barbell is greater than the negative force exerted by the Earth. The lifter does positive external work on the Earth–barbell system while the system does negative internal work on itself. (*b*) A modified free-body diagram of forces on the barbell. The net force of the lifter exceeds the net gravitational force.

**FIGURE 10-10** ■ Results of video analysis of the lifting of a 372.5 kg barbell by about a half-meter. (*a*) *y* vs. *t* of the barbell during the lift. This graph was fitted with a polynomial to allow determination of the acceleration and hence the variation over time of net force on the barbell during the lift. (*b*) A graph of net force on the barbell during the lift vs. the *y*-component of the height shows that the area under the $F_y^{\text{net}}$ vs *y* curve that defines the net work on the system is zero.

result of the lift. So, the net work-kinetic energy theorem developed in Chapter 9 tells us that

$$W_{\text{barbell}}^{\text{net}} = W_{\text{barbell}}^{\text{ext}} + W_{\text{barbell}}^{\text{grav}} = \Delta K_{\text{barbell}} = 0 \quad \text{(net work on the system).} \qquad (10\text{-}3)$$

Since the net work on the system is zero,

$$W_{\text{barbell}}^{\text{ext}} = -W_{\text{barbell}}^{\text{grav}}. \qquad (10\text{-}4)$$

As shown in Fig 10-10, an analysis of another lift, the work the lifter has to do to separate the barbell from the Earth is experimentally confirmed to be equal in magnitude to the gravitational work done by the Earth on the barbell.

In general, when an object is lifted near the surface of the Earth we often carelessly think that the work done on the lifted object is different from the work done on the system. However, as long as we calculate the work using the change in separation of the Earth and object, there is no difference.

## Defining Potential Energy Change

Alas, our weightlifter's labor did not lead to a change in kinetic energy! However, suppose the lifter dropped her barbell. The barbell would *gain* an amount of kinetic energy while falling that is just equal to the work the lifter had to do on the system to raise it. We call this increased potential for kinetic energy gain a *potential energy* change, $\Delta U$. Basically this *change* in potential energy is "stored work." The term "change" is used to allow for the possibility that the system already had some potential energy stored in it before the external work was done.

However, according to $W_{\text{sys}}^{\text{net}} = W_{\text{sys}}^{\text{ext}} + W_{\text{sys}}^{\text{int}}$ (Eq. 10-2) when the kinetic energy of the system does not change, the external work is equal to the *negative* of the internal work done by interaction forces. This leads us to a general definition of potential energy change for a conservative system (one with only conservative internal forces) in terms of the work done by internal forces on parts of the system.

> **POTENTIAL ENERGY CHANGE FOR A CONSERVATIVE SYSTEM** is defined as the negative of the internal work the system does on itself when it undergoes a reconfiguration.

Even though we used an example of a two-object system to motivate this definition, we can also apply it to many-body systems. This is discussed in more detail in Chapter 25, where we deal with the potential energy associated with electrostatic forces. Symbolically, the general definition of potential energy change for a single conservative force is

$$\Delta U \equiv -W^{\text{cons}} \quad \text{(definition of potential energy change).} \qquad (10\text{-}5)$$

Here $W^{cons}$ is the work done by a specific conservative force and $\Delta U$ is the change in potential energy associated with that force.

**Gravitational Potential Energy** When we are near the surface of the Earth, we can use the expression $W^{grav} = -mg(y_2 - y_1)$, (Eq. 9-14), to derive an expression for the change in gravitational potential energy ($\Delta U^{grav}$) of an object that is lifted from one height $y_1$ to another height $y_2$. $W^{grav}$ represents the internal work done by the system, and

$$\Delta U^{grav} = -W^{grav} = +mg(y_2 - y_1) \quad \text{(gravitational PE change near the Earth's surface).} \quad (10\text{-}6)$$

Only *changes* $\Delta U$ in gravitational potential energy (or any other type of potential energy) are physically meaningful. In an object–Earth system, there is no special separation between the center of the Earth and an object that obviously has zero potential energy. However, to simplify a calculation or a discussion, we often choose to set the gravitational potential energy value $U^{grav}$ to zero when the object is at a certain height. To do so, we rewrite Eq. 10-6 as

$$\Delta U^{grav} = U_2^{grav} - U_1^{grav} = mg(y_2 - y_1). \quad (10\text{-}7)$$

Then we take $U_1$ to be the gravitational potential energy (GPE) of the system when it is in a **reference configuration** (in which the object is at a **reference point** $y_1$). Usually we take the reference point to be $y_1 = 0$ so $U_1^{grav} = U^{grav}(y_1) = 0$. If we do this, and replace the specific point $y_2$ with the more general $y$, Eq. 10-7 becomes

$$U^{grav}(y) = mgy \quad \text{(GPE } \textit{relative} \text{ to a chosen origin).} \quad (10\text{-}8)$$

This equation tells us that:

> Near the Earth's surface, the gravitational potential energy associated with an object–Earth system depends only on the vertical position $y$ (or height) of the object relative to the reference height $y_1 = 0$, not on its horizontal location.

**Elastic (or Spring) Potential Energy** The same definition of change in potential energy (in Eq. 10-5) applies equally well to a block–spring–wall system like that shown in Fig. 10-6. So $\Delta U^{spring} = -W^{spring}$; that is,

$$W^{spring} = +\tfrac{1}{2}kx_1^2 - \tfrac{1}{2}kx_2^2, \quad \text{(Eq. 9-19)}$$

and so

$$\Delta U^{spring} = -W^{spring} = \tfrac{1}{2}kx_2^2 - \tfrac{1}{2}kx_1^2 \quad \text{(ideal spring PE change).} \quad (10\text{-}9)$$

A spring–block system has a natural zero point for potential energy when the spring is unstretched. So, to associate an elastic potential energy (EPE) value $U^{spring}$ with the block at position $x_2$, we choose the reference point to be the block's location when the spring is at its relaxed length. If we let $x_1 = 0$ at that point, then the elastic potential energy $U^{spring}$ is 0 there, and Eq. 10-9 becomes

$$\Delta U^{spring} = U_2^{spring} - U_1^{spring} = \tfrac{1}{2}kx_2^2 - 0,$$

which gives us the general expression

$$U^{spring}(x) = \tfrac{1}{2}kx^2 \quad \text{(ideal EPE relative to block location with spring relaxed).} \quad (10\text{-}10)$$

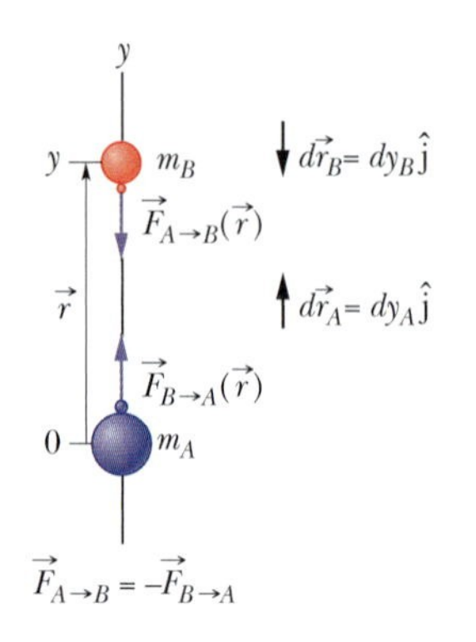

**FIGURE 10-11** ■ Two bodies interact by means of a conservative force. According to Newton's Third Law, they exert "equal and opposite" forces on each other. In general, the magnitude of these forces depends only on their separation, $|\vec{r}|$.

## Potential Energy Change for any Conservative Two-Body System

Now that we have considered two systems that can undergo potential energy change, we are ready to derive a general expression for the PE changes for any conservative two-body system. Let's start by defining a $y$ axis that passes through two interacting bodies, body $A$ and body $B$, with its origin located at body $A$ as shown in Fig. 10-11. For this choice of coordinate system, the internal interaction forces all point along the $y$ axis.

Since the change in the system's potential energy is $\Delta U = -W^{\text{cons}} - W^{\text{int}}$, the key to finding the potential energy change is to determine the internal work the particles do on each other as a result of a change in their separation. For an infinitesimally small change in separation, $dr$, the increment of internal work the system bodies do on each other, $dW^{\text{int}}$, is given by

$$dW^{\text{int}} = \vec{F}_{B\to A}(\vec{r}) \cdot d\vec{r}_A + \vec{F}_{A\to B}(\vec{r}) \cdot d\vec{r}_B.$$

Because we chose our $y$ axis along the displacement direction, we can represent the forces in terms of their $y$-components as $\vec{F}_{B\to A} = F_{A\,y}\,\hat{\mathrm{j}}$ and $\vec{F}_{A\to B} = F_{B\,y}\,\hat{\mathrm{j}}$. Here we shorten $F_{B\to A\,y}$ to $F_{A\,y}$ and $F_{A\to B\,y}$ to $F_{B\,y}$ and recall that these components can be positive or negative. This allows us to eliminate the dot products to get

$$dW^{\text{int}} = F_{A\,y}(y)\,dy_A + F_{B\,y}(y)\,dy_B.$$

Since Newton's Third Law tells us that $\vec{F}_{A\to B} = -\vec{F}_{B\to A}$ we know that the $y$-components of these vectors are related by $F_{A\,y} = -F_{B\,y}$, and we can rewrite $dW^{\text{int}}$ as

$$dW^{\text{int}} = F_{A\,y}(y)\,dy_A + F_{B\,y}(y)\,dy_B = F_{B\,y}(y)d(y_B - y_A).$$

The interaction forces always point along an axis through the two particles, so that in vector notation $\vec{r} = y\,\hat{\mathrm{j}}$. The changes in the interaction forces due to changes in separation depend only on $y$. The expression for $dW^{\text{int}}$ can be expressed in terms of only the force that particle $A$ exerts on particle $B$ and the variable $y$. This simplifies $dW^{\text{int}}$ to

$$dW^{\text{int}} = F_{B\,y}\,(y)\,dy.$$

For the most general case in which particle $B$ moves relative to particle $A$ from an initial location $y_1$ to a final location $y_2$, the internal work is given by the integral of $dW^{\text{int}}$ with respect to $y$,

$$W^{\text{int}} = \int_{y_1}^{y_2} dW^{\text{int}} = \int_{y_1}^{y_2} F_{B\,y}(y)\,dy. \qquad (10\text{-}11)$$

Substituting this into $\Delta U = -W^{\text{int}}$, we find that the change in the system's potential energy due to the change in configuration along our chosen $y$ axis is

$$\Delta U = -W^{\text{int}} = -\int_{y_1}^{y_2} F_{B\,y}(y)\,dy. \qquad (10\text{-}12)$$

We can equally well decide to place an $y'$ axis instead of a $y$ axis through the two points of interest or to develop the equation for $\Delta U$ in terms of the force of particle $B$ on particle $A$. In the absence of choosing a specific coordinate system we can substitute

$$W^{\text{int}} \equiv \int_{\vec{r}_1}^{\vec{r}_2} \vec{F}(\vec{r}) \cdot d\vec{r} \qquad (\text{Eq. 9-36})$$

for $W^{\text{int}}$ to get a general expression for the change in a conservative system's potential energy in terms of the dot product, where $\vec{r}$ is the radius vector pointing from particle $A$ to particle $B$,

$$\Delta U = -W^{\text{int}} = -\int_{r_1}^{r_2} \vec{F}_B(\vec{r}) \cdot d\vec{r}. \qquad (10\text{-}13)$$

The valuable conclusion we have reached is that

> The potential energy change of a two-body system with only conservative internal forces that depend only on the separation between the particles can be determined by considering the internal force on only *one* of the bodies.

Since $W^{\text{int}}$ is path independent, we can write $\Delta U$ as $\Delta U = U(\vec{r}_2) - U(\vec{r}_1)$. As we did earlier, we can choose a reference separation point $\vec{r}_1$ such that $U(\vec{r}_1) = 0$. Then we can express the potential energy of a particle relative to this reference for any $\vec{r}_2$ (or more generally $\vec{r}$) as $U(\vec{r})$.

**READING EXERCISE 10-3:** The net work-kinetic energy theorem predicts that the net work should be zero when the barbell is raised from its low point to its high point. Are the data in Fig. 10-10*b* consistent with this prediction? Explain. ■

**READING EXERCISE 10-4:** A particle is to move along the $x$ axis from $x_1 = 0$ to $x_2$ while a conservative internal force from a second particle force, directed along the $x$ axis, acts on the first particle. The figure shows three situations in which the $x$-component of that force varies with $x$. The force has the same maximum magnitude $F_1$ in all three situations. Rank the situations according to the change in the associated potential energy during the particle's motion, most positive first.

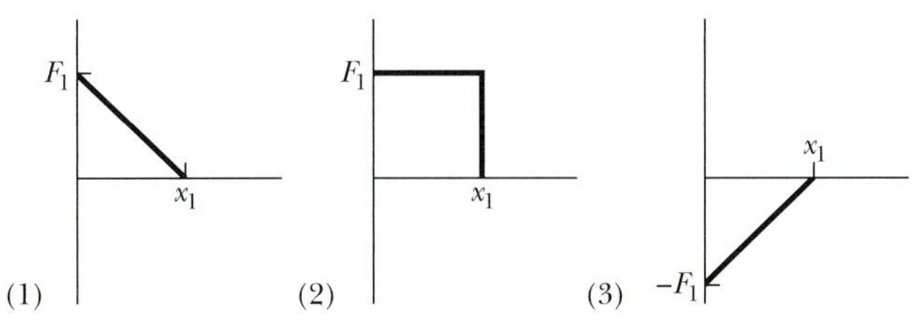


■

## TOUCHSTONE EXAMPLE 10-2: Sloth's Energy

A 2.0 kg sloth clings to a limb that is 5.0 m above the ground (Fig. 10-12).

(a) What is the gravitational potential energy $U^{\text{grav}}$ of the sloth–Earth system if we take the reference point $y_1 = 0$ to be (1) at the ground, (2) at a balcony floor that is 3.0 m above the ground, (3) at the limb, and (4) 1.0 m above the limb? Take the gravitational potential energy to be zero at $y_1 = 0$ and denote $y_2$ as $y$.

**SOLUTION** ■ The **Key Idea** here is that once we have chosen the reference point for $y_1 = 0$, we can calculate the gravitational potential energy $U^{\text{grav}}$ of the system *relative to that reference point* with Eq. 10-8. For example, for choice (1) the sloth is at $y = 5.0$ m, and

$$U^{\text{grav}} = mgy = (2.0\text{ kg})(9.8\text{ m/s}^2)(5.0\text{ m})$$
$$= 98\text{ J}. \qquad \text{(Answer)}$$

For the other choices, the values of $U^{\text{grav}}$ are

(2) $U^{\text{grav}} = mgy = mg(2.0\text{ m}) = 39\text{ J}$,

(3) $U^{\text{grav}} = mgy = mg(0) = 0\text{ J}$,

(4) $U^{\text{grav}} = mgy = mg(-1.0\text{ m}) = -19.6\text{ J} \approx -20\text{ J}$. (Answer)

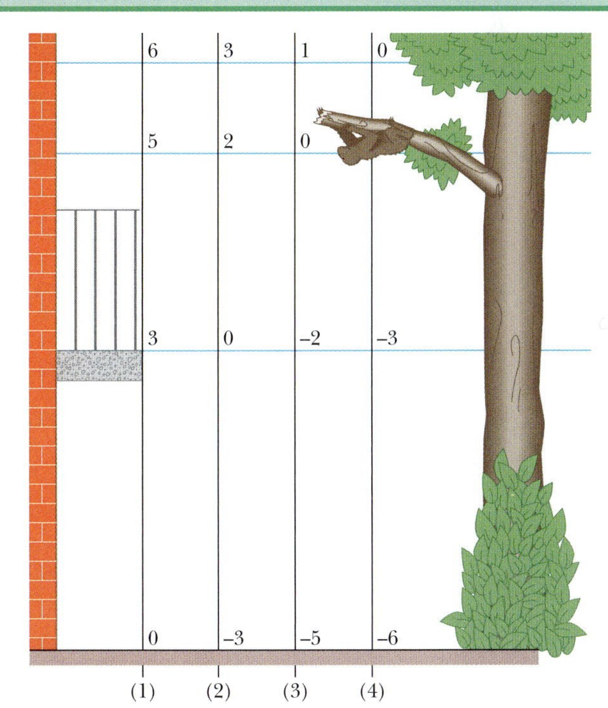


**FIGURE 10-12** ■ Four choices of reference point $y_1 = 0$. Each $y$ axis is marked in units of meters.

(b) The sloth drops to the ground. For each choice of reference point, what is the change $\Delta U^{grav}$ in the potential energy of the sloth–Earth system due to the fall?

**SOLUTION** ■ The **Key Idea** here is that the *change* in potential energy does not depend on the choice of the reference point for $y_1 = 0$; instead, it depends on the change in height $\Delta y$. For all four situations, we have the same $\Delta y = -5.0$ m. Thus, for (1) to (4), Eq. 10-6 tells us that

$$\Delta U^{grav} = mg\,\Delta y = (2.0\text{ kg})(9.8\text{ m/s}^2)(-5.0\text{ m})$$
$$= -98\text{ J}. \qquad \text{(Answer)}$$

## 10-4 Mechanical Energy Conservation

Let's consider a collection of rigid objects or particles in a system that interact only by means of conservative internal forces. Basically we are assuming that the system is *isolated* from its environment, so that no *external force* is present to do work on the system.

The system parts can have kinetic energy. For example, when a barbell is dropped the Earth–barbell system acquires kinetic energy as its gravitational potential energy decreases. This decrease in potential energy with increase in kinetic energy leads us to suspect that for isolated conservative systems, the sum of these two energies might be constant. As we saw in Chapter 7 on momentum, when a quantity is constant over time, physicists say that quantity is **conserved.** We will explore this possibility by defining a new quantity we call mechanical energy, or $E^{mec}$, that is the sum of the kinetic energy $K$ and the potential energy $U$ of a system. Symbolically we get

$$E^{mec} \equiv K + U \quad \text{(definition of mechanical energy).} \tag{10-14}$$

In this section, we examine what happens to the mechanical energy of an isolated system when all of its internal forces are conservative.

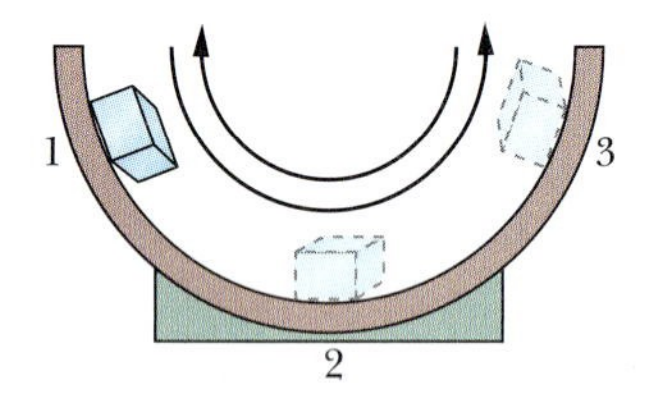


**FIGURE 10-13** ■ A small ice cube oscillates back and forth on a curved frictionless ramp constantly trading energy between potential energy and kinetic energy.

Imagine a small ice cube that is placed on a curved, frictionless ramp bolted to a table, as in Fig. 10-13. If the ice cube is released from point 1 it will oscillate back and forth. When first released it falls toward point 2 under the influence of the conservative gravitational force component parallel to the surface of the ramp. The kinetic energy of the ice cube will increase and it loses potential energy. The system's potential energy will be a minimum at point 2 when its kinetic energy is maximum. As it rises toward point 3 it loses kinetic energy and it gains potential energy.

In general, when an internal conservative force does work $W^{int}$ on an object within the system and no other objects in the system move appreciably, the system transfers energy between kinetic energy $K$ of the object and potential energy $U$ of the system. According to the net work-kinetic energy theorem, if the only work done on an object in a system is the internal work, then the change $\Delta K$ in kinetic energy is

$$\Delta K = W^{net} = W^{int}. \tag{10-15}$$

The change $\Delta U$ in the potential energy of the system is

$$\Delta U = -W^{int} \tag{10-16}$$

where $W^{int}$ is the sum of all works done by all the conservative internal forces. Combining these two equations, we find that

$$\Delta K = -\Delta U. \tag{10-17}$$

This shows that one of these energies increases exactly as much as the other decreases.

We can rewrite $\Delta K = -\Delta U$ as

$$K_2 - K_1 = -(U_2 - U_1), \tag{10-18}$$

where the subscripts refer to two different instants and thus to two different arrangements of the objects in the system. Rearranging $K_2 - K_1 = -(U_2 - U_1)$ yields

$$E^{mec} = K_1 + U_1 = K_2 + U_2 \quad \text{(conservation of mechanical energy).} \tag{10-19}$$

In words, this equation says that

> In a system where (1) no work is done on it by external forces and (2) only conservative internal forces act on the system elements, then the internal forces in the system can cause energy to be transferred between kinetic energy and potential energy, but their sum, the mechanical energy $E^{\text{mec}}$ of the system, cannot change.

This result is called the **conservation of mechanical energy.** Beware! Conservation of mechanical energy only holds under the special conditions we just outlined. Now you can see where *conservative* forces got their name. With the aid of $\Delta K = -\Delta U$ (Eq. 10-17), we can write this principle in one more form, as

$$\Delta E^{\text{mec}} = \Delta K + \Delta U = 0. \tag{10-20}$$

In cases where it holds, conservation of mechanical energy allows us to solve problems that would be quite difficult to solve using only Newton's laws:

> When the mechanical energy of a system is conserved, we can relate the sum of kinetic energy and potential energy at one instant to that at any other instant *without considering the intermediate motion* and *without finding the work done by the forces involved.*

Figure 10-14 shows an example in which the principle of conservation of mechanical energy can be applied. As a pendulum swings, the energy of the pendulum–Earth system is transferred back and forth between kinetic energy $K$ and gravitational potential energy $U$, with the sum $K + U$ being constant. If we are given the gravitational potential energy when the pendulum bob is at its highest point (Fig 10-14, stage 1), we can find the kinetic energy of the bob at the lowest point (Fig. 10-14, stage 3) using $K_2 + U_2 = K_1 + U_1$ (Eq. 10-19). The continual exchange back and forth between potential energy and kinetic energy is shown in the graph in Fig. 10-14.

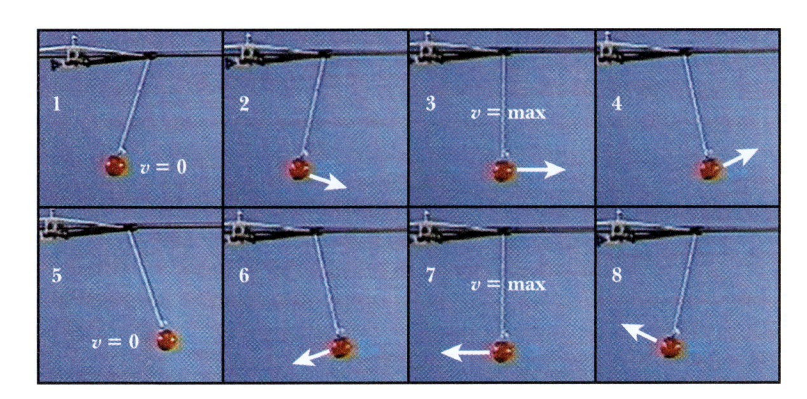


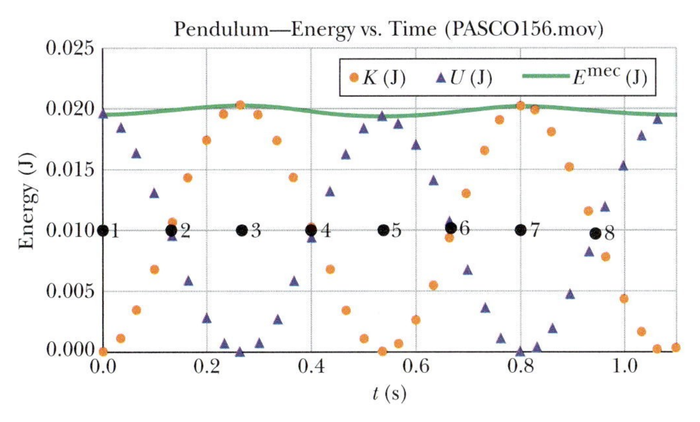


**FIGURE 10-14** ■ A pendulum with its mass of 0.10 kg concentrated in a bob at the lower end. A selection of video frames that capture its motion shows the potential and kinetic energy of the bob as it swings back and forth for one full cycle of motion. A local fit of data for angular position vs time and its first derivative was used to calculate gravitational potential energy, $U$, and kinetic energy, $K$, of the bob on a moment-by-moment basis. During the cycle, the values of the potential and kinetic energies of the pendulum–Earth system vary as the bob rises and falls. But, as shown in the graph, the total mechanical energy, $E^{\text{mec}}$, of the system remains constant within the limits of experimental uncertainty. In stages 3 and 7, all the energy is kinetic. The bob has its greatest speed while passing rapidly through its lowest point. In stages 1 and 5, all the energy is potential energy. In stages 2, 4, 6, and 8, the energy is split between potential and kinetic. The forces on the pendulum appear to be conservative when only one cycle is observed. However, the friction at the point of attachment and the presence of drag forces due to the air will cause the total mechanical energy of the system to decrease slowly with time.

For example, let us choose the lowest point of the pendulum as the reference point and set the corresponding gravitational potential energy $U_2 = 0.00$ J. Note then that the potential energy at the highest point is approximately given by $U_1 = 0.20$ J relative to the reference point. Because the bob momentarily has speed $v = 0$ at its highest point, the kinetic energy there is $K_1 = 0.00$ J. Substituting these values into $K_2 + U_2 = K_1 + U_1$ gives us the kinetic energy $K_2$ at the lowest point,

$$K_2 + 0.00\text{ J} = 0.00\text{ J} + 0.20\text{ J} \quad \text{or} \quad K_2 = 0.20\text{ J}.$$

Note that we get this result without considering the motion between the highest and lowest points (such as in Fig. 10-14, stage 7) and without finding the work done by any forces involved in the motion.

**READING EXERCISE 10-5:** The figure shows four situations—one in which an initially stationary block is dropped and three in which the block is allowed to slide down frictionless ramps. (a) Rank the situations according to the kinetic energy of the block at point $B$, greatest first. (b) Rank them according to the speed of the block at point $B$, greatest first.

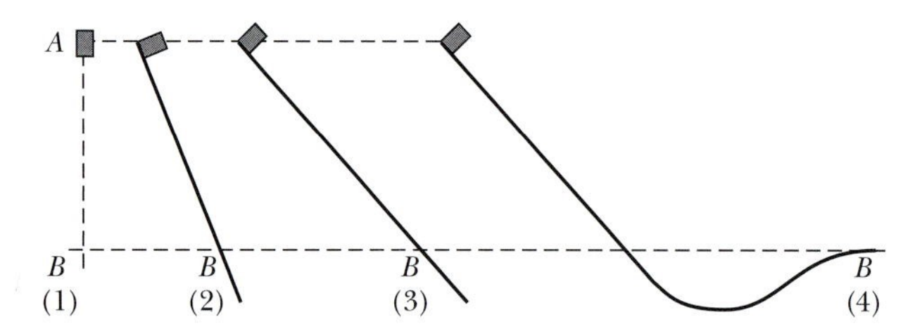


## TOUCHSTONE EXAMPLE 10-3: Bungee Jumper

A 61.0 kg bungee-cord jumper is on a bridge 45.0 m above a river. The elastic bungee cord has a relaxed length of $L =$ 25.0 m. Assume that the cord obeys Hooke's law, with a spring constant of 160 N/m. If the jumper stops before reaching the water, what is the height $h$ of her feet above the water at her lowest point?

**SOLUTION** ■ Figure 10-15 shows the jumper at the lowest point, with her feet at height $h$ and with the cord stretched by distance $d$ from its relaxed length. If we knew $d$, we could find $h$. One **Key Idea** is that perhaps we can solve for $d$ by applying the principle of conservation of mechanical energy, between her initial point (on the bridge) and her lowest point. In that case, a second **Key Idea** is that mechanical energy is conserved in an isolated system when only conservative forces cause energy transfers. Let's check.

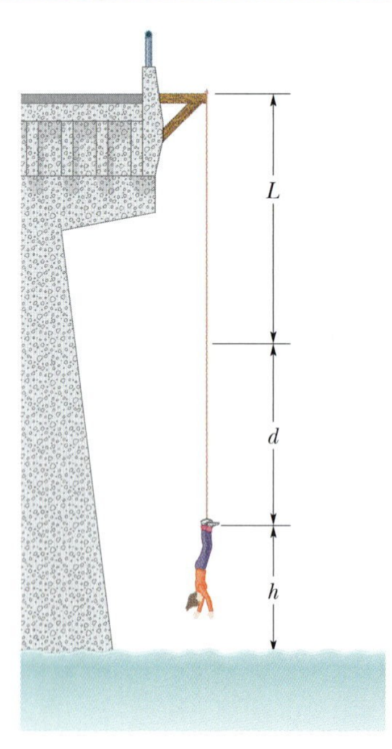


FIGURE 10-15 ■ A bungee-cord jumper at the lowest point of the jump.

*Forces:* The gravitational force does work on the jumper throughout her fall. Once the bungee cord becomes taut, the spring-like force from it does work on her, transferring energy to elastic potential energy of the cord. The force from the cord also pulls on the bridge, which is attached to Earth. The gravitational force and the spring-like force are conservative.

*System:* The jumper–Earth–cord system includes all these forces and energy transfers, and we can take it to be isolated (no work done by external forces). Thus, we *can* apply the principle of conservation of mechanical energy to the system. From Eq. 10-20, we can write the principle as

$$\Delta K + \Delta U^{\text{elas}} + \Delta U^{\text{grav}} = 0, \tag{10-21}$$

where $\Delta K$ is the change in the jumper's kinetic energy, $\Delta U^{\text{elas}}$ is the change in the elastic potential energy of the bungee cord, and $\Delta U^{\text{grav}}$ is the change in gravitational potential energy. All these changes must be computed between her initial point and her lowest point. Because she is stationary (at least momentarily) both initially and at her lowest point, $\Delta K = 0$. From Fig. 10-15 (with the bridge as origin and downward the negative $y$ direction), we see that the change $\Delta y$ in her height is $-(L + d)$, so we have

$$\Delta U^{\text{grav}} = mg\,\Delta y = -mg(L + d),$$

where $m$ is her mass. Also from Fig. 10-15, we see that the bungee cord is stretched by distance $d$. Thus, we also have

$$\Delta U^{\text{elas}} = \tfrac{1}{2}kd^2.$$

Inserting these expressions and the given data into Eq. 10-21, we obtain

$$0 + \tfrac{1}{2}kd^2 - mg(L + d) = 0,$$

or

$$\tfrac{1}{2}kd^2 - mgL - mgd = 0,$$

and then

$$\tfrac{1}{2}(160\ \text{N/m})d^2 - (61.0\ \text{kg})(9.8\ \text{m/s}^2)(25.0\ \text{m}) - (61.0\ \text{kg})(9.8\ \text{m/s}^2)d = 0.$$

Solving this quadratic equation yields

$$d = 17.9\ \text{m}.$$

The jumper's feet are then a distance of $(L + d) = 42.9$ m below their initial height. Thus,

$$h = 45.0\ \text{m} - 42.9\ \text{m} = 2.1\ \text{m}. \qquad \text{(Answer)}$$

## 10-5 Reading a Potential Energy Curve

Once again we consider a particle that is part of a system in which a conservative force acts. This time suppose that the particle is constrained to move along an $x$ axis while the conservative force does work on it. We can learn a lot about the motion of the particle from a plot of the system's potential energy $U(x)$. However, before we discuss such plots, we need one more relationship.

### Finding the Force Analytically for a Two-Body System

According to Eq. 10-12, if we choose a vertical $x$ axis passing from particle $A$ through particle $B$ (like that shown in Fig. 10-11) with its origin at particle $A$, then the change in potential energy of the system can be expressed as

$$\Delta U = -\int_{x_1}^{x_2} F_{B\,x}(x)\,dx.$$

This is the potential energy change that occurs when one of the particles, chosen as $B$, moves between $x_1$ and $x_2$ along a (vertical) $x$ axis.

Suppose we have the reverse situation. That is, suppose we happen to know $\Delta U$ and we would like to know the internal force acting on particle $B$ denoted as $F_{B\,x}(x)$. If the force on particle $B$ does not vary rapidly with $x$, the potential energy change in the system as particle $B$ moves through a distance $\Delta x$ is approximately

$$\Delta U(x) \approx -F_{B\,x}(x)\,\Delta x.$$

If we solve for $F_{B\,x}(x)$, pass to the differential limit, and drop the label $B$ (so the $x$-component of force denotes the internal force on whichever particle in the system is displaced relative to the other) we have

$$F_x^{\text{cons}}(x) = -\frac{dU(x)}{dx} \quad \text{(one-dimensional internal force).} \qquad (10\text{-}22)$$

We can check our result with $U(x) = \frac{1}{2}kx^2$, which is the elastic potential energy function for a spring force. Equation 10-22 ($F_x^{\text{cons}}(x) = -dU(x)/dx$) then yields, as expected, $F_x^{\text{cons}}(x) = F_x^{\text{spring}}(x) = -kx$, which is Hooke's law. Similarly, we can substitute $U(y) = mgy$, which is the gravitational potential energy function for a particle–Earth system, with a particle of mass $m$ at height $y$ above Earth's surface. $F_y^{\text{cons}}(y) = F_y^{\text{grav}}(y) = -dU(y)/dy$ then yields $F_y^{\text{grav}}(y) = -mg$, which is the $y$-component of gravitational force on the particle.

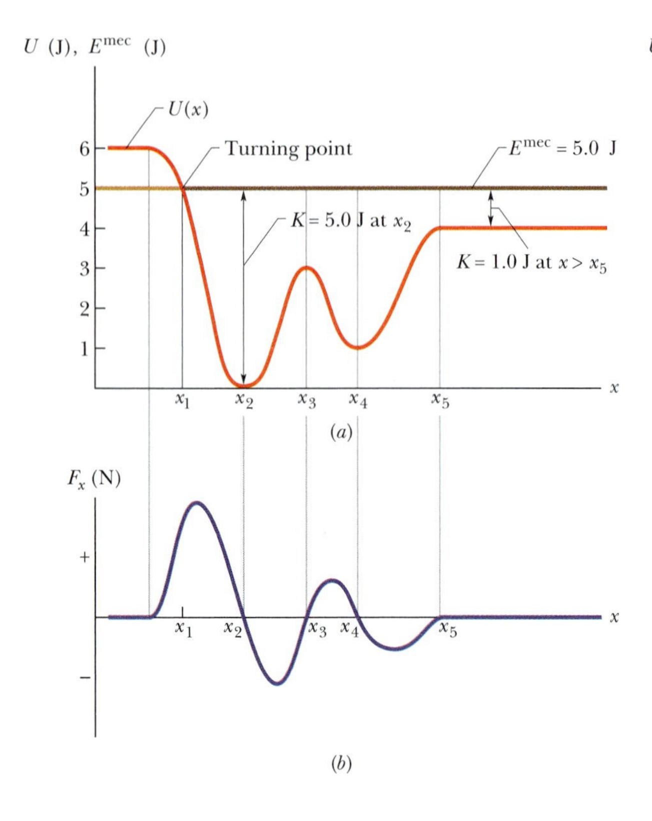


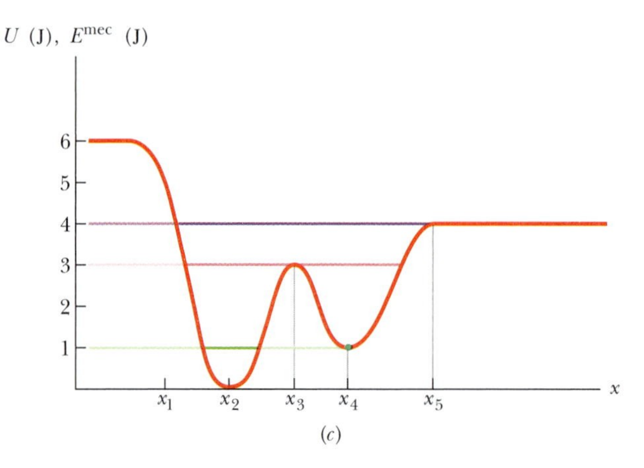


**FIGURE 10-16** ■ (*a*) A plot of $U(x)$, the potential energy function of a system containing a particle confined to move along the $x$ axis. There is no friction, so mechanical energy is conserved. (*b*) A plot of the force $F(x)$ acting on the particle, derived from the potential energy plot by taking its slope at various points. (*c*) The $U(x)$ plot of (*a*) with three different possible values of $E^{\text{mec}}$ shown.

## The Potential Energy Curve

Figure 10-16*a* is a plot of a potential energy function $U(x)$ for a system in which a particle is in one-dimensional motion while a conservative internal force $\vec{F}_x^{\text{int}}(x)$ does work on it. Since $dU(x)/dx$ in Eq. 10-22 is the slope of the $U(x)$ vs. $x$ curve, we can easily find $F_x^{\text{int}}(x)$ by (graphically) taking the slope of the $U(x)$ curve at various points and negating it. Figure 10-16*b* is a plot of $F_x^{\text{int}}(x)$ found in this way.

## Turning Points

As we discussed in Section 10-4, in the absence of a nonconservative force, the mechanical energy $E^{\text{mec}}$ of the system has a constant value given by

$$U(x) + K(x) = E^{\text{mec}}. \tag{10-23}$$

Here $K(x)$ is the *kinetic energy function* of the particle (this $K(x)$ gives the kinetic energy as a function of the particle's location $x$). We may rewrite this expression as

$$K(x) = E^{\text{mec}} - U(x). \tag{10-24}$$

Suppose that $E^{\text{mec}}$ (which has a constant value for a conservative isolated system) happens to be 5.0 J. It would be represented in Fig. 10-16*a* by a horizontal line that runs through the value 5.0 J on the energy axis. (It is, in fact, shown there.)

Equation 10-24 ($K(x) = E^{\text{mec}} - U(x)$) tells us how to determine the kinetic energy $K$ for any location $x$ of the particle: On the $U(x)$ curve, find $U$ for that location $x$ and then subtract $U$ from $E^{\text{mec}}$. For example, if the particle is at any point to the right of $x_5$, then $K = 1.0$ J. The value of $K$ is greatest (5.0 J) when the particle is at $x_2$, and least (0 J) when the particle is at $x_1$.

Since $K$ can never be negative (because $v^2$ is always positive), the particle can never move to the left of $x_1$, where $E^{\text{mec}} - U$ is negative. Instead, as the particle moves toward $x_1$ from $x_2$, $K$ decreases (the particle slows) until $K = 0$ at $x_1$ (the particle stops there).

Note that when the particle reaches $x_1$, the $x$-component of the internal force on the particle due to the rest of the system, given by

$$F_x^{\text{int}}(x) = -\frac{dU(x)}{dx},$$

is positive (because the slope $dU/dx$ is negative). This means that the particle does not remain at $x_1$ but instead begins to move to the right, opposite its earlier motion. Hence $x_1$ is a **turning point,** a place where $K = 0$ (because $U = E$) and the particle changes direction. There is no turning point (where $K = 0$) on the right side of the graph. When the particle heads to the right and $x > x_5$, there is no force on it, and it will continue indefinitely.

## Equilibrium Points

Figure 10-16*c* shows three different values for $E^{\text{mec}}$ superimposed on the plot of the same potential energy function $U(x)$. Let us see how they would change the situation. If $E^{\text{mec}} = 3.0$ J (line running through the value 3.0 J on the energy axis), there are two turning points: one is between $x_1$ and $x_2$ and the other is between $x_4$ and $x_5$. In addition, $x_3$ is a point at which $K = 0$. If the particle is located exactly there, the force on it is also zero (the slope of the curve is zero), and the particle remains stationary. However, if it is displaced even slightly in either direction, a nonzero force pushes it further in the same direction, and the particle continues to move. A particle at such a position is said to be in **unstable equilibrium.** (A marble balanced on top of a bowling ball is an example.)

Next consider the particle's behavior if $E^{\text{mec}} = 1.0$ J (line running through the value 1.0 J on the energy axis). If we place the particle at $x_4$, it is stuck there. It cannot move left or right on its own because to do so would require a negative kinetic energy. If we push it slightly left or right, a restoring force appears that moves it back to $x_4$. A particle at such a position is said to be in **stable equilibrium.** (A marble placed at the bottom of a hemispherical bowl is an example.) If we place the particle in the cup-like *potential well* centered at $x_2$, it is between two turning points. It can still move left and right somewhat, but only partway to $x_1$ or to $x_3$.

If $E^{\text{mec}} = 4.0$ J (line running through the value 4.0 J on the energy axis), the turning point shifts from $x_1$ to a point between $x_1$ and $x_2$. Also, at any point to the right of $x_5$, the system's mechanical energy is equal to its potential energy; thus, the particle has no kinetic energy and (by $F_x^{\text{int}}(x) = -dU(x)/dx$) no force acts on it. So it must be stationary. A particle at such a position is said to be in **neutral equilibrium.** (A marble placed on a horizontal tabletop is in that state.)

**READING EXERCISE 10-6:** The figure gives the potential energy function $U(x)$ for a system in which a particle is in one-dimensional motion. (a) Rank regions $AB$, $BC$, and $CD$ according to the magnitude of the force on the particle, greatest first. (b) What is the direction of the force when the particle is in region $AB$?

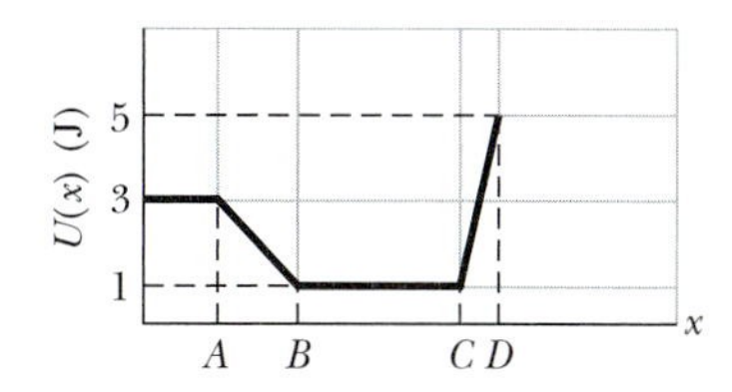

## TOUCHSTONE EXAMPLE 10-4: Shifting the Zero

Suppose that you shifted the origin of the graph shown in Fig. 10-16*a* up by 6.0 J so that the potential energy reference point from which all the values of $U(x)$ were measured was located at $x = 0$ rather than $x = x_2$.

(a) What effect would this have on the values of $U(x)$?

**SOLUTION** ■ While the plot of $U(x)$ would still have the same shape as in Fig. 10-16*a*, all of its values would be reduced by 6.0 J and so would now be negative rather than positive. For example, now $U(0) = 0$, $U(x_1) = -1$ J, $U(x_2) = -6$ J, and so on. (Answer)

(b) What effect would this have on the values of the kinetic energy?

**SOLUTION** ■ The particle's kinetic energy, $K = \frac{1}{2}mv^2$, depends only on the particle's mass and speed. Since neither of these depend on our choice of reference point from which we measure the particle's potential energy, the values of the particle's kinetic energy at each location will be the same as before. (Answer)

(c) What effect would this have on the values of $E^{\text{mec}}$?

**SOLUTION** ■ The **Key Idea** here is that $E^{\text{mec}} = U + K$. Since each value of $U$ is reduced by 6.0 J by this shift of reference point, and the values of $K$ remain the same, then $E^{\text{mec}}$ is also reduced by 6.0 J in each case. (Answer)

This change in $E^{\text{mec}}$ does *not* mean that $E^{\text{mec}}$ no longer has a constant value. It's just that it now has a *different* constant value than it had before. For example, in Fig. 10-16*a*, $E^{\text{mec}} = +5.0$ J everywhere to the right of $x = x_1$ before we shifted the potential energy's reference point. After the shift, $E^{\text{mec}} = -1.0$ J, still constant and independent of location, but now with a different value.

(d) What effect would this have on the values of $F_x(x)$, the force experienced by the particle, as pictured in Fig. 10-16*b*?

**SOLUTION** ■ The **Key Idea** here is that $F_x(x) = -dU/dx$. Subtracting a constant from $U(x)$ has no effect on its derivative since

$$\frac{d(U(x) - \text{constant})}{dx} = \frac{dU(x)}{dx} - \frac{d(\text{constant})}{dx}$$

$$= \frac{dU(x)}{dx} - 0 = \frac{dU(x)}{dx}.$$

Therefore, $F_x(x)$ will be unchanged. (Answer)

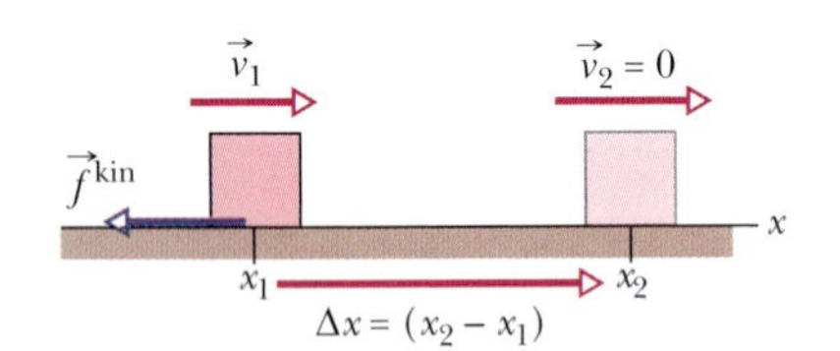


**FIGURE 10-17** ■ (*a*) A block slides across a floor while a kinetic frictional force $\vec{f}^{\text{kin}}$ $f^{\text{kin}}$ opposes the motion. The block has velocity $\vec{v}_1$ at the start of a displacement $\Delta\vec{x}$ and a velocity $\vec{v}_2 = 0$ at the end of the displacement.

## 10-6 Nonconservative Forces and Energy

We have made the claim that mechanical energy is conserved in an isolated system (no net work is done on the system by external forces) whose internal forces are conservative. Let's now consider an isolated system whose elements interact by means of nonconservative kinetic friction forces. Our example is an isolated system consisting of a sliding block and a floor (Fig. 10-17). Assume the block has an initial velocity $\vec{v}_1$. What happens to its initial kinetic energy as it slides along the floor and comes to rest so $\vec{v}_2 = 0$? According to the net work-kinetic energy theorem, the net work done on the block from the sum of all the forces acting on it will result in a kinetic energy change of the block given by

$$W^{\text{net}} = \Delta K = \tfrac{1}{2}mv_2^2 - \tfrac{1}{2}mv_1^2.$$

Since the net work is calculated from the net force, we need to write down an expression for the net force on the block. The block has a friction force, a downward gravitational force, and an upward normal force exerted on it. Since there is no motion in the vertical direction, we know that the gravitational and normal forces cancel each other out. The net force on the block is just the horizontal kinetic friction force, so $\vec{F}^{\text{net}} = \vec{f}^{\text{kin}}$. Since there are no external forces acting on the system, all the work done on the system is done by internal forces. So, $W^{\text{net}} = W^{\text{int}}$. If the block has a displacement $\Delta\vec{x}$ as it slides to rest, then

$$W^{\text{net}} = W^{\text{int}} = f_x^{\text{kin}}\Delta x = -\tfrac{1}{2}mv_1^2. \tag{10-25}$$

The product of $f_x^{\text{kin}}\Delta x$, which represents the internal work done on the system, is negative since friction forces always act in a direction opposite to an object's displacement. Since the friction force is nonconservative (the amount of work it does depends on the path taken), we cannot associate a potential energy change with it. Instead, as Eq. 10-25 indicates, the internal work done on the system has caused a *loss of kinetic energy*. This represents a loss of the only form of mechanical energy such a system can have. We can conclude that

> If the internal forces in an isolated system include nonconservative forces, then mechanical energy is not conserved.

By experimenting, we find that the block and the portion of the floor along which it slides become warmer as the block slides to a stop. If we associate the temperature of an object with a new kind of energy, thermal energy $E^{\text{thermal}}$, we may be able to continue to make use of energy conservation methods. In fact, it turns out that the kinetic energy lost in Eq. 10-25 does cause a gain in thermal energy where

$$\Delta E^{\text{thermal}} = -f_x^{\text{kin}}\Delta x \quad \text{(increase in thermal energy due to kinetic friction).} \qquad (10\text{-}26)$$

As we shall discuss in Chapter 19, the thermal energy of an object is related to temperatures and can be associated with the random motions of atoms and molecules in objects.

We define the *total energy* of the system to be the sum of its mechanical energy and other forms of energy including thermal energy, chemical energy, light, sound, and so on. Doing so, we see that we have a new principle of energy conservation for isolated systems even in the presence of nonconservative forces given by

$$\Delta E^{\text{total}} = \Delta E^{\text{mec}} + \Delta E^{\text{noncons}} = 0. \qquad (10\text{-}27)$$

Here, $\Delta E^{\text{noncons}} = \Delta E^{\text{thermal}} + \Delta E^{\text{other}}$. $\Delta E^{\text{noncons}} = \Sigma W^{\text{noncons}}$ where $\Sigma W^{\text{noncons}}$ is the sum of all work done by all nonconservative but internal forces.

**READING EXERCISE 10-7:** In three trials, a block starts with the same kinetic energy and slides across a floor that is not frictionless, as in Fig. 10-17. In all three trials, the block is allowed to slide through the same distance $\Delta x$ but has not yet come to rest. Rank the three trials according to the change in the thermal energy of the block and floor that occurs, greatest first.

| Trial | $f_x^{\text{kin}}$ | Block's Displacement $\Delta x$ |
|---|---|---|
| a | 5.0 N | 0.20 m |
| b | 7.0 N | 0.30 m |
| c | 8.0 N | 0.10 m |

## TOUCHSTONE EXAMPLE 10-5: Tamale Stops Here

In Fig. 10-18, a 2.0 kg package of tamales slides along a floor with speed $v_1 = 4.0$ m/s. It then runs into and compresses a spring, until the package momentarily stops. Its path to the initially relaxed spring is frictionless, but as it compresses the spring, a kinetic frictional force from the floor, of magnitude 15 N, acts on it. The spring constant is 10 000 N/m. By what distance $d$ is the spring compressed when the package stops?

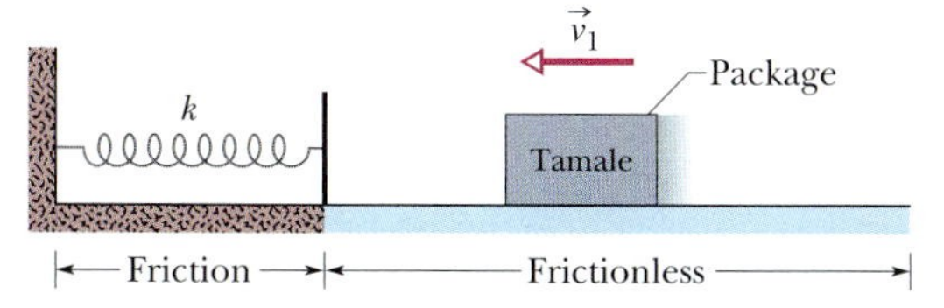


**FIGURE 10-18** ■ A package slides across a frictionless floor with velocity $\vec{v}_1$ toward a spring of spring constant $k$. When the package reaches the spring, a frictional force from the floor acts on it.

**SOLUTION** ■ A starting **Key Idea** is to examine all the forces acting on the package, and then to determine whether we have an isolated system or a system on which an external force is doing work.

*Forces:* The normal force on the package from the floor does no work on the package, because its direction is always perpendicular to that of the package's displacement. For the same reason, the gravitational force on the package does no work. As the spring is compressed, however, a spring force does work on the package, transferring energy to elastic potential energy of the spring. The spring force also pushes against a rigid wall. Because there is friction between the package and the floor, the sliding of the package across the floor increases their thermal energies.

*System:* The package–spring–floor–wall system includes all these forces and energy transfers in one isolated system. Therefore, a second **Key Idea** is that, because the system is isolated, its total energy cannot change. We can then apply the law of conservation of energy in the form of Eq. 10-27 to the system:

$$\Delta E^{\text{mec}} + \Delta E^{\text{noncons}} = 0.$$

But $\Delta E^{\text{mec}} = E_2^{\text{mec}} - E_1^{\text{mec}}$ and $\Delta E^{\text{noncons}} = \Delta E^{\text{thermal}}$,

so
$$E_2^{\text{mec}} = E_1^{\text{mec}} - \Delta E^{\text{thermal}}. \quad (10\text{-}29)$$

Let subscript 1 correspond to the initial state of the sliding package and subscript 2 correspond to the state in which the package is momentarily stopped and the spring is compressed by distance $d$. For both states the mechanical energy of the system is the sum of the package's kinetic energy ($K = \frac{1}{2}mv^2$) and the spring's potential energy ($U = \frac{1}{2}kx^2$). For state 1, $U = 0$ (because the spring is not compressed), and the package's speed is $v_1$. Thus, we have

$$E_1^{\text{mec}} = K_1 + U_1 = \tfrac{1}{2}mv_1^2 + 0.$$

For state 2, $K = 0$ (because the package is stopped), and the compression distance is $d$. Therefore, we have

$$E_2^{\text{mec}} = K_2 + U_2 = 0 + \tfrac{1}{2}kd^2.$$

Finally, by Eq. 10-26, we can substitute $(-f_x^{\text{kin}}\Delta x) = f_x^{\text{kin}}d$ for the change $\Delta E^{\text{thermal}}$ in the thermal energy of the package and the floor. We can now rewrite Eq. 10-29 as

$$\tfrac{1}{2}kd^2 = \tfrac{1}{2}mv_1^2 - f_d^{\text{kin}}d.$$

Rearranging and substituting known data give us

$$(5000\ \text{N/m})d^2 + (15\ \text{N})d - (16\ \text{J}) = 0.$$

Solving this quadratic equation yields

$$d = 0.055\ \text{m} = 5.5\ \text{cm}. \quad \text{(Answer)}$$

## 10-7 Conservation of Energy

We now have discussed several situations in which energy is transferred between objects within systems. In each situation, we assume that the energy that was involved could always be accounted for. That is, energy could not appear or disappear. In more formal language, we assumed that energy obeys a law called the **law of conservation of energy,** which is concerned with the **total energy** $E^{\text{total}}$ of a system. There are many complex situations in which it is difficult to account for all the energy. But physicists have always found that if a change in a system takes place and some energy seems to be missing, it simply has taken on a new form. This is the case with the thermal energy we talked about in the previous section. It can be accounted for by developing methods for keeping track of the kinetic energy stored in the random motions of atoms and molecules in the sliding block and floor and in the potential energy associated with the chemical bonds that hold them together.

We define total energy of a system as the sum of the system's mechanical energy, thermal energy, and other forms of energy we will not touch on here that are associated with things like sound and light. The law of conservation of energy states that:

> The total energy $E^{\text{total}}$ of a system can change only by amounts of energy that are transferred to or from the system.

When is energy transferred to or from a system? This occurs when an external force does work $W^{\text{ext}}$ on the system. The external work $W^{\text{ext}}$ done on a system is not merely a calculation procedure. It is an energy transfer process. Thus, the **law of conservation of energy** can be stated in very general terms as

$$W^{\text{ext}} = \Delta E^{\text{total}} = \Delta E^{\text{mec}} + \Delta E^{\text{noncons}}, \quad (10\text{-}28)$$

where $\Delta E^{\text{noncons}}$ is any change in thermal energy or the many other forms of energy that we have not discussed here. Included in $\Delta E^{\text{mec}}$ are changes $\Delta K$ in kinetic energy and changes $\Delta U$ in potential energy due to conservative forces such as elastic, gravitational, and electrostatic forces (which we discuss in Chapter 25).

As you may have noticed, this law of conservation of energy is *not* something we have derived from basic physics principles. It is more speculative. But in the past, whenever it has appeared to fail, scientists and engineers have been able to identify new forms of energy that allow us to hold on to the law of conservation of energy. Furthermore, each time a new form of energy has been identified, we have been able to understand whole new classes of phenomena, such as how stars shine or how radioactive atoms decay.

### TOUCHSTONE EXAMPLE 10-6: Easter Island

The giant stone statues of Easter Island were most likely moved by the prehistoric islanders by cradling each statue in a wooden sled and then pulling the sled over a "runway" consisting of almost identical logs acting as rollers. In a modern reenactment of this technique, 25 men were able to move a 9000 kg Easter Island-type statue 45 m over level ground in 2 min.

(a) Estimate the work the external force $\vec{F}^{\,ext}$ from the men did during the 45 m displacement of the statue, and determine the system on which that force did the work.

**SOLUTION** ■ One **Key Idea** is that we can calculate the work done with Eq. 9-26 ($W = |\vec{F}||\Delta\vec{r}|\cos\phi$). Here $|\Delta\vec{r}|$ is the distance 45 m, $|\vec{F}^{ext}|$ is the magnitude of the external force on the statue from the 25 men, and $\phi = 0°$. Let us estimate that each man pulled with a force magnitude equal to twice his weight, which we take to be the same value $mg$ for all the men. Thus, the magnitude of the external force was $|\vec{F}^{ext}| = (25)(2)(mg) = 50mg$. Estimating a man's mass as 80 kg, we can then write Eq. 9-26 as

$$\begin{aligned} W^{ext} &= |\vec{F}^{ext}||\Delta\vec{r}|\cos\phi = 50mgd\cos\phi \\ &= (50)(80\text{ kg})(9.8\text{ N/kg})(45\text{ m})\cos 0° \qquad \text{(Answer)} \\ &= 1.8\times 10^6\text{ J} \approx 2\text{ MJ}. \end{aligned}$$

The **Key Idea** in determining the system on which the work is done is to see which energies change. Because the statue moved, there was certainly a change $\Delta K$ in its kinetic energy during the motion. We can easily guess that there must have been considerable kinetic friction between the sled, logs, and ground, resulting in a change $\Delta E^{thermal}$ in their thermal energies. Thus, the system on which the work was done consisted of the statue, sled, logs, and ground.

(b) What was the increase $\Delta E^{thermal}$ in the thermal energy of the system during the 45 m displacement?

**SOLUTION** ■ The **Key Idea** here is that we can relate $\Delta E^{noncons} = \Delta E^{thermal}$ to the work $W^{ext}$ done by $\vec{F}^{ext}$ with the energy statement of Eq. 10-28,

$$W^{ext} = \Delta E^{mec} + \Delta E^{thermal}.$$

We know the value of $W^{ext}$ from (a). The change $\Delta E^{mec}$ in the statue's mechanical energy was zero because the statue was stationary at the beginning and end of the move and did not change in elevation. Thus, we find

$$\Delta E^{thermal} = W^{ext} = 1.8\times 10^6\text{ J} \approx 2\text{ MJ}. \qquad \text{(Answer)}$$

(c) Estimate the work that would have been done by the 25 men if they had moved the statue 10 km across level ground on Easter Island. Also estimate the total change $\Delta E^{thermal}$ that would have occurred in the statue–sled–logs–ground system.

**SOLUTION** ■ The **Key Ideas** here are the same as in (a) and (b). Thus we calculate $W^{ext}$ as in (a), but with $1\times 10^4$ m now substituted for $|\Delta\vec{r}|$. Also, we again equate $\Delta E^{thermal}$ to $W^{ext}$. We get

$$W^{ext} = \Delta E^{thermal} = 3.9\times 10^8\text{ J} \approx 400\text{ MJ}. \qquad \text{(Answer)}$$

This would have been a significant amount of energy for the men to have transferred during the movement of a statue. Still, the 25 men *could* have moved the statue 10 km, and the required energy does not suggest some mysterious source.

## 10-8 One-Dimensional Energy and Momentum Conservation

Recall that in some situations, the conservation of translational momentum allowed us to figure out what was going to happen, even when we didn't know what the forces were. (For example, when two objects collide and stick together.) Now that we have identified a second conservation law—the conservation of energy—we can figure out what is going to happen in a larger class of situations.

Consider a system of two colliding bodies. If there is to be a collision, then at least one of the bodies must be moving, so the system has a certain kinetic energy and a certain translational momentum before the collision. During the collision, the kinetic energy and translational momentum of each body are changed by the impulse from the other body. We can discuss these changes—and also the changes in the kinetic energy and translational momentum of the system as a whole—without knowing the details of the impulses that determine the changes. As was the case in Chapter 7 where we first discussed collisions, the discussion here will be limited to collisions in systems that are **closed** (no mass enters or leaves them) and **isolated** (no net external forces act on the bodies within the system).

**FIGURE 10-19** ■ Two cars after an almost head-on, almost completely inelastic collision.

## Elastic versus Inelastic Collisions

Collisions that we casually called bouncy and sticky in Chapter 7 can be classified in terms of whether or not mechanical energy is conserved. Except for a brief period during a collision, typically no potential energy is stored in a system of objects before and after the collision. So most of the time the mechanical energy in the system is equal to the total kinetic energy of the colliding objects.

**Elastic Collisions:** If the total kinetic energy of the system of two colliding bodies is unchanged by the collision, the collision is called a completely **elastic collision.** This happens if the forces between the objects during the collision are approximately conservative and spring-like. Some of the "bouncy" collisions we discussed in Chapter 7 may have been elastic collisions. However, most "bouncy" collisions are in fact not completely elastic collisions.

**Inelastic Collisions:** In everyday collisions of common bodies, such as between two cars or a ball and a bat, some energy is always transferred from kinetic energy to other forms of energy, such as thermal energy or energy of sound. Thus, the kinetic energy of the system is *not* conserved. Such a collision is defined as an **inelastic collision.** Figure 10-19 shows a dramatic example of a **completely inelastic collision.** In such collisions, the bodies always stick together and lose all their kinetic energy. Most real collisions are partially elastic and partially inelastic.

**Almost Elastic Collisions:** In some situations, we can *approximate* a collision of common bodies as elastic. Suppose that you drop a Superball onto a hard floor. If the collision between the ball and floor (or Earth) were elastic, the ball would lose no kinetic energy because of the collision and would rebound to its original height. However, the actual rebound height is somewhat short of the starting point, showing that at least some kinetic energy is lost in the collision and thus that the collision is somewhat inelastic. Still, we might choose to neglect that small loss of kinetic energy to approximate the collision as elastic.

## Distinguishing Energy and Momentum Conservation

It is easy to confuse momentum conservation with energy conservation. However, *they are not the same*. Momentum is a vector quantity defined as the product of mass and velocity. Energy is a scalar quantity that has no direction associated with it. As far as we know, *momentum is always conserved* as a result of interactions between objects in an isolated system. This is not the case for mechanical energy. *Mechanical energy is only conserved when the internal forces that do work on the system are conservative.*

Let's perform three thought experiments that illustrate some of the differences between the two conservation laws. To do this, imagine three types of collision processes described in Chapter 7 on collisions and momentum. One is a completely *inelastic collision* in which the colliding objects stick together. Another is a completely *elastic collision* in which the objects bounce off one another and the system consisting of the colliding objects loses no mechanical energy. The third is a *superelastic collision*, or explosion, in which some energy is released so the system has more mechanical energy than it did before.

In all three thought experiments, two identical carts with negligible friction are moving toward each other at the same speed (Fig. 10-20). Since they are moving on a horizontal ramp they have no change in gravitational potential energy as they move. In this special circumstance, the mechanical energy of each of the two-cart systems is the same as its kinetic energy.

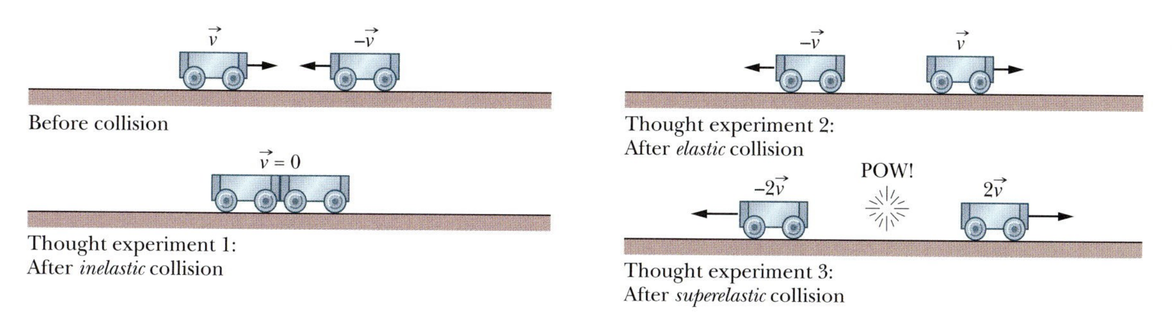


**FIGURE 10-20** ▪ Three possible outcomes of a head-on collision between two identical carts. Momentum is conserved in all three cases but not necessarily mechanical energy.

From the information in the figure, it is apparent that the momentum before and after the collision is the same in all three cases and is zero. However, in the inelastic collision shown in experiment 1, there is Velcro on the ends of the carts so they come to a dead halt when they stick together. In Section 7-6, we referred to this type of collision as "sticky." Although the temperature of the Velcro rises, there is no kinetic energy left after the collision and, hence, mechanical energy is not conserved. In experiment 2, the carts have magnets embedded in the ends that repel, causing the carts to rebound with the same speed but not the same velocity as before. In this case kinetic energy, and hence mechanical energy, is conserved. Finally, in experiment 3, wads of gunpowder glued to the cart ends ignite. Chemical potential energy is released in an explosion that causes the carts to rebound with a greater kinetic energy than they had before. Once again mechanical energy is not conserved, but translational momentum still is.

## Translational Momentum

Regardless of the details of the impulses in a collision and regardless of what happens to the total kinetic energy of the system, the total translational momentum $\vec{p}_{\text{sys}}$ of a closed, isolated system *cannot* change. The reason is that $\vec{p}_{\text{sys}}$ can be changed only by external forces (from outside the system), but the forces in the collision are internal forces (inside the system). Thus, we have this important rule:

> In a closed, isolated system in which a collision occurs, the translational momentum of each colliding body may change but the total translational momentum of the system $\vec{p}_{\text{sys}}$ cannot change, whether the collision is elastic or inelastic.

This is actually another statement of the **law of conservation of translational momentum** that we first discussed in Section 7-6. In Section 7-7, we explored translational momentum conservation for inelastic collisions—that is, "sticky collisions." In the next two sections we apply this law to elastic collisions.

## TOUCHSTONE EXAMPLE 10-7: Ballistic Pendulum

The *ballistic pendulum* was used to measure the speeds of bullets before electronic timing devices were developed. The version shown in Fig. 10-21 consists of a large block of wood of mass $M = 5.4$ kg, hanging from two long cords. A bullet of mass $m = 9.5$ g is fired into the block, coming quickly to rest. The *block + bullet* then swing upward, their centers of mass rising a vertical distance $h = 6.3$ cm before the pendulum comes momentarily to rest at the end of its arc. What is the speed of the bullet just prior to the collision?

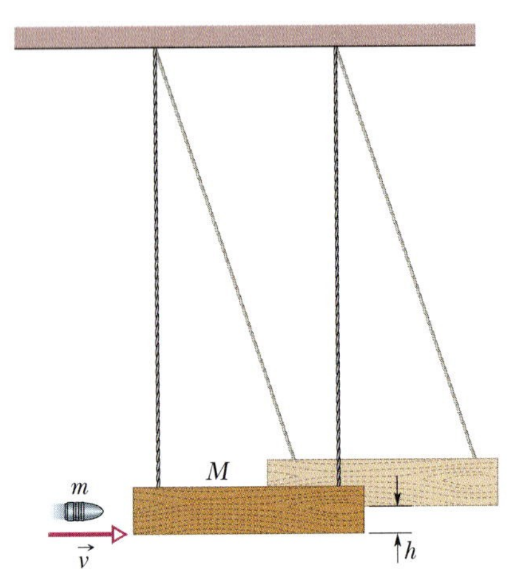


**FIGURE 10-21** ■ A ballistic pendulum, used to measure the speeds of bullets.

**SOLUTION** ■ We can see that the bullet's speed $v$ must determine the rise height $h$. However, a **Key Idea** is that we cannot use the conservation of mechanical energy to relate these two quantities because surely energy is transferred from mechanical energy to other forms as the bullet penetrates the block. Another **Key Idea** helps—we can split this complicated motion into two steps that we can separately analyze: (1) the bullet–block collision and (2) the bullet–block rise, during which mechanical energy *is* conserved.

**Step 1.** Because the collision within the bullet–block system is so brief, we can make two important assumptions: (1) During the collision, the gravitational force on the block and the force on the block from the cords are still balanced. Thus, during the collision, the net external force on the bullet–block system is zero. Therefore, the system is isolated and its total translational momentum is conserved. (2) The collision is one-dimensional in the sense that the direction of the bullet and block *just after the collision* is in the bullet's original direction of motion.

Because the collision is one-dimensional, the block is initially at rest, and the bullet sticks in the block, we use Eq. 7-21 to express the conservation of linear momentum. If the speed of the block just after the collision is $V$, we have $mv + 0 = mV + MV$ or

$$V = \frac{m}{m + M}v. \tag{10-30}$$

**Step 2.** After the "collision" between the bullet and the block is over, the bullet and block now swing up together, and the mechanical energy of the bullet–block–Earth system is conserved. (This mechanical energy is not changed by the force of the cords on the block, because that force is always directed perpendicular to the block's direction of travel.) Let's take the block's initial level as our reference level of zero gravitational potential energy. Then conservation of mechanical energy means that the system's kinetic energy at the start of the swing must equal its gravitational potential energy at the highest point of the swing. Because the speed of the bullet and block at the start of the swing is the speed $V$ immediately after the collision, we may write this conservation as

$$\tfrac{1}{2}(m + M)V^2 + 0 = 0 + (m + M)gh.$$

Substituting this result for $V$ in Eq. 10-30 leads to

$$\begin{aligned} v &= \frac{m + M}{m}\sqrt{2gh} \\ &= \left(\frac{0.0095\text{ kg} + 5.4\text{ kg}}{0.0095\text{ kg}}\right)\sqrt{(2)(9.8\text{ m/s}^2)(0.063\text{ m})} \\ &= 630\text{ m/s}. \end{aligned} \quad \text{(Answer)}$$

The ballistic pendulum is a kind of "transformer," exchanging the high speed of a light object (the bullet) for the low—and thus more easily measurable—speed of a massive object (the block).

## 10-9 One-Dimensional Elastic Collisions

### Stationary Target

As we discussed in Section 10-8, everyday collisions are inelastic but we can approximate some of them as being elastic. That is, we can assume that the total kinetic energy of the colliding bodies is approximately conserved and is not transferred to other forms of energy:

(total kinetic energy before the collision) $\approx$ (total kinetic energy after the collision).

This does not mean that the kinetic energy of each colliding body cannot change. Rather, it means this:

> In an elastic collision, the kinetic energy of each colliding body may change, but the total kinetic energy of the system is the same before the collision as it is after.

For example, consider the collision of a cue ball with an object ball of approximately the same mass in a game of pool. If the collision is head-on (the cue ball heads directly toward the object ball), the kinetic energy of the cue ball can be transferred almost entirely to the object ball. (Still, the fact that the collision makes a sound means that at least a little of the kinetic energy is transferred to the energy of the sound.)

Figure 10-22 shows two bodies whose masses are not necessarily different before and after they have a one-dimensional collision, like a head-on collision between pool balls. A projectile body of mass $m_A$ and initial velocity $\vec{v}_{A\,1}$ moves toward a target body of mass $m_B$ that is initially at rest with velocity $\vec{v}_{B\,1} = 0$. Let's assume that this two-body system is closed and isolated. Then the net linear momentum of the system is conserved, and from Eq. 7-18 we can write

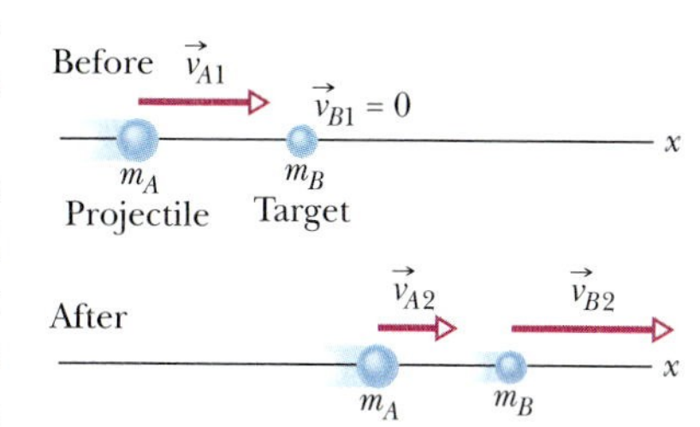


**FIGURE 10-22** ■ Body 1 moves along an $x$ axis before having an elastic collision with body 2, initially at rest. Both bodies move along that axis after the collision.

$$m_A \vec{v}_{A\,1} = m_A \vec{v}_{A\,2} + m_B \vec{v}_{B\,2} \quad \text{(linear momentum conservation)}, \tag{10-31}$$

where in general $\vec{v}_A(t) = v_{A\,x}(t)\hat{\text{i}}$ and $\vec{v}_B(t) = v_{B\,x}(t)\hat{\text{i}}$.

If the collision is also completely elastic, then the total kinetic energy is conserved and we can write

$$\tfrac{1}{2}m_A v_{A\,1}^2 = \tfrac{1}{2}m_A v_{A\,2}^2 + \tfrac{1}{2}m_B v_{B\,2}^2 \quad \text{(kinetic energy conservation)}. \tag{10-32}$$

In each of these equations, "1" signifies a time before the collision and "2" signifies a time after the collision. If we know the masses of the bodies and if we also know $\vec{v}_{A\,1}$, the initial velocity of body $A$, the only unknown quantities are $\vec{v}_{A\,2}$ and $\vec{v}_{B\,2}$, the final velocities of the two bodies. With two equations at our disposal, we should be able to find these two unknowns.

To do so, we express the velocities in terms of their $x$-components and rewrite Eq. 10-31 as

$$m_A[v_{A\,x}(t_1) - v_{A\,x}(t_2)] = m_B v_{B\,x}(t_2), \tag{10-33}$$

and Eq. 10-32 as*

$$m_A[v_{A\,x}(t_1) - v_{A\,x}(t_2)][v_{A\,x}(t_1) + v_{A\,x}(t_2)] = m_B[v_{B\,x}(t_2)]^2. \tag{10-34}$$

After dividing Eq. 10-34 by Eq. 10-33 and doing some more algebra, we obtain

$$v_{A\,x}(t_2) = \frac{m_A - m_B}{m_A + m_B} v_{A\,x}(t_1), \tag{10-35}$$

and

$$v_{B\,x}(t_2) = \frac{2m_A}{m_A + m_B} v_{A\,x}(t_1). \tag{10-36}$$

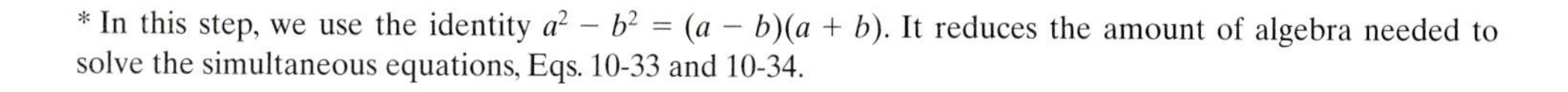

* In this step, we use the identity $a^2 - b^2 = (a - b)(a + b)$. It reduces the amount of algebra needed to solve the simultaneous equations, Eqs. 10-33 and 10-34.

We note from Eq. 10-36 that $v_{Bx}(t_2)$ is always positive (the target body with mass $m_B$ always moves forward). From Eq. 10-35 we see that $v_{Ax}(t_2)$ may be of either sign (the projectile body with mass $m_A$ moves forward if $m_A > m_B$ but rebounds if $m_A < m_B$).

Let us look at a few special situations.

1. **Equal masses.** If $m_A = m_B$, Eqs. 10-35 and 10-36 reduce to

$$v_{Ax}(t_2) = 0 \quad \text{and} \quad v_{Bx}(t_2) = v_{Ax}(t_1),$$

which we might call a pool player's result. It predicts that after a head-on collision of bodies with equal masses, body $A$ (initially moving) stops dead in its tracks and body $B$ (initially at rest) takes off with the initial speed of body $A$. In head-on collisions, bodies of equal mass simply exchange velocities. This is true even if the target particle (body $B$) is not initially at rest.

2. **A massive target.** In terms of Fig. 10-22, a massive target means that $m_B \gg m_A$. For example, we might fire a golf ball at a cannonball. Equations 10-35 and 10-36 then reduce to

$$v_{Ax}(t_2) \approx -v_{Ax}(t_1) \quad \text{and} \quad v_{Bx}(t_2) \approx \left(\frac{2m_A}{m_B}\right)v_{Ax}(t_1). \tag{10-37}$$

This tells us that body $A$ (the golf ball) simply bounces back in the same direction from which it came, its speed essentially unchanged. Body $B$ (the cannonball) moves forward at a very low speed, because the quantity in parentheses in Eq. 10-37 is much less than unity. All this is what we should expect.

3. **A massive projectile.** This is the opposite case; that is, $m_A \gg m_B$. This time, we fire a cannonball at a golf ball. Equations 10-35 and 10-36 reduce to

$$v_{Ax}(t_2) \approx v_{Ax}(t_1) \quad \text{and} \quad v_{Bx}(t_2) \approx 2v_{Ax}(t_1). \tag{10-38}$$

Equation 10-38 tells us that body $A$ (the cannonball) simply keeps on going, scarcely slowed by the collision. Body $B$ (the golf ball) charges ahead at twice the speed of the cannonball.

You may wonder: Why twice the speed? As a starting point in thinking about the matter, recall the collision described by Eq. 10-37, in which the velocity of the incident light body (the golf ball) changed from $v_{Ax}(t_1)$ to $-v_{Ax}(t_1)$, a velocity *change* of magnitude $|2v_{Ax}(t_1)|$. The same magnitude of *change* in velocity (from 0 to $|2v_{Ax}(t_1)|$) occurs in this example also.

## Moving Target

Now that we have examined the elastic collision of a projectile and a stationary target, let us examine the situation in which both bodies are moving before they undergo an elastic collision.

For the situation of Fig. 10-23, the conservation of linear momentum is written as

$$m_A\vec{v}_{A1} + m_B\vec{v}_{B1} = m_A\vec{v}_{A2} + m_B\vec{v}_{B2}, \tag{10-39}$$

FIGURE 10-23 ■ Two bodies headed for a one-dimensional elastic collision.

and the conservation of kinetic energy is written as

$$\tfrac{1}{2}m_Av_{A1}^2 + \tfrac{1}{2}m_Bv_{B1}^2 = \tfrac{1}{2}m_Av_{A2}^2 + \tfrac{1}{2}m_Bv_{B2}^2. \tag{10-40}$$

If we use similar procedures to those used in deriving Eqs. 10-35 and 10-36, we get

$$v_{A\,x}(t_2) = \frac{m_A - m_B}{m_A + m_B}\, v_{A\,x}(t_1) + \frac{2m_B}{m_A + m_B}\, v_{B\,x}(t_1), \tag{10-41}$$

and

$$v_{B\,x}(t_2) = \frac{2m_A}{m_A + m_B}\, v_{A\,x}(t_1) + \frac{m_B - m_A}{m_A + m_B}\, v_{B\,x}(t_1). \tag{10-42}$$

Note that the assignment of subscripts $A$ and $B$ to the bodies is arbitrary. If we exchange those subscripts in Fig. 10-23 and in Eqs. 10-41 and 10-42, we end up with the same set of equations. Note also that if we set $v_{B\,x}(t_1) = 0$, body $B$ becomes a stationary target, and Eqs. 10-41 and 10-42 reduce to Eqs. 10-35 and 10-36, respectively.

**READING EXERCISE 10-8:** What is the final translational momentum of the target in Fig. 10-22 if the initial translational momentum of the projectile is 6 kg·m/s and the final translational momentum of the projectile is (a) 2 kg·m/s and (b) −2 kg·m/s? (c) If the collision is elastic, what is the final kinetic energy of the target if the initial and final kinetic energies of the projectile are, respectively, 5 J and 2 J? ■

## TOUCHSTONE EXAMPLE 10-8: Colliding Pendula

Two metal spheres, suspended by vertical cords, initially just touch, as shown in Fig. 10-24. Sphere $A$, with mass $m_A = 30$ g, is pulled to the left to height $h_0 = 8.0$ cm and then released from rest. After swinging down, it undergoes an elastic collision with sphere $B$, whose mass $m_B = 75$ g. What is the velocity $\vec{v}_{A\,2}$ of sphere $A$ just after the collision?

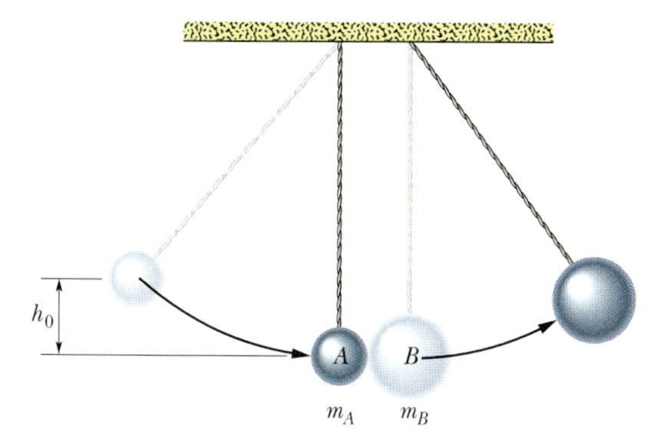


**FIGURE 10-24** ■ Two metal spheres suspended by cords just touch when they are at rest. Sphere $A$, with mass $m_A$, is pulled to the left to height $h_0$ and then released.

**SOLUTION** ■ A first **Key Idea** is that we can split this complicated motion into two steps that we can separately analyze: (1) the descent of sphere $A$ and (2) the two-sphere collision.

**Step 1.** The **Key Idea** here is that as sphere $A$ swings down, the mechanical energy of the sphere–Earth system is conserved. (The mechanical energy is not changed by the force of the cord on sphere $A$ because that force is always directed perpendicular to the sphere's direction of travel.) Let's take the lowest level as our reference level of zero gravitational potential energy. Then the kinetic energy of sphere $A$ at the lowest level must equal the gravitational potential energy of the system when sphere $A$ is at the initial height. Thus,

$$\tfrac{1}{2} m_A v_{A\,1}^2 = m_A g h_0,$$

which we solve for the speed $v_{A\,1}$ of sphere $A$ just before the collision:

$$|\vec{v}_{A\,1}| = \sqrt{2gh_0} = \sqrt{(2)(9.8\ \text{m/s}^2)(0.080\ \text{m})} = 1.252\ \text{m/s}.$$

**Step 2.** Here we can make two assumptions in addition to the assumption that the collision is elastic. First, we can assume that the collision is one-dimensional because the motions of the spheres are approximately horizontal from just before the collision ($\vec{v}_{A\,1} = v_{A\,x}(t_1)\hat{\text{i}}$) to just after it ($\vec{v}_{A\,2} = v_{A\,x}(t_2)\hat{\text{i}}$). Second, because the collision is so brief, we can assume that the two-sphere system is closed and isolated. This gives the **Key Idea** that the total translational momentum of the system is conserved. Thus, we can use Eq. 10-35 to find the velocity of sphere $A$ just after the collision:

$$v_{A\,x}(t_2) = \frac{m_A - m_B}{m_A + m_B}\, v_{A\,x}(t_1) = \frac{0.030 - 0.075\ \text{kg}}{0.030 + 0.075\ \text{kg}}\,(1.252\ \text{m/s})$$

$$= -0.537\ \text{m/s} \approx -0.54\ \text{m/s}. \qquad \text{(Answer)}$$

The minus sign tells us that sphere $A$ moves to the left just after the collision.

## 10-10 Two-Dimensional Energy and Momentum Conservation

When two bodies collide, the impulses of one on the other determine the directions in which they then travel. In particular, when the collision is not head-on, the bodies do not end up traveling along their initial axis. For such two-dimensional collisions in a closed, isolated system, the total translational momentum must still be conserved:

$$\vec{p}_{A1} + \vec{p}_{B1} = \vec{p}_{A2} + \vec{p}_{B2}. \tag{10-43}$$

If the collision is also elastic (a special case), then the total kinetic energy is also conserved

$$K_{A1} + K_{B1} = K_{A2} + K_{B2}. \tag{10-44}$$

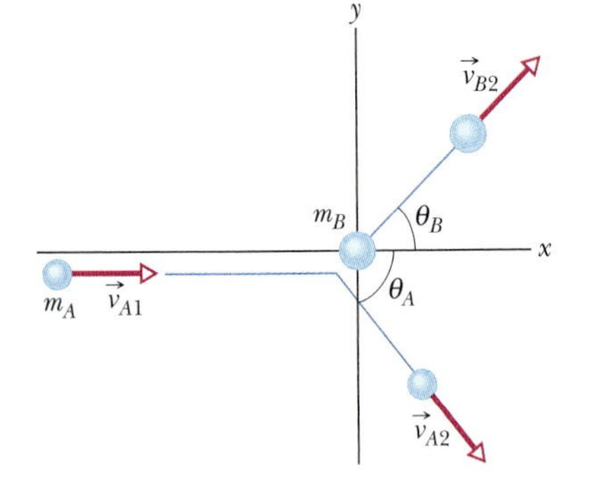


**FIGURE 10-25** ■ An elastic collision between two bodies in which the collision is not head-on. The body with mass $m_B$ (the target) is initially at rest.

Equation 10-43 is often more useful for analyzing a two-dimensional collision if we write it in terms of components on an $xy$-coordinate system. For example, let's revisit the momentum conservation analysis we did in Section 7-7 for a glancing two-dimensional collision. This time we will add the requirement that the collision be elastic so kinetic energy is conserved. Figure 10-25 shows a *glancing collision* (it is not head-on) between a projectile body and a target body initially at rest. The impulses between the bodies have sent the bodies off at angles $\theta_A$ and $\theta_B$ with respect to the $x$ axis, along which the projectile traveled initially. In this situation, we would rewrite the momentum conservation equation initially presented in Section 7-7 (Eqs. 7-24 and 7-25) in terms of components along the $x$ axis as

$$m_A|\vec{v}_{A1}|\cos 0° = m_A|\vec{v}_{A2}|\cos\theta_A + m_B|\vec{v}_{B2}|\cos\theta_B, \tag{10-45}$$

and along the $y$ axis as

$$0 = -m_A|\vec{v}_{A2}|\sin\theta_A + m_B|\vec{v}_{B2}|\sin\theta_B. \tag{10-46}$$

We can also write Eq. 10-44 for this situation as

$$\tfrac{1}{2}m_A v_{A1}^2 = \tfrac{1}{2}m_A v_{A2}^2 + \tfrac{1}{2}m_B v_{B2}^2 \quad \text{(kinetic energy target initially at rest).} \tag{10-47}$$

Equations 10-45 to 10-47 contain seven variables: two masses, $m_A$ and $m_B$; three velocity magnitudes, $v_{A1}$, $v_{A2}$, and $v_{B2}$; and two angles, $\theta_A$ and $\theta_B$. If we know any four of these quantities, we can solve the three equations for the remaining three quantities.

**READING EXERCISE 10-9:** In Fig. 10-25, suppose that the projectile has an initial $x$-component of momentum of 6 kg·m/s, a final $x$-component of momentum of 4 kg·m/s, and a final $y$-component of momentum of −3 kg·m/s. For the target, what then are (a) the final $x$-component of momentum and (b) the final $y$-component of momentum? ■

# Problems

## SEC. 10-3 ■ POTENTIAL ENERGY AS STORED WORK

**1. Spring Constant** What is the spring constant of a spring that stores 25 J of elastic potential energy when compressed by 7.5 cm from its relaxed length?

**2. Dropping a Textbook** You drop a 2.00 kg textbook to a friend who stands on the ground 10.0 m below the textbook with outstretched hands 1.50 m above the ground (Fig. 10-26). (a) How much work $W^{grav}$ is done on the textbook by the gravitational force as it drops to your friend's hands? (b) What is the change $\Delta U$ in the gravitational potential energy of the textbook–Earth system during the drop? If the gravitational potential energy $U$ of that system is taken to be zero at ground level, what is $U$ when the textbook (c) is released and (d) reaches the hands? Now take $U$ to be 100 J at ground level and again find (e) $W^{grav}$ (f) $\Delta U$, (g) $U$ at the release point, and (h) $U$ at the hands.

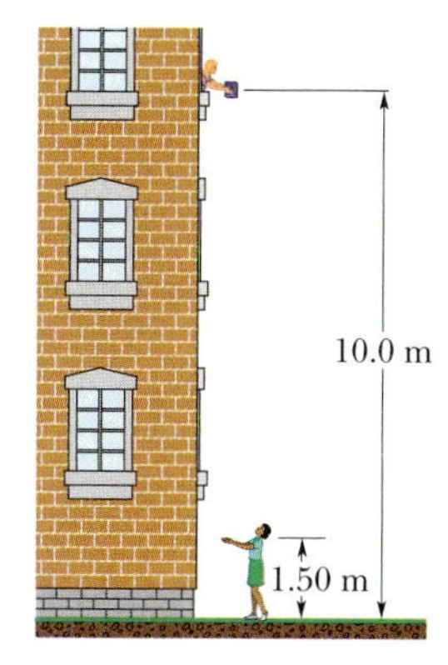

FIGURE 10-26 ■ Problems 2 and 12.

**3. Ice Flake** In Fig. 10-27, a 2.00 g ice flake is released from the edge of a hemispherical bowl whose radius $r$ is 22.0 cm. The flake–bowl contact is frictionless. (a) How much work is done on the flake by the gravitational force during the flake's descent to the bottom of the bowl? (b) What is the change in the potential energy of the flake–Earth system during that descent? (c) If that potential energy is taken to be zero at the bottom of the bowl, what is its value when the flake is released? (d) If, instead, the potential energy is taken to be zero at the release point, what is its value when the flake reaches the bottom of the bowl? (e) If the mass of the flake were doubled, would the magnitudes of the answers to (a) through (d) increase, decrease, or remain the same?

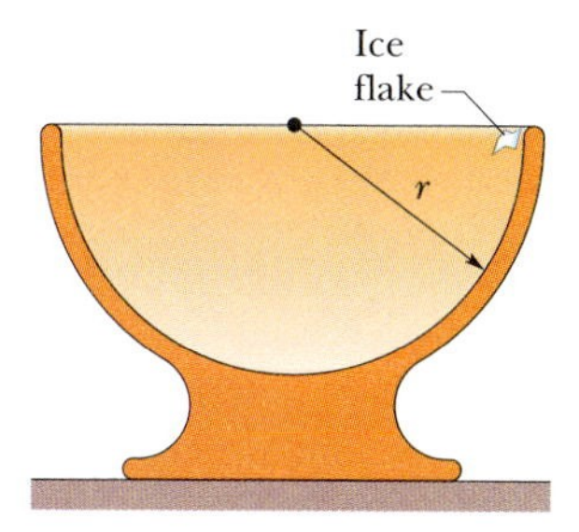

FIGURE 10-27 ■ Problems 3 and 11.

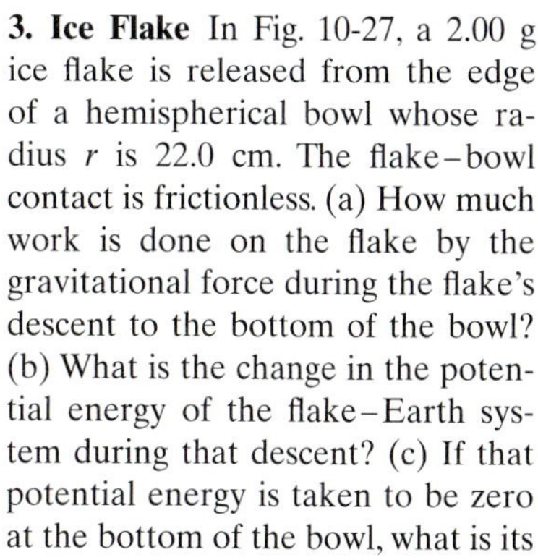

**4. Roller Coaster** In Fig. 10-28, a frictionless roller coaster of mass $m$ tops the first hill with speed $v_1$. How much work does the gravitational force do on it from that point to (a) point $A$, (b) point $B$, and (c) point $C$? If the gravitational potential energy of the coaster–Earth system is taken to be zero at point $C$, what is its value when the coaster is at (d) point $B$ and (e) point $A$? (f) If mass $m$ were doubled, would the change in the gravitational potential energy of the system between points $A$ and $B$ increase, decrease, or remain the same?

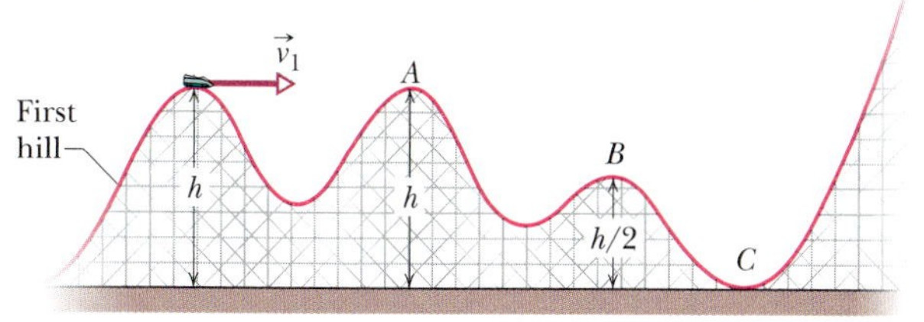

FIGURE 10-28 ■ Problems 4 and 14.

**5. Ball Attached to a Rod** Figure 10-29 shows a ball with mass $m$ attached to the end of a thin rod with length $L$ and negligible mass. The other end of the rod is pivoted so that the ball can move in a vertical circle. The rod is held in the horizontal position as shown and then given enough of a downward push to cause the ball to swing down and around and just reach the vertically upward position, with zero speed there. How much work is done on the ball by the gravitational force from the initial point to (a) the lowest point, (b) the highest point, and (c) the point on the right at which the ball is level with the initial point? If the gravitational potential energy of the ball–Earth system is taken to be zero at the initial point, what is its value when the ball reaches (d) the lowest point, (e) the highest point, and (f) the point on the right that is level with the initial point? (g) Suppose the rod were pushed harder so that the ball passed through the highest point with a nonzero speed. Would the change in the gravitational potential energy from the lowest point to the highest point then be greater, less, or the same?

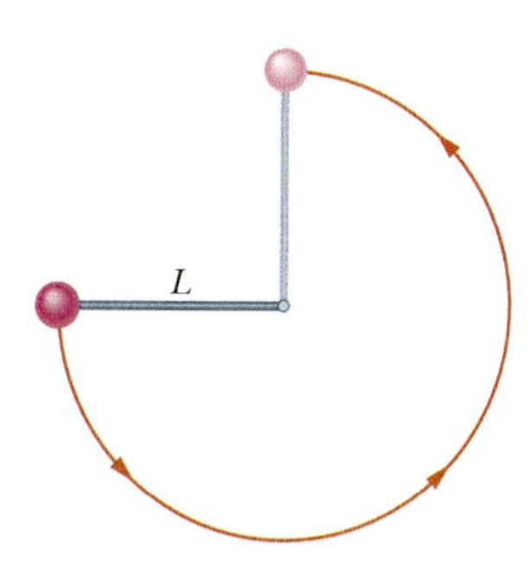

FIGURE 10-29 ■ Problems 5 and 13.

**6. Loop-the-Loop** In Fig. 10-30, a small block of mass $m$ can slide along the frictionless loop-the-loop. The block is released from rest at point $P$, at height $h = 5R$ above the bottom of the loop. How much work does the gravitational force do on the block as the block travels from point $P$ to (a) point $Q$ and (b) the top of the loop? If the gravitational potential energy of the block–Earth system is taken to be zero at the bottom of the loop, what is that potential energy when the block is (c) at point $P$, (d) at point $Q$, and (e) at the top of the loop? (f) If, instead of being released, the block is given some initial speed downward along the track, do the answers to (a) through (e) increase, decrease, or remain the same?

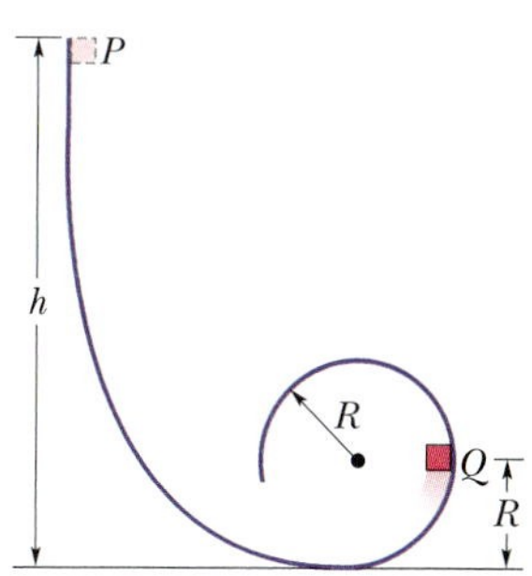

FIGURE 10-30 ■ Problems 6 and 22.

**7. Snowball** A 1.50 kg snowball is fired from a cliff 12.5 m high with an initial velocity of 14.0 m/s, directed 41.0° above the horizontal. (a) How much work is done on the snowball by the gravitational force during its flight to the flat ground below the cliff? (b) What is the change in the gravitational potential energy of the snowball–Earth system during the flight? (c) If that gravitational potential energy is taken to be zero at the height of the cliff, what is its value when the snowball reaches the ground?

**8. Thin Rod** Figure 10-31 shows a thin rod, of length $L$ and negligible mass, that can pivot about one end to rotate in a vertical circle. A heavy ball of mass $m$ is attached to the other end. The rod is pulled aside through an angle $\theta$ and released. As the ball descends to its lowest point, (a) how much work does the gravitational force do on it and (b) what is the change in the gravitational potential energy of the ball–Earth system? (c) If the gravitational potential energy is taken to be zero at the lowest point, what is its value just as the ball is released? (d) Do the magnitudes of the answers to (a) through (c) increase, decrease, or remain the same if angle $\theta$ is increased?

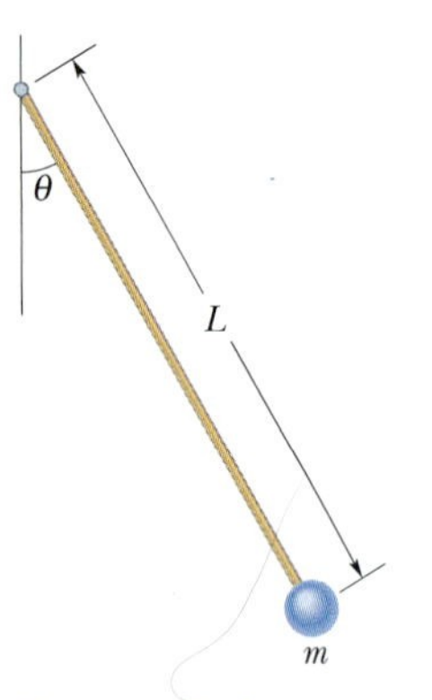


FIGURE 10-31 ■ Problems 8 and 16.

**9. Ball Thrown from Tower** At $t_1 = 0$ a 1.0 kg ball is thrown from the top of a tall tower with velocity $\vec{v}_1 = (18 \text{ m/s})\hat{i} + (24 \text{ m/s})\hat{j}$. What is the change in the potential energy of the ball–Earth system between $t_1 = 0$ and $t_2 = 6.0$ s?

## SEC. 10-4 ■ MECHANICAL ENERGY CONSERVATION

**10.** A 250 g block is dropped onto a relaxed vertical spring that has a spring constant of $k = 2.5$ N/cm (Fig. 10-32). The block becomes attached to the spring and compresses the spring 12 cm before momentarily stopping. While the spring is being compressed, what work is done on the block by (a) the gravitational force on it and (b) the spring force? (c) What is the speed of the block just before it hits the spring? (Assume that friction is negligible.) (d) If the speed at impact is doubled, what is the maximum compression of the spring?

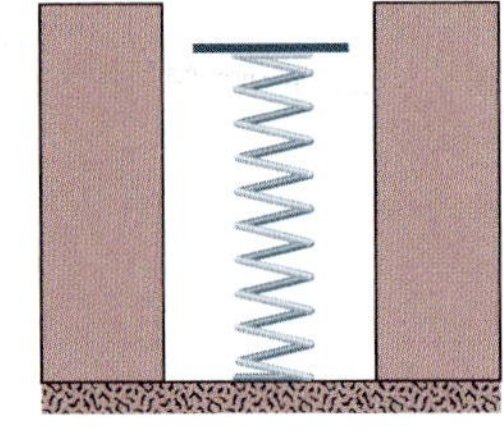

FIGURE 10-32 ■ Problem 10.

**11. Speed of Flake** (a) In Problem 3, what is the speed of the flake when it reaches the bottom of the bowl? (b) If we substituted a second flake with twice the mass, what would its speed be? (c) If, instead, we gave the flake an initial downward speed along the bowl, would the answer to (a) increase, decrease, or remain the same?

**12. Speed of Textbook** (a) In Problem 2, what is the speed of the textbook when it reaches the hands? (b) If we substituted a second textbook with twice the mass, what would its speed be? (c) If, instead, the textbook were thrown down, would the answer to (a) increase, decrease, or remain the same?

**13. Zero Speed at Vertical** (a) In Problem 5, what initial speed must be given the ball so that it reaches the vertically upward position with zero speed? What then is its speed at (b) the lowest point and (c) the point on the right at which the ball is level with the initial point? (d) If the ball's mass were doubled, would the answers to (a) through (c) increase, decrease, or remain the same?

**14. Speed of Coaster** In Problem 4, what is the speed of the coaster at (a) point $A$, (b) point $B$, and (c) point $C$? (d) How high will it go on the last hill, which is too high for it to cross? (e) If we substitute a second coaster with twice the mass, what then are the answers to (a) through (d)?

**15. Runaway Truck** In Fig. 10-33, a runaway truck with failed brakes is moving downgrade at 130 km/h just before the driver steers the truck up a frictionless emergency escape ramp with an inclination of 15°. The truck's mass is 5000 kg. (a) What minimum length $L$ must the ramp have if the truck is to stop (momentarily) along it? (Assume the truck is a particle, and justify that assumption.) Does the minimum length $L$ increase, decrease, or remain the same if (b) the truck's mass is decreased and (c) its speed is decreased?

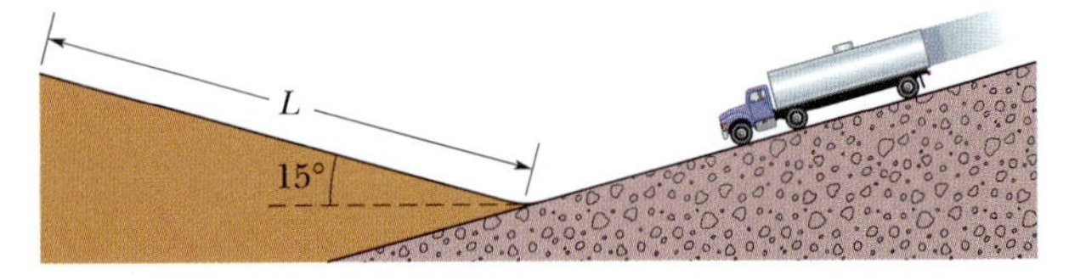


FIGURE 10-33 ■ Problem 15.

**16. Speed at Lowest Point** (a) In Problem 8, what is the speed of the ball at the lowest point if $L = 2.00$ m, $\theta = 30.0°$, and $m = 5.00$ kg? (b) Does the speed increase, decrease, or remain the same if the mass is increased?

**17. Snowball Reaches Ground** (a) In Problem 7, using energy techniques rather than the techniques of Chapter 5, find the speed of the snowball as it reaches the ground below the cliff. What is that speed (b) if the launch angle is changed to 41.0° *below* the horizontal and (c) if the mass is changed to 2.50 kg?

**18. Stone Rests on Spring** Figure 10-34 shows an 8.00 kg stone at rest on a spring. The spring is compressed 10.0 cm by the stone. (a) What is the spring constant? (b) The stone is pushed down an additional 30.0 cm and released. What is the elastic potential energy of the compressed spring just before that release? (c) What is the change in the gravitational potential energy of the stone–Earth system when the stone moves from the release point to its maximum height? (d) What is that maximum height, measured from the release point?

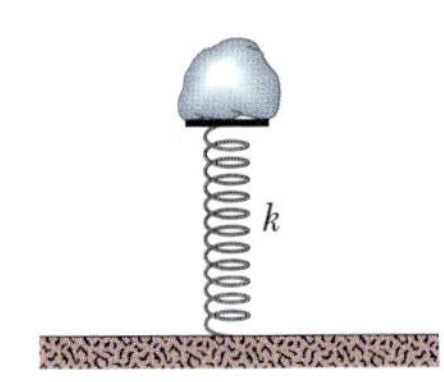


FIGURE 10-34 ■ Problem 18.

**19. Marble Fired Vertically** A 5.0 g marble is fired vertically upward using a spring gun. The spring must be compressed 8.0 cm if the marble is to just reach a target 20 m above the marble's position on the compressed spring. (a) What is the change $\Delta U^{\text{grav}}$ in the gravitational potential energy of the marble–Earth system during the 20 m ascent? (b) What is the change $\Delta U^{\text{elas}}$ in the elastic potential energy of the spring during its launch of the marble? (c) What is the spring constant of the spring?

**20. Pendulum** Figure 10-35 shows a pendulum of length $L$. Its bob (which effectively has all the mass) has speed $v_1$ when the cord makes an angle $\theta_1$ with the vertical. (a) Derive an expression for the speed of the bob when it is in its lowest position. What is the least value that $v_1$ can have if the pendulum is to swing down and then up (b) to a horizontal position, and (c) to a vertical position with the cord remaining straight? (d) Do the answers to (b) and (c) increase, decrease, or remain the same if $\theta_1$ is increased by a few degrees?

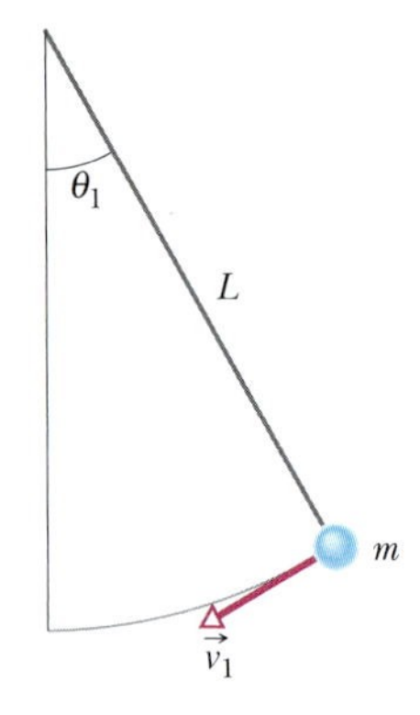


FIGURE 10-35 ■ Problem 20.

**21. Block–Spring–Incline** A 2.00 kg block is placed against a spring on a frictionless 30.0° incline (Fig. 10-36). (The block is not

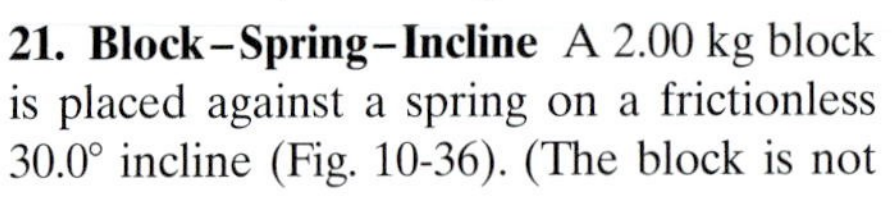

attached to the spring.) The spring, whose spring constant is 19.6 N/cm, is compressed 20.0 cm and then released. (a) What is the elastic potential energy of the compressed spring? (b) What is the change in the gravitational potential energy of the block–Earth system as the bock moves from the release point to its highest point on the incline? (c) How far along the incline is the highest point from the release point?

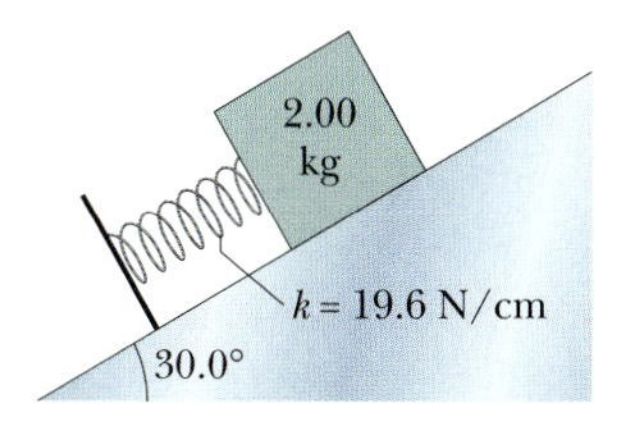


FIGURE 10-36 ■ Problem 21.

**22. Horizontal and Vertical Components** In Problem 6, what are (a) the horizontal component and (b) the vertical component of the *net* force acting on the block at point $Q$? (c) At what height $h$ should the block be released from rest so that it is on the verge of losing contact with the track at the top of the loop? (*On the verge of losing contact* means that the normal force on the block from the track has just then become zero.) (d) Graph the magnitude of the normal force on the block at the top of the loop versus initial height $h$, for the range $h = 0$ to $h = 6R$.

**23. Block on Incline Collides with Spring** In Fig. 10-37, a 12 kg block is released from rest on a 30° frictionless incline. Below the block is a spring that can be compressed 2.0 cm by a force of 270 N. The block momentarily stops when it compresses the spring by 5.5 cm. (a) How far does the block move down the incline from its rest position to this stopping point? (b) What is the speed of the block just as it touches spring?

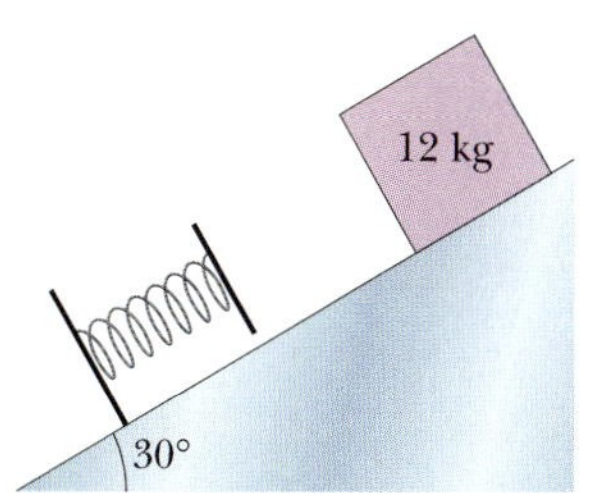


FIGURE 10-37 ■ Problem 23.

**24. Ski-Jump Ramp** A 60 kg skier starts from rest at a height of 20 m above the end of a ski-jump ramp as shown in Fig. 10-38. As the skier leaves the ramp, his velocity makes an angle of 28° with the horizontal. Neglect the effects of air resistance and assume the ramp is frictionless. (a) What is the maximum height $h$ of his jump above the end of the ramp? (b) If he increased his weight by putting on a backpack, would $h$ then be greater, less, or the same?

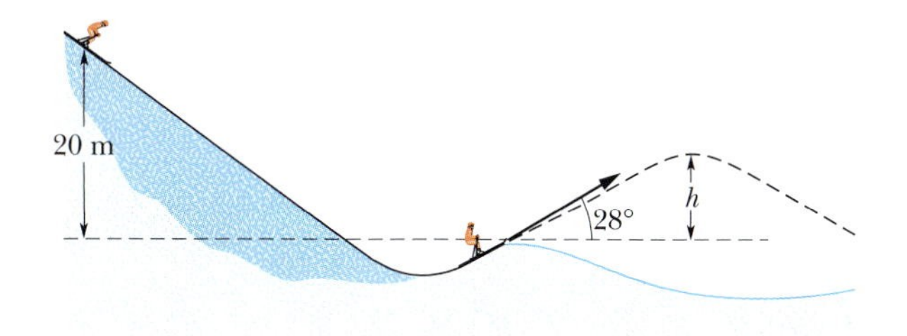


FIGURE 10-38 ■ Problem 24.

**25. Block Dropped on a Spring Two** A 2.0 kg block is dropped from a height of 40 cm onto a spring of spring constant $k = 1960$ N/m (Fig. 10-39). Find the maximum distance the spring is compressed.

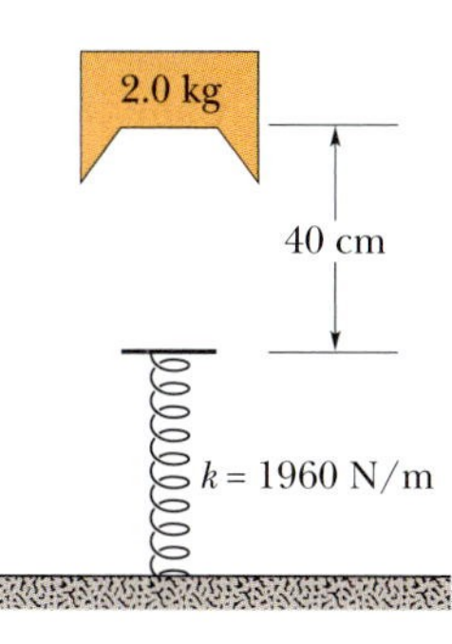


FIGURE 10-39 ■ Problem 25.

**26. Tarzan** Tarzan, who weighs 688 N, swings from a cliff at the end of a convenient vine that is 18 m long (Fig. 10-40). From the top of the cliff to the bottom of the swing, he descends by 3.2 m. The vine will break if the force on it exceeds 950 N. (a) Does the vine break? (b) If no, what is the greatest force on it during the swing? If yes, at what angle with the vertical does it break?

FIGURE 10-40 ■ Problem 26.

**27. Two Children Play** Two children are playing a game in which they try to hit a small box on the floor with a marble fired from a spring loaded gun that is mounted on a table. The target box is 2.20 m horizontally from the edge of the table; see Fig. 10-41. Bobby compresses the spring 1.10 cm, but the center of the marble falls 27.0 cm short of the center of the box. How far should Rhoda compress the spring to score a direct hit? Assume that neither the spring nor the ball encounters friction in the gun.

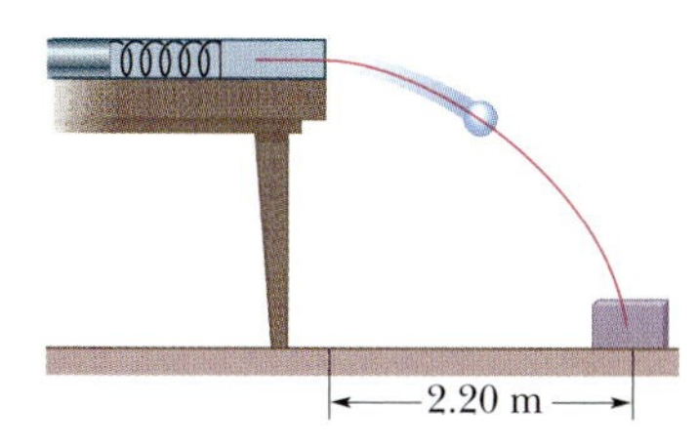


FIGURE 10-41 ■ Problem 27.

**28. Block Sticks to Spring** A 700 g block is released from rest at height $h_1$ above a vertical spring with spring constant $k = 400$ N/m and negligible mass. The block sticks to the spring and momentarily stops after compressing the spring 19.0 cm. How much work is done (a) by the block on the spring and (b) by the spring on the block? (c) What is the value of $h_1$? (d) If the block were released from height $2h_1$ above the spring, what would be the maximum compression of the spring?

**29. Complete Swing** In Fig. 10-42 show that, if the ball is to swing completely around the fixed peg, then $d > 3L/5$. (*Hint*: The ball must still be moving at the top of its swing. Do you see why?)

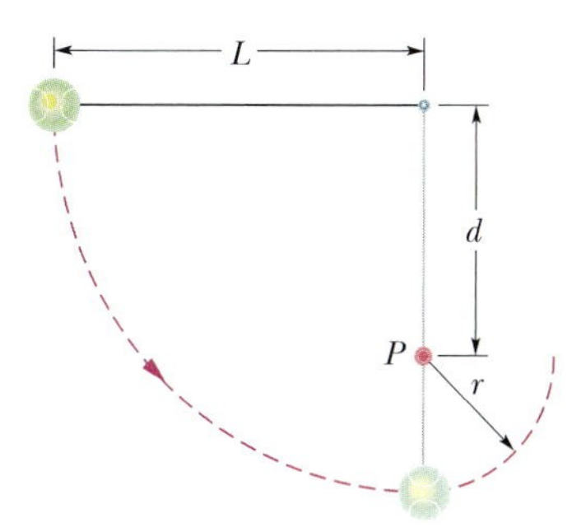


FIGURE 10-42 ■ Problems 29 and 36.

**30. To Make a Pendulum** To make a pendulum, a 300 g ball is attached to one end of a string that has a length of 1.4 m and negligible mass. (The other end of the string is fixed.) The ball is pulled to one side until the string makes an angle of 30.0° with the vertical; then (with the string taut) the ball is released from rest. Find (a) the speed of the ball when the string makes an angle of 20.0° with the vertical and (b) the maximum speed of the ball. (c) What is the angle between the string and the vertical when the speed of the ball is one-third its maximum value?

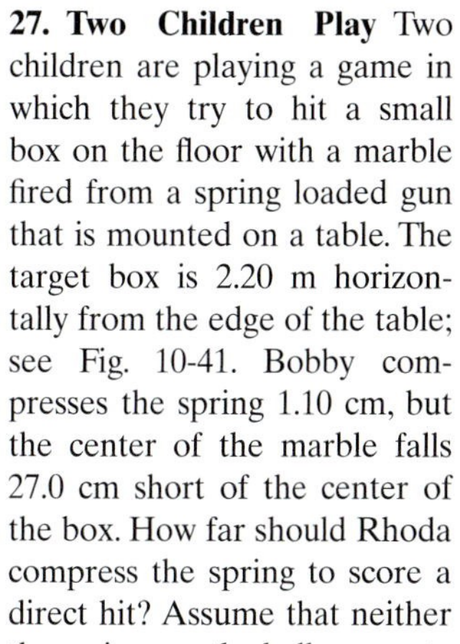

**31. Rigid Rod** A rigid rod of length $L$ and negligible mass has a ball with mass $m$ attached to one end and its other end fixed, to form a pendulum. The pendulum is inverted, with the rod straight up, and then released. At the lowest point, what are (a) the ball's speed and (b) the tension in the rod? (c) The pendulum is next released at rest from a horizontal position. At what angle from the vertical does the tension in the rod equal the weight of the ball?

**32. Spring at the Top of an Incline** In Fig. 10-43, a spring with spring constant $k = 170$ N/m is at the top of a 37.0° frictionless incline. The lower end of the incline is 1.00 m from the end of the spring, which is at its relaxed length. A 2.00 kg canister is pushed against the spring until the spring is compressed 0.200 m and released from rest. (a) What is the speed of the canister at the instant the spring returns to its relaxed length (which is when the canister loses contact with the spring)? (b) What is the speed of the canister when it reaches the lower end of the incline?

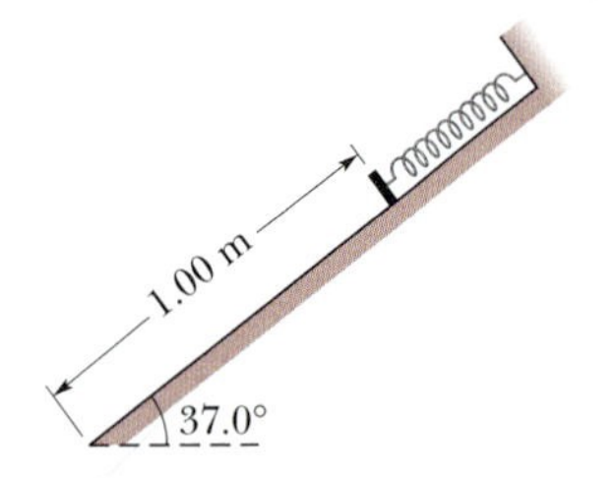


FIGURE 10-43 ■ Problem 32.

**33. Chain on Table** In Fig. 10-44, a chain is held on a frictionless table with one-fourth of its length hanging over the edge. If the chain has length $L$ and mass $m$, how much work is required to pull the hanging part back onto the table?

FIGURE 10-44 ■ Problem 33.

**34. Vertical Spring** A spring with spring constant $k = 400$ N/m is placed in a vertical orientation with its lower end supported by a horizontal surface. The upper end is depressed 25.0 cm, and a block with a weight of 40.0 N is placed (unattached) on the depressed spring. The system is then released from rest. Assume the gravitational potential energy $U^{\text{grav}}$ of the block is zero at the release point ($y_1 = 0$) and calculate the gravitational potential energy, the elastic potential energy $U^{\text{elas}}$, and the kinetic energy $K$ of the block for $y_2$ equal to (a) 0, (b) 5.00 cm, (c) 10.0 cm, (d) 15.0 cm, (e) 20.0 cm, (f) 25.0 cm, and (g) 30.0 cm, Also, (h) how far above its point of release does the block rise?

**35. Ice Mound** A boy is seated on the top of a hemispherical mound of ice (Fig. 10-45). He is given a very small push and starts sliding down the ice. Show that he leaves the ice at a point whose height is $2R/3$ if the ice is frictionless. (*Hint*: The normal force vanishes as he leaves the ice.)

FIGURE 10-45 ■ Problem 35.

**36. Ball on a String** The string in Fig. 10-42 is $L = 120$ cm long, has a ball attached to one end, and is fixed at its other end. The distance $d$ to the fixed peg at point $P$ is 75.0 cm. When the initially stationary ball is released with the string horizontal as shown, it will swing along the dashed arc. What is its speed when it reaches (a) its lowest point and (b) its highest point after the string catches on the peg?

## SEC. 10-5 ■ READING A POTENTIAL ENERGY CURVE

**37. Diatomic Molecule** The potential energy of a diatomic molecule (a two-atom system like $H_2$ or $O_2$) is given by

$$U = \frac{A}{r^{12}} - \frac{B}{r^6},$$

where $r$ is the separation of the two atoms of the molecule and $A$ and $B$ are positive constants. This potential energy is associated with the force that binds the two atoms together. (a) Find the *equilibrium separation*—that is, the distance between the atoms at which the force on each atom is zero. Is the force repulsive (the atoms are pushed apart) or attractive (they are pulled together) if their separation is (b) smaller and (c) larger than the equilibrium separation?

**38. Potential Energy Graph** A conservative force $F(x)$ acts on a 2.0 kg particle that moves along the $x$ axis. The potential energy $U(x)$ associated with $F(x)$ is graphed in Fig. 10-46. When the particle is at $x = 2.0$ m, its velocity is $-1.5$ m/s. (a) What are the magnitude and direction of $F(x)$ at this position? (b) Between what limits of $x$ does the particle move? (c) What is its speed at $x = 7.0$ m?

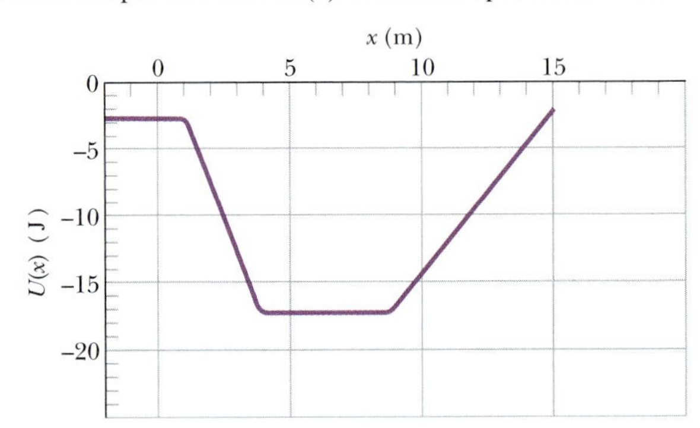


FIGURE 10-46 ■ Problem 38.

**39. Potential Energy Function** A single conservative force $F(x)$ acts on a 1.0 kg particle that moves along an $x$ axis. The potential energy $U(x)$ associated with $F(x)$ is given by

$$U(x) = -4x\,e^{-x/4}\text{ J},$$

where $x$ is in meters. At $x = 5.0$ m the particle has a kinetic energy of 2.0 J. (a) What is the mechanical energy of the system? (b) Make a plot of $U(x)$ as a function of $x$ for $0 \le x \le 10$ m, and on the same graph draw the line that represents the mechanical energy of the system. Use part (b) to determine (c) the least value of $x$ and (d) the greatest value of $x$ between which the particle can move. Use part (b) to determine (e) the maximum kinetic energy of the particle and (f) the value of $x$ at which it occurs. (g) Determine the equation for $F(x)$ as a function of $x$. (h) For what (finite) value of $x$ does $F(x) = 0$?

## SEC. 10-7 ■ CONSERVATION OF ENERGY

**40. Plastic Cube** The temperature of a plastic cube is monitored while the cube is pushed 3.0 m across a floor at constant speed by a horizontal force of 15 N. The monitoring reveals that the thermal energy of the cube increases by 20 J. What is the increase in the thermal energy of the floor along which the cube slides?

**41. Block Drawn by Rope** A 3.57 kg block is drawn at constant speed 4.06 m along a horizontal floor by a rope. The force on the block from the rope has a magnitude of 7.68 N and is directed 15.0° above the horizontal. What are (a) the work done by the rope's force, (b) the increase in thermal energy of the block–floor system, and (c) the coefficient of kinetic friction between the block and floor?

**42. Worker Pushes Block** A worker pushed a 27 kg block 9.2 m along a level floor at constant speed with a force directed 32° below the horizontal. If the coefficient of kinetic friction between block and floor was 0.20, what were (a) the work done by the worker's force and (b) the increase in thermal energy of the block–floor system?

**43. The Collie** A Collie drags its bed box across a floor by applying a horizontal force of 8.0 N. The kinetic frictional force acting on the box has magnitude 5.0 N. As the box is dragged through 0.70 m along the way, what are (a) the work done by the collie's applied force and (b) the increase in thermal energy of the bed and floor?

**44. Bullet Hits Wall** A 30 g bullet, with a horizontal velocity of 500 m/s, comes to a stop 12 cm within a solid wall. (a) What is the change in its mechanical energy? (b) What is the magnitude of the average force from the wall stopping it?

**45. Ski Jumper** A 60 kg skier leaves the end of a ski-jump ramp with a velocity of 24 m/s directed 25° above the horizontal. Suppose that as a result of air drag the skier returns to the ground with a speed of 22 m/s, landing 14 m vertically below the end of the ramp. From the launch to the return to the ground, by how much is the mechanical energy of the skier–Earth system reduced because of air drag?

**46. Frisbee** A 75 g Frisbee is thrown from a point 1.1 m above the ground with a speed of 12 m/s. When it has reached a height of 2.1 m, its speed is 10.5 m/s. What was the reduction in the mechanical energy of the Frisbee–Earth system because of air drag?

**47. Outfielder Throws** An outfielder throws a baseball with an initial speed of 81.8 mi/h. Just before an infielder catches the ball at the same level, the ball's speed is 110 ft/s. In foot-pounds, by how much is the mechanical energy of the ball–Earth system reduced because of air drag? (The weight of a baseball is 9.0 oz.)

**48. Niagara Falls** Approximately $5.5 \times 10^6$ kg of water fall 50 m over Niagara Falls each second. (a) What is the decrease in the gravitational potential energy of the water–Earth system each second? (b) If all this energy could be converted to electrical energy (it cannot be), at what rate would electrical energy be supplied? (The mass of 1 $m^3$ of water is 1000 kg.) (c) If the electrical energy were sold at 1 cent/kW · h. what would be the yearly cost?

**49. Rock Slide** During a rockslide, a 520 kg rock slides from rest down a hillside that is 500 m long and 300 m high. The coefficient of kinetic friction between the rock and the hill surface is 0.25. (a) If the gravitational potential energy $U$ of the rock–Earth system is zero at the bottom of the hill, what is the value of $U$ just before the slide? (b) How much energy is transferred to thermal energy during the slide? (c) What is the kinetic energy of the rock as it reaches the bottom of the hill? (d) What is its speed then?

**50. Block Against Horizontal Spring** You push a 2.0 kg block against a horizontal spring, compressing the spring by 15 cm. Then you release the block, and the spring sends it sliding across a tabletop. It stops 75 cm from where you released it. The spring constant is 200 N/m. What is the coefficient of kinetic friction between the block and the table?

**51. Horizontal Spring** As Fig. 10-47 shows, a 3.5 kg block is accelerated by a compressed spring whose spring constant is 640 N/m. After leaving the spring at the spring's relaxed length, the block travels over a horizontal surface, with a coefficient of kinetic friction of 0.25, for a distance of 7.8 m before stopping. (a) What is the increase in the thermal energy of the block–floor system? (b) What is the maximum kinetic energy of the block? (c) Through what distance is the spring compressed before the block begins to move?

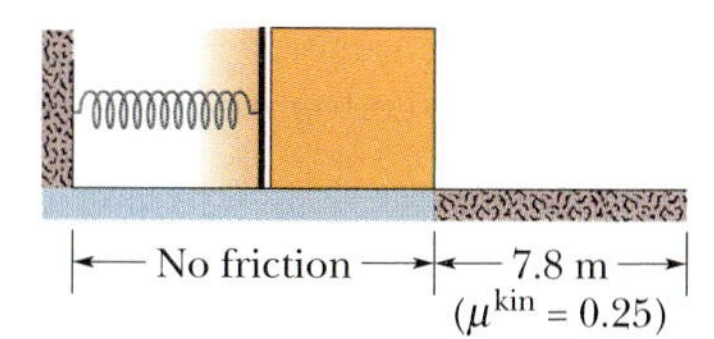


**FIGURE 10-47** ■ Problem 51.

**52. Block Slides Down an Incline** In Fig. 10-48, a block is moved down an incline a distance of 5.0 m from point $A$ to point $B$ by a force $\vec{F}$ that is parallel to the incline and has magnitude 2.0 N. The magnitude of the frictional force acting on the block is 10 N. If the kinetic energy of the block increases by 35 J between $A$ and $B$, how much work is done on the block by the gravitational force as the block moves from $A$ to $B$?

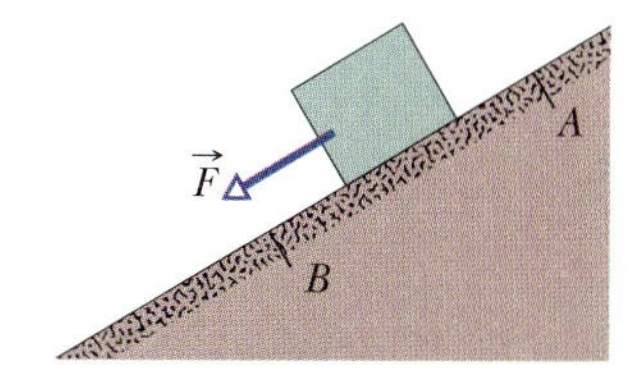


**FIGURE 10-48** ■ Problem 52.

**53. Nonconforming Spring** A certain spring is found *not* to conform to Hooke's law. The force (in newtons) it exerts when stretched a distance $x$ (in meters) is found to have magnitude $(52.8\ \text{N/m})x + (38.4\ \text{N/m}^2)x^2$ in the direction opposing the stretch. (a) Compute the work required to stretch the spring from $x_1 = 0.500$ m to $x_2 = 1.00$ m. (b) With one end of the spring fixed, a particle of mass 2.17 kg is attached to the other end of the spring when it is extended by an amount $x_2 = 1.00$ m. If the particle is then released from rest, what is its speed at the instant the spring has returned to the configuration in which the extension is $x_1 = 0.500$ m? (c) Is the force exerted by the spring conservative or nonconservative? Explain.

**54. Bundle** A 4.0 kg bundle starts up a 30° incline with 128 J of kinetic energy. How far will it slide up the incline if the coefficient of kinetic friction between bundle and incline is 0.30?

**55. Two Snowy Peaks** Two snowy peaks are 850 m and 750 m above the valley between them. A ski run extends down from the top of the higher peak and then back up to the top of the lower one, with a total length of 3.2 km and an average slope of 30° (Fig. 10-49). (a) A skier starts from rest at the top of the higher peak. At what speed will he arrive at the top of the lower peak if he coasts without using ski poles? Ignore friction. (b) Approximately what coefficient of kinetic friction between snow and skis would make him stop just at the top of the lower peak?

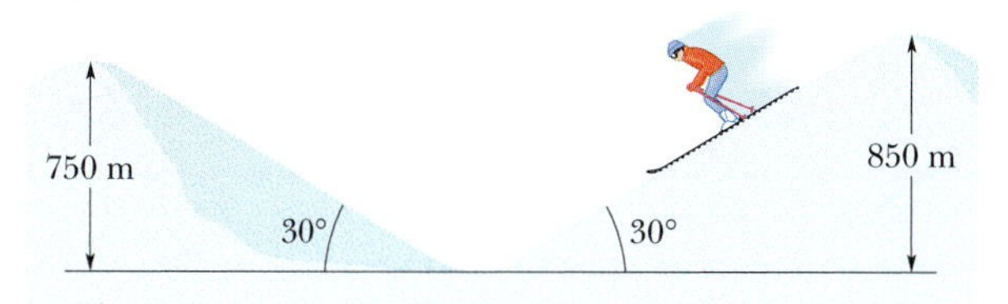


**FIGURE 10-49** ■ Problem 55.

**56. Playground Slide** A girl whose weight is 267 N slides down a 6.1 m playground slide that makes an angle of 20° with the horizontal. The coefficient of kinetic friction between slide and child is 0.10.

(a) How much energy is transferred to thermal energy? (b) If the girl starts at the top with a speed of 0.457 m/s, what is her speed at the bottom?

**57. Block and Horizontal Spring** In Fig. 10-50, a 2.5 kg block slides head on into a spring with a spring constant of 320 N/m. When the block stops, it has compressed the spring by 7.5 cm. The coefficient of kinetic friction between the block and the horizontal surface is 0.25. While the block is in contact with the spring and being brought to rest, what are (a) the work done by the spring force and (b) the increase in thermal energy of the block–floor system? (c) What is the block's speed just as the block reaches the spring?

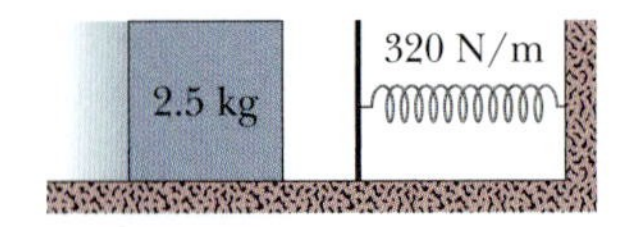


FIGURE 10-50 ■ Problem 57.

**58. Factory Worker** A factory worker accidentally releases a 180 kg crate that was being held at rest at the top of a 3.7 m-long-ramp inclined at 39° to the horizontal. The coefficient of kinetic friction between the crate and the ramp, and between the crate and the horizontal factory floor, is 0.28. (a) How fast is the crate moving as it reaches the bottom of the ramp? (b) How far will it subsequently slide across the factory floor? (Assume that the crate's kinetic energy does not change as it moves from the ramp onto the floor.) (c) Do the answers to (a) and (b) increase, decrease, or remain the same if we halve the mass of the crate?

**59. Block on a Track** In Fig. 10-51, a block slides along a track from one level to a higher level, by moving through an intermediate valley. The track is frictionless until the block reaches the higher level. There a frictional force stops the block in a distance $d$. The block's initial speed $v_1$ is 6.0 m/s; the height difference $h$ is 1.1 m; and the coefficient of kinetic friction $\mu^{kin}$ is 0.60. Find $d$.

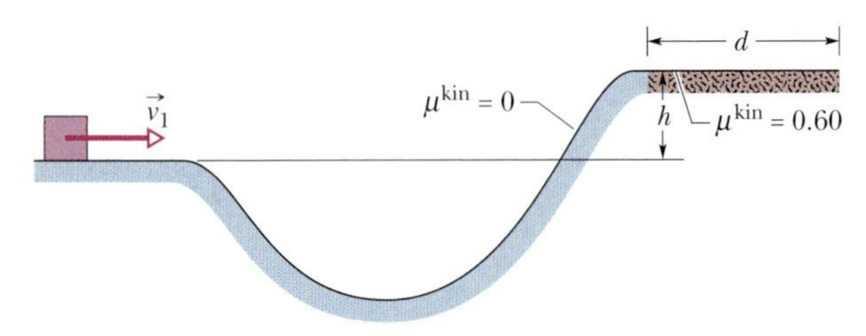


FIGURE 10-51 ■ Problem 59.

**60. Cookie Jar** A cookie jar is moving up a 40° incline. At a point 55 cm from the bottom of the incline (measured along the incline), it has a speed of 1.4 m/s. The coefficient of kinetic friction between jar and incline is 0.15. (a) How much farther up the incline will the jar move? (b) How fast will it be going when it has slid back to the bottom of the incline? (c) Do the answers to (a) and (b) increase, decrease, or remain the same if we decrease the coefficient of kinetic friction (but do not change the given speed or location)?

**61. Stone Thrown Vertically** A stone with weight $w$ is thrown vertically upward into the air from ground level with initial speed $v_1$. If a constant force $f$ due to air drag acts on the stone throughout its flight, (a) show that the maximum height reached by the stone is

$$h = \frac{v_1^2}{2g(1 + f/w)}.$$

(b) Show that the stone's speed is

$$v_2 = v_1\left(\frac{w - f}{w + f}\right)^{1/2}$$

just before impact with the ground.

**62. Playground Slide Two** A playground slide is in the form of an arc of a circle with a maximum height of 4.0 m, with a radius of 12 m, and with the ground tangent to the circle (Fig. 10-52). A 25 kg child starts from rest at the top of the slide and has a speed of 6.2 m/s at the bottom. (a) What is the length of the slide? (b) What average frictional force acts on the child over this distance? If, instead of the ground, a vertical line through the *top of the slide* is tangent to the circle, what are (c) the length of the slide and (d) the average frictional force on the child?

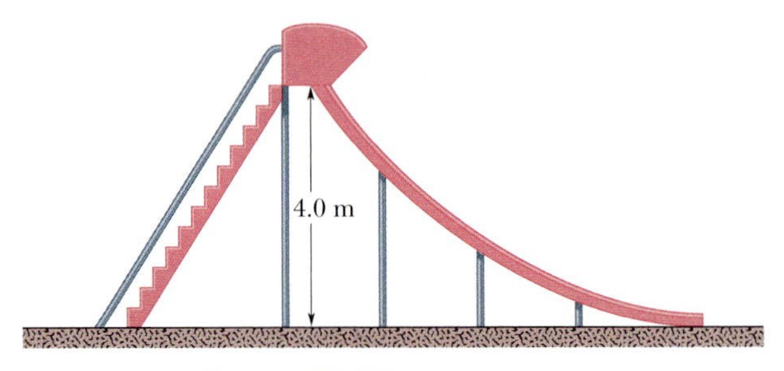


FIGURE 10-52 ■ Problem 62.

**63. Particle on a Slide** A particle can slide along a track with elevated ends and a flat central part, as shown in Fig. 10-53. The flat part has length $L$. The curved portions of the track are frictionless, but for the flat part the coefficient of kinetic friction is $\mu^{kin} = 0.20$. The particle is released from rest at point $A$, which is a height $h = L/2$ above the flat part of the track. Where does the particle finally stop?

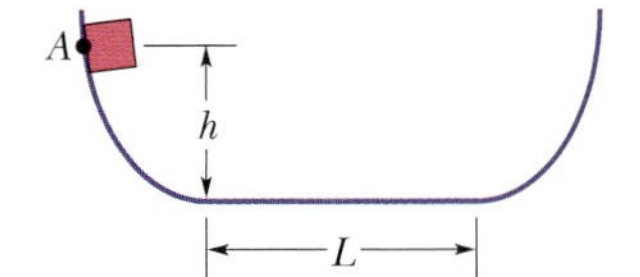


FIGURE 10-53 ■ Problem 63.

**64. Cable Breaks** The cable of the 1800 kg elevator cab in Fig. 10-54 snaps when the cab is at rest at the first floor, where the cab bottom is a distance $d = 3.7$ m above a cushioning spring whose spring constant is $k = 0.15$ MN/m. A safety device clamps the cab against guide rails so that a constant frictional force of 4.4 kN opposes the cab's motion. (a) Find the speed of the cab just before it hits the spring. (b) Find the maximum distance $x$ that the spring is compressed (the frictional force still acts during this compression). (c) Find the distance that the cab will bounce back up the shaft. (d) Using conservation of energy, find the approximate total distance that the cab will move before coming to rest. (Assume that the frictional force on the cab is negligible when the cab is stationary.)

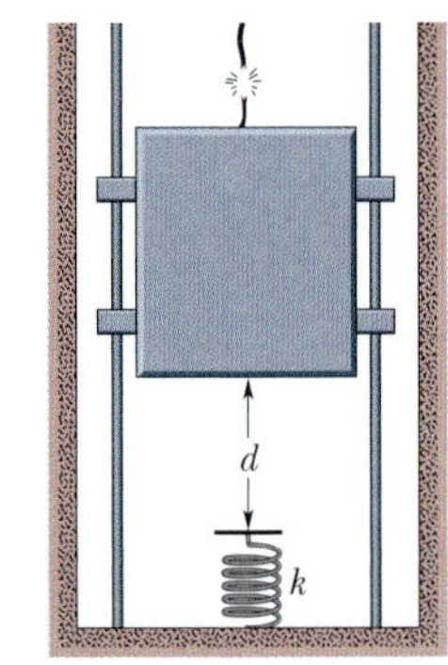


FIGURE 10-54 ■ Problem 64.

**65. At a Factory** At a certain factory, 300 kg crates are dropped vertically from a packing machine onto a conveyor belt moving at 1.20 m/s (Fig 10-55). (A motor maintains the belt's constant speed.) The coefficient of kinetic friction between the belt and each crate is 0.400. After a short time, slipping between the belt and the crate ceases, and the crate then moves along with the belt. For the period of time during which the crate is being brought to rest relative to the belt, calculate, for a coordinate system at rest in the factory, (a) the kinetic energy supplied to the crate, (b) the magnitude of the kinetic frictional force acting on the crate, and (c) the energy supplied by the motor. (d) Explain why the answers to (a) and (c) are different.

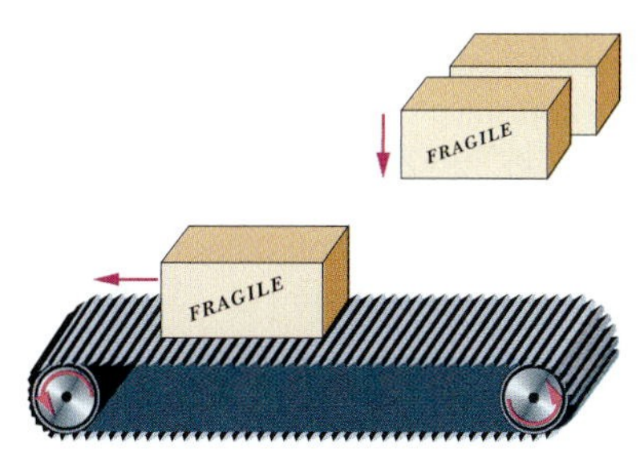


FIGURE 10-55 ■ Problem 65.

**66. A Bear Slides** A 25 kg bear slides, from rest, 12 m down a lodgepole pine tree, moving with a speed of 5.6 m/s just before hitting the ground. (a) What change occurs in the gravitational potential energy of the bear–Earth system during the slide? (b) What is the kinetic energy of the bear just before hitting the ground? (c) What is the average frictional force that acts on the sliding bear?

**67. Daniel Goodwin** In 1981, Daniel Goodwin climbed 443 m up the *exterior* of the Sears Building in Chicago using suction cups and metal clips. (a) Approximate his mass and then compute how much energy he had to transfer from biomechanical (internal) energy to the gravitational potential energy of the Earth–Goodwin system to lift his center of mass to that height. (b) How much energy would he have had to transfer if he had, instead, taken the stairs inside the building (to the same height)?

**68. Mount Everest** The summit of Mount Everest is 8850 m above sea level. (a) How much energy would a 90 kg climber expend against the gravitational force on him in climbing to the summit from sea level? (b) How many candy bars, at 1.25 MJ per bar, would supply an energy equivalent to this? Your answer should suggest that work done against the gravitational force is a very small part of the energy expended in climbing a mountain.

**69. A Woman Leaps Vertically** A 55 kg woman leaps vertically from a crouching position in which her center of mass is 40 cm above the ground. As her feet leave the floor, her center of mass is 90 cm above the ground; it rises of 120 cm at the top of her leap. (a) As she is pressing down on the ground during the leap, what is the average magnitude of the force on her from the ground? (b) What maximum speed does she attain?

**70. An Automobile with Passengers** An automobile with passengers has weight 16,400 N and is moving at 113 km/h when the driver brakes to a stop. The frictional force on the wheels from the road has a magnitude of 8230 N. Find the stopping distance.

## SECS. 10-8 TO 10-10 ■ CONSERVATION OF ENERGY AND MOMENTUM

**71. Box of Marbles** A box is put on a scale that is marked in units of mass and adjusted to read zero when the box is empty. A stream of marbles is then poured into the box from a height $h$ above its bottom at a rate of $R$ (marbles per second). Each marble has mass $m$. (a) If the collisions between the marbles and the box are completely inelastic, find the scale reading at time $t$ after the marbles begin to fill the box. (b) Determine a numerical answer when $R = 100\ \text{s}^{-1}$, $h = 7.60$ m, $m = 4.50$ g, and $t = 10.0$ s.

**72. Particle $A$ and Particle $B$** Particle $A$ and particle $B$ are held together with a compressed spring between them. When they are released, the spring pushes them apart and they then fly off in opposite directions, free of the spring. The mass of $A$ is 2.00 times the mass of $B$, and the energy stored in the spring was 60 J. Assume that the spring has negligible mass and that all its stored energy is transferred to the particles. Once that transfer is complete, what are the kinetic energies of (a) particle $A$ and (b) particle $B$?

**73. Ball and Spring Gun** In Fig. 10-56, a ball of mass $m$ is shot with speed $v_1$ into the barrel of a spring gun of mass $M$ initially at rest on a frictionless surface. The ball sticks in the barrel at the point of maximum compression of the spring. Assume that the increase in thermal energy due to friction between the ball and the barrel is negligible. (a) What is the speed of the spring gun after the ball stops in the barrel? (b) What fraction of the initial kinetic energy of the ball is stored in the spring?

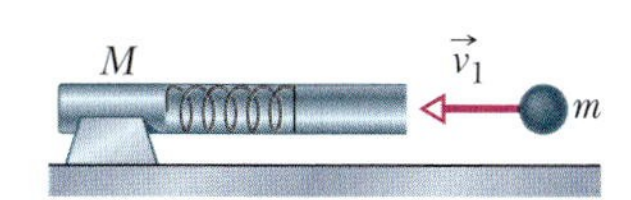


FIGURE 10-56 ■ Problem 73.

**74. Ballistic Pendulum** A bullet of mass 10 g strikes a ballistic pendulum of mass 2.0 kg. The center of mass of the pendulum rises a vertical distance of 12 cm. Assuming that the bullet remains embedded in the pendulum, calculate the bullet's initial speed.

**75. Two Blocks and a Spring** A block of mass $m_A = 2.0$ kg slides along a frictionless table with a speed of 10 m/s. Directly in front of it, and moving in the same direction, is a block of mass $m_B = 5.0$ kg moving at 3.0 m/s. A massless spring with spring constant $k = 1120$ N/m is attached to the near side of $m_B$, as shown in Fig. 10-57. When the blocks collide, what is the maximum compression of the spring? (*Hint*: At the moment of maximum compression of the spring, the two blocks move as one. Find the velocity by noting that the collision is completely inelastic at this point.)

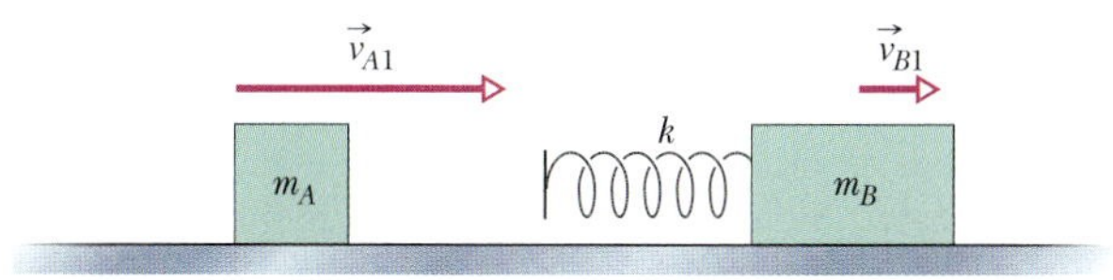


FIGURE 10-57 ■ Problem 75.

**76. Physics Book** A 4.0 kg physics book and a 6.0 kg calculus book, connected by a spring, are stationary on a horizontal frictionless surface. The spring constant is 8000 N/m. The books are pushed together, compressing the spring, and then they are released from rest. When the spring has returned to its unstretched length, the speed of the calculus book is 4.0 m/s. How much energy is stored in the spring at the instant the books are released?

**77. Neutron Scattering** Show that if a neutron is scattered through 90° in an elastic collision with an initially stationary deuteron, the neutron loses $\frac{2}{3}$ of its initial kinetic energy to the deuteron. (In atomic mass units, the mass of a neutron is 1.0 u and the mass of a deuteron is 2.0 u.)

**78. Spring Attached to Wall** A 1.0 kg block at rest on a horizontal frictionless surface is connected to an unstretched spring ($k$ = 200 N/m) whose other end is fixed (Fig. 10-58). A 2.0 kg block moving at 4.0 m/s collides with the 1.0 kg block. If the two blocks stick together after the one-dimensional collision, what maximum compression of the spring occurs when the blocks momentarily stop?

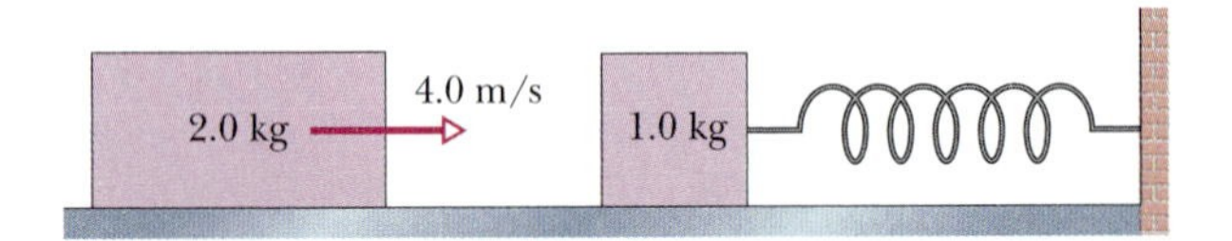


FIGURE 10-58 ■ Problem 78.

**79. Game of Pool** In a game of pool, the cue ball strikes another ball of the same mass and initially at rest. After the collision, the cue ball moves at 3.50 m/s along a line making an angle of 22.0° with its original direction of motion, and the second ball has a speed of 2.00 m/s. Find (a) the angle between the direction of motion of the second ball and the original direction of motion of the cue ball and (b) the original speed of the cue ball. (c) Is kinetic energy (of the centers of mass, don't consider the rotation) conserved?

**80. Billiard Ball** A billiard ball moving at a speed of 2.2 m/s strikes an identical stationary ball a glancing blow. After the collision, one ball is found to be moving at a speed of 1.1 m/s in a direction making a 60° angle with the original line of motion. (a) Find the velocity of the other ball. (b) Can the collision be inelastic, given these data?

**81. Three Balls** In Fig. 10-59, ball $A$ with an initial speed of 10 m/s collides elastically with stationary balls $B$ and $C$, whose centers are on a line perpendicular to the initial velocity of ball $A$ and that are initially in contact with each other. The three balls are identical. Ball $A$ is aimed directly at the contact point, and all motion is frictionless. After the collision, what are the velocities of (a) ball $B$, (b) ball $C$, and (c) ball $A$? (*Hint*: With friction absent, each impulse is directed along the line connecting the centers of the colliding balls, normal to the colliding surfaces.)

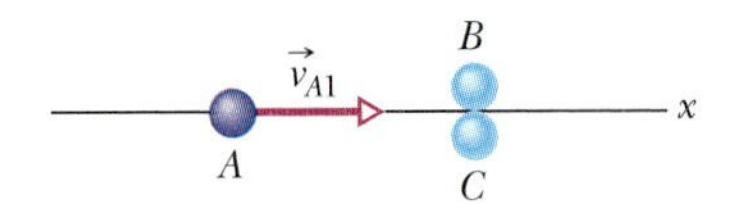


FIGURE 10-59 ■ Problem 81.

**82. Two Bodies Collide** Two 2.0 kg bodies, $A$ and $B$, collide. The velocities before the collision are $\vec{v}_{A1} = (15\text{ m/s})\hat{\text{i}} + (30\text{ m/s})\hat{\text{j}}$ and $\vec{v}_{B1} = (-10\text{ m/s})\hat{\text{i}} + (5.0\text{ m/s})\hat{\text{j}}$. After the collision, $\vec{v}_{A2} = (-5.0\text{ m/s})\hat{\text{i}} + (20\text{ m/s})\hat{\text{j}}$. All speeds are given in meters per second. (a) What is the final velocity of $B$? (b) How much kinetic energy is gained or lost in the collision?

**83. Elastic Collision of Cart** A cart with mass 340 g moving on a frictionless linear air track at an initial speed of 1.2 m/s undergoes an elastic collision with an initially stationary cart of unknown mass. After the collision, the first cart continues in its original direction at 0.66 m/s. (a) What is the mass of the second cart? (b) What is its speed after impact? (c) What is the speed of the two-cart center of mass?

**84. Electron Collision** An electron undergoes a one-dimensional elastic collision with an initially stationary hydrogen atom. What percentage of the electron's initial kinetic energy is transferred to kinetic energy of the hydrogen atom? (The mass of the hydrogen atom is 1840 times the mass of the electron.)

**85. Alpha Particle** An alpha particle (mass 4 u) experiences an elastic head-on collision with a gold nucleus (mass 197 u) that is originally at rest. (The symbol u represents the atomic mass unit.) What percentage of its original kinetic energy does the alpha particle lose?

**86. Voyager 2** Spacecraft *Voyager 2* (of mass $m$ and speed $v$ relative to the Sun) approaches the planet Jupiter (of mass $M$ and speed $V_J$ relative to the Sun) as shown in Fig. 10-60. The spacecraft rounds the planet and departs in the opposite direction. What is its speed, relative to the Sun, after this slingshot encounter, which can be analyzed as a collision? Assume $v$ = 12 km/s and $V_J$ = 13 km/s (the orbital speed of Jupiter). The mass of Jupiter is very much greater than the mass of the spacecraft ($M \gg m$).

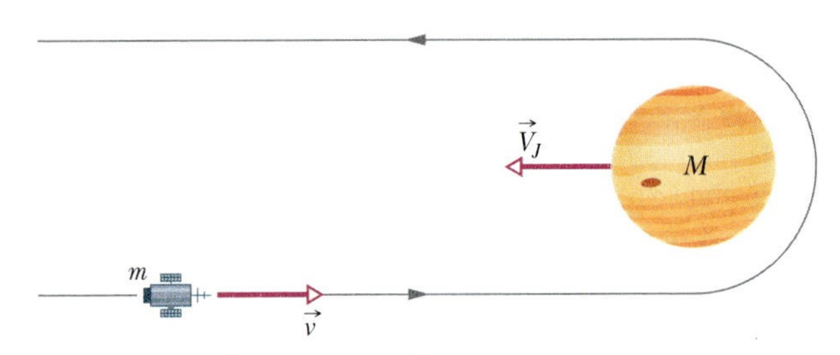


FIGURE 10-60 ■ Problem 86.

**87. Elastic Collision** A body of mass 2.0 kg makes an elastic collision with another body at rest and continues to move in the original direction but with one-fourth of its original speed. (a) What is the mass of the other body? (b) What is the speed of the two-body center of mass if the initial speed of the 2.0 kg body was 4.0 m/s?

**88. Steel Ball and Block** A steel ball of mass 0.500 kg is fastened to a cord that is 70.0 cm long and fixed at the far end. The ball is then released when the cord is horizontal (Fig. 10-61). At the bottom of its path, the ball strikes a 2.50 kg steel block initially at rest on a frictionless surface. The collision is elastic. Find (a) the speed of the ball and (b) the speed of the block, both just after the collision.

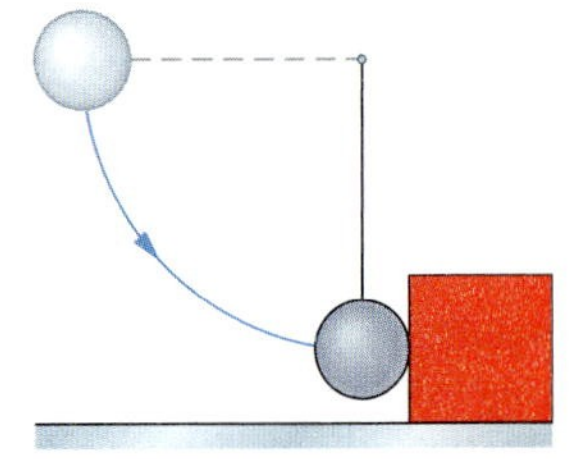

FIGURE 10-61 ■ Problem 88.

**89. Two Titanium Spheres** Two titanium spheres approach each other head-on with the same speed and collide elastically. After the collision, one of the spheres, whose mass is 300 g, remains at rest. (a) What is the mass of the other sphere? (b) What is the speed of the two-sphere center of mass if the initial speed of each sphere is 2.0 m/s?

**90. Two-Sphere Arrangement** In the two-sphere arrangement of Touchstone Example 10-8, assume that sphere $A$ has a mass of 50 g and an initial height of 9.0 cm and that sphere $B$ has a mass of 85 g. After the collision, what height is reached by (a) sphere $A$ and (b) sphere $B$? After the next (elastic) collision, what height is reached by (c) sphere $A$ and (d) sphere $B$? (*Hint*: Do not use rounded-off values.)

**91. Blocks without Friction** The blocks in Fig. 10-62 slide without friction. (a) What is the velocity $\vec{v}$ of the 1.6 kg block after the collision? (b) Is the collision elastic? (c) Suppose the initial velocity of the 2.4 kg block is the reverse of what is shown. Can the velocity $\vec{v}$ of the 1.6 kg block after the collision be in the direction shown?

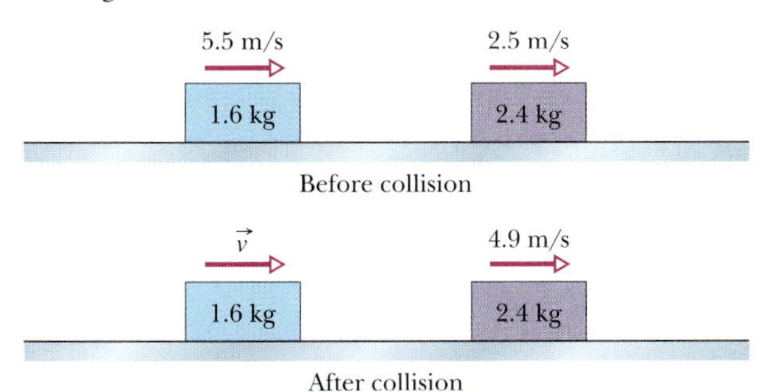


FIGURE 10-62 ■ Problem 91.

**92. Two Blocks on Frictionless Table** In Fig. 10-63, block $A$ of mass $m_A$, is at rest on a long frictionless table that is up against a wall. Block $B$ of mass $m_B$ is placed between block $A$ and the wall and sent sliding to the left, toward block $A$, with constant speed $v_{B1}$. Assuming that all collisions are elastic, find the value of $m_B$ (in terms of $m_A$) for which both blocks move with the same velocity after block $B$ has collided once with block $A$ and once with the wall. Assume the wall to have infinite mass..

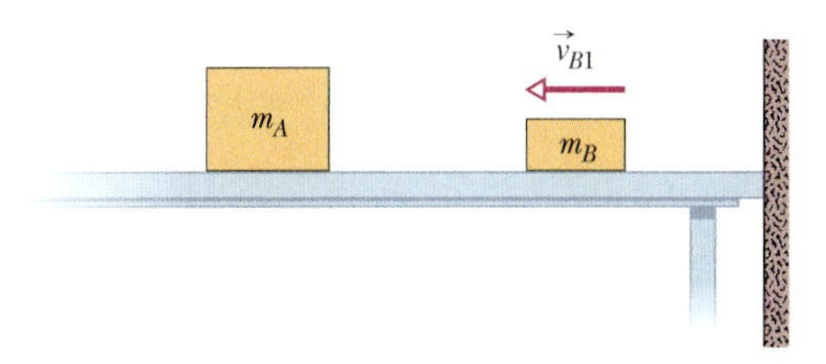


FIGURE 10-63 ■ Problem 92.

**93. Small Ball Above Larger** A small ball of mass $m$ is aligned above a larger ball of mass $M$ (with a slight separation, and the two are dropped simultaneously from $h$. (Assume the radius of each ball is negligible compared to $h$.) (a) If the larger ball rebounds elastically from the floor and then the small ball rebounds elastically from the larger ball, what ratio $m/M$ results in the larger ball stopping upon its collision with the small ball? (The answer is approximately the mass ratio of a baseball to a basketball.) (b) What height does the small ball then reach?

# Additional Problems

**94. Frictionless Ramp** In Fig. 10-64, block $A$ of mass $m_A$ slides from rest along a frictionless ramp from a height of 2.50 m and then collides with stationary block $B$, which has mass $m_B = 2.00m_A$. After the collision, block $B$ slides into a region where the coefficient of kinetic friction is 0.500 and comes to a stop in distance $d$ within that region. What is the value of distance $d$ if the collision is (a) elastic and (b) completely inelastic?

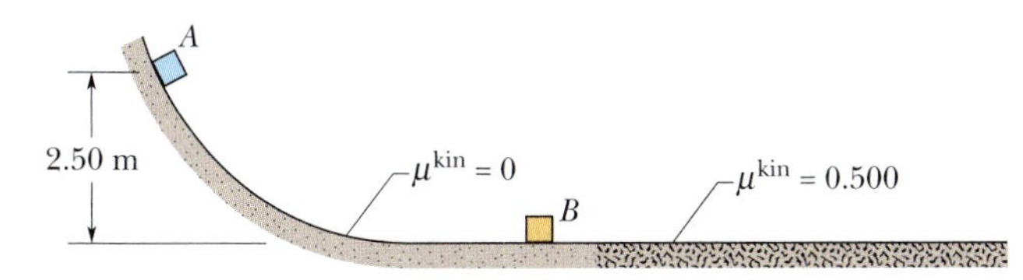


FIGURE 10-64 ■ Problem 94.

**95. Pucks on Table** In Fig. 10-65, puck $A$ of mass $m_A = 0.20$ kg is sent sliding across a frictionless lab bench, to undergo a one-dimensional elastic collision with stationary puck $B$. Puck $B$ then slides off the bench and lands a distance $d$ from the base of the bench. Puck $A$ rebounds from the collision and slides off the opposite edge of the bench, landing a distance $2d$ from the base of the bench. What is the mass of puck $B$? (*Hint*: Be careful with signs.)

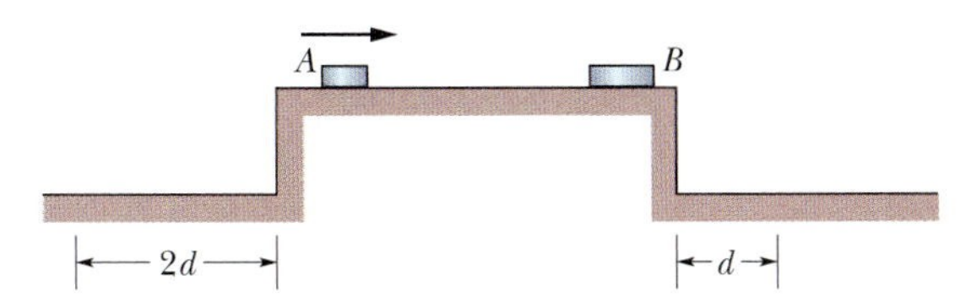


FIGURE 10-65 ■ Problem 95.

**96. Speed Amplifier** In Fig. 10-66, block $A$ of mass $m_A$ slides along an $x$ axis on a frictionless floor with a speed of $v_{A1} = 1.00$ m/s. Then it undergoes a one-dimensional elastic collision with stationary block $B$ of mass $m_B = 0.500m_A$. Next, block $B$ undergoes a one-dimensional elastic collision with stationary block $C$ of mass $m_C = 0.500m_B$. (a) What then is the speed of block $C$? Are (b) the speed, (c) the kinetic energy, and (d) the momentum of block $C$ greater than, less than, or the same as the initial values for block $A$?

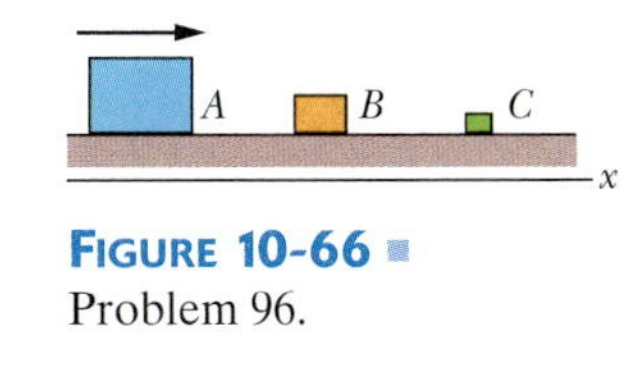


FIGURE 10-66 ■ Problem 96.

**97. Speed Amplifier Graphs** For the two-collision sequence of Problem 96, Figure 10-67*a* shows the speed $V_A$ of block $A$ plotted versus time $t$. The times for the first collision ($t_1$) and the second collision ($t_2$) are indicated. (a) On the same graph, plot the speeds of blocks $B$ and $C$. Figure 10-67*b* shows a plot of the kinetic energy of block $A$ versus time, where kinetic energy is given in terms of the initial kinetic energy $K_{A1} = 1.00$ J. (b) On the same graph, plot the kinetic energies of blocks $B$ and $C$, all in terms of $K_{A1}$. After the second collision, what percentage of the total kinetic energy do (c) block $A$, (d) block $B$, and (e) block $C$ have?

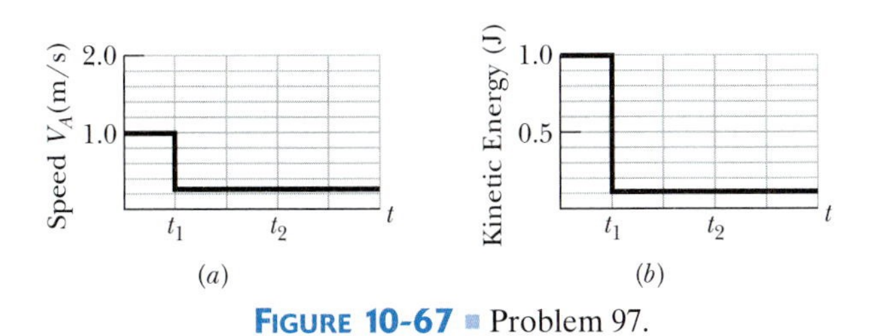


FIGURE 10-67 ■ Problem 97.

**98. The Janitor** Suppose a janitor wants to slide a trash barrel across the floor to a large trash bin. If the coefficient of kinetic friction is 0.123, determine the work done by a kinetic friction force on a 25 kg trash barrel that is pushed horizontally at a constant speed: (a) around a semicircle of diameter 2.3 m and (b) straight across the diameter.

**99. Loading Dock** Loading docks often have spring-loaded bumpers on them so that big trucks don't accidentally ruin the docks when backing up. See Fig. 10-68. Suppose a $6.45 \times 10^3$ kg truck backs into a spring-loaded dock at a speed of 2.51 m/s. If the truck compresses the dock bumper springs by 0.15 m when it slows down to zero speed, what is the effective spring constant of the bumper system? Use the correct number of significant figures.

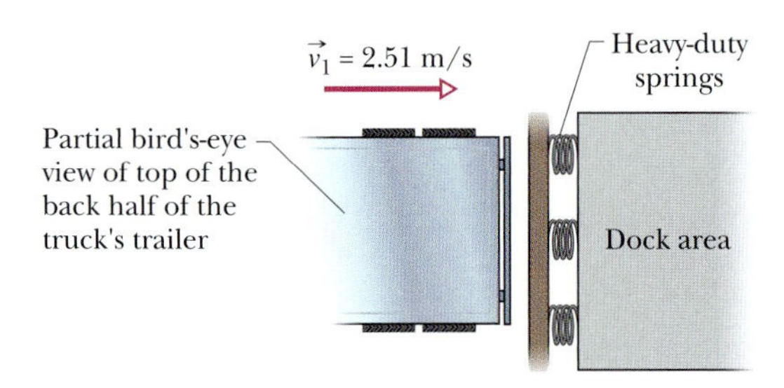


FIGURE 10-68 ■ Problem 99.

**100. Jumping into a Haystack** Tom Sawyer wanders out to the barn one fine summer's day. He notices that a haystack has recently been built just outside the barn. The barn has a second-story door into which the hay will be hauled into the barn by a crane. Tom decides it would be a neat idea to jump out of the second-story door onto the haystack. However, he knows from sad experience that if he jumps out of the second-story door onto the ground, that he is likely to break his leg. Knowing lots of physics, Tom decides to estimate whether the haystack will be able to break his fall.

He estimates the height of the haystack to be 3 meters. He presses down on top of the stack and discovers that to compress the stack by 25 cm, he has to exert a force of about 50 N. The barn door is 6 meters above the ground. Solve the problem by breaking it into pieces as follows:

**1.** Model the haystack by a spring. What is its spring constant?

**2.** Is the haystack tall enough to bring his speed to zero? (Estimate using conservation of energy.)

**3.** If he does come to a stop before he hits the ground, what will the average force exerted on him be?

**101. Closing the Door** A student is in her dorm room, sitting on her bed doing her physics homework. The door to her room is open. All of a sudden, she hears the voice of her ex-boyfriend talking to the girl in the room next door. She wants to shut the door quickly, so she throws a superball (which she keeps next to her bed for this purpose) against the door. The ball follows the path shown in Fig. 10-69. It hits the door squarely and bounces straight back.

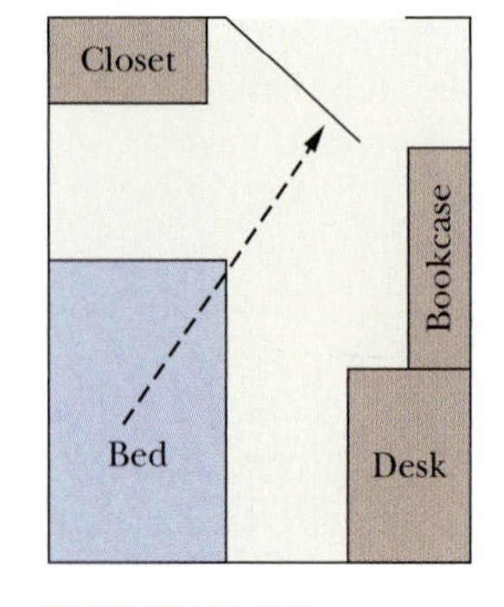


FIGURE 10-69 ■ Problem 101.

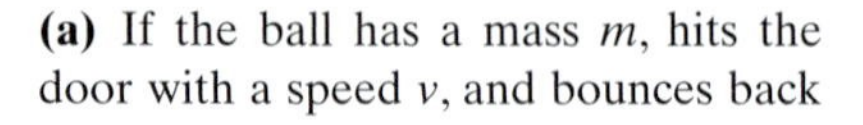

**(a)** If the ball has a mass $m$, hits the door with a speed $v$, and bounces back with a speed equal to $v$, what is the change in the ball's momentum? **(b)** If the ball was in contact with the door for a time $\Delta t$, what was the average force that the door exerted on the ball? **(c)** Would she have been better off with a clay ball of the same mass that stuck to the door? Explain your reasoning.

**102. The Astronaut and the Cream Pie*** A 77 kg astronaut, freely floating at 6 m/s, is hit by a large 36 kg lemon cream pie moving oppositely at 9 m/s. See Fig. 10-70. How much thermal energy is generated by the collision?

FIGURE 10-70 ■ Problem 102.

**103. Various Slopes** A skier wants to try different slopes of the same overall vertical height, $h$, to see which one would give him the most speed when he reaches the end of the hill (points 1, 2, and 3 in Fig. 10-71). His options are shown in the figure.

**(a)** Assuming there is no friction force between the skis and the snow, which hill would leave him with the most speed? Which would leave him with the least speed? Explain the basis for your answer.

**(b)** Assuming there is a noticeable friction force between the skis and the snow, which hill would leave him with the most speed? Which would leave him with the least speed? Explain the basis for your answer.

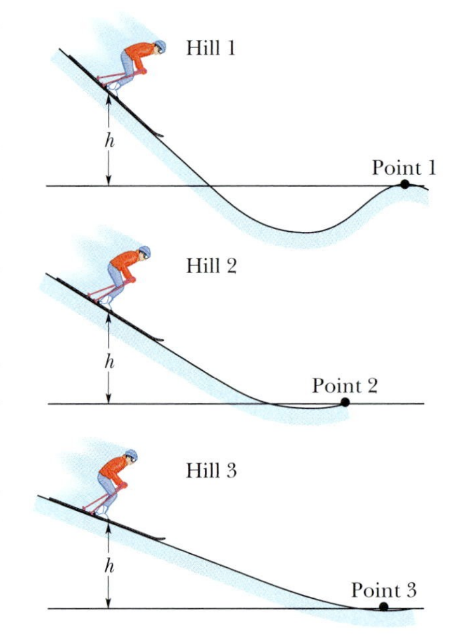


FIGURE 10-71 ■ Problem 103.

**104. Rolling Carts Down Hill**** Two carts $A$ and $B$ are identical in all respects. They roll down a hill and collide as shown in Fig. 10-72.

Figure 10-72*a*: (Case 1) cart $A$ starts from rest on a hill at a height $h$ above the ground. It rolls down and collides head-on with cart $B$, which is initially at rest on the ground. The two carts stick together.

Figure 10-72*b*: (Case 2) carts $A$ and $B$ are at rest on opposite hills at heights $h/2$ above the ground. They roll down, collide head-on with each other on the ground, and stick together.

---

* From Patrick H. Canan, *A Beginner's Guide to Classical Physics*, Corvallis, OR, School District (1982).

** Adapted from "Energy Concepts Survey" to be published in the *American Journal of Physics* by Chandralekha Singh.

Which of the following statements are true about the two-cart system *just before the carts collide* in the two cases? Give all the statements that are true. If none are true, write N.

**(a)** The kinetic energy of the system is zero in case 2.
**(b)** The kinetic energy of the system is greater in case 1 than in case 2.
**(c)** The kinetic energy of the system is the same in both cases.
**(d)** The total momentum of the system is greater in case 2 than in case 1.
**(e)** The total momentum of the system is the same in both cases.

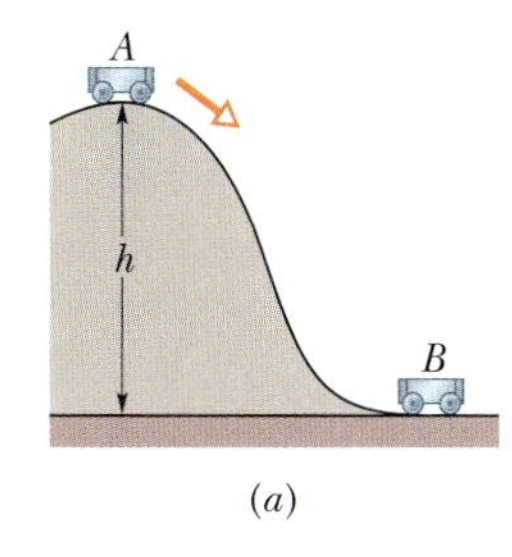


(*a*)

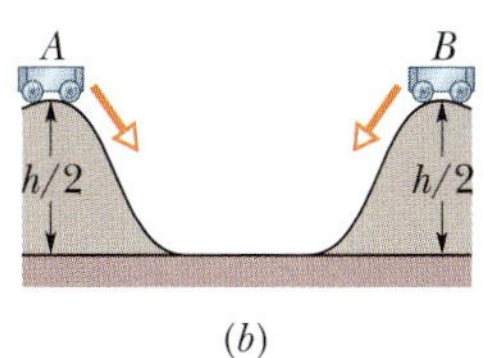


(*b*)

**FIGURE 10-72** ■ Problem 104.

Which of the following statements are true about the two-cart system *just after the carts collide* in the two cases? Give all the statements that are true. If none are true, write N.

**(f)** The kinetic energy of the system is greater in case 2 than in case 1.
**(g)** The kinetic energy of the system is the same in both cases.
**(h)** The momentum of the system is greater in case 2 than in case 1.
**(i)** The total momentum of the system is nonzero in case 1 whereas it is zero in case 2.
**(j)** The total momentum of the system is the same in both cases.

**105. Billiards Over the Edge** Two identical billiard balls are labeled $A$ and $B$ as shown in Fig. 10-73*a*. Maryland Fats places ball $A$ at the very edge of the table and ball $B$ at the other side. He strikes ball $B$ with his cue so that it flies across the table and off the edge. As it passes $A$, it just touches ball $A$ lightly, knocking it off. The balls are shown just at the instant they have left the table. Ball $B$ is moving with a speed $v_{B1}$, and ball $A$ is essentially at rest.

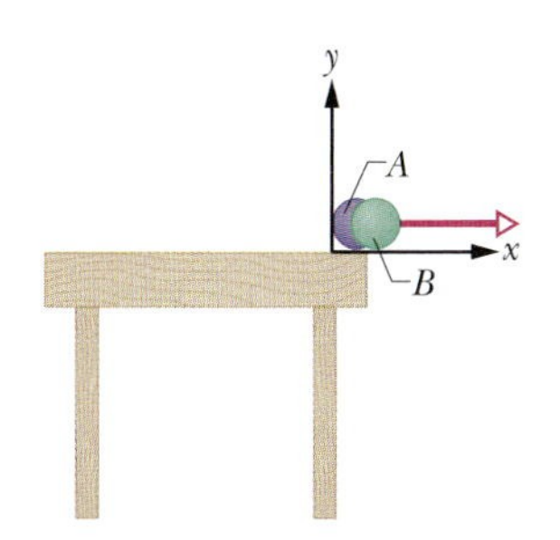


**FIGURE 10-73*a*** ■ Problem 105.

**(a)** Which ball do you think will hit the ground first? Explain your reasons for thinking so.

Fig. 10-73*b* shows a number of graphs of a quantity versus time. For each of the items below, select which graph could be a plot of that quantity vs. time. If none of the graphs are possible, write N. The time axes are taken to have $t = 0$ at the instant both balls leave the table. Use the $x$ and $y$ axes shown in Fig. 10-73*a*. For each of the following, which graph could represent:

**(b)** The $x$-component of the velocity of ball $B$?
**(c)** The $y$-component of the velocity of ball $A$?
**(d)** The $y$-component of the acceleration of ball $A$?
**(e)** The $y$-component of the force on ball $B$?
**(f)** The $y$-component of the force on ball $A$?
**(g)** The $x$-component of the velocity of ball $A$?
**(h)** The $y$-component of the acceleration of ball $B$?

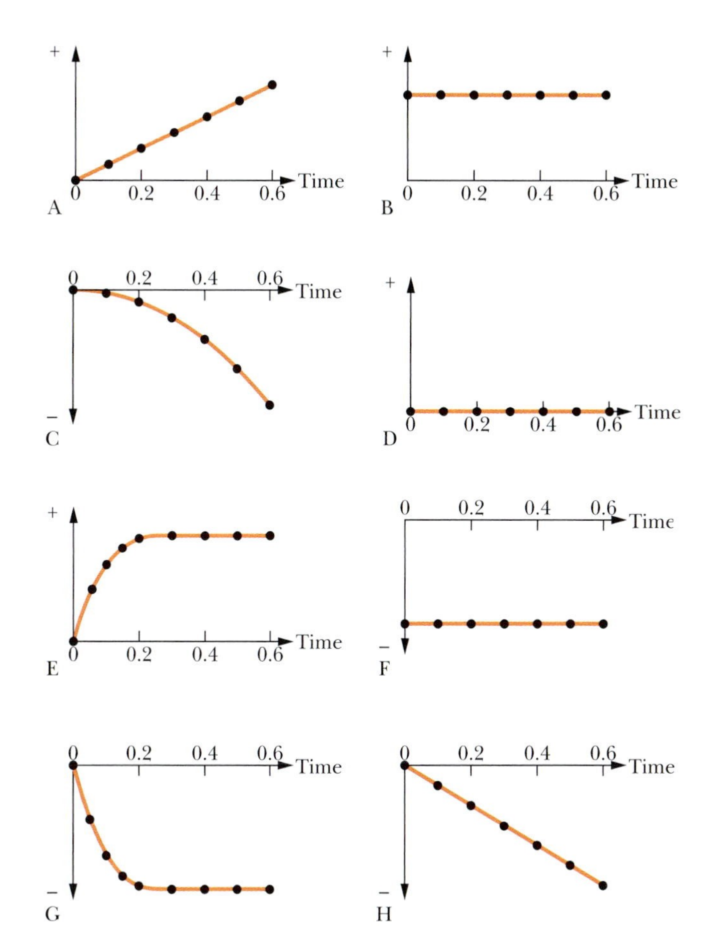


**FIGURE 10-73*b*** ■ Problem 105.

**106. When Can You Conserve Energy?** Mechanical energy conservation is sometimes a useful principle in helping us solve problems concerning the motion of objects. Suppose a single object is moving subject to a number of forces. Describe how you would know whether energy conservation would hold for the given example and in what kinds of problems you might find it appropriate to use this principle.

**107. Conserving Momentum but Not Energy?** Is it possible for a system of interacting objects to conserve momentum but not mechanical energy (kinetic plus potential)? Discuss and defend your answer, then given an example that illustrates the case you are trying to make.

**108. Momentum *and* Energy?** Is it possible for a system of interacting objects to conserve momentum and also mechanical energy (kinetic plus potential)? Discuss and defend your answer, then give an example that illustrates the case you are trying to make.

**109. Frames of Reference** Different observers can choose to use different coordinate systems. A frictionless roller coaster has been invented in which a single rider in a little cart can roll from the highest point to the lowest point, picking up kinetic energy as the cart goes downhill. The support struts for the roller coaster (shown as the grid in Fig. 10-74) are 4.00 meters apart, and the cart and rider have a combined mass of 195 kg. (a) What is the total mechanical energy of the cart-rider–Earth system according to Consuelo (an observer at the highest point on the track)? (b) What is the total mechanical energy of the cart-rider–Earth system according to

Mike (an observer at the ground level)? (c) Do Consuelo and Mike agree on the *value* of the total mechanical energy? Why or why not? (d) Do Consuelo and Mike agree that mechanical energy is conserved? Explain. (e) Assuming that mechanical energy is conserved, what is the kinetic energy of the cart and rider when it rolls over the top of the second smaller hill?

*Hint*: If you have access to the VideoPoint movie collection you may want to look at some of the roller coaster movies in the Hershey Park collection. For example, HRSY018 and HRSY019 provide similar scenarios. Although real roller coasters are not frictionless, using the VideoPoint software to find the location of a car at the top of a hill from the perspectives of two coordinate systems might be helpful.

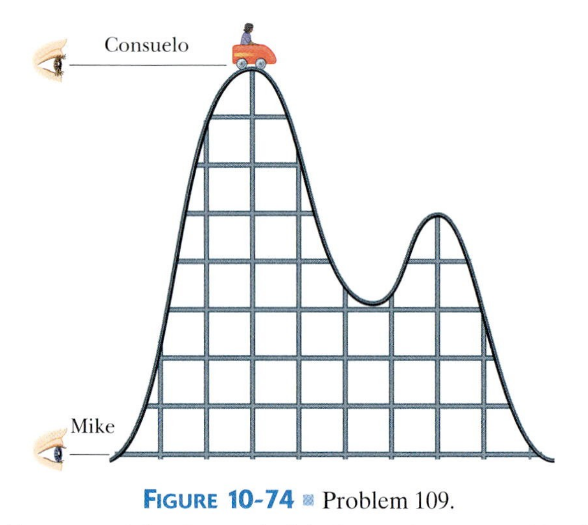


FIGURE 10-74 ■ Problem 109.

**110. Largest and Smallest** A ball is thrown from ground level with initial horizontal velocity component $v_{x\,1}$ and the initial vertical velocity component $v_{y\,1}$ and returns to ground level. Neglecting air resistance and explaining your reasoning in each instance, write expressions in terms of these two velocities for:

**(a)** The largest kinetic energy of the ball during its flight
**(b)** The smallest kinetic energy of the ball during its flight
**(c)** The maximum potential energy of the ball–earth system during the flight.
*Hint*: If you have access to the VideoPoint movie collection you may want to view the movies entitled PASCO104 and PASCO106 to remind you of the nature of a ball's path.

**111. Coffee Filter Drop** If a flat-bottomed coffee filter is dropped from rest near the surface of the Earth, it will fall more slowly than a small dense object of the same mass. You are to investigate whether or not mechanical energy is conserved during the fall of a small coffee filter using video analysis software. If you have access to the VideoPoint movie collection, you can use the movie entitled PASCO121 for this analysis. Your instructor may provide access to the movie some other way.

**(a)** If the coffee filter is dropped from rest, what is its initial velocity and kinetic energy?
**(b)** What are the final velocity and kinetic energy of the coffee filter (at the time of the last frame)? Explain how you arrived at the final velocity.
**(c)** What are the initial and final potential energies of the coffee filter?
**(d)** Is mechanical energy conserved as the coffee filter falls? Cite the evidence based on your measurements and calculations.
**(e)** How much mechanical energy, if any, is lost?
**(f)** What is the most likely source of a nonconservative force on the coffee filter? Where would missing mechanical energy probably go? Use conservation of energy concepts to explain why a paper coffee filter fall more slowly than a small dense object?

# 11 Rotation

These volleyball players leap high to block a spike. The height of their jumps would be far less if kneecaps were not a part of the human leg structure.

**How do kneecaps help people jump more effectively?**

*The answer is in this chapter.*

(a)

(b)

**FIGURE 11-1** ■ Figure skater Sarah Hughes in motion of (a) pure translation in a fixed direction and (b) pure rotation about a vertical axis.

## 11-1 Translation and Rotation

The graceful movement of figure skaters can be used to illustrate two kinds of pure motion. Figure 11-1*a* shows a skater gliding across the ice in a straight line with constant speed. Her motion is one of pure **translation.** Figure 11-1*b* shows her spinning at a constant rate about a vertical axis in a motion of pure **rotation.**

Translation is motion along a line, which we have considered in previous chapters. Rotation is turning motions, like those of wheels, gears, motors, planets, clock hands, jet engine rotors, and helicopter blades. It is our focus in this chapter. Rotational motion is everywhere around us, because most everyday objects are extended (rather than point masses) and can rotate about their centers of mass when moving freely. The characteristics of rotational motion are quite analogous to those of translational motion, and so the study of rotations will help you obtain a deeper understanding of both kinematics and the laws of translational motion. Examples of translational and rotational motion are shown in Fig. 11-2.

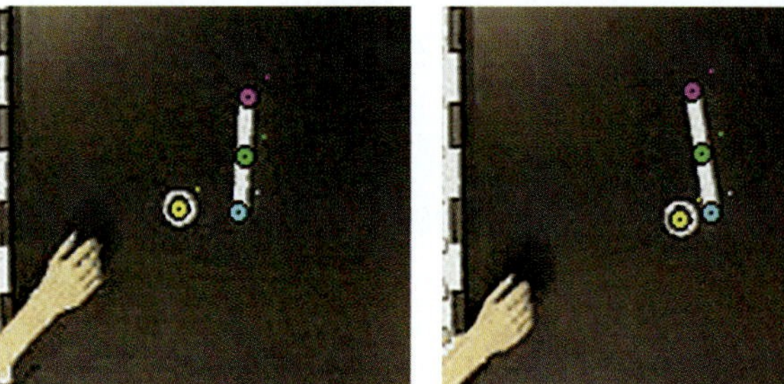

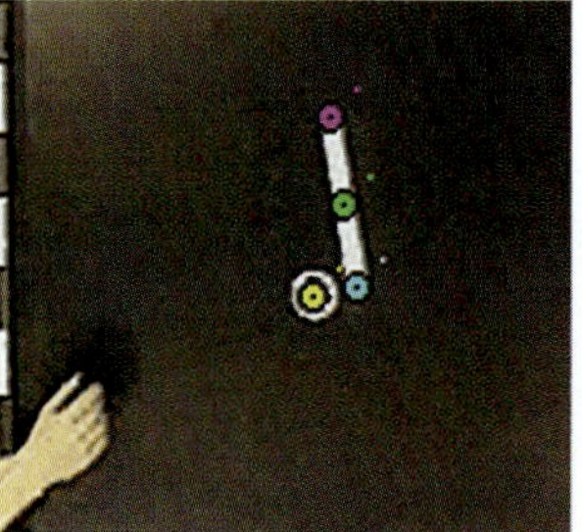

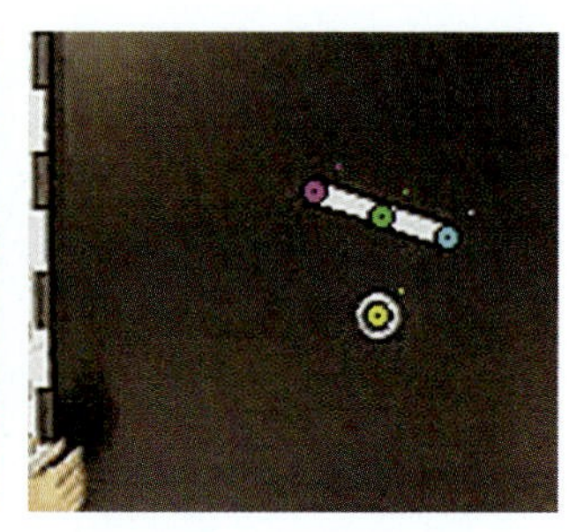

**FIGURE 11-2** ■ These video clips show a small puck colliding with a stationary rod on an air table. The puck's motion before and after the collision is purely translational. After the objects collide, the rod has a combination of rotational motion about its center of mass and translational motion of its center of mass. In Chapters 11 and 12, you will learn to use the laws of translational and rotational motion to predict the detailed outcome of collisions like this.

In this chapter we consider simple examples of rotational motion. It is the rotational analogy of motion along a line, so that we will not have to deal with rotational variables as two- or three-dimensional vectors. For example, we will limit our considerations to motions for which the axis of rotation is fixed, or at least does not accelerate, so we can always pick a frame of reference in which it doesn't move.

In Chapter 12, we will consider more complex motions involving axes of rotation that are not fixed, such as yo-yo motion. There we will also learn more about the advantages of treating rotational variables as vectors, even when the rotations are about a fixed axis. This will allow us to extend our understanding of the types of forces that can cause rotational accelerations.

## 11-2 The Rotational Variables

As is usual in physics, we like to start with a simple case so that we can make sense of the basic ideas. The big difference between what we've done before and what we are going to do now is that now we are going to consider the rotation of extended objects. This can get quite complicated if we allow the object to deform or to twist in an arbitrary way. Let's simplify by considering an object that is solid enough that we can treat it as if it has a fixed shape throughout its motion—that is, it is **perfectly rigid.** Many of these objects will have an **axis of symmetry,** a line about which the object may be turned and still look the same, like the line through the center of a cylinder or a ball. A rigid object and a nonrigid object are shown in Fig. 11-3.

In summary, in this chapter we wish to examine the rotations of rigid bodies about fixed axes. Figure 11-4 shows a rigid body in rotation about a fixed axis. The axis is called the **axis of rotation** or the **rotation axis.** This is defined as a **rigid body** because it can rotate with all its parts locked together, without any change in its shape. For example, in Fig. 11-1*b*, if the skater holds her shape while spinning, she is temporarily acting as a rigid body. But when she is moving her arms and legs relative to her body to change from one pose to another, she is not rigid. Therefore, we will not analyze the rotations of dancers and athletes except during those parts of their motions that are approximately rigid. Similarly, we will not examine the rotational motion of the Sun, because it is a ball of gas whose parts are not locked together.

**FIGURE 11-3** ■ A coffee cup serves as an example of a rigid object (upper), whereas a cloud is an example of a nonrigid object (lower).

A **fixed axis** means the rotation occurs about an axis that does not move. We also will not yet examine an object like a bowling ball rolling along a bowling alley, because the ball rotates about an axis that moves (the ball's motion is a mixture of rotation and translation).

As we know from our previous study of linear motions, in pure translational motion, every point on a body moves in a straight line. In other words, every point moves through the same *linear distance* during a particular time interval. In pure rotational motion, every point on the body moves in a circle whose center lies on the body's axis of rotation. Since the parts of a rigid body are locked together, every point moves through the same angle during a particular time interval. Hence, there are similarities and differences between translational and rotational motions. Comparisons between rotational and translational motion will appear throughout this chapter.

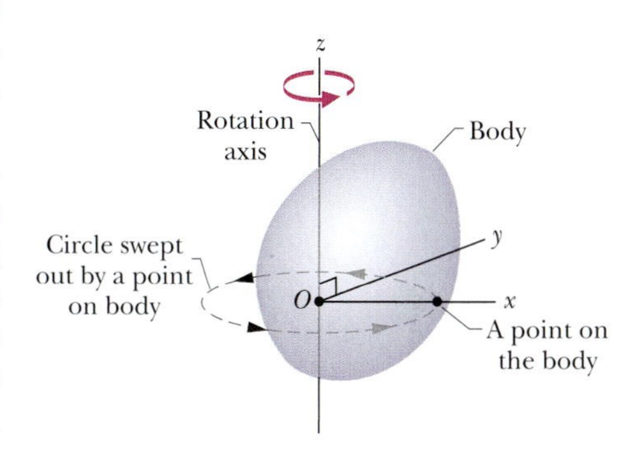


**FIGURE 11-4** ■ A rigid body of arbitrary shape in pure rotation about an axis. It is customary to choose a coordinate system in which the $z$ axis is aligned with the axis of rotation. The position of the *reference line* with respect to the rigid body is arbitrary, but it is perpendicular to the rotation axis. It is fixed in the body and rotates with the body. In this case, it must lie in the $x$-$y$ plane.

We deal now—one at a time—with specifying how an object is placed and moves rotationally. We will point out the rotational (or angular) equivalents of the translational (or linear) quantities position, displacement, velocity, and acceleration. The first step in introducing rotational quantities is to specify a coordinate system and reference line to aid in the description of motion.

Although there are many ways to specify a system for the analysis of rotational motion, the one shown in Fig. 11-4 is the most conventional. We start by choosing a rectangular coordinate system that is fixed in space. It is customary to orient the $z$ axis along the rotation axis. Next we choose a **reference line** that is perpendicular to the axis of rotation so it lies in the $x$-$y$ plane. The reference line is fixed with respect to the rotating body so that it rotates around the $z$ axis as the body rotates.

## Rotational Position

We define the rotational position $\theta$ of the body as the angle between the reference line at a given moment and the positive $x$ axis, as shown in Fig. 11-5.

For a rigid object rotating around a fixed axis, each point within the object moves in a circle around the axis of rotation. Consider a point along the reference line that is a distance $r$ from the axis. From geometry, we know that the magnitude of $\theta$ is given by

$$|\theta| = \frac{s}{r} \quad \text{(radian measure).} \tag{11-1}$$

Here $s$ is a scalar quantity that represents the length of arc (or the arc distance) between the $x$ axis (the zero rotational position) and the reference line; $r$ is the radius of that circle.

For the equation

$$|\theta| = \frac{s}{r}$$

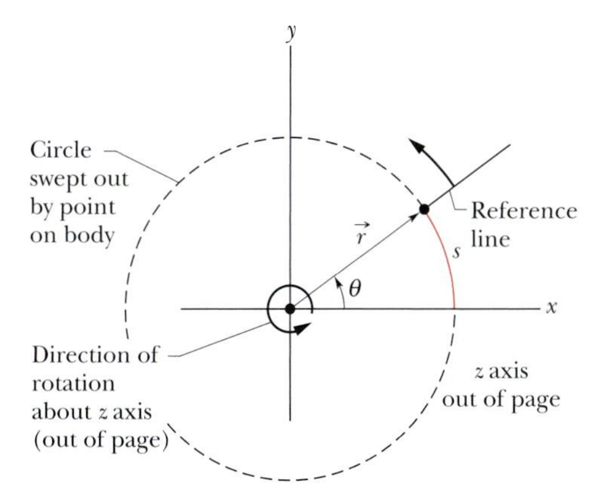


**FIGURE 11-5** ■ The rotating rigid body of Fig. 11-4 in cross section, viewed from above. The plane of the cross section is perpendicular to the rotation axis, which now extends out of the page, toward you. In this position of the body, the reference line makes an angle $\theta$ with the $x$ axis.

to be valid, the angle must be measured in radians (rad) rather than in revolutions (rev) or degrees. An angle of one radian is defined as the angle for which the length of the arc is equal to the radius of the circle. The radian, being the ratio of two lengths, is a pure number and thus has no dimensions. Although angles have no dimensions, they do have units, and it is vital to keep track of them. Because the circumference of a circle of radius $r$ is $2\pi r$, there are $2\pi$ radians in a complete circle. There are three common units used to measure angles. They are related by the equation

$$1 \text{ revolution} = 360° = \frac{2\pi r}{r} \text{ radians} = 2\pi \text{ radians.} \tag{11-2}$$

By rearranging terms algebraically, we find that

$$1 \text{ rad} = 57.3° = 0.159 \text{ rev.} \tag{11-3}$$

We do *not* reset $\theta$ to zero with each complete rotation of the reference line about the rotation axis. If we did, a smoothly rotating object would be described by a variable that jumps discontinuously. We have to keep in mind the physical, as well as the mathematical, meaning of the rotational variable. Although $\theta$, $\theta + 2\pi$, $\theta + 4\pi$, and so on, all represent the same physical position, they represent different total displacements. For example, if the reference line completes two revolutions from the zero rotational position, it is back at its starting point, but it has traveled through an angle of $\theta = 4\pi$ rad.

For pure translational motion along the $x$ direction, we can know all there is to know about a moving body if we are given $x(t)$, which is its position as a function of time. Similarly, for pure rotation, we can know all there is to know about the motion of a rigid rotating body about a fixed axis of rotation if we are given $\theta(t)$, the rotational position of the body's reference line as a function of time.

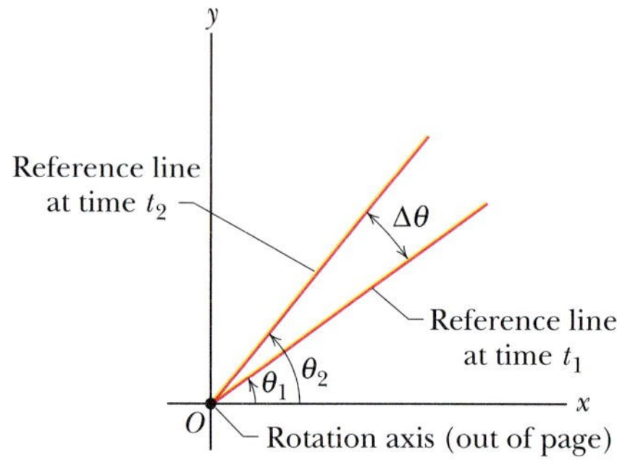


**FIGURE 11-6** ■ Bird's eye view of Fig. 11-4. The point on the reference line is at rotational position $\theta_1$ at time $t_1$, and at rotational position $\theta_2$ at a later time $t_2$. The quantity $\Delta\theta(= \theta_2 - \theta_1)$ is the rotational displacement that occurs during the interval $\Delta t\ (= t_2 - t_1)$. The body itself is not shown.

## Rotational Displacement

If the body of Fig. 11-5 rotates about the rotation axis as in Fig. 11-6, changing the rotational position of the reference line from $\theta_1$ to $\theta_2$, the body undergoes a rotational (or angular) displacement $\Delta\theta$ given by

$$\Delta\theta = \theta_2 - \theta_1. \tag{11-4}$$

This definition of rotational displacement holds not only for the rigid body as a whole, but also for *every particle within that body*, because the particles are all locked together.

If a body is in translational motion along an $x$ axis, its displacement $\Delta x$ is either positive or negative, depending on whether the body is moving in the positive or negative direction (as we have assigned them). Similarly, the rotational displacement $\Delta\theta$ of a rotating body can be either positive or negative.

Just as was the case for translational motion, the terms "positive" and "negative" are only meaningful once we have defined a coordinate system. For any situation that involves rotation about a fixed axis—for example, the rotation of the record shown in Fig. 11-7—the rotational displacement $\theta$ has a direction that is tied to the axis of rotation. Consequently, it makes sense to define a coordinate system that has one of its axes along the axis of rotation. It is standard practice to align the axis of rotation of a body along the $z$ axis of the rectangular right-handed coordinate system introduced in Section 4-5. Thus, by convention, if our right-handed rectangular coordinate system happens to be drawn so that the positive $z$ axis is out of the page along the axis of rotation of the body we are describing, then we define upward along the "vertical axis" as the positive $y$ direction and rightward along the "horizontal axis" as the positive $x$ direction.

Once we have established a coordinate system to describe a rotational motion, we can establish whether the rotational quantities of position, velocity, and acceleration are positive or negative by using a **right-hand rule,** as shown in Fig. 11-7*c*. Curl the fingers of your right hand in the direction of the rotation. If your extended thumb then points in the negative direction along the chosen axis of rotation (as is the case for the record in Fig. 11-7), we call the rotational displacement negative. If the record were to rotate in the opposite sense, the right-hand rule would tell you that the rotational displacement was positive, because your thumb would point in the opposite (positive) direction along the axis of rotation.

By using the right-hand rule, we can consider a rotational displacement to be a one-dimensional vector, where $\Delta\theta$ is its component along the axis of rotation. This assignment makes sense, since rotational displacements are meaningless unless we know what axis to relate them to. When the rotational displacement $\Delta\theta$ is positive, the object is rotating one way and when it is negative, the object is rotating the opposite way.

For the basic types of motion that we will treat in this book, the axis of rotation will not change orientation over time. In such cases, rotational displacements are said to commute. That is, the order in which you make the rotations doesn't matter. However, in more complex motions where the orientation of the axis of rotation changes direction over time, rotational displacements *do not* commute. In those cases, rotational displacements do not behave as vectors.

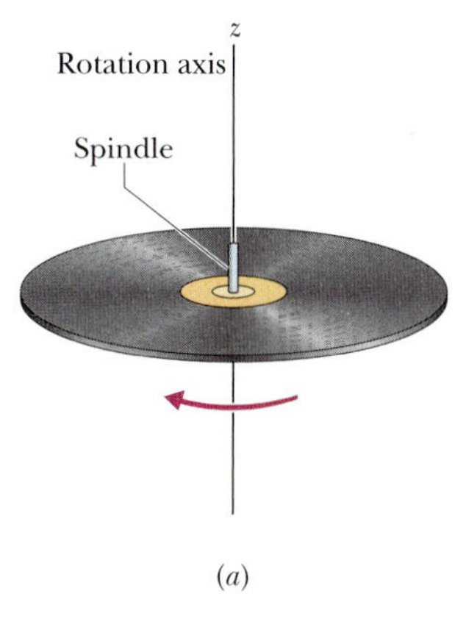


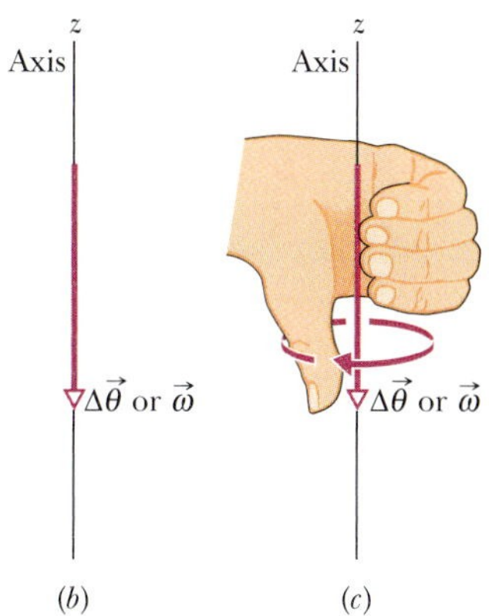


**FIGURE 11-7** ■ (*a*) A record rotating about a vertical axis that coincides with the axis of the spindle. (*b*) We establish that the rotational displacement component and the rotational velocity component are negative because our thumb points downward when using the right-hand rule. (*c*) We establish the direction of the rotational velocity vector as downward by using the right-hand rule. When the fingers of the right hand curl around the record and point the way it is moving, the extended thumb points in the direction of $\vec{\omega}$.

## The Rotational Velocity Component

Suppose (see Fig. 11-6) that our rotating body is at rotational position $\theta_1$ at time $t_1$ and at rotational position $\theta_2$ at time $t_2$. A body's **average rotational velocity** component along its axis of rotation is defined as

$$\frac{\theta_2 - \theta_1}{t_2 - t_1} = \frac{\Delta\theta}{\Delta t}, \quad \text{(average rotational velocity component)}, \tag{11-5}$$

where $\Delta\theta$ is the rotational displacement that occurs during the time interval $\Delta t = t_2 - t_1$ and $\omega$ is the lowercase Greek letter omega. Rotational velocity is often referred to as angular velocity. Note that when $z$ is the axis of rotation $\langle\omega\rangle = \langle\omega_z\rangle$.

The component $\omega_z$ of the **(instantaneous) rotational velocity** with which we shall be most concerned, is the limit of the ratio in the equation above as $\Delta t$ approaches zero. Thus,

$$\omega \equiv \lim_{\Delta t \to 0} \frac{\Delta\theta}{\Delta t} = \frac{d\theta}{dt} \quad \text{(instantaneous rotational velocity component)}. \tag{11-6}$$

Once again when $z$ is the rotation axis $\omega_z = \omega$. If we know $\theta(t)$ and it is continuous, we can find the rotational velocity component $\omega_z$ by differentiation. As is the case for the rotational displacement, the rotational velocity $\omega_z$ in this context represents the component of a one-dimensional vector along the axis of rotation relative to the coordinate system chosen to describe the motion. As a component, $\omega_z$ can be positive or negative and we do not use a vector arrow over it. Whenever the rotational position $\theta$ is becoming more positive $\omega_z$ is positive and, conversely, whenever the rotational position $\theta$ is becoming more negative $\omega_z$ is negative. Happily we will get the same result using the right-hand rule to determine whether $\omega_z$ is positive or negative.

Strictly speaking, we should always define an axis of rotation as $z$ and call $\omega_z$ the rotational velocity component. But, for the simple rotations considered in this chapter, we often refer to it casually as the rotational velocity $\omega$. Also, because the particles in a rigid body are all linked to each other, the rotational velocity is the same for *every particle in a rotating rigid body.*

**Rotational Speed** The magnitude (or absolute value) of rotational velocity is called the rotational speed. Since we have designated $\omega$ (or more correctly $\omega_z$) to be a component along the axis of rotation that can be either positive or negative, the rotational speed must be represented using an absolute value sign. Thus, to avoid confusion we always denote rotational speed as $|\omega|$ or $|\vec{\omega}|$.

**Units for Rotational Velocity** The preferred scientific unit of rotational velocity is the radian per second (rad/s). In some cases the unit revolution per second (rev/s) is used instead. Another popular measure of rotational velocity is rpm or revolutions per minute, used in automobile tachometers that measure the turning rate of engine crankshafts. The rpm is also used in conjunction with turntables used to play vinyl phonograph records.

## The Rotational Acceleration Component

If the rotational velocity of a rotating body is not constant, then the body has rotational acceleration. Let $\omega_2$ and $\omega_1$ be its rotational velocity components at times $t_2$ and $t_1$, respectively. The component of the **average rotational acceleration** along the axis of rotation of the body in the interval from $t_1$ to $t_2$ is defined as

$$\langle\alpha\rangle = \langle\alpha_z\rangle \equiv \frac{\omega_2 - \omega_1}{t_2 - t_1} = \frac{\Delta\omega}{\Delta t} \quad \text{(average rotational acceleration component),} \qquad (11\text{-}7)$$

in which $\Delta\omega$ is the component of the change in rotational velocity that occurs during the time interval $\Delta t$. The **(instantaneous) rotational acceleration** component $\alpha$, with which we shall be most concerned, is the limit of this quantity as $\Delta t$ approaches zero. Thus,

$$\alpha = \alpha_z \equiv \lim_{\Delta t \to 0} \frac{\Delta\omega}{\Delta t} = \frac{d\omega}{dt} \quad \text{(instantaneous rotational acceleration component).} \qquad (11\text{-}8)$$

Just as was the case for the rotational velocity $\omega$, these expressions for the rotational acceleration hold not only for the rotating rigid body as a whole, but also for *every particle of that body.*

Whenever the rotational velocity component $\omega$ is becoming more positive $\alpha$ is positive and, conversely, whenever the rotational velocity component $\omega$ is becoming more negative $\alpha$ is negative. Thus, the relationship between the directions of velocity and acceleration that we hold to be true for translational motion are analogous to those that apply to rotational motion.

Note that rotational acceleration, as introduced in this simple context, like rotational displacement and rotational velocity, is a component of a one-dimensional vector relative to the chosen coordinate axis aligned with the axis of rotation of the body that is rotating. Rotational acceleration is often referred to as angular acceleration.

Once again we have followed the convention of not designating what axis of rotation the acceleration component refers to. For this reason the rotational acceleration component is casually called the rotational acceleration. So for its magnitude we always denote the magnitude of rotation acceleration as $|\vec{\alpha}|$ or $|\alpha|$.

**Units for Rotational Acceleration** The preferred scientific unit of rotational acceleration is commonly the radian per second-squared (rad/s$^2$). Another common unit is the revolution per second-squared (rev/s$^2$).

**READING EXERCISE 11-1: The Sign of Rotational Velocity:** An off-center egg like the one shown in Fig. 11-4 is rotating about a $z$ axis. What is the sign of its rotational velocity component if it is rotating so (a) the angle between its reference line and the $x$ axis is increasing; (b) the angle between its reference line and the $x$ axis is decreasing? ■

**READING EXERCISE 11-2: The Sign of Rotational Acceleration**—An off-center egg like the one shown in Fig. 11-4 is rotating about a $z$ axis. What is the sign of its rotational acceleration if it is rotating so (a) the angle between its reference line and the $x$ axis is increasing and so is its speed; (b) the angle between its reference line and the $x$ axis is increasing and its speed is decreasing; (c) the angle between its reference line and the $x$ axis is decreasing but its speed is increasing; and (d) the angle between its reference line and the $x$ axis is decreasing and its speed is decreasing? ■

## TOUCHSTONE EXAMPLE 11-1: Rotating Disk

The disk in Fig. 11-8*a* is rotating about its central axis like a merry-go-round. The rotational position $\theta(t)$ of a reference line on the disk is given by

$$\theta(t) = -(1.00\text{ rad}) - (0.600\text{ rad/s})t + (0.250\text{ rad/s}^2)t^2, \quad (11\text{-}9)$$

with the zero rotational position as indicated in the figure.

(a) Graph the rotational position of the disk versus time from $t = -3.0$ s to $t = 6.0$ s. Sketch the disk and its rotational position reference line at $t = -2.0$ s, 0 s, and 4.0 s, and when the curve crosses the $t$ axis.

**SOLUTION** ■ The **Key Idea** here is that the rotational position of the disk is the rotational position $\theta(t)$ of its reference line, which is given by Eq. 11-9 as a function of time. So we graph Eq. 11-9; the result is shown in Fig. 11-8*b*.

To sketch the disk and its reference line at a particular time, we need to determine $\theta$ for that time. To do so, we substitute the time into Eq. 11-9. For $t = -2.0$ s, we get

$$\theta = -(1.00\text{ rad}) - (0.600\text{ rad/ s})(-2.0\text{ s}) + (0.250\text{ rad/s}^2)(-2.0\text{ s})^2$$

$$= 1.2\text{ rad} = 1.2\text{ rad}\,\frac{360°}{2\pi\text{ rad}} = 69°.$$

This means that at $t = -2.0$ s the reference line on the disk is rotated counterclockwise from the zero rotational position by 1.2 rad or 69° (counterclockwise because $\theta$ is positive). Sketch A in Fig. 11-8*b* shows this rotational position of the reference line. Similarly, for $t = 0$, we find $\theta = -1.00$ rad $= -57°$, which means that the reference line is rotated clockwise from the zero rotational position by 1.0 rad or 57°, as shown in sketch C. For $t = 4.0$ s, we find $\theta = 0.60$ rad $= 34°$ (sketch E). Drawing sketches for when the curve crosses the $t$ axis is easy, because then $\theta = 0$ and the reference line is momentarily aligned with the zero rotational position (sketches B and D).

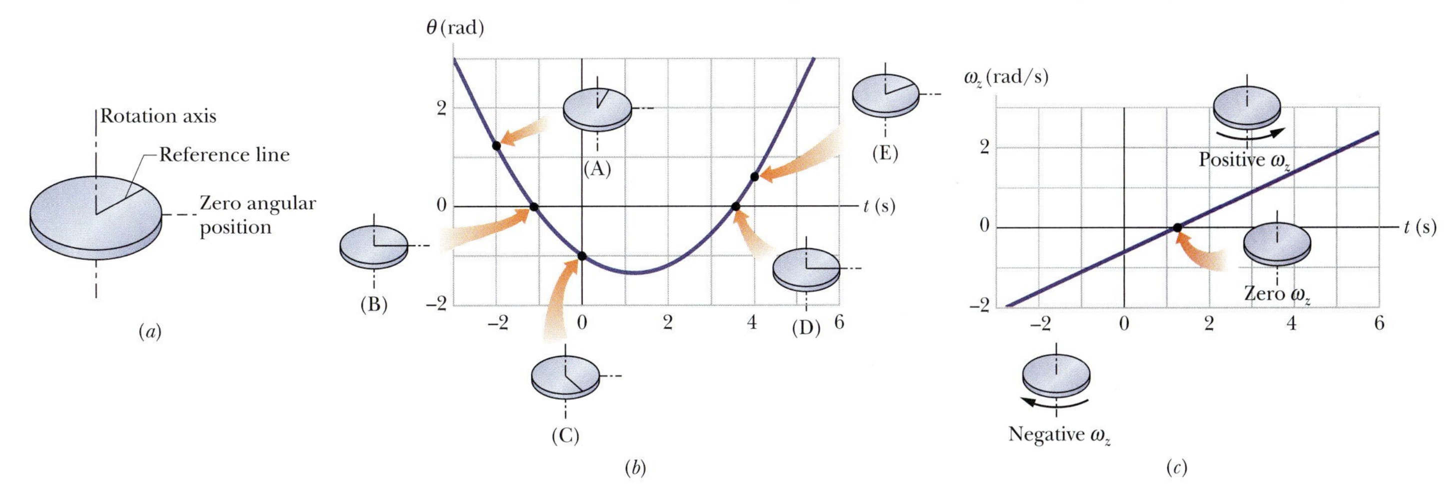


**FIGURE 11-8** ■ (*a*) A rotating disk. (*b*) A plot of the disk's rotational position $\theta(t)$. Five sketches indicate the rotational position of the reference line on the disk for five points on the curve A-E. (*c*) A plot of the $z$-component of the disk's rotational velocity $\omega_z(t)$. Positive values of $\omega_z$ correspond to counterclockwise rotation, and negative values to clockwise rotation.

(b) At what time $t^{min}$ does $\theta(t)$ reach the minimum value shown in Fig. 11-8*b*? What is that minimum value?

**SOLUTION** ■ The **Key Idea** here is that to find the extreme value (here the minimum) of a function, we take the first derivative of the function and set the result to zero. The first derivative of $\theta(t)$ is

$$\frac{d\theta}{dt} = -(.600 \text{ rad/s}) + (0.500 \text{ rad/s}^2)t. \qquad (11\text{-}10)$$

Setting this to zero and solving for $t$ give us the time at which $\theta(t)$ is minimum:

$$t^{min} = 1.20 \text{ s}. \qquad \text{(Answer)}$$

To get the minimum value of $\theta$, we next substitute $t^{min}$ into Eq. 11-9, finding

$$\theta^{min} = -1.36 \text{ rad} = -77.9°. \qquad \text{(Answer)}$$

This *minimum* of $\theta(t)$ (the bottom of the curve in Fig. 11-8*b*) corresponds to the *extreme clockwise* rotation of the disk from the zero rotational position, somewhat more than is shown in sketch C.

(c) Graph the rotational velocity $\omega$ of the disk versus time from $t = -3.0$ s to $t = 6.0$ s. Sketch the disk and indicate the direction of turning and the sign of $\omega$ at $t = -2.0$ s and 4.0 s, and also at $t^{min}$.

**SOLUTION** ■ The **Key Idea** here is that, from Eq. 11-6, the rotational velocity $\omega$ is equal to $d\theta/dt$ as given in Eq. 11-10. So, we have

$$\omega = -(.600 \text{ rad/s}) + (0.500 \text{ rad/s}^2)t. \qquad (11\text{-}11)$$

The graph of this function $\omega(t)$ is shown in Fig. 11-8*c*.

To sketch the disk at $t = -2.0$ s, we substitute that value into Eq. 11-11, obtaining

$$\omega = -1.6 \text{ rad/s} \qquad \text{(Answer)}$$

The minus sign tells us that at $t = -2.0$ s, the disk is turning clockwise as suggested by the lowest sketch in Fig. 11-8*c*.

Substituting $t = 4.0$ s into Eq. 11-11 gives us

$$\omega = 1.4 \text{ rad/s}. \qquad \text{(Answer)}$$

The implied plus sign tells us that at $t = 4.0$ s, the disk is turning counterclockwise (the highest sketch in Fig. 11-8*c*).

For $t^{min}$, we already know that $d\theta/dt = 0$. So, we must also have $\omega = 0$. That is, the disk is changing its direction of rotation when the reference line reaches the minimum value of $\theta$ in Fig. 11-8*b* as suggested by the center sketch in Fig. 11-8*c*.

(d) Use the results in parts (a) through (c) to describe the motion of the disk from $t = -3.0$ s to $t = 6.0$ s.

**SOLUTION** ■ When we first observe the disk at $t = -3.0$ s, it has a positive rotational position and is turning clockwise but slowing. It reverses its direction of rotation at rotational position $\theta = -1.36$ rad and then begins to turn counterclockwise, with its rotational position eventually becoming positive again.

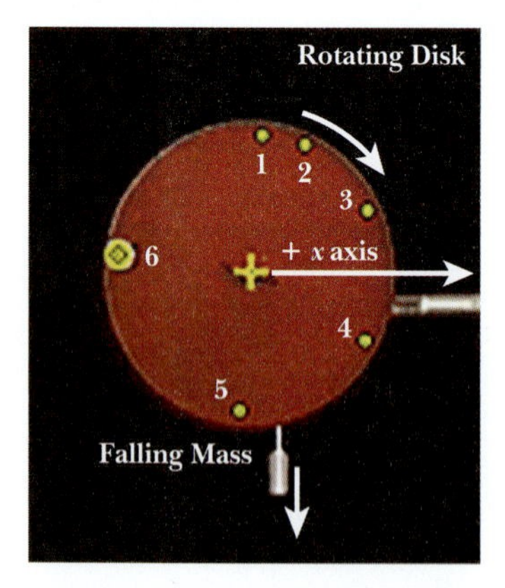


**FIGURE 11-9** ■ A falling mass is attached to the axle of a rotating disk. The mass falls with a constant translational acceleration. Video analysis shows that as the mass falls, the disk rotates from position 1 to position 6 with a constant rotational acceleration. By defining a coordinate system, both the translational and rotational accelerations can be determined using the equations in Table 11-1.

## 11-3 Rotation with Constant Rotational Acceleration

In pure translation, motion with a *constant translational acceleration* (for example, that of a falling body) is an important special case. In Table 2-1, we displayed a series of equations that hold for such motion.

Recall that in Chapter 2 we derived two primary equations $v_{2x} = v_{1x} + a_x(t_2 - t_1)$ (Eq. 2-13) and $x_2 - x_1 = v_{1x}(t_2 - t_1) + \frac{1}{2}a_x(t_2 - t_1)^2$ (Eq. 2-17) that describe velocity and position changes of an object that undergoes a constant translational acceleration. In pure rotation, the case of *constant rotational acceleration* is also important, and a parallel set of equations holds for this case also. Since the logic used to derive the analogous rotational equations is identical, we shall not derive them here. We can simply write them from the corresponding translational equations, substituting equivalent rotational quantities for the translational ones. This is done in Table 11-1. Figure 11-9 shows a situation that you can analyze using these equations. The equations for constant rotational acceleration are

$$\omega_2 = \omega_1 + \alpha(t_2 - t_1), \qquad (11\text{-}12)$$

and

$$\theta_2 - \theta_1 = \omega_1(t_2 - t_1) + \tfrac{1}{2}\alpha(t_2 - t_1)^2. \qquad (11\text{-}13)$$

Note that it is possible to derive other useful secondary equations from these two primary equations.

**TABLE 11-1**
**Equations of Motion with Constant Translational Acceleration and with Constant Rotational Acceleration**

| Equation Number | Translational Equation | Rotational Equation | Equation Number |
|---|---|---|---|
| | Primary Vector Component Equations:* | Primary Rotational Vector Component Equations | |
| (2-13) | $v_{2x} = v_{1x} + a_x(t_2 - t_1)$ | $\omega_2 = \omega_1 + \alpha(t_2 - t_1)$ | (11-12) |
| (2-17) | $x_2 - x_1 = v_{1x}(t_2 - t_1) + \frac{1}{2}a_x(t_2 - t_1)^2$ | $\theta_2 - \theta_1 = \omega_1(t_2 - t_1) + \frac{1}{2}\alpha(t_2 - t_1)^2$ | (11-13) |

**A reminder*: In cases where the initial time $t_1$ is chosen to be zero and $t_2$ is denoted as $t$, it is important to remember that whenever the term $(t_2 - t_1)$ is replaced by just $t$, then $t$ actually represents a *time period of* $\Delta t = t_2 - t_1 = t - 0$ over which the motion of interest takes place.

**READING EXERCISE 11-3:** In four situations, a rotating body has rotational position $\theta(t)$ of:

(a) $\theta(t) = 3\left[\frac{\text{rad}}{\text{s}}\right]t - 4[\text{rad}]$,

(b) $\theta(t) = -5\left[\frac{\text{rad}}{\text{s}^3}\right]t^3 + 4\left[\frac{\text{rad}}{\text{s}^2}\right]t^2 + 6[\text{rad}]$,

(c) $\theta(t) = \frac{2[\text{rad}\cdot\text{s}^2]}{t^2} - \frac{4[\text{rad}\cdot\text{s}]}{t}$, and

(d) $\theta(t) = 5\left[\frac{\text{rad}}{\text{s}^2}\right]t^2 - 3[\text{rad}]$.

To which situations do the rotational equations of Table 11-1 apply?

## TOUCHSTONE EXAMPLE 11-2: Grindstone

A grindstone (Fig. 11-10) rotates at constant rotational acceleration $\alpha = 0.35$ rad/s². At time $t_1 = 0$ s, it has a rotational velocity of $\omega_1 = -4.6$ rad/s and a reference line on it is horizontal, at the rotational position $\theta_1 = 0.0$ rad.

(a) At what time after $t = 0.0$ s is the reference line at the rotational position $\theta = 5.0$ rev?

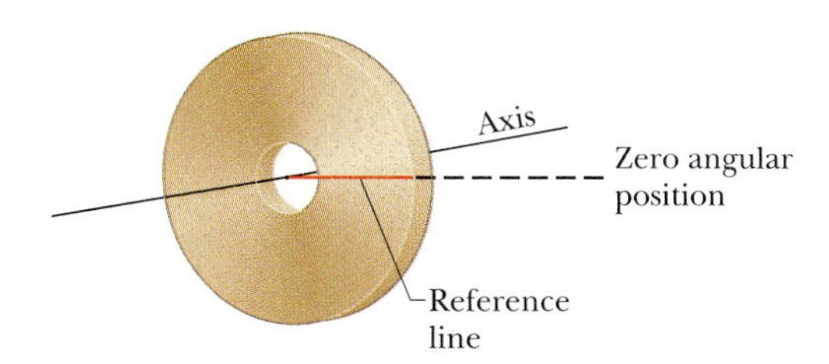


**FIGURE 11-10** A grindstone. At $t_1 = 0$ the reference line (which we imagine to be marked on the stone) is horizontal.

**SOLUTION** The **Key Idea** here is that the rotational acceleration is constant, so we can use the rotation equations of Table 11-1. We choose Eq. 11-13,

$$\theta_2 - \theta_1 = \omega_1(t_2 - t_1) + \tfrac{1}{2}\alpha(t_2 - t_1)^2,$$

because the only unknown variable it contains is the desired time $(t_2 - t_1)$. Substituting known values and setting $\theta_1 = 0.0$ rad and $\theta_2 = 5.0$ rev $= 10\pi$ rad give us

$$10\pi\text{ rad} = (-4.6\text{ rad/s})(t_2 - t_1) + \tfrac{1}{2}(0.35\text{ rad/s}^2)(t_2 - t_1)^2.$$

(We converted 5.0 rev to $10\pi$ rad to keep the units consistent.) Solving this quadratic equation for $t_2 - t_1$, we find

$$t_2 - t_1 = 32\text{ s}. \quad \text{(Answer)}$$

(b) Describe the grindstone's rotation between $t_1 = 0$ and $t_2 = 32$ s.

**SOLUTION** The wheel is initially rotating in the negative direction with rotational velocity $\omega_1 = -4.6$ rad/s, but its rotational acceleration $\alpha$ is positive. This initial opposition of the signs of rotational velocity and rotational acceleration means that the wheel slows in its rotation in the negative direction and then reverses to rotate in the positive direction. After the reference line comes back through its initial orientation of $\theta_1 = 0.0$ rad, the wheel turns an additional 5.0 rev by time $t_2 = 32$ s. (Answer)

(c) At what time $t_3$ does the grindstone change its direction of rotation?

**SOLUTION** We again go to the table of equations for constant rotational acceleration, and again we need an equation that contains only the desired unknown variable $(t_3 - t_1)$. However, now we use another **Key Idea**. The equation must also contain the variable $\omega$, so that we can set it to 0 and then solve for the corresponding time $t_3$. We choose Eq. 11-12, which yields

$$t_3 - t_1 = \frac{\omega_3 - \omega_1}{\alpha} = \frac{0.0\text{ rad/s} - (-4.6\text{ rad/s})}{0.35\text{ rad/s}^2} = 13\text{ s}. \quad \text{(Answer)}$$

## 11-4 Relating Translational and Rotational Variables

In Section 5-7, we discussed uniform circular motion, in which a particle travels at constant translational speed $v$ along a circle and around an axis of rotation. When a rigid body, such as a merry-go-round, rotates around an axis, each particle in the body moves in its own circle around that axis. Since the body is rigid, all the particles make one revolution in the same amount of time; that is, they all have the same rotational speed $|\omega|$.

However, try the following experiment. Pretend that you are a dancer or a skater. Although people are not rigid objects, they often configure their bodies into poses that are temporarily rigid. Hold out your arms like a dancer or skater and spin around in place (Fig. 11-11). What point on your body is moving fastest? Your shoulder, elbow, or fingertip?

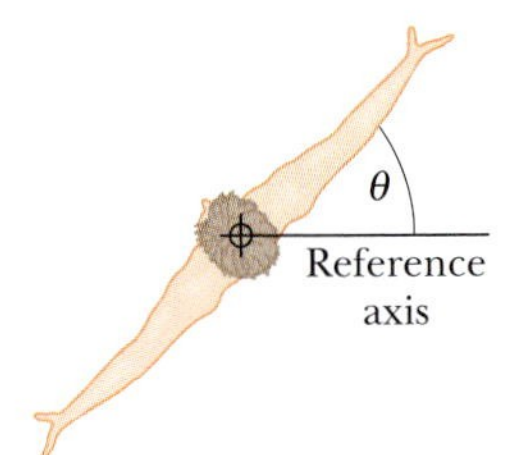


**FIGURE 11-11** ■ If you rotate your body about a fixed vertical axis with your arms extended, what moves fastest: your shoulder, your elbow, or your fingers?

As you may have gathered from the observation described above, a particle far from the axis of rotation moves at a greater translational speed $v$ than a particle close to the axis of rotation. This is because the farther the object is from the axis, the greater the circumference of the circular path the object takes in rotating about the axis. Since all the points on the object complete a revolution in the same time interval (they all have the same rotational speed $|\omega|$), those points that must travel a larger circumference must move at a higher translational speed. Hence, all points on a rotating object have the same rotational speed $|\omega|$, but not the same translational speed $|\vec{v}|$. You can also notice this on a merry-go-round. You turn with the same rotational speed $|\omega|$ regardless of your distance from the center, but your translational speed $|\vec{v}|$ increases noticeably as you move from the center to the outside edge of the merry-go-round. This is the reason we describe rotation using rotational, rather than translational, variables.

Calvin and Hobbes © 1990 Bill Watterson. (Reprinted with permission of UNIVERSAL PRESS SYNDICATE. All rights reserved.)

We often need to relate the translational variables $s$, $|\vec{v}|$, and $|\vec{a}|$ for a particular point in a rotating body to the rotational variables $|\theta|$, $|\omega|$, and $|\alpha|$ for that body. For example, we may know the rotational velocity and need to know the associated translational velocity. The two sets of variables (translational and rotational) are related by $r$, the *perpendicular distance* of the point from the rotation axis. This perpendicular distance is the distance between the point and the rotation axis, measured along a perpendicular to the axis. It is also the radius $r$ of the circle traveled by the point around the axis of rotation.

### Rotational Position and Distance Moved

If a reference line on a rigid body rotates through an angle $\theta$, a point within the body at a distance $r$ from the rotation axis moves a distance $s$ along a circular arc, where $s$ is given by

$$s = |\theta|\, r \quad \text{(for radian measure only).} \tag{11-14}$$

This is the first of our translational-rotational relations. *Caution*: The angle $\theta$ here *must be measured in radians* because $s = |\theta|\, r$ is derived from the definition of the radian.

## Relating Rotational and Translational Speed

How can we compare the translational speed $v$ of a rotating particle to its rotational speed $|\omega|$? Any small element of a rotating object that is rigid stays a fixed distance, $r$, from the axis of rotation throughout its rotation around the axis. In Section 5-7, we showed that if a rotating particle moves from one point on a circle to another, then the magnitude of its translational displacement, $|\Delta\vec{r}|$, between those points and the distance it moves along the arc of the circle, $\Delta s$, are essentially the same when the displacement is infinitesimally small. For this reason, we can find the magnitude of the instantaneous translational velocity by taking the time derivative of $s = |\theta|\, r$ (Eq. 11-14). In other words,

$$|\vec{v}| = v = \frac{|d\vec{r}\,|}{dt} = \frac{ds}{dt} = \frac{d|\theta|}{dt} r.$$

However, $d|\theta|/dt$ is the rotational speed $|\omega|$ of the rotating body, so the translational speed is given by

$$|\vec{v}| = v = |\omega|\, r \quad \text{(for radian measure only).} \tag{11-15}$$

*Caution*: Rotational speed $|\omega|$ *must be expressed in radian measure* and be denoted with an absolute value sign since $\omega$ represents a vector component.

Equation 11-15 ($v = |\omega| r$) tells us that all points within the rigid body have the same rotational speed, $|\omega|$, while points with greater radius $r$, have greater translational speed $|\vec{v}|$. This equation verifies the conclusion we already reached. Namely, that when all points on an object complete a revolution in the same time interval they all have the same rotational speed $|\omega|$. But those points that are a larger distance from the axis of rotation must travel a larger circumference and must move at a higher translational velocity. Figure 11-12*a* reminds us that the translational velocity is always tangent to the circular path of the point in question.

If the rotational speed $|\omega|$ of the rigid body is constant, then $v = |\omega|\, r$ (Eq. 11-15) tells us that the translational speed $v$ of any point within it is also constant. Thus, each point within the body undergoes uniform circular motion. We can find the period of revolution $T$ by recalling that this is the time for one revolution (which is a linear distance $2\pi r$). The rate at which that distance is traveled is equal to the circumference divided by the time needed to make one revolution. Hence,

$$v = \frac{2\pi r}{T},$$

and the period of revolution $T$, for the motion of each point and for the rigid body itself is given by

$$T = \frac{2\pi r}{v}. \tag{11-16}$$

Substituting for $v$ from $v = |\omega|\, r$ (Eq. 11-15) and canceling $r$, we find also that

$$T = \frac{2\pi}{|\omega|} \quad \text{(radian measure).} \tag{11-17}$$

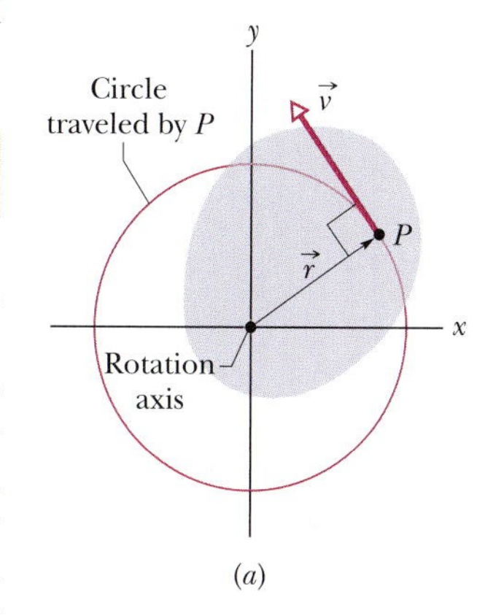


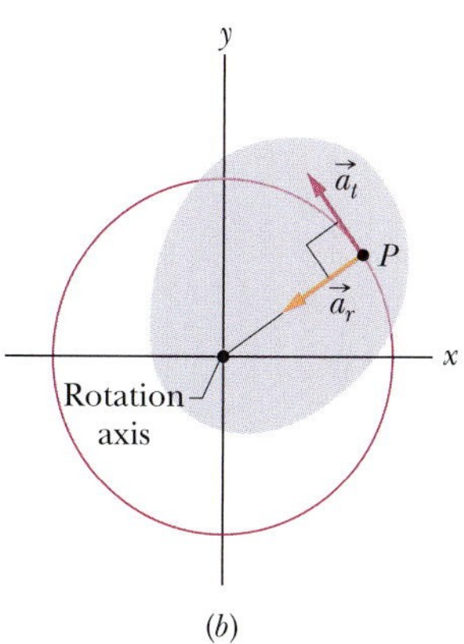


**FIGURE 11-12** ■ The rotating rigid body of Fig. 11-4, shown in cross section viewed from above. Every point of the body (such as $P$) moves in a circle around the rotation axis. (*a*) The translational velocity $v$ of every point is tangent to the circle in which the point moves. (*b*) The translational acceleration $\vec{a}$ of the point has (in general) two components, a tangential component $a_t$ and a radial component $a_r$.

## The Acceleration

Differentiating Eq. 11-15 with respect to time—again, with $r$ held constant—leads to

$$\frac{dv}{dt} = \frac{d|\omega|}{dt}r. \qquad (11\text{-}18)$$

Here we run up against a complication. In this equation, $dv/dt$ represents only the part of the magnitude of translational acceleration that is responsible for changes in the *magnitude* $|\vec{v}|$ of the translational velocity $\vec{v}$. Like $\vec{v}$, that part of the translational acceleration is tangent to the path of the point in question. We call it the *tangential component* $a_t$ of the translational acceleration of the point, and we express its magnitude

$$|\vec{a}_t| = |\alpha|r \quad \text{(radian measure)}, \qquad (11\text{-}19)$$

where the component of rotational acceleration is given by $\alpha = d\omega/dt$. *Caution:* Once again the rotational acceleration $\alpha$ in the expression $|\vec{a}_t| = |\alpha|\, r$ (Eq. 11-19) must be expressed in radian measure.

In addition, we know from our previous work that a particle (or point) moving in a circular path (even at constant velocity) has a *radial component vector* of translational acceleration, which we called the centripetal acceleration, $|\vec{a}_r| = v^2/r$ (directed radially inward), that is responsible for changes in the *direction* of the translational velocity $\vec{v}$. By substituting for $v$ from $v = |\omega|r$ (Eq. 11-15), we can write this component as

$$|\vec{a}_r| = \frac{v^2}{r} = \omega^2 r \quad \text{(radian measure)}. \qquad (11\text{-}20)$$

Thus, as Fig. 11-12*b* shows, the translational acceleration of a point on a rotating rigid body has, in general, two components. The radially inward component

$$|\vec{a}_r| = \frac{v^2}{r} = \omega^2 r,$$

is present whenever the rotational velocity of the body is not zero. That is, this component is nonzero whenever an object undergoes rotational motion. In addition, there is a tangential component $|\vec{a}_t| = |\alpha|\, r$ (Eq. 11-19) which is present whenever the rotational acceleration is nonzero. That is, this component is nonzero only if the object's rotation rate is increasing or decreasing. The total translational acceleration of a rotating rigid object is found using $|\vec{a}^{\text{tot}}|^2 = |\vec{a}_r|^2 + |\vec{a}_t|^2$.

**READING EXERCISE 11-4:** In Eq. 11-20 we did not bother to represent the squares of the magnitude of the translational and rotational speeds as $|\vec{v}|^2$ and $|\omega|^2$. Rather, we just use $v^2$ and $\omega^2$. Why is this legitimate? ■

**READING EXERCISE 11-5:** A beetle rides the rim of a rotating merry-go-round. If the rotational speed of this system (merry-go-round + beetle) is constant, does the beetle have (a) radial acceleration and (b) tangential acceleration? If the rotational speed is decreasing, does the beetle have (c) radial acceleration and (d) tangential acceleration? ■

## TOUCHSTONE EXAMPLE 11-3: Human Centrifuge

Figure 11-13 shows a centrifuge used to accustom astronaut trainees to high accelerations. The radius $r$ of the circle traveled by an astronaut is 15 m.

(a) At what constant rotational speed must the centrifuge rotate if the astronaut is to have a translational acceleration of magnitude $11g$?

**SOLUTION** ■ The **Key Idea** is this: Because the rotational speed is constant, the rotational acceleration $\alpha(= d\omega/dt)$ is zero and so is the tangential component vector of the translational acceleration ($|\vec{a}_t| = |\alpha|r$). This leaves only the radial component vector. From Eq. 11-20 ($|\vec{a}_r| = \omega^2 r$), with $|\vec{a}_r| = 11g$, we have

$$\omega = \sqrt{\frac{|\vec{a}_r|}{r}} = \sqrt{\frac{(11)\,(9.8\text{ m/s}^2)}{15\text{ m}}}$$

$$= 2.68\text{ rad/s} \approx 26\text{ rev/min.} \qquad \text{(Answer)}$$

(b) What is the tangential acceleration of the astronaut if the centrifuge accelerates at a constant rate from rest to the rotational speed found in part (a) in 120 s?

**SOLUTION** ■ The **Key Idea** here is that the magnitude of the tangential acceleration $|\vec{a}_t|$ is related to the rotational acceleration $\alpha$ by Eq. 11-19 ($|\vec{a}_t| = |\alpha|r$). Also, because the rotational acceleration is constant, we can use Eq. 11-12 ($\omega_2 = \omega_1 + \alpha(t_2 - t_1)$) from Table 11-1 to find $\alpha$ from the given rotational speeds. Putting these two equations together, we find

**FIGURE 11-13** ■ A centrifuge in Cologne, Germany, is used to accustom astronauts to the large acceleration experienced during a liftoff.

$$|\vec{a}_t| = |\alpha|r = \left|\frac{\omega_2 - \omega_1}{t_2 - t_1}\right| r$$

$$= \left|\frac{2.68\text{ rad/s} - 0}{120\text{ s}}\right|(15\text{ m}) = 0.34\text{ m/s}^2$$

$$= 0.034\,g. \qquad \text{(Answer)}$$

Although the magnitude of the final radial acceleration $|\vec{a}_r| = 11g$ is large (and alarming), the astronaut's tangential acceleration $a_t$ during the speed-up is not.

## 11-5 Kinetic Energy of Rotation

The rapidly rotating blade of a table saw certainly has kinetic energy (KE) due to that rotation. How can we express the energy? We need to treat the table saw (and any other rotating rigid body) as a collection of particles with different speeds. We can then add up the kinetic energies of all the particles $A, B, C\ldots$ to find the kinetic energy of the body as a whole. In this way we obtain, for the kinetic energy of a rigid rotating body,

$$K = \tfrac{1}{2}m_A v_A^2 + \tfrac{1}{2}m_B v_B^2 + \tfrac{1}{2}m_C v_C^2 + \cdots. \qquad (11\text{-}21)$$

The sum is taken over all the particles in the body. The problem with this sum is that the values of translational velocity are not the same for all particles. We can solve this problem by substituting for $v$ using Eq. 11-15 ($v = \omega r$) so that we have

$$\begin{aligned} K &= \tfrac{1}{2}m_A r_A^2\omega_A^2 + \tfrac{1}{2}m_B r_B^2\omega_B^2 + \tfrac{1}{2}m_C r_C^2\omega_C^2 + \cdots \\ &= \tfrac{1}{2}\omega^2\{m_A r_A^2 + m_B r_B^2 + m_C r_C^2 + \cdots\}, \end{aligned} \qquad (11\text{-}22)$$

since $\omega$ is the same for all particles.

The quantity in brackets on the right side of this equation, $\{m_A r_A^2 + m_B r_B^2 + m_C r_C^2 + \cdots\}$, tells us how the mass of the rotating body is distributed about its axis of rotation. We call that quantity the **rotational inertia** (or *moment of inertia*) $I$ of the body with respect to the axis of the rotation. It is a constant for a particular rigid body and a particular rotation axis. (That axis must always be specified if the value of $I$ is to be meaningful.)

We may write an expression defining the rotational inertia for a collection of particles as

$$I \equiv \sum m_i r_i^2 \quad \text{(rotational inertia)}, \tag{11-23}$$

where $\Sigma$ is a summation sign signifying that we sum over all the particles in the rigid rotating system. We can substitute into Eq. 11-22 ($K = \frac{1}{2}\omega^2\{m_A r_A^2 + m_B r_B^2 + m_C r_C^2 + \cdots\}$), obtaining

$$K = \tfrac{1}{2} I\omega^2 \quad \text{(rotational KE, radian measure)}, \tag{11-24}$$

as the expression we seek for the rotational kinetic energy. Because we have used the relation $v = \omega r$ in deriving $K = \frac{1}{2}I\omega^2$, $\omega$ must be expressed in radian measure. The SI unit for $I$ is the kilogram-meter-squared ($\text{kg} \cdot \text{m}^2$).

Equation 11-24 ($K = \frac{1}{2}I\omega^2$), which gives the kinetic energy of a rigid body in pure rotation, is the rotational equivalent of the formula $K = \frac{1}{2}Mv_{\text{com}}^2$, which gives the kinetic energy of a rigid body in pure translation. In both formulas, there is a factor of $\frac{1}{2}$. Where mass $M$ appears in one equation, $I$ (which involves both mass and distribution) appears in the other. Finally, each equation contains a factor of the square of a speed—translational or rotational as appropriate. The kinetic energies of translation and rotation are not different kinds of energy. They are both kinetic energy, expressed in ways that are appropriate to the motion at hand.

We noted previously that the rotational inertia of a rotating body involves not only its mass but also how that mass is distributed. Here is an example that you can literally feel. Rotate a long rod (a pole, a length of lumber, a twirling baton, or something similar), first around its central (longitudinal) axis (Fig. 11-14*a*) and then around an axis perpendicular to the rod and through the center (Fig. 11-14*b*). Both rotations involve the very same mass, but the mass of the object in the first rotation is much closer to the rotation axis. As a result, the rotational inertia of the rod is much smaller in Fig. 11-14*a* than in Fig. 11-14*b*. In general, smaller rotational inertia means easier rotation.

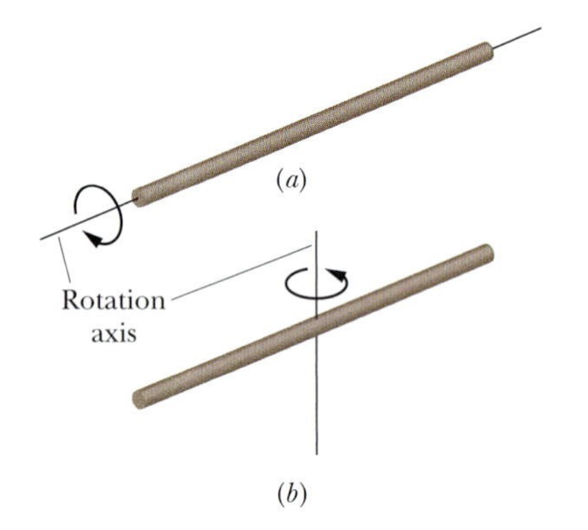


**FIGURE 11-14** ■ A long rod is much easier to rotate about (*a*) its central (longitudinal) axis through its center and perpendicular to its length because the mass is distributed closer to the rotation axis in (*a*) than in (*b*).

**READING EXERCISE 11-6:** The figure shows three small spheres that rotate about a vertical axis. The perpendicular distance between the axis and the center of each sphere is given. Rank the three spheres according to their rotational inertia about the axis, greatest first.

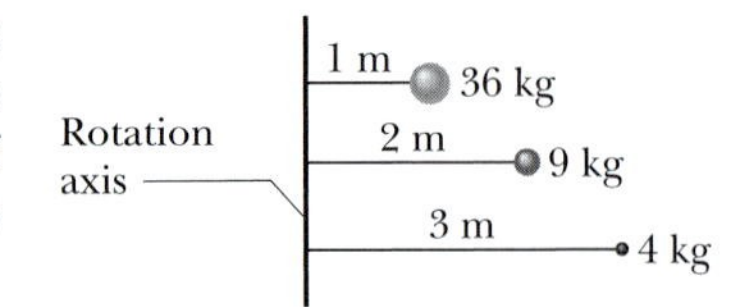


## 11-6 Calculating Rotational Inertia

If a rigid body consists of a few particles, we can calculate its rotational inertia about a given rotational axis with Eq. 11-23 ($I = \Sigma\, m_i r_i^2$). For example, consider the rotational inertia of a lump of clay (considered to be a point mass) with mass $M$ at a distance $r$ from the axis of rotation. The rotational inertia of such an object is simply $Mr^2$. Consider the rotational inertia of the object if the clay is now split into two pieces of equal mass, or eight pieces of equal mass, or even a very large number of point masses. As shown in Fig. 11-15, these pieces can be made to fashion a hoop of mass $m$ and radius $r$.

Since the total mass $M$ is divided into $n$ equal masses, we can write the rotational inertia of this hoop as the sum of the rotational inertias of each of its elements:

$$\sum_{i=1}^{i=n} m_i r_i^2 = \frac{M}{n} r_1^2 + \frac{M}{n} r_2^2 + \frac{M}{n} r_3^2 + \cdots.$$

Further, since all the point masses that make up the hoop are located the same distance $r$ away from the axis of rotation,

$$\sum_{i=1}^{i=n} m_i r_i^2 = \left(\frac{M}{n} r^2 + \frac{M}{n} r^2 + \frac{M}{n} r^2 + \cdots\right)$$

$$= r^2 \left(\frac{M}{n} + \frac{M}{n} + \frac{M}{n} + \cdots\right) \quad (11\text{-}25)$$

$$I_{\text{hoop}} = Mr^2.$$

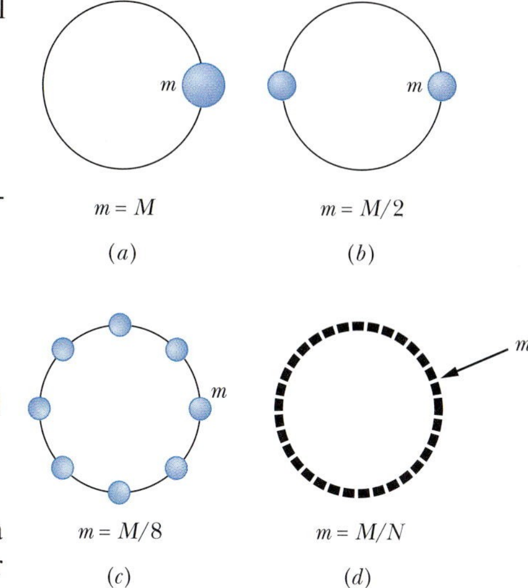


**FIGURE 11-15** ■ Imagine a thin wire that provides a "massless" circular frame for clay blobs. As a clay blob of mass $m$ is divided into more and more parts of equal mass, the distance of each smaller blob from the center of the circle is still the same. How does $I$ compare for each of the objects, $(a)$, $(b)$, $(c)$, and $(d)$?

If a rigid body consists of a great many adjacent particles (it is *continuous*, like a Frisbee), using $I = \Sigma m_i r_i^2$ would require a tedious computer calculation. Instead, for a body that has a simple geometric form, we can replace the sum $\Sigma m_i r_i^2$ with an integral, and define the rotational inertia of the body as

$$I = \int r^2\, dm \quad \text{(rotational inertia, continuous body).} \quad (11\text{-}26)$$

Table 11-2 gives the results of such integration for nine common body shapes and the indicated axes of rotation.

Studying the equations in Table 11-2 is helpful. *For example, for objects all having the same radius and mass, an object with its mass distributed very close to the axis of rotation has a smaller rotational inertia than an object with mass distributed farther out.* A case in point is the rotation of a cylinder (or equivalently a disk) about a central diameter. Table 11-2 shows that

$$I_{\text{disk}} = I_{\text{cylinder}} = \int r^2\, dm = \tfrac{1}{2} MR^2. \quad (11\text{-}27)$$

**TABLE 11-2**
**Some Rotational Inertias**

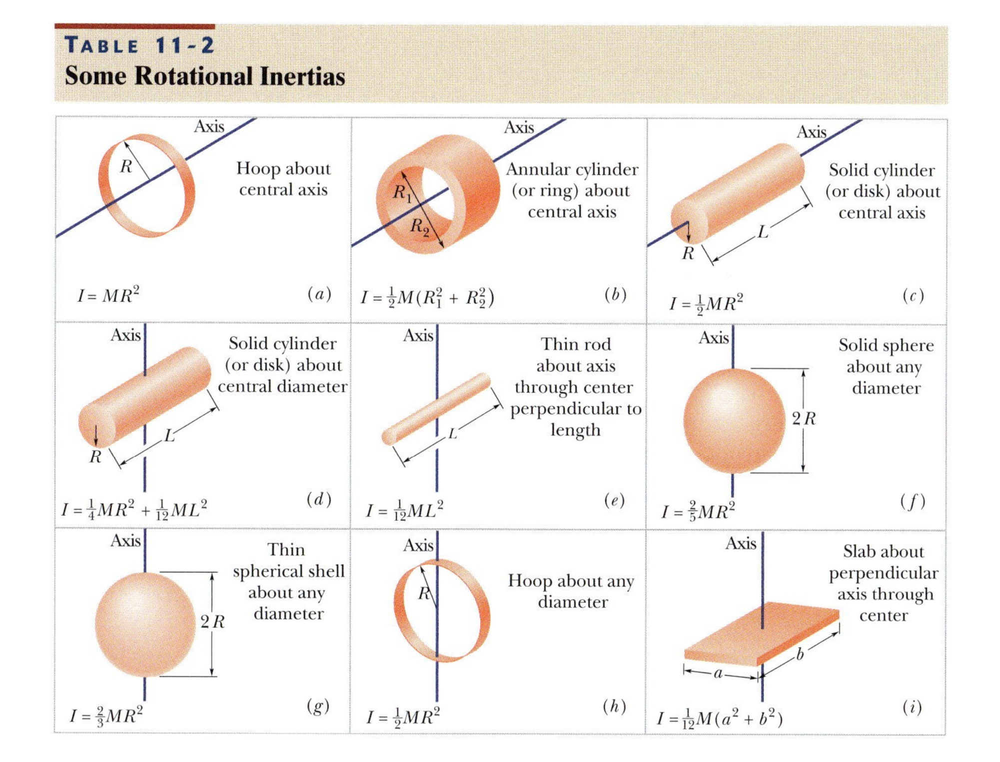

Axis
R
Hoop about central axis
I = MR²
(a)
Axis
R₁
R₂
Annular cylinder (or ring) about central axis
I = ½M(R₁² + R₂²)
(b)
Axis
R
L
Solid cylinder (or disk) about central axis
I = ½MR²
(c)
Axis
R
L
Solid cylinder (or disk) about central diameter
I = ¼MR² + 1/12 ML²
(d)
Axis
L
Thin rod about axis through center perpendicular to length
I = 1/12 ML²
(e)
Axis
2R
Solid sphere about any diameter
I = ⅖MR²
(f)
Axis
2R
Thin spherical shell about any diameter
I = ⅔MR²
(g)
Axis
R
Hoop about any diameter
I = ½MR²
(h)
Axis
a
b
Slab about perpendicular axis through center
I = 1/12 M(a² + b²)
(i)

Thus, the rotational inertia of the hoop is twice that of a disk with the same mass and radius. This is because a hoop of the same radius as the disk has *all* its mass distributed as far away from the axis of rotation as possible.

Note that an object can have more than one axis of rotation. For example, you can roll a cylinder (or disk) along a table so its axis of rotation is perpendicular to the flat face of the cylinder (as in Table 11-2(*c*)). Alternatively Table 11-2(*d*) shows a different rotational inertia equation for an axis parallel to its face. In the next subsection we present a parallel-axis theorem that will allow us to determine the rotational inertia about any rotational axis once we know its rotational inertia about another axis parallel to it.

## Parallel-Axis Theorem

Suppose we want to find the rotational inertia $I$ of a body of mass $M$ about a given axis. In principle, we can always find $I$ using integration of

$$I = \int r^2\, dm.$$

However, it is easier mathematically to find the rotational inertia of an object about an axis of symmetry that passes through the object's center of mass. Fortunately, in certain circumstances, there is a shortcut. If we know the rotational inertia of a symmetric object rotating about an axis passing through its center of mass (for example, from Table 11-2), then the rotational inertia $I$ about another parallel axis is

$$I = I_{\text{com}} + Mh^2 \quad \text{(parallel-axis theorem).} \tag{11-28}$$

Here $h$ is the perpendicular distance between the given axis and the axis through the center of mass (remember that these two axes must be parallel). The proof of this equation, known as the **parallel-axis theorem,** is fairly straightforward, because it takes advantage of the fact that the object is symmetric about its center-of-mass axis of rotation.

**READING EXERCISE 11-7:** The figure shows a book-like object (one side is longer than the other) and four choices of rotation axes, all perpendicular to the face of the object. Rank the choices according to the rotational inertia of the object about the axis, greatest first.

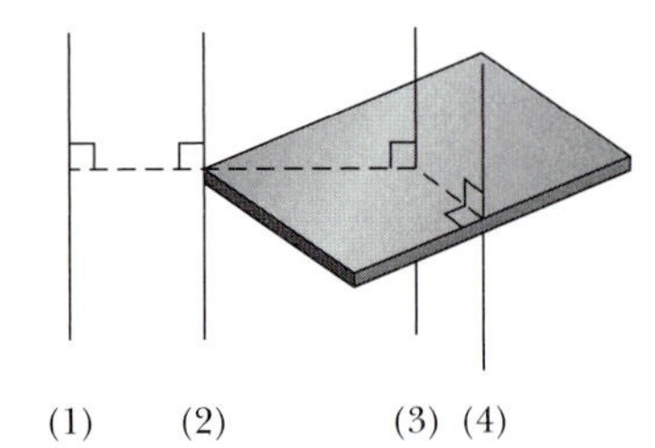


**READING EXERCISE 11-8:** Four objects having the same "radius" and mass are shown in the figure that follows. Rank the objects according to the rotational inertia about the axis shown, greatest first.

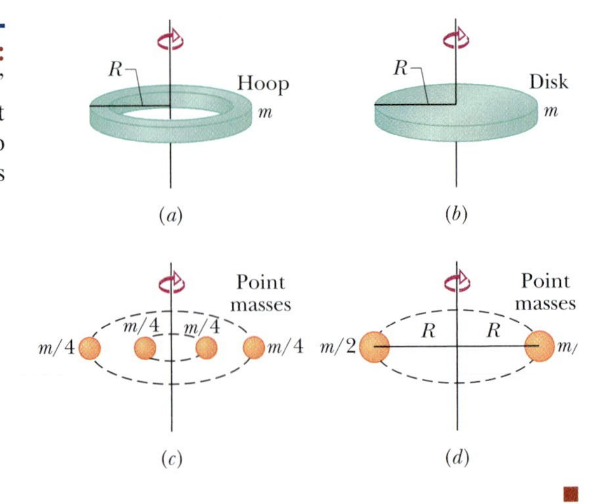

### TOUCHSTONE EXAMPLE 11-4: Rotor Failure

Large machine components that undergo prolonged, high-speed rotation are first examined for the possibility of failure in a *spin test system*. In this system, a component is *spun up* (brought up to high speed) while inside a cylindrical arrangement of lead bricks and containment liner, all within a steel shell that is closed by a lid clamped into place. If the rotation causes the component to shatter, the soft lead bricks are supposed to catch the pieces so that the failure can then be analyzed.

In early 1985, Test Devices Inc. (www.testdevices.com) was spin-testing a sample of a solid steel rotor (a disk) of mass $M = 272$ kg and radius $R = 38.0$ cm. When the sample reached a rotational speed $\omega$ of 14 000 rev/min, the test engineers heard a dull thump from the test system, which was located one floor down and one room over from them. Investigating, they found that lead bricks had been thrown out in the hallway leading to the test room, a door to the room had been hurled into the adjacent parking lot, one lead brick had shot from the test site through the wall of a neighbor's kitchen, the structural beams of the test building had been damaged, the concrete floor beneath the spin chamber had been shoved downward by about 0.5 cm, and the 900 kg lid had been blown upward through the ceiling and had then crashed back onto the test equipment (Fig. 11-16). The exploding pieces had not penetrated the room of the test engineers only by luck.

How much energy was released in the explosion of the rotor?

**FIGURE 11-16** Some of the destruction caused by the explosion of a rapidly rotating steel disk.

**SOLUTION** The **Key Idea** here is that this released energy was equal to the rotational kinetic energy $K$ of the rotor just as it reached the rotational speed of 14 000 rev/min. We can find $K$ with Eq. 11-24 ($K = \frac{1}{2}I\omega^2$), but first we need an expression for the rotational inertia $I$. Because the rotor was a disk that rotated like a merry-go-round, $I$ is given by the expression in Table 11-2($c$) ($I = \frac{1}{2}MR^2$). Thus we have

$$I = \tfrac{1}{2}MR^2 = \tfrac{1}{2}(272 \text{ kg})(0.38 \text{ m})^2 = 19.64 \text{ kg} \cdot \text{m}^2.$$

The rotational speed of the rotor was

$$\omega = (14\,000 \text{ rev/min})(2\pi \text{ rad/rev})\left(\frac{1 \text{ min}}{60 \text{ s}}\right)$$
$$= 1.466 \times 10^3 \text{ rad/s}.$$

Now we can use Eq. 11-24 to write

$$K = \tfrac{1}{2}I\omega^2 = \tfrac{1}{2}(19.64 \text{ kg} \cdot \text{m}^2)(1.466 \times 10^3 \text{ rad/s})^2$$
$$= 2.1 \times 10^7 \text{ J}. \qquad \text{(Answer)}$$

Being near this explosion was like being near an exploding bomb.

## 11-7 Torque

Now that we have defined the variables needed to describe the rotation of an object, we need to determine how forces can affect rotational motion. Because we are talking about more complex objects than point particles, we need to consider not only the forces that act on a rotating body but also the locations of those forces.

For instance, a doorknob is located as far as possible from the door's hinge line for a good reason. If you want to open a heavy door, you must certainly apply a force; that alone, however, is not enough. Where you apply that force and in what direction you push are also important. If you apply your force nearer to the hinge line than to the knob, or at any angle other than 90° to the plane of the door, you must use a greater force to move the door than if you apply the force at the knob and perpendicular to the door's plane. If you have never noticed this phenomenon, compare the force you need to open a heavy door near the hinge to that at the handle.

Figure 11-17$a$ shows a cross section of a body that is free to rotate about an axis passing through $O$ and perpendicular to the cross section. A force $\vec{F}$ is applied at point $P$, whose position relative to $O$ is defined by a position vector $\vec{r}$. The directions

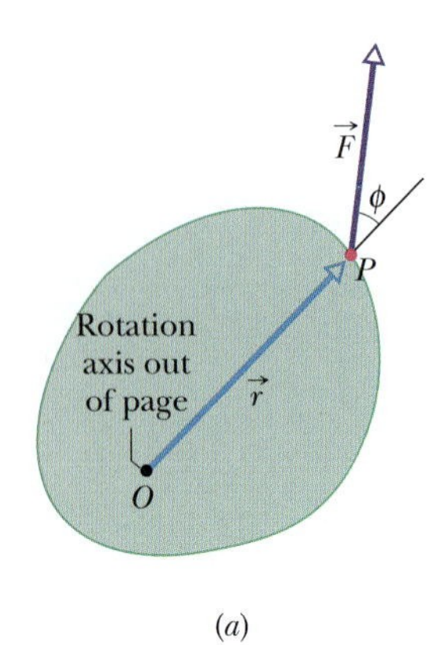


(a)

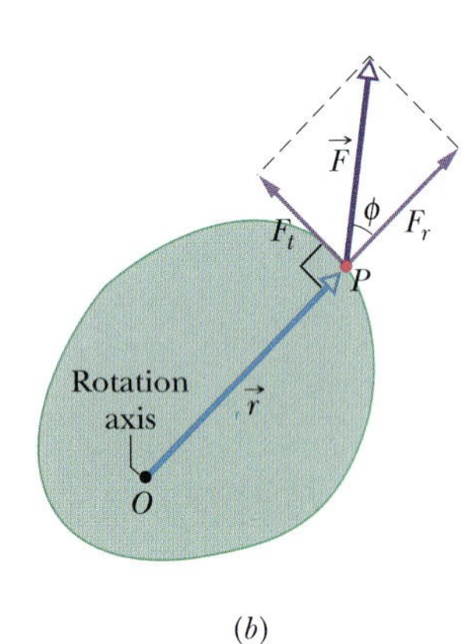


(b)

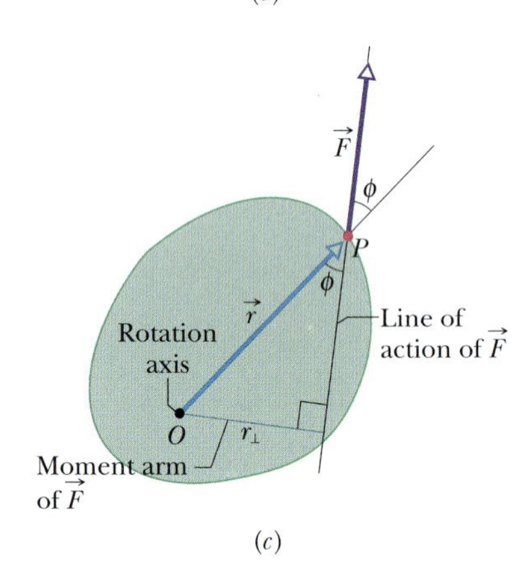


(c)

**FIGURE 11-17** ■ (*a*) A force $\vec{F}$ acts at point *P* on a rigid body that is free to rotate about an axis through *O*. The axis is perpendicular to the plane of the cross section shown here. (*b*) The torque due to this force is $|\vec{r}|\,|\vec{F}|\sin\phi$. We can also write it as $|\vec{r}|\,|\vec{F}_t|$, where $\vec{F}_t$ is the tangential component vector of $\vec{F}$. (*c*) The torque magnitude can also be written as $r_\perp|\vec{F}|$, where $r_\perp$ is the moment arm of $\vec{F}$.

of vectors $\vec{F}$ and $\vec{r}$ make an angle $\phi$ ($0 \le \phi \le 180°$) with each other. For simplicity, we consider only forces that have no component parallel to the rotation axis; in other words, $\vec{F}$ is in the plane of the page.

To determine how $\vec{F}$ results in a rotation of the body around the rotation axis, we resolve $\vec{F}$ into two components (Fig. 11-17*b*). One component, called the *radial component vector* $\vec{F}_r$, points along $\vec{r}$. This component does not cause rotation, because it acts along a line that extends through *O*. (If you pull on a door parallel to the plane of the door, you are stretching and compressing the door, but you do not cause the door to rotate.) The other vector component of $\vec{F}$, called the *tangential component vector* $\vec{F}_t$, is perpendicular to $\vec{r}$ and has magnitude $|\vec{F}_t| = |\vec{F}|\sin\phi$. This component *does* cause rotation. (If you pull on a door perpendicular to its plane, you can rotate the door.)

The ability of $\vec{F}$ to rotate the body depends not only on the magnitude of its tangential component $|\vec{F}_t|$, but also on just how far from *O* the force is applied. To include both these factors, we define a new quantity called **torque.** In general, torque is a three-dimensional vector whose direction depends on the location and direction of a net force that acts on a rigid object that can rotate. Since we are only considering fixed rotation axes in this chapter, we can represent torque here as a one-dimensional vector (as we have with the other rotational variables). For now, we will describe torque in terms of its component $\tau_z$ along a *z* axis of rotation of the body experiencing a net force.

The torque component $\tau_z$, often denoted as simply $\tau$, has either a positive or negative value, depending on the direction of rotation it would give a body initially at rest. If a body rotates so the thumb of the right hand points along the positive direction assigned to the axis of rotation, the torque component is positive. If the object rotates in the opposite way, the torque component is negative.

The magnitude of the torque can be written as the product of the magnitude of a moment arm $|\vec{r}|$ and the magnitude of the tangential component of the force $|\vec{F}_t|$. As you can see in Fig. 11-17*b*, $|\vec{F}_t| = |\vec{F}|\sin\phi$.

$$|\vec{\tau}| = |\vec{r}||\vec{F}_t| = |\vec{r}||\vec{F}|\sin\phi. \tag{11-29}$$

Two equivalent ways of computing the magnitude of torque are

$$|\vec{\tau}| = |\vec{r}|(|\vec{F}|\sin\phi) = |\vec{r}||\vec{F}_t|, \tag{11-30}$$

and

$$|\vec{\tau}| = (r\sin\phi)|\vec{F}| = r_\perp|\vec{F}|, \tag{11-31}$$

where $r_\perp$ is the perpendicular distance between the rotation axis at *O* and an extended line running through the vector $\vec{F}$ (Fig. 11-17*c*). This extended line is called the **line of action** of $\vec{F}$, and $r_\perp$ is called the **moment arm** of $\vec{F}$. Figure 11-17*c* shows that we can describe *r*, the magnitude of $\vec{r}$, as being the moment arm of the force component $F_t$.

Torque, which comes from the Latin word meaning "to twist," may be loosely identified as the turning or twisting action of the force $\vec{F}$. When you apply a force to an object—such as a screwdriver or torque wrench—with the purpose of turning that object, you are applying a torque. The SI unit of torque is the newton-meter (N · m). *Caution*: The newton-meter is also the unit of work. Torque and work, however, are quite different quantities and must not be confused. Work is often expressed in joules (1 J = 1 N · m), but torque never is.

In the next chapter, we shall discuss cases in which torque must be represented by a vector that changes direction over time.

Torques obey the superposition principle that we discussed in Chapter 3 for forces: When several torques act on a body, the **net torque** (or **resultant torque**) component is the vector sum of the individual torques. The symbol for net torque component along the axis of rotation is $\tau_z^{\text{net}}$.

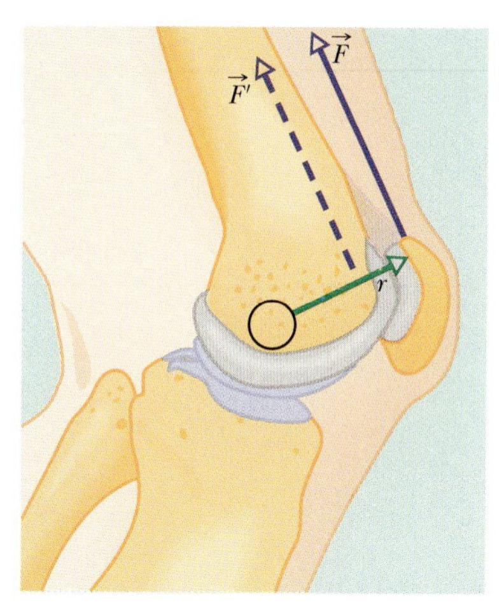


FIGURE 11-18 ■ The structure of the human knee. Note the force exerted by the quadriceps muscle, shown as $\vec{F}$, acting on the kneecap at a distance, $r$, from the pivot axis. If the kneecap were not there, the force, shown as $\vec{F}'$, would be acting along the dashed line a smaller distance $r$ from the pivot axis, which is located approximately at the small circle.

## Using Torque to Jump

So how do kneecaps allow us to jump higher? When we jump, we create a large torque in the knee joint in order to straighten the leg. If the force exerted by the strong thigh muscle (the quadriceps) is exerted along a line that is close to the pivot in the knee joint (represented by the dashed arrow in Fig. 11-18), the torque is not very great. A kneecap allows that force to be exerted farther from the pivot (represented by the solid arrow in Fig. 11-18). Recall that it is more effective to open a door by pushing far from the hinge. Similarly, the leg with a kneecap achieves more leg-straightening torque, thereby allowing for a higher jump!

**READING EXERCISE 11-9:** The figure shows an overhead view of a meter stick that can pivot about the dot at the position marked 20 (for 20 cm). All five forces on the stick have the same magnitude. Rank those forces according to the magnitude of the torque that they produce, greatest first.

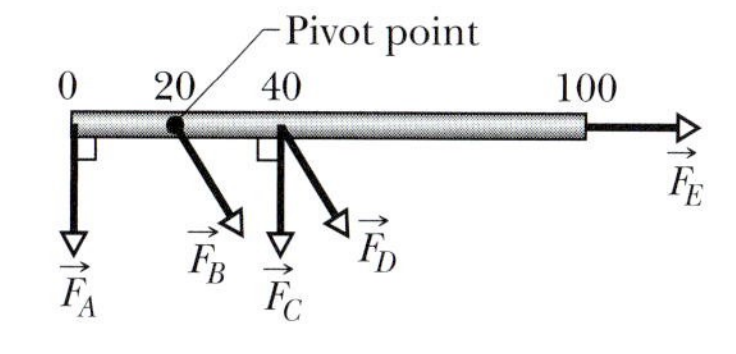


■

# 11-8 Newton's Second Law for Rotation

A torque can cause rotation of a rigid body, such as when you open a door about its hinge. Here we want to consider a special case in which a rigid body is symmetric about its axis of rotation. For this case we can relate the net torque component $\tau^{\text{net}}$ that acts on the body to the rotational acceleration component $\alpha$ the torque causes about a rotation axis. A good guess is to do so by analogy to the one-dimensional form of Newton's Second Law. If a one-dimensional net force $\vec{F}_x^{\text{net}}$ is acting along the $x$ axis, then $\vec{F}_x^{\text{net}} = m\vec{a}_x$, where $\vec{a}_x$ is the acceleration component of a body of mass $m$, due to the net force acting along the $x$ axis. For a rotation about a $z$ axis we replace $\vec{F}_x^{\text{net}}$ with $\vec{\tau}_z^{\text{net}}$, $m$ with $I$, and $a_x$ with $\alpha_z$, writing

$$\vec{\tau}_z^{\text{net}} = I\vec{\alpha}_z \quad \text{(Newton's Second Law for rotation).} \tag{11-32}$$

Remember in this context that $\vec{\tau}_z^{\text{net}}$ and $\vec{\alpha}_z$ are vector components, that we have chosen to represent as $\tau^{\text{net}}$ and $\alpha$ respectively. We can then rewrite Eq. 11-32 as

$$\tau^{\text{net}} = I\alpha \quad \text{(Newton's Second Law for symmetric rotations),} \tag{11-33}$$

where $\alpha$ must be in radian measure. This rotational analog to one-dimensional translational motion only holds when the axis of rotation does not change direction and when the body is symmetric about its axis of rotation.

## Proof of Equation 11-33

To see that Eq. 11-33 is, in fact, valid, let us see whether we can prove mathematically that $\tau^{\text{net}} = I\alpha$ by first considering the simple situation shown in Fig. 11-19. The rigid body there consists of a particle of mass $m$ on one end of a massless rod of length $r$. The rod can move only by rotating about its other end, around a rotation axis (an axle) that is perpendicular to the plane of the page. Thus, the particle can move only in a circular path that has the rotation axis at its center.

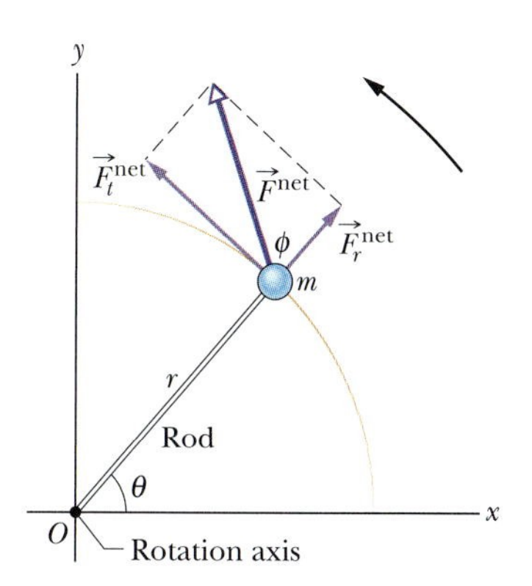


FIGURE 11-19 ■ A simple rigid body, free to rotate about an axis through $O$, consists of a particle of mass $m$ fastened to the end of a rod of length $r$ and negligible mass. An applied force $\vec{F}^{\text{net}}$ causes the body to rotate.

A force $\vec{F}^{\text{net}}$ acts on the particle. However, because the particle can move only along the circular path, only the tangential component $\vec{F}_t^{\text{net}}$ of the force (the component that is tangent to the circular path) can accelerate the particle along the path. We can relate $\vec{F}_t^{\text{net}}$ to the particle's tangential acceleration component $a_t$ along the path with Newton's Second Law, writing

$$\vec{F}_t^{\text{net}} = m\vec{a}_t.$$

So now, the magnitude of the torque acting on the particle is given by Eq. 11-30 as

$$|\vec{\tau}^{\text{net}}| = |\vec{r}|(|\vec{F}^{\text{net}}|\sin\phi) = |\vec{r}||\vec{F}_t^{\text{net}}|.$$

Note that we define a net tangential force component $\vec{F}_t^{\text{net}}$ as positive if it causes rotational and tangential accelerations that have positive components according to the right-hand rule. Conversely, $\vec{F}_t^{\text{net}}$ is negative if it leads to negative acceleration components along the axis of rotation. Since the distance $r = |\vec{r}|$ is the magnitude of a vector perpendicular to the rotation axis that points to the rotating particle, it is always positive. So the torque component can be expressed in terms of the net tangential force and acceleration components associated with a rotating body of mass $m$ as

$$\tau^{\text{net}} = F_t^{\text{net}}\, r = ma_t\, r.$$

From Eq. 11-19 ($a_t = \alpha r$), we can write this as

$$\tau^{\text{net}} = m(\alpha r)r = (mr^2)\alpha. \tag{11-34}$$

Since the quantity in parentheses on the right side of this equation is the rotational inertia, $mr^2$, of the particle about the rotation axis, Eq. 11-34 reduces to

$$\tau^{\text{net}} = I\alpha \quad \text{(radian measure)}, \qquad \text{(Eq. 11-33)}$$

which is the expression we set out to prove. We can extend this equation to any rigid body rotating about an axis of symmetry, because any such body can always be analyzed as an assembly of single particles. Both $\alpha$ and $\tau^{\text{net}}$ are vector components along the rotation axis. Since $I$ is inherently positive, $\alpha$ and $\tau^{\text{net}}$ must always have the same sign.

**READING EXERCISE 11-10:** The figure shows an overhead view of a meter stick that can pivot about a vertical axis at the point indicated, which is to the left of the stick's midpoint. Two horizontal forces, $\vec{F}_A$ and $\vec{F}_B$, are applied to the stick. Only $\vec{F}_A$ is shown. Force $\vec{F}_B$ is perpendicular to the stick and is applied at the right end. If the stick does not turn, (a) Is $\vec{F}_B$ in the same or opposite direction as $\vec{F}_A$ and (b) should the magnitude of $\vec{F}_B$ be greater than, less than, or equal to $\vec{F}_A$?

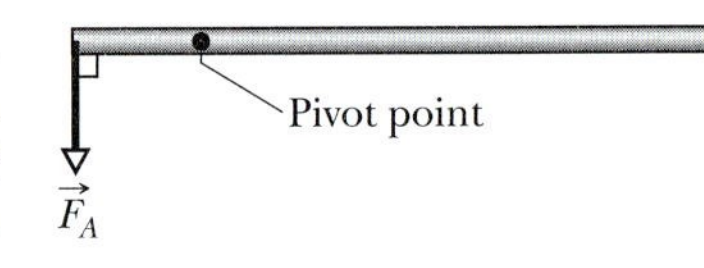


## TOUCHSTONE EXAMPLE 11-5: Accelerating a Wheel

Figure 11-20*a* shows a uniform disk, with mass $M = 2.5$ kg and radius $R = 20$ cm, mounted on a fixed horizontal axle. A block with mass $m = 1.2$ kg hangs from a massless cord that is wrapped around the rim of the disk. Find the acceleration of the falling block, the rotational acceleration of the disk, and the tension in the cord. The cord does not slip, and there is no friction at the axle.

**SOLUTION** ■ One **Key Idea** here is that, taking the block as a system, we can relate its acceleration $a$ to the forces acting on it with Newton's Second Law ($\vec{F}^{\text{net}} = m\vec{a}$). Those forces are shown in the block's free-body diagram in Fig. 11-20*b*: The force from the cord is $\vec{F}^{\text{cord}}$ and the gravitational force is $\vec{F}^{\text{grav}}$, of magnitude $mg$. We can now write Newton's Second Law for components along a vertical $y$ axis $F_y^{\text{net}} = ma_y$ as

$$|\vec{F}^{\text{cord}}| - mg = ma_y. \tag{11-35}$$

However, we cannot solve this equation for $a_y$ because it also contains the unknown $|\vec{F}^{\text{cord}}|$.

Previously, when we got stuck on the $y$ axis, we would switch to the $x$ axis. Here, we switch to the rotation of the disk and use this **Key Idea**: Taking the disk as a system, we can relate its rotational

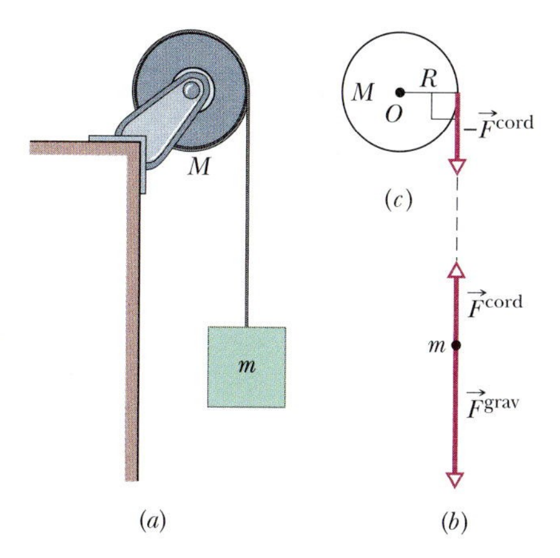


**FIGURE 11-20** ■ (*a*) The falling block causes the disk to rotate. (*b*) A free-body diagram for the block. (*c*) An incomplete free-body diagram for the disk.

acceleration $\alpha$ to the torque acting on it with Newton's Second Law for rotation ($\tau^{\text{net}} = I\alpha$). To calculate the torques and the rotational inertia $I$, we take the rotation axis to be perpendicular to the disk and through its center, at point $O$ in Fig. 11-20*c*.

The torques are then given by Eq. 11-29 ($|\vec{\tau}| = |\vec{r}||\vec{F}_t|$). The gravitational force on the disk and the force on the disk from the axle both act at the center of the disk and thus at distance $r = 0$, so their torques are zero. The force $\vec{F}^{\text{cord}}$ on the disk due to the cord acts at distance $r = R$ and is tangent to the rim of the disk. Therefore, the magnitude of its torque is $|\vec{\tau}| = R|\vec{F}^{\text{cord}}|$. From Table 11-2(*c*), the rotational inertia $I$ of the disk is $\frac{1}{2}MR^2$. Thus we can write ($\tau_z^{\text{net}} = I\alpha_z$) as

$$|\vec{\tau}| = R|\vec{F}^{\text{cord}}| = \tfrac{1}{2}MR^2|\alpha|. \qquad (11\text{-}36)$$

This equation seems equally useless because it has two unknowns, $\alpha$ and $|\vec{F}^{\text{cord}}|$, neither of which is the desired acceleration $a$. However, mustering physics courage, we can make it useful with a third **Key Idea**: Because the cord does not slip, the magnitudes of the translational acceleration $|a_y|$ of the block and of the (tangential) translational acceleration $|\vec{a}_t|$ of the rim of the disk are equal. Then, by Eq. 11-19 ($|\vec{a}_t| = |\alpha|r$) we see that here $|\alpha| = |a_y|/R$. Substituting this in Eq. 11-36 yields

$$R|\vec{F}^{\text{cord}}| = \frac{\frac{1}{2}MR^2|a_y|}{R} \quad \text{or} \quad |\vec{F}^{\text{cord}}| = \tfrac{1}{2}M|a_y|. \qquad (11\text{-}37)$$

From Fig. 11-20*a* it's apparent that $a_y$ is negative, which along with Eq. 11-37 tells us that

$$a_y = -\frac{2|\vec{F}^{\text{cord}}|}{M}. \qquad (11\text{-}38)$$

Now combining Eqs. 11-35 and 11-38 leads to

$$a_y = -g\frac{2m}{M + 2m} = -(9.8\text{ m/s}^2)\frac{(2)(1.2\text{ kg})}{2.5\text{ kg} + (2)(1.2\text{ kg})}$$
$$= -4.8\text{ m/s}^2. \qquad \text{(Answer)}$$

We then use Eq. 11-37 to find $|\vec{F}^{\text{cord}}|$:

$$|\vec{F}^{\text{cord}}| = \tfrac{1}{2}M|a_y| = \tfrac{1}{2}(2.5\text{ kg})(4.8\text{ m/s}^2) = 6.0\text{ N}. \qquad \text{(Answer)}$$

As we should expect, the magnitude of the acceleration of the falling block is less than $g$, and the tension in the cord (= 6.0 N) is less than the gravitational force on the hanging block ($= mg = 11.8$ N). We see also that the acceleration of the block and the tension depend on the mass of the disk but not on its radius. As a check, we note that the formulas derived above predict $a_y = -g$ and $T = 0$ for the case of a massless disk ($M = 0$). This is what we would expect; the block simply falls as a free body, trailing the string behind it.

From Eq. 11-19, the magnitude of the rotational acceleration of the disk is

$$|\alpha| = \frac{|a_y|}{R} = \frac{4.8\text{ m/s}^2}{0.20\text{ m}} = 24\text{ rad/s}^2. \qquad \text{(Answer)}$$

## TOUCHSTONE EXAMPLE 11-6: Judo

To throw an 80 kg opponent with a basic judo hip throw, you intend to pull his uniform with a force $\vec{F}$ and a moment arm $d_1 = 0.30$ m from a pivot point (rotation axis) on your right hip (Fig. 11-21). You wish to rotate him about the pivot point with an rotational acceleration $\alpha$ of $-6.0$ rad/s²—that is, with an rotational acceleration that is *clockwise* in the figure. Assume that his rotational inertia $I$ relative to the pivot point is 15 kg · m².

(a) What must the magnitude of $\vec{F}$ be if, before you throw him, you bend your opponent forward to bring his center of mass to your hip (Fig. 11-21*a*)?

**SOLUTION** ■ One **Key Idea** here is that we can relate your pull $\vec{F}$ on him to the given rotational acceleration $\alpha$ via Newton's Second Law for rotation ($\tau^{\text{net}} = I\alpha$). As his feet leave the floor, we can assume that only three forces act on him: your pull $\vec{F}$, a force $\vec{N}$ on him from you at the pivot point (this force is not indicated in Fig. 11-21), and the gravitational force $\vec{F}^{\text{grav}}$. To use $\tau^{\text{net}} = I\alpha$, we need the corresponding three torques, each about the pivot point.

From Eq. 11-31 ($|\tau| = r_\perp|\vec{F}|$), the torque due to your pull $\vec{F}$ is equal to $-d_1F$, where $d_1$ is the moment arm $r_\perp$ and the sign indicates the clockwise rotation this torque tends to cause. The torque due to $\vec{N}$ is zero, because $\vec{N}$ acts at the pivot point and thus has moment arm $r_\perp = 0.0$ m.

To evaluate the torque due to $\vec{F}^{\text{grav}}$, we need a **Key Idea** from Chapter 8: We can assume that $\vec{F}^{\text{grav}}$ acts at your opponent's center of mass. With the center of mass at the pivot point, $\vec{F}^{\text{grav}}$ has moment arm $r_\perp = 0.0$ m and thus the torque due to $\vec{F}^{\text{grav}}$ is zero. Thus, the only torque on your opponent is due to your pull $\vec{F}$, and we can write $\tau^{\text{net}} = I\alpha$ as

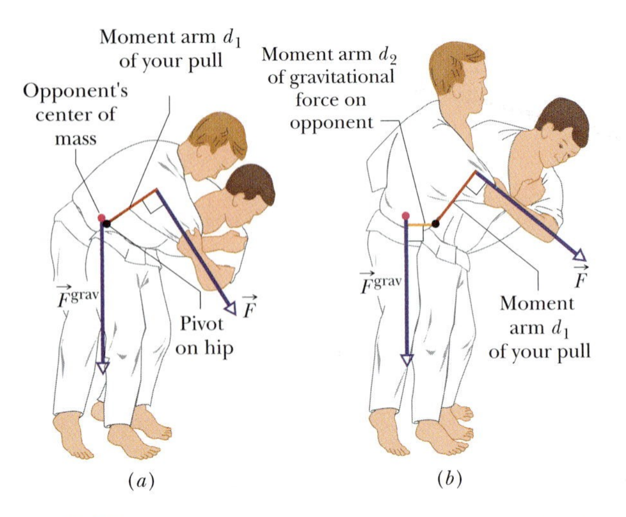


**FIGURE 11-21** ■ A judo hip throw (*a*) correctly executed and (*b*) incorrectly executed.

$$-d_1|\vec{F}| = I\alpha.$$

We then find

$$|\vec{F}| = \frac{-I\alpha}{d_1} = \frac{-(15\text{ kg}\cdot\text{m}^2)(-6.0\text{ rad/s}^2)}{0.30\text{ m}}$$

$$= 300\text{ N}. \qquad \text{(Answer)}$$

(b) What must the magnitude of $\vec{F}$ be if your opponent remains upright before you throw him, so that $\vec{F}^{grav}$ has a moment arm $d_2 = 0.12$ m from the pivot point (Fig. 11-21*b*)?

**SOLUTION** ■ The **Key Ideas** we need here are similar to those in (a) with one exception: Because the moment arm for $\vec{F}^{grav}$ is no longer zero, the torque due to $\vec{F}^{grav}$ is now equal to $d_2mg$, and is positive because the torque attempts counterclockwise rotation. Now we write $\tau^{net} = I\alpha$ as

$$-d_1|\vec{F}| + d_2mg = I\alpha,$$

which gives

$$|\vec{F}| = -\frac{I\alpha}{d_1} + \frac{d_2mg}{d_1}.$$

From (a), we know that the first term on the right is equal to 300 N. Substituting this and the given data, we have

$$|\vec{F}| = 300\text{ N} + \frac{(0.12\text{ m})(80\text{ kg})(9.8\text{ m/s}^2)}{0.30\text{ m}}$$

$$= 613.6\text{ N} \approx 610\text{ N}. \qquad \text{(Answer)}$$

The results indicate that you will have to pull much harder if you do not initially bend your opponent to bring his center of mass to your hip. A good judo fighter knows this lesson from physics. (An analysis of the physics of judo and aikido is given in "The Amateur Scientist" by J. Walker, *Scientific American*, July 1980, Vol. 243, pp. 150–161.)

## 11-9 Work and Rotational Kinetic Energy

### Net Work-Kinetic Energy Theorem for Translational Motion in One Dimension

As we discussed in Chapter 9, when a net force causes the center of mass of a rigid body of mass $m$ to accelerate along a coordinate axis, it does net work, $W^{net}$, on the body. Thus, the body's translational kinetic energy ($K = \frac{1}{2}mv^2$) can change. We can use the net work-kinetic energy theorem to relate these two quantities:

$$W^{net} = K_2 - K_1 = \Delta K \qquad \text{(work-kinetic energy theorem)}, \qquad \text{(Eq. 9-11)}$$

where $K_1$ is the kinetic energy of the object when it is located at an initial position and $K_2$ is its kinetic energy when it is displaced to a new position.

For translational motion confined to a single axis we choose to be the $x$ axis, we can calculate the net work using the expression

$$W^{net} = \int_{x_1}^{x_2} F_x^{net}(x)dx \qquad \text{(work, one-dimensional motion)}. \qquad (11\text{-}39)$$

This reduces to $W^{net} = F_x^{net}\Delta x$ when the net force is constant and the body's displacement is $\Delta x = x_2 - x_1$. The rate at which the work is done is the power, which we can find with

$$P = \frac{dW}{dt} = Fv \quad \text{(power, one-dimensional motion).} \tag{11-40}$$

## Net Work-Kinetic Energy Theorem for Rotational Motion—Fixed Axis

A similar situation exists for rotational motion. When a net torque accelerates a rigid body in rotation about a fixed axis, it also does work on the body—rotational work. Therefore, the body's rotational kinetic energy as derived in Section 11-5 as $K^{\text{rot}} = \frac{1}{2}I\omega^2$ (Eq. 11-24) can change. We can show that it is also possible to relate the change in rotational kinetic energy to the net rotational *work using the work-kinetic energy theorem* where we use rotational quantities to determine the net work and kinetic energy.

$$W^{\text{net-rot}} = \Delta K = K_2 - K_1 = \tfrac{1}{2}I\omega_2^2 - \tfrac{1}{2}I\omega_1^2 \quad \text{(rotational net work-kinetic energy theorem).} \tag{11-41}$$

Here $I$ is the rotational inertia of the body about the fixed axis and $\omega_1$ and $\omega_2$ are the rotational speeds of the body before and after the rotational work is done, respectively.

## Derivation of Rotational Work-Energy Theorem

We have already derived an expression for rotational kinetic energy as shown in Eq. 11-24. In order to derive Eq. 11-41, we need to use the definition of work to find an expression for the net rotational work $W^{\text{net-rot}}$. Then we can relate the work $W$ done on the body in Fig. 11-19 to the net torque $\tau^{\text{net}}$ (which is due to a net force $\vec{F}^{\text{net}}$ that produces it). To do this we use the relationships between rotational and translational variables. We start by considering how the net force affects a single particle located at a distance $r$ from the axis of rotation.

When a single particle moves a distance $ds$ along its circular path, only the tangential component $\vec{F}_t$ of the force accelerates the particle along the path. Therefore only $F_t$ does work on the particle. We write that infinitesimal increment of work $dW$ as $F_t\, ds$. However, we can replace $ds$ with $r\, d\theta$, where $d\theta$ is the angle through which the particle moves with respect to the $x$ axis. Thus we have

$$dW^{\text{net-rot}} = F_t^{\text{net}} r d\theta. \tag{11-42}$$

However, the product $\vec{F}_t^{\text{net}} r$ is equal to the net torque $\tau^{\text{net}}$, so we can rewrite Eq. 11-42 as

$$dW^{\text{net-rot}} = \tau^{\text{net}}\, d\theta. \tag{11-43}$$

The work done on a single rotating particle during a finite rotational displacement from $\theta_1$ to $\theta_2$ is then

$$W^{\text{net-rot}} = \int_{\theta_1}^{\theta_2} \tau^{\text{net}}\, d\theta \quad \text{(rotational work, fixed axis).} \tag{11-44}$$

If all the particles in a body rotate together, this equation for rotational work also applies to the extended body that is rigid. So we now have expressions for determining both the net rotational work and the change in rotational kinetic energy in terms of rotational variables and the same basic definitions of work and kinetic energy. This verifies that we can use the work-kinetic energy theorem to relate net work and kinetic energy change when a rigid body rotates about a fixed axis.

As is the case for work done by translational forces, rotational work is a scalar quantity that can be either positive or negative, depending on whether work is done on the rotating body or by it. The work is calculated using the product of the signed quantities torque and rotational displacement.

## Power for a Rotating Body

In addition, we can find the power $P$ associated with the rotational motion of a rigid object about a fixed axis using the equation $dW = \tau\, d\theta$ (Eq. 11-43):

$$P = \frac{dW}{dt} = \tau \frac{d\theta}{dt} = \tau\omega. \qquad (11\text{-}45)$$

The signs of both torque and rotational velocity depend on the sign of the rotation as determined by the right-hand rule.

Table 11-3 summarizes the equations that apply to the rotation of a rigid body about a fixed axis and the corresponding equations for translational motion.

**TABLE 11-3**
**Corresponding Relations for Translational and Rotational Motion**

| Pure Translation ($x$ axis) | | Pure Rotation (Symmetry about a Fixed Rotation Axis) | |
|---|---|---|---|
| Position component | $x$ | Rotational position component | $\theta$ |
| Velocity component | $v_x = dx/dt$ | Rotational velocity component | $\omega = d\theta/dt$ |
| Acceleration | $a_x = dv_x/dt$ | Rotational acceleration component | $\alpha = d\omega/dt$ |
| Mass | $m$ | Rotational inertia | $I$ |
| Newton's Second Law | $F_x^{\text{net}} = ma_x$ | Newton's Second Law | $\tau^{\text{net}} = I\alpha$ |
| Work | $W = \int F_x\, dx$ | Work | $W = \int \tau\, d\theta$ |
| Kinetic energy | $K = \frac{1}{2}mv_x^2$ | Kinetic energy | $K = \frac{1}{2}I\omega^2$ |
| Power | $P = F_x v_x$ | Power | $P = \tau\omega$ |
| Work-kinetic energy theorem | $W = \Delta K$ | Work-kinetic energy theorem | $W^{\text{rot}} = \Delta K^{\text{rot}}$ |

## TOUCHSTONE EXAMPLE 11-7: Rotating Sculpture

A rigid sculpture consists of a thin hoop (of mass $m$ and radius $R = 0.15$ m) and a thin radial rod (of mass $m$ and length $L = 2.0\,R$), arranged as shown in Fig. 11-22. The sculpture can pivot around a horizontal axis in the plane of the hoop, passing through its center.

(a) In terms of $m$ and $R$, what is the sculpture's rotational inertia $I$ about the rotation axis?

**SOLUTION** ■ A **Key Idea** here is that we can separately find the rotational inertias of the hoop and the rod and then add the results to get the sculpture's total rotational inertia $I$. From Table 11-2($h$), the hoop has rotational inertia $I_{\text{hoop}} = \frac{1}{2}mR^2$ about its diameter. From Table 11-2($e$), the rod has rotational inertia $I_{\text{com}} = mL^2/12$ about an axis through its center of mass and parallel to the sculpture's rotation axis. To find its rotational

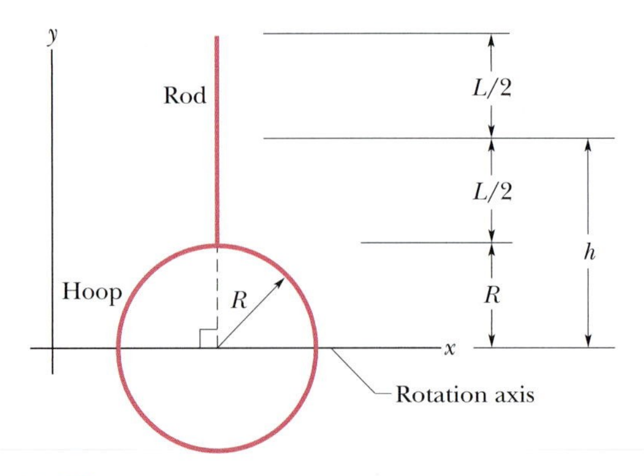


**FIGURE 11-22** ■ A rigid sculpture consisting of a hoop and two rods can rotate around a horizontal axis.

inertia $I_{rod}$ about that rotation axis, we use Eq. 11-28, the parallel-axis theorem:

$$I_{rod} = I_{com} + mh_{com}^2 = \frac{mL^2}{12} + m\left(R + \frac{L}{2}\right)^2$$
$$= 4.33mR^2,$$

where we have used the fact that $L = 2.0R$ and where the perpendicular distance between the rod's center of mass and the rotation axis $h = R + L/2$. Thus, the rotational inertia $I$ of the sculpture about the rotation axis is

$$I = I_{hoop} + I_{rod} = \tfrac{1}{2}mR^2 + 4.33mR^2$$
$$= 4.83mR^2 \approx 4.8mR^2. \qquad \text{(Answer)}$$

(b) Starting from rest, the sculpture rotates around the rotation axis from the initial upright orientation of Fig. 11-22. What is its rotational speed $\omega$ about the axis when it is inverted?

**SOLUTION** Three **Key Ideas** are required here:

1. We can relate the sculpture's speed $\omega$ to its rotational kinetic energy $K$ with Eq. 11-24 ($K = \frac{1}{2}I\omega^2$).
2. We can relate $K$ to the gravitational potential energy $U^{grav}$ of the sculpture via the conservation of the sculpture's mechanical energy $E^{mec}$ during the rotation. Thus, during the rotation, $E^{mec}$ does not change ($\Delta E^{mec} = 0$) as energy is transferred from $U^{grav}$ to $K$.
3. For the gravitational potential energy we can treat the rigid sculpture as a particle located at the center of mass, with the total mass $2m$ concentrated there.

We can write the conservation of mechanical energy ($\Delta E^{mec} = 0$) as

$$\Delta K + \Delta U^{grav} = 0. \qquad (11\text{-}46)$$

As the sculpture rotates from its initial position at rest to its inverted position, when the rotational speed is $\omega$, the change $\Delta K$ in its kinetic energy is

$$\Delta K = K_2 - K_1 = \tfrac{1}{2}I\omega^2 - 0 = \tfrac{1}{2}I\omega^2. \qquad (11\text{-}47)$$

From Eq. 10-6 ($\Delta U^{grav} = mg\Delta y$), the corresponding change $\Delta U^{grav}$ in the gravitational potential energy is

$$\Delta U^{grav} = (2m)g\,\Delta y_{com}, \qquad (11\text{-}48)$$

where $2m$ is the sculpture's total mass, and $\Delta y_{com}$ is the vertical displacement of its center of mass during the rotation.

To find $\Delta y_{com}$, we first find the initial location $y_{com}$ of the center of mass in Fig. 11-22. The hoop (with mass $m$) is centered at $y = 0$. The rod (with mass $m$) is centered at $y = R + L/2$. Thus, from Eq. 8-11, the sculpture's center of mass is at

$$y_{com} = \frac{m(0) + m(R + L/2)}{2m} = \frac{0 + m(R + 2R/2)}{2m} = R.$$

When the sculpture is inverted, the center of mass is this same distance $R$ from the rotation axis but *below* it. Therefore, the vertical displacement of the center of mass from the initial position to the inverted position is $\Delta y_{com} = -2R$.

Now let's pull these results together. Substituting Eqs. 11-47 and 11-48 into 11-46 gives us

$$\tfrac{1}{2}I\omega^2 + (2m)g\,\Delta y_{com} = 0.$$

Substituting $I = 4.83mR^2$ from (a) and $\Delta y_{com} = -2R$ from above and solving for $\omega$, we find

$$\omega = \sqrt{\frac{8g}{4.83\,R}} = \sqrt{\frac{(8)(9.8\text{ m/s}^2)}{(4.83)(0.15\text{ m})}}$$
$$= 10\text{ rad/s}. \qquad \text{(Answer)}$$

# Problems

## Sec. 11-2 ■ The Rotational Variables

**1. Flywheel** The rotational position of a flywheel on a generator is given by $\theta = (a\text{ rad/s})t + (b\text{ rad/s}^3)t^3 - (c\text{ rad/s}^4)t^4$, where $a$, $b$, and $c$ are constants. Write expressions for the wheel's (a) rotational velocity and (b) rotational acceleration.

**2. Hands of a Clock** What is the rotational speed of (a) the second hand, (b) the minute hand, and (c) the hour hand of a smoothly running analog watch? Answer in radians per second.

**3. Milky Way** Our Sun is $2.3 \times 10^4$ ly (light-years) from the center of our Milky Way galaxy and is moving in a circle around the center at a speed of 250 km/s. (a) How long does it take the Sun to make one revolution about the galactic center? (b) How many revolutions has the Sun completed since it was formed about $4.5 \times 10^9$ years ago?

**4. Rotating Wheel** The rotational position of a point on the rim of a rotating wheel is given by $\theta = (4.0\text{ rad/s})t + (3.0\text{ rad/s}^2)t^2 + (1\text{ rad/s}^3)t^3$, where $\theta$ is in radians and $t$ is in seconds. What are the rotational velocities at (a) $t_1 = 2.0$ s and (b) $t_2 = 4.0$ s? (c) What is the average rotational acceleration for the time interval that begins at $t_1 = 2.0$ s and ends at $t_2 = 4.0$ s? What are the instantaneous rotational accelerations at (d) the beginning and (e) the end of this time interval?

**5. Rotational Position** The rotational position of a point on a rotating wheel is given by $\theta = 2\text{ rad} + (4.0\text{ rad/s}^2)t^2 + (2\text{ rad/s}^3)t^3$, where $\theta$ is in radians and $t$ is in seconds. At $t_1 = 0$, what are (a) the point's rotational position and (b) its rotational velocity? (c) What is its rotational velocity at $t_3 = 4.0$ s? (d) Calculate its rotational acceleration at $t_2 = 2.0$ s. (e) Is its rotational acceleration constant?

**6. The Wheel** The wheel in Fig.11-23 has eight equally spaced spokes and a radius of 30 cm. It is mounted on a fixed axle and is spinning at 2.5 rev/s. You want to shoot a 20-cm-long arrow parallel to this axle and through the wheel without hitting any of the spokes. Assume that the arrow and the spokes are very thin. (a) What minimum speed must the arrow have? (b) Does it matter where between the axle and rim of the wheel you aim? If so, what is the best location?

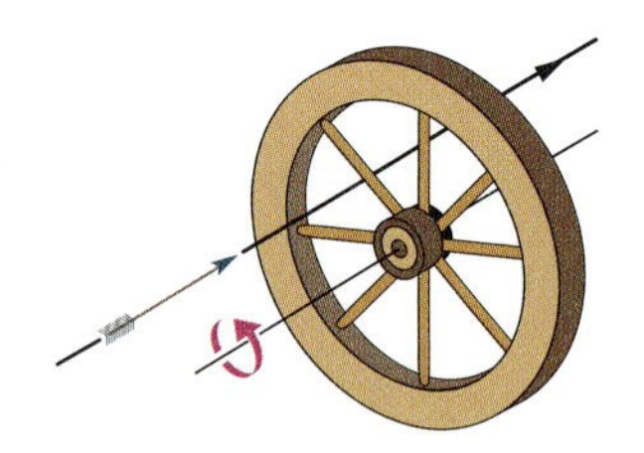

FIGURE 11-23 ■ Problem 6.

**7. A Diver** A diver makes 2.5 revolutions on the way from a 10-m-high platform to the water. Assuming zero initial vertical velocity, find the diver's average rotational velocity during a dive.

## SEC. 11-3 ■ ROTATION WITH CONSTANT ROTATIONAL ACCELERATION

**8. Automobile Engine** The rotational speed of an automobile engine is increased at a constant rate from 1200 rev/min to 3000 rev/min in 12 s. (a) What is its rotational acceleration in revolutions per minute-squared? (b) How many revolutions does the engine make during this 12 s interval?

**9. Turntable** A record turntable rotating at $33\frac{1}{3}$ rev/min slows down and stops in 30 s after the motor is turned off. (a) Find its (constant) rotational acceleration in revolutions per minute-squared. (b) How many revolutions does it make in this time?

**10. A Disk** A disk, initially rotating at 120 rad/s, is slowed down with a constant rotational acceleration of magnitude 4.0 rad/s$^2$. (a) How much time does the disk take to stop? (b) Through what angle does the disk rotate during that time?

**11. Heavy Flywheel** A heavy flywheel rotating on its central axis is slowing down because of friction in its bearings. At the end of the first minute of slowing, its rotational speed is 0.90 of its initial rotational speed of 250 rev/min. Assuming a constant rotational acceleration, find its rotational speed at the end of the second minute.

**12. A Disk Rotates** Starting from rest, a disk rotates about its central axis with constant rotational acceleration. In 5.0 s, it rotates 25 rad. During that time, what are the magnitudes of (a) the rotational acceleration and (b) the average rotational velocity? (c) What is the instantaneous rotational velocity of the disk at the end of the 5.0 s? (d) With the rotational acceleration unchanged, through what additional angle will the disk turn during the next 5.0 s?

**13. Constant Rotational Acceleration** A wheel has a constant rotational acceleration of 3.0 rad/s$^2$. During a certain 4.0 s interval, it turns through an angle of 120 rad. Assuming that the wheel starts from rest, how long is it in motion at the start of this 4.0 s interval?

**14. Starting from Rest** A wheel, starting from rest, rotates with a constant rotational acceleration of 2.00 rad/s$^2$. During a certain 3.00 s interval, it turns through 90.0 rad. (a) How long is the wheel turning before the start of the 3.00 s interval? (b) What is the rotational velocity of the wheel at the start of the 3.00 s interval?

**15. A Flywheel Has a Rotational Velocity** At $t_1 = 0$, a flywheel has a rotational velocity of 4.7 rad/s, a rotational acceleration of $-0.25$ rad/s$^2$, and a reference line at $\theta_1 = 0$. (a) Through what maximum angle $\theta^{max}$ will the reference line turn in the positive direction? For what length of time will the reference line turn in the positive direction? At what times will the reference line be at (b) $\theta = \frac{1}{2}\theta^{max}$ and (c) $\theta = -10.5$ rad (consider both positive and negative values of $t$)? (d) Graph $\theta$ versus $t$, and indicate the answers to (a), (b), and (c) on the graph.

**16. A Disk Rotates** A disk rotates about its central axis starting from rest and accelerates with constant rotational acceleration. At one time it is rotating at 10 rev/s; 60 revolutions later, its rotational speed is 15 rev/s. Calculate (a) the rotational acceleration, (b) the time required to complete the 60 revolutions, (c) the time required to reach the 10 rev/s rotational speed, and (d) the number of revolutions from rest until the time the disk reaches the 10 rev/s rotational speed.

**17. A Flywheel Turns** A flywheel turns through 40 rev as it slows from an rotational speed of 1.5 rad/s to a stop. (a) Assuming a constant rotational acceleration, find the time for it to come to rest. (b) What is its rotational acceleration? (c) How much time is required for it to complete the first 20 of the 40 revolutions?

**18. A Wheel Rotating** A wheel rotating about a fixed axis through its center has a constant rotational acceleration of 4.0 rad/s$^2$. In a certain 4.0 s interval the wheel turns through an angle of 80 rad. (a) What is the rotational velocity of the wheel at the start of the 4.0 s interval? (b) Assuming that the wheel starts from rest, how long is it in motion at the start of the 4.0 s interval?

## SEC. 11-4 ■ RELATING THE TRANSLATIONAL AND ROTATIONAL VARIABLES

**19. Record** What is the translational acceleration of a point on the rim of a 30-cm-diameter record rotating at a constant rotational speed of $33\frac{1}{3}$ rev/min?

**20. Vinyl Record** A vinyl record on a turntable rotates at $33\frac{1}{3}$ rev/min. (a) What is its rotational speed in radians per second? What is the translational speed of a point on the record at the needle when the needle is (b) 15 cm and (c) 7.4 cm from the turntable axis?

**21. Rotational Speed of Car** What is the rotational speed of car traveling at 50 km/h and rounding a circular turn of radius 110 m?

**22. Flywheel Rotating** A flywheel with a diameter of 1.20 m has a rotational speed of 200 rev/min. (a) What is the rotational speed of the flywheel in radians per second? (b) What is the translational speed of a point on the rim of the flywheel? (c) What constant rotational acceleration (in revolutions per minute-squared) will increase the wheel's rotational speed to 1000 rev/min in 60 s? (d) How many revolutions does the wheel make during that 60 s?

**23. Astronaut in Centrifuge** An astronaut is being tested in a centrifuge. The centrifuge has a radius of 10 m and, in starting, rotates according to $\theta = (0.30 \text{ rad/s}^2)t^2$, where $t$ is in seconds and $\theta$ is in radians. When $t = 5.0$ s, what are the magnitudes of the astronaut's (a) rotational velocity, (b) translational velocity, (c) tangential acceleration, and (d) radial acceleration?

**24. Spaceship** What are the magnitudes of (a) the rotational velocity, (b) the radial acceleration, and (c) the tangential acceleration of a spaceship taking a circular turn of radius 3220 km at a speed of 29 000 km/h?

**25. Speed of Light** An early method of measuring the speed of light makes use of a rotating slotted wheel. A beam of light passes through a slot at the outside edge of the wheel, as in Fig. 11-24, travels to a distant mirror, and returns to the wheel just in time to pass through the next slot in the wheel. One such slotted wheel has a radius of 5.0 cm and 500 slots at its edge. Measurements taken when the mirror is $L = 500$ m from the wheel indicate a speed of light of $3.0 \times 10^5$ km/s. (a) What is the (constant) rotational speed of the wheel? (b) What is the translational speed of a point on the edge of the wheel?

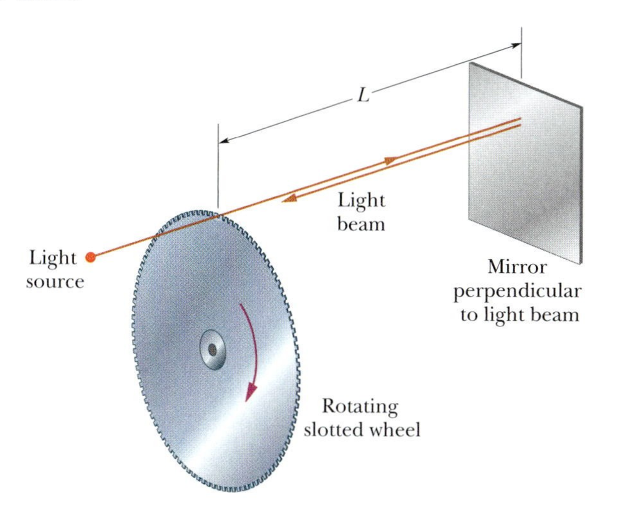


**FIGURE 11-24** ■ Problem 25.

**26. Steam Engine** The flywheel of a steam engine runs with a constant rotational velocity of 150 rev/min. When steam is shut off, the friction of the bearings and of the air stops the wheel in 2.2 h. (a) What is the constant rotational acceleration, in revolutions per minute-squared, of the wheel during the slowdown? (b) How many rotations does the wheel make during the slowdown? (b) How many rotations does the wheel make before stopping? (c) At the instant the flywheel is turning at 75 rev/min, what is the tangential component of the translational acceleration of a flywheel particle that is 50 cm from the axis of rotation? (d) What is the magnitude of the net translational acceleration of the particle in (c)?

**27. Polar Axis of Earth** (a) What is the rotational speed $\omega$ about the polar axis of a point on Earth's surface at a latitude of 40° N? (Earth rotates about that axis.) (b) What is the translational speed $v$ of the point? What are (c) $\omega$ and (d) $v$ for a point at the equator?

**28. Gyroscope** A gyroscope flywheel of radius 2.83 cm is accelerated from rest at 14.2 rad/s$^2$ until its rotational speed is 2760 rev/min. (a) What is the tangential acceleration of a point on the rim of the flywheel during this spin-up process? (b) What is the radial acceleration of this point when the flywheel is spinning at full speed? (c) Through what distance does a point on the rim move during the spin-up?

**29. Coupled Wheels** In Fig. 11-25, wheel $A$ of radius $r_A = 10$ cm is coupled by belt $B$ to wheel $C$ of radius $r_C = 25$ cm. The rotational speed of wheel $A$ is increased from rest at a constant rate of 1.6 rad/s$^2$. Find the time for wheel $C$ to reach a rotational speed of 100 rev/min, assuming the belt does not slip. (*Hint*: If the belt does not slip, the translational speeds at the rims of the two wheels must be equal.)

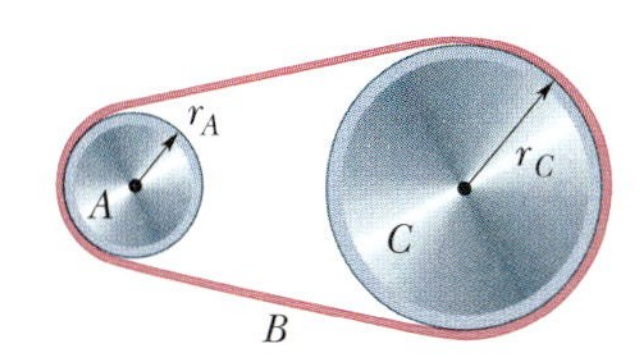


**FIGURE 11-25** ■ Problem 29.

**30. Fixed Axis** An object rotates about a fixed axis, and the rotational position of a reference line on the object is given by $\theta = (0.40 \text{ rad}) e^{(2.05^{-1})t}$. Consider a point on the object that is 4.0 cm from the axis of rotation. At $t = 0$, what are the magnitudes of the point's (a) tangential component of acceleration and (b) radial component of acceleration?

**31. Pulsar** A pulsar is a rapidly rotating neutron star that emits a radio beam like a lighthouse emits a light beam. We receive a radio pulse for each rotation of the star. The period $T$ of rotation is found by measuring the time between pulses. The pulsar in the Crab nebula (Fig. 11-26) has a period of rotation of $T = 0.033$ s that is increasing at the rate of $1.26 \times 10^{-5}$ s/y. (a) What is the pulsar's rotational acceleration? (b) If its rotational acceleration is constant, how many years from now will the pulsar stop rotating? (c) The pulsar originated in a supernova explosion seen in the year 1054. What was the intial $T$ of the pulsar? (Assume constant rotational acceleration since the pulsar originated.)

**FIGURE 11-26** ■ Problem 31. The Crab nebula resulted from a star whose explosion was seen in 1054. In addition to the gaseous debris seen here, the explosion left a spinning neutron star at its center. The star has a diameter of only 30 km.

**32. Turntable Two** A record turntable is rotating at $33\frac{1}{3}$ rev/min. A watermelon seed is on the turntable 6.0 cm from the axis of rotation. (a) Calculate the translational acceleration of the seed, assuming that it does not slip. (b) What is the minimum value of the coefficient of static friction, $\mu^{\text{stat}}$, between the seed and the turntable if the seed is not to slip? (c) Suppose that the turntable achieves its rotational speed by starting from rest and undergoing a constant rotational acceleration for 0.25 s. Calculate the minimum $\mu^{\text{stat}}$ required for the seed not to slip during the acceleration period.

## SEC. 11-5 ■ KINETIC ENERGY OF ROTATION

**33. Rotational Inertia of Wheel** Calculate the rotational inertia of a wheel that has a kinetic energy of 24 400 J when rotating at 602 rev/min.

**34. Oxygen Molecule** The oxygen molecule $O_2$ has a mass of $5.30 \times 10^{-26}$ kg and a rotational inertia of $1.94 \times 10^{-46}$ kg · m$^2$ about an axis through the center of the line joining the atoms and perpendicular to that line. Suppose the center of mass of an $O_2$ molecule in a gas has a translational speed of 500 m/s and the molecule has a rotational kinetic energy that is $\frac{2}{3}$ of the translational kinetic energy of its center of mass. What then is the molecule's rotational speed about the center of mass?

## SEC. 11-6 ■ CALCULATING ROTATIONAL INERTIA

**35. Two Solid Cylinders** Two uniform solid cylinders, each rotating about its central (longitudinal) axis, have the same mass of 1.25 kg and rotate with the same rotational speed of 235 rad/s, but they differ in radius. What is the rotational kinetic energy of (a) the smaller cylinder, of radius 0.25 m, and (b) the larger cylinder, of radius 0.75 m?

**36. Communications Satellite** A communications satellite is a solid cylinder with mass 1210 kg, diameter 1.21 m, and length 1.75 m. Prior to launching from the shuttle cargo bay, it is set spinning at 1.52 rev/s about the cylinder axis (Fig. 11-27). Calculate the satellite's (a) rotational inertia about the rotation axis and (b) rotational kinetic energy.

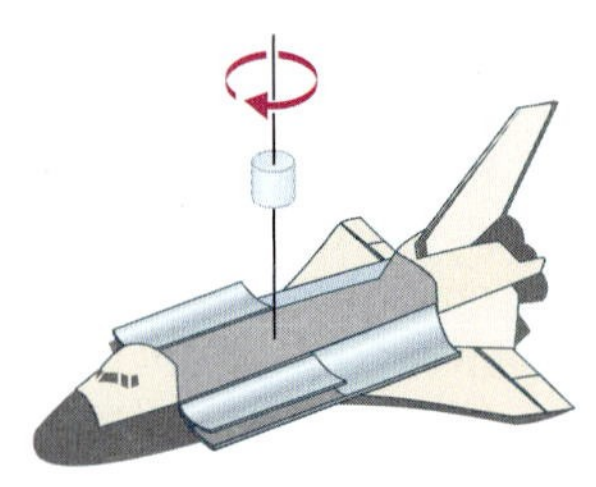

FIGURE 11-27 ■ Problem 36.

**37. Two Particles** In Fig. 11-28, two particles, each with mass $m$, are fastened to each other, and to a rotation axis at $O$, by two thin rods, each with length $d$ and mass $M$. The combination rotates around the rotation axis with rotational velocity $\omega$. In terms of these symbols, and measured about $O$, what are the combination's (a) rotational inertia and (b) kinetic energy?

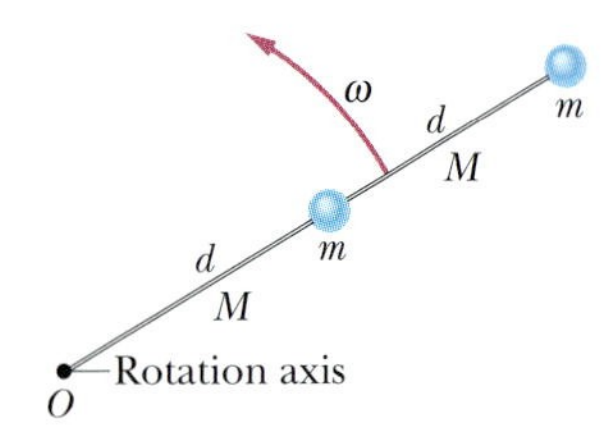


FIGURE 11-28 ■ Problem 37.

**38. Helicopter Blades** Each of the three helicopter rotor blades shown in Fig. 11-29 is 5.20 m long and has a mass of 240 kg. The rotor is rotating at 350 rev/min. (a) What is the rotational inertia of the rotor assembly about the axis of rotation? (Each blade can be considered to be a thin rod rotated about one end.) (b) What is the total kinetic energy of rotation?

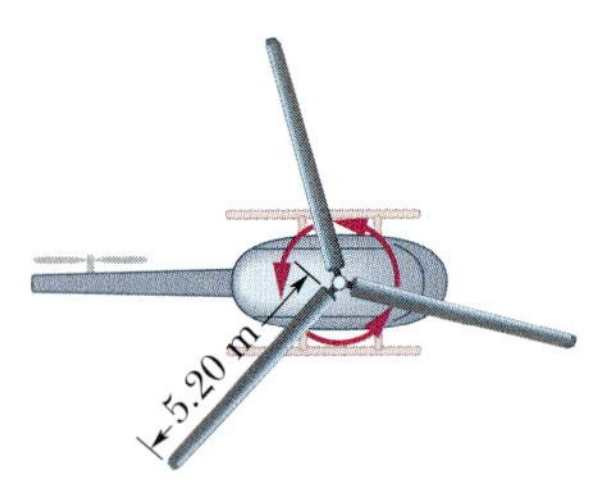


FIGURE 11-29 ■ Problem 38.

**39. Meter Stick** Calculate the rotational inertia of a meter stick, with mass 0.56 kg, about an axis perpendicular to the stick and located at the 20 cm mark. (Treat the stick as a thin rod.)

**40. Four Identical Particles** Four identical particles of mass 0.50 kg each are placed at the vertices of a 2.0 m × 2.0 m square and held there by four massless rods, which form the sides of the square. What is the rotational inertia of this rigid body about an axis that (a) passes through the midpoints of opposite sides and lies in the plane of the square, (b) passes through the midpoint of one of the sides and is perpendicular to the plane of the square, and (c) lies in the plane of the square and passes through two diagonally opposite particles?

**41. Uniform Solid Block** The uniform solid block in Fig. 11-30 has mass $M$ and edge lengths $a$, $b$, and $c$. Calculate its rotational inertia about an axis through one corner and perpendicular to the large faces.

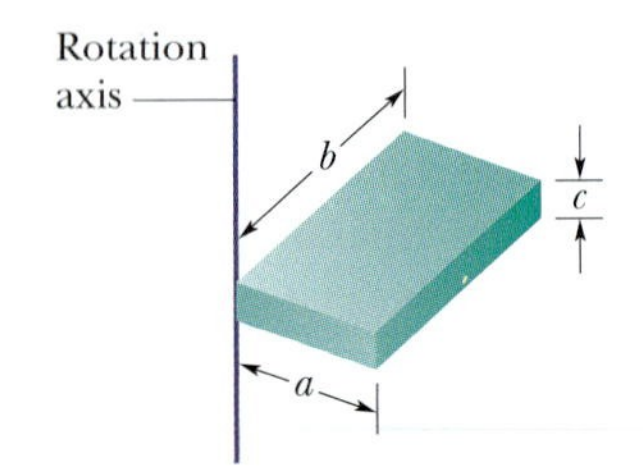


FIGURE 11-30 ■ Problem 41.

**42. Masses and Coordinates** The masses and coordinates of four particles are as follows: 50 g, $x = 2.0$ cm, $y = 2.0$ cm; 25 g, $x = 0$, $y = 4.0$ cm; 25 g, $x = -3.0$ cm, $y = -3.0$ cm; 30 g, $x = -2.0$ cm, $y = 4.0$ cm. What are the rotational inertias of this collection about the (a) $x$, (b) $y$, and (c) $z$ axes? (d) Suppose the answers to (a) and (b) are $A$ and $B$, respectively. Then what is the answer to *(c)* in terms of $A$ and $B$?

**43. Solid Cylinder—Thin Hoop** (a) Show that the rotational inertia of a solid cylinder of mass $M$ and radius $R$ about its central axis is equal to the rotational inertia of a thin hoop of mass $M$ and radius $R/\sqrt{2}$ about its central axis. (b) Show that the rotational inertia $I$ of any given body of mass $M$ about any given axis is equal to the rotational inertia of an *equivalent hoop* about that axis, if the hoop has the same mass $M$ and a radius $k$ given by

$$k = \sqrt{\frac{I}{M}}.$$

The radius $k$ of the equivalent hoop is called the *radius of gyration* of the given body.

**44. Delivery Trucks** Delivery trucks that operate by making use of energy stored in a rotating flywheel have been used in Europe. The trucks are charged by using an electric motor to get the flywheel up to its top speed of $200\pi$ rad/s. One such flywheel is a solid, uniform cylinder with a mass of 500 kg and a radius of 1.0 m. (a) What is the kinetic energy of the flywheel after charging? (b) If the truck operates with an average power requirement of 8.0 kW, for how many minutes can it operate between chargings?

## SEC. 11-7 ■ TORQUE

**45. Small Ball** A small ball of mass 0.75 kg is attached to one end of a 1.25-m-long massless rod, and the other end of the rod is hung from a pivot. When the resulting pendulum is 30° from the vertical, what is the magnitude of the torque about the pivot?

**46. Bicycle Pedal Arm** The length of a bicycle pedal arm is 0.152 m, and a downward force of 111 N is applied to the pedal by the rider's foot. What is the magnitude of the torque about the pedal arm's pivot point when the arm makes an angle of (a) 30°, (b) 90°, and (c) 180° with the vertical?

**47. Pivoted at *O*** The body in Fig. 11-31 is pivoted at $O$, and two forces act on it as shown. (a) Find an expression for the net torque on the body about the pivot. (b) If $r_A = 1.30$ m, $r_B = 2.15$ m, $F_A = 4.20$ N, $F_B = 4.90$ N, $\theta_A = 75.0°$, and $\theta_B = 60.0°$, what is the net torque about the pivot?

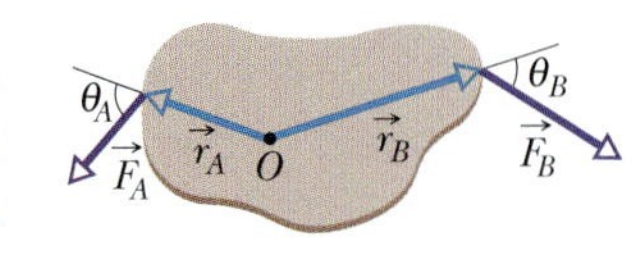


FIGURE 11-31 ■ Problem 47.

**48. Three Force** The body in Fig. 11-32 is pivoted at $O$. Three forces act on it in the directions shown: $F_A = 10$ N at point $A$, 8.0 m from $O$; $F_B = 16$ N at point $B$, 4.0 m from $O$; and $F_C = 19$ N at point $C$, 3.0 m from $O$. What is the net torque about $O$?

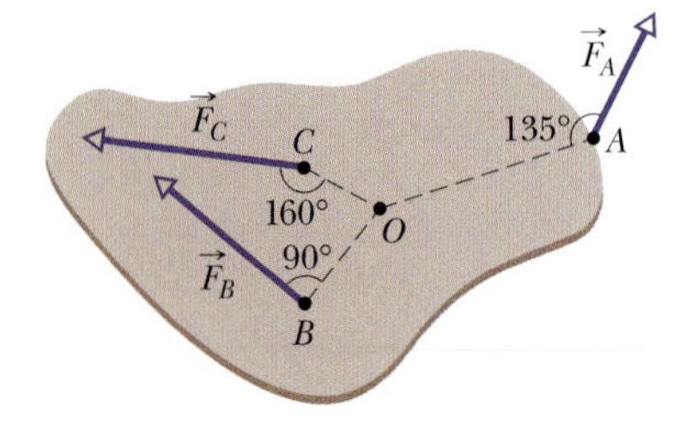


FIGURE 11-32 ■ Problem 48.

## SEC. 11-8 ■ NEWTON'S SECOND LAW FOR ROTATION

**49. Diver's Launch** During the launch from a board, a diver's rotational speed about her center of mass changes from zero to 6.20 rad/s in 220 ms. Her rotational inertia about her center of mass is 12.0 kg · m$^2$. During the launch, what are the magnitudes of (a) her average rotational acceleration and (b) the average external torque on her from the board?

**50. Torque on a Certain Wheel** A torque of 32.0 N · m on a certain wheel causes a rotational acceleration of 25.0 rad/s$^2$. What is the wheel's rotational inertia?

**51. Thin Spherical Shell** A thin spherical shell has a radius of 1.90 m. An applied torque of 960 N · m gives the shell a rotational acceleration of 6.20 rad/s$^2$ about an axis through the center of the shell. What are (a) the rotational inertia of the shell about that axis and (b) the mass of the shell?

**52. Cylinder Having Mass** In Fig. 11-33, a cylinder having a mass of 2.0 kg can rotate about its central axis through point $O$. Forces are applied as shown: $F_A = 6.0$ N, $F_B = 4.0$ N, $F_C = 2.0$ N, and $F_D = 5.0$ N. Also, $R_1 = 5.0$ cm and $R_2 = 12$ cm. Find the magnitude and direction of the rotational acceleration of the cylinder. (During the rotation, the forces maintain their same angles relative to the cylinder.)

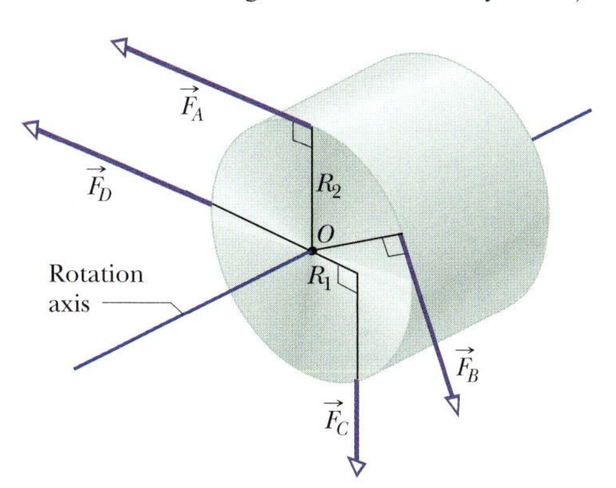


FIGURE 11-33 ■ Problem 52.

**53. Lawrence Livermore Door** Figure 11-34 shows the massive shield door at a neutron test facility at Lawrence Livermore Laboratory; this is the world's heaviest hinged door. The door has a mass of 44,000 kg, a rotational inertia about a vertical axis through its huge hinges of $8.7 \times 10^4$ kg · m$^2$, and a (front) face width of 2.4 m. Neglecting friction, what steady force, applied at its outer edge and perpendicular to the plane of the door can move it from rest through an angle of 90° in 30 s?

FIGURE 11-34 ■ Problem 53.

**54. Wheel on a Frictionless Axis** A wheel of radius 0.20 m is mounted on a frictionless horizontal axis. The rotational inertia of the wheel about the axis is 0.050 kg · m$^2$. A massless cord wrapped around the wheel is attached to a 2.0 kg block that slides on a horizontal frictionless surface. If a horizontal force of magnitude $P$ = 3.0 N is applied to the block as shown in Fig. 11-35, what is the magnitude of the rotational acceleration of the wheel? Assume that the string does not slip on the wheel.

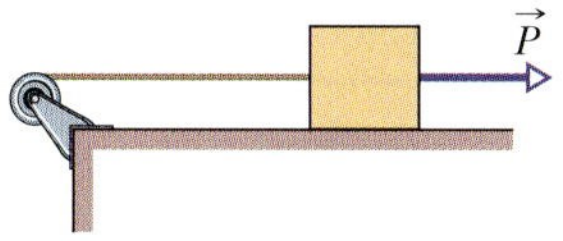

FIGURE 11-35 ■ Problem 54.

**55. Two Blocks on a Pulley** In Fig. 11-36, one block has mass $M$ = 500 g, the other has mass $m$ = 460 g, and the pulley, which is mounted in horizontal frictionless bearings, has a radius of 5.00 cm. When released from rest, the heavier block falls 75.0 cm in 5.00 s (without the cord slipping on the pulley). (a) What is the magnitude of the blocks' acceleration? What is the tension in the part of the cord that supports (b) the heavier block and (c) the lighter block? (d) What is the magnitude of the pulley's rotational acceleration? (e) What is its rotational inertia?

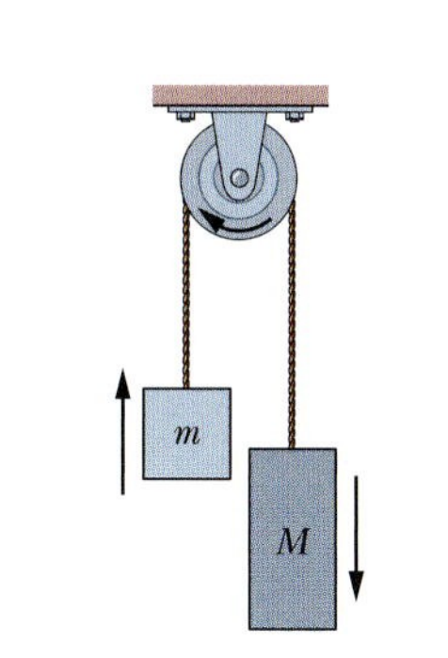


FIGURE 11-36 ■ Problem 55.

**56. A Pulley** A pulley, with a rotational inertia of $1.0 \times 10^{-3}$ kg · m$^2$ about its axle and a radius of 10 cm, is acted on by a force applied tangentially at its rim. The force magnitude varies in time as $F = (0.50\text{ N/s})t + (0.30\text{ N/s}^2)t^2$, with $F$ in newtons and $t$ in seconds. The pulley is initially at rest. At $t$ = 3.0 s what are (a) its rotational acceleration and (b) its rotational speed?

**57. Two Blocks on a Rod** Figure 11-37 shows two blocks, each of mass $m$, suspended from the ends of a rigid massless rod of length $L_1 + L_2$, with $L_1$ = 20 cm and $L_2$ = 80 cm. The rod is held horizontally on the fulcrum and then released. What are the magnitudes of the initial accelerations of (a) the block closer to the fulcrum and (b) the other block?

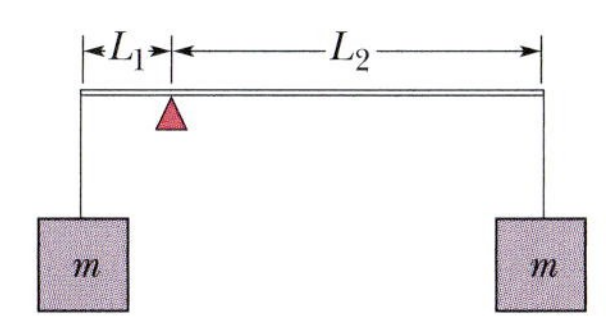


FIGURE 11-37 ■ Problem 57.

## SEC. 11-9 ■ WORK AND ROTATIONAL KINETIC ENERGY

**58. Speed of the Block** (a) If $R$ = 12 cm, $M$ = 400 g, and $m$ = 50 g in Fig. 11-20, find the speed of the block after it has descended 50 cm starting from rest. Solve the problem using energy conservation principles. (b) Repeat (a) with $R$ = 5.0 cm.

**59. Crankshaft** An automobile crankshaft transfers energy from the engine to the axle at the rate of 100 hp (=74.6 kW) when rotating at a speed of 1800 rev/min. What torque (in newton-meters) does the crankshaft deliver?

**60. Thin Hoop** A 32.0 kg wheel, essentially a thin hoop with radius 1.20 m, is rotating at 280 rev/min. It must be brought to a stop in 15.0 s. (a) How much work must be done to stop it? (b) What is the required average power?

**61. Thin Rod of Length $L$** A thin rod of length $L$ and mass $m$ is suspended freely from one end. It is pulled to one side and then allowed to swing like a pendulum, passing through its lowest position with rotational speed $\omega$. In terms of these symbols and $g$, and neglecting friction and air resistance, find (a) the rod's kinetic energy at its lowest position and (b) how far above that position the center of mass rises.

**62. Accelerating the Earth** Calculate (a) the torque, (b) the energy, and (c) the average power required to accelerate Earth in 1 day from rest to its present rotational speed about its axis.

**63. Meter Stick Held Vertically** A meter stick is held vertically with one end on the floor and is then allowed to fall. Find the speed of the other end when it hits the floor, assuming that the end on the floor does not slip. (*Hint*: Consider the stick to be a thin rod and use the conservation of energy principle.)

**64. Cylinder Rotates about Horizontal** A uniform cylinder of radius 10 cm and mass 20 kg is mounted so as to rotate freely about a horizontal axis that is parallel to and 5.0 cm from the central longitudinal axis of the cylinder. (a) What is the rotational inertia of the cylinder about the axis of rotation? (b) If the cylinder is released from rest with its central longitudinal axis at the same height as the axis about which the cylinder rotates, what is the rotational speed of the cylinder as it passes through its lowest position?

**65. The Letter *H*** A rigid body is made of three identical thin rods, each with length $L$, fastened together in the form of a letter H (Fig. 11-38). The body is free to rotate about a horizontal axis that runs along the length of one of the legs of the H. The body is allowed to fall from rest from a position in which the plane of the H is horizontal. What is the rotational speed of the body when the plane of the H is vertical?

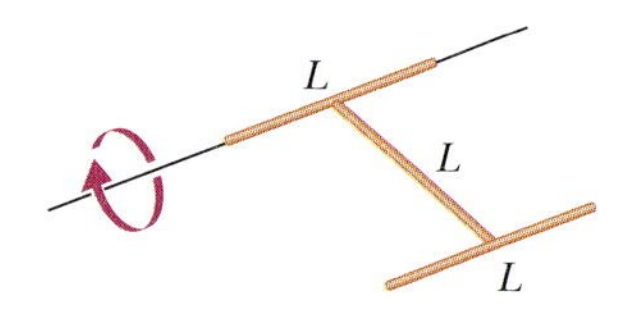


FIGURE 11-38 ■ Problem 65.

**66. Uniform Spherical Shell** A uniform spherical shell of mass $M$ and radius $R$ rotates about a vertical axis on frictionless bearings (Fig. 11-39). A massless cord passes around the equator of the shell, over a pulley of rotational inertia $I$ and radius $r$, and is attached to a small object of mass $m$. There is no friction on the pulley's axle; the cord does not slip on the pulley. What is the speed of the object after it falls a distance $h$ from rest? Use energy considerations.

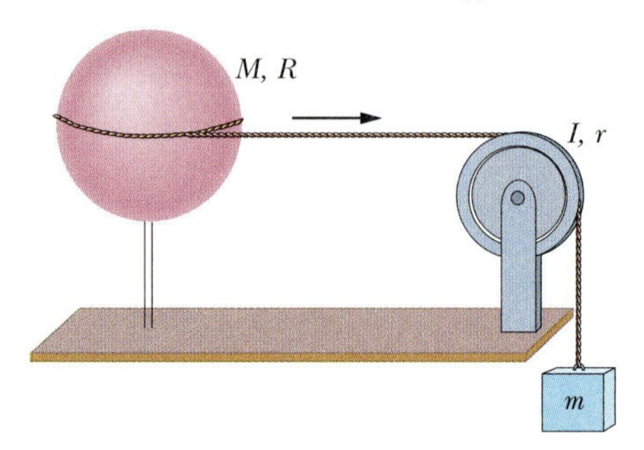


FIGURE 11-39 ■ Problem 66.

**67. Tall Cylinder-Shaped Chimney** A tall, cylinder-shaped chimney falls over when its base is ruptured. Treat the chimney as a thin rod of length $H$, and let $\theta$ be the angle the chimney makes with the vertical. In terms of these symbols and $g$, express the following: (a) the rotational speed of the chimney, (b) the radial acceleration of the chimney's top, and (c) the tangential acceleration of the top. (*Hint*: Use energy considerations, not a torque. In part (c) recall that $\alpha = d\omega/dt$.) (d) At what angle $\theta$ does the tangential acceleration equal $g$?

# Additional Problems

**68. Judo** In a judo foot-sweep move, you sweep your opponent's left foot out from under him while pulling on his gi (uniform) toward that side. As a result, your opponent rotates around his right foot and onto the mat. Figure 11-40 shows a simplified diagram of your opponent as you face him, with his left foot swept out. The rotational axis is through point $O$. The gravitational force $\vec{F}^{\,grav}$ on him effectively acts at his center of mass, which is a horizontal distance of $d = 28$ cm from point $O$. His mass is 70 kg, and his rotational inertia about point $O$ is 65 kg · m$^2$. What is the magnitude of his initial rotational acceleration about point $O$ if your pull $\vec{F}^{\,app}$ on his gi is (a) negligible and (b) horizontal with a magnitude of 300 N and applied at height $h = 1.4$ m?

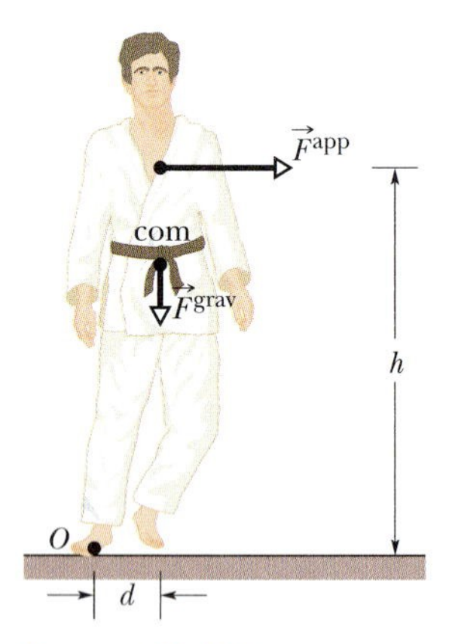


FIGURE 11-40 ■ Problem 68.

**69. Disk Rod** Figure 11-41 shows an arrangement of 15 identical disks that have been glued together in a rod-like shape of length $L$ and (total) mass $M$. The arrangement can rotate about a perpendicular axis through its central disk at point $O$. (a) What is the rotational inertia of the arrangement about that axis? (b) If we approximated the arrangement as being a uniform rod of mass $M$ and length $L$, what percentage error would we make in using the formula in Table 11-2*e* to calculate the rotational inertia?

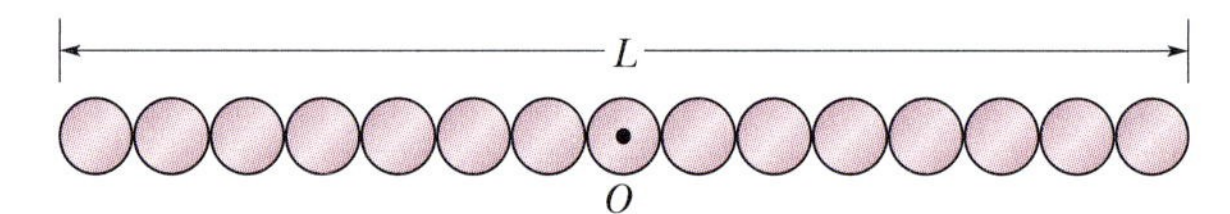


FIGURE 11-41 ■ Problem 69.

**70. Summing Up to Estimate Rotational Inertia.** By performing an integration it can be shown that the general equation for the rotational inertia of a thin rod of length $L$ and mass $M$ about an axis through one end of the rod that is perpendicular to its length is given by

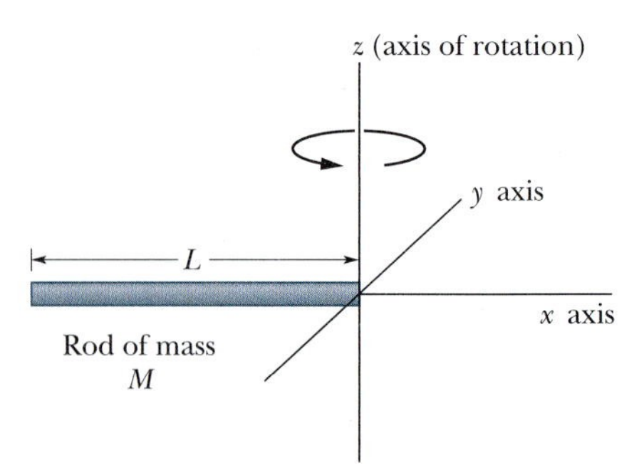


FIGURE 11-42 ■ Problem 70.

$$I = \tfrac{1}{3}ML^2.$$

Consider a rod of length $L = 0.50$ m that has a mass of $M = 1.2$ kg rotating as shown in Fig. 11-42.

**(a)** Calculate the theoretical value of the rotational inertia.

**(b)** Estimate the rotational inertia of the rod by breaking it into 50 small point masses each having a mass of $M/50$, with the first point mass being 0.01 m from the axis of rotation, the second mass being 0.02 m from the axis of rotation, and so on. Use a spreadsheet to do your estimated calculations of the rotational inertia of the rod.
**(c)** Compare the theoretically calculated value with the estimated value. Are they similar?

**71. Calculation of Torque (Angle Method)** Before the finger holes are drilled, a uniform bowling ball of radius 0.120 m has a net gravitational force of 65 N exerted on it by the Earth. Assume that this force acts through the center of mass of the bowling ball. Determine the magnitude and direction of this net force and the resulting torques on the bowling ball about four axes that are *perpendicular* to the plane of the paper passing through points $A$, $B$, $C$, and $D$ as shown in Fig. 11-43. *Hint*: You can use the $|\vec{\tau}| = |\vec{r}||\vec{F}|\sin\theta$ form of the torque equation.

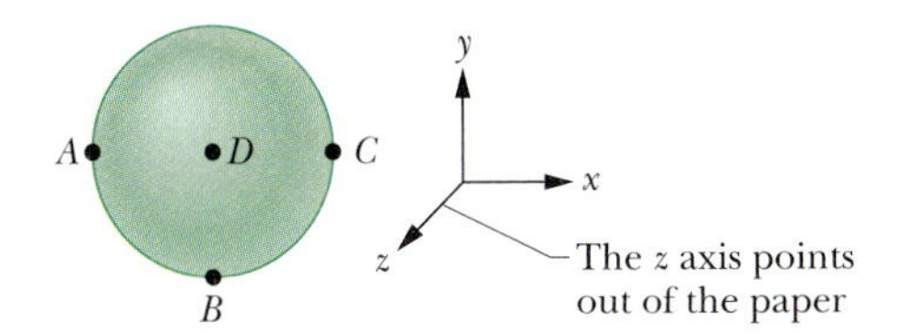


FIGURE 11-43 ■ Problem 71.

**72. Simple Yo-Yo** Consider a "yo-yo" consisting of a disk fixed to an axle that has two strings wrapped around it. As the axle rolls off the strings, the disk and the axle fall as shown in Fig. 11-44.

**(a)** If the disk has fallen through a vertical distance of $d = 30$ cm and the radius of the string and axle is given by $r = 50$ mm, how many revolutions has the disk gone through?
**(b)** If the disk is rotating faster and faster with a constant rotational acceleration and takes 25 s to fall through the distance $d$ from rest, what is the magnitude of its rotational acceleration $\alpha$?
**(c)** What is the magnitude of its rotational velocity $\omega$ after the 25 seconds have elapsed? *Hint*: Use the rotational kinematic equations.

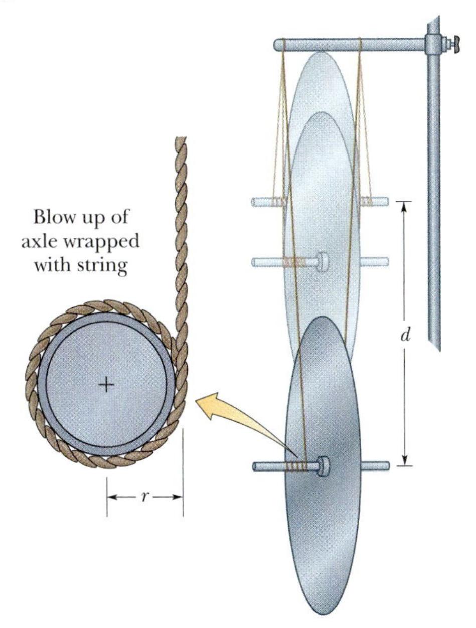


FIGURE 11-44 ■ Problem 72.

**73. Buying Wire** Wire is often delivered wrapped on a large cylindrical spool. Suppose such a spool is supported by resting on a horizontal metal rod pushed through a hole that runs through the center of the spool. A worker is pulling some wire off the spool by exerting a force on it as shown in Fig. 11-45. (If you have ever bought wire in a hardware store, this is the way they usually store and dispense it.)

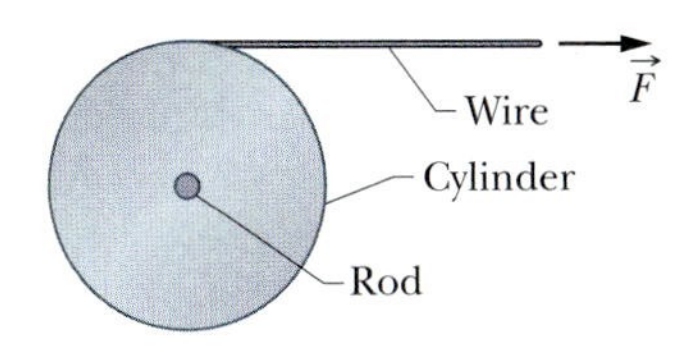


FIGURE 11-45 ■ Problem 73.

Suppose the spool rotates on the rod essentially without friction. The spool is approximately a uniform cylinder with a mass of 50 kg and a radius of 30 cm. The worker pulls on the wire for 2 s with a force of 30 N. At the end of the 2 s he immediately clamps on a brake that very quickly stops the spool's rotation. Just before he puts the brake on, how fast is the spool rotating? How much wire does he pull off the spool?

**74. Fly on an LP** An old-fashioned record player spins a disk at approximately a constant angular velocity, $\omega$. A fly of mass $m$ is sitting on the disk as it turns, at a point a distance $R$ from the center.

**(a)** What force keeps the fly from sliding off the rotating disk? What direction does the force point? How big is it? For the last question, express your answer in terms of the symbols given in the description above.

**(b)** If the fly has a mass of 0.5 grams, is sitting 10 cm from the center of the disk, and the disk is turning at a rate of 33 rev/min, what is the coefficient of friction?

**75. Rotational Inertia and Rotational Acceleration** A small spool of radius $r_s$ and a large Lucite disk of radius $r_d$ are connected by an axle that is free to rotate in an almost frictionless manner inside of a bearing as shown in Fig. 11-46. A string is wrapped around the spool and a mass, $m$, which is attached to the string, is allowed to fall.

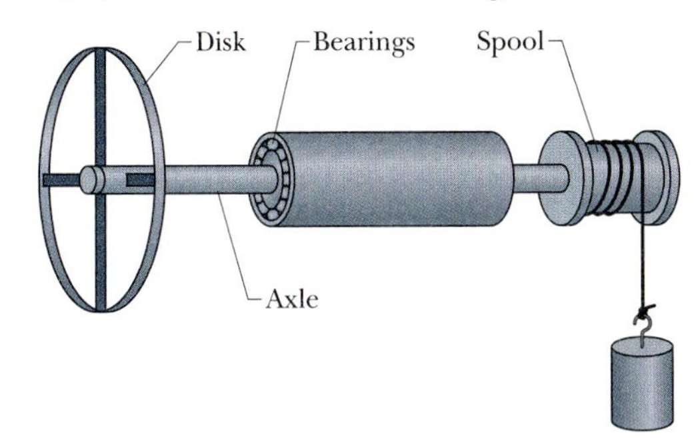


FIGURE 11-46 ■ Problem 75.

**(a)** Draw a free-body diagram showing the forces on the falling mass, $m$, in terms of $m$, $g$, and $\vec{F}^{\text{tension}}$.
**(b)** If the magnitude of the translational acceleration of the mass is measured to be $a$, what is the equation that should be used to calculate, $|\vec{F}^{\text{tension}}|$ in the string? In other words what equation relates $m$, $g$, $|\vec{F}^{\text{tension}}|$ and $a$? *Note*: In a system where $|\vec{F}^{\text{tension}}| - mg = ma$, if $a \ll g$ then $|\vec{F}^{\text{tension}}| \approx mg$.
**(c)** What is the magnitude of torque, $\tau$, on the spool–axle–disk system as a result of the tension in the string, $|\vec{F}^{\text{tension}}|$, acting on the spool?
**(d)** What is the magnitude of the rotational acceleration, $\alpha$, of the rotating system as a function of the translational acceleration, $a$, of the falling mass, and the radius, $r_s$, of the spool?

**(e)** The rotational inertias of the axle and the spool are so small compared to the rotational inertia of the disk the they can be neglected. If only the rotational inertia, $I_d$, of the large disk of radius $r_d$ is considered, what is the equation that can be used to predict the value of $I_d$ as a function of the torque on the system, $\tau$, and the magnitude of the rotational acceleration, $\alpha$, of the disk?
**(f)** What is the theoretical value of the rotational inertia, $I_d$, of a disk of mass $M$ and radius $r_d$ in terms of $M_d$ and $r_d$?

**76. Round and Round** Little Jay is enjoying his first ride on a merry-go-round. (He is riding a stationary horse rather than one that goes up and down.) A schematic view of the merry-go-round as seen from above is shown in Fig. 11-47*a* with a convenient coordinate system. A bit after the merry-go-round has started and is going around uniformly, we start our clock. Little Jay's position and velocity at time $t_1 = 0$ are shown as a dot and arrow. At $t_1 = 0$ is the net force acting on Jay equal to zero? If it is, write "Yes" and give a reason why you think so. If it isn't, write "No" and specify the type of force and the object responsible for exerting it.

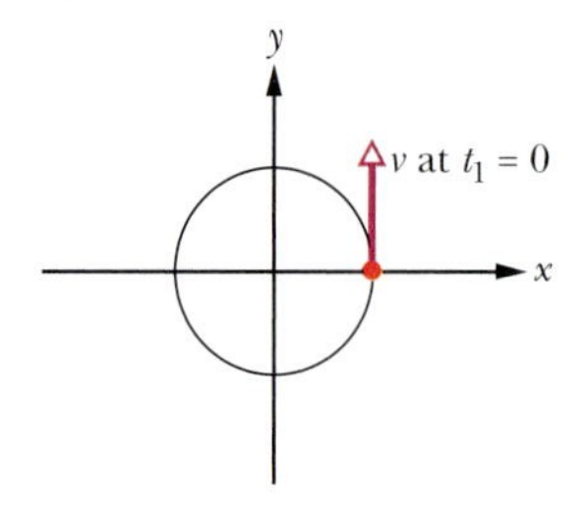

FIGURE 11-47*a* ■ Problem 76.

For the next six parts, specify which of the graphs shown in Fig. 11-47*b* could represent the indicated variable for Jay's motion. If none of the graphs work, write "N."

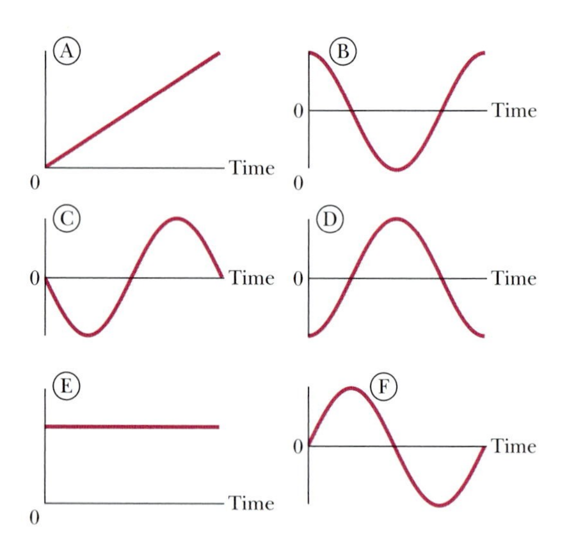

FIGURE 11-47*b* ■ Problem 76.

**(a)** The $x$-component of Jay's velocity
**(b)** The angle Jay's position vector makes with the $x$ axis
**(c)** The $y$-component of the force keeping Jay moving in a circle
**(d)** Jay's rotational velocity
**(e)** Jay's translational speed
**(f)** The $x$-component of Jay's position

**77. Comparing Rotational Inertias** If all three of the objects shown in Fig. 11-48 have the same radius and mass, which one has the most rotational inertia about its indicated axis of rotation? Which one has the least rotational inertia? Explain the reasons for your answer. *Hint*: Consider which one has its mass distributed farthest from the axis of rotation.

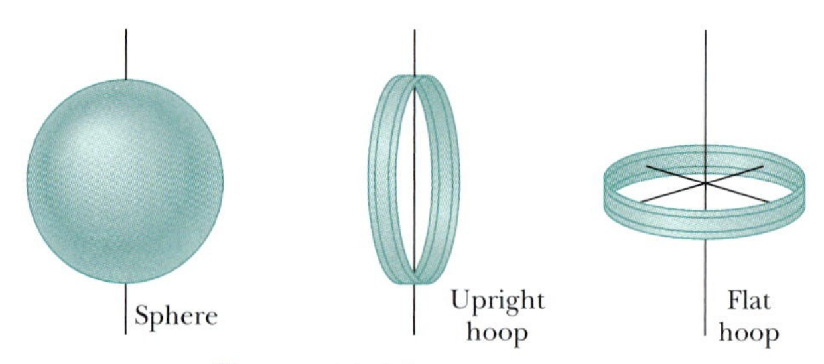

FIGURE 11-48 ■ Problem 77.

**78. Rotational Vs. Translational Energy of Motion**
**(a)** Describe how a solid ball can move so that

i. Its total kinetic energy is just the energy of motion of its center of mass
ii. Its total kinetic energy is the energy of its motion relative to its center of mass

**(b)** Two bowling balls are moving down a bowling alley so that their centers of mass have the same velocity, but one just slides down the alley, while the other rolls down the alley. Which ball has more energy? Explain your reasoning.

**79. Closing the Door** A student is in her dorm room, sitting on her bed doing her physics homework. The door to her room is open. Suddenly she hears the voice of her ex-boyfriend down the hall, talking to the girl in the room next door. She wants to shut the door quickly, so she throws a superball (which she keeps next to her bed for this purpose) against the door. The ball follows the path shown in Fig. 11-49. It hits the door squarely and bounces straight back. Does the ball's effectiveness in closing the door depend on where on the door the ball hits? If it does, where should it hit to be most effective? Explain your reasoning.

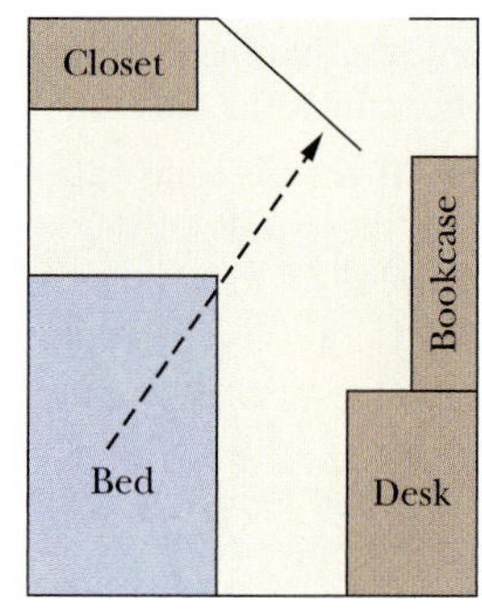

FIGURE 11-49 ■ Problem 79.

**80. Cleaning Up with Flywheels** One proposal for reducing air pollution is the flywheel-driven automobile. Instead of an engine, the car contains a large steel disk, or flywheel, which is mounted to rotate about a vertical axis. It is set spinning at a high rotational velocity in the early morning using electric power (from plugging it into the wall). If the car is to be about the same size as a typical car today, estimate the amount of energy that could be stored in a rotating steel flywheel that fits under the car's hood. You may find some of the following numbers useful:

- density of steel = 6 g/cm$^3$
- mass of a typical car = 1000 kg
- maximum speed of flywheel = 1000 revolutions/minute
- fraction of carbon monoxide pollution produced by vehicles = 60%

**81. Spinning with the Earth** Because the earth is spinning about its axis once a day, you are also spinning about the earth's axis once a day. Estimate the rotational kinetic energy you have as a result of this motion.

**82. Keep the Dust Off Your Hard Drive!** A current-generation hard drive in a computer spins at a rate of 7000 rpm. The disk in the drive is about the same size as a floppy disk. Estimate the coefficient of friction that would permit the frictional force to keep a speck of dust sitting on the disk from sliding off. Assume that the speck has a mass of 50 mg. Discuss the implications of your result.

**83. Kinetic Energy of a Bicycle Wheel** Estimate the rotational kinetic energy of a bicycle wheel as the bicycle it is a part of is being ridden down the street.

**84. Ferris Wheel** Use a video analysis software program to analyze the motion of a Ferris wheel. If you have access to the VideoPoint movie collection, use the movie with filename HRSY001. This is a movie of the Cyclops Ferris wheel at Hershey Park. A sample frame is shown in Fig. 11-50.

FIGURE 11-50 ■ Problem 84.

**(a)** What is the nature of the rotational speed of a Ferris wheel as a function of time? Is it increasing, decreasing, or remaining constant? Cite the evidence for your answer.
**(b)** At $t = 0.1000$ s, what is the translational speed of a point on the inner circle?
**(c)** At $t = 0.1000$ s, what is the translational speed of a point on the outer circle?

**85. Falling Mass Turns Disk** Use a video analysis software program to analyze the motion of a disk that is attached to a spool. A falling mass attached by a string to the spool causes the spool and disk to undergo a rotational acceleration. If you have access to the VideoPoint movie collection, use the movie with filename DSON014 and analyze the first 12 frames.

**(a)** Is the acceleration of the disk constant? Explain what you did and cite the evidence for your conclusions.
**(b)** Describe the nature of the rotational acceleration. Does it increase, decrease, or stay the same? If you concluded that the rotational acceleration is constant, then determine what its value is in rad/s$^2$. Explain how you arrived at your conclusions. Show relevant data and graphs.
**(c)** What is the equation that describes the angle through which the disk has moved as a function of time? Explain how you determined this equation.
**(d)** What is the equation that describes the rotational velocity of the disk as a function of time? Explain how you derived this equation.

# 12 Complex Rotations

Image courtesy Ringling Brothers and Barnum & Bailey® THE GREATEST SHOW ON EARTH

In 1897, a European "aerialist" made the first triple somersault during the flight from a swinging trapeze to the hands of a partner. For the next 85 years aerialists attempted to complete a *quadruple* somersault, but not until 1982 was it done before an audience. Miguel Vazquez of the Ringling Bros. and Barnum & Bailey Circus rotated his body in four complete circles in midair before his brother Juan caught him. Both were stunned by their success.

**Why was the feat so difficult, and what feature of physics made it (finally) possible?**

*The answer is in this chapter.*

## 12-1 About Complex Rotations

This chapter presents an extension of the concepts in rotational motion that we began discussing in the last chapter. In Chapter 11, we studied the relationships between translational and rotational quantities like position and angle, translational velocity and rotational velocity, translational acceleration and rotational acceleration, and force and torque. We limited our discussion to the rotation of rigid bodies with a constant rotational inertia $I$ (constant mass distribution) about a fixed axis. Furthermore, because the rotation took place about a fixed axis, we were able to treat rotational quantities as one-dimensional components along the axis of rotation. In this chapter, we extend our study to more complex motion in which either the axis of rotation does not stay fixed in space, or the rotational inertia of the rotating body changes over time.

There are several distinct types of complex motion of interest to scientists and engineers that we will consider:

1. Rotations about an axis of rotation that moves but does not change direction. This type of motion is a combination of rotational motion and translational motion. Examples include the motion of yo-yos, wheels, and bowling balls (Fig. 12-1).
2. Rotations about an axis of rotation which changes direction. The axes of rotation for Frisbees and boomerangs change direction as a result of interactions both with air molecules and the Earth's gravity. The axis of a spinning top changes direction as it loses energy. A simple example of this type of motion is someone flipping the axis of a spinning wheel (Fig. 12-2).
3. Rotating objects that have fixed axes of rotation but undergo changes in rotational inertia while spinning. For example, skaters who pull in their arms are reducing their rotational inertia (Fig. 12-3). Stellar matter does the same thing when collapsing into a neutron star.

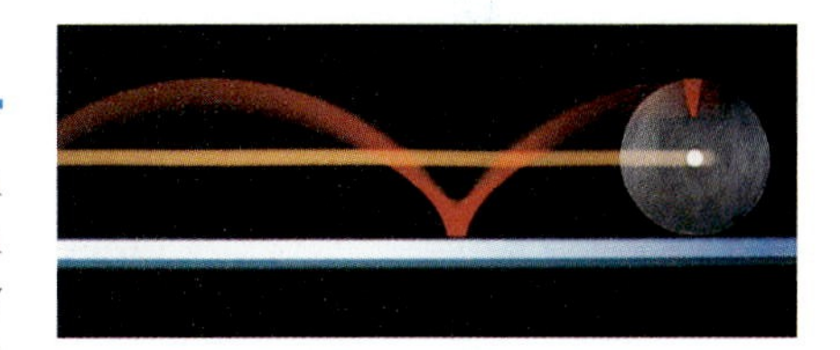

**FIGURE 12-1** ■ A time exposure photograph of a rolling disk. Small lights have been attached to the disk, one at its center and one at its edge. The latter traces out a curve called a *cycloid*.

**FIGURE 12-2** ■ A student applies a torque to the axis of a rotating bicycle wheel in order to change the direction of the wheel's axis of rotation. (Photo courtesy of PASCO scientific.)

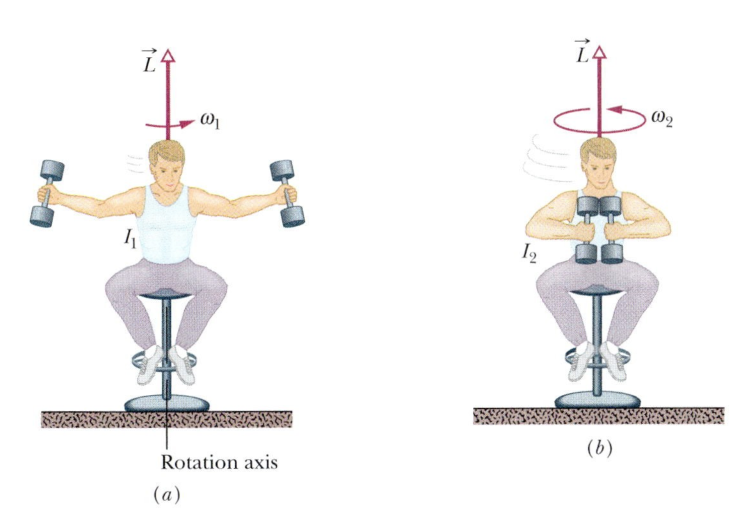


**FIGURE 12-3** ■ A student reduces his rotational inertia while rotating by pulling in his arms.

We start our study of complex rotations by considering the kinetic energy associated with combined translational and rotational motions. We will then apply Newton's Second Law in both a translational form and a rotational form to the motion of a yo-yo traveling up and down a string. This will provide us with a model of how one might find an expression for the translational acceleration in a complex motion. Before moving on to the task of analyzing motions involving changes in axis direction and rotational inertia, we will develop the mathematical tools we will require in order to treat torque and other rotational quantities as three-dimensional vectors.

Finally, because the law of conservation of translational momentum in cases of zero net force is such a powerful analysis tool, we will explore the concept of rotational momentum as a rotational corollary of translational momentum. We can then recast Newton's Second Law of rotation so it relates changes in rotational momentum to the net applied torque, showing that rotational momentum is conserved when the net torque acting on the system is zero. In the last sections of the chapter, we bring all of this together, using Newton's laws in their rotational form, the new concept of conservation of rotational momentum, and the vector mathematics we developed to explain the complex motions of aerialists, divers, spacecraft navigation, and neutron star rotation.

## 12-2 Combining Translations with Simple Rotations

We begin our discussion of complex rotations by considering motions that are combinations of translation and rotation. For example, when a bicycle moves along a straight track, the center of each wheel moves forward with a translational speed $v_{com}$. At any given instant if the wheel is rolling without slipping, the top point on the wheel is moving forward at twice $v_{com}$ relative to the track, and the bottom point on the wheel is not moving. However, every point on the wheel also rotates about the center with rotational speed $\omega$. Hence, the rolling motion of a wheel is a combination of purely translational and purely rotational motions.

A yo-yo is another example of this type of motion. As a yo-yo rolls down a string, it undergoes rotational motion. However, it also undergoes translational motion as it falls. One way to view such motion is as rotation about an axis that is moving (translating) downward. We will more carefully consider this type of motion by analyzing the forces and torques at work in the case of the falling yo-yo. But first, let's consider energy issues involved in motions that combine rotation with translation.

### Energy Considerations

If a yo-yo rolls down its string for a distance $h$, the yo-yo-Earth system loses potential energy in the amount of $mgh$ but gains kinetic energy in both translational ($\frac{1}{2}mv_{com}^2$) and rotational ($\frac{1}{2}I_{com}\omega^2$) forms. As the yo-yo climbs back up, the system loses kinetic energy and regains potential energy.

> An object that undergoes combined rotational and translation motion has two types of kinetic energy: a rotational kinetic energy ($\frac{1}{2}I_{com}\omega^2$) due to its rotation about its center of mass and a translational kinetic energy ($\frac{1}{2}Mv_{com}^2$) due to translation of its center of mass. The total kinetic energy of the object is the sum of these two.

In a modern yo-yo, the string is not tied to the axle but is looped around it. When the yo-yo "hits" the bottom of its string, an upward force on the axle from the string stops the descent. The yo-yo then spins about its axle inside the loop and has only rotational kinetic energy. The yo-yo keeps spinning ("sleeping") until you "wake it" by jerking on the string, causing the string to catch on the axle and the yo-yo to climb back up. The rotational kinetic energy of the yo-yo at the bottom of its string (and thus the sleeping time) can be considerably increased by throwing the yo-yo downward so it starts down the string with initial speeds $v_{com}$ and $\omega$ instead of rolling down from rest.

### The Forces of Rolling

The simultaneous application of Newton's Second Law in both its translational and rotational forms allows us to calculate the acceleration of an object in situations where the motion combines rotation and translation. As an example of this technique, let's attempt to find an expression for the translational acceleration $a_{com}$ of a yo-yo rolling down a string. We will use Newton's Second Law, noting the following points:

1. The yo-yo rolls down a string that makes angle $\theta = 90°$ with the horizontal.
2. The yo-yo rolls on an axle of radius $R_0$ (Fig. 12-4*a*).
3. The yo-yo is slowed by the tension force, $\vec{T}$, exerted on it by the string (Fig. 12-4*b*).

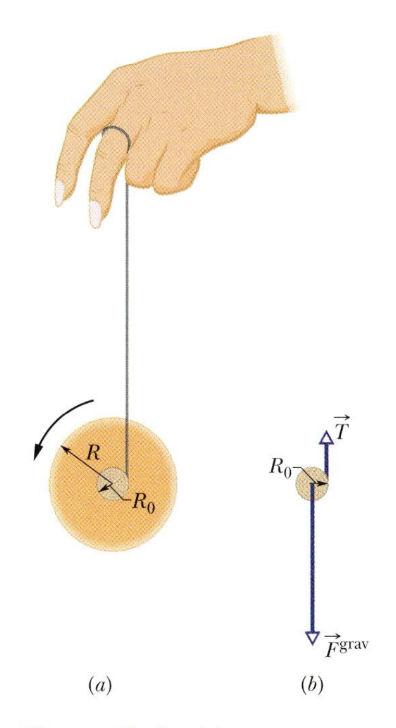


**FIGURE 12-4** ■ (*a*) A yo-yo, shown in cross section. The string, of negligible thickness, is wound around an axle of radius $R_0$. (*b*) A free-body diagram for the falling yo-yo. Only the axle is shown.

The net force acting on the yo-yo is the vector sum of the gravitational force of the Earth on the yo-yo and the tension in the string. This net force causes the yo-yo to speed up or slow down. That is, the net force causes a translational acceleration $\vec{a}_{com}$ of the center of mass along the direction of travel. The net force also causes the yo-yo to rotate faster or slower, which means it causes a rotational acceleration $\alpha$ about the center of mass. From Chapter 11, we know that we can relate the magnitudes of the translational acceleration $\vec{a}_{com}$ and the rotational acceleration $\alpha$ by

$$a_{com} = \alpha R_0 \quad \text{(smooth rolling motion).} \tag{12-1}$$

If we want to find an expression for the yo-yo's acceleration $a_{com\,y}$ down the string, we can do this by using Newton's Second Law in the component form of both its translational version ($F_y^{net} = Ma_y$) and its rotational version ($\tau^{net} = I\alpha$).

We start by drawing the forces on the body as shown in Fig. 12-4:

1. The gravitational force $\vec{F}^{grav}$ on the body is directed downward. It acts at the center of mass of the yo-yo.
2. The tension in the string is directed upward. It acts at the point of contact outside of the yo-yo's central axis.

We can write Newton's Second Law for components along the $y$ axis in Fig. 12-4 ($F_y^{net} = ma_y$) as

$$T_y - Mg = -Ma_{com\,y}. \tag{12-2}$$

Here $M$ is the mass of the yo-yo. This equation contains two unknowns: the positive tension force component ($T_y = +|\vec{T}|$) and the component describing the vertical acceleration of the center of mass ($a_{com\,y}$).

Now we can use Newton's Second Law in rotational form to analyze the yo-yo's rotation about its center of mass (which coincides with its central axis). First, we shall use $|\vec{\tau}| = r_\perp|\vec{F}|$ (Eq. 11-31) to determine the magnitude of torque on the yo-yo about that point. The perpendicular distance from the rotation axis to the tension force (or moment arm) is $R_0$. So, the magnitude of torque that causes the yo-yo to rotate is given by $|\vec{T}||\vec{R}_0| = T_y R_0$. By the right-hand rule that we learned in Chapter 11, the resulting rotational acceleration would be positive (out of the page). Since the rotational acceleration is positive, we know the torque that produced the rotational acceleration is also positive.

The other force acting on the yo-yo, the gravitational force $\vec{F}^{grav}$, acts at the center of mass of the yo-yo. That is the center of the object itself, and so the gravitational force has a zero moment arm ($r_\perp = 0$) about the center of mass. Thus, the gravitational force produces zero torque. So we can write the rotational version of Newton's Second Law in component form ($\tau_z^{net} = I\alpha_z$) about an axis through the body's center of mass as

$$T_y R_0 = I_{com}\alpha_z. \tag{12-3}$$

As was the case for the equation resulting from the application of Newton's Second Law in its translational form, this equation contains two unknowns, $T_y$ and $\alpha_z$.

However, there is a relationship between $a_{\text{com } y}$ and $\alpha$. We can use that relationship ($a_{\text{com}} = \alpha_z R_0$) to tie together the rotational and translational expressions of Newton's Second Law. Thus, we substitute $a_{\text{com } y}/R_0$ for $\alpha_z$ in the expression above (Eq. 12-3) and solve for the magnitude of the tension force, $T = T_y$, to obtain

$$T_y = I_{\text{com}} \frac{a_{\text{com } y}}{R_0^2}. \qquad (12\text{-}4)$$

Substituting the right side of the equation above for $T_y$ in the relationship we derived based on the translational motion of the yo-yo,

$$T_y - Mg = -Ma_{\text{com } y}, \qquad (12\text{-}5)$$

we then find

$$a_{\text{com } y} = -\frac{g}{1 + I_{\text{com}}/MR_0^2}, \qquad (12\text{-}6)$$

where $I_{\text{com}}$ is the yo-yo's rotational inertia about its center, $R_0$ is its axle radius, and $M$ is its mass. A yo-yo has the same downward acceleration when it is climbing back up the string, because the forces on it are still those shown in Fig. 12-4*b*.

## TOUCHSTONE EXAMPLE 12-1: Hoop, Disk, Sphere

Consider a hoop, a disk, and a sphere, each of mass $M$ and radius $R$, that roll smoothly along a horizontal table. For each, what fraction of its kinetic energy is associated with the translation of its center of mass?

**SOLUTION** ■ The **Key Idea** is that the kinetic energy of a smoothly rolling body is the sum of its translational kinetic energy ($\frac{1}{2}Mv_{\text{com}}^2$) and its rotational kinetic energy ($\frac{1}{2}I_{\text{com}}\omega^2$). Therefore, the fraction of the kinetic energy associated with translation is

$$\text{frac} = \frac{\frac{1}{2}Mv_{\text{com}}^2}{\frac{1}{2}Mv_{\text{com}}^2 + \frac{1}{2}I_{\text{com}}\omega^2}. \qquad (12\text{-}7)$$

We can greatly simplify the right side of Eq. 12-7 by substituting $v_{\text{com}}/R$ for $\omega$ (Eq. 11-15) and realizing that the expressions for rotational inertia in Table 11-2 are all of the form $\beta MR^2$, where $\beta$ is a numerical coefficient (the "front number"). Here $\beta$ is 1 for a hoop, $\frac{1}{2}$ for a disk, and $\frac{2}{5}$ for a sphere. Thus, we can substitute $\beta MR^2$ for $I_{\text{com}}$ in Eq. 12-7.

After these substitutions and some cancellations, Eq. 12-7 becomes

$$\text{frac} = \frac{1}{1 + \beta}. \qquad (12\text{-}8)$$

Now, substituting the $\beta$ values for the hoop, disk, and sphere, we can generate Table 12-1 to show the fractional splits of translational and rotational kinetic energy. For example, 0.67 of the kinetic energy of the disk is associated with the translation.

The relative split between translational and rotational energy depends on the relative size of the rotational inertia of the rolling object. As Table 12-1 shows, the rolling object (the hoop) that has its mass farthest from the central axis of rotation (and so has the largest rotational inertia) has the largest share of its kinetic energy in rotational motion. The object (the sphere) that has its mass closest to the central axis of rotation (and so has the smallest rotational inertia) has the smallest share in rotational motion.

**TABLE 12-1**
**The Relative Splits between Rotational and Translational Energy for Rolling Objects**

| Object | Rotational Inertia $I_{\text{com}}$ | Fraction of Energy in Translation | Fraction of Energy in Rotation |
|---|---|---|---|
| Hoop | $1MR^2$ | 0.50 | 0.50 |
| Disk | $\frac{1}{2}MR^2$ | 0.67 | 0.33 |
| Sphere | $\frac{2}{5}MR^2$ | 0.71 | 0.29 |
| General[a] | $\beta MR^2$ | $\frac{1}{1+\beta}$ | $\frac{\beta}{1+\beta}$ |

[a] $\beta$ may be computed for any rolling object as $I_{\text{com}}/MR^2$.

## TOUCHSTONE EXAMPLE 12-2: Racing Down a Ramp

A uniform hoop, disk, and sphere, with the same mass $M$ and same radius $R$, are released simultaneously from rest at the top of a ramp of length $L = 2.5$ m and angle $\theta = 12°$ with the horizontal. The objects roll without slipping down the ramp. No appreciable energy is lost to friction.

(a) Which object wins the race down to the bottom of the ramp?

**SOLUTION** ■ Two **Key Ideas** are these: First, the objects begin with the same mechanical energy $E^{\text{mec}}$, because they start from rest and the same height. Second, $E^{\text{mec}}$ is conserved during the race to the bottom, because the only force doing work on the object-ramp-Earth system is the gravitational force. (The normal force on them from the ramp and the frictional force at their point of contact with the ramp do not cause energy transfers). Further, at any given point along the ramp, the objects must have the same kinetic energy $K$ because the same amount of energy has been transferred from gravitational potential energy to kinetic energy.

If the objects were sliding down the ramp, this means they would have the same speed. However, another **Key Idea** is that they do not have the same speed $v_{\text{com}}$ because each object shares its kinetic energy between its translational motion down the ramp and its rotational motion around its center of mass. As we saw in Touchstone Example 12-1 and Table 12-1, the sphere has the greatest fraction (0.71) as translational energy, so it has the greatest $v_{\text{com}}$ and wins the race. Figure 12-5 shows the order of the objects during the race.

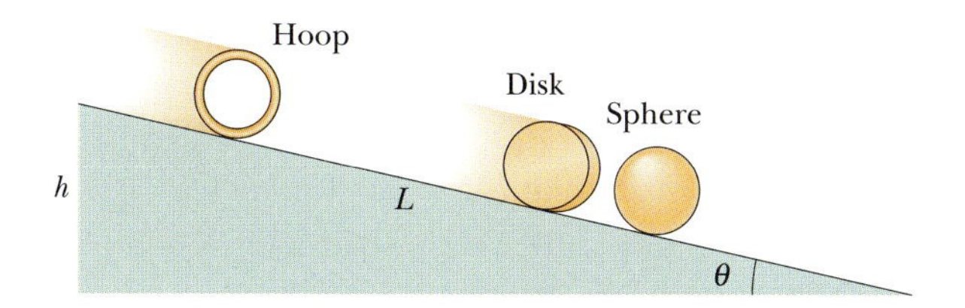


**FIGURE 12-5** ■ A hoop, a disk, and a sphere roll smoothly from rest down the last segment of a very long ramp of angle $\theta$.

(b) What is $v_{\text{com}}$ for each object at the bottom of the ramp?

**SOLUTION** ■ Again, the **Key Idea** here is that mechanical energy is conserved. Let us choose the bottom of the ramp as our reference height for zero gravitational potential energy, so at the finish each object-ramp-Earth system has $U_2 = 0$. The initial kinetic energy for all three objects is $K_1 = 0$. The initial potential energy is $U_1 = Mgh = Mg(L\sin\theta)$. Now we can write the conservation of mechanical energy $E_2^{\text{mec}} = E_1^{\text{mec}}$ as

$$K_2 + U_2 = K_1 + U_1$$

or

$$(\tfrac{1}{2}I_{\text{com}}\omega^2 + \tfrac{1}{2}Mv_{\text{com}}^2) + 0 = 0 + Mg(L\sin\theta).$$

Substituting $\omega = v_{\text{com}}/R$ and solving for $v_{\text{com}}$ give us

$$v_{\text{com}} = \sqrt{\frac{2gL\sin\theta}{1 + I_{\text{com}}/MR^2}}, \quad \text{(Answer)} \qquad (12\text{-}9)$$

which is the symbolic answer to the question.

Note that the speed depends not on the mass or the radius of the rolling object but only on the distribution of its mass about its central axis, which enters through the term $I_{\text{com}}/MR^2$. A marble and a bowling ball will have the same speed at the bottom of the ramp and will thus roll down the ramp in the same time. A bowling ball will beat a disk of any mass or radius, and almost anything that rolls will beat a hoop.

For the rolling hoop (see the hoop listing in Table 12-1) we have $I_{\text{com}}/MR^2 = 1$, so Eq. 12-9 yields

$$v_{\text{com}} = \sqrt{\frac{2gL\sin\theta}{1 + I_{\text{com}}/MR^2}} = \sqrt{\frac{(2)(9.8\ \text{m/s}^2)(2.5\ \text{m})(\sin 12°)}{1+1}} = 2.3\ \text{m/s}. \quad \text{(Answer)}$$

From a similar calculation, we obtain $v_{\text{com}} = 2.6$ m/s for the disk ($I_{\text{com}}/MR^2 = \frac{1}{2}$) and 2.7 m/s for the sphere ($I_{\text{com}}/MR^2 = \frac{2}{5}$).

## 12-3 Rotational Variables as Vectors

In the previous chapter, we considered only rotations that are about a fixed axis. We used the right-hand rule to determine whether the alignments for rotational displacement and velocity, representing the direction of rotation, are positive or negative. By assigning a standard coordinate system with the $z$ axis along the axis of rotation, we treated the variables $\Delta\theta$ and $\omega$ as components along the $z$ axis. Since rotational acceleration is defined in terms of changes of rotational velocity over time, the variable $\alpha$ could also be treated as a component along the axis of rotation. Thus, we developed a useful foundation for treating rotational quantities as vectors.

How can we work with rotational variables mathematically in cases where the axis of rotation is changing direction? For example, as a spinning top loses energy, its

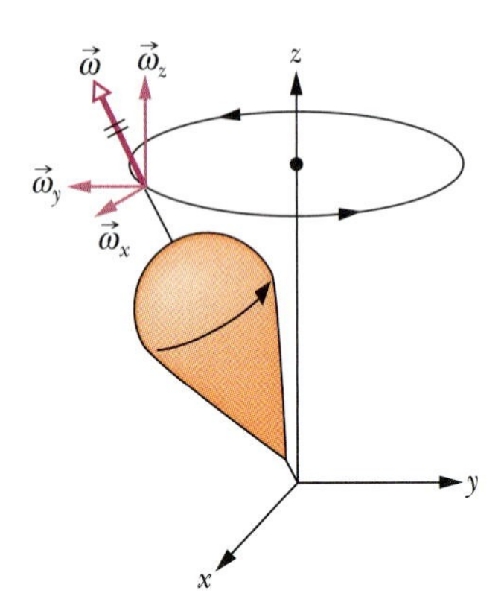


FIGURE 12-6 ■ The rotational velocity of a top rotating about an axis of symmetry always points along its axis of rotation. In the case where the rotational velocity changes direction, it must be described as a three-dimensional vector. Its components at one moment in time are shown relative to a right-handed coordinate system in which the $z$ axis points up in the vertical direction.

axis of rotation begins turning (in technical terms, "precessing") around a vertical axis as shown in Fig. 12-6. In this example, and many others like it, it seems logical to explore the feasibility of expressing rotational variables as three-dimensional vectors. In Fig. 12-6, we can then choose to define a right-hand coordinate system with the $z$ axis vertical. At any particular moment, the rotational displacement or velocity can be thought of as a vector pointing along the axis of rotation of the top.

In such a system, the rotational displacement must be described as a three-dimensional vector. Although we have not worked with three-dimensional vectors very much, we did introduce the decomposition of vectors into rectangular components in Section 4-4. In order to decompose a vector into components, we use unit vectors $\hat{i}$, $\hat{j}$, and $\hat{k}$ (discussed in Section 4-5) that point, respectively, in the positive directions of the $x$, $y$, and $z$ axes shown in Fig. 12-6.

This method of using unit vectors enables us to decompose a rotational variable in terms of vector components in the familiar way. Using the rotational velocity vector as an example, we get

$$\vec{\omega} = \vec{\omega}_x + \vec{\omega}_y + \vec{\omega}_z = \omega_x\hat{i} + \omega_y\hat{j} + \omega_z\hat{k}. \tag{12-10}$$

## Do Rotational Displacements and Velocities Behave Like Vectors?

It is not easy to get used to the way in which rotational quantities are represented as vectors. We instinctively expect that something should be moving *along* the direction of a vector. That is not the case when we attempt to use vectors to describe rotations. In the world of pure rotation, a vector defines an axis of rotation, not a direction in which something moves. Instead, a single particle or the many particles that make up a rigid body rotate *around* the direction of the vector. Nonetheless, a vector can be used to describe a rotational motion if it obeys the rules for vector manipulation discussed in Chapters 2 and 4. In particular, we stated in Chapter 2 that a vector is a mathematical entity that has both magnitude and direction, and that can be added, subtracted, multiplied, and transformed according to well-accepted mathematical rules. We have established that rotational variables seem to have both magnitude and direction. But we were vague about what the "well-accepted mathematical rules" for vector operations really are. One of these rules, used when vector addition was defined in Chapter 4, requires that the order of vector addition not matter, so that, for instance,

$$\vec{a} + \vec{b} = \vec{b} + \vec{a}.$$

Now for a caution: It turns out that *large rotational displacements cannot* be treated as vectors. Why not? We can certainly give them both magnitude and direction, as we did for the rotational velocity vector in Fig. 12-6. However, to be represented as a vector, a quantity must *also* obey the rules of vector addition. Rotational displacements fail this test.

Figure 12-7 shows an example of how large rotational displacements can fail the test. A book that is horizontal is given two 90° rotational displacements, first in the order shown in Fig. 12-7*a* and then in the order shown in Fig. 12-7*b*. Although each of the two rotational displacements are identical, the order in which they are applied is not. The book ends up with different orientations. Thus, the addition of the two *large* rotational displacements depends on their order and they cannot be vectors.

Fortunately, it can be shown mathematically that for small displacements, the order of the rotations does not matter. Since instantaneous rotational velocity is defined as

$$\vec{\omega} = \lim_{\Delta t \to 0} \frac{\Delta\vec{\theta}}{\Delta t} = \frac{d\vec{\theta}}{dt}, \tag{Eq. 11-6}$$

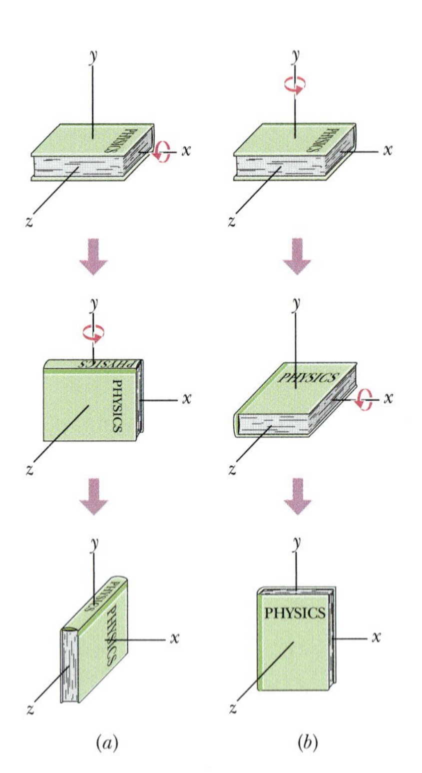


FIGURE 12-7 ■ (*a*) From its initial position, at the top in the figure, the book is given two successive 90° rotations, first about the (horizontal) $x$ axis and then about the (vertical) $y$ axis. (*b*) The book is given the same rotations, but in the reverse order.

it is made up of infinitesimally small displacements. Thus, it appears that any series of small rotational displacements, as well as instantaneous rotational velocities, behave like vectors. Since rotational acceleration is constructed as a vector difference between rotational velocity vectors, it should behave like a vector also. Thus, we conclude that the basic rotational velocity and acceleration variables behave like vectors so long as they are determined by using small rotational displacements $\Delta\vec{\theta}$.

## Can Torque Be Described as a Vector?

Recall that torque is a kind of "turning force" that can cause rotational accelerations about an axis. It was constructed mathematically for a very simple situation by combining the force acting on a single particle and the distance between that force and the particle's rotational axis. To see whether it is feasible to define torque as a three-dimensional vector, let's revisit the simple situation presented in Chapter 11.

In Section 11-8, we considered a single force acting on a particle that is attached to a "massless" rigid rod, which is, in turn, connected to a point (we'll call that point the origin of a coordinate system). We find that this force can cause the particle to rotate in a circle, but only when the force has a component that is tangent to the circle. As shown in Fig. 11-19, *this circle lies in the same plane as the force*. This means that the direction of the particle's rotational velocity and acceleration will be along its axis of rotation.

In Sections 11-7 and 11-8 we explored the relationship between torque ($\tau$) and rotational acceleration ($\alpha$) for special cases where the axis of rotation of a symmetric body is aligned with an axis of symmetry. We used the definition of rotational inertia ($I$) and Newton's Second Law for translational motion to show that for a single force acting on a particle,

$$\tau = I\alpha, \qquad \text{(Eq. 11-33)}$$

provided that we define the magnitude of torque ($\tau$) in this situation to be given by

$$|\tau| = |\vec{r}||\vec{F}|\sin\phi, \qquad \text{(Eq. 11-29)}$$

where $\phi$ is the *smaller* of the two angles between the vectors $\vec{r}$ and $\vec{F}$.

In fact, by regarding the rotational inertia $I$ as a scalar, and $\alpha$ and $\tau$ as components of a vector along the axis of rotation of the particle, we presented $\tau_z = I\alpha_z$ as the one-dimensional rotational analog to the expression $F_x = ma_x$ that describes motion along a straight line. We assume that for both expressions the acceleration that results from the application of a torque (or force) is in the same direction as the force (or torque). If we generalize this analogy between the translational and rotational laws of motion to three dimensions, then we expect that if $\vec{F}^{\text{net}} = m\vec{a}$, then

$$\vec{\tau}^{\text{net}} = I\vec{\alpha}, \qquad \text{(12-11)}$$

where $\vec{\alpha}$ and $\vec{\tau}^{\text{net}}$ are three-dimensional vectors that point in the *same direction*. If this is the case, then $\vec{\tau}$ is a vector that must be *perpendicular* to both the applied force $\vec{F}$ and the position vector $\vec{r}$ that extends from the axis of rotation to the particle experiencing the force. The torque vector must also have a magnitude given by $|\tau| = |\vec{r}||\vec{F}|\sin\phi$ (Eq. 11-29).

In Section 9-8, we discussed the fact that there are two different methods defined by mathematicians for multiplying vectors. One, known as the scalar (or dot) product, is used to define the amount of work, $W$, done on an object that undergoes a translational displacement $\vec{d}$ under the influence of a constant force $\vec{F}$. Work is a scalar quantity that is invariant to coordinate rotations and is given by

$$W = \vec{F}\cdot\vec{d} = |\vec{F}||\vec{d}|\cos\theta.$$

The other type of vector multiplication is known as the **vector** (or **cross**) **product.** The vector product $\vec{c}$ of two vectors $\vec{a}$ and $\vec{b}$ is given by

$$\vec{c} = \vec{a} \times \vec{b}.$$

As its name suggests, the **vector product** of two vectors is itself a vector. It is not hard to convince yourself that any two vectors determine a plane. We define the vector that results when a vector product is calculated *to be perpendicular to the plane determined by the vectors being multiplied.*

Recall that the plane a particle rotates in is perpendicular to the axis of rotation along which we expect the torque vector to point. This suggests that we may be able to express torque as a vector product. It also turns out that the magnitude of a vector product is equal to the product of the magnitudes of the two vectors being multiplied times the sine of the angle between them. This is also how the magnitude of a torque about a fixed axis is determined.

It appears it may be valid to define torque as the vector product of the position vector $\vec{r}$ and the force vector $\vec{F}$ so that

$$\vec{\tau} = \vec{r} \times \vec{F} \quad \text{(tentative definition of torque).}$$

In the next section, we discuss the mathematical properties of the vector product.

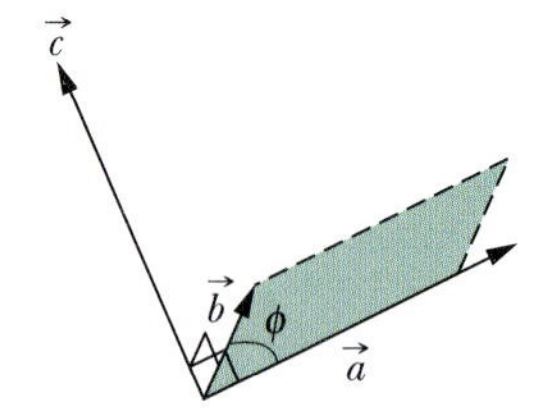


**FIGURE 12-8** ■ A parallelogram of area $A = |\vec{a}||\vec{b}|\sin\phi$ with $\vec{c}$ being perpendicular to $\vec{a}$ and $\vec{b}$. $\vec{a}$ and $\vec{b}$ lie in the same plane.

## 12-4 The Vector or Cross Product

Is there a natural way to associate a vector with the product of a pair of vectors? If we think about a pair of vectors in three-dimensional space, we see that they have two directions (unless they point in the same direction). There are only three mutually perpendicular directions, so we could choose the direction not used—the one perpendicular to the plane determined by the two vectors we are trying to multiply—as the direction of the vector product. Here's one way to think about it. Consider two vectors of lengths $|\vec{a}|$ and $|\vec{b}|$ pointing in different directions with $\phi$ being the *smaller angle* between them. The two vectors can be considered to be two sides of a parallelogram of area $A = |\vec{a}||\vec{b}|\sin\phi$.

This area has a direction, though we don't often think of area that way. The same area can be turned and oriented in different ways in space. We can choose to describe its orientation by an arrow perpendicular to the area. This suggests that we create a vector product $\vec{a} \times \vec{b} = \vec{c}$ that has the magnitude equal to the size of the area, $A = |\vec{a}||\vec{b}|\sin\phi$, and a direction perpendicular to the two vectors $\vec{a}$ and $\vec{b}$ (Fig. 12-8).

We should point out, though, that the area could actually have two different directions associated with it, with one direction pointing perpendicular to one side of the area and one direction pointing perpendicular to the other side of the area. These two vectors point in opposite directions, so they are just the negative of each other. Since the choice between these two directions is arbitrary, we will use the right-hand rule to choose which direction to associate with the product. Applying the right-hand rule to this vector product means if we point our straightened fingers on our right hand in the direction of the first vector so we can curl them to the direction of the second vector, then the direction of our extended thumb will be the direction associated with the vector product as shown in Fig. 12-9.

This discussion also implies that the vector (or cross) product of two vectors that point in the same direction must be zero. We know this because the area created by two such vectors is zero. Furthermore, we can't know in what direction a zero area would point!

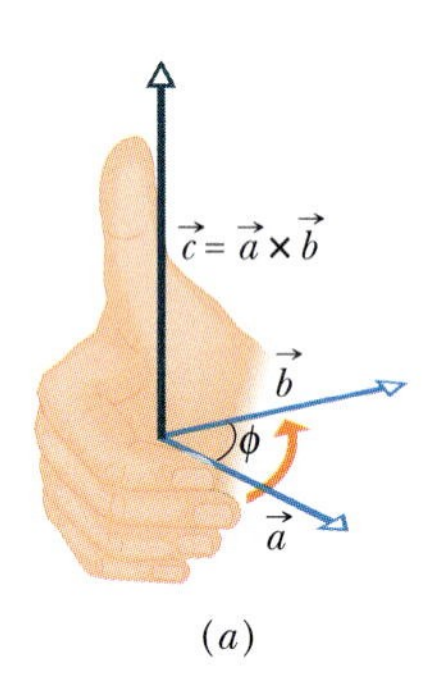


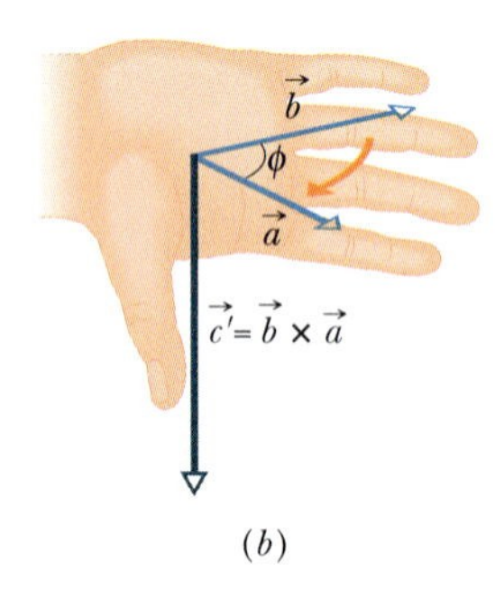


**FIGURE 12-9** ■ Illustration of the right-hand rule for vector products. (*a*) Sweep vector $\vec{a}$ into vector $\vec{b}$ with the fingers of your right hand. Your outstretched thumb shows the direction of vector $\vec{c} = \vec{a} \times \vec{b}$. (*b*) Showing that $\vec{a} \times \vec{b}$ is the reverse of $\vec{b} \times \vec{a}$.

If the vectors $\vec{a}$ and $\vec{b}$ are not actually length vectors (but have some other unit), we can just generalize the discussion above as if we were working with an area. So, in general, the vector product of any two vectors $\vec{a}$ and $\vec{b}$, written $\vec{a} \times \vec{b}$, produces a third vector $\vec{c}$ whose magnitude is

$$|\vec{c}| = |\vec{a}||\vec{b}|\sin\phi, \tag{12-12}$$

where $\phi$ (phi) is the *smaller* of the two angles between $\vec{a}$ and $\vec{b}$. (You must use the smaller of the two angles between the vectors because $\sin\phi$ and $\sin(360° - \phi)$ differ in algebraic sign.) Because of the notation $\vec{a} \times \vec{b}$, the vector product is known as the **cross product** of $\vec{a}$ and $\vec{b}$ or, more simply, "a cross b."

> If $\vec{a}$ and $\vec{b}$ are parallel or antiparallel, $\vec{a} \times \vec{b} = 0$. The magnitude of $\vec{a} \times \vec{b}$, which can be written as $|\vec{a} \times \vec{b}|$, is maximum when $\vec{a}$ and $\vec{b}$ are perpendicular to each other.

Remember that the order of the vector multiplication is important. In Fig. 12-9*b*, we are determining the direction of $\vec{c}\,' = \vec{b} \times \vec{a}$, so the fingers are placed to sweep $\vec{b}$ into $\vec{a}$ through the smaller angle. The thumb ends up in the opposite direction from before, and so it must be that $\vec{c}\,' = -\vec{c}$ or $\vec{a} \times \vec{b} = -\vec{b} \times \vec{a}$.

In unit-vector notation, we can write

$$\vec{a} \times \vec{b} = (a_x\hat{\text{i}} + a_y\hat{\text{j}} + a_z\hat{\text{k}}) \times (b_x\hat{\text{i}} + b_y\hat{\text{j}} + b_z\hat{\text{k}}), \tag{12-13}$$

which can be expanded according to the distributive law. That is, each component of the first vector is to be crossed with each component of the second vector. The cross products of unit vectors are given in Appendix E (see Products of Vectors). For example, in the expansion of the equation above, we have

$$a_x\hat{\text{i}} \times b_x\hat{\text{i}} = a_xb_x(\hat{\text{i}} \times \hat{\text{i}}) = 0, \tag{12-14}$$

because the two unit vectors $\hat{\text{i}}$ and $\hat{\text{i}}$ are parallel and thus have a zero cross product. Similarly, we have

$$a_x\hat{\text{i}} \times b_y\hat{\text{j}} = a_xb_y(\hat{\text{i}} \times \hat{\text{j}}) = a_xb_y\hat{\text{k}}. \tag{12-15}$$

In the last step, we used Eq. 12-12 to evaluate the magnitude of $\hat{\text{i}} \times \hat{\text{j}}$ as unity (one). (The vectors $\hat{\text{i}}$ and $\hat{\text{j}}$ each have a dimensionless magnitude of unity, and the angle between them is 90°.) Also, we used the right-hand rule to get the direction of $\hat{\text{i}} \times \hat{\text{j}}$ as being in the positive direction of the $z$ axis (thus in the direction of $\hat{\text{k}}$).

Continuing to expand Eq. 12-13, we can show that

$$\vec{a} \times \vec{b} = (a_yb_z - b_ya_z)\hat{\text{i}} + (a_zb_x - b_za_x)\hat{\text{j}} + (a_xb_y - b_xa_y)\hat{\text{k}}. \tag{12-16}$$

We can also evaluate a cross product by setting up and evaluating a determinant (as shown in Appendix E) or by using a vector-capable calculator.

To check whether any $xyz$ coordinate system is a right-handed coordinate system, use the right-hand rule shown in Fig. 12-9 for the cross product $\hat{\text{i}} \times \hat{\text{j}} = \hat{\text{k}}$ with that system. If your fingers sweep $\hat{\text{i}}$ (positive direction of $x$) into $\hat{\text{j}}$ (positive direction of $y$) with the outstretched thumb pointing in the positive direction of $z$, then the system is right-handed.

**READING EXERCISE 12-1:** Vectors $\vec{c}$ and $\vec{d}$ have magnitudes of 3 units and 4 units, respectively. What is the angle between the directions of $\vec{c}$ and $\vec{d}$ if the magnitude of the vector product $\vec{c} \times \vec{d}$ is (a) zero, (b) 12 units, (c) 6 units? ■

**TOUCHSTONE EXAMPLE 12-3: Vector Product**

In Fig. 12-10, vector $\vec{a}$ lies in the $xy$ plane, has a magnitude of 18 units, and points in a direction 250° from the positive direction of $x$. Also, vector $\vec{b}$ has a magnitude of 12 units and points along the positive direction of $z$. What is the vector product $\vec{c} = \vec{a} \times \vec{b}$?

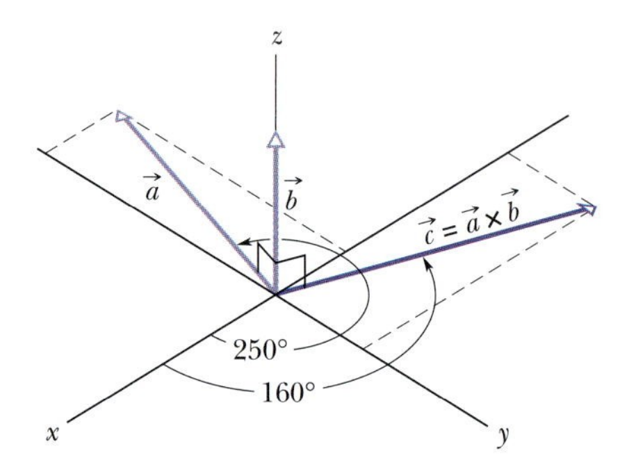


**FIGURE 12-10** ■ Vector $\vec{c}$ (in $xy$ plane) is the vector (or cross) product of vectors $\vec{a}$ and $\vec{b}$.

**SOLUTION** ■ One **Key Idea** is that when we have two vectors in magnitude-angle notation, we find the magnitude of their cross product (that is, the vector that results from taking their cross product) with Eq. 12-12. Here that means the magnitude of $\vec{c}$ is

$$|\vec{c}| = |\vec{a}||\vec{b}|\sin\phi = (18)(12)(\sin 90°) = 216. \quad \text{(Answer)}$$

A second **Key Idea** is that with two vectors in magnitude-angle notation, we find the direction of their cross product with the right-hand rule of Fig. 12-9. In Fig. 12-10, imagine placing the fingers of your right hand around a line perpendicular to the plane of $\vec{a}$ and $\vec{b}$ (the line on which $\vec{c}$ is shown) such that your fingers sweep $\vec{a}$ into $\vec{b}$. Your outstretched thumb then gives the direction of $\vec{c}$. Thus, as shown in Fig. 12-10, $\vec{c}$ lies in the $xy$ plane. Because its direction is perpendicular to the direction of $\vec{a}$, it is at an angle of

$$250° - 90° = 160° \quad \text{(Answer)}$$

from the positive direction of $x$.

## 12-5 Torque as a Vector Product

In Chapter 11, we defined the torque component, $\tau_z$, for a rigid body that can rotate around a fixed axis. In that case, each particle in the body was forced to move in a path that is a circle about that axis. We now use the vector product to expand the definition of torque to apply it to an individual particle that moves along *any* path relative to a fixed *point* (rather than a fixed axis). The path need no longer be a circle, and we must write the torque as a vector $\vec{\tau}$ that may have any direction.

Figure 12-11*a* shows a particle at point $A$ in the $xy$ plane. A single force $\vec{F}$ in that plane acts on the particle. The particle's position relative to the origin $O$ is given by position vector $\vec{r}$. The torque $\vec{\tau}$ acting on the particle relative to the fixed point $O$ is a vector quantity defined as the vector product of $\vec{r}$ and $\vec{F}$ so that

$$\vec{\tau} = \vec{r} \times \vec{F} \quad \text{(torque defined).} \quad (12\text{-}17)$$

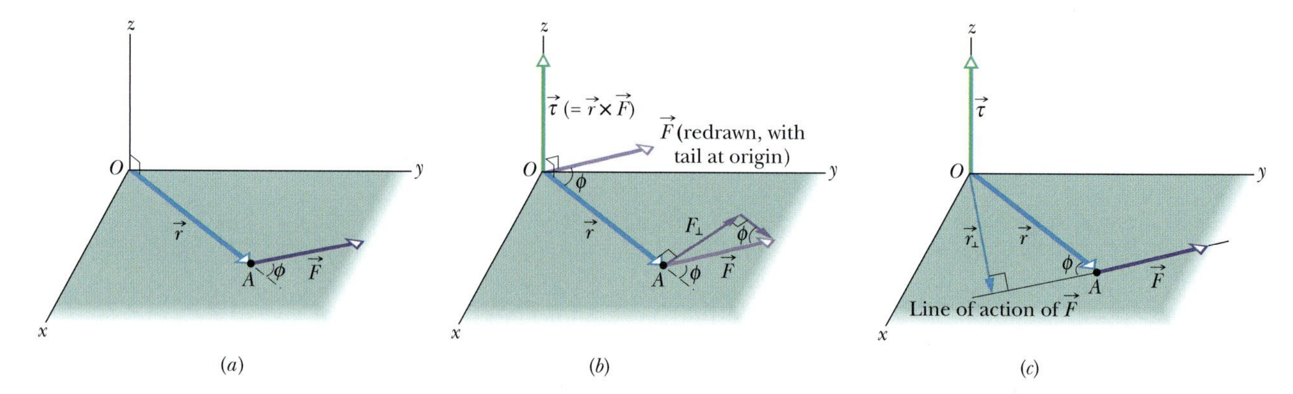


**FIGURE 12-11** ■ Defining torque. (*a*) A force $\vec{F}$, lying in the $xy$ plane, acts on a particle at point $A$. (*b*) This force produces a torque $\vec{\tau}(= \vec{r} \times \vec{F})$ on the particle with respect to the origin $O$. By the right-hand rule for vector (cross) products, the torque vector points in the positive direction of $z$. Its magnitude is equivalently given by $rF_\perp$ in (*b*) and by $r_\perp F$ in (*c*).

We can evaluate the vector (or cross) product in this definition of $\vec{\tau}$ by using the rules for such products given above. To find the direction of $\vec{\tau}$, we slide the vector $\vec{F}$ (without changing its direction) until its tail is at the origin $O$, so that the two vectors in the vector product are tail to tail as in Fig. 12-11*b*. We then use the right-hand rule for vector products in Fig. 12-9, sweeping the fingers of the right hand from $\vec{r}$ (the first vector in the product) into $\vec{F}$ (the second vector). The outstretched right thumb then gives the direction of $\vec{\tau}$. In Fig. 12-11*c*, the direction of $\vec{\tau}$ is again shown to be in the positive direction of the $z$ axis.

When drawing diagrams of three-dimensional vectors, we often need a way to show that a vector points into or out of the plane of the page.

> The symbol ⊗ is used to denote a vector that points into the plane of the page. The symbol ⊙ denotes a vector pointing out of the page.

## TOUCHSTONE EXAMPLE 12-4: Three Torques

In Fig. 12-12*a*, three forces, each of magnitude 2.0 N, act on a particle. The particle is in the $xz$ plane at point $a$ given by position vector $\vec{r}$, where $r = 3.0$ m and $\theta = 30°$. Force $\vec{F}_A$ is antiparallel to the $x$ axis, force $\vec{F}_B$ is antiparallel to the $z$ axis, and force $\vec{F}_C$ is antiparallel to the $y$ axis. What is the torque, with respect to the origin $O$, due to each force?

**SOLUTION** ■ The **Key Idea** here is that, because the three force vectors do not lie in a plane, we cannot evaluate their torques as in Chapter 11. Instead, we must use vector (or cross) products, given by Eq. 12-17 ($\vec{\tau} = \vec{r} \times \vec{F}$) with their directions given by the right-hand rule for vector products.

Because we want the torques with respect to the origin $O$, the vector $\vec{r}$ required for each cross product is the given position vector. To determine the angle $\phi$ between the direction of $\vec{r}$ and the direction of each force, we shift the force vectors of Fig. 12-12*a*, each in turn, so that their tails are at the origin. Figures 12-12*b*, *c*, and *d*, which are direct views of the $xz$ plane, show the shifted force vectors $\vec{F}_A$, $\vec{F}_B$, and $\vec{F}_C$, respectively. (Note how much easier the angles are to see.) In Fig. 12-12*d*, the angle between the directions of $\vec{r}$ and $\vec{F}_C$ is 90° and the symbol ⊗ means $\vec{F}_C$ is directed into the page.

Now, applying Eq. 12-17 for each force, we find the magnitudes of the torques to be

$$\tau_A = rF_A \sin\phi_A = (3.0 \text{ m})(2.0 \text{ N})(\sin 150°) = 3.0 \text{ N} \cdot \text{m},$$

$$\tau_B = rF_B \sin\phi_B = (3.0 \text{ m})(2.0 \text{ N})(\sin 120°) = 5.2 \text{ N} \cdot \text{m},$$

and $$\tau_C = rF_C \sin\phi_C = (3.0 \text{ m})(2.0 \text{ N})(\sin 90°) = 6.0 \text{ N} \cdot \text{m}.$$

(Answer)

To find the directions of these torques, we use the right-hand rule, placing the fingers of the right hand so as to rotate $\vec{r}$ into $\vec{F}$ through the *smaller* of the two angles between their directions. The thumb points in the direction of the torque. Thus $\vec{\tau}_A$ is directed into the page in Fig. 12-12*b*, $\vec{\tau}_B$ is directed out of the page in Fig. 12-12*c*, and $\vec{\tau}_C$ is directed as shown in Fig. 12-12*d*. All three torque vectors are shown in Fig. 12-12*e*

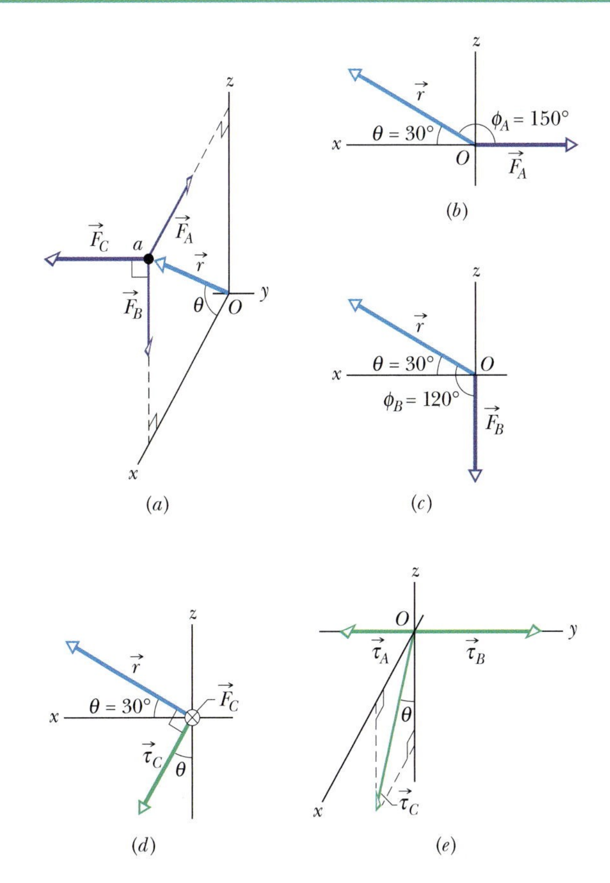


**FIGURE 12-12** ■ (*a*) A particle at point $a$ is acted on by three forces, each antiparallel to a coordinate axis. The angle $\phi$ (used in finding torque) is shown (*b*) for $\vec{F}_A$ and (*c*) for $\vec{F}_B$. (*d*) Torque $\vec{\tau}_C$ is perpendicular to both $\vec{r}$ and $\vec{F}_C$ (force $\vec{F}_C$ is directed into the plane of the figure). (*e*) The torques (relative to the origin $O$) acting on the particle.

## 12-6 Rotational Form of Newton's Second Law

Recall that the concept of translational momentum $\vec{p}$ and the principle of conservation of momentum are extremely powerful tools. They allow us to predict the outcome of, say, a collision between two cars without knowing the details of what goes on during the collision. Here we begin a discussion of the rotational counterpart of $\vec{p}$. In Chapter 7, we found that we could write Newton's Second Law in the form

$$\vec{F}^{\text{net}} = m\vec{a} = \frac{d\vec{p}}{dt} \quad \text{(single particle).} \tag{12-18}$$

This relationship expresses the close relation between force and translational momentum for a single particle. It can be generalized to extended bodies. It also leads directly to the powerful idea that translational momentum is conserved in the absence of a net external force.

We have seen enough of the parallelism between translational and rotational quantities to be hopeful that there is a rotational corollary to $\vec{F}^{\text{net}} = d\vec{p}/dt$. In search of the equivalent expression, we start with

$$\vec{\tau}^{\text{net}} = \vec{r} \times \vec{F}^{\text{net}}, \tag{12-19}$$

and replace the force vector with $m\vec{a}$. This gives us

$$\vec{\tau}^{\text{net}} = \vec{r} \times m\vec{a} \quad \text{or} \quad \vec{\tau}^{\text{net}} = \vec{r} \times m\frac{d\vec{v}}{dt}. \tag{12-20}$$

For a constant mass, the expression $\vec{r} \times m\,d\vec{v}/dt$ above can be replaced with $d(\vec{r} \times m\vec{v})/dt$.

The equality of these two expressions is more clearly seen in reverse. Namely,

$$\frac{d(\vec{r} \times m\vec{v})}{dt} = \frac{md(\vec{r} \times \vec{v})}{dt} \quad \text{(for constant mass).}$$

Then applying the product rule of derivatives

$$\frac{d(\vec{r} \times m\vec{v})}{dt} = m\left(\vec{r} \times \frac{d\vec{v}}{dt} + \frac{d\vec{r}}{dt} \times \vec{v}\right). \tag{12-21}$$

However, $d\vec{r}/dt$ is the object's velocity $\vec{v}$, and $\vec{v} \times \vec{v} = 0$. Thus, we can rewrite the equation above as

$$\frac{d(\vec{r} \times m\vec{v})}{dt} = m\left(\vec{r} \times \frac{d\vec{v}}{dt}\right),$$

or

$$\frac{d(\vec{r} \times m\vec{v})}{dt} = \vec{r} \times m\frac{d\vec{v}}{dt}. \tag{12-22}$$

So, from Eq. 12-20 above,

$$\vec{\tau}^{\text{net}} = \vec{r} \times m\frac{d\vec{v}}{dt} = \frac{d(\vec{r} \times m\vec{v})}{dt},$$

or

$$\vec{\tau}^{\text{net}} = \frac{d(\vec{r} \times \vec{p})}{dt}. \tag{12-23}$$

Comparing this expression to $\vec{F}^{\text{net}} = d\vec{p}/dt$, we see that if we choose to define the rotational momentum, $\vec{\ell}$, as the rotational corollary of translational momentum, then

$\vec{\ell} = \vec{r} \times \vec{p}$. We now have an equivalent expression for rotations as we do for translations. Namely,

$$\vec{\tau}^{\,\text{net}} = \frac{d\vec{\ell}}{dt} \quad \text{(single particle)}. \tag{12-24}$$

In words,

> The (vector) sum of all the torques acting on a particle is equal to the time rate of change of the rotational momentum of that particle.

Be careful, though: $\vec{\tau}^{\,\text{net}} = d\vec{\ell}/dt$ has no meaning unless the net torque $\vec{\tau}^{\,\text{net}}$, and the rotational momentum $\vec{\ell}$, are defined with respect to the same origin. Many texts refer to rotational momentum as angular momentum.

**READING EXERCISE 12-2:** The figure shows the position vector $\vec{r}$ of a particle at a certain instant, and four choices for the direction of a force that is to accelerate the particle. All four choices lie in the $xy$ plane. Rank the choices according to the magnitude of the time rate of change ($d\vec{\ell}/dt$) they produce in the rotational momentum of the particle about point $O$, greatest first.

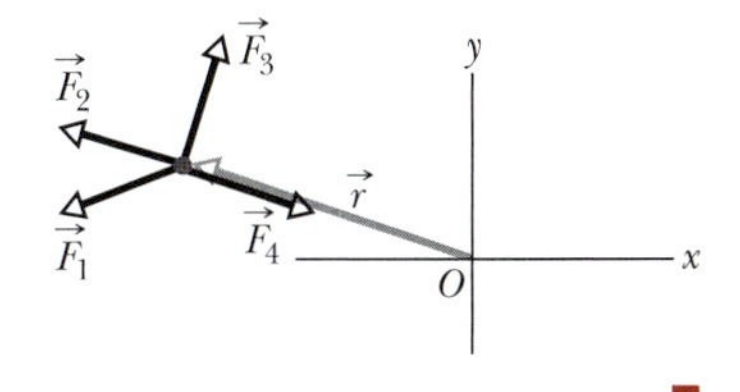


## 12-7 Rotational Momentum

Figure 12-13 shows a particle of mass $m$ with translational momentum $\vec{p} = m\vec{v}$ as it passes through point $A$ in the $xy$ plane. The **rotational momentum** $\vec{\ell}$ of this particle with respect to the origin $O$ is a vector quantity defined as

$$\vec{\ell} \equiv \vec{r} \times \vec{p} = m(\vec{r} \times \vec{v}) \quad \text{(rotational momentum defined)}, \tag{12-25}$$

where $\vec{r}$ is the position vector of the particle with respect to $O$. Note carefully that to have rotational momentum about $O$, the particle does *not* have to rotate around $O$. Comparison of $\vec{\tau} = \vec{r} \times \vec{F}$ (Eq. 12-17) and $\vec{\ell} = \vec{r} \times \vec{p}$ (Eq. 12-25) shows that rotational momentum bears the same relation to translational momentum as torque does to force. The SI unit of rotational momentum is the kilogram-meter-squared per second ($\text{kg} \cdot \text{m}^2/\text{s}$), equivalent to the joule-second ($\text{J} \cdot \text{s}$).

To find the direction of the rotational momentum vector $\vec{\ell}$ in Fig. 12-13, we slide the vector $\vec{p}$ until its tail is at the origin $O$. Then we use the right-hand rule for vector products, sweeping our right-hand fingers from $\vec{r}$ into $\vec{p}$. The outstretched thumb then shows that the direction of $\vec{\ell}$ is in the positive (upward) direction of the $z$ axis in Fig. 12-13. To find the magnitude of $\vec{\ell}$, we use the general definition of a cross product to write

$$|\vec{\ell}| = |\vec{r}||mv|\sin\phi \quad \text{or} \quad \ell = rmv\sin\phi, \tag{12-26}$$

where $\phi$ is the smaller angle between $\vec{r}$ and $\vec{p}$. From Fig. 12-13$a$, we see that $\ell = rmv\sin\phi$ can be rewritten as

$$\ell = rp_{\perp} = rmv_{\perp}, \tag{12-27}$$

where $p_{\perp}$ is the component of $\vec{p}$ perpendicular to $\vec{r}$, $v_{\perp}$ is the component of $\vec{v}$ perpendicular to $\vec{r}$ and $r$ is the magnitude of $\vec{r}$. From Fig. 12-13$b$, we see that $\ell = rmv\sin\phi$. So, $\vec{\ell} = \vec{r} \times \vec{p}$ (Eq. 12-25) can also be rewritten as

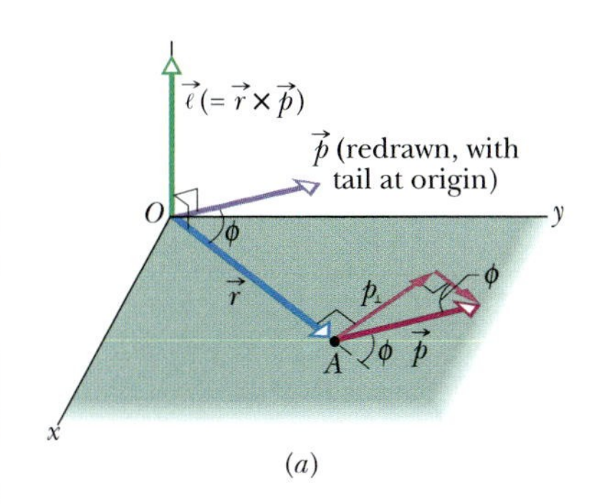


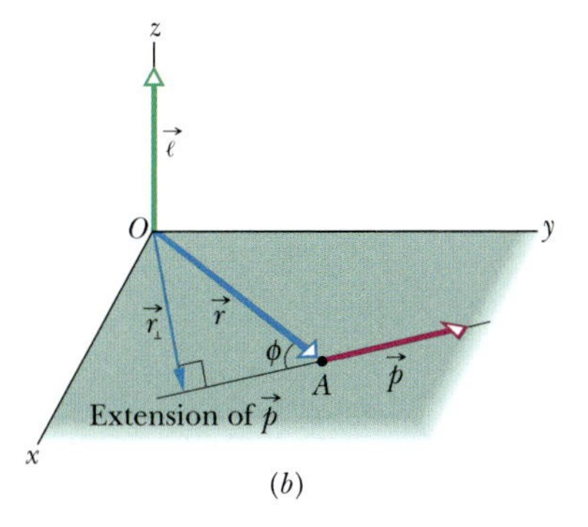


**FIGURE 12-13** Defining rotational momentum. A particle passing through point $A$ has translational momentum $\vec{p} = m\vec{v}$, with the vector $\vec{p}$ lying in the $xy$ plane. The particle has rotational momentum $\vec{\ell} = \vec{r} \times \vec{p}$ with respect to the origin $O$. By the right-hand rule, the rotational momentum vector points in the positive direction of $z$. ($a$) The magnitude of $\vec{\ell}$ is given by $\ell = rp_{\perp} = rmv_{\perp}$. ($b$) The magnitude of $\vec{\ell}$ is also given by $\ell = r_{\perp}p = r_{\perp}mv$.

$$\ell = r_\perp p = r_\perp mv, \tag{12-28}$$

where $r_\perp$ is the perpendicular distance between $O$ and the extension of $\vec{p}$.

Just as is true for torque, rotational momentum has meaning only with respect to a specified origin. Moreover, if the particle in Fig. 12-13 did not lie in the $xy$ plane, or if the translational momentum $\vec{p}$ of the particle did not also lie in that plane, the rotational momentum $\vec{\ell}$ would not be parallel to the $z$ axis. The direction of the rotational momentum vector is always perpendicular to the plane formed by the position and translational momentum vectors $\vec{r}$ and $\vec{p}$.

**READING EXERCISE 12-3:** In the diagrams below there is an axis of rotation perpendicular to the page that intersects the page at point $O$. Figure (*a*) shows particles 1 and 2 moving around point $O$ in opposite rotational directions, in circles with radii 2 m and 4 m. Figure (*b*) shows particles 3 and 4 traveling in the same direction, along straight lines at perpendicular distances of 2 m and 4 m from point $O$. Particle 5 moves directly away from $O$. All five particles have the same mass and the same constant speed. (a) Rank the particles according to the magnitudes of their rotational momentum about point $O$, greatest first. (b) Which particles have rotational momentum about point $O$ that is directed into the page?

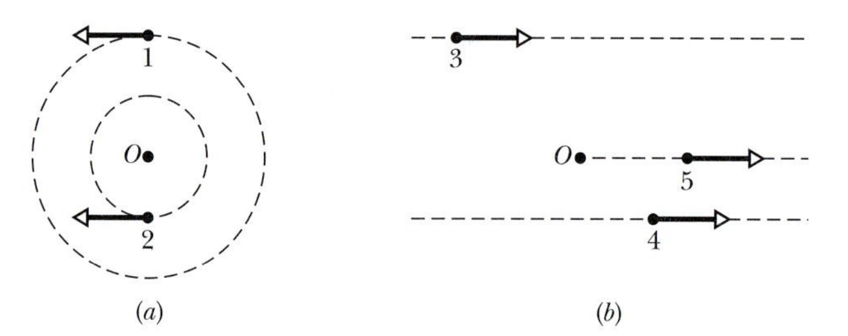


## 12-8 The Rotational Momentum of a System of Particles

Having a rotational equivalent of translational momentum is interesting, but what we would really like to do with such a quantity is to use it to understand the rotational motion of complex objects in dynamic situations. This is what made translational momentum so useful. For example, why is it that a skater spins faster when she pulls in her arms? How do we steer spaceships? Why do neutron stars spin so much faster than other stars? To understand these and other real-world situations, we must develop an expression for the rotational momentum of a system of particles.

Just as we did for translational momentum, we can use a principle of superposition for rotational momentum. We define the total rotational momentum $\vec{L}$ of a system of particles to be the vector sum of the rotational momenta $\vec{\ell}$ of the individual particles

$$\vec{L} = \vec{\ell}_A + \vec{\ell}_B + \vec{\ell}_C + \cdots + \vec{\ell}_n = \sum_{i=A}^{n} \vec{\ell}_i, \tag{12-29}$$

in which $i$ $(A,B,C,\ldots)$ labels the particles. With time, the rotational momenta of individual particles may change, either because of interactions within the system (between the individual particles) or because of influences that may act on the system from the outside.

We can find the change in $\vec{L}$ as these changes take place by taking the time derivative of

$$\vec{L} = \sum_{i=A}^{n} \vec{\ell}_i.$$

Thus,

$$\frac{d\vec{L}}{dt} = \sum_{i=A}^{n} \frac{d\vec{\ell}_i}{dt}. \tag{12-30}$$

From $\vec{\tau}^{\,\text{net}} = d\vec{\ell}/dt$ (Eq. 12-24), we see that

$$\frac{d\vec{L}}{dt} = \sum_{i=1}^{n} \vec{\tau}_i^{\,\text{net}}. \tag{12-31}$$

In the equation above, the right side is the sum of the torques acting on the particles that make up the system. This sum includes torques that result from all the forces acting on the system, whether they originate from within the system (internal forces) or outside of it (external forces). However, the internal torques sum to zero, as did the internal forces in the analogous expression $\vec{F}^{\,\text{net}} = d\vec{P}/dt$.

In general,

$$\vec{\tau}^{\,\text{net}} = \frac{d\vec{L}}{dt} \quad \text{(system of particles)}, \tag{12-32}$$

where $\vec{\tau}^{\,\text{net}}$ is the net torque acting on the system. In practice, this is just the vector sum of all external torques on all particles in the system, since the internal torques sum to zero.

This equation is Newton's Second Law in rotational form, for a system of particles. It says:

> The net (external) torque $\vec{\tau}^{\,\text{net}}$ acting on a system of particles is equal to the time rate of change of the system's total rotational momentum $\vec{L}$.

$\vec{\tau}^{\,\text{net}} = d\vec{L}/dt$ (Eq. 12-32) is analogous to $\vec{F}^{\,\text{net}} = d\vec{P}/dt$. However, it requires extra caution: Torques and the system's rotational momentum must be measured relative to the same origin.

## 12-9 The Rotational Momentum of a Rigid Body Rotating About a Fixed Axis

We next evaluate the rotational momentum of an extended system of particles that form a rigid body that rotates about a fixed axis. Figure 12-14 shows such a body. In Chapter 8, when we discussed the translational motion of extended systems, we derived an expression for the translational momentum of the object in terms of the velocity of its center of mass,

$$\vec{p}_{\text{sys}} = M_{\text{sys}}\vec{v}_{\text{com}} \quad \text{(translational momentum, system of particles)}. \tag{12-33}$$

We can develop an analogous expression for rotational motion. Let's start our development with a single mass element $\Delta m_i$ that rotates with a rotational velocity whose component along the axis of rotation is $\omega$. In Fig. 12-14 we see that the mass element has a translational momentum $\vec{p}_i$ and a position vector $\vec{r}_i$ relative to the axis of rotation. These vectors change constantly as the mass element rotates in a circle about its

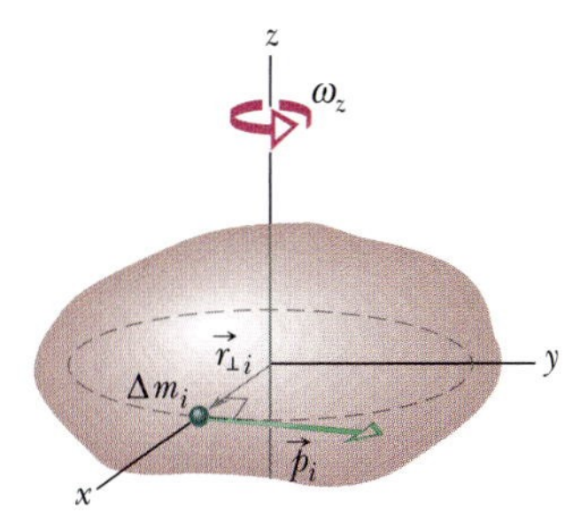


**FIGURE 12-14** ■ A rigid body rotates about the $z$ axis with rotational speed $\omega$. A mass element of mass $\Delta m_i$ within the body moves about the $z$ axis in a circle with radius $r$. The translational momentum $\vec{p}_i$ and the position vector $\vec{r}_i$ of the mass element relative to the axis of rotation change constantly as the mass element rotates.

axis of rotation. However, the translational momentum and position vectors are always perpendicular to each other and lie in the $x$-$y$ plane. This means that the rotational momentum vector only has a component in the $z$ direction. Since rotational momentum is defined as $\vec{\ell} \equiv \vec{r} \times \vec{p} = m(\vec{r} \times \vec{v})$ (Eq. 12-25) we see that for our simple situation

$$\vec{\ell}_i = \vec{r}_i \times \vec{p}_i = \ell_{iz}\hat{\mathrm{k}} = r_i p_i \hat{\mathrm{k}} = r_i(\Delta m_i v_i)\hat{\mathrm{k}}.$$

If we replace the translational speed $v_i$ with $\omega r_i$, where the rotational velocity component $\omega = \omega_z$ does not depend on which mass element we are considering (that is, the entire object moves as one), this equation reduces to

$$\ell_{iz} = r_i(\Delta m_i \omega_i r_i) = \Delta m_i r_i^2 \omega.$$

Aha! The term $\Delta m_i r_i^2$ is just the rotational inertia, $\Delta I_i$, of the $i$th mass element. So we can sum over all the mass elements to get a total rotational momentum of the rotating body given by

$$L_z = \sum \ell_{iz} = \sum \Delta m_i r_i^2 \omega_z = \sum \Delta I_i \omega = I\omega_z.$$

Recalling that the rotational analogy of mass is rotational inertia $I$, and that all points in a rigid rotating body move with the same rotational velocity $\omega = \omega_z$, we write the analogous expression for the rotational momentum of an extended object for an arbitrary choice of coordinate axes as

$$\vec{L} = I\vec{\omega} \quad \text{(rigid symmetric body, fixed axis through com).} \quad (12\text{-}34)$$

As you will see in the next section, this expression is very useful in situations where rotational momentum is conserved. It allows us to explain why rotating objects that change from one shape to another (such as a spinning ice skater) can speed up or slow down the rate of turn. However, you must remember that the rotational momentum $\vec{L}$, can only be expressed as $I\vec{\omega}$ when the rotational momentum and the rotational inertia, $I$, are taken about the same axis.

If an extended body is not symmetric with respect to its axis of rotation and its rotation axis does not pass through its center of mass, calculation of rotational inertia and momentum can become quite complex. For example, you can get different values of $I$ when the object rotates about different axes. (Compare, for example, a long rod rotating about its central axis and about one end.) Furthermore, in some cases, the rotational momentum is not aligned along the axis of rotation. These more complicated cases require the mathematics of "tensors" to handle them correctly; which is beyond the scope of this book.

**READING EXERCISE 12-4:** In the figure, a disk, a hoop, and a solid sphere are made to spin about fixed central axes (like a top) by means of strings wrapped around them, with the strings producing the same constant tangential force $\vec{F}$ on all three objects. The three objects have the same mass and radius, and they are initially stationary. Rank the objects according to (a) their rotational momentum about their central axes and (b) their rotational speed, greatest first, when the strings have been pulled for a certain time $t$.

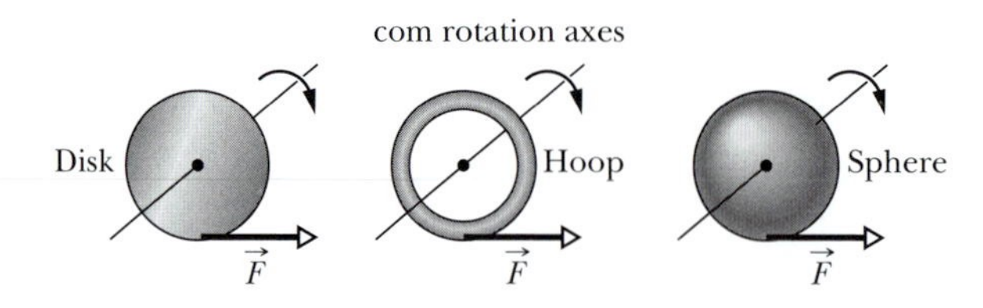

## TOUCHSTONE EXAMPLE 12-5: First Ferris Wheel

George Washington Gale Ferris, Jr., a civil engineering graduate from Rensselaer Polytechnic Institute, built the original Ferris wheel (Fig. 12-15) for the 1893 World's Columbian Exposition in Chicago. The wheel, an astounding engineering construction at the time, carried 36 wooden cars, each holding as many as 60 passengers, around a circle of radius $R = 38$ m. The mass of each car was about $1.1 \times 10^4$ kg. The mass of the wheel's structure was about $6 \times 10^5$ kg, which was mostly in the circular grid at the rim of the wheel from which the cars were suspended. The cars were loaded 6 at a time, and once all 36 cars were full, the wheel made a complete rotation at a constant rotational speed $\omega_F$ in about 2 min.

**FIGURE 12-15** ■ The original Ferris wheel, built in 1893 near the University of Chicago, towered over the surrounding buildings.

(a) Estimate the magnitude $L$ of the rotational momentum of the wheel and its passengers while the wheel rotated at $\omega_F$.

**SOLUTION** ■ The **Key Idea** here is that we can treat the wheel, cars, and passengers as a rigid object rotating about a fixed axis, at the wheel's axle. Then Eq. 12-34 ($\vec{L} = I\vec{\omega}$) gives the magnitude of the rotational momentum of that object. We need to find the rotational inertia $I$ of this object and the rotational speed $\omega_F$.

To find $I$, let us start with the loaded cars. Because we can treat them as particles, at distance $R$ from the axis of rotation, we know from Eq. 11-23 that their rotational inertia is $I_{pc} = M_{pc}R^2$, where $M_{pc}$ is their total mass. Let us assume that the 36 cars are each filled with 60 passengers, each of mass 70 kg. Then their total mass is

$$M_{pc} = 36[1.1 \times 10^4 \text{ kg} + 60(70 \text{ kg})] = 5.47 \times 10^5 \text{ kg}$$

and their rotational inertia is

$$I_{pc} = M_{pc}R^2 = (5.47 \times 10^5 \text{ kg})(38 \text{ m})^2$$
$$= 7.90 \times 10^8 \text{ kg} \cdot \text{m}^2.$$

Next we consider the structure of the wheel. Let us assume that the rotational inertia of the structure is due mainly to the circular grid suspending the cars. Further, let us assume that the grid forms a hoop of radius $R$, with a mass $M_{hoop}$ of $3 \times 10^5$ kg (half the wheel's mass). From Table 11-2(*a*), the rotational inertia of the hoop is

$$I_{hoop} = M_{hoop}R^2 = (3.0 \times 10^5 \text{ kg})(38 \text{ m})^2$$
$$= 4.33 \times 10^8 \text{ kg} \cdot \text{m}^2.$$

The combined rotational inertia $I$ of the cars, passengers, and hoop is then

$$I = I_{pc} + I_{hoop} = 7.90 \times 10^8 \text{ kg} \cdot \text{m}^2 + 4.33 \times 10^8 \text{ kg} \cdot \text{m}^2$$
$$= 1.22 \times 10^9 \text{ kg} \cdot \text{m}^2.$$

To find the rotational speed $\omega_F$, we use Eq. 11-5 ($\langle\omega_z\rangle = \Delta\theta/\Delta t$). Here the wheel goes through a rotational displacement of $\Delta\theta = 2\pi$ rad in a time period $\Delta t = 2$ min. Thus, we have

$$|\langle\omega_F\rangle| = |\omega_F| = \frac{2\pi \text{ rad}}{(2 \text{ min})(60 \text{ s/min})} = 0.0524 \text{ rad/s},$$

since at constant rotational speed $\langle\omega_F\rangle = \omega_F$. Now we can find the magnitude $L$ of the rotational momentum with Eq. 12-34:

$$|\vec{L}| = I|\vec{\omega}_F| = (1.22 \times 10^9 \text{ kg} \cdot \text{m}^2)(0.0524 \text{ rad/s})$$
$$= 6.39 \times 10^7 \text{ kg} \cdot \text{m}^2/\text{s} \approx 6.4 \times 10^7 \text{ kg} \cdot \text{m}^2/\text{s}. \quad \text{(Answer)}$$

(b) Assume that the fully loaded wheel is rotated from rest to $\omega_F$ in a time period $\Delta t = 5.0$ s. What is the magnitude $|\langle\tau\rangle|$ of the average net external torque acting on it during $\Delta t$?

**SOLUTION** ■ The **Key Idea** here is that the average net external torque is related to the rate of change in the rotational momentum of the loaded wheel by Eq. 12-32 ($\vec{\tau}^{\,net} = d\vec{L}/dt$). The wheel rotates about a fixed axis to reach rotational speed $\omega_F$ in time period $\Delta t$ and the change $\Delta L$ is from zero to the answer for part (a). Thus, we have

$$|\langle\vec{\tau}^{\,net}\rangle| = \left|\frac{\Delta\vec{L}}{\Delta t_1}\right| = \frac{6.39 \times 10^7 \text{ kg} \cdot \text{m}^2/\text{s} - 0}{5.0 \text{ s}}$$
$$\approx 1.3 \times 10^7 \text{ N} \cdot \text{m}. \quad \text{(Answer)}$$

## 12-10 Conservation of Rotational Momentum

So far we have discussed two powerful conservation laws, the conservation of energy and the conservation of translational momentum. Now we meet a third law of this type, involving the conservation of rotational momentum. We start from Eq. 12-32 ($\vec{\tau}^{\,\text{net}} = d\vec{L}/dt$), which is Newton's Second Law in rotational form. If no net external torque acts on the system, this equation becomes $d\vec{L}/dt = 0$, or

$$\vec{L} = \text{a constant} \qquad (\vec{\tau}^{\,\text{net}} = 0). \tag{12-35}$$

This result, called the **law of conservation of rotational momentum,** can also be written as

$$\left\{\begin{matrix}\text{Net rotational momentum}\\ \text{at some initial time } t_1\end{matrix}\right\} = \left\{\begin{matrix}\text{Net rotational momentum}\\ \text{at some later time } t_2\end{matrix}\right\}$$

or

$$\vec{L}_1 = \vec{L}_2 \qquad (\vec{\tau}^{\,\text{net}} = 0). \tag{12-36}$$

Equation 12-35 ($\vec{L} =$ a constant) and Eq. 12-36 ($\vec{L}_1 = \vec{L}_2$) tell us:

> If the net (external) torque acting on a system is zero, the rotational momentum $\vec{L}$ of the system remains constant, no matter what changes take place within the system.

Equations 12-32 ($\vec{\tau}^{\,\text{net}} = d\vec{L}/dt$) and 12-36 ($\vec{L}_1 = \vec{L}_2$) are vector equations. As such, they are equivalent to three component equations corresponding to the conservation of rotational momentum in three mutually perpendicular directions. Depending on the torques acting on a system, the rotational momentum of the system might be conserved in only one or two directions but not in all three:

> If the component of the net *external* torque on a system along a certain axis is zero, then the component of the rotational momentum of the system along that axis cannot change, no matter what changes take place within the system.

We can apply this law to the isolated body in Fig. 12-14, which rotates around the $z$ axis. Suppose that the initially rigid body somehow redistributes its mass relative to that rotation axis, changing its rotational inertia about that axis. Equation 12-35 ($\vec{L} =$ a constant) and Eq. 12-36 ($\vec{L}_1 = \vec{L}_2$) state that the rotational momentum of the body cannot change in the absence of a net external torque. Substituting $\vec{L} = I\vec{\omega}$ (Eq. 12-34) for the rotational momentum along the rotational axis into Eq. 12-36, we write this conservation law as

$$I_1\vec{\omega}_1 = I_2\vec{\omega}_2 \qquad (\vec{\tau}^{\,\text{net}} = 0). \tag{12-37}$$

Here the subscripts refer to the values of the rotational inertia $I$ and rotational speed $\omega$ before and after the redistribution of mass.

Like the other two conservation laws that we have discussed, $\vec{L} =$ a constant and $\vec{L}_1 = \vec{L}_2$ hold beyond the limitations of Newtonian mechanics. They hold for particles whose speeds approach that of light (where the theory of special relativity reigns), and they remain true in the world of subatomic particles (where quantum physics reigns). No exceptions to the law of conservation of rotational momentum have ever been found.

We now discuss four examples involving this law.

1. ***The spinning volunteer.*** Figure 12-16 shows a student seated on a stool that can rotate freely about a vertical axis. The student, who has been set into rotation at a modest initial rotational speed $\omega_1$, holds two dumbbells in his outstretched hands. His rotational momentum vector $\vec{L}$ lies along the vertical rotation axis, pointing upward.

   The instructor now asks the student to pull in his arms. This action reduces his rotational inertia from its initial value $I_1$ to a smaller value $I_2$ because he moves mass closer to the rotation axis. His rate of rotation increases markedly, from $\vec{\omega}_1$ to $\vec{\omega}_2$. The student can then slow down by extending his arms once more.

   No net external torque acts along the vertical axis of the system consisting of the student, stool, and dumbbells. Thus, the rotational momentum of that system about the rotation axis must remain constant. In Fig. 12-16*a*, the student's rotational speed $|\vec{\omega}_1|$ is relatively low and his rotational inertia $I_1$ is relatively high. According to Eq. 12-37, ($I_1\vec{\omega}_1 = I_2\vec{\omega}_2$) his rotational speed in Fig. 12-16*b* must be greater to compensate for the decreased rotational inertia.

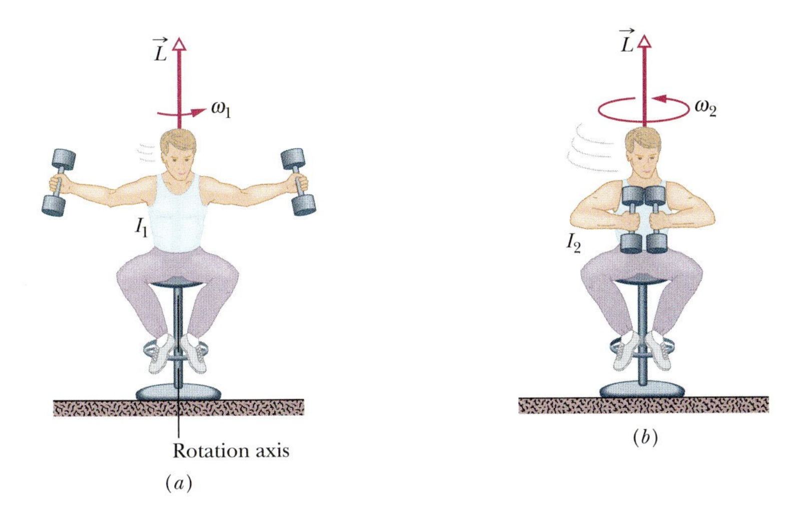


**FIGURE 12-16** ■ (*a*) The student has a relatively large rotational inertia and a relatively small rotational speed. (*b*) By decreasing his rotational inertia, the student automatically increases his rotational speed. The rotational momentum $\vec{L}$ of the rotating system remains unchanged.

2. ***The springboard diver.*** Figure 12-17 shows a diver doing a forward one-and-a-half-somersault dive. As you should expect from our discussion in Chapter 8, her center of mass follows a parabolic path. She leaves the springboard with a definite rotational momentum $\vec{L}$ about an axis through her center of mass, represented by a vector pointing into the plane of Fig. 12-17, perpendicular to the page. When she is in the air, no net external torque acts on her about her center of mass (assuming air drag is negligible). So, her rotational momentum about her center of mass cannot change. By pulling her arms and legs into the *closed pike* position (in the fourth image), she reduces her rotational inertia about the same axis and thus, according to $I_1\vec{\omega}_1 = I_2\vec{\omega}_2$ (Eq. 12-37), increases her rotational speed. Pulling out of the closed pike position (and back into the *open layout position*) at the end of the dive increases her rotational inertia. This slows her rotation rate so she can enter the water with little splash. Even in a more complicated dive involving both twisting and somersaulting, the rotational momentum of the diver must be conserved, in both magnitude *and* direction, throughout the dive.

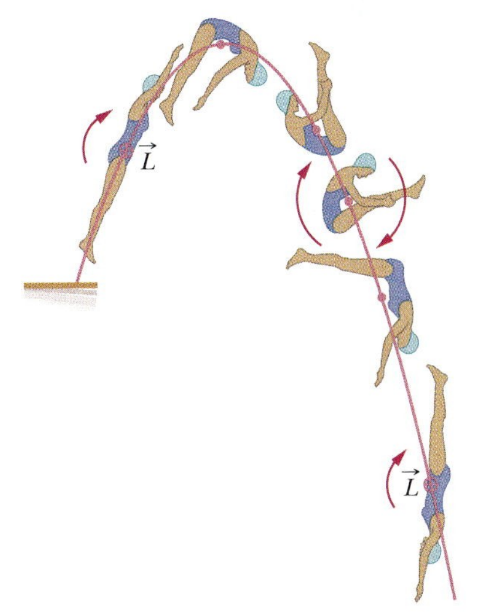


**FIGURE 12-17** ■ A diver rotates about her center of mass as she falls. Since she has no net torque relative to her center of mass, her rotational momentum is constant throughout the dive. Note also that her center of mass (see the dots) follows a parabolic path as she falls.

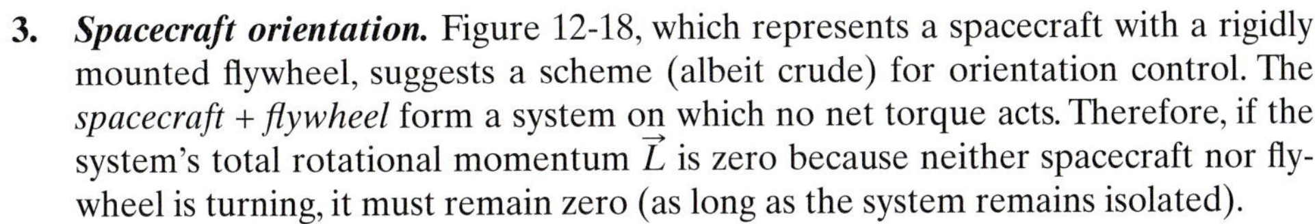

3. ***Spacecraft orientation.*** Figure 12-18, which represents a spacecraft with a rigidly mounted flywheel, suggests a scheme (albeit crude) for orientation control. The *spacecraft + flywheel* form a system on which no net torque acts. Therefore, if the system's total rotational momentum $\vec{L}$ is zero because neither spacecraft nor flywheel is turning, it must remain zero (as long as the system remains isolated).

   To change the orientation of the spacecraft, the flywheel is made to rotate (Fig. 12-18*a*). The spacecraft will start to rotate in the opposite direction to maintain the system's rotational momentum at zero. When the flywheel is then brought to rest, the spacecraft will also stop rotating but will have changed its orientation (Fig. 12-18*b*). Throughout, the rotational momentum of the system *spacecraft + flywheel* never differs from zero.

Interestingly, the spacecraft *Voyager 2*, on its 1986 flyby of the planet Uranus, was set into unwanted rotation by this flywheel effect every time its tape recorder was turned on at high speed. The ground staff at the Jet Propulsion Laboratory had to program the on-board computer to turn on counteracting thruster jets every time the tape recorder was turned on.

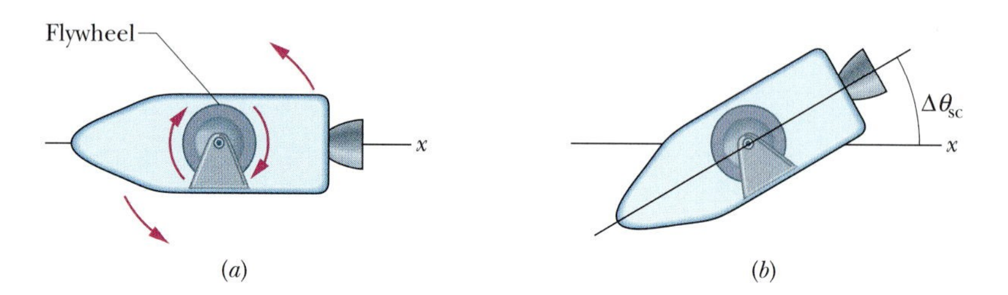


**FIGURE 12-18** ■ (*a*) An idealized spacecraft containing a flywheel. If the flywheel is made to rotate clockwise as shown, the spacecraft itself will rotate counterclockwise because the total rotational momentum must remain zero. (*b*) When the flywheel is braked to a stop, the spacecraft will also stop rotating but will have reoriented its axis by the angle $\Delta\theta_{sc}$.

4. ***The incredible shrinking star.*** When the nuclear fire in the core of a star burns low, the star may eventually begin to collapse, building up pressure in its interior. The collapse may go so far as to reduce the radius of the star from something like that of the Sun to the incredibly small value of a few kilometers. The star then becomes a *neutron star*—its material has been compressed to an extremely dense gas of neutrons.

   During this shrinking process, the star is an isolated system and its rotational momentum $\vec{L}$ cannot change. Because its rotational inertia is greatly reduced, its rotational speed is correspondingly greatly increased, to as much as 600 to 800 revolutions per *second.* For comparison, the Sun, a typical star, rotates at about one revolution per month.

## Summary of Rotational vs. Translational Equations

Table 12-2 supplements Table 11-3 with some of the new equations developed in this chapter. It extends our list of corresponding translational and rotational relations.

**TABLE 12-2**
**More Corresponding Relations for Translational and Rotational Motion[a]**

| Translational | | Rotational | |
|---|---|---|---|
| Force | $\vec{F}$ | Torque | $\vec{\tau} = \vec{r} \times \vec{F}$ |
| Translational momentum | $\vec{p}^{\,sys}$ | Rotational momentum | $\vec{\ell} = \vec{r} \times \vec{p}$ |
| Translational momentum[b] | $\vec{p}^{\,sys} = \Sigma \vec{p}_i$ | Rotational momentum[b] | $\vec{L} = \Sigma \vec{\ell}_i$ |
| Translational momentum[b] | $\vec{p}^{\,sys} = M\vec{v}_{com}$ | Rotational momentum[c] | $\vec{L} = I\vec{\omega}$ |
| Newton's Second Law[b] | $\Sigma \vec{F}^{\,ext} = \dfrac{d\vec{p}^{\,sys}}{dt}$ | Newton's Second Law[b] | $\Sigma \vec{\tau}^{\,ext} = \dfrac{d\vec{L}}{dt}$ |
| Conservation law[d] | $\vec{p}^{\,sys} =$ a constant | Conservation law[d] | $\vec{L} =$ a constant |

[a] See also Table 11-3.
[b] For systems of particles, including rigid bodies.
[c] For a rigid body about a fixed axis, with $L$ being the component along that axis.
[d] For a closed, isolated system ($\vec{F}^{\,net} = 0$, $\vec{\tau}^{\,net} = 0$).

**READING EXERCISE 12-5:** A rhinoceros beetle rides the rim of a small disk that rotates like a merry-go-round. If the beetle crawls toward the center of the disk, do the following (each relative to the central axis) increase, decrease, or remain the same: (a) the rotational inertia of the beetle–disk system, (b) the rotational momentum of the system, and (c) the rotational speed of the beetle and disk? ■

## TOUCHSTONE EXAMPLE 12-6: Student with a Wheel

Figure 12-19*a* shows a student sitting on a stool that can rotate freely about a vertical axis. The student, initially at rest, is holding a bicycle wheel whose rim is loaded with lead and whose rotational inertia $I_{\rm wh}$ about its central axis is 1.2 kg · m². The wheel is rotating at a rotational speed $\omega_{\rm wh}$ of 3.9 rev/s; as seen from overhead, the rotation is counterclockwise. The axis of the wheel is vertical, and the rotational momentum $\vec{L}_{\rm wh}$ of the wheel points vertically upward. The student now inverts the wheel (Fig. 12-19*b*) so that, as seen from overhead, it is rotating clockwise. Its rotational momentum is then $-\vec{L}_{\rm wh}$. The inversion results in the student, the stool, and the wheel's center rotating together as a composite rigid body about the stool's rotation axis, with rotational inertia $I_{\rm b} = 6.8\ {\rm kg\cdot m^2}$. (The fact that the wheel is also rotating about its center does not affect the mass distribution of this composite body; thus, $I_{\rm b}$ has the same value whether or not the wheel rotates.) With what rotational speed $\omega_{\rm b}$ and in what direction does the composite body rotate after the inversion of the wheel?

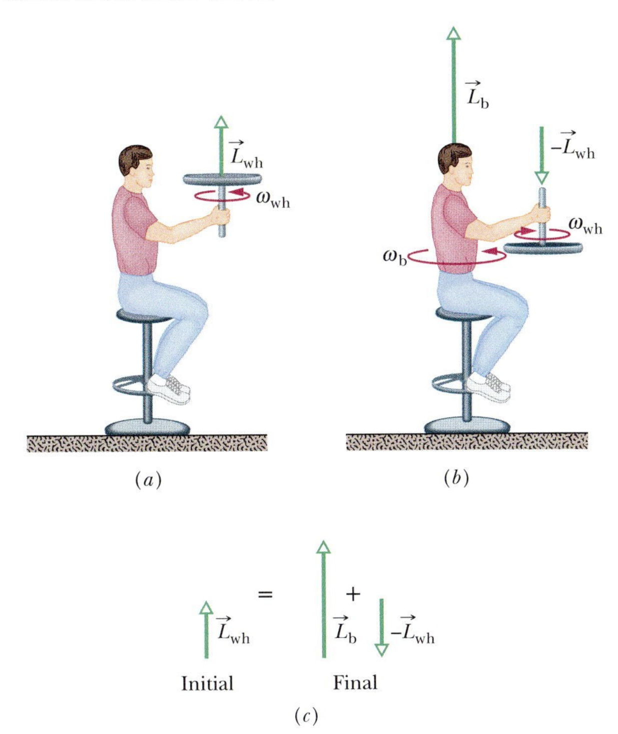


**FIGURE 12-19** ■ (*a*) A student holds a bicycle wheel rotating around the vertical. (*b*) The student inverts the wheel, setting himself into rotation. (*c*) The net angular momentum of the system must remain the same in spite of the inversion.

**SOLUTION** ■ The **Key Ideas** here are these:

1. The rotational speed $\omega_{\rm b}$ we seek is related to the final rotational momentum $\vec{L}_{\rm b}$ of the composite body about the stool's rotation axis by Eq. 12-34 ($\vec{L} = I\vec{\omega}$).
2. The initial rotational speed $\omega_{\rm wh}$ of the wheel is related to the rotational momentum $\vec{L}_{\rm wh}$ of the wheel's rotation about its center by the same equation.
3. The vector addition of $\vec{L}_{\rm b}$ and $\vec{L}_{\rm wh}$ gives the total rotational momentum $\vec{L}^{\rm tot}$ of the system of student, stool, and wheel.
4. As the wheel is inverted, no net *external* torque acts on that system to change $\vec{L}^{\rm tot}$ about any vertical axis. (Torques due to forces between the student and the wheel as the student inverts the wheel are *internal* to the system.) So, the system's total rotational momentum is conserved about any vertical axis.

The conservation of $\vec{L}^{\rm tot}$ is represented with vectors in Fig. 12-19*c*. We can also write it in terms of components along a vertical axis as

$$L_{{\rm b}\,y}(t_2) + L_{{\rm wh}\,y}(t_2) = L_{{\rm b}\,y}(t_1) + L_{{\rm wh}\,y}(t_1), \tag{12-38}$$

where $t_1$ and $t_2$ refer to the initial state (before inversion of the wheel) and the final state (after inversion). Because inversion of the wheel inverted the wheel's rotational momentum vector, we substitute $-L_{{\rm wh}\,y}(t_1)$ for $L_{{\rm wh}\,y}(t_2)$. Then, if we set $L_{{\rm b}\,y}(t_1) = 0$ (because the student, the stool, and the wheel's center were initially at rest), Eq. 12-38 yields

$$L_{{\rm b}\,y}(t_2) = 2L_{{\rm wh}\,y}(t_1).$$

We next substitute $I_{\rm b}\omega_{{\rm b}\,y}$ for $L_{{\rm b}\,y}$ and $I_{\rm wh}\omega_{{\rm wh}\,y}$ for $L_{{\rm wh}\,y}$ and solve for $\omega_{\rm b}$, finding

$$\omega_{{\rm b}\,y}\hat{\rm j} = \frac{2I_{\rm wh}}{I_{\rm b}}\omega_{{\rm wh}\,y}\hat{\rm j}$$

$$= \frac{(2)(1.2\ {\rm kg\cdot m^2})(3.9\ {\rm rev/s})}{6.8\ {\rm kg\cdot m^2}}\hat{\rm j} = (1.4\ {\rm rev/s})\,\hat{\rm j}.$$

(Answer)

The fact that this final rotational velocity points upward tells us that the student rotates counterclockwise about the stool axis as seen from overhead. If the student wishes to stop rotating, he has only to invert the wheel once more.

## TOUCHSTONE EXAMPLE 12-7: Quadruple Somersault

During a jump to his partner, an aerialist is to make a quadruple somersault lasting a time $t = 1.87$ s. For the first and last quarter-revolution, he is in the extended orientation shown in Fig. 12-20, with rotational inertia $I_1 = 19.9 \text{ kg}\cdot\text{m}^2$ around his center of mass (the dot). During the rest of the flight he is in a tight tuck, with rotational inertia $I_2 = 3.93 \text{ kg}\cdot\text{m}^2$. What must be his rotational speed $\omega_2$ around his center of mass during the tuck?

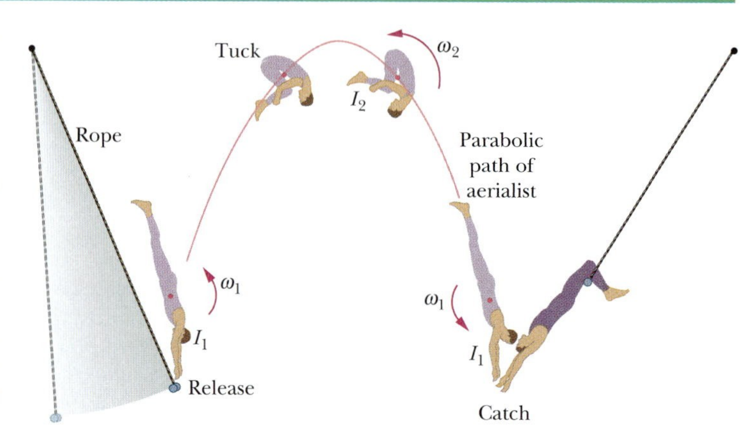


**FIGURE 12-20** ■ An aerialist performing a multiple somersault.

**SOLUTION** ■ Obviously he must turn fast enough to complete the 4.0 rev required for a quadruple somersault in the given 1.87 s. To do so, he increases his rotational speed to $\omega_2$ by tucking. We can relate $\omega_2$ to his initial rotational speed $\omega_1$ with this **Key Idea:** His rotational momentum about his center of mass is conserved throughout the free flight because there is no net external torque about his center of mass to change it. From Eq. 12-37, we can write the conservation of rotational momentum ($\vec{L}_1 = \vec{L}_2$) as

$$I_1\vec{\omega}_1 = I_2\vec{\omega}_2,$$

or

$$\vec{\omega}_1 = \frac{I_2}{I_1}\vec{\omega}_2. \tag{12-39}$$

A second **Key Idea** is that these rotational speeds are related to the angles through which he must rotate and the time available to do so. At the start and at the end, he must rotate in the extended orientation for a total angle of $\Delta\theta_1 = 0.500$ rev (two quarter-turns) in a time we shall call $\Delta t_1$. In the tuck, he must rotate through an angle of $\Delta\theta_2 = 3.50$ rev in a time $\Delta t_2$. From Eq. 11-5 ($\langle\omega\rangle = \Delta\theta/\Delta t$), we can write

$$\Delta t_1 = \frac{\Delta\theta_1}{\omega_1} \quad \text{and} \quad \Delta t_2 = \frac{\Delta\theta_2}{\omega_2}.$$

Thus, his total flight time is

$$\Delta t = \Delta t_1 + \Delta t_2 = \frac{\Delta\theta_1}{\omega_1} + \frac{\Delta\theta_2}{\omega_2},$$

which we know to be 1.87 s. Now substituting from Eq. 12-39 yields for $\omega_1$

$$\Delta t = \frac{(\Delta\theta_1)I_1}{\omega_2 I_2} + \frac{\Delta\theta_2}{\omega_2} = \frac{1}{\omega_2}\left(\Delta\theta_1\frac{I_1}{I_2} + \Delta\theta_2\right).$$

Inserting the known data, we obtain

$$1.87 \text{ s} = \frac{1}{\omega_2}\left((0.500 \text{ rev})\frac{19.9 \text{ kg}\cdot\text{m}^2}{3.93 \text{ kg}\cdot\text{m}^2} + 3.50 \text{ rev}\right),$$

which gives us

$$\omega_2 = 3.23 \text{ rev/s}. \qquad \text{(Answer)}$$

This rotational speed is so fast that the aerialist cannot clearly see his surroundings or fine-tune his rotation by adjusting his tuck. The possibility of an aerialist making a four-and-a-half-somersault flight, which would require a greater value of $\omega_2$ and thus a smaller $I_2$ via a tighter tuck, seems very small.

## TOUCHSTONE EXAMPLE 12-8: Turnstile Takes a Hit

(This touchstone example is long and challenging, but it is helpful because it pulls together many ideas of Chapters 11 and 12.) In the overhead view of Fig. 12-21, four thin, uniform rods, each of mass $M$ and length $d = 0.50$ m, are rigidly connected to a vertical axle to form a turnstile. The turnstile rotates clockwise about the axle, which is attached to a floor, with initial rotational velocity $\vec{\omega}_1 = (-2.0 \text{ rad/s})\hat{j}$. A mud ball of mass $m = \frac{1}{3}M$ and initial speed $v_1 = 12$ m/s is thrown along the path shown and sticks to the end of one rod. What is the final rotational velocity $\vec{\omega}_2$ of the ball–turnstile system?

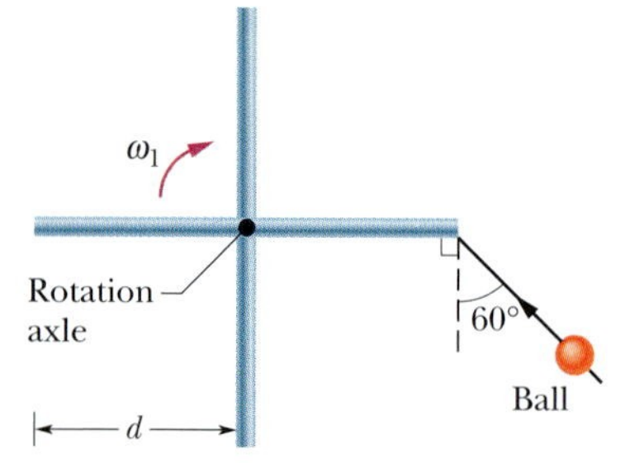


**FIGURE 12-21** ■ An overhead view of four rigidly connected rods rotating freely around a central axle, and the path a mud ball takes to stick onto one of the rods.

**SOLUTION** ■ A **Key Idea** here can be stated in a question-and-answer format. The question is this: Does the system have a quantity that is conserved during the collision and that involves rotational velocity, so that we can solve for $\vec{\omega}_2$? To answer, let us check the conservation possibilities:

1. The total kinetic energy $K$ is *not* conserved, because the collision between ball and rod is completely inelastic (the ball sticks). So, some energy must be transferred from kinetic energy to other types of energy (such as thermal energy). For the same reason, total mechanical energy is not conserved.
2. The total translational momentum $\vec{P}$ is also *not* conserved, because during the collision an external force acts on the turnstile at the attachment of the axle to the floor. (This is the force that keeps the turnstile from moving across the floor when it is hit by the mud ball.)
3. The total rotational momentum $\vec{L} = L_y\hat{j}$ of the system about the axle *is* conserved because there is no net external torque to change $\vec{L}$. (The forces in the collision produce only internal torques; the external force on the turnstile acts at the axle, has zero moment arm, and thus does not produce an external torque.)

We can write the conservation of the system's total rotational momentum ($\vec{L}_2 = \vec{L}_1$) about the axle as

$$L_{\text{ts}\,y}(t_2) + L_{\text{ball}\,y}(t_2) = L_{\text{ts}\,y}(t_1) + L_{\text{ball}\,y}(t_1), \qquad (12\text{-}40)$$

where ts stands for turnstile. The final rotational velocity $\vec{\omega}_2$ is contained in the terms $L_{\text{ts}\,y}(t_2)$ and $L_{\text{ball}\,y}(t_2)$ because those final rotational momenta depend on how fast the turnstile and ball are rotating. To find $\vec{\omega}_2$, we consider first the turnstile and then the ball, and then we return to Eq. 12-40.

*Turnstile*: The **Key Idea** here is that, because the turnstile is a rotating rigid object, Eq. 12-34 ($\vec{L} = I\vec{\omega}$) gives its rotational momentum. Thus we can write its final and initial rotational momenta about the axle as

$$L_{\text{ts}\,y}(t_2) = I_{\text{ts}}\omega_{2\,y} \quad \text{and} \quad L_{\text{ts}\,y}(t_1) = I_{\text{ts}}\omega_{1\,y}. \qquad (12\text{-}41)$$

Because the turnstile consists of four rods, each rotating around an end, the rotational inertia $I_{\text{ts}}$ of the turnstile is four times the rotational inertia $I_{\text{rod}}$ of each rod about its end. From Table 11-2(*e*), we know that the rotational inertia $I_{\text{com}}$ of a rod about its center is $\frac{1}{12}Md^2$, where $M$ is its mass and $d$ is its length. To get $I_{\text{rod}}$, we use the parallel-axis theorem of Eq. 11-28 ($I = I_{\text{com}} + Mh^2$). Here perpendicular distance $h$ is $d/2$. Thus, we find

$$I_{\text{rod}} = \tfrac{1}{12}Md^2 + M\left(\frac{d}{2}\right)^2 = \tfrac{1}{3}Md^2.$$

With four rods in the turnstile, we then have

$$I_{\text{ts}} = \tfrac{4}{3}Md^2. \qquad (12\text{-}42)$$

*Ball*: Before the collision, the ball is like a particle moving along a straight line, as in Fig. 12-13. So, to find the ball's initial rotational momentum $L_{\text{ball}\,y}(t_1)$ about the axle, we can use any of Eqs. 12-25 through 12-28, but Eq. 12-27 ($\ell = rmv_\perp$) is easiest. Here $\ell$ is $L_{\text{ball}\,y}(t_1)$. Just before the ball hits, its radial distance $r$ from the axle is $d$, and the component $v_\perp$ of the ball's velocity perpendicular to $r$ is $v_1 \cos 60°$.

To give a sign to this rotational momentum, we mentally draw a position vector from the turnstile's axle to the ball. As the ball approaches the turnstile, this position vector rotates counterclockwise about the axle, so the ball's rotational momentum is a positive quantity. We can now rewrite $\ell = rmv_\perp$ as

$$L_{\text{ball}\,y}(t_1) = mdv_1 \cos 60°. \qquad (12\text{-}43)$$

After the collision, the ball is like a particle rotating in a circle of radius $d$. So, from Eq. 11-23 ($I = \Sigma m_i r_i^{\,2}$), we have $I_{\text{ball}} = md^2$ about the axle. Then from Eq. 12-34 ($\vec{L} = I\vec{\omega}$), we can write the final rotational momentum of the ball about the axle as

$$L_{\text{ball}\,y}(t_2) = I_{\text{ball}}\omega_{2y} = md^2\omega_{2y}. \qquad (12\text{-}44)$$

*Return to Eq. 12-40:* Substituting from Eqs. 12-41 through 12-44 into Eq. 12-40, we have

$$\tfrac{4}{3}Md^2\omega_{2\,y} + md^2\omega_{2\,y} = \tfrac{4}{3}Md^2\omega_{1\,y} + mdv_1\cos 60°.$$

Substituting $M = 3m$ and solving for $\omega_{2\,y}$, we find

$$\begin{aligned}\omega_{2\,y} &= \frac{1}{5d}(4d\omega_{1\,y} + v_1\cos 60°)\\ &= \frac{1}{5(0.50\text{ m})}[4(0.50\text{ m})(-2.0\text{ rad/s}) + (12\text{ m/s})(\cos 60°)]\\ &= 0.80\text{ rad/s}. \qquad \text{(Answer)}\end{aligned}$$

Thus, the turnstile is now turning counterclockwise.

# Problems

## SEC. 12-2 ■ COMBINING TRANSLATIONS WITH SIMPLE ROTATIONS

***Unless otherwise noted, rolling occurs without slipping.***

**1. An Automobile Traveling** An automobile traveling 80.0 km/h has tires of 75.0 cm diameter. (a) What is the rotational speed of the tires about their axles? (b) If the car is brought to a stop uniformly in 30.0 complete turns of the tires (without skidding), what is the magnitude of the rotational acceleration of the wheels? (c) How far does the car move during the braking?

**2. Car's Tire** Consider a 66-cm-diameter tire on a car traveling at 80 km/h on a level road in the positive direction of an $x$ axis. Relative to a woman in the car, what are (a) the translational velocity $\vec{v}$ and (b) the magnitude $a$ of the translational acceleration of the center of the wheel? What are (c) $\vec{v}$ and (d) $a$ for a point at the top of the tire? What are (e) $\vec{v}$ and (f) $a$ for a point at the bottom of the tire?

Now repeat the questions relative to a hitchhiker sitting near the road: What are (g) $\vec{v}$ at the wheel's center, (h) $a$ at the wheel's center, (i) $\vec{v}$ at the tire top, (j) $a$ at the tire top, (k) $\vec{v}$ at the tire bottom, and (l) $a$ at the tire bottom?

**3. A Hoop Rolls** A 140 kg hoop rolls along a horizontal floor so that its center of mass has a speed of 0.150 m/s. How much work must be done on the hoop to stop it?

**4. Thin-Walled Pipe** A thin-walled pipe rolls along the floor. What is the ratio of its translational kinetic energy to its rotational kinetic energy about an axis parallel to its length and through its center of mass?

**5. Car Has Four Wheels** A 1000 kg car has four 10 kg wheels. When the car is moving, what fraction of the total kinetic energy of the car is due to rotation of the wheels about their axles? Assume that the wheels have the same rotational inertia as uniform disks of the same mass and size. Why do you not need the radius of the wheels?

**6. A Body of Radius *R*** A body of radius $R$ and mass $m$ is rolling smoothly with speed $v$ on a horizontal surface. It then rolls up a hill to a maximum height $h$. (a) If $h = 3v^2/4g$, what is the body's rotational inertia about the rotational axis through its center of mass? (b) What might the body be?

**7. A Uniform Solid Sphere** A uniform solid sphere rolls down an incline. (a) What must be the incline angle if the translational acceleration of the center of the sphere is to have a magnitude of $0.10g$? (b) If a frictionless block were to slide down the incline at that angle, would its acceleration magnitude be more than, less than, or equal to $0.10g$? Why?

**8. A Hollow Sphere** A hollow sphere of radius 0.15 m, with rotational inertia $I = 0.040\ \text{kg}\cdot\text{m}^2$ about a line through its center of mass, rolls without slipping up a surface inclined at 30° to the horizontal. At a certain initial position, the sphere's total kinetic energy is 20 J. (a) How much of this initial kinetic energy is rotational? (b) What is the speed of the center of mass of the sphere at the initial position? What are (c) the total kinetic energy of the sphere and (d) the speed of its center of mass after it has moved 1.0 m up along the incline from its initial position?

**9. Yo-Yo's Inertia** A yo-yo has a rotational inertia of $950\ \text{g}\cdot\text{cm}^2$ and a mass of 120 g. Its axle radius is 3.2 mm, and its string is 120 cm long. The yo-yo rolls from rest down to the end of the string. (a) What is the magnitude of its translational acceleration? (b) How long does it take to reach the end of the string? As it reaches the end of the string, what are its (c) translational speed, (d) translational kinetic energy, (e) rotational kinetic energy, and (f) rotational speed?

**10. Instead of Rolling** Suppose that the yo-yo in Problem 9, instead of rolling from rest, is thrown so that its initial speed down the string is 1.3 m/s. (a) How long does the yo-yo take to reach the end of the string? As it reaches the end of the string, what are its (b) total kinetic energy, (c) translational speed, (d) translational kinetic energy, (e) rotational speed, and (f) rotational kinetic energy?

## SEC. 12-4 ■ THE VECTOR OR CROSS PRODUCT

**11. Area of Triangle** Show that the area of the triangle contained between $\vec{a}$ and $\vec{b}$ and the solid line connecting their tips in Fig. 12-22 is $\frac{1}{2}|\vec{a} \times \vec{b}|$.

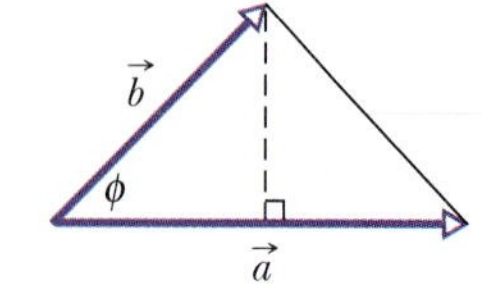


FIGURE 12-22 ■ Problem 11.

**12. The Product** In the product $\vec{F} = q\vec{v} \times \vec{B}$, take $q = 2$,

$$\vec{v} = 2.0\,\hat{\text{i}} + 4.0\,\hat{\text{j}} + 6.0\,\hat{\text{k}}$$

and

$$\vec{F} = 4.0\,\hat{\text{i}} - 20\,\hat{\text{j}} + 12\,\hat{\text{k}}.$$

What then is $\vec{B}$ in unit-vector notation if $B_x = B_y$?

**13. Show That** (a) Show that $\vec{a}\cdot(\vec{b} \times \vec{a})$ is zero for all vectors $\vec{a}$ and $\vec{b}$. (b) What is the magnitude of $\vec{a} \times (\vec{b} \times \vec{a})$ if there is an angle $\phi$ between the directions of $\vec{a}$ and $\vec{b}$?

**14. For the Following** For the following three vectors, what is $3\vec{C}\cdot(2\vec{A} \times \vec{B})$?

$$\vec{A} = 2.00\hat{\text{i}} + 3.00\hat{\text{j}} - 4.00\hat{\text{k}}$$

$$\vec{B} = -3.00\hat{\text{i}} + 4.00\hat{\text{j}} + 2.00\hat{\text{k}}$$

$$\vec{C} = 7.00\hat{\text{i}} - 8.00\hat{\text{j}}$$

## SEC. 12-5 ■ TORQUE AS A VECTOR PRODUCT

**15. In a Given Plane** Show that, if $\vec{r}$ and $\vec{F}$ lie in a given plane, the torque $\vec{\tau} = \vec{r} \times \vec{F}$ has no component in that plane.

**16. A Plum** What are the magnitude and direction of the torque about the origin on a plum located at coordinates (−2.0, 0.0, 4.0) m due to force $\vec{F}$ whose only component is (a) $F_x = 6.0$ N, (b) $F_x = -6.0$ N, (c) $F_z = 6.0$ N, and (d) $F_z = -6.0$ N?

**17. Particle Located at** What are the magnitude and direction of the torque about the origin on a particle located at coordinates (0.0, −4.0, 3.0) m due to (a) force $\vec{F}_A$ with components $F_{Ax} = 2.0$ N and $F_{Ay} = F_{Az} = 0$, and (b) force $\vec{F}_B$ with components $F_{Bx} = 0$, $F_{By} = 2.0$ N, and $F_{Bz} = 4.0$ N?

**18. Pebble** Force $\vec{F} = (2.0\ \text{N})\hat{\text{i}} - (3.0\ \text{N})\hat{\text{k}}$ acts on a pebble with position vector $\vec{r} = (0.50\ \text{m})\hat{\text{j}} - (2.0\ \text{m})\hat{\text{k}}$, relative to the origin. What is the resulting torque acting on the pebble about (a) the origin and (b) a point with coordinates (2.0, 0.0, −3.0) m?

**19. Particle at Origin** Force $\vec{F} = (-8.0\ \text{N})\hat{\text{i}} + (6.0\ \text{N})\hat{\text{j}}$ acts on a particle with position vector $\vec{r} = (3.0\ \text{m})\hat{\text{i}} + (4.0\ \text{m})\hat{\text{j}}$. What are (a) the torque on the particle about the origin and (b) the angle between the directions of $\vec{r}$ and $\vec{F}$?

**20. Jar of Jalapeños** What is the torque about the origin on a jar of jalapeño peppers located at coordinates (3.0 m, −2.0 m, 4.0 m) due to (a) force $\vec{F}_A = (3.0\ \text{N})\hat{\text{i}} - (4.0\ \text{N})\hat{\text{j}} + (5.0\ \text{N})\hat{\text{k}}$, (b) force $\vec{F}_B = (-3.0\ \text{N})\hat{\text{i}} - (4.0\ \text{N})\hat{\text{j}} - (5.0\ \text{N})\hat{\text{k}}$, and (c) the vector sum of $\vec{F}_A$ and $\vec{F}_B$? (d) Repeat part (c) about a point with coordinates (3.0 m, 2.0 m, 4.0 m) instead of about the origin.

## SEC. 12-6 ■ ROTATIONAL FORM OF NEWTON'S SECOND LAW

**21. A Particle with Velocity** A 3.0 kg particle with velocity $\vec{v} = (5.0\ \text{m/s})\hat{\text{i}} - (6.0\ \text{m/s})\hat{\text{j}}$ is at $x = 3.0$ m, $y = 8.0$ m. It is pulled by a 7.0 N force in the negative $x$ direction. (a) What is the rotational momentum of the particle about the origin? (b) What torque about the origin acts on the particle? (c) At what rate is the rotational momentum of the particle changing with time?

**22. Acted on by Two Torques** A particle is acted on by two torques about the origin: $\vec{\tau}_1$ has a magnitude of 2.0 N · m and is directed in the positive direction of the $x$ axis, and $\vec{\tau}_2$ has a magnitude of

4.0 N · m and is directed in the negative direction of the $y$ axis. What are the magnitude and direction of $d\vec{\ell}/dt$, where $\vec{\ell}$ is the rotational momentum of the particle about the origin?

**23. Torque About the Origin** What torque about the origin acts on a particle moving in the $xy$ plane, clockwise about the origin, if the particle has the following magnitudes of rotational momentum about the origin:

**(a)** 4.0 kg · m²/s,
**(b)** $(4.0 \frac{1}{s^2})t^2$ kg · m²/s,
**(c)** $(4.0 \frac{1}{s^{1/2}})\sqrt{t}$ kg · m²/s,
**(d)** $(4.0\ s^2)/t^2$ kg · m²/s?

**24. At Time $t$** At time $t = 0$, a 2.0 kg particle has position vector $\vec{r} = (4.0\text{ m})\hat{i} - (2.0\text{ m})\hat{j}$ relative to the origin. Its velocity just then is given by $\vec{v} = (-6.0\text{ m/s}^3)t^2\ \hat{i}$. About the origin and for $t > 0$, what are (a) the particle's rotational momentum and (b) the torque acting on the particle? (c) Repeat (a) and (b) about a point with coordinates (−2.0, −3.0, 0.0) m instead of about the origin.

## Sec. 12-7 ■ Rotational Momentum

**25. Two Objects** Two objects are moving as shown in Fig. 12-23. What is their total rotational momentum about point $O$?

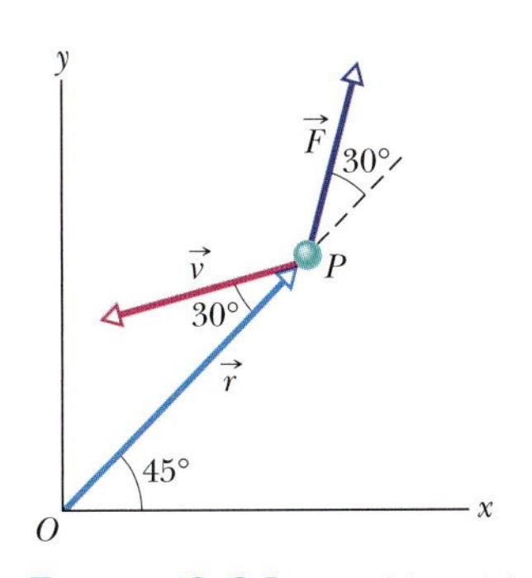


FIGURE 12-23 ■ Problem 25.

**26. A Particle $P$** In Fig. 12-24, a particle $P$ with mass 2.0 kg has position vector $\vec{r}$ of magnitude 3.0 m and velocity $\vec{v}$ of magnitude 4.0 m/s. A force $\vec{F}$ of magnitude 2.0 N acts on the particle. All three vectors lie in the $xy$ plane oriented as shown. About the origin, what are (a) the rotational momentum of the particle and (b) the torque acting on the particle?



FIGURE 12-24 ■ Problem 26.

**27. At a Certain Time** At a certain time, a 0.25 kg object has a position vector $\vec{r} = (2.0\text{ m})\hat{i} + (-2.0\text{ m})\hat{k}$ in meters. At that instant, its velocity in meters per second is $\vec{v} = (-5.0\text{ m/s})\hat{i} + (5.0\text{ m/s})\hat{j}$ and the force in newtons acting on it is $\vec{F} = (4.0\text{ N})\hat{j}$. (a) What is the rotational momentum of the object about the origin? (b) What torque acts on it?

**28. Particle-Like Object** A 2.0 kg particle-like object moves in a plane with velocity components $v_x = 30$ m/s and $v_y = 60$ m/s as it passes through the point with $(x, y)$ coordinates of (3.0, −4.0) m. Just then, what is its rotational momentum relative to (a) the origin and (b) the point (−2.0, −2.0) m?

**29. Two Particles of Mass $m$** Two particles, each of mass $m$ and speed $v$, travel in opposite directions along parallel lines separated by a distance $d$. (a) In terms of $m$, $v$, and $d$, find an expression for the magnitude $L$ of the rotational momentum of the two-particle system around a point midway between the two lines. (b) Does the expression change if the point about which $L$ is calculated is not midway between the lines? (c) Now reverse the direction of travel for one of the particles and repeat (a) and (b).

**30. At the Instant** A 4.0 kg particle moves in an $xy$ plane. At the instant when the particle's position and velocity are $\vec{r} = (2.0\text{ m})\hat{i} + (4.0\text{ m})\hat{j}$ and $\vec{v} = (-4.0\text{ m/s})\hat{j}$, the force on the particle is $\vec{F} = (-3.0\text{ N})\hat{i}$. At this instant, determine (a) the particle's rotational momentum about the origin, (b) the particle's rotational momentum about the point $x = 0$, $y = 4.0$ m, (c) the torque acting on the particle about the origin, and (d) the torque acting on the particle about the point $x = 0.0$ m, $y = 4.0$ m.

## Sec. 12-9 ■ The Rotational Momentum of a Rigid Body Rotating About a Fixed Axis

**31. Flywheel** The rotational momentum of a flywheel having a rotational inertia of 0.140 kg · m² about its central axis decreases from 3.00 to 0.800 kg · m²/s in 1.50 s. (a) What is the magnitude of the average torque acting on the flywheel about its central axis during this period? (b) Assuming a constant rotational acceleration, through what angle does the flywheel turn? (c) How much work is done on the wheel? (d) What is the average power of the flywheel?

**32. Sanding Disk** A sanding disk with rotational inertia $1.2 \times 10^{-3}$ kg · m² is attached to an electric drill whose motor delivers a torque of 16 N · m. Find (a) the rotational momentum of the disk about its central axis and (b) the rotational speed of the disk 33 ms after the motor is turned on.

**33. $d$ Apart** Three particles, each of mass $m$, are fastened to each other and to a rotation axis at $O$ by three massless strings, each with length $d$ as shown in Fig. 12-25. The combination rotates around the rotational axis with rotational velocity $\omega$ in such a way that the particles remain in a straight line. In terms of $m$, $d$, and $\omega$, and relative to point $O$, what are (a) the rotational inertia of the combination, (b) the rotational momentum of the middle particle, and (c) the total rotational momentum of the three particles?

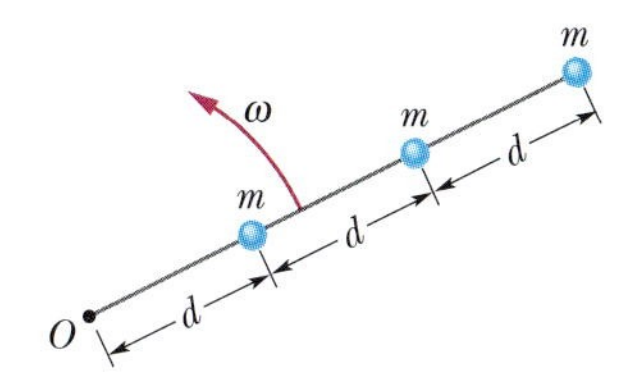


FIGURE 12-25 ■ Problem 33.

**34. Impulsive Force** An impulsive force $\vec{F}(t) = F_x(t)\hat{i}$ acts for a short time $\Delta t$ on a rotating rigid body constrained to rotate about the $z$ axis with rotational inertia $I$. Show that

$$\left(\int \tau_z\, dt\right)\hat{k} = (|\langle\vec{F}\rangle| R\, \Delta t)\hat{k} = I(\omega_{2\,z} - \omega_{1\,z})\hat{k},$$

where $\tau_z\hat{k}$ is the torque due to the force, $R$ is the moment arm of the force, $\langle\vec{F}\rangle$ is the average value of the force during the time it acts on the body, and $\omega_{1\,z}\hat{k}$ and $\omega_{2\,z}\hat{k}$ are the rotational velocities of the body just before and just after the force acts. (The quantity $(\int \tau_z\, dt)\hat{k} = (|\langle\vec{F}\rangle| R\, \Delta t)\hat{k}$ is called the rotational *impulse,* in analogy with $\langle\vec{F}\rangle\,\Delta t$, the translational impulse.)

**35. Two Cylinders** Two cylinders having radii $R_A$ and $R_B$ and rotational inertias $I_A$ and $I_B$ about their central axes are supported by axles perpendicular to the plane of Fig. 12-26. The large cylinder is initially rotating clockwise with rotational velocity $\vec{\omega}_1$.

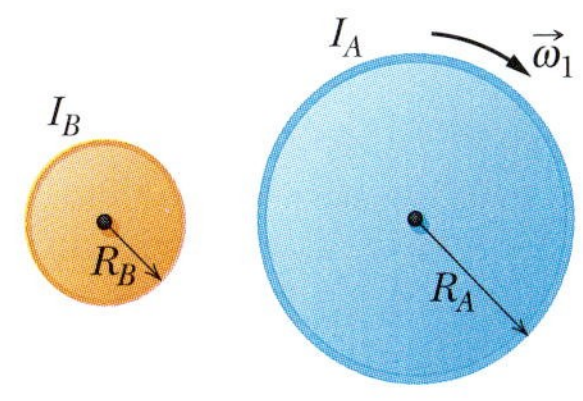


FIGURE 12-26 ■ Problem 35.

The small cylinder is moved to the right until it touches the large cylinder and is caused to rotate by the frictional force between the two. Eventually, slipping ceases, and the two cylinders rotate at constant rates in opposite directions. Find the final rotational velocity $\vec{\omega}_2$ of the small cylinder in terms of $I_A$, $I_B$, $R_A$, $R_B$, and $\vec{\omega}_1$. (*Hint*: Neither rotational momentum nor kinetic energy is conserved. Apply the rotational impulse equation of Problem 34.)

**36. Rigid Structure** Figure 12-27 shows a rigid structure consisting of a circular hoop of radius $R$ and mass $m$, and a square made of four thin bars, each of length $R$ and mass $m$. The rigid structure rotates at a constant speed about a vertical axis, with a period of rotation of 2.5 s. Assuming $R = 0.50$ $m$ and $m = 2.0$ kg, calculate (a) the structure's rotational inertia about the axis of rotation and (b) its rotational momentum about that axis.

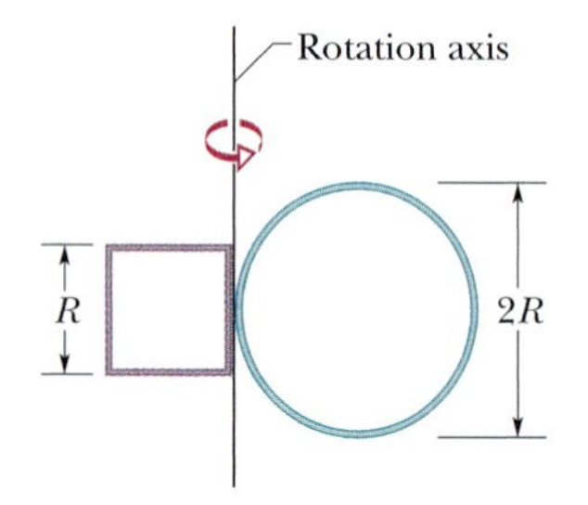


FIGURE 12-27 ■ Problem 36.

## SEC. 12-10 ■ CONSERVATION OF ROTATIONAL MOMENTUM

**37. A Man Stands on a Platform** A man stands on a platform that is rotating (without friction) with a rotational speed of 1.2 rev/s; his arms are outstretched and he holds a brick in each hand. The rotational inertia of the system consisting of the man, bricks, and platform about the central axis is 6.0 kg · m$^2$. If by moving the bricks the man decreases the rotational inertia of the system to 2.0 kg · m$^2$, (a) what is the resulting rotational speed of the platform and (b) what is the ratio of the new kinetic energy of the system to the original kinetic energy? (c) What provided the added kinetic energy?

**38. Rotor** The rotor of an electric motor has rotational inertia $I_m = 2.0 \times 10^{-3}$ kg · m$^2$ about its central axis. The motor is used to change the orientation of the space probe in which it is mounted. The motor axis is mounted parallel to the axis of the probe, which has rotational inertia $I_p = 12$ kg · m$^2$ about its axis. Calculate the number of revolutions of the rotor required to turn the probe through 30° about its axis.

**39. Wheel is Rotating** A wheel is rotating freely at rotational speed 800 rev/min on a shaft whose rotational inertia is negligible. A second wheel, initially at rest and with twice the rotational inertia of the first, is suddenly coupled to the same shaft. (a) What is the rotational speed of the resultant combination of the shaft and two wheels? (b) What fraction of the original rotational kinetic energy is lost?

**40. Two Disks** Two disks are mounted on low-friction bearings on the same axle and can be brought together so that they couple and rotate as one unit. (a) The first disk, with rotational inertia 3.3 kg · m$^2$ about its central axis, is set spinning at 450 rev/min. The second disk, with rotational inertia 6.6 kg · m$^2$ about its central axis, is set spinning at 900 rev/min in the same direction as the first. They then couple together. What is their rotational speed after coupling? (b) If instead the second disk is set spinning at 900 rev/min in the direction opposite the first disk's rotation, what is their rotational speed and direction of rotation after coupling?

**41. Playground** In a playground, there is a small merry-go-round of radius 1.20 m and mass 180 kg. Its radius of gyration (see Problem 43 of Chapter 11) is 91.0 cm. A child of mass 44.0 kg runs at a speed of 3.00 m/s along a path that is tangent to the rim of the initially stationary merry-go-round and then jumps on. Neglect friction between the bearings and the shaft of the merry-go-round. Calculate (a) the rotational inertia of the merry-go-round about its axis of rotation, (b) the magnitude of the rotational momentum of the running child about the axis of rotation of the merry-go-round, and (c) the rotational speed of the merry-go-round and child after the child has jumped on.

**42. Collapsing Spinning Star** The rotational inertia of a collapsing spinning star changes to $\frac{1}{3}$ its initial value. What is the ratio of the new rotational kinetic energy to the initial rotational kinetic energy?

**43. Track on a Wheel** A track is mounted on a large wheel that is free to turn with negligible friction about a vertical axis (Fig. 12-28). A toy train of mass $m$ is placed on the track and, with the system initially at rest, the electrical power is turned on. The train reaches a steady speed $v$ with respect to the track. What is the rotational speed of the wheel if its mass is $M$ and its radius is $R$? (Treat the wheel as a hoop, and neglect the mass of the spokes and hub.)

FIGURE 12-28 ■ Problem 43.

**44. Two Skaters** In Fig. 12-29, two skaters, each of mass 50 kg, approach each other along parallel paths separated by 3.0 m. They have opposite velocities of 1.4 m/s each. One skater carries one end of a long pole with negligible mass, and the other skater grabs the other end of it as she passes. Assume frictionless ice. (a) Describe quantitatively the motion of the skaters after they have become connected by the pole. (b) What is the kinetic energy of the two-skater system?

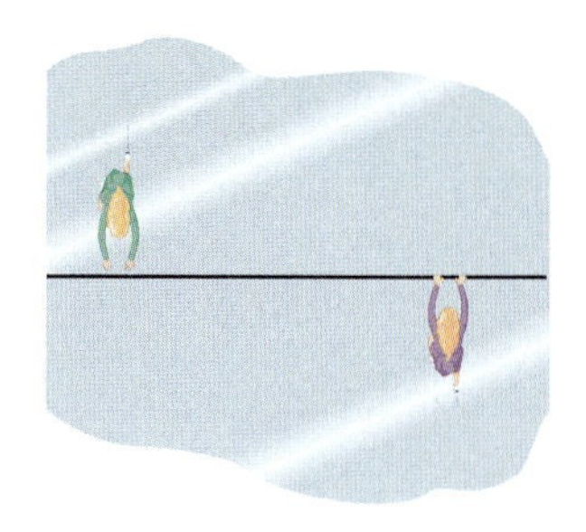

FIGURE 12-29 ■ Problem 44.

Next, the skaters each pull along the pole so as to reduce their separation to 1.0 m. What then are (c) their rotational speed and (d) the kinetic energy of the system? (e) Explain the source of the increased kinetic energy.

**45. A Cockroach** A cockroach of mass $m$ runs counterclockwise around the rim of a lazy Susan (a circular dish mounted on a vertical axle) of radius $R$ and rotational inertia $I$ and having frictionless bearings. The cockroach's speed (relative to the ground) is $v$, whereas the lazy Susan turns clockwise with rotational speed $\omega_1$. The cockroach finds a bread crumb on the rim and, of course, stops. (a) What is the rotational speed of the lazy Susan after the cockroach stops? (b) Is mechanical energy conserved?

**46. Girl on a Merry-go-Round** A girl of mass $M$ stands on the rim of a frictionless merry-go-round of radius $R$ and rotational inertia $I$ that is not moving. She throws a rock of mass $m$ horizontally in a direction that is tangent to the outer edge of the merry-go-round. The speed of the rock, relative to the ground, is $v$. Afterward, what are (a) the rotational speed of the merry-go-round and (b) the translational speed of the girl?

**47. Vinyl Record** A horizontal vinyl record of mass 0.10 kg and radius 0.10 m rotates freely about a vertical axis through its center with a rotational speed of 4.7 rad/s. The rotational inertia of the record about its axis of rotation is $5.0 \times 10^{-4}$ kg · m$^2$. A wad of wet putty of mass 0.020 kg drops vertically onto the record from above and sticks to the edge of the record. What is the rotational speed of the record immediately after the putty sticks to it?

**48. Uniform Thin Rod** A uniform thin rod of length 0.50 m and mass 4.0 kg can rotate in a horizontal plane about a vertical axis through its center. The rod is at rest when a 3.0 g bullet traveling in the horizontal plane of the rod is fired into one end of the rod. As viewed from above, the direction of the bullet's velocity makes an angle of 60° with the rod (Fig. 12-30). If the bullet lodges in the rod and the rotational velocity of the rod is 10 rad/s immediately after the collision, what is the bullet's speed just before impact?

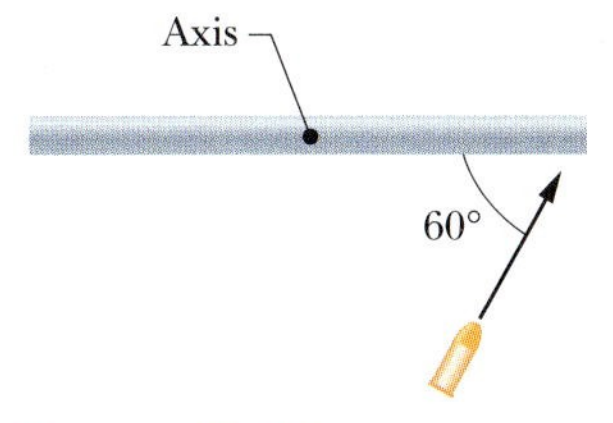


**FIGURE 12-30** ■ Problem 48.

**49. Putty Wad** Two 2.00 kg balls are attached to the ends of a thin rod of negligible mass, 50.0 cm long. The rod is free to rotate in a vertical plane without friction about a horizontal axis through its center. With the rod initially horizontal (Fig. 12-31), a 50.0 g wad of wet putty drops onto one of the balls, hitting it with a speed of 3.00 m/s and then sticking to it. (a) What is the rotational speed of the system just after the putty wad hits? (b) What is the ratio of the kinetic energy of the entire system after the collision to that of the putty wad just before? (c) Through what angle will the system rotate until it momentarily stops?

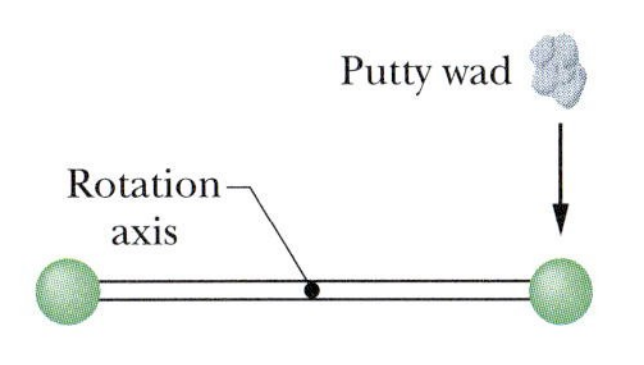


**FIGURE 12-31** ■ Problem 49.

**50. Cockroach on a Disk** A cockroach of mass $m$ lies on the rim of a uniform disk of mass $10.0m$ that can rotate freely about its center like a merry-go-round. Initially the cockroach and disk rotate together with a rotational velocity of $\omega_1$. Then the cockroach walks halfway to the center of the disk. (a) What is the change $\Delta\omega$ in the rotational velocity of the cockroach–disk system? (b) What is the ratio $K_2/K_1$ of the new kinetic energy of the system to its initial kinetic energy? (c) What accounts for the change in the kinetic energy?

**51. Earth's Polar Ice Caps** If Earth's polar ice caps fully melted and the water returned to the oceans, the oceans would be deeper by about 30 m. What effect would this have on Earth's rotation? Make an estimate of the resulting change in the length of the day. (Concern has been expressed that warming of the atmosphere resulting from industrial pollution could cause the ice caps to melt.)

**52. Horizontal Platform** A horizontal platform in the shape of a circular disk rotates on a frictionless bearing about a vertical axle through the center of the disk. The platform has a mass of 150 kg, a radius of 2.0 m, and a rotational inertia of 300 kg · m$^2$ about the axis of rotation. A 60 kg student walks slowly from the rim of the platform toward the center. If the rotational speed of the system is 1.5 rad/s when the student starts at the rim, what is the rotational speed when she is 0.50 m from the center?

**53. Uniform Disk** A uniform disk of mass $10m$ and radius $3.0r$ can rotate freely about its fixed center like a merry-go-round. A smaller uniform disk of mass $m$ and radius $r$ lies on top of the larger disk, concentric with it. Initially the two disks rotate together with a rotational velocity of 20 rad/s. Then a slight disturbance causes the smaller disk to slide outward across the larger disk, until the outer edge of the smaller disk catches on the outer edge of the larger disk. Afterward, the two disks again rotate together (without further sliding). (a) What then is their rotational velocity about the center of the larger disk? (b) What is the ratio $K_2/K_1$ of the new kinetic energy of the two-disk system to the system's initial kinetic energy?

**54. A Child Stands** A 30 kg child stands on the edge of a stationary merry-go-round of mass 100 kg and radius 2.0 m. The rotational inertia of the merry-go-round about its axis of rotation is 150 kg · m$^2$. The child catches a ball of mass 1.0 kg thrown by a friend. Just before the ball is caught, it has a horizontal velocity of 12 m/s that makes an angle of 37° with a line tangent to the outer edge of the merry-go-round, as shown in the overhead view of Fig. 12-32. What is the rotational speed of the merry-go-round just after the ball is caught?

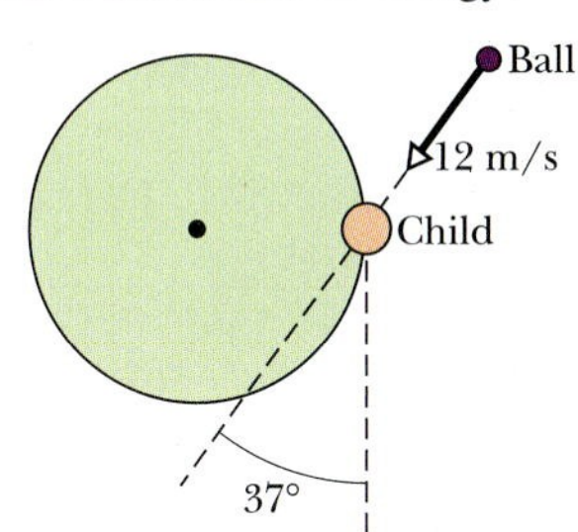


**FIGURE 12-32** ■ Problem 54.

**55. Bullet Hits Block** In Fig. 12-33, a 1.0 g bullet is fired into a 0.50 kg block that is mounted on the end of a 0.60 m nonuniform rod of mass 0.50 kg. The block–rod–bullet system then rotates about a fixed axis at point $A$. The rotational inertia of the rod alone about $A$ is 0.060 kg · m$^2$. Assume the block is small enough to treat as a particle on the end of the rod. (a) What is the rotational inertia of the block–rod–bullet system about point $A$? (b) If the rotational speed of the system about $A$ just after the bullet's impact is 4.5 rad/s, what is the speed of the bullet just before the impact?

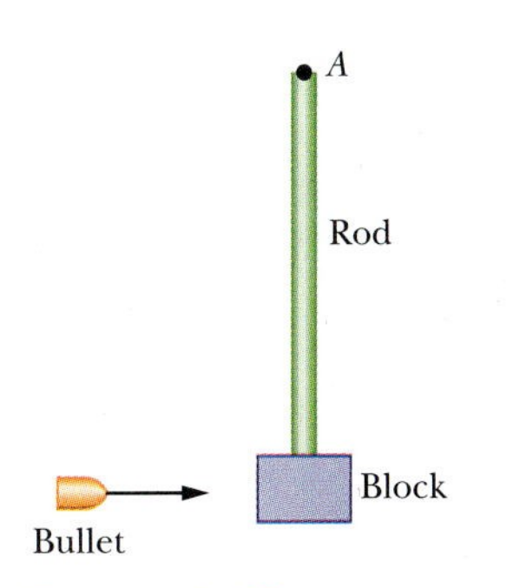


**FIGURE 12-33** ■ Problem 55.

**56. Uniform Rod** In Fig. 12-34, a uniform rod (length = 0.60 m, mass 1.0 kg) rotates about an axis through one end, with a rotational inertia of 0.12 kg · m$^2$. As the rod swings through its lowest position, the end of the rod collides with a small 0.20 kg putty wad that sticks to the end of the rod. If the rotational speed of the rod just before the collision is 2.4 rad/s, what is the rotational speed of the rod–putty system immediately after the collision?

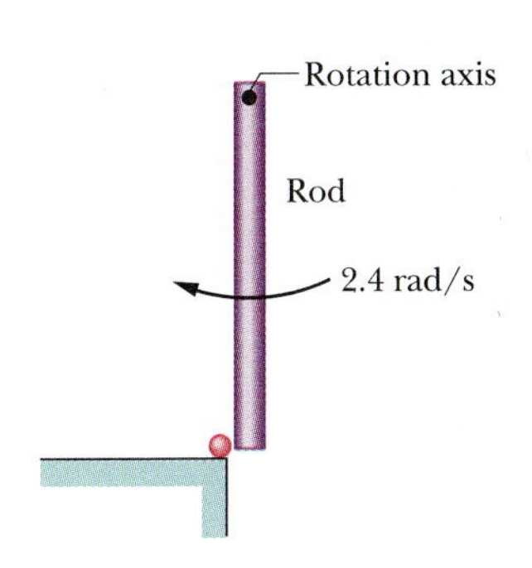


**FIGURE 12-34** ■ Problem 56.

**57. Particle on a Slide** The particle of mass $m$ in Fig. 12-35 slides down the frictionless surface through height $h$ and collides with the uniform vertical rod (of mass $M$ and length $d$), sticking to it. The rod pivots about point $O$ through the angle $\theta$ before momentarily stopping. Find $\theta$.

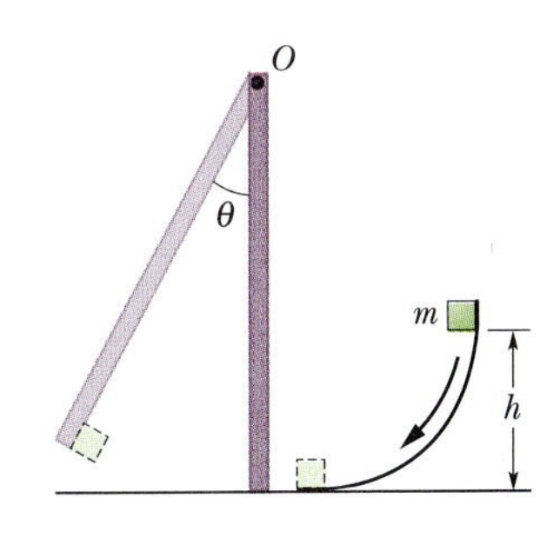


**FIGURE 12-35** ■ Problem 57.

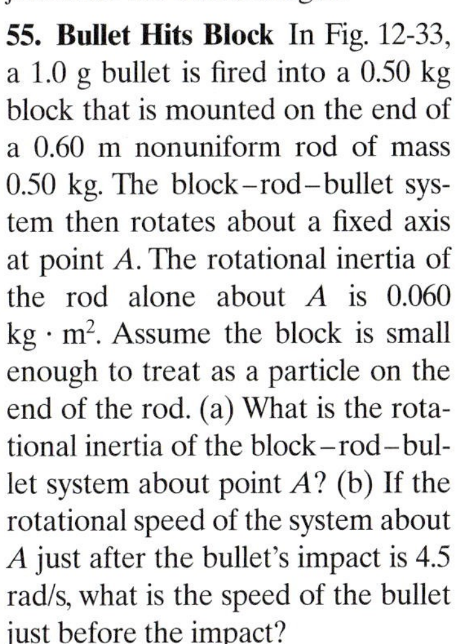

# Additional Problems

**58. Finding a Mistake Using Dimensional Analysis** As part of an examination a few years ago, a student went through the algebraic manipulations on an exam shown in Fig. 12-36. At this point you don't know what the symbols mean, but given the information about the dimensions associated with each symbol, decide the following:

**(a)** Is it possible that the final equation in Fig. 12-36 is correct? Justify your answer.
**(b)** If the final equation is not correct, does that mean that the starting equation is necessarily wrong? Explain.
**(c)** If the final equation is not correct and the starting equation is not wrong, can you find the error using dimensional analysis? If so, do so. If not, explain why.

$[M] = \mathrm{M}$

$[g] = \mathrm{L/T^2}$

$[h] = \mathrm{L}$

$[\omega] = \mathrm{1/T}$

$[v] = \mathrm{L/T}$

$[R] = \mathrm{L}$

$[I] = \mathrm{ML^2}$

$$Mgh = \tfrac{1}{2}Mv^2 + \tfrac{1}{2}I\omega^2$$

$$Mgh = \tfrac{1}{2}Mv^2 + \tfrac{1}{2}(MR^2)\omega^2$$

$$Mgh = \tfrac{1}{2}Mv^2 + \tfrac{1}{2}(MR^2)\left(\frac{v^2}{R}\right)^2$$

$$gh = \tfrac{1}{2}v^2 + \tfrac{1}{2}v^4$$

FIGURE 12-36 ■ Problem 58.

Note: M stands for a mass unit, L is for length unit, and T is for a time unit.

**59. Comparing Conserved Quantities** The four objects in Fig. 12-37 are moving as indicated by the arrows. A curved arrow indicates rolling without slipping in the direction. For object (*a*), use the coordinates shown. For the others, take the origin at the center of the circle. Use the directions associated with the coordinate axes shown for object (*a*). Construct a table with the values of the magnitudes total translational momentum, total rotational momentum, and total energy of motion at the instant shown for each case. Express your answers in terms of $m$, $v$, and $R$. (Include an indicator of the direction where appropriate.) Which system has the largest and smallest of each of the quantities? Explain your reasoning.

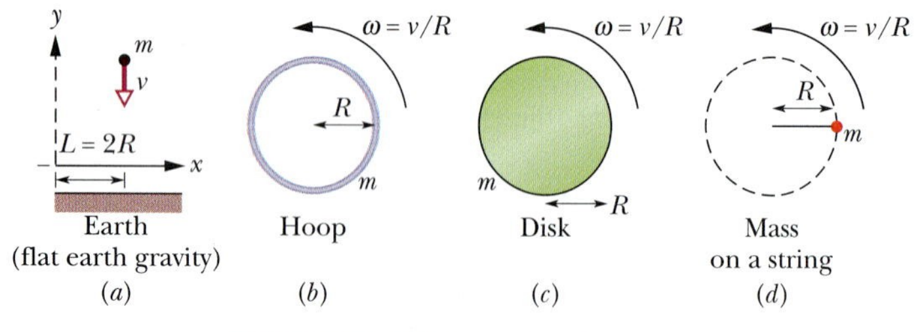


FIGURE 12-37 ■ Problem 59.

**60. Designing a Yo-Yo** In testing a design for a yo-yo, an engineer begins by constructing a simple prototype—a string wound about the rim of a wooden disk. She puts an axle riding on nearly frictionless ball bearings through the axis of the wooden disk and fixes the ends of the axle. See Fig. 12-38. In order to measure the moment of inertia of the disk, she attaches a weight of mass $m$ to the string and measures how long it takes to fall a given distance. (a) Assuming the rotational inertia of the disk is given by $I$, and the radius of the disk is $R$, find the time for the mass to fall a distance $h$ starting from rest. (b) The engineer doesn't have a very accurate stopwatch but wants to get a measurement good to a few percent. She decides that a fall time of 2 seconds would work. How big a mass should she use? Imagine you were setting up this experiment, and make reasonable estimates of the parameters you need.

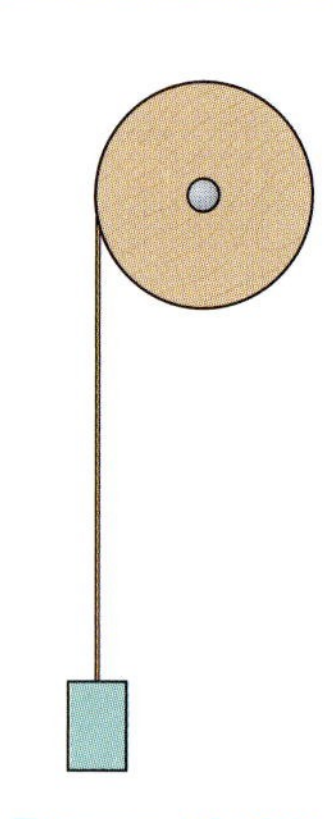

FIGURE 12-38 ■ Problem 60.

**61. Approximating Atwood** Figure 12-39 shows an Atwood's machine with two unequal masses attached by a massless string. The pulley has a mass of 20 g and a radius of 2 cm. (a) State three approximations that you can make to simplify your calculation of the motion of the blocks. ("Making an approximation" is the process of ignoring a physical effect because you expect it to be small and have little effect on your result if you only care about a few significant figures. If you want more significant figures, you may have to include those effects.) (b) Using your approximations, find the acceleration of block $A$. (c) What happens to your result if the two masses are equal? Is the result what you expect? Explain. (d) If you have ignored the rotational inertia of the pulley in your calculation in part (a) of this problem, set up the equations that would allow you to solve for the acceleration when it is included (but don't solve them).

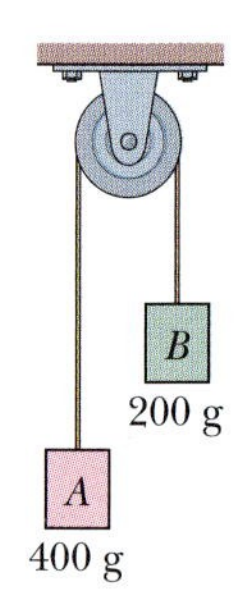


FIGURE 12-39 ■ Problem 61.

**62. The Refrigerator Door** A refrigerator has separate shelves on the door for storing bottles. Thin plastic straps keep the bottles from falling off the door. Someone in the house slams the door with a bit too much vigor and a heavy bottle breaks the strap. Do you think the bottle would be more likely to break the plastic strap if it is close to the hinge? Close to the handle? Or doesn't it matter? Explain your answer in terms of the physics we have learned.

APPENDIX A

# The International System of Units (SI)*

## 1 SI Base Units

**1. The SI Base Units**

| Quantity | Name | Symbol | Definition |
|---|---|---|---|
| length | meter | m | "... the length of the path traveled by light in vacuum in 1/299 792 458 of a second." (1983) |
| mass | kilogram | kg | "... this prototype [a certain platinum–iridium cylinder] shall henceforth be considered to be the unit of mass." (1889) |
| time | second | s | "... the duration of 9 192 631 770 periods of the radiation corresponding to the transition between the two hyperfine levels of the ground state of the cesium-133 atom." (1967) |
| electric current | ampere | A | "... that constant current which, if maintained in two straight parallel conductors of infinite length, of negligible circular cross section, and placed 1 meter apart in vacuum, would produce between these conductors a force equal to $2 \times 10^{-7}$ newton per meter of length." (1946) |
| thermodynamic temperature | kelvin | K | "... the fraction 1/273.16 of the thermodynamic temperature of the triple point of water." (1967) |
| amount of substance | mole | mol | "... the amount of substance of a system which contains as many elementary entities as there are atoms in 0.012 kilogram of carbon-12." (1971) |
| luminous intensity | candela | cd | "... the luminous intensity, in a given direction, of a source that emits monochromatic radiation of frequency $540 \times 10^{12}$ hertz and that has a radiant intensity in that direction of 1/683 watt per steradian." (1979) |

## 2 The SI Supplementary Units

**2. The SI Supplementary Units**

| Quantity | Name of Unit | Symbol |
|---|---|---|
| plane angle | radian | rad |
| solid angle | steradian | sr |

*Adapted from "The International System of Units (SI)," National Bureau of Standards Special Publication 330, 2001 edition. The definitions above were adopted by the General Conference of Weights and Measures, an international body, on the dates shown. In this book we do not use the candela.

## 3 Some SI Derivations

**3. Some SI Derived Units**

| Quantity | Name of Unit | Symbol | In Terms of other SI Units |
|---|---|---|---|
| area | square meter | $m^2$ | |
| volume | cubic meter | $m^3$ | |
| frequency | hertz | Hz | $s^{-1}$ |
| mass density (density) | kilogram per cubic meter | $kg/m^3$ | |
| speed, velocity | meter per second | m/s | |
| rotational velocity | radian per second | rad/s | |
| acceleration | meter per second per second | $m/s^2$ | |
| rotational acceleration | radian per second per second | $rad/s^2$ | |
| force | newton | N | $kg \cdot m/s^2$ |
| pressure | pascal | Pa | $N/m^2$ |
| work, energy, quantity of heat | joule | J | $N \cdot m$ |
| power | watt | W | J/s |
| quantity of electric charge | coulomb | C | $A \cdot s$ |
| potential difference, electromotive force | volt | V | W/A |
| electric field strength | volt per meter (or newton per coulomb) | V/m | N/C |
| electric resistance | ohm | $\Omega$ | V/A |
| capacitance | farad | F | $A \cdot s/V$ |
| magnetic flux | weber | Wb | $V \cdot s$ |
| inductance | henry | H | $V \cdot s/A$ |
| magnetic flux density | tesla | T | $Wb/m^2$ |
| magnetic field strength | ampere per meter | A/m | |
| entropy | joule per kelvin | J/K | |
| specific heat | joule per kilogram kelvin | $J/(kg \cdot K)$ | |
| thermal conductivity | watt per meter kelvin | $W/(m \cdot K)$ | |
| radiant intensity | watt per steradian | W/sr | |

## 4 Mathematical Notation

Poorly chosen mathematical notation can be a source of considerable confusion to those trying to learn and to do physics. For example, ambiguity in the meaning of a mathematical symbol can prevent a reader from understanding the meaning of a crucial relationship. It is also difficult to solve problems when the symbols used ot represent different quantities are not distinctive. In this text we have taken special care to use mathematical notation in ways that allow important distinctions to be easily visible both on the printed page and in handwritten work.

An excellent starting point for clear mathematical notation is the U.S. National Institute of Standard and Technology's Special Publication 811 (SP 811), *Guide for the Use of the International System of Units (SI)*, available at http://physics.nist.gov/cuu/Units/bibliography.html. In addition to following the National Institute guidelines, we have made a number of systematic choices to facilitate the translation of printed notation into handwritten mathematics. For example:

- Instead of making vectors bold, vector quantities (even in one dimension) are denoted by an arrow above the symbol. So printed equations look like handwritten equations. Example: $\vec{v}$ rather than $\boldsymbol{v}$ is used to denote an instantaneous velocity.
- In general, each vector component has an explicit subscript denoting that it represents the component along a chosen coordinate axis. The one exception is the position vector, $\vec{r}$, whose components are simply written as $x$, $y$, and $z$. For example, $\vec{r} = x\hat{\imath} + y\hat{\jmath} + z\hat{k}$, whereas, $\vec{v} = v_x\hat{\imath} + v_y\hat{\jmath} + v_z\hat{k}$.
- To emphasize the distinction between a vector's components and its magnitude, we write the magnitude of a vector, such as $\vec{F}$, as $|\vec{F}|$. However, when it is obvious that a magnitude is being described, we use the plain symbol (such as $F$ with no coordinate subscript) to denote a vector's magnitude.
- We often choose to spell out the names of objects that are associated with mathematical variables—writing, for example, $\vec{v}_{\text{ball}}$ and not $\vec{v}_{\text{b}}$ for the velocity of a ball.
- Numerical subscripts most commonly denote sequential times, positions, velocities, and so on. For example, $x_1$ is the $x$-component of the position of some object at time $t_1$, whereas $x_2$ is the value of that parameter at some later time $t_2$. We have avoided using the subscript zero to denote initial values, as in $x_0$ to denote "the initial position along the $x$ axis," to emphasize that *any* time can be chosen as the initial time for consideration of the subsequent time evolution of a system.
- To avoid confusing the numerical time sequence labels with object labels, we prefer to use capital letters as object labels. For example, we would label two particles A and B rather than 1 and 2. Thus, $\vec{p}_{\text{A}\,1}$ and $\vec{p}_{\text{B}\,1}$ would represent the translational momenta of two particles before a collision whereas $\vec{p}_{\text{A}\,2}$ and $\vec{p}_{\text{B}\,2}$ would be their momenta after a collision.
- To avoid excessively long strings of subscripts, we have made the unconventional choice to write all adjectival labels as *super*scripts. Thus, Newton's Second Law is written $\vec{F}^{\text{net}} = m\vec{a}$ whereas the sum of the forces acting on a certain object might be written as $\vec{F}^{\text{net}} = \vec{F}^{\text{grav}} + \vec{F}^{\text{app}}$. To avoid confusion with mathematical exponents, an adjectival label is never a single letter.
- Following a usage common in contemporary physics, the time average of a variable $\vec{v}$ is denoted as $\langle\vec{v}\rangle$ and not as $\vec{v}_{\text{avg}}$.
- Physical constants such as $e$, $c$, $g$, $G$, are all **positive** scalar quantities.

## 5 Significant Figures and the Precision of Numerical Results

Quoting the result of a calculation or a measurement to the correct number of significant figures is merely a way of telling your reader roughly how precise you believe the result to be. Quoting too many significant figures overstates the precision of your result and quoting too few implies less precision than the result may actually possess. So how many significant figures should you quote when reporting your result.

### Determining Significant Figures

Before answering the question of how many significant figures to quote, we need to have a clear method for determining how many significant figures a reported number has. The standard method is quite simple:

> **METHOD FOR COUNTING SIGNIFICANT FIGURES:** Read the number from left to right, and count the first nonzero digit and all the digits (zero or not) to the right of it as significant.

Using this rule, 350 mm, 0.000350 km, and 0.350 m each has *three* significant figures. In fact, each of these numbers merely represents the same distance, expressed in different units. As you can see from this example, the number of *decimal places* that a number has is *not* the same as its number of *significant figures*. The first of these distances has zero decimal places, the second has six decimal places, and the third has three, yet all three of these numbers have three significant figures.

One consequence of this method is especially worth noting. Trailing zeros count as significant figures. For example, 2700 m/s has four significant figures. If you really meant it to have only three significant figures, you would have to write it either as 2.70 km/s (changing the unit) or $2.70 \times 10^3$ m/s (using scientific notation.)

## A Simple Rule for Reporting Significant Figures in a Calculated Result

Now that you know how to count significant figures, how many should the result of a calculation have? A simple rule that will work in most calculations is:

> **SIGNIFICANT FIGURES IN A CALCULATED RESULT:** The common practice is to quote the result of a calculation to the number of significant figures of the *least* precise number used in the calculation.

Although this simple rule will often either understate or (less frequently) overstate the precision of a result, it still serves as a good rule-of-thumb for everyday numerical work. In introductory physics you will only rarely encounter data that are known to better than two, three, or four significant figures. This simple rule then tells you that you can't go very far wrong if you round off all your final results to three significant figures.

There are two situations in which the simple rule should *not* be applied to a calculation. One is when an exact number is involved in the calculation and another is when a calculation is done in parts so that intermediate results are used.

1. ***Using Exact Data*** There are some obvious situations in which a number used in a calculation is exact. Numbers based on counting items are exact. For example, if you are told that there are 5 people on an elevator, there are exactly 5 people, not 4.7 or 5.1. Another situation arises when a number is exact by definition. For example, the conversion factor 2.54 cm/inch does *not* have three significant figures because the inch is *defined* to be exactly 2.5400000 . . . cm. *Data that are known exactly should not be included when deciding which of the original data has the fewest significant figures.*
2. ***Significant Figures in Intermediate Results*** Only the final result at the end of your calculation should be rounded using the simple rule. Intermediate results should never be rounded. Spreadsheet software takes care of this for you, as does your calculator if you store your intermediate results in its memory rather than writing them down and then rekeying them. If you must write down intermediate results, keep a few more significant figures than your final result will have.

## Understanding and Refining the Simple Significant Figure Rule

Quoting the result of a calculation or measurement to the correct number of significant figures is a way of indicating its precision. You need to understand what limits the precision of data before you fully understand how to use the simple rule or its exceptions.

**Absolute Precision** There are two ways of talking about precision. First there is *absolute precision*, which tells you explicitly the smallest scale division of the measurement. It's always quoted in the same units as the measured quantity. For example, saying "I measured the length of the table to the nearest centimeter" states the absolute precision of the measurement. The absolute precision tells you how many *decimal places* the measurement has; it alone does not determine the number of significant figures. Example: if a table is 235 cm long, then 1 cm of absolute precision translates into three significant figures. On the other hand, if a table is for a doll's house and is only 8 cm long, then the same 1 cm of absolute precision has only one significant figure.

**Relative Precision** Because of this problem with absolute precision, scientists often prefer to describe the precision of data *relative* to the size of the quantity being measured. To use the previous examples, the *relative precision* of the length of the real table in the previous example is 1 cm out of 235 cm. This is usually stated as a ratio (1 part in 235) or as a percentage ($1/235 = 0.004255 \approx 0.4\%$). In the case of the toy table, the same 1 cm of absolute precision yields a relative precision of only 1 part in 8 or $1/8 = 0.125 = 12.5\%$.

**Inconsistencies between Significant Figures and Relative Precision** There is an inconsistency that goes with using a certain number of significant figures to express relative precision. Quoted to the same number of significant figures, the relative precision of results can be quite different. For example, 13 cm and 94 cm both have two significant figures. Yet the first is specified to only 1 part in 13 or $1/13 \approx 10\%$, whereas the second is known to 1 part in 94 or $1/94 \approx 1\%$. This bias toward greater relative precision for results with larger first significant figures is one weakness of using significant figures to track the precision of calculated results. You can partially address this problem, by including one more significant figure than the simple rule suggests, when the final result of a calculation has a 1 as its first significant figure.

**Multiplying and Dividing** When multiplying or dividing numbers, the *relative* precision of the result cannot exceed that of the least precise number used. Since the number of significant figures in the result tells us its relative precision, the simple rule is all that you need when you multiply or divide. For example, the area of a strip of paper of measured size is 280 cm by 2.5 cm would be correctly reported, according to the simple rule, as $7.0 \times 10^2$ $cm^2$. This result has only two significant figures since the less precise measurement, 2.5 cm, that went into the calculation had only two significant figures. Reporting this result as 700 $cm^2$ would not be correct since this result has three significant figures, exceeding the relative precision of the 2.5 cm measurement.

**Addition and Subtraction** When adding or subtracting, you line up the decimal points before you add or subtract. This means that it's the *absolute* precision of the least precise number that limits the precision of the sum or the difference. This can lead to some exceptions to the simple rule. For example, adding 957 cm and 878 cm yields 1835 cm. Here the result is reliable to an absolute precision of about 1 cm since both of the original distances had this reliability. But the result then has four significant figures whereas each of the original numbers had only three. If, on the other hand, you take the difference between these two distances you get 79 cm. The difference is still reliable to about 1 cm, but that absolute precision now translates into only two significant figures worth of relative precision. So, you should be careful when adding or subtracting, since addition can actually increase the relative precision of your result and, more important, subtraction can reduce it.

**Evaluating Functions** What about the evaluation of functions? For example, how many significant figures does the sin(88.2°) have? You can use your calculator to answer this question. First use your calculator to note that sin(88.2°) = 0.999506. Now add 1 to the least significant decimal place of the argument of the function and evaluate it again. Here this gives sin(88.3°) = 0.999559. Take the last significant figure in the result to be *the first one from the left that changed* when you repeated the calculation. In this example the first digit that changed was the 0; it became a 5 (the second 5) in the recalculation. So, using the empirical approach gives you five significant figures.

# APPENDIX B Some Fundamental Constants of Physics*

| Constant | Symbol | Computational Value | Best (1998) Value | |
|---|---|---|---|---|
| | | | Value[a] | Uncertainty[b] |
| Speed of light in a vacuum | $c$ | $3.00 \times 10^8$ m/s | 2.997 924 58 | exact |
| Elementary charge | $e$ | $1.60 \times 10^{-19}$ C | 1.602 176 462 | 0.039 |
| Gravitational constant | $G$ | $6.67 \times 10^{-11}$ m$^3$/s$^2 \cdot$ kg | 6.673 | 1500 |
| Universal gas constant | $R$ | 8.31 J/mol $\cdot$ K | 8.314 472 | 1.7 |
| Avogadro constant | $N_A$ | $6.02 \times 10^{23}$ mol$^{-1}$ | 6.022 141 99 | 0.079 |
| Boltzmann constant | $k_B$ | $1.38 \times 10^{-23}$ J/K | 1.380 650 3 | 1.7 |
| Stefan–Boltzmann constant | $\sigma$ | $5.67 \times 10^{-8}$ W/m$^2 \cdot$ K$^4$ | 5.670 400 | 7.0 |
| Molar volume of ideal gas at STP[d] | $V_m$ | $2.27 \times 10^{-2}$ m$^3$/mol | 2.271 098 1 | 1.7 |
| Electric constant (permittivity) | $\epsilon_0$ | $8.85 \times 10^{-12}$ C$^2$/N $\cdot$ m$^2$ | 8.854 187 817 62 | exact |
| Coulomb constant | $k = 1/4\pi\epsilon_0$ | $8.99 \times 10^9$ N $\cdot$ m$^2$/C$^2$ | 8.987 551 787 | $5 \times 10^{-10}$ |
| Magnetic constant (permeability) | $\mu_0$ | $1.26 \times 10^{-6}$ N/A$^2$ | 1.256 637 061 43 | exact |
| Planck constant | $h$ | $6.63 \times 10^{-34}$ J $\cdot$ s | 6.626 068 76 | 0.078 |
| Electron mass[c] | $m_e$ | $9.11 \times 10^{-31}$ kg | 9.109 381 88 | 0.079 |
| | | $5.49 \times 10^{-4}$ u | 5.485 799 110 | 0.0021 |
| Proton mass[c] | $m_p$ | $1.67 \times 10^{-27}$ kg | 1.672 621 58 | 0.079 |
| | | 1.0073 u | 1.007 276 466 88 | $1.3 \times .10^{-4}$ |
| Ratio of proton mass to electron mass | $m_p/m_e$ | 1840 | 1836.152 667 5 | 0.0021 |
| Electron charge-to-mass ratio | $e/m_e$ | $1.76 \times 10^{11}$ C/kg | 1.758 820 174 | 0.040 |
| Neutron mass[c] | $m_n$ | $1.68 \times 10^{-27}$ kg | 1.674 927 16 | 0.079 |
| | | 1.0087 u | 1.008 664 915 78 | $5.4 \times 10^{-4}$ |
| Hydrogen atom mass[c] | $m_{1H}$ | 1.0078 u | 1.007 825 031 6 | 0.0005 |
| Deuterium atom mass[c] | $m_{2H}$ | 2.0141 u | 2.014 101 777 9 | 0.0005 |
| Helium atom mass[c] | $m_{4He}$ | 4.0026 u | 4.002 603 2 | 0.067 |
| Muon mass | $m_\mu$ | $1.88 \times 10^{-28}$ kg | 1.883 531 09 | 0.084 |
| Electron magnetic moment | $\mu_e$ | $9.28 \times 10^{-24}$ J/T | 9.284 763 62 | 0.040 |
| Proton magnetic moment | $\mu_p$ | $1.41 \times 10^{-26}$ J/T | 1.410 606 663 | 0.041 |
| Bohr magneton | $\mu_B$ | $9.27 \times 10^{-24}$ J/T | 9.274 008 99 | 0.040 |
| Nuclear magneton | $\mu_N$ | $5.05 \times 10^{-27}$ J/T | 5.050 783 17 | 0.040 |
| Bohr radius | $r_B$ | $5.29 \times 10^{-11}$ m | 5.291 772 083 | 0.0037 |
| Rydberg constant | $R$ | $1.10 \times 10^7$ m$^{-1}$ | 1.097 373 156 854 8 | $7.6 \times 10^{-6}$ |
| Electron Compton wavelength | $\lambda_C$ | $2.43 \times 10^{-12}$ m | 2.426 310 215 | 0.0073 |

[a]Values given in this column should be given the same unit and power of 10 as the computational value.
[b]Parts per million.
[c]Masses given in u are in unified atomic mass units, where 1 u = $1.660\,538\,73 \times 10^{-27}$ kg.
[d]STP means standard temperature and pressure: 0°C and 1.0 atm (0.1 MPa).

*The values in this table were selected from the 1998 CODATA recommended values (www.physics.nist.gov).

APPENDIX C

# Some Astronomical Data

**Some Distances from Earth**

| | | | |
|---|---|---|---|
| To the Moon* | $3.82 \times 10^8$ m | To the center of our galaxy | $2.2 \times 10^{20}$ m |
| To the Sun* | $1.50 \times 10^{11}$ m | To the Andromeda Galaxy | $2.1 \times 10^{22}$ m |
| To the nearest star (Proxima Centauri) | $4.04 \times 10^{16}$ m | To the edge of the observable universe | $\sim 10^{26}$ m |

* Mean distance.

**The Sun, Earth, and the Moon**

| Property | Unit | Sun | | Earth | Moon |
|---|---|---|---|---|---|
| Mass | kg | $1.99 \times 10^{30}$ | | $5.98 \times 10^{24}$ | $7.36 \times 10^{22}$ |
| Mean radius | m | $6.96 \times 10^8$ | | $6.37 \times 10^6$ | $1.74 \times 10^6$ |
| Mean density | $kg/m^3$ | 1410 | | 5520 | 3340 |
| Free-fall acceleration at the surface | $m/s^2$ | 274 | | 9.81 | 1.67 |
| Escape velocity | km/s | 618 | | 11.2 | 2.38 |
| Period of rotation[a] | — | 37 d at poles[b] | 26 d at equator[b] | 23 h 56 min | 27.3 d |
| Radiation power[c] | W | $3.90 \times 10^{26}$ | | | |

[a] Measured with respect to the distant stars.
[b] The Sun, a ball of gas, does not rotate as a rigid body.
[c] Just outside Earth's atmosphere solar energy is received, assuming normal incidence, at the rate of 1340 $W/m^2$.

## Some Properties of the Planets

| | Mercury | Venus | Earth | Mars | Jupiter | Saturn | Uranus | Neptune | Pluto |
|---|---|---|---|---|---|---|---|---|---|
| Mean distance from Sun, $10^6$ km | 57.9 | 108 | 150 | 228 | 778 | 1430 | 2870 | 4500 | 5900 |
| Period of revolution, y | 0.241 | 0.615 | 1.00 | 1.88 | 11.9 | 29.5 | 84.0 | 165 | 248 |
| Period of rotation,[a] d | 58.7 | $-243$[b] | 0.997 | 1.03 | 0.409 | 0.426 | $-0.451$[b] | 0.658 | 6.39 |
| Orbital speed, km/s | 47.9 | 35.0 | 29.8 | 24.1 | 13.1 | 9.64 | 6.81 | 5.43 | 4.74 |
| Inclination of axis to orbit | $<28°$ | $\approx 3°$ | 23.4° | 25.0° | 3.08° | 26.7° | 97.9° | 29.6° | 57.5° |
| Inclination of orbit to Earth's orbit | 7.00° | 3.39° | | 1.85° | 1.30° | 2.49° | 0.77° | 1.77° | 17.2° |
| Eccentricity of orbit | 0.206 | 0.0068 | 0.0167 | 0.0934 | 0.0485 | 0.0556 | 0.0472 | 0.0086 | 0.250 |
| Equatorial diameter, km | 4880 | 12 100 | 12 800 | 6790 | 143 000 | 120 000 | 51 800 | 49 500 | 2300 |
| Mass (Earth = 1) | 0.0558 | 0.815 | 1.000 | 0.107 | 318 | 95.1 | 14.5 | 17.2 | 0.002 |
| Density (water = 1) | 5.60 | 5.20 | 5.52 | 3.95 | 1.31 | 0.704 | 1.21 | 1.67 | 2.03 |
| Surface value of $g$,[c] $m/s^2$ | 3.78 | 8.60 | 9.78 | 3.72 | 22.9 | 9.05 | 7.77 | 11.0 | 0.5 |
| Escape velocity,[c] km/s | 4.3 | 10.3 | 11.2 | 5.0 | 59.5 | 35.6 | 21.2 | 23.6 | 1.1 |
| Known satellites | 0 | 0 | 1 | 2 | 16 + ring | 18 + rings | 17 + rings | 8 + rings | 1 |

[a] Measured with respect to the distant stars.
[b] Venus and Uranus rotate opposite their orbital motion.
[c] Gravitational acceleration measured at the planet's equator.

# APPENDIX D Conversion Factors

Conversion factors may be read directly from these tables. For example, 1 degree = $2.778 \times 10^{-3}$ revolutions, so $16.7° = 16.7 \times 2.778 \times 10^{-3}$ rev. The SI units are fully capitalized. Adapted in part from G. Shortley and D. Williams, *Elements of Physics*, 1971, Prentice-Hall, Englewood Cliffs, N.J.

## Plane Angle

| | ° | ′ | ″ | RADIAN | rev |
|---|---|---|---|---|---|
| 1 degree = | 1 | 60 | 3600 | $1.745 \times 10^{-2}$ | $2.778 \times 10^{-3}$ |
| 1 minute = | $1.667 \times 10^{-2}$ | 1 | 60 | $2.909 \times 10^{-4}$ | $4.630 \times 10^{-5}$ |
| 1 second = | $2.778 \times 10^{-4}$ | $1.667 \times 10^{-2}$ | 1 | $4.848 \times 10^{-6}$ | $7.716 \times 10^{-7}$ |
| 1 RADIAN = | 57.30 | 3438 | $2.063 \times 10^{5}$ | 1 | 0.1592 |
| 1 revolution = | 360 | $2.16 \times 10^{4}$ | $1.296 \times 10^{6}$ | 6.283 | 1 |

## Solid Angle

1 sphere = $4\pi$ steradians = 12.57 steradians

## Length

| | cm | METER | km | in. | ft | mi |
|---|---|---|---|---|---|---|
| 1 centimeter = | 1 | $10^{-2}$ | $10^{-5}$ | 0.3937 | $3.281 \times 10^{-2}$ | $6.214 \times 10^{-6}$ |
| 1 METER = | 100 | 1 | $10^{-3}$ | 39.37 | 3.281 | $6.214 \times 10^{-4}$ |
| 1 kilometer = | $10^{5}$ | 1000 | 1 | $3.937 \times 10^{4}$ | 3281 | 0.6214 |
| 1 inch = | 2.540 | $2.540 \times 10^{-2}$ | $2.540 \times 10^{-5}$ | 1 | $8.333 \times 10^{-2}$ | $1.578 \times 10^{-5}$ |
| 1 foot = | 30.48 | 0.3048 | $3.048 \times 10^{-4}$ | 12 | 1 | $1.894 \times 10^{-4}$ |
| 1 mile = | $1.609 \times 10^{5}$ | 1609 | 1.609 | $6.336 \times 10^{4}$ | 5280 | 1 |

1 angström = $10^{-10}$ m
1 nautical mile = 1852 m
= 1.151 miles = 6076 ft
1 fermi = $10^{-15}$ m
1 light-year = $9.460 \times 10^{12}$ km
1 parsec = $3.084 \times 10^{13}$ km
1 fathom = 6 ft
1 Bohr radius = $5.292 \times 10^{-11}$ m
1 yard = 3 ft
1 rod = 16.5 ft
1 mil = $10^{-3}$ in.
1 nm = $10^{-9}$ m

## Area

| METER$^2$ | cm$^2$ | ft$^2$ | in.$^2$ |
|---|---|---|---|
| 1 SQUARE METER = 1 | $10^4$ | 10.76 | 1550 |
| 1 square centimeter = $10^{-4}$ | 1 | $1.076 \times 10^{-3}$ | 0.1550 |
| 1 square foot = $9.290 \times 10^{-2}$ | 929.0 | 1 | 144 |
| 1 square inch = $6.452 \times 10^{-4}$ | 6.452 | $6.944 \times 10^{-3}$ | 1 |

1 square mile = $2.788 \times 10^7$ ft$^2$ = 640 acres
1 barn = $10^{-28}$ m$^2$
1 acre = 43 560 ft$^2$
1 hectare = $10^4$ m$^2$ = 2.471 acres

## Volume

| METER$^3$ | cm$^3$ | L | ft$^3$ | in.$^3$ |
|---|---|---|---|---|
| 1 CUBIC METER = 1 | $10^6$ | 1000 | 35.31 | $6.102 \times 10^4$ |
| 1 cubic centimeter = $10^{-6}$ | 1 | $1.000 \times 10^{-3}$ | $3.531 \times 10^{-5}$ | $6.102 \times 10^{-2}$ |
| 1 liter = $1.000 \times 10^{-3}$ | 1000 | 1 | $3.531 \times 10^{-2}$ | 61.02 |
| 1 cubic foot = $2.832 \times 10^{-2}$ | $2.832 \times 10^4$ | 28.32 | 1 | 1728 |
| 1 cubic inch = $1.639 \times 10^{-5}$ | 16.39 | $1.639 \times 10^{-2}$ | $5.787 \times 10^{-4}$ | 1 |

1 U.S. fluid gallon = 4 U.S. fluid quarts = 8 U.S. pints = 128 U.S. fluid ounces = 231 in.$^3$
1 British imperial gallon = 277.4 in.$^3$ = 1.201 U.S. fluid gallons

## Mass

Quantities in the colored areas are not mass units but are often used as such. When we write, for example, 1 kg "=" 2.205 lb, this means that a kilogram is a *mass* that *weighs* 2.205 pounds at a location where $g$ has the standard value of 9.80665 m/s$^2$.

| g | KILOGRAM | slug | u | oz | lb | ton |
|---|---|---|---|---|---|---|
| 1 gram = 1 | 0.001 | $6.852 \times 10^{-5}$ | $6.022 \times 10^{23}$ | $3.527 \times 10^{-2}$ | $2.205 \times 10^{-3}$ | $1.102 \times 10^{-6}$ |
| 1 KILOGRAM = 1000 | 1 | $6.852 \times 10^{-2}$ | $6.022 \times 10^{26}$ | 35.27 | 2.205 | $1.102 \times 10^{-3}$ |
| 1 slug = $1.459 \times 10^4$ | 14.59 | 1 | $8.786 \times 10^{27}$ | 514.8 | 32.17 | $1.609 \times 10^{-2}$ |
| 1 atomic mass unit = $1.661 \times 10^{-24}$ | $1.661 \times 10^{-27}$ | $1.138 \times 10^{-28}$ | 1 | $5.857 \times 10^{-26}$ | $3.662 \times 10^{-27}$ | $1.830 \times 10^{-30}$ |
| 1 ounce = 28.35 | $2.835 \times 10^{-2}$ | $1.943 \times 10^{-3}$ | $1.718 \times 10^{25}$ | 1 | $6.250 \times 10^{-2}$ | $3.125 \times 10^{-5}$ |
| 1 pound = 453.6 | 0.4536 | $3.108 \times 10^{-2}$ | $2.732 \times 10^{26}$ | 16 | 1 | 0.0005 |
| 1 ton = $9.072 \times 10^5$ | 907.2 | 62.16 | $5.463 \times 10^{29}$ | $3.2 \times 10^4$ | 2000 | 1 |

1 metric ton = 1000 kg

## Density

Quantities in the colored areas are weight densities and, as such, are dimensionally different from mass densities. See note for mass table.

| slug/ft$^3$ | KILOGRAM/ METER$^3$ | g/cm$^3$ | lb/ft$^3$ | lb/in.$^3$ |
|---|---|---|---|---|
| 1 slug per foot$^3$ = 1 | 515.4 | 0.5154 | 32.17 | $1.862 \times 10^{-2}$ |
| 1 KILOGRAM per METER$^3$ = $1.940 \times 10^{-3}$ | 1 | 0.001 | $6.243 \times 10^{-2}$ | $3.613 \times 10^{-5}$ |
| 1 gram per centimeter$^3$ = 1.940 | 1000 | 1 | 62.43 | $3.613 \times 10^{-2}$ |
| 1 pound per foot$^3$ = $3.108 \times 10^{-2}$ | 16.02 | $16.02 \times 10^{-2}$ | 1 | $5.787 \times 10^{-4}$ |
| 1 pound per inch$^3$ = 53.71 | $2.768 \times 10^4$ | 27.68 | 1728 | 1 |

## Time

| y | d | h | min | SECOND |
|---|---|---|---|---|
| 1 year = 1 | 365.25 | $8.766 \times 10^3$ | $5.259 \times 10^5$ | $3.156 \times 10^7$ |
| 1 day = $2.738 \times 10^{-3}$ | 1 | 24 | 1440 | $8.640 \times 10^4$ |
| 1 hour = $1.141 \times 10^{-4}$ | $4.167 \times 10^{-2}$ | 1 | 60 | 3600 |
| 1 minute = $1.901 \times 10^{-6}$ | $6.944 \times 10^{-4}$ | $1.667 \times 10^{-2}$ | 1 | 60 |
| 1 SECOND = $3.169 \times 10^{-8}$ | $1.157 \times 10^{-5}$ | $2.778 \times 10^{-4}$ | $1.667 \times 10^{-2}$ | 1 |

## Speed

| ft/s | km/h | METER/SECOND | mi/h | cm/s |
|---|---|---|---|---|
| 1 foot per second = 1 | 1.097 | 0.3048 | 0.6818 | 30.48 |
| 1 kilometer per hour = 0.9113 | 1 | 0.2778 | 0.6214 | 27.78 |
| 1 METER per SECOND = 3.281 | 3.6 | 1 | 2.237 | 100 |
| 1 mile per hour = 1.467 | 1.609 | 0.4470 | 1 | 44.70 |
| 1 centimeter per second = $3.281 \times 10^{-2}$ | $3.6 \times 10^{-2}$ | 0.01 | $2.237 \times 10^{-2}$ | 1 |

1 knot = 1 nautical mi/h = 1.688 ft/s     1 mi/min = 88.00 ft/s = 60.00 mi/h

## Force

Force units in the colored areas are now little used. To clarify: 1 gram-force (= 1 gf) is the force of gravity that would act on an object whose mass is 1 gram at a location where $g$ has the standard value of 9.80665 m/s$^2$.

| dyne | NEWTON | lb | pdl | gf | kgf |
|---|---|---|---|---|---|
| 1 dyne = 1 | $10^{-5}$ | $2.248 \times 10^{-6}$ | $7.233 \times 10^{-5}$ | $1.020 \times 10^{-3}$ | $1.020 \times 10^{-6}$ |
| 1 NEWTON = $10^5$ | 1 | 0.2248 | 7.233 | 102.0 | 0.1020 |
| 1 pound = $4.448 \times 10^5$ | 4.448 | 1 | 32.17 | 453.6 | 0.4536 |
| 1 poundal = $1.383 \times 10^4$ | 0.1383 | $3.108 \times 10^{-2}$ | 1 | 14.10 | $1.410 \times 10^2$ |
| 1 gram-force = 980.7 | $9.807 \times 10^{-3}$ | $2.205 \times 10^{-3}$ | $7.093 \times 10^{-2}$ | 1 | 0.001 |
| 1 kilogram-force = $9.807 \times 10^5$ | 9.807 | 2.205 | 70.93 | 1000 | 1 |

1 ton = 2000 lb

## Pressure

| | atm | dyne/cm$^2$ | inch of water | cm Hg | PASCAL | lb/in.$^2$ | lb/ft$^2$ |
|---|---|---|---|---|---|---|---|
| 1 atmosphere = | 1 | $1.013 \times 10^6$ | 406.8 | 76 | $1.013 \times 10^5$ | 14.70 | 2116 |
| 1 dyne per centimeter$^2$ = | $9.869 \times 10^{-7}$ | 1 | $4.015 \times 10^{-4}$ | $7.501 \times 10^{-5}$ | 0.1 | $1.405 \times 10^{-5}$ | $2.089 \times 10^{-3}$ |
| 1 inch of water[a] at 4°C = | $2.458 \times 10^{-3}$ | 2491 | 1 | 0.1868 | 249.1 | $3.613 \times 10^{-2}$ | 5.202 |
| 1 centimeter of mercury[a] at 0°C = | $1.316 \times 10^{-2}$ | $1.333 \times 10^4$ | 5.353 | 1 | 1333 | 0.1934 | 27.85 |
| 1 PASCAL = | $9.869 \times 10^{-6}$ | 10 | $4.015 \times 10^{-3}$ | $7.501 \times 10^{-4}$ | 1 | $1.450 \times 10^{-4}$ | $2.089 \times 10^{-2}$ |
| 1 pound per inch$^2$ = | $6.805 \times 10^{-2}$ | $6.895 \times 10^4$ | 27.68 | 5.171 | $6.895 \times 10^3$ | 1 | 144 |
| 1 pound per foot$^2$ = | $4.725 \times 10^{-4}$ | 478.8 | 0.1922 | $3.591 \times 10^{-2}$ | 47.88 | $6.944 \times 10^{-3}$ | 1 |

[a] Where the acceleration of gravity has the standard value of 9.80665 m/s$^2$.

1 bar = $10^6$ dyne/cm$^2$ = 0.1 MPa  1 millibar = $10^3$ dyne/cm$^2$ = $10^2$ Pa  1 torr = 1 mm Hg

## Energy, Work, Heat

Quantities in the colored areas are not energy units but are included for convenience. They arise from the relativistic mass–energy equivalence formula $E = mc^2$ and represent the energy released if a kilogram or unified atomic mass unit (u) is completely converted to energy (bottom two rows) or the mass that would be completely converted to one unit of energy (rightmost two columns).

| | Btu | erg | ft·lb | hp·h | JOULE | cal | kW·h | eV | MeV | kg | u |
|---|---|---|---|---|---|---|---|---|---|---|---|
| 1 British thermal unit = | 1 | $1.055 \times 10^{10}$ | 777.9 | $3.929 \times 10^{-4}$ | 1055 | 252.0 | $2.930 \times 10^{-4}$ | $6.585 \times 10^{21}$ | $6.585 \times 10^{15}$ | $1.174 \times 10^{-14}$ | $7.070 \times 10^{12}$ |
| 1 erg = | $9.481 \times 10^{-11}$ | 1 | $7.376 \times 10^{-8}$ | $3.725 \times 10^{-14}$ | $10^{-7}$ | $2.389 \times 10^{-8}$ | $2.778 \times 10^{-14}$ | $6.242 \times 10^{11}$ | $6.242 \times 10^5$ | $1.113 \times 10^{-24}$ | 670.2 |
| 1 foot-pound = | $1.285 \times 10^{-3}$ | $1.356 \times 10^7$ | 1 | $5.051 \times 10^{-7}$ | 1.356 | 0.3238 | $3.766 \times 10^{-7}$ | $8.464 \times 10^{18}$ | $8.464 \times 10^{12}$ | $1.509 \times 10^{-17}$ | $9.037 \times 10^9$ |
| 1 horsepower-hour = | 2545 | $2.685 \times 10^{13}$ | $1.980 \times 10^6$ | 1 | $2.685 \times 10^6$ | $6.413 \times 10^5$ | 0.7457 | $1.676 \times 10^{25}$ | $1.676 \times 10^{19}$ | $2.988 \times 10^{-11}$ | $1.799 \times 10^{16}$ |
| 1 JOULE = | $9.481 \times 10^{-4}$ | $10^7$ | 0.7376 | $3.725 \times 10^{-7}$ | 1 | 0.2389 | $2.778 \times 10^{-7}$ | $6.242 \times 10^{18}$ | $6.242 \times 10^{12}$ | $1.113 \times 10^{-17}$ | $6.702 \times 10^9$ |
| 1 calorie = | $3.969 \times 10^{-3}$ | $4.186 \times 10^7$ | 3.088 | $1.560 \times 10^{-6}$ | 4.186 | 1 | $1.163 \times 10^{-6}$ | $2.613 \times 10^{19}$ | $2.613 \times 10^{13}$ | $4.660 \times 10^{-17}$ | $2.806 \times 10^{10}$ |
| 1 kilowatt hour = | 3413 | $3.600 \times 10^{13}$ | $2.655 \times 10^6$ | 1.341 | $3.600 \times 10^6$ | $8.600 \times 10^5$ | 1 | $2.247 \times 10^{25}$ | $2.247 \times 10^{19}$ | $4.007 \times 10^{-11}$ | $2.413 \times 10^{16}$ |
| 1 electron-volt = | $1.519 \times 10^{-22}$ | $1.602 \times 10^{-12}$ | $1.182 \times 10^{-19}$ | $5.967 \times 10^{-26}$ | $1.602 \times 10^{-19}$ | $3.827 \times 10^{-20}$ | $4.450 \times 10^{-26}$ | 1 | $10^{-6}$ | $1.783 \times 10^{-36}$ | $1.074 \times 10^{-9}$ |
| 1 million electron-volts = | $1.519 \times 10^{-16}$ | $1.602 \times 10^{-6}$ | $1.182 \times 10^{-13}$ | $5.967 \times 10^{-20}$ | $1.602 \times 10^{-13}$ | $3.827 \times 10^{-14}$ | $4.450 \times 10^{-20}$ | $10^{-6}$ | 1 | $1.783 \times 10^{-30}$ | $1.074 \times 10^{-3}$ |
| 1 kilogram = | $8.521 \times 10^{13}$ | $8.987 \times 10^{23}$ | $6.629 \times 10^{16}$ | $3.348 \times 10^{10}$ | $8.987 \times 10^{16}$ | $2.146 \times 10^{16}$ | $2.497 \times 10^{10}$ | $5.610 \times 10^{35}$ | $5.610 \times 10^{29}$ | 1 | $6.022 \times 10^{26}$ |
| 1 unified atomic mass unit = | $1.415 \times 10^{-13}$ | $1.492 \times 10^{-3}$ | $1.101 \times 10^{-10}$ | $5.559 \times 10^{-17}$ | $1.492 \times 10^{-10}$ | $3.564 \times 10^{-11}$ | $4.146 \times 10^{-17}$ | $9.320 \times 10^8$ | 932.0 | $1.661 \times 10^{-27}$ | 1 |

### Power

| Btu/h | ft · lb/s | hp | cal/s | kW | WATT |
|---|---|---|---|---|---|
| 1 British thermal unit per hour = 1 | 0.2161 | $3.929 \times 10^{-4}$ | $6.998 \times 10^{-2}$ | $2.930 \times 10^{-4}$ | 0.2930 |
| 1 foot-pound per second = 4.628 | 1 | $1.818 \times 10^{-3}$ | 0.3239 | $1.356 \times 10^{-3}$ | 1.356 |
| 1 horsepower = 2545 | 550 | 1 | 178.1 | 0.7457 | 745.7 |
| 1 calorie per second = 14.29 | 3.088 | $5.615 \times 10^{-3}$ | 1 | $4.186 \times 10^{-3}$ | 4.186 |
| 1 kilowatt = 3413 | 737.6 | 1.341 | 238.9 | 1 | 1000 |
| 1 WATT = 3.413 | 0.7376 | $1.341 \times 10^{-3}$ | 0.2389 | 0.001 | 1 |

### Magnetic Field

| gauss | TESLA | milligauss |
|---|---|---|
| 1 gauss = 1 | $10^{-4}$ | 1000 |
| 1 TESLA = $10^4$ | 1 | $10^7$ |
| 1 milligauss = 0.001 | $10^{-7}$ | 1 |

1 tesla = 1 weber/meter$^2$

### Magnetic Flux

| maxwell | WEBER |
|---|---|
| 1 maxwell = 1 | $10^{-8}$ |
| 1 WEBER = $10^8$ | 1 |

APPENDIX

# E Mathematical Formulas

## Geometry

Circle of radius $r$: circumference $= 2\pi r$; area $= \pi r^2$.
Sphere of radius $r$: area $= 4\pi r^2$; volume $= \frac{4}{3}\pi r^3$.
Right circular cylinder of radius $r$ and height $h$:
area $= 2\pi r^2 + 2\pi rh$; volume $= \pi r^2 h$.
Triangle of base $a$ and altitude $h$: area $= \frac{1}{2}ah$.

## Quadratic Formula

If $ax^2 + bx + c = 0$, then $x = \dfrac{-b \pm \sqrt{b^2 - 4ac}}{2a}$.

## Trigonometric Functions of Angle $\theta$

$\sin\theta = \dfrac{y}{r}$ $\cos\theta = \dfrac{x}{r}$

$\tan\theta = \dfrac{y}{x}$ $\cot\theta = \dfrac{x}{y}$

$\sec\theta = \dfrac{r}{x}$ $\csc\theta = \dfrac{r}{y}$

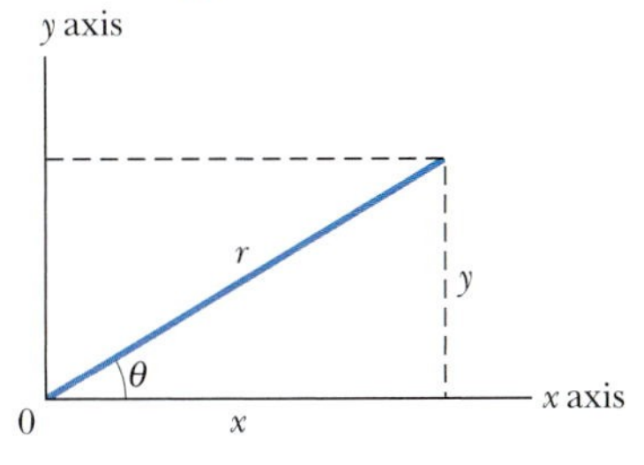


## Pythagorean Theorem

In this right triangle,
$a^2 + b^2 = c^2$

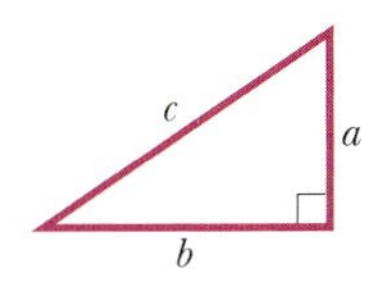


## Triangles

Angles are $A, B, C$
Opposite sides are $a, b, c$
Angles $A + B + C = 180°$
$\dfrac{\sin A}{a} = \dfrac{\sin B}{b} = \dfrac{\sin C}{c}$
$c^2 = a^2 + b^2 - 2ab\cos C$
Exterior angle $D = A + C$

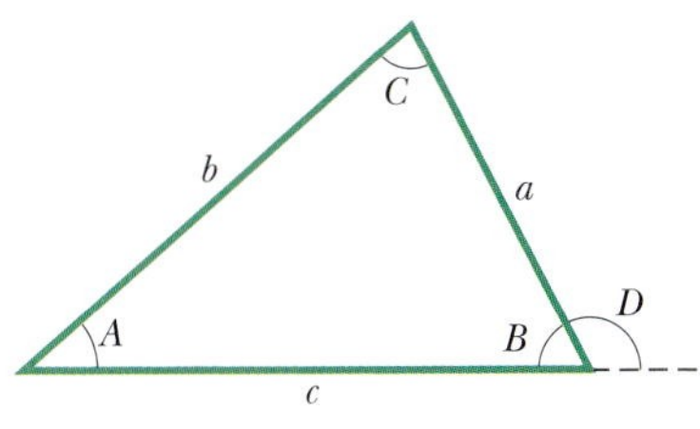


## Mathematical Signs and Symbols

$=$ equals
$\approx$ equals approximately
$\sim$ is the order of magnitude of
$\neq$ is not equal to
$\equiv$ is identical to, is defined as
$>$ is greater than ($\gg$ is much greater than)
$<$ is less than ($\ll$ is much less than)
$\geq$ is greater than or equal to (or, is no less than)
$\leq$ is less than or equal to (or, is no more than)
$\pm$ plus or minus
$\propto$ is proportional to
$\Sigma$ the sum of
$\langle x \rangle$ the average value of $x$

## Trigonometric Identities

$\sin(90° - \theta) = \cos\theta$
$\cos(90° - \theta) = \sin\theta$
$\sin\theta/\cos\theta = \tan\theta$
$\sin^2\theta + \cos^2\theta = 1$
$\sec^2\theta - \tan^2\theta = 1$
$\csc^2\theta - \cot^2\theta = 1$
$\sin 2\theta = 2\sin\theta\cos\theta$
$\cos 2\theta = \cos^2\theta - \sin^2\theta = 2\cos^2\theta - 1 = 1 - 2\sin^2\theta$
$\sin(\alpha \pm \beta) = \sin\alpha\cos\beta \pm \cos\alpha\sin\beta$
$\cos(\alpha \pm \beta) = \cos\alpha\cos\beta \mp \sin\alpha\sin\beta$
$\tan(\alpha \pm \beta) = \dfrac{\tan\alpha \pm \tan\beta}{1 \mp \tan\alpha\tan\beta}$
$\sin\alpha \pm \sin\beta = 2\sin\frac{1}{2}(\alpha \pm \beta)\cos\frac{1}{2}(\alpha \mp \beta)$
$\cos\alpha + \cos\beta = 2\cos\frac{1}{2}(\alpha + \beta)\cos\frac{1}{2}(\alpha - \beta)$
$\cos\alpha - \cos\beta = -2\sin\frac{1}{2}(\alpha + \beta)\sin\frac{1}{2}(\alpha - \beta)$

## Binomial Theorem

$$(1 + x)^n = 1 + \frac{nx}{1!} + \frac{n(n-1)x^2}{2!} + \cdots \qquad (x^2 < 1)$$

## Exponential Expansion

$$e^x = 1 + x + \frac{x^2}{2!} + \frac{x^3}{3!} + \cdots$$

## Logarithmic Expansion

$$\ln(1 + x) = x - \tfrac{1}{2}x^2 + \tfrac{1}{3}x^3 - \cdots \qquad (|x| < 1)$$

## Trigonometric Expansions ($\theta$ in radians)

$$\sin \theta = \theta - \frac{\theta^3}{3!} + \frac{\theta^5}{5!} - \cdots$$

$$\cos \theta = 1 - \frac{\theta^2}{2!} + \frac{\theta^4}{4!} - \cdots$$

$$\tan \theta = \theta + \frac{\theta^3}{3} + \frac{2\theta^5}{15} + \cdots$$

## Cramer's Rule

Two simultaneous equations in unknowns $x$ and $y$,

$$a_1x + b_1y = c_1 \quad \text{and} \quad a_2x + b_2y = c_2,$$

have the solutions

$$x = \frac{\begin{vmatrix} c_1 & b_1 \\ c_2 & b_2 \end{vmatrix}}{\begin{vmatrix} a_1 & b_1 \\ a_2 & b_2 \end{vmatrix}} = \frac{c_1b_2 - c_2b_1}{a_1b_2 - a_2b_1}$$

and

$$y = \frac{\begin{vmatrix} a_1 & c_1 \\ a_2 & c_2 \end{vmatrix}}{\begin{vmatrix} a_1 & b_1 \\ a_2 & b_2 \end{vmatrix}} = \frac{a_1c_2 - a_2c_1}{a_1b_2 - a_2b_1}.$$

## Products of Vectors

Let $\hat{\text{i}}$, $\hat{\text{j}}$, and $\hat{\text{k}}$ and be unit vectors in the $x$, $y$, and $z$ directions. Then

$$\hat{\text{i}} \cdot \hat{\text{i}} = \hat{\text{j}} \cdot \hat{\text{j}} = \hat{\text{k}} \cdot \hat{\text{k}} = 1, \qquad \hat{\text{i}} \cdot \hat{\text{j}} = \hat{\text{j}} \cdot \hat{\text{k}} = \hat{\text{k}} \cdot \hat{\text{i}} = 0,$$

$$\hat{\text{i}} \times \hat{\text{i}} = \hat{\text{j}} \times \hat{\text{j}} = \hat{\text{k}} \times \hat{\text{k}} = 0,$$

$$\hat{\text{i}} \times \hat{\text{j}} = \hat{\text{k}}, \qquad \hat{\text{j}} \times \hat{\text{k}} = \hat{\text{i}}, \qquad \hat{\text{k}} \times \hat{\text{i}} = \hat{\text{j}}.$$

Any vector $\vec{a}$ with components $a_x$, $a_y$, and $a_z$ along the $x$, $y$, and $z$ axes can be written as

$$\vec{a} = a_x\hat{\text{i}} + a_y\hat{\text{j}} + a_z\hat{\text{k}}.$$

Let $\vec{a}$, $\vec{b}$, and $\vec{c}$ be arbitrary vectors with magnitudes $a$, $b$, and $c$. Then

$$\vec{a} \times (\vec{b} + \vec{c}) = (\vec{a} \times \vec{b}) + (\vec{a} \times \vec{c})$$

$$(s\vec{a}) \times \vec{b} = \vec{a} \times (s\vec{b}) = s(\vec{a} \times \vec{b}) \quad (s = \text{a scalar}).$$

Let $\theta$ be the smaller of the two angles between $\vec{a}$ and $\vec{b}$. Then

$$\vec{a} \cdot \vec{b} = \vec{b} \cdot \vec{a} = a_xb_x + a_yb_y + a_zb_z = ab\cos\theta$$

$$\vec{a} \times \vec{b} = -\vec{b} \times \vec{a} = \begin{vmatrix} \hat{\text{i}} & \hat{\text{j}} & \hat{\text{k}} \\ a_x & a_y & a_z \\ b_x & b_y & b_z \end{vmatrix}$$

$$= \hat{\text{i}}\begin{vmatrix} a_y & a_z \\ b_y & b_z \end{vmatrix} - \hat{\text{j}}\begin{vmatrix} a_x & a_z \\ b_x & b_z \end{vmatrix} + \hat{\text{k}}\begin{vmatrix} a_x & a_y \\ b_x & b_y \end{vmatrix}$$

$$= (a_yb_z - b_ya_z)\hat{\text{i}} + (a_zb_x - b_za_x)\hat{\text{j}} + (a_xb_y - b_xa_y)\hat{\text{k}}$$

$$|\vec{a} \times \vec{b}| = ab\sin\theta$$

$$\vec{a} \cdot (\vec{b} \times \vec{c}) = \vec{b} \cdot (\vec{c} \times \vec{a}) = \vec{c} \cdot (\vec{a} \times \vec{b})$$

$$\vec{a} \times (\vec{b} \times \vec{c}) = (\vec{a} \cdot \vec{c})\vec{b} - (\vec{a} \cdot \vec{b})\vec{c}$$

## Derivatives and Integrals

In what follows, the letters $u$ and $v$ stand for any functions of $x$, and $a$ and $m$ are constants. To each of the indefinite integrals should be added an arbitrary constant of integration. The *Handbook of Chemistry and Physics* (CRC Press Inc.) gives a more extensive tabulation.

### Derivatives

**1.** $\dfrac{dx}{dx} = 1$

**2.** $\dfrac{d}{dx}(au) = a\dfrac{du}{dx}$

**3.** $\dfrac{d}{dx}(u + v) = \dfrac{du}{dx} + \dfrac{dv}{dx}$

**4.** $\dfrac{d}{dx}x^m = mx^{m-1}$

**5.** $\dfrac{d}{dx}\ln x = \dfrac{1}{x}$

**6.** $\dfrac{d}{dx}(uv) = u\dfrac{dv}{dx} + v\dfrac{du}{dx}$

**7.** $\dfrac{d}{dx}e^x = e^x$

**8.** $\dfrac{d}{dx}\sin x = \cos x$

9. $\frac{d}{dx}\cos x = -\sin x$

10. $\frac{d}{dx}\tan x = \sec^2 x$

11. $\frac{d}{dx}\cot x = -\csc^2 x$

12. $\frac{d}{dx}\sec x = \tan x \sec x$

13. $\frac{d}{dx}\csc x = -\cot x \csc x$

14. $\frac{d}{dx}e^u = e^u \frac{du}{dx}$

15. $\frac{d}{dx}\sin u = \cos u \frac{du}{dx}$

16. $\frac{d}{dx}\cos u = -\sin u \frac{du}{dx}$

**Integrals**

1. $\int dx = x$

2. $\int au\, dx = a\int u\, dx$

3. $\int (u + v)\, dx = \int u\, dx + \int v\, dx$

4. $\int x^m dx = \frac{x^{m+1}}{m+1} \quad (m \neq -1)$

5. $\int \frac{dx}{x} = \ln|x|$

6. $\int u\frac{dv}{dx}\, dx = uv - \int v\frac{du}{dx}\, dx$

7. $\int e^x\, dx = e^x$

8. $\int \sin x\, dx = -\cos x$

9. $\int \cos x\, dx = \sin x$

10. $\int \tan x\, dx = \ln|\sec x|$

11. $\int \sin^2 x\, dx = \frac{1}{2}x - \frac{1}{4}\sin 2x$

12. $\int e^{-ax}\, dx = -\frac{1}{a}e^{-ax}$

13. $\int xe^{-ax}\, dx = -\frac{1}{a^2}(ax + 1)e^{-ax}$

14. $\int x^2 e^{-ax}\, dx = -\frac{1}{a^3}(a^2x^2 + 2ax + 2)e^{-ax}$

15. $\int_0^\infty x^n e^{-ax}\, dx = \frac{n!}{a^{n+1}}$

16. $\int_0^\infty x^{2n} e^{-ax^2}\, dx = \frac{1 \cdot 3 \cdot 5 \cdots (2n - 1)}{2^{n+1}a^n}\sqrt{\frac{\pi}{a}}$

17. $\int \frac{dx}{\sqrt{x^2 + a^2}} = \ln(x + \sqrt{x^2 + a^2})$

18. $\int \frac{x\, dx}{(x^2 + a^2)^{3/2}} = -\frac{1}{(x^2 + a^2)^{1/2}}$

19. $\int \frac{dx}{(x^2 + a^2)^{3/2}} = \frac{x}{a^2(x^2 + a^2)^{1/2}}$

20. $\int_0^\infty x^{2n+1} e^{-ax^2}\, dx = \frac{n!}{2a^{n+1}} \quad (a > 0)$

21. $\int \frac{x\, dx}{x + d} = x - d\ln(x + d)$

APPENDIX F

# Properties of the Elements

All physical properties are for a pressure of 1 atm unless otherwise specified.

| Element | Symbol | Atomic Number Z | Molar Mass, g/mol | Density, $g/cm^3$ at 20°C | Melting Point, °C | Boiling Point, °C | Specific Heat, J/(g·°C) at 25°C |
|---|---|---|---|---|---|---|---|
| Actinium | Ac | 89 | (227) | 10.06 | 1323 | (3473) | 0.092 |
| Aluminum | Al | 13 | 26.9815 | 2.699 | 660 | 2450 | 0.900 |
| Americium | Am | 95 | (243) | 13.67 | 1541 | — | — |
| Antimony | Sb | 51 | 121.75 | 6.691 | 630.5 | 1380 | 0.205 |
| Argon | Ar | 18 | 39.948 | $1.6626 \times 10^{-3}$ | −189.4 | −185.8 | 0.523 |
| Arsenic | As | 33 | 74.9216 | 5.78 | 817 (28 atm) | 613 | 0.331 |
| Astatine | At | 85 | (210) | — | (302) | — | — |
| Barium | Ba | 56 | 137.34 | 3.594 | 729 | 1640 | 0.205 |
| Berkelium | Bk | 97 | (247) | 14.79 | — | — | — |
| Beryllium | Be | 4 | 9.0122 | 1.848 | 1287 | 2770 | 1.83 |
| Bismuth | Bi | 83 | 208.980 | 9.747 | 271.37 | 1560 | 0.122 |
| Bohrium | Bh | 107 | 262.12 | — | — | — | — |
| Boron | B | 5 | 10.811 | 2.34 | 2030 | — | 1.11 |
| Bromine | Br | 35 | 79.909 | 3.12 (liquid) | −7.2 | 58 | 0.293 |
| Cadmium | Cd | 48 | 112.40 | 8.65 | 321.03 | 765 | 0.226 |
| Calcium | Ca | 20 | 40.08 | 1.55 | 838 | 1440 | 0.624 |
| Californium | Cf | 98 | (251) | — | — | — | — |
| Carbon | C | 6 | 12.01115 | 2.26 | 3727 | 4830 | 0.691 |
| Cerium | Ce | 58 | 140.12 | 6.768 | 804 | 3470 | 0.188 |
| Cesium | Cs | 55 | 132.905 | 1.873 | 28.40 | 690 | 0.243 |
| Chlorine | Cl | 17 | 35.453 | $3.214 \times 10^{-3}$ (0°C) | −101 | −34.7 | 0.486 |
| Chromium | Cr | 24 | 51.996 | 7.19 | 1857 | 2665 | 0.448 |
| Cobalt | Co | 27 | 58.9332 | 8.85 | 1495 | 2900 | 0.423 |
| Copper | Cu | 29 | 63.54 | 8.96 | 1083.40 | 2595 | 0.385 |
| Curium | Cm | 96 | (247) | 13.3 | — | — | — |
| Darmstadtium | Ds | 110 | (271) | — | — | — | — |
| Dubnium | Db | 105 | 262.114 | — | — | — | — |
| Dysprosium | Dy | 66 | 162.50 | 8.55 | 1409 | 2330 | 0.172 |
| Einsteinium | Es | 99 | (254) | — | — | — | — |
| Erbium | Er | 68 | 167.26 | 9.15 | 1522 | 2630 | 0.167 |
| Europium | Eu | 63 | 151.96 | 5.243 | 817 | 1490 | 0.163 |
| Fermium | Fm | 100 | (237) | — | — | — | — |
| Fluorine | F | 9 | 18.9984 | $1.696 \times 10^{-3}$(0°C) | −219.6 | −188.2 | 0.753 |

| Element | Symbol | Atomic Number Z | Molar Mass, g/mol | Density, g/cm³ at 20°C | Melting Point, °C | Boiling Point, °C | Specific Heat, J/(g·°C) at 25°C |
|---|---|---|---|---|---|---|---|
| Francium | Fr | 87 | (223) | — | (27) | — | — |
| Gadolinium | Gd | 64 | 157.25 | 7.90 | 1312 | 2730 | 0.234 |
| Gallium | Ga | 31 | 69.72 | 5.907 | 29.75 | 2237 | 0.377 |
| Germanium | Ge | 32 | 72.59 | 5.323 | 937.25 | 2830 | 0.322 |
| Gold | Au | 79 | 196.967 | 19.32 | 1064.43 | 2970 | 0.131 |
| Hafnium | Hf | 72 | 178.49 | 13.31 | 2227 | 5400 | 0.144 |
| Hassium | Hs | 108 | (265) | — | — | — | — |
| Helium | He | 2 | 4.0026 | $0.1664 \times 10^{-3}$ | −269.7 | −268.9 | 5.23 |
| Holmium | Ho | 67 | 164.930 | 8.79 | 1470 | 2330 | 0.165 |
| Hydrogen | H | 1 | 1.00797 | $0.08375 \times 10^{-3}$ | −259.19 | −252.7 | 14.4 |
| Indium | In | 49 | 114.82 | 7.31 | 156.634 | 2000 | 0.233 |
| Iodine | I | 53 | 126.9044 | 4.93 | 113.7 | 183 | 0.218 |
| Iridium | Ir | 77 | 192.2 | 22.5 | 2447 | (5300) | 0.130 |
| Iron | Fe | 26 | 55.847 | 7.874 | 1536.5 | 3000 | 0.447 |
| Krypton | Kr | 36 | 83.80 | $3.488 \times 10^{-3}$ | −157.37 | −152 | 0.247 |
| Lanthanum | La | 57 | 138.91 | 6.189 | 920 | 3470 | 0.195 |
| Lawrencium | Lr | 103 | (257) | — | — | — | — |
| Lead | Pb | 82 | 207.19 | 11.35 | 327.45 | 1725 | 0.129 |
| Lithium | Li | 3 | 6.939 | 0.534 | 180.55 | 1300 | 3.58 |
| Lutetium | Lu | 71 | 174.97 | 9.849 | 1663 | 1930 | 0.155 |
| Magnesium | Mg | 12 | 24.312 | 1.738 | 650 | 1107 | 1.03 |
| Manganese | Mn | 25 | 54.9380 | 7.44 | 1244 | 2150 | 0.481 |
| Meitnerium | Mt | 109 | (266) | — | — | — | — |
| Mendelevium | Md | 101 | (256) | — | — | — | — |
| Mercury | Hg | 80 | 200.59 | 13.55 | −38.87 | 357 | 0.138 |
| Molybdenum | Mo | 42 | 95.94 | 10.22 | 2617 | 5560 | 0.251 |
| Neodymium | Nd | 60 | 144.24 | 7.007 | 1016 | 3180 | 0.188 |
| Neon | Ne | 10 | 20.183 | $0.8387 \times 10^{-3}$ | −248.597 | −246.0 | 1.03 |
| Neptunium | Np | 93 | (237) | 20.25 | 637 | — | 1.26 |
| Nickel | Ni | 28 | 58.71 | 8.902 | 1453 | 2730 | 0.444 |
| Niobium | Nb | 41 | 92.906 | 8.57 | 2468 | 4927 | 0.264 |
| Nitrogen | N | 7 | 14.0067 | $1.1649 \times 10^{-3}$ | −210 | −195.8 | 1.03 |
| Nobelium | No | 102 | (255) | — | — | — | — |
| Osmium | Os | 76 | 190.2 | 22.59 | 3027 | 5500 | 0.130 |
| Oxygen | O | 8 | 15.9994 | $1.3318 \times 10^{-3}$ | −218.80 | −183.0 | 0.913 |
| Palladium | Pd | 46 | 106.4 | 12.02 | 1552 | 3980 | 0.243 |
| Phosphorus | P | 15 | 30.9738 | 1.83 | 44.25 | 280 | 0.741 |
| Platinum | Pt | 78 | 195.09 | 21.45 | 1769 | 4530 | 0.134 |
| Plutonium | Pu | 94 | (244) | 19.8 | 640 | 3235 | 0.130 |
| Polonium | Po | 84 | (210) | 9.32 | 254 | — | — |
| Potassium | K | 19 | 39.102 | 0.862 | 63.20 | 760 | 0.758 |
| Praseodymium | Pr | 59 | 140.907 | 6.773 | 931 | 3020 | 0.197 |
| Promethium | Pm | 61 | (145) | 7.22 | (1027) | — | — |
| Protactinium | Pa | 91 | (231) | 15.37 (estimated) | (1230) | — | — |

| Element | Symbol | Atomic Number Z | Molar Mass, g/mol | Density, $g/cm^3$ at 20°C | Melting Point, °C | Boiling Point, °C | Specific Heat, J/(g·°C) at 25°C |
|---|---|---|---|---|---|---|---|
| Radium | Ra | 88 | (226) | 5.0 | 700 | — | — |
| Radon | Rn | 86 | (222) | $9.96 \times 10^{-3}$ (0°C) | (−71) | −61.8 | 0.092 |
| Rhenium | Re | 75 | 186.2 | 21.02 | 3180 | 5900 | 0.134 |
| Rhodium | Rh | 45 | 102.905 | 12.41 | 1963 | 4500 | 0.243 |
| Rubidium | Rb | 37 | 85.47 | 1.532 | 39.49 | 688 | 0.364 |
| Ruthenium | Ru | 44 | 101.107 | 12.37 | 2250 | 4900 | 0.239 |
| Rutherfordium | Rf | 104 | 261.11 | — | — | — | — |
| Samarium | Sm | 62 | 150.35 | 7.52 | 1072 | 1630 | 0.197 |
| Scandium | Sc | 21 | 44.956 | 2.99 | 1539 | 2730 | 0.569 |
| Seaborgium | Sg | 106 | 263.118 | — | — | — | — |
| Selenium | Se | 34 | 78.96 | 4.79 | 221 | 685 | 0.318 |
| Silicon | Si | 14 | 28.086 | 2.33 | 1412 | 2680 | 0.712 |
| Silver | Ag | 47 | 107.870 | 10.49 | 960.8 | 2210 | 0.234 |
| Sodium | Na | 11 | 22.9898 | 0.9712 | 97.85 | 892 | 1.23 |
| Strontium | Sr | 38 | 87.62 | 2.54 | 768 | 1380 | 0.737 |
| Sulfur | S | 16 | 32.064 | 2.07 | 119.0 | 444.6 | 0.707 |
| Tantalum | Ta | 73 | 180.948 | 16.6 | 3014 | 5425 | 0.138 |
| Technetium | Tc | 43 | (99) | 11.46 | 2200 | — | 0.209 |
| Tellurium | Te | 52 | 127.60 | 6.24 | 449.5 | 990 | 0.201 |
| Terbium | Tb | 65 | 158.924 | 8.229 | 1357 | 2530 | 0.180 |
| Thallium | Tl | 81 | 204.37 | 11.85 | 304 | 1457 | 0.130 |
| Thorium | Th | 90 | (232) | 11.72 | 1755 | (3850) | 0.117 |
| Thulium | Tm | 69 | 168.934 | 9.32 | 1545 | 1720 | 0.159 |
| Tin | Sn | 50 | 118.69 | 7.2984 | 231.868 | 2270 | 0.226 |
| Titanium | Ti | 22 | 47.90 | 4.54 | 1670 | 3260 | 0.523 |
| Tungsten | W | 74 | 183.85 | 19.3 | 3380 | 5930 | 0.134 |
| *Unununium | Uuu | 111 | (272) | — | — | — | — |
| *Unbium | Uub | 112 | (285) | — | — | — | — |
| Ununquadium | Uuq | 114 | (285) | — | — | — | — |
| Uranium | U | 92 | (238) | 18.95 | 1132 | 3818 | 0.117 |
| Vanadium | V | 23 | 50.942 | 6.11 | 1902 | 3400 | 0.490 |
| Xenon | Xe | 54 | 131.30 | $5.495 \times 10^{-3}$ | −111.79 | −108 | 0.159 |
| Ytterbium | Yb | 70 | 173.04 | 6.965 | 824 | 1530 | 0.155 |
| Yttrium | Y | 39 | 88.905 | 4.469 | 1526 | 3030 | 0.297 |
| Zinc | Zn | 30 | 65.37 | 7.133 | 419.58 | 906 | 0.389 |
| Zirconium | Zr | 40 | 91.22 | 6.506 | 1852 | 3580 | 0.276 |

The values in parentheses in the column of molar masses are the mass numbers of the longest-lived isotopes of those elements that are radioactive. Melting points and boiling points in parentheses are uncertain.

The data for gases are valid only when these are in their usual molecular state, such as $H_2$, He, $O_2$, Ne, etc. The specific heats of the gases are the values at constant pressure.

*Primary source*: Adapted fron J. Emsley, *The Elements*, 3rd ed., 1998, Clarendon Press, Oxford (www.webelements.com). Data on newest elements are current.

*Newest elements: As of May 2003 in the WebElements Periodic Table.

APPENDIX G

# Periodic Table of the Elements

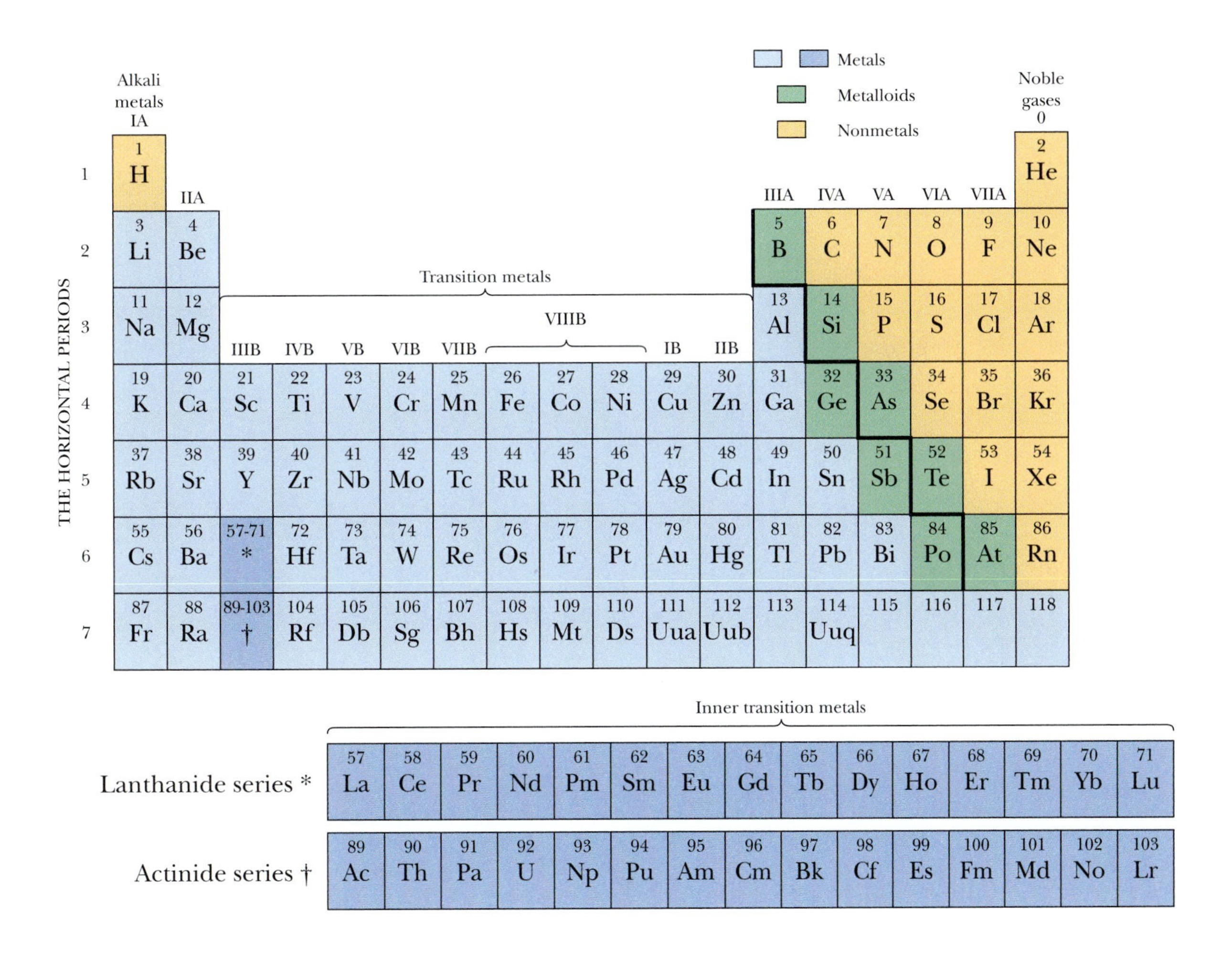

Metals
Metalloids
Nonmetals
Alkali metals
IA
Noble gases
0
IIA
IIIA IVA VA VIA VIIA
Transition metals
VIIIB
IIIB IVB VB VIB VIIB IB IIB
THE HORIZONTAL PERIODS
1
1 H
2 He
2
3 Li 4 Be 5 B 6 C 7 N 8 O 9 F 10 Ne
3
11 Na 12 Mg 13 Al 14 Si 15 P 16 S 17 Cl 18 Ar
4
19 K 20 Ca 21 Sc 22 Ti 23 V 24 Cr 25 Mn 26 Fe 27 Co 28 Ni 29 Cu 30 Zn 31 Ga 32 Ge 33 As 34 Se 35 Br 36 Kr
5
37 Rb 38 Sr 39 Y 40 Zr 41 Nb 42 Mo 43 Tc 44 Ru 45 Rh 46 Pd 47 Ag 48 Cd 49 In 50 Sn 51 Sb 52 Te 53 I 54 Xe
6
55 Cs 56 Ba 57-71 * 72 Hf 73 Ta 74 W 75 Re 76 Os 77 Ir 78 Pt 79 Au 80 Hg 81 Tl 82 Pb 83 Bi 84 Po 85 At 86 Rn
7
87 Fr 88 Ra 89-103 † 104 Rf 105 Db 106 Sg 107 Bh 108 Hs 109 Mt 110 Ds 111 Uua 112 Uub 113 114 Uuq 115 116 117 118
Inner transition metals
Lanthanide series *
57 La 58 Ce 59 Pr 60 Nd 61 Pm 62 Sm 63 Eu 64 Gd 65 Tb 66 Dy 67 Ho 68 Er 69 Tm 70 Yb 71 Lu
Actinide series †
89 Ac 90 Th 91 Pa 92 U 93 Np 94 Pu 95 Am 96 Cm 97 Bk 98 Cf 99 Es 100 Fm 101 Md 102 No 103 Lr


The names of elements 104 through 109 (Rutherfordium, Dubnium, Seaborgium, Bohrium, Hassium, and Meitnerium, respectively) were adopted by the International Union of Pure and Applied Chemistry (IUPAC) in 1997. As of May 2003, elements 110, 111, 112, and 114 have been discovered. See www.webelements.com for the latest information and newest elements.

# Answers to Reading Exercises and Odd-Numbered Problems

**(Answers that involve a proof, graph, or otherwise lengthy solution are not included.)**

## Chapter 1

**RE 1-1:** Examples include second or hour, meter or inch, and gram or kilogram.

**RE 1-2:** A 12-inch ruler would more likely change less over time than your foot, especially if you are still growing.

**RE 1-3:** The length of one day or the time it takes for the earth to rotate 360° about its own axis is not constant, because the speed of the earth's rotation is slowly decreasing with time.

**RE 1-4:** (a) Since 24 h of time occurs for each 360° of rotation or 4 min for each degree of longitude or 240 s for each degree of longitude, 20 min and 13 s will relate to a rotation or longitude change of (1213 s)/(240 s/deg) = 5.05 degrees of longitude change. (b) If the clock is off by 2 min or 120 s, the longitude will be off by (120 s)/(240 s/degree) = 0.5 degrees of longitude. (c) 360° or one revolution relates to one circumference of length. Therefore 0.5°/360° = $x$/(24 000 nautical miles), or $x = 33.3$ nautical miles off course. Sailor beware!

**RE 1-5:** (a) If your watches are synchronized, you should measure the same time for the flash. For the same duration of time between the flash and thunder you both should have accurate watches and be located close to one another. (b) No, the 12 h (smaller) clock shows a time of 7:44 or a total elapsed time of 464 min since 12 o'clock. This is 464 min/(1440 min/day) = .322 day elapsed. The 10 h (larger) clock shows a time of 8.23 hours elapsed since 10 o'clock (12 o'clock on the other scale) or 8.23/(20 hr/day) = .412 day elapsed.

**RE 1-6:** One of many possible procedures would be to use the balance to determine the amount of clay equal to 1 kg. Divide the clay into 1000 equal volume pieces. Assuming the density of the clay is uniform, each clay piece now has a mass of 1 gram. Use these pieces with the balance and the object whose mass is to be determined to find its mass.

**RE 1-7:** (a) It is correct to write 1 min/60 s = 1 because 1 minute and 60 seconds are the same *length* of time. It is meaningless to say 1/60 = 1 when no units are specified. These numbers are not the same in the absence of the context of the units. (b) In terms of conversion factors and chain-link conversions, the number of minutes in a day is given by

$$1\ \text{d} = (1\ \text{d})\left(\frac{24\ \text{h}}{1\ \text{d}}\right)\left(\frac{60\ \text{min}}{1\ \text{h}}\right) = 1440\ \text{m}.$$

**RE 1-8:** (a) 2. (b) Exact, if the cows were counted. (c) 6. Remember that the leading zeros don't count. (d) 7. Trailing zeros do count. (e) Exact, by definition.

**RE 1-9:** (a) 11. (b) Probably 3, we can't be sure. (c) $2.09 \times 10^{10}$ ft. (d) $10^{10}$ ft (ten to the tenth feet).

**RE 1-10:** (a) You should keep all digits for intermediate results; thus you should use $A = 1.96\ \text{cm}^2$ for calculating $V$. (b) 2.7 cm$^3$; in this situation the answer can be to no more significant figures than the original data. (c) 2.8 cm$^3$.

**RE 1-11:** (a) 27; (b) 198.0; (c) 0.6; (d) 0.9986, see *Evaluating Functions* in Appendix A, Section 5. (e) Since five is an exact number, the four significant numbers in the average length limits the answer to 10.67 m.

**RE 1-12:** (a) 0.01 s; (b) .01 s out of 1.78 s or .01/1.78 = 0.00562, or about 0.6%.

## Problems

1. (a) 0.98 ft/ns; (b) 0.30 mm/ps. 3. C, D, A, B, E; the important criterion is the constancy of the daily variation, not its magnitude. 5. 0.12 AU/min 7. 2.1 h. 9. $1.21 \times 10^{12}$ $\mu$s. 11. (a) 160 rods; (b) 40 chains. 13. (a) $4.00 \times 10^4$ km; (b) $5.10 \times 10^8$ km$^2$. 15. $1.9 \times 10^{22}$ cm$^3$. 17. $1.1 \times 10^3$ acre-feet. 19. $9.0 \times 10^{49}$. 21. (a) $10^3$ kg; (b) 158 kg/s. 23. (a) $1.18 \times 10^{-29}$ m$^3$. 25. 3.8 mg/s. 27. $8 \times 10^2$ km. 29. $6.0 \times 10^{26}$. 31. (a) 60.8 W; (b) 43.3 Z. 33. 89 km. 35. $\approx 1 \times 10^{36}$. 37. 700 to 1500. 39. (a) 293 U.S. bushels; (b) $3.81 \times 10^3$ U.S. bushels. 41. $9.4 \times 10^{-3}$. 43. 5.95 km. 45. $1.9 \times 10^5$ kg. 47. $2 \times 10^4$ to $4 \times 10^4$. 49. 10.7. 59. (a) 13 597 kg; (b) 4917 L; (c) 6172 kg; (d) 20 075 L; (e) 45%

## Chapter 2

**RE 2-1:** (b), (c), and (d).

**RE 2-2:** Correct order: (c), (b), and (a).

**RE 2-3:** Yes, the displacement can be positive as long as the particle moves to a less negative position.

**RE 2-4:** (a) Average velocity is the displacement divided by the total time $\langle v_x \rangle = 10$ mi/30 min $= 0.33$ mi/min due east. (b) Average speed is the total distance traveled divided by the total time $\langle s \rangle = 30$ mi/30 min $=$ 1 mi/min. (c) The answers are different because the displacement is different from the total distance traveled in the 30 minute time period.

**RE 2-5:** Instantaneous speed. The speedometer only tells you the speed at which you are currently driving, not your acceleration or direction.

**RE 2-6:** (a) Remember that the velocity is the time derivative of the position equation. The velocity will be constant if it has no time dependence. Position equations 1 and 4 give a constant velocity. (b) The velocity is negative in equations 2 and 3.

**RE 2-7:** In returning to $x_1$ the total displacement $\Delta x = x_1 - x_1$ is zero. Since $\langle v_x \rangle = \Delta x / \Delta t$, the average velocity is also zero.

**RE 2-8:** (a) +, (b) −, (c) −, (d) +; remember that $\vec{a}$ will have the same direction as $\Delta \vec{v}$ or $\vec{v}_2 - \vec{v}_1$.

**RE 2-9:** The equations of Table 2-1 apply when $a_x$ is constant. Take the second derivative of $x$ with respect to $t$ to find $a_x$. Only equations 1, 3 and 4 give a constant $a_x$ ($a_x = 0$ is a constant).

## Problems

1. 414 ms. 3. (a) +40 km/h; (b) 40 km/h. 5. (a) 73 km/h; (b) 68 km/h; (c) 70 km/h; (d) 0. 7. (a) 0, −2, 0, 12 m; (b) +12 m; (c) +7 m/s. 9. 1.4 m. 11. (a) −6 m/s; (b) negative $x$ direction; (c) 6 m/s; (d) first smaller, then zero, and then larger; (e) yes ($t = 2$s); (f) no. 13. 100 m. 15. (a) velocity squared; (b) acceleration; (c) $m^2/s^2$, $m/s^2$. 17. 20 $m/s^2$, in the direction opposite to its initial velocity. 19. (a) $m/s^2$, $m/s^3$; (b) 1.0 s; (c) 82 m; (d) −80 m; (e) 0, −12, −36, −72 m/s; (f) −6, −18, −30, −42 $m/s^2$. 21. 0.10 m. 23. (a) 1.6 m/s; (b) 18 m/s. 25. (a) $3.1 \times 10^6$ s = 1.2 months; (b) $4.6 \times 10^{13}$ m. 27. $1.62 \times 10^{15}$ $m/s^2$. 29. 2.5 s. 31. (a) 3.56 $m/s^2$; (b) 8.43 m/s. 33. (a) 5.00 m/s; (b) 1.67 $m/s^2$; (c) 7.50 m. 35. (a) 0.74 s; (b) −6.2 $m/s^2$. 37. (a) 10.6 m; (b) 41.5 s. 39. (a) 30 s; (b) 300 m. 41. (a) 54 m, 18 m/s, −12 $m/s^2$; (b) 64 m at $t = 4.0$ s; (c) 24 m/s at $t = 2.0$ s; (d) −24 $m/s^2$; (e) 18 m/s. 49. (a) 0.75 s; (b) 50 m. 57. Since there is some latitude in what might be considered "the right answer" here, we have elected to mention some Web sites (current as of May 2002) where graphs for model rocket kinematics are shown: http://www.rocket-roar.com/rap/alt.html; http://mks.niobrara.com/altitude.html; http://www.boilerbay.com/rockets/; 59. 40 m.

## Chapter 3

**RE 3-1:** (a) The velocity of the cart on the carpet goes to zero at $t = 1.1$ s. (b) The velocity of the cart on the track at $t = 1.1$ s is approximately 0.65 m/s, so it still has (0.65 m/s/0.80 m/s) or 81% of its initial speed.

**RE 3-2:** (a) An elevator or car starting or stopping, or a merry-go-round moving at a constant speed. (b) The person feels heavy during startup and light during stopping. Objects, such as a marble, start to move with no apparent reason on the merry-go-round floor.

**RE 3-3:** (a) No acceleration: Sliding a block along a table with a small steady force or shoving on a huge object like a desk or car, etc., can result in either constant velocity motion or an inability to move the object (desk or car). (b) Acceleration: Pushing hard on a sliding block, pushing on a rolling ball, pushing or pulling someone on a vehicle with wheels, etc.

**RE 3-4:** You would attach one end of the rubber band to a post and hook the other end of the rubber band to a calibrated spring scale. Then you would record the unstretched length of the rubber band and the fact that the force on it is 0 N. Next you would pull on the rubber band with the spring scale until it reads 1 N and record the new length of the rubber band. Then you would repeat the process as the spring scale reads 2 N, 3 N, etc., recording the rubber-band length each time. In that way you can generate either a look-up table or a graph of force vs. rubber-band length. If greater precision is needed, you could take data for many more force-scale readings.

**RE 3-5:** (a) $\vec{F} = (-26\text{ N})\hat{i}$, $\vec{a} = (-0.42\text{ m/s}^2)\hat{i}$; (b) $m = F/a = 62$ kg; (c) 62 kg

**RE 3-6:** (a) The mass measurement in part (b) above uses the ratio of the force to acceleration and hence is the inertial mass. (b) We assumed that the student is on the surface of planet Earth and that the bathroom scale was calibrated for the same planet.

**RE 3-7:** In both cases (a) and (b) the acceleration is zero, therefore the net force must also be zero. This will require all three forces to add to zero as vectors. (a) This requires $\vec{F}_C$ to point to the left in the diagram with a magnitude of 2 N so $\vec{F}_C = (-2\text{ N})\hat{i}$. (b) Since the acceleration is also zero in this case, we still have $\vec{F}_C = (-2\text{ N})\hat{i}$.

**RE 3-8:** (a) Bottom right cart has a net force of −5 N, top left has +4 N, top right has −1 N, and bottom left has a net force of zero. (b) Since the acceleration and net force are directly proportional, the accelerations rank in the same order.

**RE 3-9:** In the chosen coordinate system, all the accelerations in the $v$ vs. $t$ graphs shown is Fig. 3-2 are negative since the slopes are negative. (a) The box on carpet acceleration is about −3.9 $m/s^2$ as determined by calculating the slope of the $v$ vs. $t$ graph. Slope = (0.00 − 0.90)(m/s)/(0.23 − 0.00)(s). (b) The cart on track acceleration is about −0.15 $m/s^2$ as determined by calculating the slope of the $v$ vs. $t$ graph. Slope = (0.62 − 0.80)(m/s)/(1.2 − 0.0)(s).

**RE 3-10:** (a) There appear to be no other horizontal forces on the moving objects except friction. Thus, we can assume that the net force on each object is due to a friction force. This friction force seems to be constant since the acceleration is constant and we assume that $F_x^{net} = ma_x$. (b) Box on carpet $F_x^{fric} = ma_x =$ 0.5 kg × (− 3.9 $m/s^2$) = −2 N. It points to the left. (c) Cart on track $F_x^{fric} = ma_x = 0.5$ kg × (−0.15 $m/s^2$) = −0.08 N. It also points to the left.

**RE 3-11:** (a) A tossed object is changing its velocity at all times. Just before it reaches the top of its flight it has a positive velocity and just after it has a negative velocity. Since acceleration is rate of change of velocity over time, even the instantaneous acceleration doesn't go to zero over an infinitesimal time interval. (b) The Fig. 3-22 graph of velocity vs. time is linear with a constant negative slope. Since slope of a $v_y$ vs. $t$ graph represents the acceleration component $a_y$, then $a_y$ = constant so $\vec{a} = a_y\hat{j}$ is constant.

**RE 3-12:** Change every $x$ in the two equations in Table 2-1 to a $y$. Then replace $a_y$ (previously $a_x$) with $-g$.

**RE 3-13:** (a) The unmagnetized paperclip will be attracted to the magnet and, in turn, the magnet will be attracted toward the paperclip. Newton's Third Law tells us that these attractive forces will be equal in magnitude to one another but opposite in direction; the force on the magnet will be to the left and the force on the paperclip will be to the right. (b) Newton's Third Law applies to all forces of interaction of which this is just one example.

## Problems

1. 16 N. 3. (a) 0.02 m/s$^2$; (b) $8 \times 10^4$ km; (c) $2 \times 10^3$ m/s. 5. $1.2 \times 10^5$ N. 7. (a) $4.9 \times 10^5$ N; (b) $1.5 \times 10^6$ N. 9. (a) 245 m/s$^2$; (b) 20.4 kN. 11. (a) 8.0 m/s; (b) $+x$ direction. 13. 8.0 cm/s$^2$. 15. $1.8 \times 10^4$ N. 17. (a) 31.3 kN; (b) 24.4 kN. 19. $2Ma/(a + g)$. 21. 2.4 N. 23. (a) 1.23 N; (b) 2.46 N; (c) 3.69 N; (d) 4.92 N; (e) 6.15 N; (f) 0.25 N. 25. (a) 3.2 s; (b) 1.3 s. 27. (a) 3.70 m/s; (b) 1.74 m/s; (c) 0.154 m. 29. 4.0 m/s. 31. 22 cm and 89 cm below the nozzle. 33. (a) 5.4 s; (b) 41 m/s. 35. (a) 1.23 cm; (b) 4 times, 9 times, 16 times, 25 times. 37. (a) 29.4 m; (b) 2.45 s. 39. (a) 3260 N (b) $2.7 \times 10^3$ kg; (c) 1.2 m/s 41. (a) 17 s; (b) 290 m. 43. (a) 11 N; (b) 2.2 kg; (c) 0; (d) 2.2 kg. 45. (a) 494 N, up; (b) 494 N, down. 47. (a) 1.1 N. 49. 5.1 m/s. 51. (a) 466 N; (b) 527 N.

# Chapter 4

**RE 4-1:** Displacement (1) is identical as the ball ends up going a net distance of 6 meters north and 3 meters west. Displacement (2) is different. It actually has an equal magnitude but the ball has moved in the opposite direction. *Note*: Displacement does not depend on where something starts or ends, but only on how much and in what direction its position has changed relative to where it started.

**RE 4-2:** (a) The maximum magnitude occurs when the two vectors point in the same direction. This gives a magnitude for vector $\vec{c}$ of 3 m + 4 m = 7 m. (This answer is not correct without a unit attached.) (b) The minimum magnitude occurs when the two vectors point in the opposite directions. This gives a magnitude for vector $\vec{c}$ of 4 m − 3 m = 1 m.

**RE 4-3:** Methods (c), (d), and (f) work since the parallelogram methods (c) and (d) show that the same correct resultant can be obtained regardless of the order in which components are added. Method (f) shows an equivalent construction using components. All the other vectors point in the wrong directions.

**RE 4-4:** The vectors in figures (b) and (d) have the same components as the standard vector.

**RE 4-5:** Compare Figs 4-12 and 4-13.

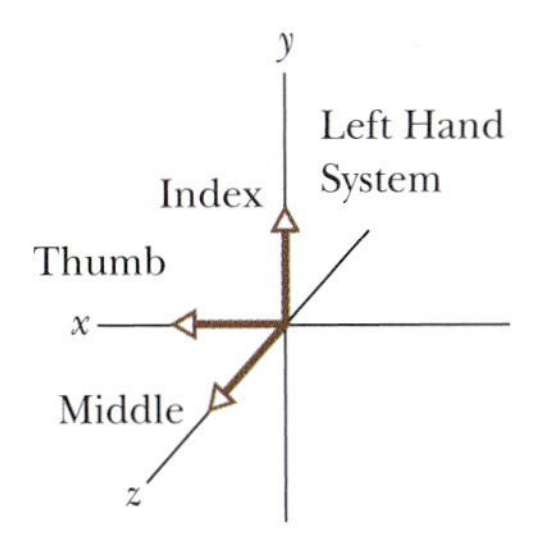


**RE 4-6:** (a & b). The $x$- and $y$-components of $\vec{d}_1$ are both positive. The $x$-component of $\vec{d}_2$ is positive but the $y$-component points down in a negative direction. (c) Using the parallelogram method to get the vector sum of $\vec{d}_1$ and $\vec{d}_2$ results in a vector that has both $x$- and $y$-components that are positive.

**RE 4-7:** This is a kind of artificial question since units of force and acceleration are different as are units of displacement and velocity. However, if the scalars (mass and time respectively) act as compressors or stretchers, then the simplistic answers would be (a) The force vector would point in the same direction as the acceleration vector but be three times as long. (b) The velocity vector would point off in the same direction as the displacement vector and be twice as long since the displacement was divided by 0.5 s.

**RE 4-8:**

(a) $\vec{F} = m\vec{a} = 3.0\,\text{kg}[(1.8\,\text{m/s}^2)\hat{\text{i}} + (1.0\,\text{m/s}^2)\hat{\text{j}}] = (5.4\,\text{N})\hat{\text{i}} + (3.0\,\text{N})\hat{\text{j}}$.

(b) $\langle\vec{v}\rangle = \dfrac{\Delta\vec{r}}{\Delta t} = \dfrac{(3.2\,\text{m})\hat{\text{i}} + (-0.8\,\text{m})\hat{\text{j}}}{0.5\,\text{s}} = (6.4\,\text{m/s})\hat{\text{i}} + (-1.6\,\text{m/s})\hat{\text{j}}$.

## Problems

1. The displacements should be (a) parallel, (b) antiparallel, (c) perpendicular. 3. (a) 5; (b) 1; (c) 7. 5. (a) −2.5 m; (b) −6.9 m. 7. (a) 47.2 m; (b) 122°. 9. (a) 168 cm; (b) 32.5° above the floor. 11. (a) 6.42 m; (b) no; (c) yes; (d) yes; (e) a possible answer: $(4.30\,\text{m})\hat{\text{i}} + (3.70\,\text{m})\hat{\text{j}} + (3.00\,\text{m})\hat{\text{k}}$; (f) 7.96 m. 13. (a) 370 m; (b) 36° north of east; (c) 425 m; (d) the distance. 15. (a) $(-9\,\text{m})\hat{\text{i}} + (10\,\text{m})\hat{\text{j}}$; (b) 13 m; (c) + 132°. 17. (a) 4.2 m; (b) 40° east of north; (c) 8.0 m; (d) 24° north of west. 19. (a) $(3.0\,\text{m})\hat{\text{i}} - (2.0\,\text{m})\hat{\text{j}} + (5.0\,\text{m})\hat{\text{k}}$; (b) $(5.0\,\text{m})\hat{\text{i}} - (4.0\,\text{m})\hat{\text{j}} - (3.0\,\text{m})\hat{\text{k}}$; (c) $(-5.0\,\text{m})\hat{\text{i}} + (4.0\,\text{m})\hat{\text{j}} + (3.0\,\text{m})\hat{\text{k}}$. 21. (a) 38 m; (b) 320°; (c) 130 m; (d) 1.2°; (e) 62 m; (f) 130°. 23. (a) 1.59 m; (b) 12.1 m; (c) 12.2 m; (d) 82.5°. 29. (a) Put axes along cube edges, with the origin at one corner. Diagonals are $a\hat{\text{i}} + a\hat{\text{j}} + a\hat{\text{k}}$, $a\hat{\text{i}} + a\hat{\text{j}} - a\hat{\text{k}}$, $a\hat{\text{i}} - a\hat{\text{j}} - a\hat{\text{k}}$, $a\hat{\text{i}} - a\hat{\text{j}} + a\hat{\text{k}}$; (b) 54.7°; (c) $\sqrt{3}\,a$. 31. 4.1. 33. (a) 103 km; (b) 60.9° north of due west. 35. (a) 15 m; (b) south; (c) 6.0 m; (d) north. 37. 5.0 km, 4.3° south of due west. 39. 5.39 m at 21.8° left of forward. 41. (a) 4.28 m; (b) 11.7 m. 43. (a) −80 m; (b) 110 m; (c) 143 m; (d) +168° (counterclockwise). 45. 3.6 m. 47. (a) 1.84 m; (b) 69° north of east. 49. (a) 9.51 m; (b) 14.1 m; (c) 13.4 m; (d) 10.5 m. 51. (a) $9.19\hat{\text{i}} + 7.71\hat{\text{j}}$; (b) $14.0\hat{\text{i}} + 3.41\hat{\text{j}}$

# Chapter 5

**RE 5-1:** (a) No, because in Fig. 5-5 the vertical positions of the ball on the right are the same as those of the ball on the left. (b) No. The horizontal positions of the ball on the right are equally spaced, indicating that horizontal velocity of the ball is constant and unaffected by the falling.

**RE 5-2:** The skateboarder's vertical motion is independent of his horizontal velocity. This is why the skateboarder lands back on his skateboard after his jump.

**RE 5-3:** (a) At each of the three points, the force vector points straight down and has a constant magnitude and (b) the same is true for the three acceleration vectors. (c) The horizontal component of each of the three velocity vectors points to the right and has a constant size. The vertical component of the velocity at the left point is directed straight upward and is slightly larger than the common size of the horizontal velocity components. The vertical component of the velocity at the center point is zero, while at the right point it is directed downward and is smaller in size than the horizontal velocity component.

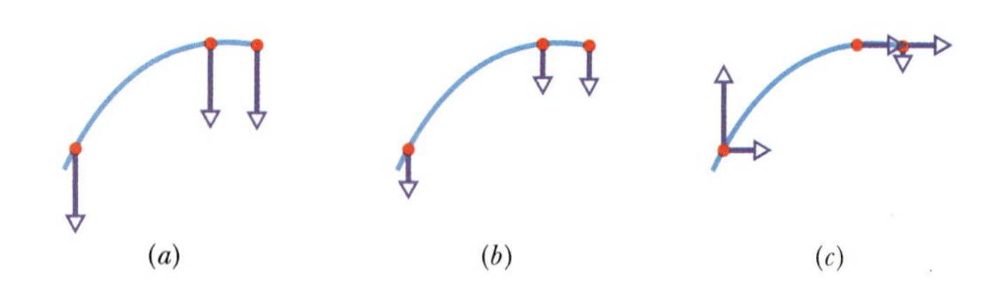


**RE 5-4:** (a) The $x$-component of velocity is not changing and is the slope of Fig. 5-9. From the data in the figures, the slope is about 2.3 m/s. The initial $y$-component of velocity is the initial slope of Fig 5-10, which is about 3.5 m/s. The launch angle will be the inverse tangent of 3.5/2.3 or about 57°. (b) Using a protractor about 57°, too.

**RE 5-5:** (a) The horizontal component of velocity remains constant. (b) The vertical component of velocity is changing constantly as there is a vertical acceleration. (c) The horizontal component of its acceleration is zero. The only force (gravity) acting is in the vertical direction. (d) The vertical component of its acceleration is constant (9.8 m/s$^2$ downward).

**RE 5-6:** (a) Using Eq. 5-15 and noting that $\Delta x = 8$ m and $\Delta y = -6$ m gives a displacement of $\Delta\vec{r} = (8\text{ m})\hat{\imath} + (-6\text{ m})\hat{\jmath}$. (b) No, since it has components along both axes.

**RE 5-7:** (a) When traveling clockwise, the $x$-component of the particle's velocity is positive when it is in the I and II quadrant, and its $y$-component is negative in the I quadrant, so the particle is now in the I quadrant. (b) When traveling counterclockwise, the $x$-component of the particle's velocity is positive when it is in the III and IV quadrant, and it's $y$-component is negative in the III quadrant, so the particle is then in the III quadrant.

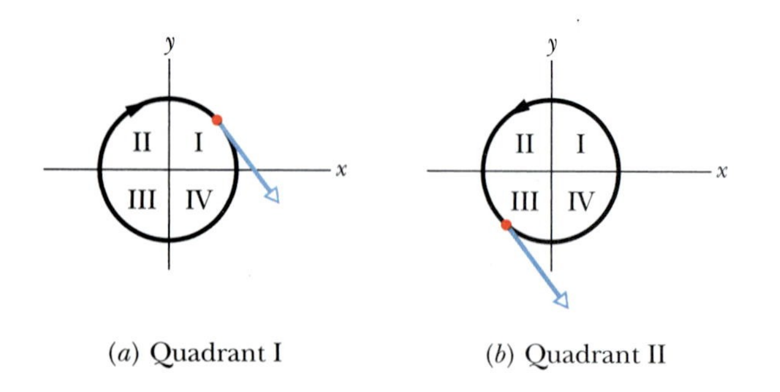


(a) Quadrant I (b) Quadrant II

**RE 5-8:** Remember that the $x$-component of acceleration will be in the direction of the change in the $x$-component of velocity and the $y$-component of acceleration will be in the direction of the change in the $y$-component of velocity. Just knowing the trajectory or path of the particle does not give you the direction of the acceleration. You also need to know how the velocity is changing as the particle travels along its trajectory. Therefore, if the change in the velocity vector is in the direction of the path of the particle, $\vec{a}$ will be tangent to the trajectory. However, you will study other situations (Section 5-7) where $\vec{a}$ is actually perpendicular to the trajectory, that is, the change in the velocity is perpendicular to the trajectory.

**RE 5-9:** The centripetal force is always inward toward the center of the curve. According to Newton's First Law, the passenger wants to travel in a straight line unless acted upon by a force. The centripetal force acts on the passenger through the friction between the passenger and the car seat. If that frictional force is not strong enough, the passenger tends to travel in a straight line and slides to the outside edge of the seat, where both the seat and the side of the car can provide the centripetal force needed to move your body in a curved path.

## Problems

1. (a) 62 ms; (b) 480 m/s. 3. (a) 0.205 s; (b) 0.205 s; (c) 20.5 cm; (d) 61.5 cm. 5. (a) 2.00 ns; (b) 2.00 mm; (c) $1.00 \times 10^7$ m/s; (d) $2.00 \times 10^6$ m/s. 7. (a) 16.9 m; (b) 8.21 m; (c) 27.6 m; (d) 7.26 m; (e) 40.2 m; (f) 0. 9. 4.8 cm. 13. (a) 11 m; (b) 23 m; (c). 17 m/s; (d) 63° below the horizontal. 15. (a) 24 m/s; (b) 65° above the horizontal. 17. (a) 10 s; (b) 897 m. 19. the third. 21. (a) 202 m/s; (b) 806 m; (c) 161 m/s; (d) −171 m/s. 23. (a) yes; (b) 2.56 m. 25. between the angles 31° and 63° above the horizontal. 27. (a) $(-5.0\text{ m})\hat{\imath} + (8.0\text{ m})\hat{\jmath}$; (b) 9.4 m; (c) 122°; (e) $(8\text{ m})\hat{\imath} + (-8\text{ m})\hat{\jmath}$; (f) 11 m; (g) −45°. 29. (a) $(-7.0\text{ m})\hat{\imath} + (12\text{ m})\hat{\jmath}$; (b) $xy$ plane. 31. 8.43 m at −129°. 33. 7.59 km/h, 22.5° east of north; 35. (a) $(3.00\text{ m/s})\hat{\imath} + (-8.00\text{ m/s}^2)t\,\hat{\jmath}$; (b) $(3.00\text{ m/s})\hat{\imath} + (-16.00\text{ m/s})\hat{\jmath}$; (c) 16.3 m/s; (d) −79.4° 37. 0.421 m/s at 3.1° west of due north. 39. (a) $(6.00\text{ m})\hat{\imath} + (-106\text{ m})\hat{\jmath}$; (b) $(19.0\text{ m/s})\hat{\imath} + (-224\text{ m/s})\hat{\jmath}$; (c) $(24.0\text{ m/s}^2)\hat{\imath} + (-336\text{ m/s}^2)\hat{\jmath}$; (d) −85.2° to $+x$. 41. (a) $(-1.5\text{ m/s})\hat{\jmath}$; (b) $(4.5\text{ m})\hat{\imath} + (-2.25\text{ m})\hat{\jmath}$. 43. (a). 45 m; (b) 22 m/s. 45. (a) $(8\text{ m/s}^2)t\,\hat{\jmath}$; (b) $(8\text{ m/s}^2)\hat{\jmath}$. 47. (a) 22 m; (b) 15 s. 49. (a) 7.49 km/s; (b) 8.00 m/s$^2$. 51. (a) 19 m/s; (b) 35 rev/min; (c) 1.7 s. 53. (a) 0.034 m/s$^2$; (b) 84 min. 55. (a) 12 s; (b) 4.1 m/s$^2$, down; (c) 4.1 m/s$^2$, up. 57. 160 m/s$^2$. 59. (4.00 m, 6.00 m)

## Chapter 6

**RE 6-1:** If you gather the tails of the three vectors shown in the helicopter diagram, you get the free-body diagram shown in (c).

**RE 6-2:** Use the balance in Fig. 3-9 and place one object on the left pan and the other object on the right pan. If the two objects have the same mass they will balance one another. They would have the same weight if they both gave the same reading on the spring scale. Also you could realize that if they have the same mass, they have the same weight since $W =$ mg and $g$ is a constant. The weight and mass are not the same. The weight is a force, and the mass is mass. Yes, since the weight and mass are proportional, the ratios are the same.

**RE 6-3:** It's true that the planet is yanking down on the patient but this is a force equal to his weight. However, since the normal force from the floor is equal and opposite, there is no net force and hence no acceleration.

**RE 6-4:** (a) In this case, at constant speed $a$ equals zero and thus the net force must equal zero, requiring $\vec{N}$ and $\vec{F}^{\text{grav}}$ to be equal in magnitude and opposite in direction. (b) Since the only two forces acting on the block are $\vec{N}$ and $\vec{F}^{\text{grav}}$, to have an upward acceleration we must have a net upward force, meaning that the magnitude of $\vec{N}$ is now larger than that of $\vec{F}^{\text{grav}}$. (c) Slowing down means an acceleration or net force in the downward direction, requiring the magnitude of $\vec{F}^{\text{grav}}$ to be larger than that of $\vec{N}$. What do you think would happen to $\vec{N}$ if the elevator cable broke and the block fell freely with $a = g$?

**RE 6-5:** In both answers to follow we are assuming the only forces acting on the block in the horizontal direction are the friction force and the pull of the cord, which is what the force sensor is measuring. Since in both cases there is no acceleration, these two forces must be equal and opposite, allowing us to equate the force sensor reading to the frictional force. (a) From the graph it looks like the block breaks free when the force is about 9.5 N. The total mass is 0.7956 kg, and the normal force that equals the weight is $mg$; therefore using Eq. 6-11, we find $\mu^{\text{stat}} = 9.5/(0.7956 \times 9.8) = 1.22$. Notice that the coefficient of friction has no units. (b) From the graph, the force needed to keep the block moving at a constant speed is about 3.0 N. Using Eq. 6-10, $\mu^{\text{kin}} = 3.0/(0.7956 \times 9.8) = 0.38$.

**RE 6-6:** (a) Zero; (b) 5 N; (c) No; (d) Yes, there is now a net force of 2 N on the block causing it to accelerate; (e) 8 N.

**RE 6-7:** It is true that friction has both a bad side and a good side. Friction always tries to retard motion. If you desire that motion then friction is bad—for example, the pistons in your car engine—and we do everything we can (lubricants) to eliminate it. However, there are other times when we don't want motion (slippage) to occur, as when we are walking or riding a bike, and the force of friction allows us to do these activities.

**RE 6-8:** Think of the cord as an object with a mass you are trying to accelerate with only two forces—the one at one end from the hand and the other at the other end from the block. We will assume that the length of the cord hanging down on each side is the same so we can ignore the force of gravity on the cord. (a) If the cord is not accelerating then the magnitudes of the two forces are equal and cancel. (b) If the block is accelerating then so is the cord and the force of the hand on the cord is greater than that of the block. (c). In this case the acceleration is opposite to b and the pull force of the hand is less than the pull force due to the block.

**RE 6-9:** Look at Eq. 6-25. The only things in this equation that will change with the size of the drops are the mass, $m$, and the cross-sectional area $A$. So for this exercise $v_t^2$ is proportional to $m/A$. How will this ratio change with the size of the drops? $A$ changes as $r^2$ and $m$ changes as $(\rho_{\text{water}})$(volume) and since volume goes as $r^3$, we finally determine that $m/A$ and hence $v_t^2$ goes as $r$. Therefore, large drops have greater speeds than small drops.

## Problems

1. (a) $F_x = 1.88$ N; (b) $F_y = 0.684$ N; (c) $(1.88\text{ N})\hat{i} + (0.684\text{ N})\hat{j}$. 3. 2.9 m/s$^2$. 5. $(3\text{ N})\hat{i} + (-11\text{ N})\hat{j}$. 7. (a) $(-32\text{ N})\hat{i} + (-21\text{ N})\hat{j}$; (b) 38 N; (c) 213° from $+x$. 9. (a) 108 N; (b) 108 N; (c) 108 N. 11. (a) 200 N; (b) 120 N. 13. 0.61. 15. (a) 190 N; (b) 0.56 m/s$^2$. 17. (a) 0.13 N; (b) 0.12. 19. (a) no; (b) $(-12\text{ N})\hat{i} + (5\text{ N})\hat{j}$. 23. (a) 300 N; (b) 1.3 m/s$^2$. 25. (a) 66 N; (b) 2.3 m/s$^2$. 27. (b) $3.0 \times 10^7$ N. 29. 100 N. 31. (a) 0; (b) 3.9 m/s$^2$ down the incline; (c) 1.0 m/s$^2$ down the incline. 33. (a) 3.5 m/s$^2$; (b) 0.21 N; (c) blocks move independently. 35. 490 N 37. (a) 6.1 m/s$^2$, leftward; (b) 0.98 m/s$^2$, leftward. 39. $g(\sin\theta - \sqrt{2}\mu^{\text{kin}}\cos\theta)$. 41. 9.9 s. 43. 6200 N. 45. 2.3. 47. 1.5 mm. 49. (a) 68 N (b) 73 N. 51. (a) $2.2 \times 10^{-3}$ N; (b) $3.7 \times 10^{-3}$ N. 53. (a) $4.6 \times 10^3$ N for each bolt; (b) $5.8 \times 10^3$ N. 55. (a) 180 N; (b) 640 N. 57. (a) 3.1 N; (b) 14.7 N. 59. (a) $6.8 \times 10^3$ N (b) $-21°$ or 159°. 61. (b) $F/(m + M)$; (c) $MF/(m + M)$; (d) $F(m + 2M)/2\ (m + M)$. 63. $1.8 \times 10^4$ N. 65. about 48 km/h. 67. 21 m. 69. $\sqrt{Mgr/m}$. 71. (a) light; (b) 778 N; (c) 223 N. 73. 2.2 km. 75. (b) 8.74 N; (c) 37.9 N, radially inward; (d) 6.45 m/s. 77. (a) $\sqrt{Rg\tan(\theta + \tan^{-1}(\mu^{\text{stat}}))}$; (b) graph; (c) 41.3 m/s; (d) 21.2 m/s. 81. (a) 3.0 N, up the incline; (b) 3.0 N, up the incline; (c) 1.6 N, up the incline; (d) 4.4 N, up the incline; (e) 1.0 N, down the incline. 83. 0.54

# Chapter 7

**RE 7-1:** (a) The 60 s encounter between the *Titanic* and an iceberg was a collision. (b) A tennis ball encountering a racket for 2 s is not a collision.

**RE 7-2:** (a) $|\vec{F}_1| > |\vec{F}_3| > |\vec{F}_2| = |\vec{F}_4| = 0$. Since the slopes represent $\Delta\vec{p}/\Delta t$ the magnitude is greatest where the slope is steepest. Thus, ranking is by steepness of slope. (b) Since the momentum is initially positive, the particle speeds up in region 1, drifts in region 2, and *slows down in region 3,* where its momentum is becoming less positive (and hence more negative).

**RE 7-3:** The change in the egg's momentum is $m\vec{v}_2 - m\vec{v}_1$, and since $\vec{v}_2$ is zero the change is just $m\vec{v}_1$. The time you take in catching the egg does not affect the momentum change since the initial and final velocities are still the same. However the time taken in the catch will affect the average force the egg experiences. Since the change in momentum equals the impulse, which equals the average force times the time the force acts, making the time of the catch longer makes the average force on the egg less and hence a greater likelihood of a successful catch. In order to make $\Delta t$ as large as possible, you move your hands and body backwards once the catch is made in order to bring the egg to zero speed over the largest time interval possible.

**RE 7-4:** (a) $p_{1x}$ is to the right and $+$, $p_{2x}$ is to the left and $-$, therefore $\Delta p_x$ is $-$. Remember that $\Delta$ is always final minus initial, and here we have a negative number minus a positive number giving a negative result. (b) $\Delta p_y$ is zero since the $y$ component of the momentum does not change in the bounce. (c) The direction of $\Delta\vec{p}$ is left. To see this, draw the two momentum vectors and subtract the initial from the final. Remember: To subtract vectors add the negative of the second to the first.

**RE 7-5:** (1) Assuming the carts are frictionless, the system consisting of the firecracker and the two carts is an isolated system and momentum should be conserved. In fact, if the firecracker is initially at rest and explodes symmetrically, then the carts should move off at the same speed in opposite directions. (2) Assuming the carts are not frictionless, then the track and the table and the Earth become part of the system. We might not see the carts come off with the same speeds in opposite directions. Instead the Earth might move (imperceptibly) to make up the difference. However, momentum is always conserved, so it should be so for our new system.

**RE 7-6:** (a) Zero, since no external forces are acting and hence the total momentum is conserved. (b) No, since the $y$-component of momentum must also be conserved. (c) The second piece must be moving in the negative direction on the $x$ axis so that the total momentum after the explosion is zero.

**RE 7-7:** We need a mass for the grapefruit—let's say 1.0 kg. The grapefruit's momentum starts at zero and goes to (1 kg)(2 m/s) = 2 kg · m/s, therefore $\Delta p$ = 2 kg · m/s. The change in the Earth's momentum will be equal and opposite, therefore the change in the Earth's speed will be 2 kg · m/s divided by the mass of the Earth. If you look at the inside front cover of this text, you find $m_{\text{Earth}} = 5.98 \times 10^{24}$ kg. Dividing, you get $v_{\text{Earth}} = 3.3 \times 10^{-25}$ m/s. Did you feel the Earth move?

## Problems

1. 24 km/h. 3. (a) $(-4.0 \times 10^4$ kg · m/s $)\hat{\text{i}}$; (b) west. 5. (a) 30°; (b) $(-0.572$ kg · m/s$)\hat{\text{j}}$. 7. 2.5 m/s. 9. 3000 N. 11. 67 m/s, in opposite direction. 13. (a) 42 N · s; (b) 2100 N. 15. (a) $(7.4 \times 10^3$ N · s $)\hat{\text{i}} + (-7.4 \times 10^3$ N · s $)\hat{\text{j}}$; (b) $(-7.4 \times 10^3$ N · s $)\hat{\text{i}}$; (c) $2.3 \times 10^3$ N; (d) $2.1 \times 10^4$ N; (e) −45°. 17. 10 m/s. 19. (a) 1.0 kg · m/s; (b) 10 N; (c) 1700 N; (d) the answer for (b) includes time between pellet collisions. 21. 41.7 cm/s. 23. (a) 46 N; (b) none. 25. ≈2 mm/y. 27. 3.0 mm/s, away from the stone. 29. (a) 4.6 m/s; (b) 3.9 m/s; (c) 7.5 m/s. 31. increases by 4.4 m/s. 33. 190 m/s. 35. (a) $\{m_A/(m_A + m_B)\}v_{A\,1}$. 37. (a) 7290 m/s; (b) 8200 m/s. 39. 4400 km/h. 41. 8.1 m/s at 38° south of east. 43. (a) 11.4 m/s; (b) 95.1° clockwise from $+x$. 45. (a) 61.7 km/h; (b) 63.4° south of west. 47. (a) 2.5 m/s. 49. 1.0 m/s north. 51. (a) $1.4 \times 10^{-22}$ kg · m/s; (b) 150°; (c) 120°. 53. 14 m/s, 135° from the other pieces. 55. 3.0 m/s. 57. 120°. 59. (a) $4.15 \times 10^5$ m/s; (b) $4.84 \times 10^5$ m/s. 61. (a) 41°; (b) 4.76 m/s; (c) no. 63. 2.0 m/s, $-x$ direction. 65. 108 m/s. 67. (a) $1.57 \times 10^6$ N; (b) $1.35 \times 10^5$ kg; (c) 2.08 km/s. 69. $2.2 \times 10^{-3}$

# Chapter 8

**RE 8-1:** (a) At the center; (b) in the lower right quadrant; (c) on the negative $y$ axis; (d) at the center; (e) in the lower left quadrant; (f) at the center

**RE 8-2:** (a) The spacing between successive halfway points is the same, which suggests that the velocity represented by these points is constant.

(b) $v = |\Delta\vec{r}|/\Delta t = 0.41\,\text{m}/[(12/15)\text{s}] = 0.51$ m/s

**RE 8-3:** Since there are no outside forces on the system, the center of mass of the system will not change. Thus, the skaters will end up meeting at the origin of the original coordinate system in all three situations (a), (b), and (c). The only difference is that in case (a) Ethel will be holding one end of the "massless" pole at the end, in case (b) Fred will be holding an end of the "massless" pole, and in case (c) one-third of the "massless" pole will be sticking out behind Fred and two-thirds will be sticking out behind Ethel.

## Problems

1. (a) −4.5 m; (b) −5.5 m. 3. (a) 4600 km; (b) $0.73R_e$. 5. (a) 1.1 m; (b) 1.3 m; (c) shifts toward topmost particle. 7. (a) −0.25 m; (b) 0. 9. $6.8 \times 10^{-12}$ m from the nitrogen atom, along axis of symmetry. 11. (a) $H/2$; (b) $H/2$; (c) descends to lowest point and then ascends to $H/2$; (d) $\dfrac{HM}{m}\left(\sqrt{1 + \dfrac{m}{M}} - 1\right)$. 13. $x_{\text{com}} = B/2$ and $y_{\text{com}} = H/3$. 15. $x_{\text{com}} = B/2$ and $y_{\text{com}} = 4R/(3\pi)$. 17. (a) 0,0; (b) 0. 19. (−1.50 m, −1.43 m). 21. 29 m. 23. 72 km/h. 25. (a) 28 cm; (b) 2.3 m/s. 27. 53 m. 29. (a) halfway between the containers; (b) 26 mm toward the heavier container; (c) down; (d) $-1.6 \times 10^{-2}$ m/s². 31. 4.2 m. 33. 12 m/s, 132° counterclockwise from east. 37. (a) 33 m/s; (b) 8.7 m/s. 39. (a) 540 m/s; (b) 40.4°. 41. (a) $0.2000v^{\text{rel}}$; (b) $0.2103v^{\text{rel}}$; (c) $0.2095v^{\text{rel}}$. 43. (a) 1.0 m/s north; (b) 3 m north

# Chapter 9

**RE 9-1:** (a) Decreases. (b) Remains the same. Remember that the kinetic energy is a scalar and depends on the velocity squared, so −2 m/s and 2 m/s give the same kinetic energy. (c) Negative for situation (a) and zero for situation (b). Situation (b) is interesting. How can the net work done be zero? Try breaking the velocity change into two changes: first from −2 m/s to zero, then from zero to 2 m/s. For the first change the work is negative and for the second change the work is positive. When we add the two works together, we get zero for the total.

**RE 9-2:** $c > a > b = d$

**RE 9-3:** Use Eq. 9-19: (a) positive; (b) negative; (c) zero. Think through your calculated answers. Do they make sense? For example, in (a) as the block moves from −3 cm to the origin, the spring force and displacement are in the same direction giving a positive work; from the origin to 2 cm the spring force and displacement are in opposite directions giving a negative work, but the positive work is larger because the displacement is larger giving a net positive work.

**RE 9-4:** $d > c > b > a$

**RE 9-5:** The power is zero at all times since $\vec{F}$ and $\vec{v}$ are always perpendicular in uniform circular motion.

## Problems

1. $1.2 \times 10^6$ m/s. 3. (a) 3610 J; (b) 1900 J; (c) $1.1 \times 10^{10}$ J. 5. (a) $2.9 \times 10^7$ m/s; (b) $2.1 \times 10^{-13}$ J. 7. (a) $7.5 \times 10^4$ J; (b) $3.8 \times 10^4$ kg · m/s; (c) 38° south of east. 9. $1.18 \times 10^4$ kg. 11. (a) 3.7 m/s; (b) 1.3 N·s; (c) $1.8 \times 10^2$ N. 13. (a) 42 J; (b) 30 J; (c) 12 J; (d) 6.48 m/s, positive direction of $x$ axis; (e) 5.48 m/s, positive direction of $x$ axis; (f) 3.46 m/s, positive direction of $x$ axis. 15. $AB$: +, $BC$: 0, $CD$: −, $DE$: +. 17. (a) 170 N; (b) 340 m; (c) $-5.8 \times 10^4$ J; (d) 340 N; (e) 170 m; (f) $-5.8 \times 10^4$ J. 19. 800 J. 21. (a) 98 N; (b) 4.0 cm; (c) 3.9 J; (d) −3.9 J. 23. 0, by both methods. 25. (a) −0.043 J; (b) −0.13 J. 27. (a) 6.0 N; (b) − 2.5 N; (c) 15 N 29. 15.3 J. 31. (a) 590 J; (b) 0; (c) 0; (d) 590 J. 33. 6.8 J. 35. (a) 1.20 J; (b) 1.10 m/s. 37. (a) 1.50 J; (b) increases. 39. (a) $1.2 \times 10^4$ J; (b) $-1.1 \times 10^4$ J; (c) 1100 J; (d) 5.4 m/s. 41. (a) $-3Mgd/4$; (b) $Mgd$; (c) $Mgd/4$; (d) $\sqrt{gd/2}$. 43. 20 J. 45. (a) $8.84 \times 10^3$; (b) $7.84 \times 10^3$ J; (c) $6.84 \times 10^3$ J. 47. (a) 2.3 J; (b) 2.6 J. 49. 490 W. 51. (a) 0.83 J; (b) 2.5 J; (c) 4.2 J; (d) 5.0 W. 53. 740 W. 55. 68 kW. 57. (a) $1.8 \times 10^5$ ft · lb; (b) 0.55 hp. 59. (a) 8.8 m/s; (b) 2600 J; (c) 1.6 kW. 61. 24 W 63. (a) $2.1 \times 10^6$ kg; (b) $\sqrt{100 + 1.5t}$ m/s; (c) $(1.5 \times 10^6)/\sqrt{100 + 1.5t}$ N; (d) 6.7 km 65. (a) ≈ $1 \times 10^5$ megatons; (b) ≈ ten million bombs

# Chapter 10

**RE 10-1:** No, for the force to be conservative the work done in going between two points must not depend on the path taken. Also, if you go from 2 to 1 instead of 1 to 2 the work will change sign. Therefore, for the force in the exercise to be conservative the work for the bottom path should have a negative sign.

**RE 10-2:** A Hot Wheels® car that traverses path $b$ should lose more kinetic energy than one that traverses path $a$. This is because path $b$ is longer so the friction forces have more distance to act on path $b$.

**RE 10-3:** The kinetic energy of the barbell is zero before the lift and zero after the lift, as evidenced by the fact that $y$ vs. $t$ is a constant at $t = 0.0$ s and at $t = 2.0$ s. Since the kinetic energy change $\Delta K = 0.0$ J, then the net work on the barbells should be zero. An examination of graph 10-10$b$ shows that the positive work is approximately given by the area under the $F^{net}$ vs. $y$ curve. $W^+ =$ area under the positive portion of the curve $= (0.5)(116\text{ N})(.15\text{ m}) = +8.7$ J and $W^- =$ area under the negative portion of the curve $(0.5)(58\text{ J})(.45 - .15)\text{ m} = -8.7$ J. So $W^{net} = W^+ + W^- = 0.0$ J.

**RE 10-4:** Use Eq. 10-13. Note that the change in the potential energy is the negative of the area under the curves in the figure. The most positive will be (3) and the least positive (2).

**RE 10-5:** Without friction, the decrease in the potential energy will equal the increase in the kinetic energy. (a) Therefore, since all four blocks are losing the same amount of potential energy, they will all have the same kinetic energy at point $B$. (b) Since the kinetic energies are the same, the speeds are the same.

**RE 10-6:** Use the equation $F_x^{int}(x) = -dU(x)/dx$. The force is the negative of the slope of the $U$ vs. $x$ curve. (a) Ranking *magnitudes* with the greatest first: $CD$, $AB$, $BC$. (b) The slope is negative, hence the force is in the positive $x$ direction.

**RE 10-7:** $b > a > c$ as determined by the equation $\Delta E^{thermal} = f_x^{kin}\Delta x$.

**RE 10-8:** (a) 4 kg · m/s; (b) 8 kg · m/s; (c) assuming an elastic collision, 3 J.

**RE 10-9:** (a) 2 kg · m/s. (b) Since the initial $y$-component is zero, the final must be zero. Therefore, the final $y$-component of momentum for the target is 3 kg · m/s.

## Problems

1. 89 N/cm. 3. (a) 4.31 mJ; (b) −4.31 mJ; (c) 4.31 mJ; (d) −4.31 mJ; (e) all increase. 5. (a) $mgL$; (b) $-mgL$; (c) 0; (d) $-mgL$; (e) $mgL$; (f) 0; (g) same. 7. (a) 184 J; (b) −184 J; (c) −184 J. 9. −320 J 11. (a) 2.08 m/s; (b) 2.08 m/s; (c) increase. 13. (a) $\sqrt{2gL}$; (b) $2\sqrt{gL}$; (c) $\sqrt{2gL}$; (d) all the same. 15. (a) 260 m; (b) same; (c) decrease. 17. (a) 21.0 m/s; (b) 21.0 m/s; (c) 21.0 m/s. 19. (a) 0.98 J; (b) −0.98 J; (c) 3.1 N/cm. 21. (a) 39.2 J; (b) 39.2 J; (c) 4.00 m. 23. (a) 35 cm; (b) 1.7 m/s. 25. 10 cm. 27. 1.25 cm. 31. (a) $2\sqrt{gL}$; (b) $5mg$; (c) 71°. 33. $mgL/32$. 37. (a) $1.12(A/B)^{1/6}$; (b) repulsive; (c) attractive. 39. (a) −3.7 J; (c) 1.29 m; (d) 9.12 m; (e) 2.16 J ; (f) 4.0 m; (g) $(4 - x)e^{-x/4}$ N; (h) 4 m. 41. (a) 30.1 J; (b) 30.1 J; (c) 0.22. 43. (a) 5.6 J; (b) 3.5 J. 45. 11 kJ. 47. 20 ft · lb. 49. (a) 1.5 MJ; (b) 0.51 MJ; (c) 1.0 MJ; (d) 63 m/s. 51. (a) 67 J; (b) 67 J; (c) 46 cm. 53. (a) 31.0 J; (b) 5.35 m/s; (c) conservative. 55. (a) 44 m/s; (b) 0.036. 57. (a) −0.90 J; (b) 0.46 J; (c) 1.0 m/s. 59. 1.2 m. 63. in the center of the flat part. 65. (a) 216 J; (b) 1180 N; (c) 432 J; (d) motor also supplies thermal energy to crate and belt. 67. (a) 0.2 to 0.3 MJ; (b) same amount. 69. (a) 860 N; (b) 2.4 m/s. 71. (a) $mR(\sqrt{2gh} + gt)$; (b) 5.06 kg. 73. (a) $mv_1/(m + M)$; (b) $M/(m + M)$. 75. 25 cm. 79. (a) 41°; (b) 4.76 m/s; (c) no. 81. (a) 6.9 m/s, 30° to $+x$ direction; (b) 6.9 m/s, −30° to $+x$ direction; (c) 2.0 m/s, $-x$ direction. 83. (a) 99 g; (b) 1.9 m/s; (c) 0.93 m/s. 85. 7.8%. 87. (a) 1.2 kg; (b) 2.5 m/s. 89. (a) 100 g; (b) 1.0 m/s. 91. (a) 1.9 m/s, to the right; (b) yes; (c) no, total kinetic energy would have increased. 93. (a) 1/3; (b) $4h$. 95. 1.0 kg. 97. (c) 11%; (d) 10%; (e) 79%

# Chapter 11

**RE 11-1:** (a) Positive, since $\theta$ is increasing. (b) Negative, since $\theta$ is decreasing.

**RE 11-2:** (a) Positive; (b) negative; (c) negative; (d) positive

**RE 11-3:** Find the angular acceleration, $\alpha$, by taking the second derivative of $\theta$ with respect to $t$. The accelerations for (a) and (d) do not depend on $t$ and are therefore constant, and hence the equations of Table 11-1 apply.

**RE 11-4:** Since the speeds are being squared, $v^2$ and $\omega^2$ will always be positive quantities.

**RE 11-5:** (a) Yes, the centripetal acceleration; (b) no, since $\alpha$ is zero; (c) yes; (d) yes, since $\alpha$ is no longer zero.

**RE 11-6:** Calculate $mr^2$ for each, and you'll find they are all the same.

**RE 11-7:** (1) > (2) > (4) > (3). Remember that $I$ depends not only on the mass but also on how far that mass is from the chosen axis.

**RE 11-8:** $I_a = I_d = mr^2$, $I_b = \frac{1}{2}mr^2$, $I_c = \frac{5}{8}mr^2$, so $a = d > c > b$.

**RE 11-9:** $A = C > D > B = E =$ zero. For $A$ and $C$, $\phi$ is 90°; for $D$, $\phi$ is between zero and 90°; for $E$, $\phi$ is zero; and for $C$, $r$ is zero.

**RE 11-10:** (a) Same direction. (b) Less.

## Problems

1. (a) $a + 3bt^2 - 4ct^3$; (b) $6bt - 12ct^2$. 3. (a) $5.5 \times 10^{15}$ s; (b) 26. 5. (a) 2 rad; (b) 0; (c) 130 rad/s; (d) 32 rad/s²; (e) no. 7. 11 rad/s. 9. (a) −67 rev/min²; (b) 8.3 rev. 11. 200 rev/min. 13. 8.0 s. 15. (a) 44 rad; (b) 5.5 s, 32 s; (c) −2.1 s, 40 s. 17. (a) 340 s; (b) $-4.5 \times 10^{-3}$ rad/s²; (c) 98 s. 19. 1.8 m/s², toward the center. 21. 0.13 rad/s. 23. (a) 3.0 rad/s; (b) 30 m/s; (c) 6.0 m/s²; (d) 90 m/s². 25. (a) $3.8 \times 10^3$ rad/s; (b) 190 m/s. 27. (a) $7.3 \times 10^{-5}$ rad/s; (b) 350 m/s; (c) $7.3 \times 10^{-5}$ rad/s; (d) 460 m/s. 29. 16 s. 31. (a) $-2.3 \times 10^{-9}$ rad/s²; (b) 2600 y; (c) 24 ms. 33. 12.3 kg · m². 35. (a) 1100 J; (b) 9700 J. 37. (a) $5md^2 + 8/3Md^2$; (b) $(5/2m + 4/3M)d^2\omega^2$. 39. 0.097 kg · m². 41. $\frac{1}{3}M(a^2 + b^2)$. 45. 4.6 N · m. 47. (a) $r_1F_A \sin\theta_1 - r_2F_B \sin\theta_2$; (b) −3.8 N · m. 49. (a) 28.2 rad/s²; (b) 338 N · m. 51. (a) 155 kg · m²; (b) 64.4 kg. 53. 130 N. 55. (a) 6.00 cm/s²; (b) 4.87 N; (c) 4.54 N; (d) 1.20 rad/s²; (e) 0.0138 kg · m². 57. (a) 1.73 m/s²; (b) 6.92 m/s². 59. 396 N · m. 61. (a) $mL^2\omega^2/6$; (b) $L^2\omega^2/6g$. 63. 5.42 m/s 65. $\frac{3}{2}\sqrt{\frac{g}{L}}$. 67. (a) $[(3g/H)(1 - \cos\theta)]^{0.5}$; (b) $3g(1 - \cos\theta)$; (c) $3/2g \sin\theta$; (d) 41.8°. 69. (a) $0.083519ML^2 \approx 0.084ML^2$; (b) low by (only) 0.22%

## Chapter 12

**RE 12-1:** (a) When is the sin of the angle between the vectors zero? Sin is zero for 0° and 180°. (b) Here the sin needs to equal ±1. This occurs at 90° and 270°. (c) Here $|\vec{c}||\vec{d}|\sin\phi = 3 \cdot 4 \sin\phi = 6$ so $\phi = \sin^{-1}(6/12)$ so $\phi = 30°$ or 150°.

**RE 12-2:** The time rate of change of the rotational momentum is equal to the net torque. $3 > 1 > 2 = 4 =$ zero.

**RE 12-3:** (a) $1 = 3 > 2 = 4 > 5 =$ zero, since $r_\perp$ is 4 m for both 1 and 3 and 2 m for both 2 and 4 and zero for 5. (b) Particles 2 and 3 have negative rotational momentum about $o$, since $\vec{\ell} = \vec{r} \times \vec{p}$ points into the page for each of them.

**RE 12-4:** (a) Since the rate of change of the rotational momentum is equal to the applied torque, which is the same for all three cases, all three objects increase their rotational momentum at the same rate; and assuming all three started from rest, they will all have the same rotational momentum at any given time. (b) Look at Table 11-2 (Some Rotational Inertias). Note that $I_{hoop} > I_{disk} > I_{sphere}$. Since $L = I\omega$ and they all have the same $L$, the object with the biggest $I$ will have the smallest $\omega$; $\omega_{sphere} > \omega_{disk} > \omega_{hoop}$.

**RE 12-5:** (a) Decrease, since although the total mass of the system has not changed, it is distributed closer to the axis of rotation. (b) Remain the same, since there is no net external torque. (c) If $I$ decreases and $L$ is constant, then $\omega$ must increase.

## Problems

1. (a) 59.3 rad/s; (b) 9.31 rad/s$^2$; (c) 70.7 m. 3. −3.15 J. 5. 1/50 7. (a) 8.0°; (b) more. 9. (a) 13 cm/s$^2$; (b) 4.4 s; (c) 55 cm/s; (d) $1.8 \times 10^{-2}$ J; (e) 1.4 J; (f) 27 rev/s. 11. (a) 10 s; (b) 897 m. 13. the third. 17. (a) 10 N · m, parallel to $yz$ plane, at 53° to $+y$; (b) 22 N · m, $-x$. 19. (a) (50 N · m)$\hat{k}$; (b) 90°. 21. (a) (−170 kg · m$^2$/s)$\hat{k}$; (b) (+56 N · m)$\hat{k}$; (c) (+56 kg · m$^2$/s$^2$)$\hat{k}$. 23. (a) 0; (b) $8t$ N · m, in $-z$ direction; (c) $2/\sqrt{t}$ N · m, $-z$; (d) $8/t^3$ N · m, $+z$. 25. 9.8 kg · m$^2$/s. 27. (a) 0; (b) (8.0 N · m)$\hat{i}$ + (8.0 N · m)$\hat{k}$. 29. (a) $mvd$; (b) no; (c) 0, yes. 31. (a) −1.47 N · m; (b) 20.4 rad; (c) −29.9 J; (d) 19.9 W. 33. (a) $14md^2$; (b) $4md^2\omega$; (c) $14md^2\omega$. 35. $\omega_1 R_A R_B I_A/(I_A R_B^2 + I_B R_A^2)$. 37. (a) 3.6 rev/s; (b) 3.0; (c) in moving the bricks in, the forces on them from the man transferred energy from internal energy of the man to kinetic energy. 39. (a) 267 rev/min; (b) 2/3. 41. (a) 149 kg · m$^2$; (b) 158 kg · m$^2$/s; (c) 0.746 rad/s 43. $\dfrac{m}{M+m}\left(\dfrac{v}{R}\right)$ 45. (a) $(mRv - I\omega_1)/(I + mR^2)$; (b) no, energy transferred to internal energy of cockroach. 47. 3.4 rad/s. 49. (a) 0.148 rad/s; (b) 0.0123; (c) 181°. 51. The day would be longer by about 0.8 s. 53. (a) 18 rad/s; (b) 0.92 55. (a) 0.24 kg · m$^2$; (b) 1800 m/s 57. $\theta = \cos^{-1}\left[1 - \dfrac{6m^2h}{d(2m+M)(3m+M)}\right]$. 59. 11.0 m/s 61. (a) 0.180 m ; (b) clockwise

# Photo Credits

## Dedication

Photo courtesy Jean Arons.

## Introduction

Opener: Courtesy of the Archives, California Institute of Technology. Reproduced with permission of the Estate of Richard Feynman. Figure I-1: FOX-TROT ©1999 Bill Amend. Reprinted with permission of UNIVERSAL PRESS SYNDICATE. All rights reserved.

## Preface

Figures P-1, P-2 and P-3: Courtesy Priscilla Laws.

## Chapter 1

Opener: Larry Bray/Taxi/Getty Images. Figure 1-2: Detlev van Ravenswaay/Photo Researchers. Figure 1-3: Courtesy Randall Feenstra, Carnegie Mellon University. Figure 1-4: ©Steven Pitkin. Figure 1-5: Courtesy National Institute of Standards and Technology. Figure 1-7: Courtesy Bureau International des Poids et Mesures, France. Figure 1-8: Courtesy Fisher Scientific. Figures 1-9, 1-10 and 1-11 (left): Courtesy Vernier Software and Technology. Figure 1-11 (right): Courtesy Pasco Scientific. Figure 1-12: Courtesy Ron Thornton. Figures 1-13 and 1-14: Courtesy Priscilla Laws. Figure 1-15: Getty Images News and Sport Services. Figure 1-20: Worldscapes/Age Fotostock America, Inc. Figure 1-21: Lynda Richardson/Corbis Images. Figure 1-22: Corbis Digital Stock.

## Chapter 2

Opener: Niagara Gazette/Corbis Sygma. Figures 2-6, 2-7 and 2-9: Courtesy Pat Cooney. Figure 2-10: Courtesy Priscilla Laws. Figure 2-13: Courtesy U.S. Air Force. Figure 2-16: Courtesy Ron Thornton. Figures 2-18, 2-19, 2-20 and 2-21b: Courtesy Priscilla Laws. Figure 2-30: Courtesy Vernier Software and Technology. Figure 2-42: Courtesy Priscilla Laws.

## Chapter 3

Opener: Nicole Duplaix/Corbis Images. Figure 3-1: American Institute of Physics/Photo Researchers. Figures 3-2, 3-3 and 3-5: Courtesy Priscilla Laws. Figure 3-8: Courtesy Vernier Software and Technology. Figure 3-10: Courtesy Ohaus Corporation. Figures 3-12 and 3-13: Courtesy Priscilla Laws. Page 67: ©AP/Wide World Photos. Figure 3-19a: Courtesy Priscilla Laws. Figure 3-20: James A. Sugar/Corbis Images. Figures 3-22 and 3-28: Courtesy Priscilla Laws. Figure 3-34: Clive Newton/Corbis Images. Figures 3-45, 3-55 and 3-56: Courtesy Priscilla Laws.

## Chapter 4

Opener: Courtesy David des Marais, Cave Research Foundation. Figure 4-2: FOX-TROT ©1999 Bill Amend. Reprinted with permission of UNIVERSAL PRESS SYNDICATE. All rights reserved.

## Chapter 5

Opener: ©AP/Wide World Photos. Figure 5-1: Photo by Andrew Davidhazy/RIT. Figure 5-2: Bettmann/Corbis Images. Figure 5-4: Courtesy Priscilla Laws. Figure 5-5: Richard Megna/Fundamental Photographs. Figure 5-6: Courtesy Priscilla Laws. Figure 5-7: Jammie Budge/Gamma-Presse, Inc. Figures 5-9 and 5-10: Courtesy Priscilla Laws. Figure 5-23: Adam Woolfitt/Corbis Images. Figure 5-30, Problem 6: Allsport/Getty Images.

## Chapter 6

Opener: © Natural History Photographic Agency. Page 145: ©2003 Tom Thaves. Reprinted with permission. Figure 6-14: Courtesy Priscilla Laws. Figure 6-16: Courtesy Pasco Scientific. Figure 6-17: Courtesy Priscilla Laws. Figure 6-26: Jean Y. Ruszniewki/Stone/Getty Images. Figure 6-28: Joe McBride/Stone/Getty Images. Figure 6-29: Courtesy Priscilla Laws. Figure 6-41: Jerry Schad/Photo Researchers. Figure 6-47: Susan Copen Oken/Dot, Inc. Figure 6-73: Peter Turnley/Corbis Images. Figure 6-84: Courtesy Joe Redish. Figure 6-92a: Courtesy Priscilla Laws.

## Chapter 7

Opener: Terje Rakke/The Image Bank/Getty Images. Figure 7-1a: Charles and Josette Lenars/Corbis Images. Figure 7-1b: Science Photo Library/Photo Researchers. Figure 7-1c: Photo by Andrew Davidhazy/RIT. Figure 7-5: Courtesy Robert Teese. Figure 7-9: Courtesy Priscilla Laws. Figure 7-10: Courtesy PASCO scientific. Figures 7-11, 7-12, 7-13 and 7-14: Courtesy Priscilla Laws. Figure 7-17: Courtesy NASA. Page 198: FOXTROT ©2000 Bill Amend. Reprinted with permission of UNIVERSAL PRESS SYNDICATE.. All rights reserved. Figure 7-19: Courtesy NASA. Figure 7-21: George Long/Sports Illustrated/Time, Inc. Picture Collection. Figure 7-22: Superman #48, October 1990. ©D.C. Comics. All rights reserved. Reprinted with permission. Figures 7-33 and 7-38: Courtesy Priscilla Laws.

## Chapter 8

Opener: Lois Greenfield. Figure 8-1a: Richard Megna/Fundamental Photographs. Figures 8-6, 8-10, 8-11 and 8-12: Courtesy Priscilla Laws. Figure 8-23: Adam Crowley/PhotoDisc, Inc./Getty Images.

## Chapter 9

Opener and Figure 9-22: ©AP/Wide World Photos. Figure 9-21: ©Photri.

## Chapter 10

Opener: Malcolm S. Kirk/Peter Arnold, Inc. Figure 10-1: Dimitri Lundt/Corbis Images. Figure 10-7: Photo provided courtesy of Mattel, Inc. Figure 10-9a: ©AP/Wide World Photos. Figure 10-14: Courtesy Priscilla Laws. Figure 10-19: Courtesy Mercedes-Benz of North America.

## Chapter 11

Opener: Arthur Tilley/Stone/Getty Images. Figure 11-1a: Doug Pensinger/Getty Images News and Sport Services. Figure 11-1b: Duomo/Corbis Images. Figures 11-2: Courtesy PASCO scientific and Priscilla Laws. Figure 11-9: Courtesy Priscilla Laws. Page 308: Calvin and Hobbes ©1990 Bill Watterson. Reprinted with permission of UNIVERSAL PRESS SYNDICATE. All rights reserved. Figure 11-13: Roger Ressmeyer/Corbis Images. Figure 11-16: Courtesy Test Devices, Inc. Figure 11-26: Courtesy Lick Observatory. Figure 11-34: Courtesy Lawrence Livermore Laboratory, University of California. Figure 11-50: Courtesy Mark Luetzelschwab.

## Chapter 12

Opener: Image courtesy Ringling Brothers and Barnum & Bailey®, THE GREATEST SHOW ON EARTH. Figure 12-1: Richard Megna/Fundamental Photographs. Figure 12-2: Courtesy PASCO Scientific. Figure 12-15: From *Shepp's World's Fair* Photographed by James W. Shepp and Daniel P. Shepp, Globe Publishing Co., Chicago and Philadelphia, 1893. Photo provided courtesy of Jeffery Howe.

# Index

## T



## U



## V



## W

## R



## S

## X

## Y



## Z

## Mathematical Formulas*

### Quadratic Formula

If $ax^2 + bx + c = 0$, then $x = \dfrac{-b \pm \sqrt{b^2 - 4ac}}{2a}$

### Binomial Theorem

$$(1 + x)^n = 1 + \frac{nx}{1!} + \frac{n(n-1)x^2}{2!} + \cdots \qquad (x^2 < 1)$$

### Products of Vectors

Let $\theta$ be the smaller of the two angles between $\vec{a}$ and $\vec{b}$. Then

$$\vec{a} \cdot \vec{b} = \vec{b} \cdot \vec{a} = a_x b_x + a_y b_y + a_z b_z = |\vec{a}||\vec{b}| \cos\theta$$

$$\vec{a} \times \vec{b} = -\vec{b} \times \vec{a} = \begin{vmatrix} \hat{\mathrm{i}} & \hat{\mathrm{j}} & \hat{\mathrm{k}} \\ a_x & a_y & a_z \\ b_x & b_y & b_z \end{vmatrix}$$

$$= \hat{\mathrm{i}} \begin{vmatrix} a_y & a_z \\ b_y & b_z \end{vmatrix} - \hat{\mathrm{j}} \begin{vmatrix} a_x & a_z \\ b_x & b_z \end{vmatrix} + \hat{\mathrm{k}} \begin{vmatrix} a_x & a_y \\ b_x & b_y \end{vmatrix}$$

$$= (a_y b_z - b_y a_z)\hat{\mathrm{i}} + (a_z b_x - b_z a_x)\hat{\mathrm{j}} + (a_x b_y - b_x a_y)\hat{\mathrm{k}}$$

$$|\vec{a} \times \vec{b}| = |\vec{a}||\vec{b}| \sin\theta$$

### Trigonometric Identities

$$\sin\alpha \pm \sin\beta = 2\sin\tfrac{1}{2}(\alpha \pm \beta)\cos\tfrac{1}{2}(\alpha \mp \beta)$$

$$\cos\alpha + \cos\beta = 2\cos\tfrac{1}{2}(\alpha + \beta)\cos\tfrac{1}{2}(\alpha - \beta)$$

### Derivatives and Integrals

$$\frac{d}{dx}\sin x = \cos x \qquad \int \sin x\,dx = -\cos x$$

$$\frac{d}{dx}\cos x = -\sin x \qquad \int \cos x\,dx = \sin x$$

$$\frac{d}{dx}e^x = e^x \qquad \int e^x\,dx = e^x$$

$$\int \frac{dx}{\sqrt{x^2 + a^2}} = \ln(x + \sqrt{x^2 + a^2})$$

$$\int \frac{x\,dx}{(x^2 + a^2)^{3/2}} = -\frac{1}{(x^2 + a^2)^{1/2}}$$

$$\int \frac{dx}{(x^2 + a^2)^{3/2}} = \frac{x}{a^2(x^2 + a^2)^{1/2}}$$

### Cramer's Rule

Two simultaneous equations in unknowns $x$ and $y$,

$$a_1 x + b_1 y = c_1 \qquad \text{and} \qquad a_2 x + b_2 y = c_2,$$

have the solutions

$$x = \frac{\begin{vmatrix} c_1 & b_1 \\ c_2 & b_2 \end{vmatrix}}{\begin{vmatrix} a_1 & b_1 \\ a_2 & b_2 \end{vmatrix}} = \frac{c_1 b_2 - c_2 b_1}{a_1 b_2 - a_2 b_1}$$

and

$$y = \frac{\begin{vmatrix} a_1 & c_1 \\ a_2 & c_2 \end{vmatrix}}{\begin{vmatrix} a_1 & b_1 \\ a_2 & b_2 \end{vmatrix}} = \frac{a_1 c_2 - a_2 c_1}{a_1 b_2 - a_2 b_1}.$$

* See Appendix E for a more complete list.

## The Greek Alphabet

| | | | | | | | | |
|---|---|---|---|---|---|---|---|---|
| Alpha | A | $\alpha$ | Iota | I | $\iota$ | Rho | P | $\rho$ |
| Beta | B | $\beta$ | Kappa | K | $\kappa$ | Sigma | $\Sigma$ | $\sigma$ |
| Gamma | $\Gamma$ | $\gamma$ | Lambda | $\Lambda$ | $\lambda$ | Tau | T | $\tau$ |
| Delta | $\Delta$ | $\delta$ | Mu | M | $\mu$ | Upsilon | $\Upsilon$ | $\upsilon$ |
| Epsilon | E | $\epsilon$ | Nu | N | $\nu$ | Phi | $\Phi$ | $\phi$, $\varphi$ |
| Zeta | Z | $\zeta$ | Xi | $\Xi$ | $\xi$ | Chi | X | $\chi$ |
| Eta | H | $\eta$ | Omicron | O | $o$ | Psi | $\Psi$ | $\psi$ |
| Theta | $\Theta$ | $\theta$ | Pi | $\Pi$ | $\pi$ | Omega | $\Omega$ | $\omega$ |